# Therapeutics of Neural Stimulation for Neurological Disorders

Yuping Wang
Editor

# Therapeutics of Neural Stimulation for Neurological Disorders

Springer

*Editor*
Yuping Wang
Department of neurology
Xuanwu Hospital, Capital Medical University
Beijing, China

ISBN 978-981-99-4537-5    ISBN 978-981-99-4538-2 (eBook)
https://doi.org/10.1007/978-981-99-4538-2



This Springer imprint is published by the registered company Springer Nature Singapore Pte Ltd.
The registered company address is: 152 Beach Road, #21-01/04 Gateway East, Singapore 189721, Singapore

# Contents

## Contributors

**Tianhao Bao** Department of Geriatrics, Yunnan Province Mental Health Center, Kunming Medical University, Kunming, Yunnan, China

**Jia Chen** Department of Neurology, Xuanwu Hospital, Capital Medical University, Beijing, China

**Runze Chen** Department of Pediatric, Xuanwu Hospital, Capital Medical University, Beijing, China

**Weibi Chen** Department of Neurology, Xuanwu Hospital, Capital Medical University, Beijing, China

**Bo Cui** Department of Neurology, Xuanwu Hospital, Capital Medical University, Beijing, China

**Xiaona Dai** Department of Neurology, Xuanwu Hospital, Capital Medical University, Beijing, China

**Yuanyuan Dang** Department of Neurosurgery, The First Medical Center, Chinese PLA General Hospital, Beijing, China

**Yan Ding** Department of Neurology, Xuanwu Hospital, Capital Medical University, Beijing, China

**Hongfeng Duan** Beijing PsycheArk Science and Technology Development Co., Ltd., Beijing, China

**Yiran Duan** Department of Neurology, Beijing Friendship Hospital, Capital Medical University, Beijing, China

**Jialin Du** Department of Pharmacy Phase I Clinical Trial Center, Xuanwu Hospital, Capital Medical University, Beijing, China

**Yuan Geng** Brain Aging and Cognitive Neuroscience Laboratory of Hebei Province; Neuromedical Technology Innovation Center of Hebei Province; Department of Neurology, Hebei Hospital of Xuanwu Hospital, Capital Medical University, the First Hospital of Hebei Medical University, Shijiazhuang, Hebei, China

**Yuzhou Guan** Department of Neurology, Peking Union Medical College Hospital, Beijing, China

**Bingqi Guo** Department of Neurology, Xuanwu Hospital, Capital Medical University, Beijing, China

**Xin Guo** Department of Neurology, Neuromedical Technology Innovation Center of Hebei Province, Brain Aging and Cognitive Neuroscience Laboratory of Hebei Province, Hebei Hospital, Xuanwu Hospital of Capital Medical University, The First Hospital of Hebei Medical University, Shijiazhuang, Hebei, China

**Jiaqi Han** Department of Neurology, Xuanwu Hospital, Capital Medical University, Beijing, China

**Tao Han** Department of Neurology, Shandong Provincial Hospital Affiliated to Shandong First Medical University, Jinan, Shandong, China

**Wensi Hao** Department of Neurology, Xuanwu Hospital, Capital Medical University, Beijing, China

**Jianghong He** Department of Neurosurgery, Beijing Tiantan Hospital, Capital Medical University, Beijing, China

**Liu He** Department of Functional Neurosurgery, Beijing Institute of Functional Neurosurgery, Xuanwu Hospital, Capital Medical University, Beijing, China

**Yue Hou** Department of Neurology, Xuanwu Hospital, Capital Medical University, Beijing, China

**Jiaxi Huang** Center for Neurological Function Test and Neuromodulation, West China Xiamen Hospital, Sichuan University, Xiamen, China

**Zhaoyang Huang** Department of Neurology, Xuanwu Hospital, Capital Medical University, Beijing, China

**Xiaofei Jia** Department of Functional Neurosurgery, Beijing Institute of Functional Neurosurgery, Xuanwu Hospital, Capital Medical University, Beijing, China

**Yu Jia** Department of Neurology, Xuanwu Hospital, Capital Medical University, Beijing, China

**Gongjun Ji** The School of Mental Health and Psychological Sciences, Anhui Medical University, Hefei, China

**Hua Lin** Department of Neurology, Xuanwu Hospital, Capital Medical University, Beijing, China

**Yicong Lin** Department of Neurology, Xuanwu Hospital, Capital Medical University, Beijing, China

**Siran Li** Department of Neurology, Xuanwu Hospital, Capital Medical University, Beijing, China

**Chunyan Liu** Department of Neurology, Xuanwu Hospital, Capital Medical University, Beijing, China

**Jianghong Liu** Department of Neurology, Xuanwu Hospital, Capital Medical University, Beijing, China

**Sitong Liu** Department of Neurology, Beijing Friendship Hospital, Capital Medical University, Beijing, China

**Yingxuan Li** Department of Neurology, Xuanwu Hospital, Capital Medical University, Beijing, China

**Qinying Ma** Brain Aging and Cognitive Neuroscience Laboratory of Hebei Province; Neuromedical Technology Innovation Center of Hebei Province; Department of Neurology, Hebei Hospital of Xuanwu Hospital, Capital Medical University, the First Hospital of Hebei Medical University, Shijiazhuang, Hebei, China

**Bing Ni** Department of Functional Neurosurgery, Beijing Institute of Functional Neurosurgery, Xuanwu Hospital, Capital Medical University, Beijing, China

**Lili Niu** Shenzhen Institutes of Advanced Technology, Chinese Academy of Sciences, Shenzhen, China

**Yulian Niu** Department of Neurology, Daxing District People's Hospital, Beijing, China

**Na Clara Pan** Department of Neurology, Xuanwu Hospital, Capital Medical University, Beijing, China

**Wenwen Shen** Department of Pediatric, Xuanwu Hospital, Capital Medical University, Beijing, China

**Zhiyuan Shen** Department of Neurology, Neuromedical Technology Innovation Center of Hebei Province, Brain Aging and Cognitive Neuroscience Laboratory of Hebei Province, Hebei Hospital, Xuanwu Hospital of Capital Medical University, The First Hospital of Hebei Medical University, Shijiazhuang, Hebei, China

**Haiqing Song** Department of Neurology, Xuanwu Hospital, Capital Medical University, Beijing, China

**Penghui Song** Department of Neurology, Xuanwu Hospital, Capital Medical University, Beijing, China

**Weiqun Song** Department of Rehabilitation, Xuanwu Hospital, Capital Medical University, Beijing, China

**Xueli Song** Department of Neurology, Yuquan Hospital, Tsinghua University (Tsinghua University Hospital of Integrated Traditional Chinese and Western Medicine), Beijing, China

**Ying Sun** Department of Neurology, Xuanwu Hospital, Capital Medical University, Beijing, China

**Shujuan Tian** Department of Neurology, Neuromedical Technology Innovation Center of Hebei Province, Brain Aging and Cognitive Neuroscience Laboratory of Hebei Province, Hebei Hospital, Xuanwu Hospital of Capital Medical University, The First Hospital of Hebei Medical University, Shijiazhuang, Hebei, China

**Changming Wang** Department of Neurosurgery, Xuanwu Hospital, Capital Medical University, Beijing, China

**Di Wang** School of Engineering Medicine, Beihang University, Beijing, China

**Huicong Wang** Department of Neurology, Xuanwu Hospital, Capital Medical University, Beijing, China

**Jing Wang** Department of Neurobiology, School of Basic Medical Sciences, Capital Medical University, Beijing, China

**Kai Wang** The School of Mental Health and Psychological Sciences, Anhui Medical University, Hefei, China

**Kun Wang** Department of Neurology, Beijing Puren Hospital, Beijing, China

**Li Wang** Department of Neurology, Xuanwu Hospital, Capital Medical University, Beijing, China

School of Life Sciences, Beijing Institute of Technology, Beijing, China

**Mingwei Wang** Brain Aging and Cognitive Neuroscience Laboratory of Hebei Province; Neuromedical Technology Innovation Center of Hebei Province; Department of Neurology, Hebei Hospital of Xuanwu Hospital, Capital Medical University, the First Hospital of Hebei Medical University, Shijiazhuang, Hebei, China

**Shiyu Wang** Department of Pediatric, Xuanwu Hospital, Capital Medical University, Beijing, China

**Wei Wang** Department of Neurology, Xuanwu Hospital, Capital Medical University, Beijing, China

**Weimin Wang** School of Electronics, Peking University, Beijing, China

**Xueyuan Wang** Department of Functional Neurosurgery, Beijing Institute of Functional Neurosurgery, Xuanwu Hospital, Capital Medical University, Beijing, China

**Zu Wang** Department of Neurology, Xuanwu Hospital, Capital Medical University, Beijing, China

**Hua Wei** Department of Neurology, Xuanwu Hospital, Capital Medical University, Beijing, China

**Ruicai Xiong** Department of Psychiatry, The Fourth People's Hospital (Mental Health Centre) of Pengzhou City, Chengdu, Sichuan, China

**Cuiping Xu** Department of Functional Neurosurgery, Beijing Institute of Functional Neurosurgery, Xuanwu Hospital, Capital Medical University, Beijing, China

**Qing Xue** Department of Neurology, Xuanwu Hospital, Capital Medical University, Beijing, China

**Zhexue Xu** Department of Neurology, Xuanwu Hospital, Capital Medical University, Beijing, China

Department of Neurology, Dongfang Hospital Beijing University of Chinese Medicine, Beijing, China

**Dongju Yang** Department of Neurology, Xuanwu Hospital, Capital Medical University, Beijing, China

**Junlan Yang** Department of Geriatrics, The Fourth People's Hospital (Mental Health Centre) of Pengzhou City, Chengdu, Sichuan, China

**Nuo Yang** Department of Neurology, Xuanwu Hospital, Capital Medical University, Beijing, China

**Xiaotong Yang** Department of Neurology, The First Hospital of Hebei Medical University, Shijiazhuang, Hebei, China

**Yingxue Yang** Department of Neurology, Xuanwu Hospital, Capital Medical University, Beijing, China

**Tao Yu** Department of Functional Neurosurgery, Beijing Institute of Functional Neurosurgery, Xuanwu Hospital, Capital Medical University, Beijing, China

**Jingwen Zhang** Department of Neurology, Xuanwu Hospital, Capital Medical University, Beijing, China

**Tingting Zhang** Department of Neurology, Xuanwu Hospital, Capital Medical University, Beijing, China

**Xiating Zhang** Department of Neurology, Xuanwu Hospital, Capital Medical University, Beijing, China

**Yuqing Zhang** Department of Functional Neurosurgery, Beijing Institute of Functional Neurosurgery, Xuanwu Hospital, Capital Medical University, Beijing, China

**Hairong Zheng** Shenzhen Institutes of Advanced Technology, Chinese Academy of Sciences, Shenzhen, China

**Zhong Zheng** Neurobiological Laboratory, West China Hospital, Sichuan University, Chengdu, Sichuan, China

Center for Neurological Function Test and Neuromodulation, West China Xiamen Hospital, Sichuan University, Xiamen, China

**Jingshu Zhou** Center for Neurological Function Test and Neuromodulation, West China Xiamen Hospital, Sichuan University, Xiamen, China

**Qihui Zhou** Department of Neurology, Beijing Chaoyang Hospital, Capital Medical University, Beijing, China

**Qilin Zhou** Department of Neurology, Fuwai Hospital, National Center for Cardiovascular Diseases, State Key Laboratory of Cardiovascular Disease, Chinese Academy of Medical Sciences and Peking Union Medical College, Beijing, China

**Hongwei Zhu** Department of Functional Neurosurgery, Beijing Institute of Functional Neurosurgery, Xuanwu Hospital, Capital Medical University, Beijing, China

**Ke Zou** Neurobiological Laboratory, West China Hospital, Sichuan University, Chengdu, Sichuan, China

# Abbreviations

| | |
|---|---|
| 1C-Leu | Leucine |
| AAN | American Academy of Neurology |
| ACC | Anterior cingulate cortex |
| AC-DC | Alternating current to direct current |
| Ach | Acetylcholine |
| AD | Alzheimer's Disease |
| ADAS | AD Assessment Scale |
| ADAS-cog | AD Assessment scale-cognitive subscale |
| ADHD | Attention-deficit hyperactivity disorder |
| ADL | Activities of daily living |
| ADTF | Adaptive directed transfer function |
| AIDS | Acquired immune deficiency syndrome |
| ALIC | Anterior limb of the internal capsule |
| AMG | Amygdala |
| amGPi | Anteromedial globus pallidus interna |
| AMT | Active motor threshold |
| ANOVA | One-way analysis of variance |
| ANS | Autonomic nervous system |
| ANSI | American National Standards Institute |
| AOS | Apraxia of speech |
| AP | Acoustic pressure |
| ASD | Autism spectrum disorder |
| ASEX | Arizona sexual experience scale |
| ASH | Amphid sheath glia |
| atDCS | Anodal transcranial direct current stimulation |
| ATP | Adenosine triphosphate |
| AVH | Auditory-verbal hallucinations |
| AVLT | Auditory verbal learning test |
| AWR | Abdominal withdrawal reflex |
| BAD | Barry-Albright Dystonia |
| BAI | Beck Anxiety Inventory |
| BDAE | Boston Diagnostic Aphasia Examination |
| BDI | Beck Depression Inventory |
| BDI-II | Beck Depression Inventory-II |
| BDNF | Brain-derived neurotrophic factor |
| BFMDRS | Burke-Fahn-Marsden Dystonia Rating Scale |
| BPRS | Brief Psychiatric Rating Scale |
| BRMS | Bech-Rafaelsen Mania Scale |
| BTX-A | Botulinum toxin A |
| cAMP | Cyclic adenosine monophosphate |
| CAPSIT-PD | Core Assessment Program for Surgical Interventional Therapies in Parkinson's Disease |

| | |
|---|---|
| CBF | Cerebral blood flow |
| CBI | Cerebellar brain inhibition |
| CBT | Cognitive behavioral therapy |
| CBT-I | Cognitive behavioral therapy for Insomnia |
| CCEP | Cortico-cortical evoked potential |
| CCO | Cytochrome C oxidase |
| CDR | Clinical Dementia Rating |
| CEN | Central executive network |
| CGI-I | Clinical Global Impressions Improvement |
| CGI-S | Clinical global impression rating scale |
| CGRP | Calcitonin gene-related peptide |
| CHART | Craig handicap assessment and reporting technique |
| cHRNA7 | α7-cholinergic receptor gene |
| CIMT | Constraint-induced movement therapy |
| CIQ | Community integration questionnaire |
| CM | Centromedian nucleus |
| CMA | Cingulate motor area |
| CMAP | Compound muscle action potential |
| CMCT | Central motor conduction time |
| C-MET | 11C-labeled methionine |
| CMP | Chronic musculoskeletal pain |
| CM-pf | Centromedian-parafascicular nuclei complex |
| cMUT | Capacitive micromachined ultrasound transducer |
| CNS | Central nervous system |
| CPAP | Continuous positive airway pressure |
| Cr | Creatine |
| CRPS | Complex regional pain syndrome |
| CRS-R | The JFK Coma Recovery Scale—Revised |
| CSD | Current source density |
| CSP | Cortical silent period |
| C-SSRS | Columbia-Suicide Severity Rating Scale |
| CST | Corticospinal tract |
| CSTC | Cortico-striatal-thalamo-cortical |
| CT | Computerized tomography |
| cTBS | Continuous theta burst stimulation |
| CTD | Chronic Tic Disorder |
| ctDCS | Cathodal transcranial direct current stimulation |
| CVP | Chronic visceral pain |
| CVSS | Chinese vegetative state scale |
| DBS | Deep brain stimulation |
| DBS-ST | Deep brain stimulation of subthalamus |
| DC | Direct current |
| DC | Duty cycle |
| DC-DC | Direct current to direct current |
| dcMEP | Direct cortical motor evoked potential |
| DCR | Direct cortical responses |
| DLB | Dementia with Lewy bodies |
| DLPFC | Dorsal lateral prefrontal cortex |
| DLT | Dichotic listening test |
| DMN | Default Mode Network |
| DN4 | Douleur neuropathique 4 |
| DoC | Disorders of consciousness |
| DOSS | Dysphagia outcome and severity scale |

| | |
|---|---|
| DPM | Descending pain modulation |
| DPN | Diabetic peripheral neuropathy |
| DRG | Dorsal root ganglion |
| DSM | Diagnostic and statistical manual of mental disorders |
| DSM-5 | Diagnostic and Statistical Manual of Mental Disorders (fifth edition) |
| DTF | Direct Transfer Function |
| DTI | Diffusion tensor imaging |
| dTMS | Deep transcranial magnetic stimulation |
| DWI | Diffusion-weighted imaging |
| ECT | Electroconvulsive therapy |
| EDS | Excessive daytime somnolence |
| EEG | Electroencephalogram |
| EMG | Electromyogram |
| EMSE | Epidemiology-based mortality score in SE |
| ENS | Enteric nervous system |
| EPDS | Edinburg postnatal depression scale |
| ERD/ERS | Event-related desynchronization/synchronization |
| ERPs | Event-related potentials |
| ESQ | Esophageal Symptoms Questionnaire |
| ESS | Epworth Sleep Score |
| eTNS | External trigeminal nerve stimulation |
| FBSS | Failed back surgery syndrome |
| FDA | Food and Drug Administration |
| FEDSS | Fiberoptic endoscopic dysphagia severity scale |
| fEPSP | Field excitatory postsynaptic potential |
| FF | Fundamental frequency |
| FIM | Functional independence measure |
| fMRI | Functional magnetic resonance imaging |
| FOG | Freezing of gait |
| FOUR | Full outline of unresponsiveness scale |
| FPGA | Field Programmable Gate Array |
| FPS | Faces pain scale |
| FST | Forced swim test |
| FSTT | Finger sequence tapping tasks |
| FTM-TRS | Fahn-Tolosa-Marin Tremor Rating Scale |
| F-Tyr | F-labeled tyrosine |
| FUS | Focused ultrasound |
| GABA | γ-aminobutyric acid |
| GAD-7 | Generalized anxiety disorder-7 |
| GCS | Glasgow Coma Scale |
| GCS-P | Glasgow-Pittsburgh Coma Scale |
| GDS | Global Deterioration Scale |
| GMS | Grading Muscle Strength |
| GOS-E | Glasgow Outcome Scale-Extended |
| GPDs | Generalized periodic discharges |
| GPe | Globus pallidus externus |
| GPi | Globus pallidus internus |
| GWT | Global-workspace theory |
| HADS | Hospital Anxiety and Depression Scale |
| HAMA | Hamilton Anxiety Scale |
| HAMD | Hamilton Depression Scale |
| Hb | Habenula |
| HD | Huntington's disease |

| | |
|---|---|
| hd-EEG | High-density electroencephalography |
| HD-MST | High-dose magnetic seizure therapy |
| HFA | High-frequency activity |
| HFO | High-frequency oscillations |
| HF-rTMS | High frequency repetitive transcranial magnetic stimulation |
| HIS | Hachinski Ischemic Score |
| HMVTS | Hopkins Motor and Vocal Tic Scale |
| HPA | Hypothalamic-pituitary-adrenal |
| HPC | Hippocampus |
| IADL | Instrumental Activities of Daily Living |
| IAPS | International Affective Picture System |
| IBS | Irritable bowel syndrome |
| ICARS | International Cooperative Ataxia Rating Scale |
| ICD | International Classification of Diseases |
| ICD-11 | International Classification of Diseases (eleventh edition) |
| ICD-3 | Classification of Sleep Disorders |
| ICF | Intracortical facilitation |
| ICS | Innsbruck Coma Scale |
| IFC | Inferior frontal cortex |
| IFCN | International Federation of Clinical Neurophysiology |
| IFG | Inferior frontal gyrus |
| ILAE | International League Against Epilepsy |
| IPG | Implantable pulse generator |
| IPL | Inferior parietal lobe |
| iRBD | Idiopathic Rapid Eye Movement Sleep Behavior Disorder |
| IRLS-RS | International RLS rating scale |
| ISI | Insomnia severity index |
| ISI | Interstimulus interval |
| ISP | Ipsilateral silent period |
| ISPPA | Spatial peak pulse average intensity |
| ISPTA | Spatial peak temporal average intensity |
| iTBS | Intermittent theta burst stimulation |
| ITP | Inferior thalamic peduncle |
| LANSS | Leeds assessment of neuropathic symptoms and signs |
| LBT | Line bisection task |
| LC | Induction capacitor |
| LC | Locus coeruleus |
| LD | Light diode |
| LEA | Left ear advantage |
| LED | Light emitting diode |
| LEPs | Laser-evoked potentials |
| LFP | Local field potential |
| LF-rTMS | Low-frequency repetitive transcranial magnetic stimulation |
| LH | Hypothalamus |
| LH | Left hemispheric |
| LHb | Lateral habenula |
| LIFU/LIFUS | Low-intensity focused ultrasound |
| LPDs | Lateral periodic discharges |
| LSSS | Liverpool Seizure Severity Scale |
| LTC | Lateral temporal cortex |
| LTD | Long-term depression |
| LTP | Long-term potentiation |
| M1 | Primary motor cortex |

| | |
|---|---|
| MADRS | Montgomery Depression Rating Scale |
| MARS | Medication adherence rating scale |
| MC | Motor cortex |
| MCE | Motor cortex excitability |
| MCID | Minimal clinically important difference |
| MCS | Minimally conscious state |
| MDQ | Mayo Dysphagia Questionnaire |
| MDS-UPDRS | Movement Disorder Society-sponsored revision of the Unified Parkinson's Disease Rating Scale |
| MECT | Modified electric convulsive therapy |
| MEG | Magnetoencephalography |
| MEP | Motor evoked potential |
| MFB | Medial forebrain bundle |
| MMSE | Mini-Mental State Examination |
| MOS | Medical Outcome Study |
| mPFC | Medial prefrontal cortex |
| MPQ | McGill pain questionnaire |
| MPR | Multiplanar Reformation |
| MQ | Memory quotient |
| MRgFUS | MR-guided focused ultrasound |
| MRI | Magnetic Resonance Imaging |
| MSA | Multiple system atrophy |
| MScL | Escherichia coli mechanosensitive channel of large conductance |
| MSI | Magnetic source imaging |
| MST | Magnetic seizure therapy |
| MT | Motor threshold |
| MVAAR | Multi-variable Adaptive Auto Regression |
| NAA | N-acetyl aspartic acid |
| NAc | Nucleus accumbens |
| NADH | Nicotinamide adenine dinucleotide |
| NCP | Nociceptive pain |
| NCSE | Nonconvulsive status epilepticus |
| NDDI-E | Neurological Disorders Depression Inventory for Epilepsy |
| NHS3 | National Hospital Seizure Severity Scale |
| NIBS | Noninvasive brain stimulation |
| NMDA | N-methyl-D-aspartic acid |
| NMES | Neuromuscular electrical stimulation |
| NMQ | Nordic Musculoskeletal Questionnaire |
| NO | Nitric oxide |
| NPI | Neuropsychiatric Inventory |
| NPP | Neuropathic pain |
| NPQ | Nerve Pain Questionnaire |
| NREM | Nonrapid eye movement sleep |
| NRS | Numeric rating scale |
| NSAIDs | Nonsteroidal Anti-Inflammatory Drugs |
| NTS | Nucleus Tractus Solitaries |
| OAB | Overactive bladder |
| OCD | Obsessive-compulsive disorder |
| OFC | Orbitofrontal cortex |
| OLT | Open-label trial |
| OSAS | Obstructive Sleep Apnea Syndrome |
| otDCS | Oscillatory transcranial direct current stimulation |
| PAG | Periaqueductal gray matter |

| | |
|---|---|
| PANSS | Positive and Negative Symptoms Scale |
| PAS | Paired associative stimulation |
| PBM | Photobiomodulation |
| PBN | Parabrachial Nuclei |
| PD | Parkinson's disease |
| PDC | Partial directed coherence |
| pDoC | Prolonged disorders of consciousness |
| PDSS | Postpartum depression screen scale |
| PFC | Prefrontal cortex |
| PGI | Patient Global Impression scale |
| PHQ-9 | Brief Patient Health Questionnaire |
| PL | Phase-locking |
| PLM | Periodic limb movements |
| PLV | Phase locking value |
| PMC | Premotor cortex |
| PMS | Peripheral magnetic stimulation |
| PNS | Peripheral nerve stimulation |
| PPN | Pedunculopontine nucleus |
| ppTMS | Paired-pulse transcranial magnetic stimulation |
| pre-SMA | Pre-supplementary motor area |
| PRF | Pulse repetition frequency |
| PSD | Post stroke depression |
| PSG | Polysomnogram |
| PSMS | Physical-self maintenance scale |
| PSQI | Pittsburgh Sleep Quality Index |
| PTP | Post-tetanic potentiation |
| PTSD | Post-traumatic stress disorder |
| pvGPi | Posteroventral globus pallidus interna |
| QIDS-SR | Quick Inventory of Depressive Symptomatology Self-Rated |
| QoL | Quality of life |
| QOLIE-31 | Quality of Life in Epilepsy Inventory-31 |
| QPS | Quadri-pulse stimulation |
| QST | Quantitative sensory testing |
| RAS | Reticular activating system |
| RBD | Rapid Eye Movement Sleep Behavior Disorder |
| rCBF | Regional cerebral blood flow |
| RCT | Randomized controlled trial |
| RE | Nucleus reuniens of the midline thalamus |
| REA | Right ear advantage |
| REN | Remote electrical neuromodulation |
| RF | Radio frequency |
| RH | Right hemisphere |
| RLC | Resistance inductance capacitance |
| RLS | Restless legs syndrome |
| rLSMS | Repetitive lumbosacral nerve magnetic stimulation |
| RMT | Resting motor threshold |
| RNA | Ribonucleic acid |
| rTMS | Repetitive transcranial magnetic stimulation |
| RVM | Rostral ventromedial medulla |
| SAI | Short interval afferent inhibition |
| SAINT | Stanford accelerated intelligent neuromodulation therapy |
| SARA | Scale for the Assessment and Rating of Ataxia |
| SAS | Self-Anxiety Scale |

| | |
|---|---|
| SAS | Self-Rating Anxiety Scale |
| SCC | Subcallosal cingulate cortex |
| SCD | Significant current density |
| SCL-90 | 90 Items Symptom Checklist (Symptom Check List 90) |
| SCS | Significant current scattering |
| SCS | Spinal cord stimulation |
| SCWT | Stroop Color and Word Test |
| SCZ | Schizophrenia |
| SD | Sonication duration |
| SDMT | Symbol Digit Modalities Test |
| SDS | Self-Rating Depression Scale |
| SE | Status epilepticus |
| SEEG | Stereo electroencephalography |
| SF-MPQ-2 | Short-form McGill pain questionnaire-2 |
| SG | Sensory gating |
| sgACC | Subgenual anterior cingulate cortex |
| SIB | Severe Impairment Battery |
| SICI | Short interval intracortical inhibition |
| slMFB | Supero-lateral branch of the medial forebrain bundle |
| SMA | Supplementary motor area |
| SN | Salience Network |
| SNr | Substantia nigra pars reticulata |
| SNRIs | Selective norepinephrine reuptake inhibitors |
| SNS | Sacral nerve stimulation |
| SOL | Sleep onset latency |
| so-tDCS | Slow oscillatory transcranial direct current stimulation |
| SPES | Single-pulse electrical stimulation |
| spTMS | Single-pulse TMS |
| SQUID | Superconducting quantum interference device |
| SRTT | Serial reaction time task |
| SSRI | Selective serotonin reuptake inhibitors |
| SSS | Stanford Sleepiness Scale |
| STDP | Spike timing dependent plasticity |
| sTMS | Single-pulse transcranial magnetic stimulation |
| STN | Subthalamic nucleus |
| STP | Short-term potentiation |
| tACS | Transcranial alternating current stimulation |
| TAES | Transcutaneous acupoint electrical stimulation |
| taVNS | Transcutaneous Auricular Vagus Nerve Stimulation |
| TBI | Traumatic Brain Injury |
| TBS | Theta burst stimulation |
| tCS | Transcranial current stimulation |
| tcVNS | Transcutaneous Cervical Vagus Nerve Stimulation |
| tDCS | Transcranial direct current stimulation |
| tDCS-TMS | Transcranial direct current stimulation-transcranial magnetic stimulation |
| TENS | Transcutaneous electrical nerve stimulation |
| TEP | TMS-evoked potential |
| tES | Transcranial electrical stimulation |
| tESMEP | Transcranial electrical stimulation motor evoked potential |
| tFUS | Transcranial focused ultrasound stimulation |
| TIA | Transient Ischemic Attack |
| TIS | Tic Disorder Score |
| TMS | Transcranial magnetic stimulation |

| | |
|---|---|
| TMS-EEG | Transcranial magnetic stimulation-electroencephalography |
| TMS-fMR | Transcranial magnetic stimulation-functional magnetic resonance |
| TMS-MEP | Transcranial magnetic stimulation-motor evoked potential |
| tNIRS | Transcranial near-infrared stimulation |
| TPC | Temporoparietal cortex |
| tPCS | Transcranial pulsed current stimulation |
| TRD | Treatment-resistant depression |
| tRNS | Transcranial random noise stimulation |
| TRPA1 | Transient receptor potential ankyrin 1 |
| TRS | Treatment-resistant schizophrenia |
| TS | Tourette Syndrome |
| tsDCS | Transcutaneous spinal direct current stimulation |
| TST | Total sleep time |
| TTD | Provisional Tic Disorder |
| t-VNS | Transcutaneous Vagus Nerve Stimulation |
| TWSTR | Toronto Western Spasmodic Torticollis Rating Scale |
| UDRS | Unified Dystonia Rating Scale |
| UHDRS | Unified Huntington's Disease Rating Scale |
| UPDRS | Unified Parkinson's Disease Rating Scale |
| UWS | Unresponsive wakefulness syndrome |
| VAS | Visual analogue scale |
| VC/VS | Ventral capsule and ventral striatum |
| VDCCs | Voltage-dependent calcium channels |
| VDS | Verbal descriptor scale |
| VEPs | Viscoelastic properties |
| VFSS | Video-fluoroscopic swallowing study |
| VIN | Ventral intermediate nucleus |
| VMPFC | Ventromedial prefrontal cortex |
| VNS | Vagus nerve stimulation |
| VP | Ventral pallidum |
| VRS | Verbal rating scale |
| VS | Vegetative state |
| VSEP | Vagus somatosensory evoked potentials |
| vSub | Ventral subiculum of the hippocampus |
| VTA | Ventral tegmental area |
| WAIS | Wechsler Adult Intelligence Scale |
| WAIS-RC | Wechsler Adult Intelligence Scale-Revised by China |
| WHIM | Wessex head injury matrix |
| WST | Water Swallowing Test |
| Y-BOCS | Yale-Brown obsessive-compulsive scale |
| YGTSS | Yale Global Tic Severity Scale |
| α7-AchR | α7-nicotinic receptor |

# Functional Neuroanatomy Basics

1

Bo Cui, Yue Hou, Yu Jia, and Qilin Zhou

## 1 Brain Structure

### 1.1 Overall Structure of Hemispheres

Cerebral fissures, as anatomical boundaries, separate left and right hemispheres. Each hemisphere is separated by the lateral sulcus and central sulcus into three lobes: the temporal lobe inferior to the lateral fissure, the frontal lobe anterior to the central sulcus, and the parietal lobe deep to the lateral fissure. The last part of the lateral side is the occipital lobe, and the connecting line from the upper boundary of the sulcus parietooccipitalis in the medial surface of the hemisphere to the lower notch (preoccipital notch) constitutes the lateral boundary among occipital, parietal, and temporal lobes (Fig. 1.1).

There is a parallel sulcus precentralis in front of the frontal sulcus centralis, and the gyrus precentralis is located between the sulcus centralis and the sulcus precentralis. The front of the frontal sulcus precentralis is subdivided into gyrus frontalis superior, gyrus frontalis medius, and gyrus frontalis inferior by two transverse sulci running back and forth.

The temporal lobe is subdivided into gyrus temporalis superior, gyrus temporalis medius, and gyrus temporalis inferior by two transverse sulci. Below the lateral fissure, the transverse gyrus runs on the anterolateral side above the gyrus temporalis superior. It is the anterior boundary of the primary auditory area and the planum temporal or superior temporal plane, and in addition, the anatomy of the individual may determine or be determined by function.

B. Cui (✉) · Y. Hou · Y. Jia
Department of Neurology, Xuanwu Hospital, Capital Medical University, Beijing, China

Q. Zhou
Department of Neurology, Fuwai Hospital, National Center for Cardiovascular Diseases, State Key Laboratory of Cardiovascular Disease, Chinese Academy of Medical Sciences and Peking Union Medical College, Beijing, China

Posterior to the gyrus sulcus postcentralis of the parietal lobe, separated by the sulcus postcentralis from the rest of the convex surface of the parietal lobe. There is the primary somatosensory cortex in the gyrus postcentralis. The lowest bound of the gyrus postcentralis adjacent to the lateral fissure contains the secondary somatosensory cortex. The transverse sulcus divides the rest of the parietal lobe into lobulus parietalis superior and lobulus parietalis inferior. The lobulus parietalis inferior, anterior part is the gyrus supramarginalis encircles around posterior boundary of lateral fissure, and posterior part is the gyrus angularis encircles posterior boundary of superior temporal sulcus.

Sensory cortex: The first somatosensory area is located in the posterior part of the gyrus postcentralis and the paracentral lobule. It receives the contralateral hemisomatic pain, temperature, touch, and pressure from the nucleus ventralis posterolateralis thalami, as well as the sense of position and motion perception. The characteristics of projection of various parts of the body in this area are: (1) Upside down, but the head is straight; (2) Left-right crossing; (3) The projection of each part of the body in this area depends on the sensitivity of this part' s feelings. For example, the receptors of fingers and lips are the densest, with the largest projection range in the sensory area (Stokard et al. 1977).

Motor cortex: The first somatomotor cortex is located in the gyrus precentralis and the anterior paracentral lobule, and this center has a certain local corresponding positioning relationship for the management of skeletal muscle movement, which is characterized by (1) Upside down, but the head is straight; (2) Left-right crossing; (3) The size of projection area of each part of the body depends on the importance and complexity of function. This region simultaneously receives fibers from the nucleus anterior ventralis thalami, the nucleus ventralis lateralis thalami, and the nucleus ventralis posterolateralis thalami, and emits fibers to form pyramidal tract, reaching the brainstem motor nuclei and anterior horn of spinal cord.



Y. Wang (ed.), *Therapeutics of Neural Stimulation for Neurological Disorders*, https://doi.org/10.1007/978-981-99-4538-2_1

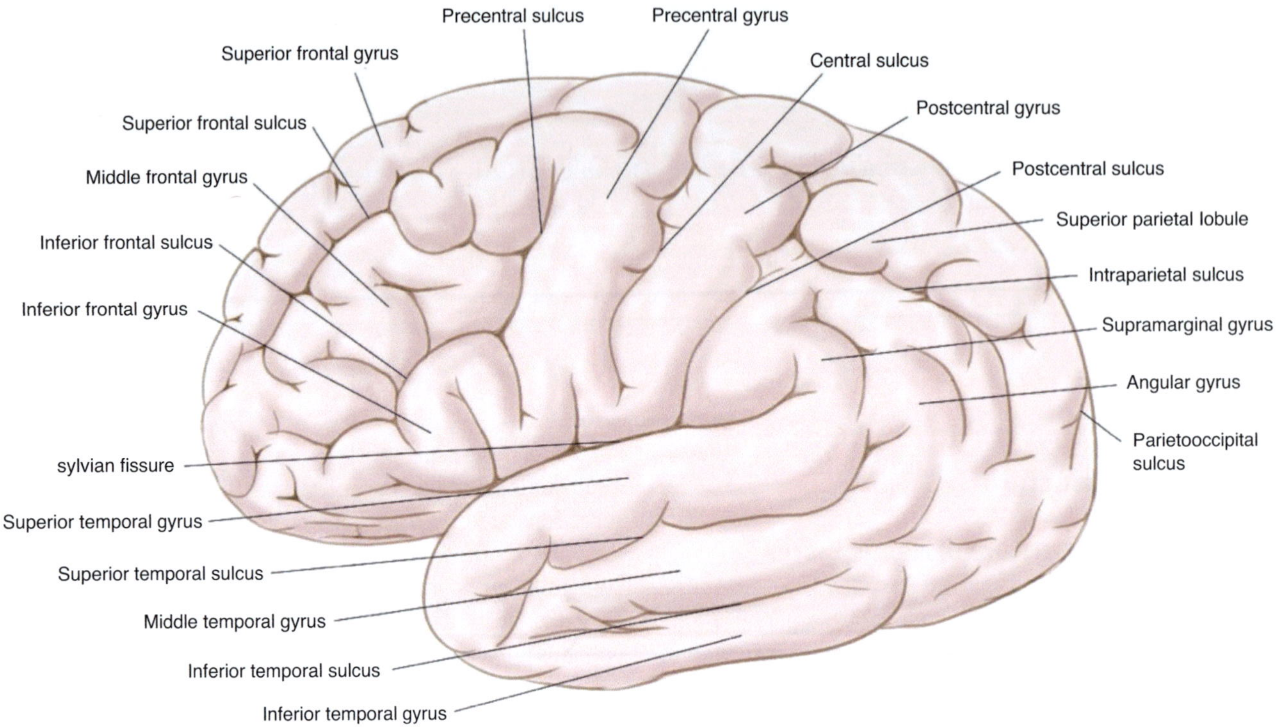


**Fig. 1.1** Lateral surface of hemisphere

Auditory cortex: It is located in the transverse temporal gyrus; it receives the fibers from the medial geniculate body. The auditory center on each side receives the impulse from both ears, so damaging the auditory center on one side will not cause total deafness.

Visual cortex: It is located in the occipital cortex above and below the sulcus calcarinus, that is, the superior cuneus and the inferior gyrus lingualis, receiving fibers from the lateral geniculate body.

Speech center: There are corresponding speech centers in the dominant hemisphere of human cerebral cortex (left hemisphere of right handedness and some left handedness): (1) Motor language center (Broca area), located at the posterior part of gyrus frontalis inferior; (2) Writing center is located at the posterior part of the gyrus frontalis medius; (3) Auditory language center is located in the posterior part of gyrus temporalis superior; (4) Visual language center is located in the gyrus angularis of lobulus parietalis inferior.

The medial hemisphere (Fig. 1.2) and the median plane surround the corpus callosum and merge successively with the lower surface of the hemisphere at the lower rear part. Among the main sulci on the medial surface, three run radially, while one runs parallel to the corpus callosum. The latter is called the sulcus cinguli, which separates the gyrus cinguli from the first gyrus frontalis, which is centripetal and the medial surface of the lobulus paracentralis. The medial surface of the paracentral gyrus of the frontal lobe and parietal lobe is separated by one of three radial sulci, that is, the marginal sulcus from the cingulate sulcus and the medial surface of the parietal lobe, the other two radial sulci are slightly backward. The sulcus parietooccipitalis separates the precuneus of the lobulus parietalis from the cuneus of medial occipital lobe and is bounded downward by the sulcus calcarinus. The two sulci meet at the anterior part and add to the posterior bound of the sulcus cinguli because it surrounds the splenium of corpus callosum, it demarcates the isthmus of the cingulate cortex at the dorsal side. As the gyrus cinguli travels anteriorly and inferiorly around the splenium, it merges with the subiculum hippocampi on the innermost part of the temporal lobe. Hidden in the recess between the temporal horn of the lateral ventricle and the lateral surface of the mesencephalon, the hippocampal gyrus travels anteriorly and posteriorly to the subiculum hippocampi, and they are separated by the hippocampal sulcus. They converge forward to a small nodule containing amygdaloid complex.

The inferior surface of the hemisphere (Fig. 1.3) consists of the orbital frontal lobe and the inferior medial surface of the occipital lobe and temporal lobe. Some irregular orbital gyri and the medial gyrus rectus are located on the medial side of the olfactory bulb and the olfactory tract, which constitute the orbitofrontal surface. The border between the temporal lobe and occipital lobe is unclear. The anterolateral fusiform gyrus or temporal occipital gyrus, and the postero-

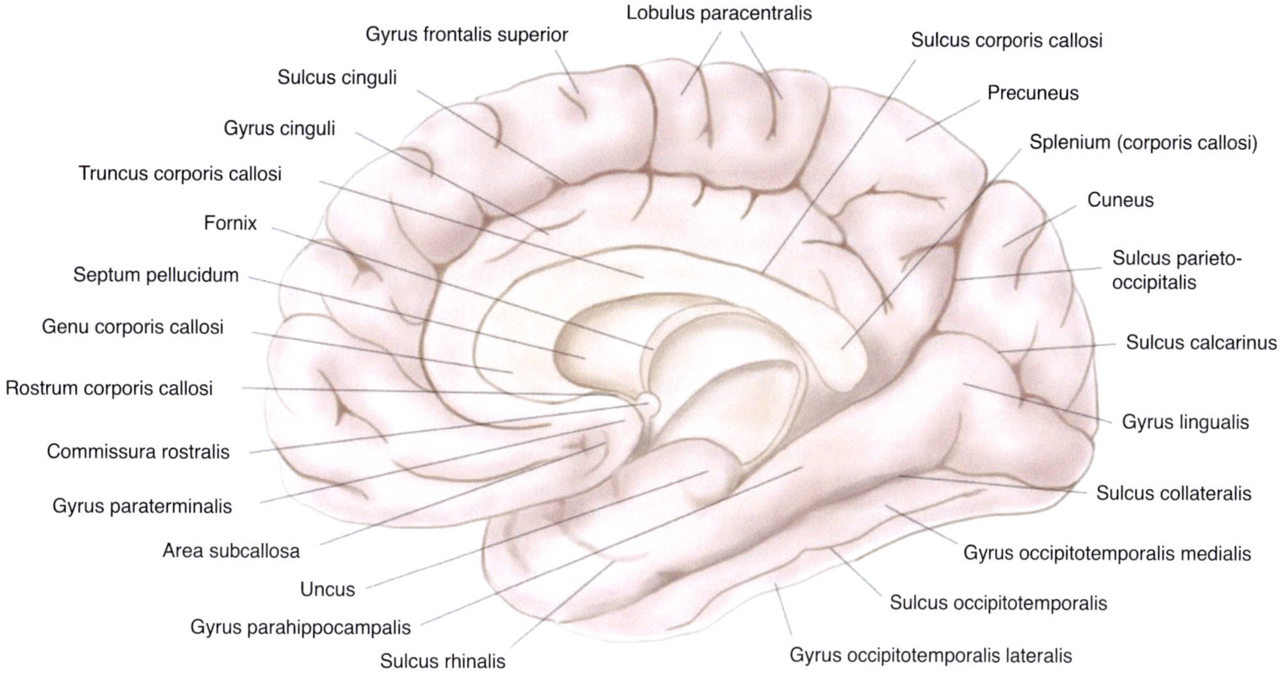


**Fig. 1.2** Medial surface of hemisphere

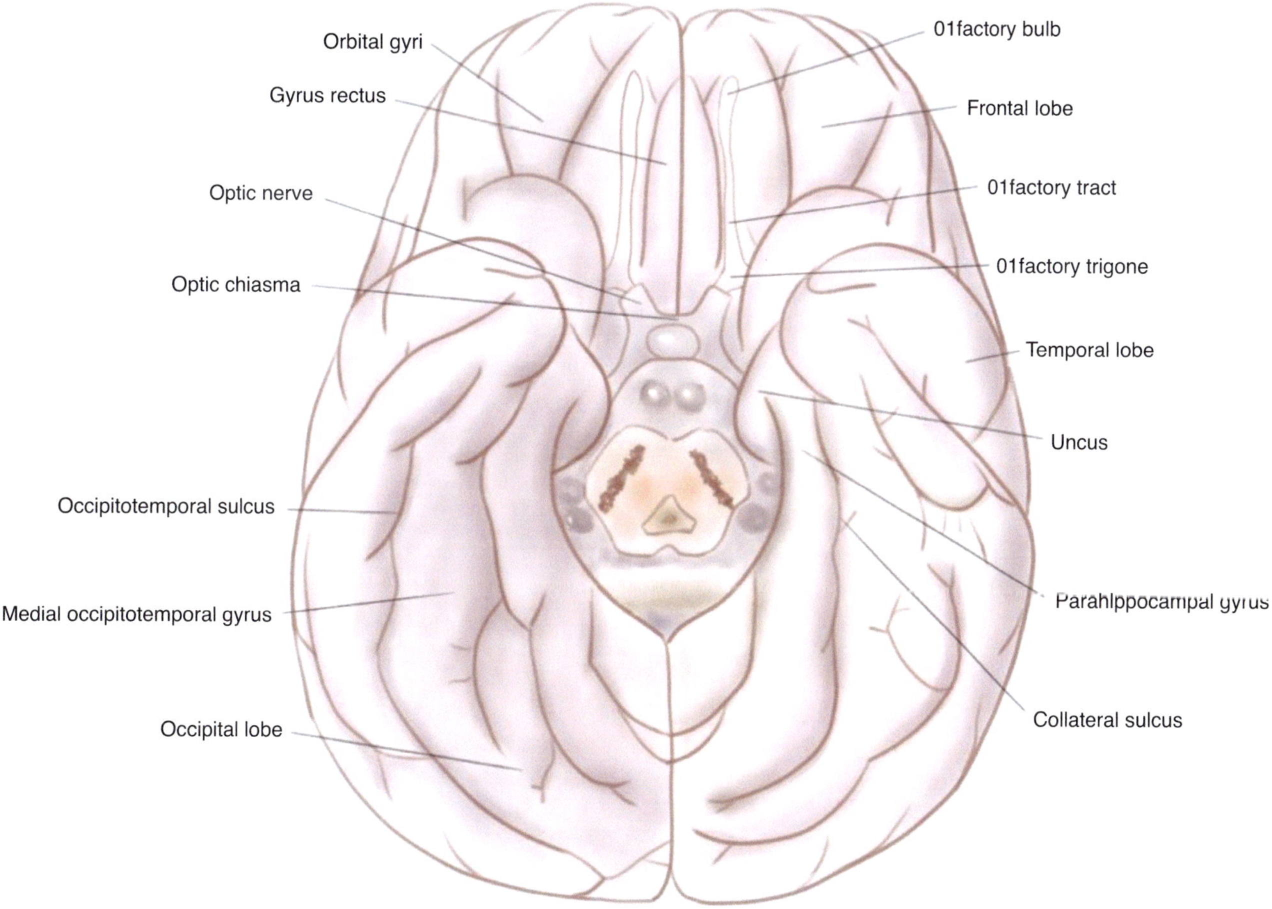


**Fig. 1.3** Inferior view of brain

medial gyrus lingualis can be distinguished by the scanline method between the collateral sulcus (lateral to the subiculum hippocampi) and the gyrus temporalis inferior.

## 1.2 Microstructure of Neocortex

The neocortex of primates can be divided into six layers from outside to inside (Fig. 1.4), which are called: the first molecular layer; the second external granular layer consisting of small round or stellate neurons; the third external pyramidal layer, containing medium-sized pyramidal neurons, with their larger apical dendrites pointing to the surface; the fourth granular layer, except the small round active ingredients, containing thick clumps of fibers in the horizontal direction; the fifth internal pyramidal layer or lamina ganglionaris, which is composed of larger pyramidal neurons; and the sixth polymorphic layer, which is composed of spindle neurons. There is also a certain heterogeneity in function. The granular layer is larger than the pyramidal layer in the area receiving a large number of sensory projections. However, the pyramidal layer is larger than the granular layer in the area emitting larger motor projections to the brainstem and spinal cord.

The hemisphere processes information from inside and outside of the individual. Most of the afferent information outside the body reaches the primary cortical areas through thalamus. Information on homeostasis originates from brainstem and hypothalamus, which mainly reaches the pericallosum, the medial temporal lobe, insular lobe and orbital cortex through the medial thalamus. In order to function, both systems need to be "activated" by the reticular formation of the brainstem (Kahle and Frotscher 2015a, b).

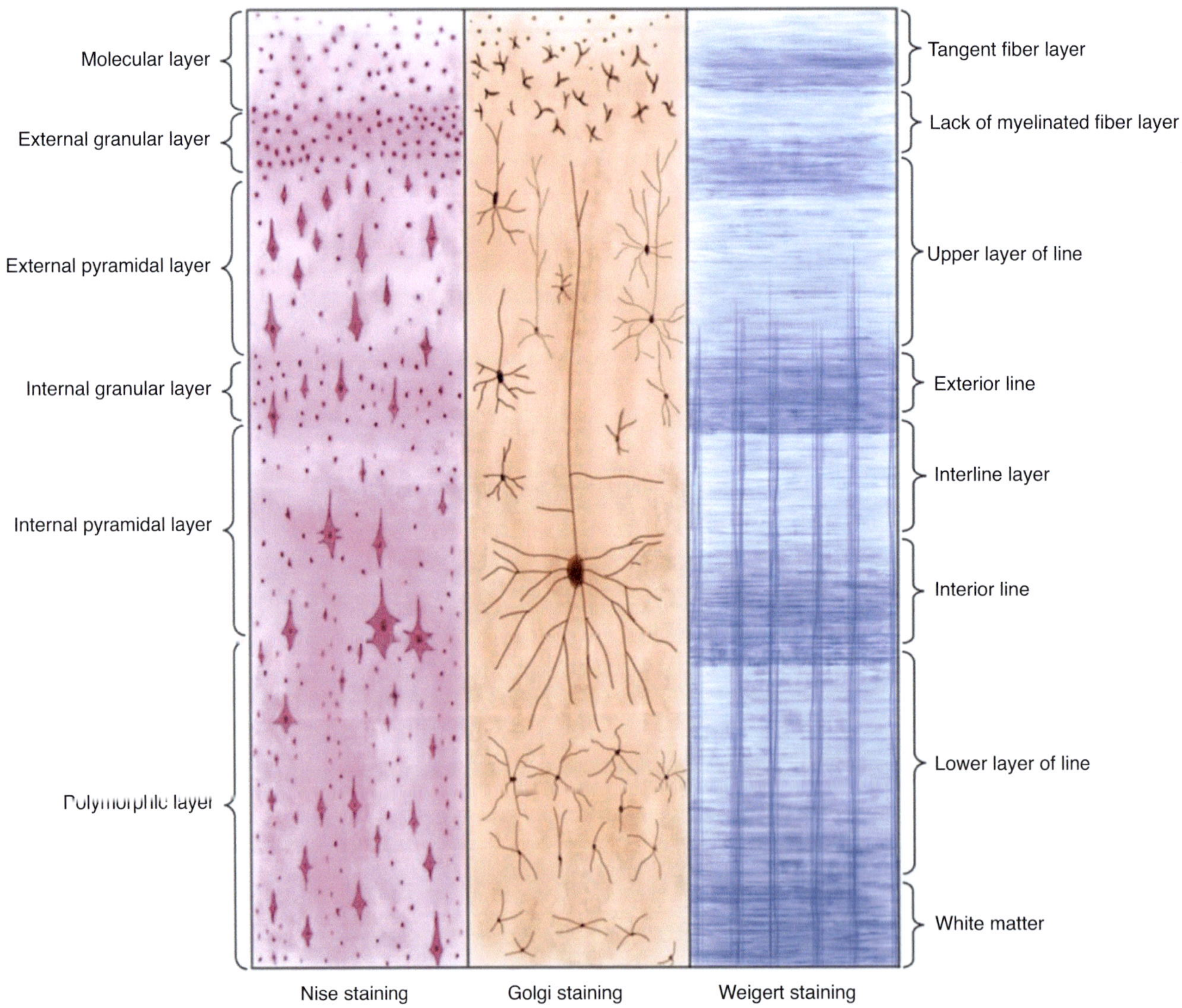


**Fig. 1.4** Cell stratification and fiber stratification in human neocortex

## 1.3 Function of Hemispheres

The posterior part to central sulcus of each hemisphere is mainly involved in the processing of sensory information about the external world. The anterior part of each hemisphere is involved in the execution of motor behaviors by individuals. These two aspects, especially the latter, require the integration of sensory information in different forms such as visual and somatosensory. The lesions of the "primary" sensory area cause the loss of specific sensory patterns. The cortex surrounding the primary sensory area processes the information of a specific pattern and integrates it with the information from other sensory organs as well as the physiological environment information of the individual.

In short, the cortex (secondary sensory area) adjacent to the primary sensory area processes a single form of sensory information, often maintaining an integrally located organizational structure, while the third cortical sensory cortex, located between different secondary sensory areas, integrates multimodal sensory information. For example, the somatosensory information reaches the gyrus postcentralis via the medulla oblongata-tritocerebrum pathway in a somatotopical localization manner, and it is projected to the lobulus parietalis superior (upper and lower limbs) and the lobulus parietalis inferior (head) according to the somatotopical localization. The somatotopical localization information is integrated with the auditory information in the posterior inferior part of the lobulus parietalis inferior (gyrus angularis). Visual and sensory loss occurs with the posterior medial part of the parietal lobe (precuneus) lesion. Multimodal-related area lesions result in lack of multimodal integration in motor execution. For example, bilateral lesions in the posterior part of the lobulus parietalis superior cause hand dyskinesia under visual guidance.

The activation of primary, secondary, and multimodal sensory areas is highly stimulus-specific and task-specific. For example, the posterior part of the right gyrus temporalis superior is activated by the auditory changes of spoken and non-spoken sounds, while the left gyrus supramarginalis of the multimodal association area is more specifically engaged in detecting the changes of learning units for phonology. It is a multimodal task. Attention to complex or nuanced sensory tasks, such as the detection of specific target syllables, requires the enhancement of sensory efficacy and a greater degree of cortical activation. Such activation is probably realized through an attention network, which is described in more detail under the topic of disturbance in attention below. Areas around the intraparietal sulcus and in the temporal-occipital region "regulates" the activation of the primary and secondary sensory area. The prefrontal, insular and cingulate cortex also regulates activity in the sensory area (Gao 2009; Jia and Chen 2018).

The anterior part to the central gyrus of the hemisphere contains procedures related to the planning, initiation, and execution of movement. The medial surface of frontal cortex (gyrus cinguli and supplementary motor cortex) is closely connected with reticular activating system and limbic lobe. It seems to regulate and drive the shift to a meaningful direction, which is called cortical attention. Although the patient seems to be alert, patients with bilateral large lesions in this area are still unable to move and mute "akinetic mutism." Information from the limbic system about past events and their perception of the well-being and health of the individual reaches the anterior part of the frontal lobe, where it forms the best course of action with the multimodal association area from thalamus and hemisphere. The frontal cortex at the rostral to gyrus precentralis is involved in the complex motor procedure, which is triggered by the "instruction" of the medial area of the frontal lobe and is executed through the subcortical nucleus (basal ganglia and brainstem nucleus) and the primary motor cortex. The primary motor cortex plays an important role in performing fine distal movements, while axial movements, such as walking, are mediated by subcortical structures to a large extent. The sensory nuclei in the cerebellum and brainstem, including the vestibular nuclear complex, provide the necessary feedback information for the motor system. The different function of different cortical area lies the theory basis of scalp electrode location (Fig. 1.5).

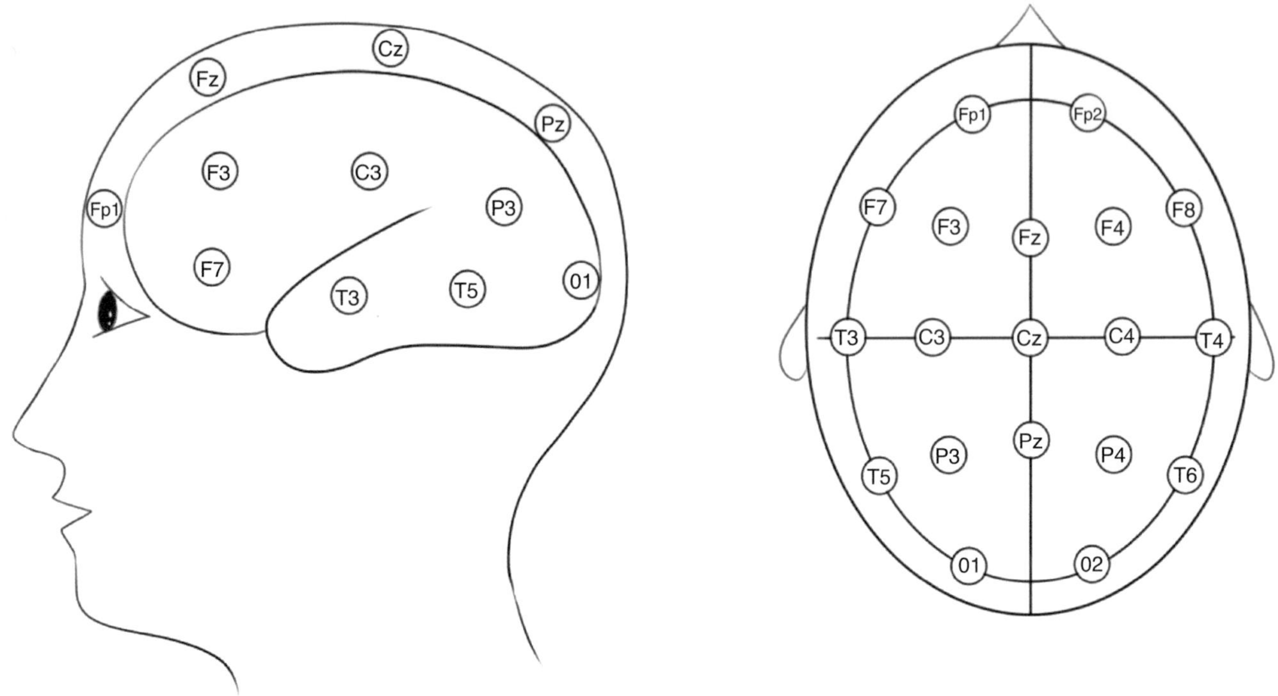


**Fig. 1.5** Location of scalp electrodes

## 2 Visual Pathway and Pupillary Light Reflex Pathway

### 2.1 Visual Pathway

The visual pathway is composed of three-order neurons, which starts from the retinal ganglion cell axon, passes through the optic nerve, optic chiasma, optic tract, lateral geniculate body, and visual radiation, and finally reaches the occipital visual cortex (Fig. 1.6) (Mattle and Mumenthaler 2017). The first-order neurons are bipolar cells located in the medial layers of the retina, the second-order neurons are ganglion cells located in the innermost layer of the retina, and the cell bodies of the third-order neurons are located in the lateral geniculate body of the posterior thalamus. Optic radiation is composed of fibers emitted from the lateral geniculate nucleus, which is projected to the visual areas (striated cortex) on both sides of the sulcus calcarinus of the telencephalon via the posterior limb of the internal capsule, resulting in vision.

Among them, the retina is composed of three layers of cells, with the outermost layers being cone and rod cells, the middle layer being bipolar cells and the inner layer being ganglion cells. The axons of ganglion cells converge to the optic papilla to form the optic nerve. The optic nerve from the optic papilla to the optic chiasma, has a length of about 5 cm, of which 3.5 cm is located in the orbit and 1.5 cm in the optic canal and cranial cavity (Brazis et al. 2012). The bilateral optic nerves merge backward at the basal cistern above the sella turcica to form optic chiasma, and the optic chiasma extends posterolaterally to form bilateral optic tracts. The optic nerve fibers from the nasal retina cross each other to the contralateral optic tract, while the optic nerve fibers from the temporal retina do not cross and enter the ipsilateral optic tract backward via the outer side of the optic chiasma. Therefore, the left optic tract contains fibers from the left half of the retina of both eyes, and the right optic tract contains fibers from the right half of the retina of both eyes. The bilateral visual tracts bypass the lower part of the cerebral peduncle backwards from the optic chiasma. When reaching the posterolateral thalamus, each visual tract is further divided into two branches with different sizes, and the larger lateral branch is the visual fiber, which enters the lateral geniculate body; the smaller medial branch is the centripetal fiber of pupillary reflex, reaching the pretectal area and the superior colliculus of the corpora quadrigemina. The fibers from the superior colliculus constitute the tectospinal tract which travels downward to the spinal cord to complete the visual reflex. The pretectal area is a part of the pupillary light reflex pathway. The lateral geniculate body is located at the inferolateral part of the thalamus, outside of the cerebral peduncle. The fibers of the optic tract reach the cells here, where the peripheral neurons of the visual pathway terminate, and then the fibers of the optic radiation are emitted by the cells of the lateral geniculate body. Optic radiation fibers arise from the lateral geniculate body, pass backwards

**Fig. 1.6** Visual pathway

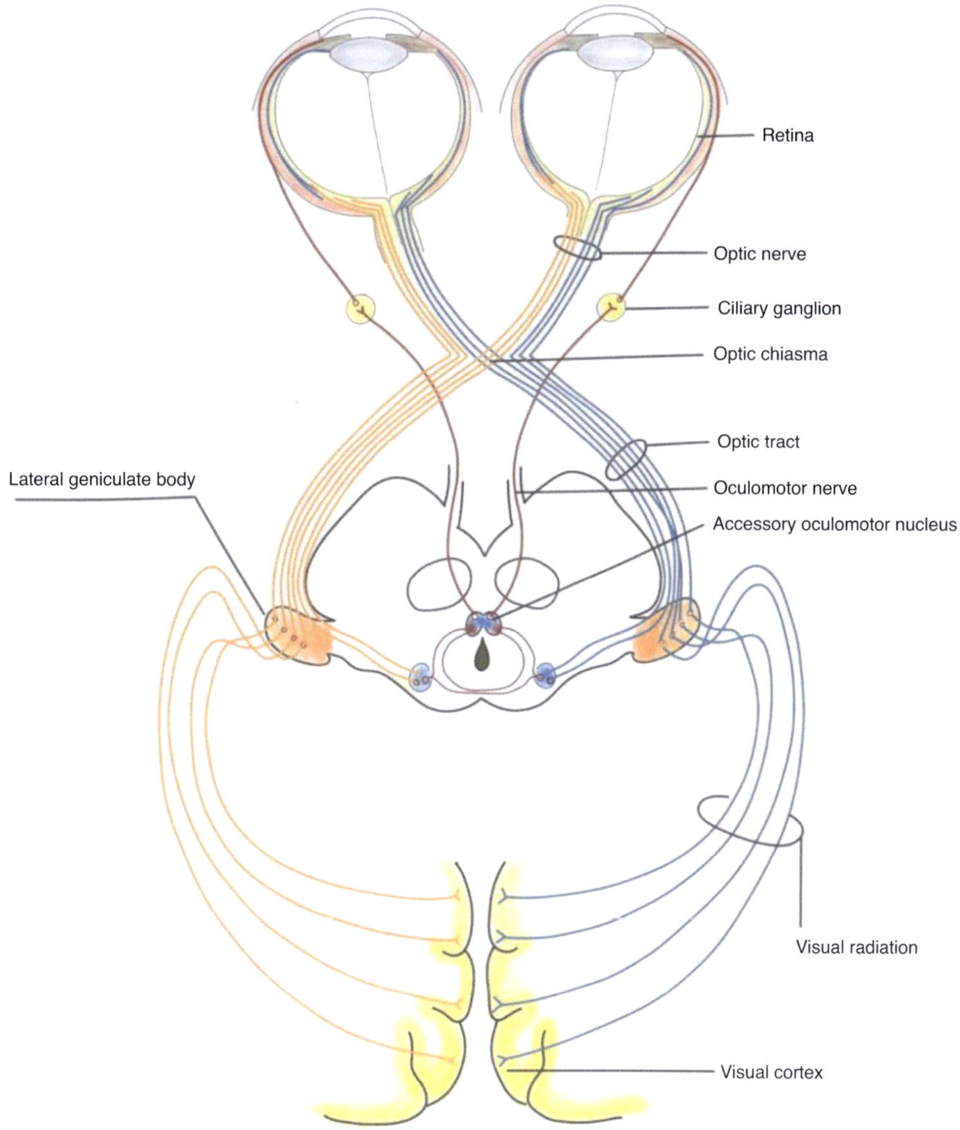


through the posterior limbs of the internal capsule and behind the sensory fibers, and go to the occipital lobe. The fibers from the upper retina pass through the deep part of parietal lobe and occipital lobe, bypass the posterior horn of lateral ventricle from the lateral side, and terminate at the cuneus above the sulcus calcarinus; the fibers from the lower retina pass deep in the temporal lobe, bypass outwardly the inferior horn of the lateral ventricle, and terminate at the lingual gyrus below the sulcus calcarinus. The striate areas which located above and below the sulcus calcarinus of occipital lobe on the medial surface of both cerebral hemispheres are the visual cortex centers, and the striate areas on each side connect with the retina of the ipsilateral half of both eyes.

Visual pathway runs through almost the whole brain, and lesions in various parts of the brain can cause damage to different parts of the visual pathway, showing certain symptoms and signs. According to these manifestations, it can help to determine the location of brain lesions. The main lesions are also divided from the optic papilla to the optic nerve, optic chiasma, optic tract, lateral geniculate body, optic radiation, and finally to the occipital visual cortex. For the examination of optic nerve function, pupillary light reflex, visual acuity, visual field, and fundus condition are mainly used for evaluation.

Pupillary light reflex is described below. First, unilateral visual field refers to the entire spatial range that a monocular eyeball can see when fixed to look at the front horizontally, which involves the functions of retina, visual pathway, and visual cortex. The visual field can be divided into temporal and nasal half visual fields. Due to the refraction of light by the eyeball refractive device, the image in the nasal half visual field is projected onto the temporal half retina, that in the temporal half visual field is projected onto the nasal half retina, that in the upper half visual field is projected onto the lower half retina, and that in the lower half visual field is projected onto the upper half retina (Brazis et al. 2012).

Different visual field defects can be caused when different parts of the visual pathway are damaged. (1) The unilateral optic nerve injury can cause total blindness of the visual field of the affected eye; (2) The injury in the media part of the optic chiasm (cross fiber) (pituitary tumor compression) can cause temporal hemianopsia in the visual field of both eyes; (3) The injury in the lateral part of unilateral optic chiasma (non-crossed fiber) (compression of internal carotid artery aneurysm) can cause nasal hemianopsia in the affected side's visual field; (4) The injury in the unilateral optic tract and its later parts (visual radiation and visual cortex) can cause homonymous hemianopsia of contralateral visual field of both eyes (e.g., the injury in the right side can cause the nasal hemianopsia of right visual field and the temporal hemianopsia of left visual field) (Brazis et al. 2012).

## 2.2 Pupillary Light Reflex Pathway

Pupillary light reflex refers to the miosis of both eyes caused by illumination on one pupil. Among them, the miosis on the illuminated side is called direct light reflex, and that on the opposite side is called indirect light reflex. The light reflex pathway is as follows: the visual impulse produced by retina is conducted via optic nerve, optic chiasma and optic tract, and a few fibers in optic tract pass through brachium of superior colliculus to pretectal area, and form synapses with cells in pretectal area. Fibers from pretectal area (light reflex center) connect with bilateral accessory oculomotor nucleus, which emits parasympathetic preganglionic fibers to reach ciliary ganglion via oculomotor nerve. After changing neurons, postganglionic fibers are emitted and distributed to the sphincter pupillae, leading to miosis (Brazis et al. 2012). (Retina → optic nerve → optic chiasma → bilateral optic tract → brachium of superior colliculus → pretectal area → bilateral accessory oculomotor nucleus → oculomotor nerve → ciliary ganglion → postganglionic fiber → sphincter pupillae contraction → bilateral miosis.)

Pupillary light reflex is clinically important, and its disappearance may indicate a critical illness. However, the injury to optic nerve or oculomotor nerve can also cause the change of pupillary light reflex. For example, when the unilateral optic nerve is damaged, due to the interruption of information transmission, the bilateral pupillary light reflex disappears when the pupil of the affected side is illuminated; however, when the pupil of the unaffected side is illuminated, the bilateral pupillary light reflexes exist. Clinical manifestations show that the direct light reflex disappears and the indirect light reflex exists on the affected side. For example, when the unilateral oculomotor nerve is damaged, due to the interruption of information transmission, no matter which side of the pupil is illuminated, the light reflex of the affected side disappears (the direct and indirect light reflexes of the affected side disappear), but the direct and indirect light reflexes of the unaffected side exist.

# 3 Auditory Pathway

The auditory pathway (Fig. 1.7) can be conceptualized as a four-relay neuronal network, as follows: (1) The auditory nerve (cochlea) extends from the spiral organ (also known as the Corti's organ) to the cochlear nucleus; (2) The fibers from the cochlear nucleus cross to the contralateral inferior colliculus; (3) The fibers from the inferior colliculus extend to the medial geniculate body; and (4) The fibers from the medial geniculate body are projected to the auditory cortex of the gyrus temporalis superior (Brazis et al. 2012, Kahle et al. 2015).

## 3.1 First-Order Neurons

Auditory receptors are hair cells of neuroepithelium of spiral organ. This is the structure of the cochlea. The hair cells located at the top of the cochlea are stimulated by low-frequency tones, while those located at the base are stimulated by high-frequency tones. The cell body of first-order neuron of auditory pathway is on the spiral ganglion of cochlear nerve, which is located in Rosenthal's canal (cochlear spiral canal) at the base of bony spiral plate. The afferent components of these cells contact with cochlear hair cells, and most of them converge on the internal hair cells, while a small part separates and communicates with the external hair cells. The cochlear nerve is formed by the central processes of the bipolar neurons of the cochlear ganglion (the first neurons of the auditory pathway); it exits from the petrous bone at the internal acoustic meatus, travels a short distance in the subarachnoid space, and enters the brainstem in the cerebellopontine angle (Rohkamm 2014).

## 3.2 Second-Order Neurons

The afferent fibers of cochlear nerve enter the inferior brainstem at the junction between medulla oblongata and pons and are divided into dorsal cochlear nuclear, anteroventral nuclear, and posteroventral nuclear, which dominate the cochlear complex. The dorsal side of these nuclei receives nerve fibers dominated by "high frequency" (basal) hair cells, while the ventral side receives fibers from "low frequency" (apical) hair cells. The dorsal and ventral cochlear nuclei contain second-order neurons and emit some to project to the contralateral brainstem, whose ascending fibers

**Fig. 1.7** Auditory pathway

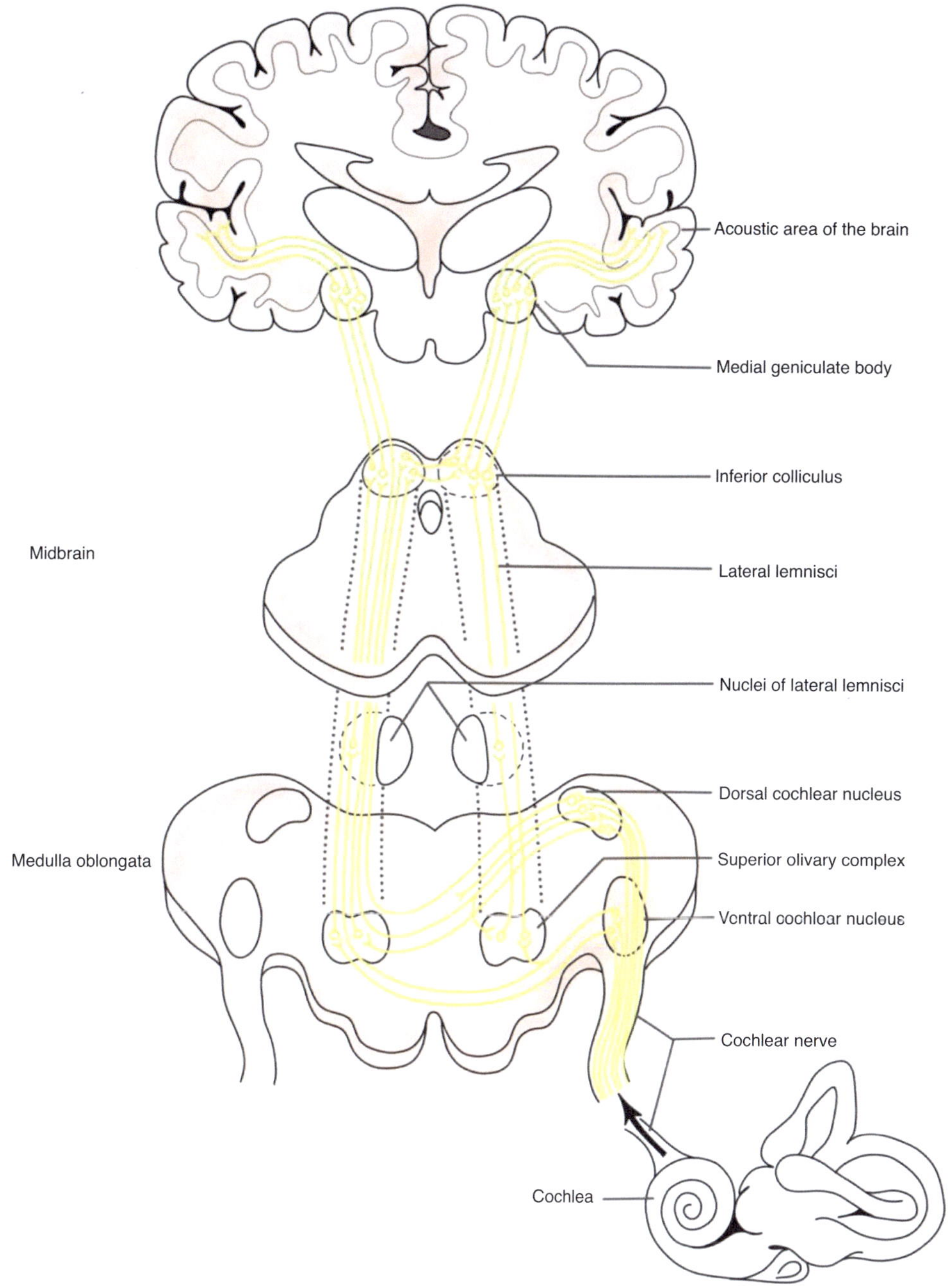


become the lateral lemniscus of fiber bundles projected to the central nucleus of the inferior colliculus. These projections include dorsal acoustic striae (from dorsal cochlear nucleus), intermediate acoustic striae (from the back of ventral cochlear nucleus), and ventral acoustic striae (from ventral cochlear nucleus) that is part of rhombus. Cross fibers of rhombus and lateral lemniscus ascending to the opposite side in the contralateral superior olivary complex. The fibers of the lateral lemniscus ascend with some fibers that terminate in the lateral lemniscus nuclei along this path and terminate in the inferior colliculus.

## 3.3 Third-Order Neurons

The inferior colliculus, located at the caudal end of the superior colliculus of the tectum mesencephali, contains third-order neurons and serves as an important relay nucleus for receiving ascending and descending afferents in the auditory pathway. Generally, all ascending auditory pathways terminate in the inferior colliculus. The fibers from the lateral lemniscus terminate in the important central nucleus of the inferior colliculus. Projections from the inferior colliculus terminate in the medial geniculate body, low-frequency fibers terminate in the parietal-lateral region, and high-frequency fibers terminate in the medial part of this nucleus.

### 3.4 Fourth-Order Neurons

The medial geniculate body is the auditory relay nucleus of thalamus and mainly emits temporal lobe fiber of geniculate body or auditory radiation. The auditory radiation passes through the dense bundle from the outside, which partially penetrates the ventral and lateral parts of the posterior half of the putamen, and partially travels in the white matter below it. Most of the fibers terminate in the primary auditory cortex layer IV (AI, Brodmann area 41) located in the transverse temporal gyrus of Heschl, but some of them terminate in the auditory cortex association area (AII, Brodmann area 42) (Brazis et al. 2012). The termination of primary auditory cortex adapts it to the mode of tension theory, high-frequency tones terminate at the inside and low-frequency tones terminate at the outside. Brodmann area 41 connects with ventral branch interactively, while Brodmann area 42 connects with dorsal branch of medial geniculate body. Through a large fiber bundle of the corpus callosum, the auditory cortex of each side also connects with the interactive area of the other cerebral hemisphere.

There are many commissural connections along the auditory pathway; however, not at the medial geniculate body level. Unilateral brainstem lesion sometimes shows bilateral manifestations. The neurons of the superior olivary complex receive the afferent from both ears. There is connection between two cochlear nuclei, namely, between two dorsal nuclei of the lateral lemniscus via the commissure of Probst, between the inferior colliculus on each side via the commissure of the inferior colliculus, and between the central nucleus of the inferior colliculus and the contralateral medial geniculate body via the brachium of inferior colliculus. Another important consideration is that there are two different projection systems from the inferior colliculus upward. The first one (including the central nucleus of inferior colliculus, medial geniculate body, and primary auditory cortex) is called the core system. The other (including the central peripheral area of the inferior colliculus, the non-laminated portion of the medial geniculate body, and the secondary auditory cortex) is known as the belt projection, which serves as a multimodal system for receiving bilateral auditory and non-auditory information.

There are also some descending auditory pathways parallel to ascending fibers, which are integrated in the feedback control of auditory afferent (Brazis et al. 2012). They include cortical geniculate fiber, cortical inferior colliculus fiber, inferior colliculus geniculate fiber, inferior colliculus efferent and efferent cochlear bundle from superior olivary complex to hair cells of spiral organ.

The blood supply to the auditory nuclei of the cochlea and brainstem comes from the internal auditory (labyrinthine) artery, which is usually a branch of the anterior inferior cerebellar artery. In internal acoustic meatus, internal auditory artery supplies ganglion cells, auditory nerves, dura mater, and arachnoid, and subsequently divides into common cochlear artery and anterior vestibular artery. The superior olivary complex and the lateral lemniscus are supplied by the circumflex branch of the basilar artery, the inferior colliculus is supplied by the branch of the superior cerebellar artery and the corpora quadrigemina artery, and the medial geniculate body receives the blood supply from the thalamogeniculate artery. Branches of the middle cerebral artery supply the primary auditory cortex and cortex association area.

## 4 Somatosensory System

The human brain connects with the internal and external environment through a variety of receptors and related neuronal chains. Somatosensory system refers to all sensations except visual, auditory, taste, olfactory sensations, and vestibular balance. Different somatic sensations depend on different receptors, neural pathways, and brain activities. The somatosensory system includes multiple important neural pathways, which are mainly responsible for the information conduction of pain and temperature sensation, rough tactile sensation, and proprioception. The systemic pathway is generally composed of three-order neurons: the first-order neurons are pseudounipolar or bipolar neurons, with cell bodies located in the sensory ganglion; the peripheral branch distributed on various receptors; the central branch connects to the brain or spinal cord. The second-order neurons are multipolar neurons. The cell body is located in the spinal cord or brainstem; axons project upward to the diencephalon. The third-order neurons are multipolar neurons. The cell body is generally located in the diencephalon. According to the difference in the running location and function of each pathway, this section focuses on the introduction of proprioceptive pathways, pain and temperature sensation, and rough tactile sensation pathways.

### 4.1 Proprioceptive Pathway

Proprioception, also known as deep sensation, refers to the position, motion, and vibration senses of muscles, tendons, and joints. Proprioceptive pathway includes conscious and unconscious proprioceptive pathways.

## 4.2 Conscious Proprioceptive Pathway

Conscious proprioceptive pathway is composed of three-order neurons.

First-order neuron: The cell body is located in the spinal ganglion. The peripheral process distributes in proprioceptors and fine tactile receptors of muscles, tendons, and joints via spinal nerves. The central process enters posterior funiculus of spinal cord via medial dorsal roots. The ascending branch from segments below the fifth thoracic segment forms a fasciculus gracilis at medial posterior funiculus. The ascending branch from segments above the fourth thoracic segment forms a fasciculus cuneatus at lateral posterior funiculus. The two fasciculi ascend to the medulla oblongata and terminate in the gracile nucleus and the cuneate nucleus, respectively.

Second-order neuron: Cell bodies are gracile nucleus and cuneate nucleus. The emitted arcuate fibers forward bypass the ventral central gray matter, and cross left and right at the midline to form decussation of medial lemniscus. After the decussation, the fibers run behind the pyramids on both sides of the medial medulla and fold upward, which is called medial lemniscus. The medial lemniscus is located at anterior margin of pontine tegmentum, at the posterolateral red nucleus in the tegmentum mesencephali, and terminates upward at the ventral posterolateral nucleus.

Third-order neuron: The cell body is ventral posterolateral nucleus. The fibers constitute central thalamic radiations, which are projected to the middle and upper parts of the gyrus postcentralis and the rear of the lobulus paracentralis via the posterior limb of internal capsule, and some of the fibers are projected to the gyrus precentralis.

For the injuries above the decussation of medial lemniscus, the symptoms are manifested as the disappearance of proprioception and fine tactile sensation in the distribution range below the contralateral lesion site of the injury; for the injury below the decussation of medial lemniscus, the symptoms are manifested at the ipsilateral side of injury.

## 4.3 Unconscious Proprioceptive Pathway

Unconscious proprioceptive pathway is composed of two-order neurons.

First-order neuron: The cell body is located in the spinal ganglion. The peripheral process distributes at proprioceptors in muscles, tendons, and joints. The central process enters the spinal cord via posterior root of spinal nerve.

Second-order neuron: The cell bodies are located in the dorsal nucleus of spinal cord C8-L3, the spinal border cells of L2-S3 and the nuclei cuneatus accessorius of medulla oblongata. Fibers from dorsal nucleus form fasciculus cerebellospinalis at the ipsilateral lateral funiculus, which enters paleocerebellum cortex via inferior cerebellar peduncle. Most of fibers emitted from marginal cells cross to contralateral lateral funiculus via anterior white commissure to form anterior spinocerebellar tract, which enters paleocerebellum cortex via superior cerebellar peduncle. The fibers from nuclei cuneatus accessorius of medulla oblongata constitute cuneocerebellar tract and enter paleocerebellum cortex via inferior cerebellar peduncle.

## 4.4 Pain-Temperature Sensation and Rough Tactile Sensation Pathway

Pain-temperature sensation and rough tactile sensation pathway, also known as superficial sensory pathway, is composed of three-order neurons. According to different distributions, it can be divided into the exteroceptive sensory pathway of trunk and limbs, and face.

## 4.5 Exteroceptive Sensory Pathways of Trunk and Limbs

The exteroceptive sensory pathway of trunk and limbs is composed of three-order neurons.

First-order neurons: The cell body is located in spinal ganglion. The peripheral process acts as the receptors of pain, temperature, and tactile in the skin of the trunk and limbs via spinal nerves. The central process enters the spinal cord via the lateral part (fine fibers, conducting pain and temperature sensation) and the medial part (conducting rough tactile and pressure sensation) of dorsal roots. It ascends 1–2 segments in dorsolateral fasciculus and terminates in dorsal horn cells.

Second-order neurons: Cell bodies are mainly located in the posterior horn (layers I, IV, and V) of the spinal cord gray matter. The fibers ascend obliquely 1–2 spinal segments via anterior white commissure, cross to the contralateral lateral funiculus and anterior funiculus, and ascend to form lateral spinothalamic tract (conducting pain and temperature sensation) and anterior spinothalamic tract (also containing a small part of non-crossing fibers, conducting rough tactile and pressure sensation). After entering the brainstem, the fibers combine and ascend, which is called spinal lemniscus. It passes through the dorsolateral inferior olivary nucleus of medulla oblongata, the lateral medial lemniscus of pons and mesencephalon, and terminates upward in the ventral posterolateral nucleus.

Third-order neurons: The cell body is located in the ventral posterolateral nucleus. The fibers constitute central thalamic radiations, which is projected to the middle and upper parts of the gyrus postcentralis and the rear of the lobulus paracentralis via the posterior limb of internal capsule.

Unilateral spinothalamic tract damaging will lead to the disorder of pain and temperature sensation in the contralateral trunk and limbs.

### 4.6 Exteroceptive Sensory Pathways of Head and Face

The exteroceptive sensory pathway of head and face is composed of three-order neurons.

First-order neurons: The cell body is trigeminal ganglion. The peripheral processes distribute in the skin of the head and face, and the receptors of oral and nasal mucosa via the branches of the trigeminal nerve. The central process enters the pons via the root of trigeminal nerve. Among them, the fibers conducting pain and temperature sensation descend to form spinal tract of trigeminal nerve, which terminates in the spinal trigeminal nucleus; the fibers conducting tactile and pressure sensation ascend and terminate in the pontine nuclei of trigeminal nerve.

Second-order neurons: The cell bodies are trigeminal spinal nucleus and pontine nucleus of trigeminal nerve. The fibers cross to the opposite side to form trigeminal lemniscus, which terminate in ventral posteromedial nucleus of thalamus.

Third-order neurons: The cell body is located in the ventral posteromedial nucleus of thalamus. The fibers constitute central thalamic radiations, which is projected to the lower parts of the gyrus postcentralis via the posterior limb of internal capsule.

For example, if the trigeminal lemniscus or above is damaged, the disorder of pain and temperature sensation may occur in the contralateral head and face.

## 5 Motor System

Motor system underlies all human behaviors. It is mainly composed of bone, joint and skeletal muscle, which cooperate to complete the commanded actions under the regulation of motor pathways. According to the difference of morphology and function, we divided motor pathways into two systems, namely, the pyramidal system and the extrapyramidal system.

### 5.1 Pyramidal System

Pyramidal system, the most important motor nervous system, modulates the voluntary movement of skeletal muscle, which is composed of upper motor neuron and lower motor neuron.

Upper motor neuron: Giant pyramidal cells (Betz cells) and other types of pyramidal cells with cell bodies located in layer V of gyrus precentralis and anterior paracentral lobule (areas 4 and 6). Axons are pyramidal tracts. The fiber bundle of the pyramidal tract descending to the anterior horn cells is the corticospinal tract. The fiber bundle descending to the brainstem neuromotor nucleus is the corticobulbar tract.

Lower motor neuron: The cell bodies are located in the anterior horn cells, and the cranial nerve motor nucleus of brainstem. Axons are somatic motor fibers of cranial nerves and spinal nerves. Cell bodies and axons of lower motor neurons constitute the last road to conduct motor impulses.

### 5.2 Corticospinal Tract

The cell bodies are located in pyramidal cells of primary motor cortex, premotor cortex, supplementary motor cortex, and primary somatic sensory area of cerebral cortex. The fibers pass through the anterior posterior limb of internal capsule to the lateral part of 3/5 of mesencephalic crus cerebri, and then to the basilar pons and finally to the medullary pyramid. About 90% of fibers cross to the contralateral side to form the decussation of pyramids, and then run in the back of the opposite lateral funiculus of spinal cord to form the lateral corticospinal tract. About 10% of non-crossed fibers run on the innermost side of the anterior funiculus to form the anterior corticospinal tract.

The lateral corticospinal tract terminates in the lateral nucleus of anterior horn cells section by segment during descending, which dominates the motor function of limb muscles. During descending, most fibers of the anterior corticospinal tract cross to the contralateral side via anterior white commissure section by section, terminate in the medial nucleus of contralateral anterior horn cells, and a few non-crossed fibers terminate in the medial nucleus of ipsilateral anterior horn cells, dominating the motor function of trunk muscles. The limb muscles are dominated by the contralateral cerebral cortex, while the trunk muscles are dominated by the bilateral cerebral cortex. When the unilateral corticospinal tract is damaged above the decussation of pyramids, it causes motor dysfunction of the contralateral limb, but has no obvious effect on the movement of trunk muscles.

### 5.3 Corticobulbar Tract

The cell bodies are located in pyramidal cells of head and face projection areas of primary motor cortex and primary somatosensory area in cerebral cortex.

The fibers pass through the genu of internal capsule to the medial part of 3/5 of the mesencephalic crus cerebri, and then to the basilar pons and the medullary pyramid. The corticobulbar tract is accompanied by the corticospinal tract.

The lower motor neurons controlled by the corticobulbar tract include cranial nerve motor nuclei and special visceral motor nuclei. Most fibers terminate in bilateral motor nuclei of cranial nerves, including oculomotor nucleus, trochlear nucleus, and abductor nucleus; trigeminal motor nucleus, superior facial nucleus, nucleus ambiguus, and accessory nucleus. A few fibers completely cross to the contralateral side and terminate in the inferior facial nucleus and the hypoglossal nucleus.

The corticobulbar tract dominates bilateral extraocular muscles, masticatory muscles, upper facial expression muscles, throat muscles, sternocleidomastoid muscles, and trapezius muscles, as well as contralateral lower facial expression muscles and lingualis. Unilateral corticobulbar tract injury results in dysfunction of the inferior part of contralateral facial nucleus and the innervated area of hypoglossal nucleus.

The pyramidal system regulates and controls the excitation and inhibition of lower neurons, thus ensuring the coordination and unity of general movements. In case of systemic lesion and dysfunction, the lower motor neurons will be in a released and out-of-control state, and fine voluntary movement dysfunction may occur. The physical examination shows muscular hypertonia (clasp-knife phenomenon), tendon hyperreflexia and positive pathological reflex, inconsistent increase of muscle tension for extensor and flexor of limbs, dominant for increase of muscle tension of flexor of upper limbs, and extensor of lower limbs and no involuntary movement.

## 5.4 Extrapyramidal System

Extrapyramidal system is an integral part of motor system, which refers to the pathway beyond pyramidal system that influences and controls somatic motor. The main structures of extrapyramidal system include cerebral cortex, striatum, cerebellum, thalamus, subthalamic nucleus, red nucleus, substantia nigra, pontine nucleus, vestibular nucleus, and reticular formation of brainstem, etc. The main function of the system is to adjust muscle tension, coordinate muscle movement, maintain posture, complete habitual and rhythmic movements, etc. In this section, we will introduce the extrapyramidal system from two parts: striatal circuit and cerebellar circuit.

## 5.5 Striatal Circuit

### 5.5.1 Cortico-Striatal-Cortical Circuit

The corticostriatal fibers from the cerebral cortex terminate in neostriatum (caudoputamen) via internal capsule. Fibers are emitted and reach the globus pallidus interior and projected to nucleus anterior ventralis thalami and nucleus ventralis lateralis thalami, and back to the cerebral cortex motor cortex via thalamocortical radiation. The specific circuit plays an important role in feedback regulation of the cortical motor cortex that emits pyramidal tract.

### 5.5.2 Striatum-Substantia Nigra Circuit

There is a back-and-forth fiber connection between neostriatum and substantia nigra pars compacta, which is classified according to its function. The striatum-substantia nigra circuit fibers belong to dopaminergic fibers, which rely on dopaminergic neurotransmitters for information transmission. This circuit is involved in regulating the balance among striatum, thalamus, and cerebral motor cortex.

### 5.5.3 Globus Pallidus-Subthalamic Nucleus Circuit

There is also a close back-and-forth fiber connection between globus pallidus and subthalamic nucleus. The main function of subthalamic nucleus is to inhibit globus pallidus, which is of great significance in integrating the efferent information of basal ganglia and controlling motor behavior.

## 5.6 Cerebellar Circuit

### 5.6.1 Cortico-Pontine-Cerebellar-Cortical Circuit

Fibers from frontal, parietal, occipital, and temporal lobes of cerebral cortex descend through both sides of internal capsule and crus cerebri, enter pons and terminate in ipsilateral pontine nucleus (constituting frontal, parietal, occipital, and temporal pontine tracts). Fibers from the pontine nucleus pass the midline, enter the cerebellum through the contralateral middle cerebellar peduncle, and terminate in the neocerebellar cortex. Fibers from the cerebellar cortex terminate in the dentate nucleus. Fibers from the posterior dentate nucleus pass through the superior cerebellar peduncle to the contralateral red nucleus and ventromedial and ventral anterior nuclei of dorsal thalamus via decussationes tegmenti. Fibers from red nucleus constitute rubrospinal tract via ventral tegmental decussation, which descends and terminates in the motor cells of anterior horn of spinal cord, and the released nerve impulses are finally transmitted to skeletal muscle via spinal nerves. Fibers from ventralis intermedius nucleus and ventral anterior nucleus reach the motor cortex of cerebral cortex, forming the cortico-pontine-cerebellar-cortical circuit. This circuit connects the cerebellum with the cerebrum back and forth, and the cerebellum integrates the afferent information of cerebral cortex and spinal cord, which plays an important role in the coordination movement of muscles.

If the extrapyramidal system is damaged, paralysis will not occur, but muscle tone, muscle coordination, and posture disorder will occur. Abnormalities such as Parkinson's syndrome, choreiform, tic and dystonia may occur. The physical examination indicates alternated muscle tone and negative pathological reflex, and the performance of involuntary movements such as resting tremor and athetosis may occur, with easily variable and unstable signs.

## 6 Cognitive Anatomical Basis of Cerebrum

### 6.1 Memory Circuit

Memory refers to the ability to acquire new information and maintain it over time. A large number of studies have shown that memory is involved by multiple neural circuits that support different aspects of memory, such as short-term memory, working memory, procedural memory, semantic memory, and episodic memory.

The memory circuit mainly includes: (1) Medial temporal lobe structure, which is mainly involved in the formation and consolidation of new episodic memory; (2) Prefrontal cortex, which participates in information encoding and extraction possibly based on properties of processed materials; (3) Temporal lobe cortex, which is mainly involved in the storage of episodic and semantic knowledge.

The medial structure of temporal lobe includes hippocampus, parahippocampal gyrus, amygdala, etc., among which the hippocampus plays an important role in memory. The classical memory circuit of hippocampus is that different neocortex areas project to one or more parahippocampal gyrus including periolfactory cortex, parahippocampal cortex, and entorhinal cortex. These parahippocampal areas connect with each other and project to different hippocampal areas, including dentate gyrus, hippocampal areas CA1 and CA3 and inferior horn of hippocampus. Therefore, parahippocampal areas is the condense part of various cortical inputs. Moreover, parahippocampal areas transmit information from cortex to hippocampus. After being processed by hippocampus, information can be fed back to the same cortical region for information input via parahippocampal areas. A series of studies in animals and patients with brain injury have found that the medial temporal lobe structure is a key component in organizing and consolidating long-term memory, which is permanently stored in the neocortex in a distributed manner.

Not all brain regions have the same potential to store information although a wide range of brain regions cooperate in learning and memory to form circuits that support and produce specific memory processes. At the cellular level, synaptic plasticity between neurons of the neural network in medial temporal lobe, neocortex, cerebellum, and other regions is the most likely mechanism of learning and memory.

### 6.2 Emotion Circuit

In cognitive neuroscience research, it is of great significance to identify and understand the neural network underlying various emotional states and emotional processing (Pessoa 2016). The emotion circuit theory proposed by James Papez in the 1930s suggested that emotional responses involved a circuit composed of hypothalamus, anterior thalamus, cingulate gyrus, and hippocampus. Paul Maclean later named the emotion circuit Papez's circuit, and further expanded it by adding some other structures like amygdala, orbitofrontal gyrus, and part of the basal ganglia. The expanded emotion circuit is named as the limbic system.

In early works in identifying the neural circuit of emotion, investigators tend to regard emotion as an independent component, which can be located on a specific neural circuit (such as the limbic system). In recent years, with the development of technology and in-depth study, we have recognized that emotion is a multifaceted behavior, which involves different neural circuits according to different tasks or situations.

Studies have found that the amygdala is an important brain region for emotional participation, which is particularly important in the fear response. The orbitofrontal gyrus is associated with explicit emotion recognition of anger. The anterior insular cortex is related to aversion experience. Anterior cingulate gyrus receives nerve projections from amygdala, orbitofrontal gyrus, and anterior insular cortex, which is essential for generalized emotional processing.

### 6.3 Motor Circuit

The cerebral cortex can directly or indirectly control the activity of spinal cord neurons. The direct connection is provided by the corticospinal tract. The corticospinal tract mainly originates from the primary motor cortex or BA4. Corticospinal fibers can also be found in part of the somatomotor cortex and in BA6. BA6 contains the secondary motor cortex, which is divided into lateral area (premotor cortex) and medial area (supplementary motor cortex). It is further divided into the ventral premotor cortex, dorsal premotor cortex, supplementary motor cortex, and presupplementary motor cortex.

Many other association areas in the cortex are also related to motor function. The Broca area, insula (deep in Broca area), and posterior inferior frontal gyrus are all involved in the production of speech acts. BA8 includes the frontal eye field, which controls eye movement. The posterior parietal lobe area participates in the planning and control of movements, and along the intraparietal sulcus, it can be subdivided into different areas related to eye movements, arm movements, and hand movements.

Although the cerebral cortex is directly connected with the structure of spinal cord through the corticospinal tract, it can also regulate motor function through four other ways. First, the motor cortex and the premotor cortex receive the afferent fibers from most regions of the cortex via the cortico-cortical tract. Second, some neuronal axons of the motor cortex terminate in the brainstem, thus regulating the function of the extrapyramidal system. Third, there are a large number of fiber connections between motor cortex and cerebellum and basal ganglia. Fourth, the corticomedullary tract contains cortical fibers that terminate in cranial nerves.

The anatomical structure of the motor cortex follows two principles. First, there is organized and orderly characterization of somatic-specific areas in both motor cortex and premotor cortex. The characterization of a somatic-specific area does not match its actual size. The size of the cortical area characterizing a somatic-specific area reflects the importance of this somatic-specific area in movement and the control level required to manipulate this area. The characterization of a somatic-specific area is strictly limited to one side of the body. Each cerebral hemisphere mainly controls the movement of the opposite side of the body. The second basic principle is related to the connection between motor cortex areas. Each component constitutes several movement control levels. That at the lowest level is the spinal cord, which not only provides the connection point between the nervous system and muscles, but also controls the simple reflex movement at this level. Those at the highest level are premotor cortex and association area, and the processing of these areas plays a key role in motor planning based on current perceptual input, experience, and future goals. With the help of cerebellum and basal ganglia, the motor cortex and brainstem structures convert action command into movement.

Viewing the motor system as a hierarchical structure can make us realize that motor control is a distributed processing. The highest level may not care about the details of a motion but instead, provide the possibility for the low level to convert instructions into motion. The cortex can regulate low-level movement by additional means. This ability makes the movement of organisms more flexible. We can produce many movements in response to an input of a sensory signal. The cortical structure can also make our movement depend on exogenous cues to the minimum.

## 6.4 Language Circuit

Language is unique in psychological function because only human beings have a real language system. As early as more than a century ago, we knew that the peripheral region of the lateral fissure in the left hemisphere (dominant hemisphere) was involved in the process of speech understanding and production. Sound comprehension involves the superior temporal gyrus, and the injury in this area will lead to pure word deafness. Written word processing takes place in the occipital-temporal area of the left hemisphere, and the injury in this area will lead to pure dyslexia. Broca aphasic patients have injury in the corresponding brain area (Broca area) that controls the production of spoken language, so they show significant difficulty in producing speech. Wernicke aphasia is caused by the injury in the brain region responsible for speech comprehension (the Wernicke area is in the back third of superior temporal gyrus). Aphasia can also be caused by injury in the connecting fibers (arcuate fascicles) between Wernicke and Broca areas, and conduction aphasia is caused by this injury. Patients with such aphasia have difficulty in producing spontaneous speech and retelling speech. Oral apraxia or difficulty in pronunciation of words is caused by insula injury.

However, the classical model is insufficient to explain the calculation process of supporting languages. Peter Hagoort put forward a new linguistic neural model, the memory-integration-control model, which integrates recent brain and language study results. The temporal lobe is particularly important for storing and extracting word characterization. The phonetic and phonemic features of words are stored in the brain region that extends to the superior temporal sulcus around the posterior superior temporal gyrus (including Wernicke area), while the semantic information is distributed in different regions of the left middle temporal gyrus and inferior temporal gyrus. The process of combining and integrating (unifying) phonetic, lexical-semantic, and syntactic information involves many regions of the frontal lobe, including Broca area or left inferior frontal gyrus. At present, it seems that the left inferior frontal gyrus involves all three kinds of integration processing: BA47 and BA45 are involved in semantic integration, BA45 and BA44 are involved in syntactic integration, and BA44 and part of BA6 are involved in phonetic integration. When people need to alternately speak in actual communication, such as conversation, the control component of this model is particularly important. There are not many studies on cognitive control in language comprehension, but some brain regions that are involved in cognitive control in other tasks, such as anterior cingulate gyrus and dorsolateral prefrontal cortex (i.e., BA46/9), also play a role in cognitive control in language comprehension.

## 6.5 Awareness Circuit

Awareness refers to a person's awareness of his own perception and feeling to his own self-awareness, that is, the perception that he regards oneself as a subject with intentionality and free will. The first meaning of awareness is phenomenological awareness.

At present, the most recognized consciousness perception model is the global-workspace theory (GWT). Based on previous professional cognitive theory, GWT assumes that in order to make sensory information enter to consciousness, it must be provided to many different cognitive systems (such as working memory) through the global neuron workspace composed of repeated activation between sensory cortex and frontoparietal cortex. This model believes that consciousness is inseparable from the function provided by the global workspace and makes a strong prediction in several aspects, that is, (1) at any moment, only one content or object can dominate in consciousness; (2) without a certain degree of attention, there will be no conscious perception; (3) the perception of consciousness necessarily involves the "ignition" of neuronal activity in the frontoparietal workspace (large-scale nonlinear amplification).

Another recognized model of consciousness perception is the theoretical model of neuronal alliances, in which each alliance represents a different object or content, and the winner of the competition determines the content of perception at that moment. The areas where such alliances might be placed were initially unknown; recent attention has focused on the advanced sensory area of the "back of the brain," the non-primary visual cortex. Importantly, this theory predicts the one-to-one mapping between some neuronal activities and contents of conscious perception, so that the activation of these neurons will lead to the perception of the content; otherwise, these contents will be excluded from conscious perception.

In addition, there is a local recurrence theoretical model. This conscious perception model puts forward the importance of local and repeated processing between early and higher order sensory (i.e., visual) areas. An important feature of the local recurrence model is that conscious perception is considered to be completely separated from attention, memory, and other cognitive mechanisms required for perceptual reporting because it only originates from repeated activities in the sensory area. In turn, this leads to the prediction of some phenomenology of the conscious state—which is not only inaccessible, but also unable to be reported because of its inaccessibility, making it controversial. Note that the difference between feedforward processing (not assumed to be conscious) and cyclic processing (assumed to be conscious) is also derived from the global workspace theory model and the winning alliance theory. Obviously, the cyclic processing of cortex plays an important role in many modern theories of consciousness, and the cyclic processing between the cortex and thalamus is also possible.

Other theoretical models of consciousness perception also include reverse hierarchy theory model, information integration theory model, self-sustaining signal regeneration theory model, intermediate level theory model, higher order or "metacognition" theory model, fixed ring theory model, attention mode theory model and classification theory model.

## References

Brazis PW, Joseph C et al (2012) Localization in clinical neurology. People's Medical Publishing House, Beijing

Gao X (2009) Human anatomy. Peking University Press, pp 338–342

Jia J, Chen S (2018) Neurology. People's Medical Public House, pp 3–61

Kahle W, Frotscher M (2015a) Sensory organs. In: Kahle W, Frotscher M (eds) Color atlas of human anatomy, Vol. 3: nervous system and sensory organs, 7th edn. Thieme

Kahle W, Frotscher M (2015b) Color atlas of human anatomy: Vol. 3 nervous system and sensory organs, 7th edn. Thieme

Mattle H, Mumenthaler M (2017) Fundamentals of neurology, 2nd edn. Thieme

Pessoa L (2016) Chapter 29 - the emotional brain, Conn's translational neuroscience. Elsevier, Amsterdam

Rohkamm R (2014) Color atlas of neurology, 2nd edn. Thieme Medical Publishers

Stokard JJ, Stokard JE, Sharbrough FW (1977) Detection and localization of occult lesions with brainstem auditory responses. Mayo Clin Proc 52:761–769

# Basics of Neurophysiology and Brain Network

# 2

Na Clara Pan and Qihui Zhou

There are about 100 billion neurons in the human brain. Each neuron has 2000 to tens of thousands of dendrites, which enable connections with other neurons (100–10,000 synapses in higher animals), thus forming complex neural networks. Therefore, the formation and remodeling of brain neural networks depend on the morphogenesis and dynamic changes of neurons. This chapter focuses on the basics of neurophysiology and brain networks.

## 1 Structure of Cell Membrane

The structure and function of a cell are closely related to the cell membrane. The cell membrane isolates the intracellular environment from the surrounding environment and also enables independence between components (including nucleus and organelle) in a eukaryotic cell. The cell membrane is an elastic semipermeable membrane composed of phospholipids. Any cell membrane is similar in its structure and construction, which consists of a phospholipid bilayer with proteins. These membrane proteins have diverse specific functions, including acting as receptors in response to extracellular signals, enabling selective transport of molecules across membranes, and participating in charge transport and oxidative phosphorylation. In addition, membrane proteins control the interactions among intracellular organelles. Given this, these membrane structures underlie a variety of biological processes and specific membrane functions, which will be discussed in detail later.

N. C. Pan (✉)
Department of Neurology, Xuanwu Hospital, Capital Medical University, Beijing, China
e-mail: panna@xwhosp.org

Q. Zhou
Department of Neurology, Beijing Chaoyang Hospital, Capital Medical University, Beijing, China

### 1.1 Lipid Bilayer

Phospholipids, as fundamental building blocks of cell membrane, are amphiphilic molecules and composed of two hydrophobic fatty acid chains attached to a phosphate-containing hydrophilic group. Due to the poor water solubility of its fatty acid tail, a phospholipid spontaneously forms a bilayer membrane in an aqueous solution, with the hydrophobic tail buried in the membrane, and the polar groups exposed on both sides of the membrane and in contact with water (Fig. 2.1). Such a phospholipid bilayer membrane forms a stable barrier between the internal and external water environment and is the basic structure of any bio-membrane.

Lipids account for about 50% of cell membrane mass, which varies with the type of membrane. For example, about 50% of the content of cytoplasmic membrane is lipid and 50% is protein; the mitochondrial inner membrane contains

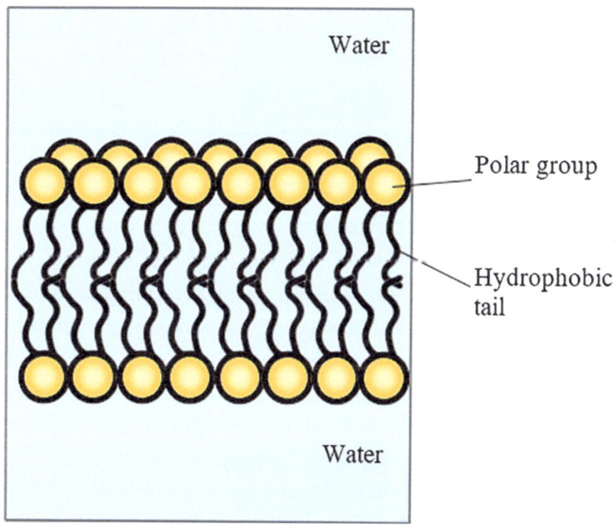


**Fig. 2.1** Lipid bilayer phospholipids can spontaneously form a stable dual membrane with polar groups exposed to water and hydrophobic tails buried inside the membrane



Y. Wang (ed.), *Therapeutics of Neural Stimulation for Neurological Disorders*, https://doi.org/10.1007/978-981-99-4538-2_2

abnormally high content of proteins (about 75%), which are involved in charge transfer and oxidative phosphorylation. The lipid components of different cell membranes are also different. The cytoplasmic membrane of *Escherichia coli* is mainly composed of phosphatidylethanolamine, accounting for 80% of the total lipid. The cytoplasmic membrane of mammals is more complex and contains four main phospholipids (phosphatidylcholine, phosphatidylserine, phosphatidylethanolamine, and sphingomyelin), which together constitute 50–60% of the total membrane lipid. In addition to phospholipids, the cytoplasmic membrane of animal cells also contains glycolipids and cholesterol, generally accounting for about 40% of the total lipid molecules.

An important property of membrane lipid bilayer is characterized by a two-dimensional fluid. Individual molecules (including lipids and proteins) can rotate freely and move laterally. This fluidity is a key property of the membrane, determined by both temperature and lipid composition. For example, the interaction between shorter fatty acid chains is weaker than that between longer fatty acid chains. Therefore, the cell membrane containing shorter fatty acid chains is less rigid and can still remain fluid state at lower temperatures. In addition, due to the action of double bonds, kinks are introduced into the fatty acid chain, making it more difficult for them to gather together. Therefore, lipids containing unsaturated fatty acids can increase the fluidity of the membrane.

Cholesterol plays a unique role in determining membrane fluidity due to its unique hydrocarbon ring structure. The cholesterol molecule is inserted into the bilayer through its polar hydroxyl group close to the hydrophilic group of the phospholipid, and the rigid hydrocarbon ring of cholesterol interacts with the fatty acid chain region near the phospholipid, thereby reducing the fluidity outside the fatty acid chain and increasing the stiffness of the membrane. On the other hand, cholesterol inserted into phospholipids will interfere with the interaction between fatty acid chains, so that the cell membrane has a certain fluidity at lower temperatures.

## 1.2 Membrane Protein

Protein is another main component of the cell membrane, accounting for 25–75% of mass of various cell membranes. The membrane structure model was first proposed by Jonathan Singer and Garth Nicolson in 1972, which regarded the cell membrane as a fluid mosaic of proteins inserted into the lipid bilayer. Phospholipids are the basic structural tissues of the cell membrane, and membrane proteins fulfill specific functions of different cell membranes. These proteins can be divided into two categories according to their relationship with cell membrane: transmembrane proteins that directly embed in the lipid bilayer; peripheral membrane proteins that do not directly insert into the cell membrane but interact with the cell membrane by interacting with membrane proteins.

Many transmembrane proteins span the lipid bilayer and are partially exposed on either side of the membrane. The transmembrane portion of these proteins is usually a helical region composed of 20–25 non-polar amino acids. The hydrophobic side chains of these amino acids interact with the fatty acid chains of membrane lipids. Like phospholipids, transmembrane protein is an amphiphilic molecule, with hydrophilic portion exposed to the water environment on either side of the membrane. Some transmembrane proteins span the membrane only once, while some transmembrane proteins have multiple transmembrane regions. Transmembrane proteins of most eukaryotic cytoplasmic membranes are modified by carbohydrates and exposed to the cell surface, which may involve the interaction between cells. Proteins can also be anchored to cell membranes by lipids covalently linked to polypeptide chains. Different lipid modifications anchor proteins to the cytoplasmic and extracellular surfaces of the plasma membrane. Proteins can be anchored to the cytoplasmic surface of the membrane by adding an amino acid containing a 14-carbon fatty acid (myristic acid) or a 16-carbon fatty acid (palmitic acid) or adding 15–20 carbophenyls to the side chain of cysteine residues. In addition, the protein is fixed on the extracellular surface of the plasma membrane by adding glycolipid at its carboxyl end.

## 1.3 Transmembrane Transportation

The selective permeability of the biological cell membranes to small molecules enables cells to control and maintain their internal components. Only small uncharged molecules can diffuse freely through the phospholipid bilayer membrane. Small non-polar molecules, such as $O_2$ and $CO_2$, are soluble in the lipid bilayer and therefore easily cross the cell membrane. Small uncharged polar molecules, such as $H_2O$, can also diffuse through the membrane, but larger uncharged polar molecules, such as glucose, cannot pass through the membrane. Charged ions, regardless of size, cannot diffuse through the phospholipid bilayer, and even hydrogen ions cannot freely diffuse through the lipid bilayer.

Many uncharged polar molecules (such as glucose) pass through the cell membrane via the action of specific transmembrane proteins, which act as transporters. Such transporters determine the selective permeability of the cell membrane and play a vital role in the function of the cell membrane. They contain multiple transmembrane regions, forming channels through lipid bilayers, allowing polar or charged molecules to pass through the membrane via protein pores without interacting with the hydrophobic fatty acid chains of membrane phospholipids.

There are two categories of membrane transporters: channel proteins that can form permeability pore structures and carrier proteins that release selectively bound small molecules to the other side of the membrane by conformational changes. Channel proteins form open pores in the cell membrane, allowing molecules of appropriate size to pass freely. For example, ion channels allow inorganic ions such as $Na^+$, $K^+$, $Ca^{2+}$, and $Cl^-$ to pass through the cell membrane. The pores formed by these channel proteins are not permanently open. On the contrary, they can selectively open and close according to extracellular signals, thus allowing cells to control the motion of ions across the cell membrane. Such regulated ion channels mediate the transmission of electrochemical signals in nerve and muscle cells. Unlike channel proteins, carrier proteins selectively bind and transport specific small molecules, such as glucose. Instead of forming open channels, the carrier protein acts like an enzyme to promote specific molecules across the cell membrane. In particular, the carrier protein binds to specific molecules, and then carries out conformational changes, thus opening the channel, so that the transported molecules can pass through the membrane and be released on the other side.

The kinetic transport direction of molecules transported via channels or carrier proteins across the membrane, i.e., concentration and electrochemical gradient, is called passive transport. However, carrier proteins also provide a mechanism by which changes in energy for transporting molecules across membranes can be combined with other forms of metabolic energy, just as enzyme reactions can be combined with hydrolysis or synthesis of adenosine triphosphate (ATP). For example, if the molecular transporting in this direction is coupled with the hydrolysis of ATP as an energy source—a process known as active transport, the molecules can cross the membrane in the direction opposite to the energy gradient (e.g., opposite to the concentration gradient). Therefore, the free energy stored as ATP can be used to control the internal composition of cells and drive the biosynthesis of cell components.

## 2 Resting Potential and Action Potential

The action of ion transporters creates substantial transmembrane gradients for most ions. The ion concentration directly measured in the abnormally large neurons found in the squid nervous system in previous work showed that the internal $K^+$ of neurons was much higher than the external, and the external $Na^+$ was much higher than the internal. Similar concentration gradients also occur in neurons of most animals, including humans. However, since the ionic strength in mammalian blood is lower than that of marine animals such as squid, the concentration of each ion in mammalian blood is several times lower. These transporter-dependent concentration gradients are the indirect source of resting neuron membrane potential and action potential.

### 2.1 Ionic Basis of Cell Membrane Resting Potential

Once the ion concentration gradients on different neuronal membranes are known, the equilibrium potentials of $K^+$ and other main ions can be calculated using the Nernst equation (Eq. 2.1). Because the resting membrane potential of squid neurons is about −65 mV, $K^+$ is the ion closest to electrochemical equilibrium when the cells are in resting state. This indicates that the permeability of the resting membrane to $K^+$ is greater than other ions, and this permeability is the source of the resting potential.

$$E = E_0 - \left(\frac{RT}{nF}\right)\ln Q \tag{2.1}$$

where $E$ represents the non-equilibrium potential, $E_0$ represents the equilibrium potential, $R$ represents the gas constant, $T$ represents the temperature, $n$ represents the number of electrons exchanged in the electrochemical reaction, $F$ represents the Faraday constant, and $Q$ represents the reaction coefficient.

When Hodgkin and Katz performed experiments on live squid neurons, they found that the resting membrane potential did change when the external $K^+$ concentration changed, and the negativity decreased when the external $K^+$ concentration increased. When the external $K^+$ concentration increased to be equal to the internal $K^+$ concentration of neurons, making the $K^+$ equilibrium potential 0 mV, the resting membrane potential was also about 0 mV. Briefly, the resting membrane potential changed with the logarithm of $K^+$ concentration, and its slope was close to 58 mV per 10 times of $K^+$ concentration. Since other ions such as $Cl^-$ and $Na^+$ also had slight permeability and little effect on the resting potential, the values obtained were not exactly 58 mV. In general, however, adjusting the external concentration of these other ions had only a small effect, and $K^+$ permeability was indeed the primary source of resting membrane potential.

### 2.2 Mechanism of Resting Potential

In typical multicellular animals, the concentration of intracellular $K^+$ ions is about 10 times of $K^+$ concentration of extracellular fluid, while the concentrations of extracellular $Na^+$ and $Cl^-$ ions are much higher than those of intracellular ones. These concentration gradients are maintained by $Na^+$/

$K^+$ ATPase and cell energy consumption. The cytoplasmic membrane contains a large number of opened "static" $K^+$ channels, which only allow $K^+$ to pass through. The resting potential is mainly determined by the motion of $K^+$: When a $K^+$ crosses the cell membrane along the concentration gradient, it leaves an extra negative charge on the cytoplasm surface and deposits a positive charge on the outer surface. According to Nernst equation calculation and typical external and cytoplasmic $K^+$ concentrations, the resting potential is usually about −60 mV, which is smaller than the $K^+$ equilibrium potential $E_K$. When the $K^+$ concentration around the resting cells changes, the measured membrane potential presents a new value, and closes to the calculated value of $E_K$ again, which indicates that the resting potential is mainly caused by the motion of $K^+$ through the opened $K^+$ channel on the cytoplasmic membrane.

The intracellular environment is complex because there are some opened $Na^+$ and $Cl^-$ channels in the resting cell membrane. Of course, cells also contain other ions, such as $HPO_4^{2-}$, $SO_4^{2-}$, and $Mg^{2+}$, but few channels can accept these ions. In addition, the membrane potential of excitable cells such as neurons and muscle cells are mainly affected by the opening and closing of $K^+$, $Na^+$, and $Cl^-$ channels. Therefore, the three kinds of ions are the only ones we need to consider here.

In order to calculate the membrane potentials of different ion concentrations, the permeability constant $P$ of each ion is defined as a measure of how easy it is for ions to pass through the membrane per unit area (1 $cm^2$) driven by 1 m of concentration difference; it is proportional to the number of open ion channels and the number of ions that each channel can conduct per second (channel conductivity). Therefore, $P_K$, $P_{Na}$, and $P_{Cl}$ are indicators measuring the "leak performance" of the membrane per unit area for these ions. Permeability is generally not measured directly. On the contrary, the permeability of membrane to a given ion is the product of the number of open ion channels and the conductivity of each channel, both of which can be measured by different techniques. Since the conductivities of different ion channels are nearly identical, differences in membrane permeability to $Na^+$, $K^+$, and $Cl^-$ largely reflect differences in the number of open channels specific to each ion. What matters is not the absolute magnitude of the permeability of each ion but the ratio of $Na^+$ and $Cl^-$ to $K^+$ permeability. The potential $E$ (in millivolts) across the cell surface membrane is given by a more complex Nernst equation, in which the weight of ion concentrations is proportional to the relative magnitude of their permeability constants as shown in Eq. (2.2):

$$E = 59\log_{10}\frac{[K_o]+[Na_o]\frac{P_{Na}}{P_K}+[Cl_i]\frac{P_{Cl}}{P_K}}{[K_i]+[Na_i]\frac{P_{Na}}{P_K}+[Cl_o]\frac{P_{Cl}}{P_K}} \tag{2.2}$$

where the "$o$" and "$i$" subscripts indicate the concentration of ions outside and inside the cell. $[K_o]$ and $[Na_o]$ are placed on the numerator because of the opposite charge ($Z$ value in the Nernst equation), while $[Cl_o]$ is placed on the denominator. On the contrary, $[K_i]$ and $[Na_i]$ are on the denominator, while $[Cl_i]$ is on the numerator. If the corresponding ion concentration and permeability are known, the membrane potential at any time and any position in the neuron can be calculated using the equation.

In resting neurons, the permeability of the cell membrane to $Na^+$ or $Cl^-$ is about one-tenth to $K^+$ ($P_{Na}/P_K = P_{Cl}/P_K = 0.1$). That is, there are ten times more $K^+$ channels than $Na^+$ or $Cl^-$ channels. By substituting the typical ion concentration and these permeability ratios into Eq. (2.2), we can calculate a membrane potential of −52.9 mV, which is closer to $E_K$ (−91.1 mV) than to $E_{Na}$ (+64.7 mV). Although $E_{Cl}$ (−87.2 mV) is close to $E_K$, since there are few open $Cl^-$ channels, their contribution to the resting potential is small. The resting potential is not equal to $E_K$ because the membrane also contains some open $Na^+$ channels; the inflow of $Na^+$ along the concentration gradient increases the positive charge inside the cell membrane to make the cell membrane potential rise (or fall).

## 2.3 Change of Membrane Potential

It can be known from Eq. (2.2) that the change in the permeability of cell membrane to various ions will cause the change of membrane potential. Here, we summarize the effects of opening and closing of various channels on the change of membrane potential:

1. The opening of $Na^+$ channel (increased $P_{Na}$) causes depolarization of the membrane. The membrane potential shifts positively, and if $P_{Na}$ increases sufficiently, the membrane potential will be close to $E_{Na}$. Intuitively, $Na^+$ tends to flow from the outside to the inside of the cell, reducing the concentration gradient, leaving excess negative ions outside the membrane and more positive ions on the surface of cytoplasm. On the contrary, closing $Na^+$ channel and reducing $P_{Na}$ result in the hyperpolarization of the membrane tending to a more negative potential.
2. $K^+$ channel opening (increased $P_K$) results in repolarization of the cell membrane. The membrane potential becomes more negative, close to $E_K$. Intuitively, this is because more $K^+$ ions flow from inside to outside of the

cell, reducing their concentration gradient, leaving excess negative ions inside the cell and more positive ions outside the cell membrane. On the contrary, $K^+$ channel closing and $P_K$ reduction result in membrane depolarization and the potential tending to the positive direction.

3. The opening of "non-specific" cation channels makes $Na^+$ and $K^+$ equal, which also causes membrane depolarization. Such channels allow $K^+$ to flow outward from the cytoplasm and $Na^+$ to flow inward; the net effect is to drive the membrane potential towards zero.
4. $Cl^-$ channel opening (increased $P_{Cl}$) causes hyperpolarization of the cell membrane and may close to $E_{Cl}$. Intuitively, $Cl^-$ tends to flow from the outside to the inside of the cell, reducing the concentration gradient, leaving excess positive ions outside the membrane and more negative ions inside the cells. In muscle cells, the resting $Cl^-$ channel, rather than the resting $K^+$ channel, is the main determinant of the internal negative resting potential. On the contrary, the closing of $Cl^-$ channel and reduction of $P_{Cl}$ will cause depolarization with the potential tending to the positive direction.

At the resting potential, the voltage-gated ion channels close and no ions pass through them. However, when a region of the cell membrane is slightly depolarized, the voltage-gated sodium channel opens for a period of time, allowing $Na^+$ to flow in, causing sudden and transient depolarization related to action potentials. Under the action of action potential, with the opening (and closing) of the voltage-gated $Na^+$ channel, the transient opening of the voltage-gated $K^+$ channel causes the membrane potential to return to the resting state and even become more negative (hyperpolarization) in a short time. Thus, the ability of axons to conduct action potentials for long distances without attenuation depends on the controlled opening and closing of the voltage-gated $Na^+$ and $K^+$ channels. Binding of neurotransmitters to ligand-gated ion channels in postsynaptic cells triggers the changes in postsynaptic membrane potential during synaptic impulse transmission.

## 2.4 Ionic Basis of Action Potential

What causes membrane potential depolarization of neurons and finally induces action potentials? The conventional answer is that increased $Na^+$ permeability leads to the release of action potentials, but the details are worth exploring. The equilibrium potential of $Na^+$ ($E_{Na}$) in neurons is positive, and in fact it is also true in most cells. Therefore, if the permeability of the membrane to $Na^+$ is high, the membrane potential will be close to $E_{Na}$. Based on these considerations, Hodgkin and Katz hypothesized that the action potential was generated because the neuron membrane was temporarily permeable to $Na^+$. Hodgkin and Katz tested the role of $Na^+$ in the generation of action potential using the same method for studying resting potential by ion substitution test and studied what happened to action potential when $Na^+$ was removed from extracellular components. They found that decreasing the external $Na^+$ concentration would reduce the rising rate and peak amplitude of action potential, and there was a specific linear relationship between the amplitude of action potential and the logarithm of external $Na^+$ concentration. In contrast, decreasing $Na^+$ concentration had little effect on resting membrane potential. Therefore, when the cell membrane of resting neurons had only a slight permeability to $Na^+$, the permeability of cell membrane to $Na^+$ became very strong during rising phase and overshoot phase of action potential. The transient increase of $Na^+$ permeability was due to the opening of $Na^+$ selective channels, which were basically closed at resting state. The membrane pump maintained a large electrochemical gradient for $Na^+$, and the concentration of $Na^+$ outside the neuron was much higher than that inside the neuron. When the $Na^+$ channel opened, $Na^+$ flowed into the neurons, causing the membrane potential to depolarize and close to the $E_{Na}$.

During the overshoot phase of action potential, the membrane potential stayed near $E_{Na}$ (about +58 mV) for a short time because the increase of permeability of the membrane to $Na^+$ itself was transient. The membrane potential rapidly repolarized to a resting level, followed by even a transient hyperpolarization. After the action potential overshot, the permeability of the cell membrane to $Na^+$ was inactivated while the permeability to $K^+$ increased, during transient hyperpolarization, because $K^+$ permeability became greater than that at rest. When the enhanced $K^+$ permeability decreased, the action potential ended and the membrane potential thus returned to the normal resting level. In conclusion, the ionic substitution test performed by Hodgkin and

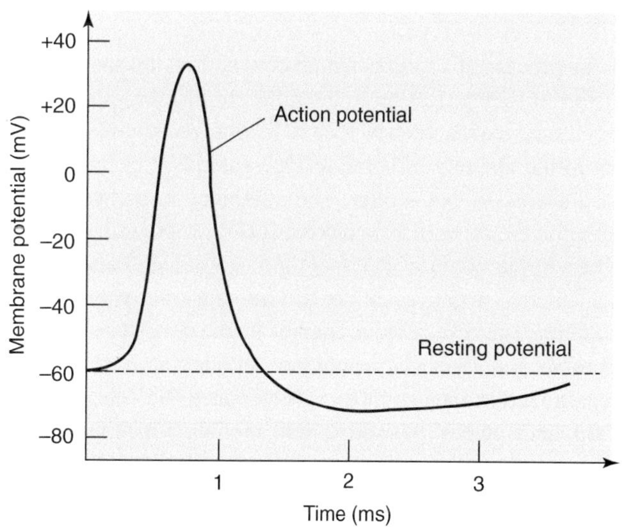


**Fig. 2.2** Dynamic changes of membrane potential of squid giant axon during action potential

Katz provided compelling evidence that the resting membrane potential was caused by the high permeability of the resting membrane to $K^+$, while the depolarization of the action potential was caused by the transient increase of the membrane $Na^+$ permeability (Fig. 2.2).

After stimulation at time 0, the membrane potential rapidly became more positive, approaching the $E_{Na}$ value, and then became more negative. A transient increase in $Na^+$ permeability due to the transient opening of the voltage-gated $Na^+$ channel allowed the $Na^+$ to flow in, causing the membrane to become depolarized. Before this, the opening of voltage-gated $K^+$ channels and the resulting $K^+$ outflow caused the membrane to become hyperpolarized in a short time.

## 2.5 Voltage-Gated Cation Channel

The action potential is a cycle of membrane depolarization, repolarization, hyperpolarization, and return to resting value. This cycle lasts for 1–2 ms and can occur hundreds of times per second. These periodic changes in membrane potential are caused by transient increases in permeability of a region of the membrane, first $Na^+$ and then $K^+$. More specifically, these electrical changes are due to the opening and closing of voltage-gated $Na^+$ and $K^+$ channels with the change of membrane potential. The role of these channels in action potential generation and conduction has been elucidated in the classical studies of squid giant axons, in which multiple microelectrodes can be inserted without damaging the integrity of the plasma membrane. However, all neurons use the same basic mechanism.

## 2.6 Voltage-Gated $Na^+$ Channel

In the process of action potential conduction, the passive diffusion of depolarization to the adjacent distal region of the membrane makes the new region depolarized slightly, resulting in the opening of some voltage-gated $Na^+$ channels and the increase of $Na^+$ inflow. The combination of two forces acting in the same direction drives $Na^+$ into the cell: the concentration gradient of $Na^+$ and the resting membrane potential (internal negative) attract $Na^+$ into the cell. As more $Na^+$ ions enter the cell, positive charges inside the cell membrane increase and the cell membrane depolarizes further. This depolarization causes more voltage-gated $Na^+$ channels to open, causing $Na^+$ to explode into motion, which is accomplished in less than a millisecond. In less than a millisecond, at the peak of the action potential, $Na^+$ of the permeable membrane in this region becomes much higher than $K^+$ and $Cl^-$, and the membrane potential method of $E_{Na}$, the equilibrium potential of membrane permeation is only $Na^+$. When the membrane potential approaches $E_{Na}$, further net inward movement of $Na^+$ stops because the concentration gradient of $Na^+$ larger than the external is balanced (internal positive) by the membrane potential $E_{Na}$. The action potential reaches the maximum point. The measurement peak of squid giant axon action potential is 35 mV, which is similar to the calculated value of $E_{Na}$ (55 mV) at 440 mM in vitro and 50 mM in vivo of $Na^+$ concentration. The relationship between the size of the action potential and the concentration of $Na^+$ ions inside and outside the cell is confirmed in the experiment. For example, if the concentration of $Na^+$ in the solution soaking the squid axon decreases to one-third of the normal concentration, the magnitude of depolarization decreases by 40 mV, almost as predicted.

Like all channel proteins, voltage-gated $Na^+$ channels contain an aqueous pore through which ions flow. Entering the channel from the outside, encountering a wide vestibule first, then a narrow aperture that selects the type of ions allowed to pass through. This aperture leads to a large internal vestibule. At the end of the cytoplasm, there is a segment of protein, the "gate," which closes the aperture in a resting state. In the resting state, the voltage-gated $Na^+$ channel is closed, but if the membrane is depolarized, it can be opened. The greater the depolarization, the greater the chance of any channel opening. The opening is triggered by the movement of the helical structure induced by the voltage in response to membrane depolarization, causing a small conformational change on the grid that opens the channel and allows the ion flow. Once opened, the channels remain open for about 1 ms, during this period, about 6000 $Na^+$ ions pass through them. $Na^+$ inflow is further prevented by moving the channel inactivation segment to the channel opening. The channel is inactivated and cannot be reopened as long as the membrane remains depolarized. As we will discuss later, the refractory period of $Na^+$ channel is important in determining the unidirectionality of the action potential. A few milliseconds after the internal negative resting potential, channels are reestablished return to a closed resting state, and "start" again, are ready to open by depolarization.

## 2.7 Voltage-Gated $K^+$ Channel

The voltage-gated $K^+$ channel protein opens when the voltage-gated $Na^+$ channel closes and fewer $Na^+$ ions enter the cell. This results in the observed increases in $K^+$ permeability and $K^+$ outflow from the cytoplasm, and $K^+$ refluxes to its resting potential from cytoplasm. In fact, at a transient moment, the membrane becomes hyperpolarized, and the potential is close to $E_K$, which is more negative than the resting potential.

The opening of the voltage-gated $K^+$ channel is caused by the membrane depolarization of the action potential. Unlike voltage-gated $Na^+$ channels, most types of voltage-gated $K^+$ channels remain open as long as the membrane depolarizes, and close only when the membrane potential returns to an internal negative value. Since voltage-gated $K^+$ channels open approximately 1 ms after initial depolarization, they are also known as delayed $K^+$ channels. Eventually all voltage-gated $K^+$ and $Na^+$ channels close. The only open channel is a non-voltage-gated $K^+$ channel, which produces the internal negative potential characteristic of resting-state; as a result, the membrane potential returns to its resting value.

## 2.8 Action Potential and Electrical Impulse Conduction

The conduction, transmission, and reception abilities of the characteristic electrical activities of neurons are caused by the opening and closing of specific ion channel proteins in the neuronal plasma membrane. Each open channel allows only a small number of ions to move from one side of the membrane to the other; however, the movement of these ions results in significant changes of the membrane potential. In order to understand how voltage-gated $Na^+$ channels and $K^+$ channels allow the action potential to conduct downward along one direction of axons, we first need to check how the plasma membrane with only $K^+$ channels depolarize. In electrical properties, neurons with only $K^+$ channels are like a long underwater telegram line. It consists of insulator cell membrane with poor conductivity property, which separates two media (cytoplasm and extracellular fluid) with high conductivity to ions.

Suppose a microelectrode is inserted in the axon and the electrode is connected to a current source (e.g., a battery), so that the potential at that point suddenly depolarizes and remains at the potential. There will be a relatively excessive positive charges inside the membrane, mainly $K^+$. These ions will tend to move away from the initial depolarization position and thus depolarize adjacent parts of the membrane. This is called passive diffusion of depolarization. In contrast to the action potential, passive diffusion occurs evenly in both directions. Furthermore, the magnitude of depolarization decreases with distance from the initial depolarized position because some excess cations leak back from the membrane via the static cation channels. Only a fraction of excess cations is carried long distances longitudinally along the axon. The degree of passive diffusion of depolarization depends on two properties of neurons: the permeability of the cell membrane to ions and the conductivity of the cytoplasm.

The passive diffusion of depolarization is larger for large-diameter neurons than for small-diameter neurons because the conductivity of the cytoplasm of neurons depends on its cross-sectional area. The larger the area of conducting current, the more the number of ions (in length per unit neuron). Thus, $K^+$ can, on average, travel farther along large axons than small axons before "leaking" back to axon through the membrane. As a result, large-diameter neurons passively depolarize, faster and farther away. However, the depolarized membrane can only passively transmit a short distance, from 0.1 mm to about 5 mm. The depolarization of dendrites and cell bodies usually diffuses in this way although some dendrites have action potentials. Neurons with very short axons also undergo axonal depolarization through passive diffusion. However, passive transmission does not allow electrical signals to transmit over long distances.

At the peak of action potential, the passive diffusion of membrane depolarization is sufficient to depolarize the "downstream" portion of the membrane. This results in the opening of few $Na^+$ channels in the region, thus increasing the degree of depolarization in the region, resulting in the exposed opening of more $Na^+$ channels. In this way, it is ensured that the transmission of action potential does not decrease. Since the voltage-gated $Na^+$ channels remain inactive within a few milliseconds after opening, these $Na^+$ channels immediately "lag" behind the action potential and cannot reopen even if the potential in this segment depolarizes due to passive diffusion. $Na^+$ channel cannot open during the refractory period, which ensures the unidirectional transmission of action potentials from the cell body to the axonal terminal and limits the number of action potentials that can be conducted by neurons per second. The reopening of the sodium channel "behind" the action potential is also prevented by the membrane hyperpolarization, which is caused by the opening of voltage-gated $K^+$ channel.

## 2.9 Myelin Sheath

The cell bodies of motor neurons innervating the muscles in human legs are located in the spinal cord, with axons about 1 m long. A neuron's axon is covered by a myelin sheath, which accelerates the nerve-impulse transmission; as a result, it takes only 10 ms for the action potential to pass through the axon and stimulate muscle contraction. The action potential conduction velocity may vary in the range of 10–100 m/s, depending on the type of myelinated neuron. Without myelin, the velocity will be approximately equal to 1 m/s, and coordination exercises, such as running, will be impossible.

Myelin sheath is a pile of specialized plasma membrane produced by glial cells, which wraps itself on axons. In the peripheral nervous system, these glial cells are called Schwann cells; in the central nervous system, they are called oligodendrocytes. Usually, several axons are surrounded by a single glial cell. In vertebrates and some invertebrates, axons are accompanied by neuroglia cells, but the specialization of neuroglia cells to form myelin sheath mainly occurs in vertebrates. On the surface of vertebrate neuroglia cells, myelinated protein-related glycoprotein and other protein bind to adjacent axons and trigger the formation of myelin sheath. The myelin membrane, like all membranes, contains phospholipid bilayer membrane, but unlike many other membranes, it contains only a few proteins. The major myelin protein in the peripheral nervous system is PO, which causes adjacent serosa to stack closely together. Myelin in nervous centralis contains cytoplasmic protein and membrane protein, which are called myelin and proteolipid, respectively, with functions similar to PO.

Myelin sheath around axons is formed by many glial cells. Each myelinated region formed by a single glial cell is separated from the next region, an unmyelinated region called the Ranvier node (or simply node). Since myelin prevents the transfer of ions between the axonal cytoplasm and extracellular fluid, all electrical activity in the axon is confined to Ranvier node, where ions can flow through the axonal membrane. Glial cells secrete protein hormone, which triggers the accumulation of $Na^+$ channels at the node in some way. Therefore, the nodal region contains high-density voltage-gated $Na^+$ channels (axonal primary membrane $\approx$10,000 per square micron), while there is few $Na^+$ channel in the axonal membrane region between nodes. $Na^+/K^+$ ATPases maintaining the axonal ion gradient are also located at the node. The connexins of fibrocytoskeletal proteins bind to these proteins and keep them in the nodal membrane.

The length of myelinated nerve is several millimeters, which is used for passive depolarization diffusion, because ions can only cross the axonal membrane at the unmyelinated nodes. Therefore, during membrane depolarization, the action potential-related excess cytoplasmic positive ions generated at one node passively diffuse to the next node via the axonal cytoplasm with little loss or attenuation, causing the depolarization of one node to quickly spread to the next node. This allows the action potential to actually "jump" from one node to another. As expected, myelinated nerves are used in signaling circuits where speed is important. The evolution of myelin also allows more rapidly conducting axons to occupy less space, which is clearly crucial for the evolution of vertebrate brains.

## 3 Synaptic Transmission

Despite certain variability in morphology among different types of neurons, each neuron consists of four regions with distinct functions: cell body, dendrite, axon, and axon terminal. The cell body contains the nucleus, where the synthesis of nearly all neuronal proteins and cell membranes occurs. The synthesis of some proteins occurs in dendrites; the synthesis of proteins that are required for the renewal of axons and nerve terminals occurs in the cell body, where they are assembled into membranous vesicles or multiprotein particles. In anterograde transport, these substances travel along axonal microtubules to the terminal, where they intercalate into the plasma membrane or other organelles. The axonal microtubules are also a path for the movement of injured membranes and organelles towards the cell body, which is called retrograde transportation. Almost every neuron has an axon, which ranges in diameter from a micron of some nerves in the human brain to a millimeter of the squid giant fibers. Axons are especially responsible for the conduction of electrical impulses encoded by action potentials from the cell body outward to the axon terminal. These properties distinguish the action potential from other types of potential changes across the plasma membrane and allow the action potential to move along the axon without attenuation.

Understanding synaptic transmission is the key to understanding the basic operation of the nervous system at the cellular level. Without transmission, there is no direct communication between cells. The whole effect of the nervous system is to control and coordinate the functions of the body, so that the body can respond and act on the environment. Synaptic transmission is the key process of nervous system integration. It refers to the process in which chemical or electrical signals on synapses are released from one neuron and spread to other neurons or target cells, where signals that stimulate, inhibit, or regulate cell activities are generated. Through synaptic transmission, the electrical signal in one neuron travels from the end of the axon into another cell, where it begins to have pulses with properties different from its own.

### 3.1 Chemical Synapses and Their Neurotransmitters

Synapses generally transmit signals only in one direction: an axon terminal from the presynaptic cell sends signals that are picked up by the postsynaptic cell. There are generally two types of synapses: chemical synapses and electrical synapses.

In 1921, Otto Loewi proposed that synaptic transmission was chemical. He believed that stimulating the vague nerve, which dominates the heart, would cause the heart to beat more slowly, and that if a chemical transmitter was released at the end of the vague nerve, it could be collected. Then, if the collected solution was applied to another heart, and the solution contained a chemical similar to vague nerve stimulation, the solution should be able to slow down the heart's beating. He did this experiment and found that the vague nerve did release a substance that could slow the heart rate. This experiment provided the first clear evidence that synaptic transmission was chemical. Later, it was found that this substance was acetylcholine (ACh), which had been suspected to be a transmitter of this synapse. Subsequently, experiments conducted in many laboratories showed that transmission was completely chemical in most synapses of the mammalian nervous system.

Many small molecules synthesized in the cytoplasm of axonal terminals act as neurotransmitters in various chemical synapses. "Classical" neurotransmitters are stored in synaptic vesicles, with uniform-sized organelles, diameters of 40–50 nm. In addition to ACh, classical neurotransmitters in Fig. 2.3 are amino acids or derivatives of amino acids. Nucleotides such as ATP and their corresponding nucleosides also act as neurotransmitters although lacking phosphate groups. Generally, each neuron produces only one typical neurotransmitter. Neurotransmitters enter the synaptic cleft from synaptic vesicles and then bind to specific receptors on the postsynaptic cytoplasmic membrane, resulting in changes in its permeability to ions. Many neurons secrete neuropeptides, a set of different signaling molecules, including endorphin, vasopressin, oxytocin, and gastrin. Compared with traditional neurotransmitters, neuropeptides are stored in different types of vesicles. The extracellular secretion of two types of transmitters is triggered by the local rise of intracellular $Ca^{2+}$, but neuropeptides are released outside the synaptic region. The effects of neuropeptide transmitters are various and usually last a long time (from hours to days).

In addition to ACh, all of these are amino acids (glycine and glutamic acid) or are derived from specific amino acids. Three neurotransmitters synthesized from tyrosine, in which contain catechol moiety (blue highlight), are called catecholamines.

The axonal terminals of chemical presynaptic cells contain vesicles filled with specific neurotransmitters. The postsynaptic cell may be the dendrite or cell body of another neuron, muscle or gland cell, or another axon. When the action potential of presynaptic cells reaches the axonal terminals, it induces a local rise of cytoplasmic $Ca^{2+}$ levels. Furthermore, it causes some vesicles to fuse with plasma membranes, releasing their contents into synaptic cleft, that is, narrow spaces between cells. Neurotransmitters diffuse in the synaptic cleft; it takes about 0.5 ms for them to bind to receptors on postsynaptic cells. The binding of neurotransmitters leads to the change of ion permeability of postsynaptic cell membrane, and then changes the potential of cell membrane at this time. If the postsynaptic cell is a neuron, it may be sufficient to induce action potentials. If the postsynaptic cell is a muscle, it may cause contraction; if it is a gland cell, it can induce hormone secretion. In some cases, enzymes within the cytoskeleton will destroy the neurotransmitter after it acts; in other cases, the signal terminates when the neurotransmitter diffuses or is transported back to presynaptic cells. Some postsynaptic neurons also send signals to presynaptic neurons. Such retrograde signals can be gases such as nitric oxide and carbon monoxide, or peptide hormone. This type of signal changes the ability of presynaptic cells to send signals to postsynaptic cells and is considered to be important in many types of learning.

One way to classify synapses is whether the activity of neurotransmitters tends to promote or inhibit the production

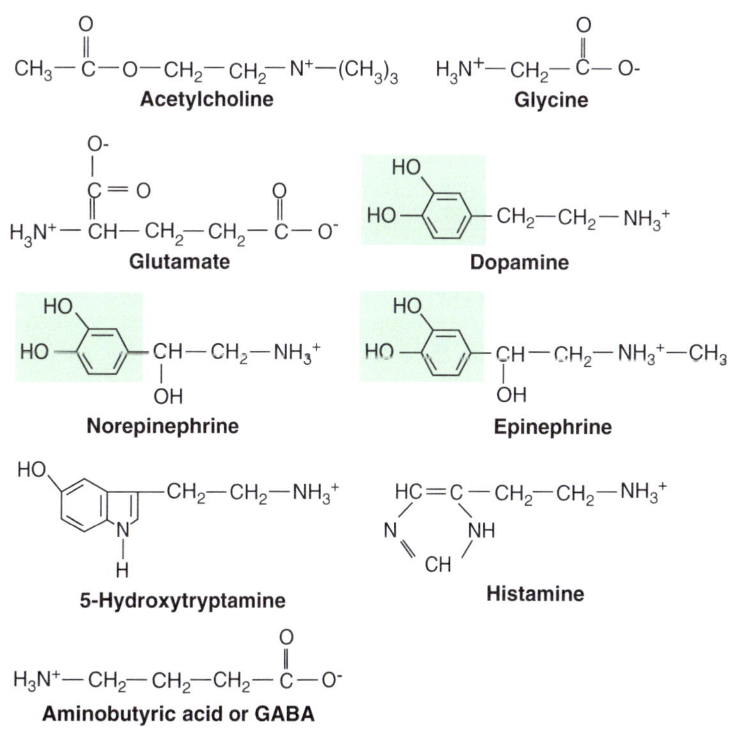


**Fig. 2.3** Chemical structure of several small molecular neurotransmitters

of action potentials in postsynaptic cells. Neurotransmitters bind to excitatory receptors to open $Na^+$ or $K^+$ channels. These non-voltage-gated ion channels can be part of the receptor protein or a single protein, which opens when the activated receptor generates cytoplasmic signals. Channel opening leads to depolarization of postsynaptic plasma membrane and promotes the generation of action potentials. On the contrary, the binding of neurotransmitters to inhibitory receptors on postsynaptic cells leads to the opening of $K^+$ or $Cl^-$ channels, and the resultant membrane hyperpolarization inhibits the generation of action potentials in postsynaptic cells. The same neurotransmitter (such as ACh) can produce excitatory response in some postsynaptic cells and inhibitory response in other cells. Many neuro-neuronal and most neuro-muscular chemical synapses are excitatory. Neurotransmitter receptors in postsynaptic plasma membrane can also be divided into two categories: ligand-gated ion channels and G protein-coupled receptors. Synapses containing both types can be either excitatory or inhibitory, but the response rates of the two types vary widely.

### 3.2 Electrical Synapse

In an electrical synapse, ions move directly from one neuron to another neuron via gap junctions. The membrane depolarization related to action potential in presynaptic cells leads to the depolarization of postsynaptic cells via gap junctions, thus producing action potential. Such batteries are called electrically coupled. In the heart, for example, electrical synapses allow muscle cells to contract synchronously. Gap junctions also connect non-excitable cells, allowing small molecules such as cAMP and amino acids to be transferred from one cell to another.

Electrical synapses also have the dominance of speed and certainty; the direct transmission of pulses avoids a delay of about 0.5 ms, which is the characteristic of chemical synapses. In some cases, this dominance of a fraction of a millisecond may mean the difference between life and death. For example, electrical synapses in a goldfish's brain regulate a tail flick reflex, thus enabling fish to escape from predators. As another example, the electrical coupling in the cell body and dendrites would ensure simultaneous depolarization of a whole cell. The large number of electrical synapses in many cold-blooded fishes indicate that they might be an adaptation to low temperature because the rate of cell metabolism in cold environment reduces the impulse transmission rate between chemical synapses.

Chemical synapses have two important dominances over electronic synapses in the pulse transmission of presynaptic cells. The first is signal amplification, which is common in neuromuscular synapses. The action potential of a single presynaptic motor neuron can cause the contraction of multiple muscle cells because the release of relatively few signal molecules at one synapse can stimulate the contraction of postsynaptic cells. The second dominance is signal calculation, which is common in synapses involving interneurons, especially in the central nervous system. A single neuron can be affected simultaneously by the signals received by multiple excitatory and inhibitory synapses. Neurons constantly average these signals and decide whether to produce action potentials. In this process, various depolarizations and hyperpolarizations produced by synapses passively spread from dendrites to cell bodies through cell membranes, and then spread to axon hillocks, where they gather. When the membrane on axon hillock depolarizes to a certain voltage called threshold potential, action potential will be produced. Whether a neuron produces an action potential in the axon hillock depends on the balance of the time, amplitude, and location of various inputs it receives; the signal calculation is different among various kinds of neurons. In a sense, each neuron is a microcomputer, which averages all receptor activation and electrical interference on the cell membrane and determines whether to trigger action potentials and transmit them to the axon. In any specific neuron, the action potential always has the same size. The frequency of action potential in a specific neuron is an important parameter of its ability to send signals to other cells.

## 4 Neuroplasticity and Neural Connection

Neuroplasticity refers to the ability of the nervous system to respond functionally and structurally to various internal and external events or damages, which is a key component of nerve development and normal function of the nervous system, and also a response to changing environment, aging, or pathological injury. Although it has been assumed that the plasticity of the nervous system is only limited to the development of the nervous system, there is now sufficient evidence that the adult brain has the ability to maintain reorganization or plasticity, especially in the critical period when neurons compete for synaptic positions, at this time the body is very sensitive to external influences. Plasticity changes also include the ability of neurons to alter their function, and the amount and type of neurotransmitters they produce. One indicator of neuroplasticity is synaptic plasticity regulation, which is closely related to learning and memory. There are two main modes of synaptic plasticity: long-term potentiation (LTP) and long-term depression (LTD).

## 4.1 Long-Term Potentiation

In the 1970s, Tim Bliss and Terje Lomo first reported the phenomenon of LTP, i.e., the improvement of synaptic efficacy after synaptic activity (Bliss and Gardner-Medwin 1973). After that, LTP has attracted attention as a potential memory-related mechanism. LTP can show many characteristics of synaptic mechanisms of associative memory, such as rapid induction, synaptic specificity, associative interaction, persistence, and dependence on synaptic-related activities, which may be the basis of memory formation. Therefore, these characteristics of LTP make it the key to participate in synaptic-related changes during the storage of neural information.

An important characteristic of LTP is that it can be induced rapidly. LTP appears within minutes shortly after stimulation induction (which usually includes postsynaptic depolarization caused by high-frequency stimulation of a sufficient number of afferent fibers). In some cases, the actual development of LTP is masked by short-term potentiation (STP), i.e., the synaptic strength increases exponentially, which also involves the N-methyl-D-aspartic acid (NMDA) receptor, and then gradually attenuates within 5–20 min. LTP is not always expressed so quickly and shows incremental growth in 10–20 minutes. This increment can only be observed in some specific cases, and the exact reason is unclear. The method used to induce LTP may determine the size of the initial LTP. For example, rapid LTP induction is achieved by directly stimulating the connection and execution pathways input to the CA3. However, when LTP is induced in a combined manner, a slow development and increment of LTP is usually observed, although LTP increases relatively quickly, it will take some time to grow completely. It should be noted that the development of LTP does not require STP, that is, LTP is also often observed even in the absence of STP. STP is a phenomenon, like rapidly induced LTP. For example, STP can be observed 0.5–1 s after 400 Hz burst pulse stimulation but is rarely observed in simulated theta rhythm (or the so-called theta pulse stimulation) stimulation emitted at 200 ms intervals. Therefore, it remains to be verified whether STP reflects processes related to endogenous synaptic activity patterns under normal circumstances, and whether it plays a role in learning or memory.

Another characteristic of LTP is its associativity. If a set of high-frequency stimulated afferent nerves induce LTP, a single active synapse can also express LTP, provided that the synapses are synergetic within the limited window. Associativity of LTP can be achieved by activating NMDA receptors, especially glutamate and postsynaptic depolarization, which are essential to alleviate the magnesium ion blockade of NMDA channels. Associativity is essentially the same as the nature of synergy. LTP has a specific threshold, which can be induced only when a certain number of active afferent fibers are achieved. LTP can be induced in "cooperative" synapses, indicating that the associativity is an inherent attribute of this phenomenon. Although it is thought that the term associativity refers to a cooperative LTP involving different afferent systems, it should be noted that the first demonstration of cooperativity uses two different afferent systems of the dentate gyrus: the medial and lateral perforator pathways, since the interaction between a critical number of afferent fibers may be the basis of this effect.

Another important characteristic of LTP is input specificity, which exists in almost all forms of LTP. It means that LTP has synaptic specificity and is only limited to synapses of activated afferent nerves. This is in contrast to non-associativity phenomena, for example, the more general response promotion is observed in other afferent inputs during sensitization without specific stimulation (or afferent input). Obviously, input specificity is an important feature, because when the plasticity of a single synapse is regulated, the information storage capacity will increase.

Another characteristic of LTP is its persistence. Before it was discovered, the only activity-dependent electrophysiological change with a duration close to LTP was post-tetanic potentiation (PTP), a presynaptic phenomenon mainly lasting from seconds to minutes. In contrast, LTP of hippocampal formation can last from hours to weeks or months, depending on specific stimulation parameters. In animals, LTP decreases gradually, and usually declines within 1–2 weeks. Although this period of time is clearly too short for long-term memory storage, a few points should be noted regarding the lifetime of LTP. First, LTP of hippocampus is not necessarily permanent. The current study results support the view that the hippocampus has a time-limited role in memory; persistent long-term memory is gradually consolidated in the neocortex. This view suggests that the memory formed in the hippocampus may be transferred to the neocortex and consolidated in the neocortex under slow-wave sleep or rapid eye movement sleep. Therefore, if the hippocampus is indeed a temporary information repository, LTP may not last long. In fact, long-term retention of any information beyond the normal weeks in the hippocampus may even damage hippocampal memory, thereby interfering with previously stored patterns of synaptic activity. How long can LTP last? It has been reported that LTP decay is an active process mediated by NMDA receptors, and blocking these receptors can prevent LTP decay. Although it is unclear at present, many forms of long-term depression (LTD), like LTP, are long-term changes in synaptic strength, which will reduce synaptic strength. Like LTP, LTD usually needs to activate NMDA receptor. Therefore, LTD may mediate LTP decay. However, since the LTP decay is an active process that requires NMDA receptor activation (or LTD induction), LTP is theoretically permanent. As long as it is not erased by the subsequent synaptic activity and NMDAR activation, it can exist as long as the synapse itself.

### 4.2 Long-Term Depression

It is particularly noteworthy that the rapid induction of LTP follows the principle of presynaptic and postsynaptic association formally proposed by Donald Hebb. However, a mechanism of increased synaptic strength cannot operate alone; otherwise, the synaptic strength can only continue to increase, eventually reaching a saturation point. The existence of other mechanisms that allow LTP inversion is necessary. This phenomenon is also observed on synapses showing LTP, which is called LTD. In the 1990s, with the viewpoint that "any substance as a temporary information repository must have some way to reduce synaptic strength" put forward, LTD gradually became the focus of many studies.

Compared with LTP, researchers have found that different forms of LTD have different induction mechanisms. Homosynaptic LTD refers to LTD with synaptic activity, which is usually induced by repeated low-frequency (0.5–5 Hz) stimulation. In most synapses, like LTP, homosynaptic LTD has input specificity, which depends on the activation and binding of NMDA receptor, and also requires the participation of calcium ions. However, the level of calcium influx required for LTD induction may be lower than that for LTP. To some extent, this reflects the regulation of phosphatase related to calcium-induced LTD, with much smaller changes in calcium concentrations.

LTD can also be observed when synaptic activity or LTP occur at adjacent synapses. This type of LTD is called a heterosynaptic LTD because it is observed on unenergized (or even inactive) synapses. Heterosynapses usually project most strongly to the odontoid process, and induction of LTP in one set of afferent fibers (such as the medial perforant pathway) can induce a heterosynaptic LTD response induced by a single set of inactive afferent fibers (the lateral perforant path), and vice versa. LTD induction is sensitive to both NMDA receptors and voltage-dependent calcium channels (VDCCs), indicating that the low level of calcium required for LTD may be provided by VDCCs activated in response to NMDA receptors, possibly through different NMDA receptors at different extra-synaptic locations (such as NR2B variants of NMDA receptors or similar receptors in dendrites and spinal bases). In addition, a different form of LTD in CA1 is also reported, involving metabotropic glutamate receptors, especially type I metabotropic receptors except NMDA receptors. This type of LTD is seen after the application of mGluR1/5 agonists and blocked by mGluR1/5 antagonists. Similarly, in the mossy fiber pathway, LTD appears to depend on postsynaptic factors, including postsynaptic calcium ions, while the other depends on the activation of presynaptic metabotropic glutamate receptors.

Different types or forms of LTD induction mechanism may reflect the different effects of these plastic forms in hippocampal function and memory. As mentioned earlier, one mechanism that can enhance synaptic strength will be accompanied by other mechanisms that reverse LTP or weaken synapses; otherwise, synaptic plasticity will always be disturbed in the process of gradual saturation. Since many nervous systems are characterized by gradual degeneration rather than continuous interference, synaptic enhancement in hippocampus is strictly regulated, and an activity-dependent mechanism may be used to weaken synaptic strength.

Therefore, LTD may play a role in reversing LTP (also called mitigation). However, LTD may also play a role in normalizing synaptic strength, which will ensure that the net excitatory input of neurons remains unchanged. This is a very important aspect of neuronal homeostasis: maintaining the dynamic range of neuronal output. Notably, LTP induced in one neuron's synapses might lead to a "net" decrease in the strength of other inactive synapses on the same neuron. LTD can also play a role in a small number of coding processes, ensuring that only the most active synapses increase strength in response to a given input. Other less active synapses are depressed, retaining the coding necessary for a distributed storage system using the Hebb rule.

## 5 Critical Brain Functional Network

The nervous system is a complex network of neurons and glia that are structurally interconnected and dynamically linked in complex patterns. Both structural and functional datasets of the brain can be presented as complex networks, facilitating network modeling and analysis, which can be performed across scales, from the microscale of individual neurons to the macroscale of whole-brain recordings. Understanding the role of connectivity in brain function is a long-term goal for cellular and systems neuroscience. Understanding the fundamental role of brain functional connectivity has been one of the core topics throughout its history. Scientists have prompted to establish a complete wiring diagram of the *C. elegans* nervous system and to map the interregional projections in the mammalian cerebral cortex. These early network diagrams sparked a flurry of theoretical and computational studies aimed at exploiting connectivity data to define spatial layouts, clusters and small-world properties, etc.

The brain network is a collection of nodes (neuron elements) and edges (their interconnections). The brain network is constructed by measuring the structural or functional relationships between pairs of neurons or brain regions. These pairwise relationships are summarized in the form of connectivity matrices that describe the relationships between nodes and edges (i.e., network topology). Techniques for extracting brain network data from structure or function are constantly evolving. Current approaches include reconstructing the morphology and connectivity of single-cell neurons using electron or light microscopy, improvements in large-scale optical recordings and noninvasive imaging techniques, and several new methods of labeling and tracking.

An important distinction lies in the differences between structural and brain functional networks. Structural networks are derived from anatomical datasets and represent the physical synaptic connections between neural elements, while functional networks are derived from neural recordings and represent their statistical relationships, such as covariances or cross-correlations. Structural networks are usually sparse (most possible structural connections do not exist) and relatively stable over time. In contrast, functional networks undergo rapid changes during both spontaneous and task-evoked activities and can be configured from vast quantities of time-series analysis measures. Importantly, "connectivity" in functional networks is only about the similarity or coherence of neural event processes, which may depend on, but do not directly correspond to, structural connections.

## 5.1 Structural Network

The structural network, also known as brain connectivity topology, refers to a network that is composed of individual neurons, groups of neurons with similar functions, nuclei and brain regions (as nodes) and the signaling pathways between them as edges. The network is characterized by dynamically evolving features across space and time and is an essential element for exploring brain functions. Researchers have been working to study the structure of the brain at different levels, from individual neurons to brain regions. Brain connectivity topology may be elucidated at three scales, i.e., microscale, mesoscale, and macroscale (Sporns 2016).

## 5.2 Microscale

Brain connectivity topology at microscale refers to the connections between neurons. Microscale studies rely on the development of automation histology (sophisticated electron or light microscopies) and reconstruction techniques with both sensitivity and scalability. Although it is impossible to map whole-brain network diagrams of complex organisms using these techniques, they have successfully mapped specific circuits in the nervous system of invertebrates and vertebrates.

Microscale connectivity studies in three model organisms (*C. elegans*, Drosophila, and mice) have increased our understanding of structural connectivity at the cellular level. The wiring diagram of the posterior nervous system of *C. elegans* adult male was reconstructed from serial electron micrograph sections. The neural network was found to have many features: multiple parallel pathways connecting sensory neurons to effector neurons, recurrent and reciprocal connectivity among sensory neurons, and modular substructure. The connectivity may be associated with specific aspects of sensorimotor processing and behavior. Studies of microscale wiring patterns in Drosophila have shown that the topology of specific subcircuits can be explained based on the mechanisms of wiring minimization and capacity mutual exclusion, both of which are aimed at saving space. The circuit reconstruction technique reveals the specific pattern of interneuron connectivity.

## 5.3 Mesoscale

Mesoscale brain network topology is found in many organisms. Some of the most important findings were investigated in Drosophila and mouse models. Some researchers collected high-resolution 3D images of about 16,000 single neurons in the Drosophila brain and used these images to assemble the whole-brain connection matrix. The individual neuron images are aggregated into functional partitions, the so-called local processing unit, to produce a mesoscale connectome composed of 41 nodes and their weighted interconnections. Cluster analysis reveals different network communities or modules, which functionally specialize in visual, olfactory, auditory, and motor processing. Some researchers obtained the rat brain network connection diagram through high-resolution optical imaging and tracking, in which the directional and weighted anatomical connections among 295 gray matter areas were plotted. Network analysis of this map showed that there were multiple densely connected clusters and network connection nodes in the rat brain. Due to the existence of network modules, different modules are strongly associated with different functions or behaviors, the connection modes of various regions are different, and the potential function contributions are also different, so high clustering phenomenon is common.

## 5.4 Macroscale

The macroscale mainly focuses on interregional or large-range projections in the primate cerebral cortex. Henry Kennedy et al. revealed the connection anatomy of macaque cerebral cortex (Kennedy et al. 2013). Retrograde tracer was injected at 29 cortical areas, and then the labeling density of the whole cortex was strictly quantified, and varying degrees of unknown connections among some areas were found. At present, the human brain connectome mainly relies on magnetic resonance diffusion imaging and tractography to reconstruct structural connectivity. The unique feature of noninvasive imaging methods is that they allow data to be acquired from a large number of individuals, thus opening up opportunities to measure individual variability and the relationship of connection characteristics to behavioral and cognitive performance, which is a step towards "population neuroscience".

A large number of studies have generated network diagrams of human connectomes. The network studies of the human structural connectivity pattern mainly lie in the generalized degree distribution study, including some nodes keeping a very high number of connections. Due to resolution constraints and problems associated with node division, the exact shape of the degree distribution is still somewhat uncertain, and most studies have shown that the degree distribution of nodes follows truncated exponential power-law distribution. Another common feature encountered in most, if not all, studies of human connectome networks is the "small world property", i.e., the presence of high clustering coefficients and short characteristic path lengths. This is important because the existence of small-world property is consistent with anatomical and functional separation (obtained by high clustering) and has the ability of global integration at the same time (obtained by short paths). The development of diffusion spectrum imaging technique has solved the problem of cross-fibers in the scale of single MRI voxel. In 2008, Hagmann established a brain network including 998 regions of interest using diffusion spectrum imaging technique, analyzed the degree of nodes, connection strength and centrality, and found that the nodes of the network were mainly distributed in the medial parietal lobe (Hagmann et al. 2008).

## 5.5 Functional Network

The structural network helps us understand the relationship among various brain regions, which is the structural basis for the explanation of neurophysiological activities. However, due to the complexity of brain network structure, the specific relationship between structure and function is unclear.

In the connection of brain functional network, there are functional connectivity and effective connectivity among neural activity signals recorded by different nodes. Functional connectivity refers to the time consistency of spatially separated neurophysiological events and whether electrophysiological signals of different brain regions are related or coherent, which can be described by functional connectivity. Effective connectivity refers to the causal relationship between two different neural units, which is a directed connectivity. Functional connectivity indicators include the simplest correlation, mutual information, Phase Locking Value (PLV), and Phase lag index (PLI), etc. Effective connectivity indicators mainly include Direct Transfer Function (DTF), Partial directed coherence (PDC), etc.

At present, studies on brain functional networks are mainly carried out on the macroscale. After the establishment of the brain functional connectivity network using techniques such as functional magnetic resonance imaging (fMRI), electroencephalogram (EEG), magnetoencephalography (MEG), and multi-electrode array, N*N functional/effective connectivity matrixes (N represents the number of cortical ROI brain regions) can be constructed by using the functional connectivity or effective connectivity indicators. With such a large-scale network, the method of graph theory can be used to analyze some specific and objective indicators, degree, clustering coefficient, characteristic path length, global efficiency, local efficiency, modularization of the network, etc., in combination with complex network analytical procedures, revealing its topological principles, so as to understand the working mechanism inside the brain.

fMRI is a special MRI, which can reflect the changes of neurons activity related to hemodynamic response in the brain by measuring the changes of blood flow oxygen content in the brain at resting or during activity, providing an important way for studying the function of the human brain. Brain fMRI based on blood oxygenation level has higher spatial resolution but relatively low temporal resolution. Scalp EEG is a test for detecting the electrical activity of the brain through a small metal plate (electrode) attached to the scalp. EEG has a temporal resolution of milliseconds and a high sampling rate, but the spatial resolution of scalp EEG is poor. MEG is a noninvasive technique used to study human brain activity. It can measure sustained brain activity on a millisecond basis and show where brain activity occurs.

These techniques complement each other, and fMRI signals show relative neuron activity by measuring the oxygenation of blood flow near active neurons, which means that fMRI signal analysis is always compared with reference neuron activity, indirectly reflecting brain activity. EEG and MEG signals are directly derived from the electrical activity of neurons, which can show absolute neuronal activity. MEG provides more accurate spatial localization of neural activity than EEG, and EEG is a supplementary method to record brain activity.

In 2005, Salvador et al. constructed the brain functional network of normal subjects at resting-state for the first time (Salvador et al. 2005). Based on the ALL brain atlas, they took 90 regions with different brain components as brain network nodes, calculated the partial correlation coefficients among various nodes of each subject, and obtained their respective functional connectivity. Finally, the functional networks of all subjects were analyzed, and it was found that their brain networks had "small world" property, optimized connectivity structure and high topological stability. Eguíluz measured the functional connectivity between voxels activated by any two specific tasks using the correlation coefficient at the voxel level for the first time and found that the voxel-based brain functional network is a scale-free "small world" network (Eguíluz et al. 2005). Many subsequent study results have consistently shown that the human brain is an efficient "small world" network (Sunaga et al. 2020). Several common brain network systems in the brain are as follows:

## 5.6 Default Mode Network

The default mode network (DMN) is a set of brain regions that are preferentially active when individuals are in a passive environment (heedless to the external environment) and diverting their attention from external stimuli. It mainly consists of the medial prefrontal cortex, anterior cingulate cortex, posterior cingulate cortex and precuneus, as well as bilateral inferior parietal lobules, etc. DMN works in a contradictory manner compared to other brain systems for external attention and sensory processing. When DMN is most active, the brain systems for focusing external attention are less active, and vice versa. Studies have shown that DMN may play a role in monitoring the external environment or may represent an individual's response to the external environment. Some studies have also hypothesized that DMN may be constructed on the basis of an individual's past experience; it is involved in the construction of self-relevant mental simulations, which are widely used in functions including memorizing, thinking about the future and inferring other people's viewpoints and ideas. DMN is related to a wide range of neuropsychiatric diseases including autism, schizophrenia, depression, obsessive-compulsive disorder, attention deficit/hyperactivity disorder, and Alzheimer's disease. Studies have found that decreased activity in DMN is associated with autism, while hyperactivity is observed in patients with schizophrenia. In patients with Alzheimer's disease, DMN is endangered before others, by amyloid deposition caused by the course of disease.

## 5.7 Frontal Parietal Control Network

The frontal parietal control network is mainly involved in goal-directed behavior, which is highly activated under the stimulation of external environment, participates in many advanced cognitive tasks, and plays an important role in adaptive cognitive control. The frontal parietal control network mainly includes brain regions such as the dorsolateral prefrontal cortex, parietal sulcus, dorsal anterior cingulate cortex, and anterior insula. Frontal parietal control network is primarily responsible for focusing attention on the associated stimulation, as behavioral choices are weighed against changing conditions, steady-state demand of background, and environment. In order to achieve this degree of response flexibility, the brain must control the performance of the posterior sensorimotor and keep relevant data in the brain until the action is activated and selected. Among them, the dorsolateral prefrontal cortex and the intraparietal sulcus are considered to be its core regions. Studies have found that the dorsolateral prefrontal cortex showed continuous activation in the activity of goal-directed tasks, and intraparietal sulcus was considered to play an important role in top-down attention control. The results showed that the frontal parietal control network could initiate and adjust controls during the task to stabilize and control goal-directed behavior. The intraparietal sulcus cortex might send stimulation conflict signals to the dorsolateral prefrontal cortex, which then adjusted control parameters accordingly to stabilize goal-directed behavior.

## 5.8 Salience Network

The salience network (SN) is mainly responsible for tracking brain internal events and external input information, evaluating and selecting various information, and making adaptive behavioral responses. The brain regions mainly included are bilateral anterior insula and dorsal anterior cingulate cortex, which are closely connected to structures such as the subcortical and limbic systems. The nervous system is constantly stimulated by internal and external stimuli. It is very important that how to find out the stimuli most related to steady state from this large amount of input information, preliminarily integrate the information and mark them accordingly. Only after evaluation can an organism decide what to do (or not to do) next. The regions such as the anterior insula and the dorsal anterior cingulate cortex in the salience network can simultaneously show activation in response to different forms of significant responses, including emotional dimension of pain, empathy for pain, hunger or pleasure, different feelings brought by various pressures or music, etc. The salience network is closely related to neuropsychiatric diseases such as schizophrenia, depression, and obesity.

## 5.9 Dorsal Attention Network

The dorsal attention network is mainly involved in top-down directed attention. The dorsal system is bilateral and consists of the junction of the intraparietal sulcus and the connecting area (frontal eye field) of precentral gyrus and superior frontal sulcus of each hemisphere. Attention is not a single function, it includes the limitation of resources and the selection arises from different processing levels and different cognitive fields, including perception, action, language, and memory. The dorsal attention network involves spontaneous (top-down) directed attention. The studies have found that the dorsal attention network is significantly activated when the subject's attention is guided by cues prompting direction, time, or topic during task operation, which represents the highly concentrated attention and efficient state of the brain to ensure the task is successfully completed. The dorsal attention network is in a state of opposition to the default mode network described above to some extent. When the default network is activated, our attention is often in a state of distraction, while the activation of the dorsal attention network can inhibit the default mode network to regain concentration. The dorsal attention network is closely related to neuropsychiatric diseases such as depression and attention deficit.

## 5.10 Ventral Attention Network

The ventral attention network is one of the two sensory orientation systems in the human brain, and the other is the dorsal attention network. The ventral attention network is considered to be mostly involved in involuntary behavior. Usually, the ventral attention network can be activated by infrequent or unexpected events related to behaviors. Therefore, this network is related to stimulus-driven attention control. The neural network is lateralized the right hemisphere, including the right temporoparietal junction and the right prefrontal cortex. The system shows increased activity when significant targets are identified, especially when they are in unexpected locations. After the sensory stimulation changes, increased activity of the ventral system can also be observed at the beginning, during the task cancellation, and at the end of the test. During top-down guided attention processing, such as visual search, or under high visual short-term memory, the activity in the ventral region (e.g., the temporoparietal junction) is inhibited. This is explained as a filtering mechanism during the concentration state, in order to protect the goal-driven behavior and visual short-term memory content from the interference of irrelevant interferents. These top-down signals most likely originate in the dorsal regions, such as the IPS and the frontal eye field, which are active during visual searching. Interestingly, when the significant non-target in the visual searching sequence carries the target stimulus information, the inactivation of the temporoparietal junction turns into activation. Furthermore, the analysis of effective connectivity indicates that this behavior depends to a large extent on dorsal-ventral interactions.

## References


Bliss T, Gardner-Medwin A (1973) Long-lasting potentiation of synaptic transmission in the dentate area of the unanesthetized rabbit following stimulation of the perforant path. J Physiol 232:357

Eguíluz VM, Chialvo DR, Cecchi GA et al (2005) Scale-free brain functional networks. Phys Rev Lett 94(1):018102

Hagmann P, Cammoun L, Gigandet X et al (2008) Mapping the structural core of human cerebral cortex. PLoS Biol 6(7):e159

Kennedy H, Knoblauch K, Toroczkai Z (2013) Why data coherence and quality is critical for understanding interareal cortical networks. NeuroImage 80:37–45

Salvador R, Suckling J, Coleman MR et al (2005) Neurophysiological architecture of functional magnetic resonance images of human brain. Cereb Cortex 15(9):1332–1342

Sporns O (2016) Connectome networks: from cells to systems. In: Kennedy H, Van Essen DC, Christe Y (eds) Micro-, meso- and macro-connectomics of the brain. Springer, Cham

Sunaga M, Takei Y, Kato Y et al (2020) Frequency-specific resting connectome in bipolar disorder: an MEG study. Front Psych 11:597

# Neuropsychiatric Function Evaluation

# 3

Liu He, Jiaqi Han, Wei Wang, Yan Ding, Yulian Niu, Shiyu Wang, and Weibi Chen

## 1 Clinical Evaluation of Motor Function

The evaluation of motor function is usually carried out in the following order: muscle volume, muscle tone, muscle strength, coordination, involuntary movement, posture and gait.

### 1.1 Muscle Volume

#### 1.1.1 Definition

Whether there is atrophy or hypertrophy in muscle volume and contour.

#### 1.1.2 Examination

Due to congenital and developmental difference, there are individual variances. For example, the muscle volume of the dominant limb is often larger than that of the contralateral limb; there may be smaller muscle volume in malnourished, sedentary, and elderly people, and athletes' muscles may show physiological hypertrophy. The evaluation of muscle volume should be organically combined with other examination of motor systems, especially the evaluation of muscle strength and muscle tone.

The volume and contour of muscle can be evaluated by visual examination, palpation and measurement. Visual examination: whether there is flattening, hollowing or bulging of muscle masses, the following parts shall be specially examined: face, shoulder, pelvic and distal limbs, especially the palm, thenar and interossei. Palpation is used to evaluate muscle volume, contour and texture. Normal muscle has certain elasticity under pressure, and can quickly return to its original state after being released. In true hypertrophy, the muscle becomes hard and strong; in pseudohypertrophy, the muscle becomes larger in appearance but feels like a dough or elastic rubber-like toy; atrophied muscles usually feel soft. Measurement is helpful for fine judgment, and the limb circumference can be measured with a tape measure for left-right comparison and follow-up observation. The symmetrical point should be selected in left and right limbs for measurement of the circumference to avoid measurement errors. If muscle atrophy or hypertrophy is found, its location, distribution and scope shall be recorded to determine whether it is generalized, lateralized, symmetrical or focal, whether it is limited to the innervation area of a peripheral nerve or the range of motion of a joint. If possible, the specific affected muscle or muscle group should be identified.

### 1.2 Muscle Tone

#### 1.2.1 Definition

Muscle tone refers to the tension of a muscle in a relaxed state, or the resistance of a completely relaxed muscle in passive movement. Under normal circumstances, the muscle has a slight, continuous, constant resistance when being passively moved in the relaxed state, by which the muscle maintains its position in the resting state, this is very important for maintaining a standing posture.

L. He
Department of Functional Neurosurgery, Beijing Institute of Functional Neurosurgery, Xuanwu Hospital, Capital Medical University, Beijing, China

J. Han · W. Wang (✉) · Y. Ding · W. Chen
Department of Neurology, Xuanwu Hospital, Capital Medical University, Beijing, China

Y. Niu
Department of Neurology, Daxing District People's Hospital, Beijing, China

S. Wang
Department of Pediatric, Xuanwu Hospital, Capital Medical University, Beijing, China



Y. Wang (ed.), *Therapeutics of Neural Stimulation for Neurological Disorders*, https://doi.org/10.1007/978-981-99-4538-2_3

### 1.2.2 Examination of Muscle Tone

It is a certain difficult to evaluate muscle tone. There is a certain subjectivity in the judgment of muscle tone, and the judgment result may be different for different examiners. At present, there is no quantitative method to judge muscle tone, and the measurement mainly depends on the examiner's judgment through clinical examination. In order to judge muscle tone accurately, rich clinical experience is necessary. When there is only a slight increase in muscle tone, it is difficult to distinguish it from the increased muscle tone in patients with tension or anxiety.

In the muscle tone examination, the most important is to judge muscular resistance by passively moving the limbs in the state of complete relaxation of the muscle, and evaluate the extent of its extension and flexion and range of motion at the same time. Abnormalities of tone are more easily found in the limbs than in the trunk. When passively moving the limbs, first inspect slowly in the range of motion of the whole limb, and then perform passive motion at different speeds. The judgment by comparing the same muscles on both sides during the examination is more helpful to find the changes of muscle tone. In addition, there are some special inspection techniques (Miller et al. 2009):

#### 1.2.2.1 Babinski Tonus Test

The patient is instructed to abduct the upper arm at the shoulder, and then check the passive flexibility of the forearm at the elbow. When the muscle tone decreases, the degree of flexion and range of motion of the forearm will increase, and the angle between forearm and upper arm during elbow flexion is smaller than normal. When muscle tone increases, the amplitude and range of motion of elbow flexion will decrease. In passive flexion of elbow, the angle between forearm and upper arm is an obtuse angle.

#### 1.2.2.2 Head-Dropping Test

The patient lies supine with pillow removed and completely relaxes with eyes closed and attention diverted. The examiner places one hand under the patient's occiput, lifts the head with the other hand, then lets the head drop. The normal head drops quickly on the examiner's protective hand; in patients with extrapyramidal disorder rigidity, the head dropping will be delayed to a gentle drop, which is caused by the rigidity affecting the flexors of the neck. In meningitis, flexion of the neck may cause neck resistance and neck pain.

#### 1.2.2.3 Shoulder Shaking Test

The examiner places the hands on the patient's shoulders, and then quickly shakes them, observing the back-and-forth movement of the patient's upper limbs. In patients with extrapyramidal disorders, the swing range of affected limbs reduces; in patients with decreased muscle tone, especially in patients with cerebellar diseases, the amplitude of arm swing is greater than that in normal people.

#### 1.2.2.4 Hand Posture

Patients with decreased muscle tone, especially those with cerebellar disease or Sydenham's chorea, may develop a specific posture of the hand, hyperextension of upper limbs and hands, wrist flexion, and hyperextension of fingers ("scoop sign") with mild pronation. When the patient's upper limbs raise above the head, the pronated posture is more pronounced, with palmar eversion.

#### 1.2.2.5 Activation Maneuver Test

During the examination, the patient keeps relaxed. The examiner flexes and straightens the patient's fingers and wrists while talking with him/her. If the basal ganglia disease is suspected, the examiner should rotate one wrist of the patient and let the patient repeatedly open or hold the other hand.

## 1.3 Muscle Strength

### 1.3.1 Definition

The ability of a muscle to resist resistance during contraction.

### 1.3.2 Examination of Muscle Strength

Muscle strength tests require muscles to be in a position where their major movements can be evaluated. The examiner asks the patient to make specific movements to show the movement of a certain muscle and maintain this movement to fully resist the examiner. Before muscle strength tests, patients should be informed how to cooperate with tests. If there is a decrease in muscle strength, attention should be paid to whether the weakness is constant or variable, and whether the muscle strength is improved after rest. Semi-quantitative results of muscle strength can be obtained by physical examination. If it is necessary to make quantitative judgment on muscle strength, various ergometries, muscle strength meters and dynamometers can be used.

When examining muscle strength, each muscle is scored separately, usually using the Grading Muscle Strength (GMS) scale, which divides muscle strength into 5 grades and 6 levels, as shown in Table 3.1.

The strength of each person is different, which is related to many factors such as height, weight, gender, physique, and age. When examining muscle strength, these differences will affect the examiner's judgment. At the same time, different examiners may have inconsistent results for the same patient because of these factors. In order to avoid the influence of above factors on results, the muscles that are less affected by individual factors should be checked as far as possible. The muscles with small muscle volume can be checked, for finger-to-finger movement, it is best to judge by antagonizing the corresponding muscles of the tested person, for example, the doctor uses the abductor pollicis brevis to resist the patient's abductor pollicis brevis. When examining the very powerful gastrocnemius muscle, it is difficult to find slight abnormalities when using the upper limb to resist the patient's foot movements. The examiner can make the patient walk on toes and heels, or make the patient jump on one foot, which is often easier to find abnormalities. Another is to make full use of the lever principle, for example, in the examination of the deltoid muscle, the patient extends his/her upper arm horizontally, if you place your hand on the middle and upper arms and press it down, it is better to press it down at the elbow, this is to use the lever principle, the delicate examiner can also fight against the strong male (William 2013). Examine according to the motion of each joint, see Table 3.2 for details. The examination of muscle strength of each muscle needs to test the corresponding specific movement strength, the specific method is shown in Table 3.3. No muscle strength of all muscles should be tested for every patient, and the key muscle groups should be selected for examination according to the condition.

**Table 3.1** Grading of muscle strength

| Grade | Muscle contraction |
|---|---|
| 0/5 | No muscle contraction is seen or identified with palpation; paralysis. |
| 1/5 | Muscle contraction is seen or identified with palpation, but it is insufficient to produce joint motion even with elimination of gravity. |
| 2/5 | The muscles can move joints against gravity to complete the whole range of motion, only when the force of gravity is eliminated. |
| 3/5 | The muscles can move joints against gravity to complete the whole range of motion, but not against any resistance. |
| 4/5 | The muscles can move joints against gravity to complete the whole range of motion, and against moderate resistance applied by the examiner. |
| 5/5 | The muscles can move joints against gravity to complete the whole range of motion, and against full resistance applied by the examiner. |

Notes:
1. The limit level of range of motion caused by contracture shall be evaluated according to the range of motion that may be completed, and shall be recorded in detail
2. Attention shall be paid to the influence of pain on activity, thus affecting the judgment of muscle strength grading
3. The following methods are commonly used to judge subtle differences in muscle strength: 4+/5 muscle strength can resist resistance, but there is obvious weakness. 5−/5 muscle strength is almost normal, but there is some slight weakness

**Table 3.2** Examination of muscle strength of muscle groups

| | |
|---|---|
| Shoulder | Abduction, adduction |
| Elbow | Flexion, extension |
| Wrist | Flexion, extension |
| Finger | Flexion, extension |
| Hip | Flexion, extension, abduction, adduction |
| Knee | Flexion, extension |
| Ankle | Dorsiflexion, plantar-flexion |
| Toe | Dorsiflexion, plantar-flexion |
| Trunk | Without the help of upper limb activities, the contractility of abdominal muscles is tested by raising the head and shoulders in supine position; the contractility of paraspinal muscles is tested by raising the head and shoulders in prone position. |

**Table 3.3** Test methods of muscle strength of skeletal muscles

| Muscle | Function | Test Methods |
|---|---|---|
| Supraspinatus | Upper arm abduction | Abduction of upper arm from vertical position with resistance applied by the examiner. |
| Infraspinatus | Upper arm external rotation | The upper arm is flexed 90° vertically, the upper arm is rotated outwards forcibly, and the examiner pushes the patient's forearm inward |
| Serratus anterior | Subscapular angle abduction and forward | Extend the arm and push forward with resistance applied by the examiner. The scapula at paralyzed side leaves the chest wall and presents as winged scapula. |
| Latissimus dorsi | Upper arm adduction, extension and internal rotation | The upper arm forces downward from the horizontal abduction with resistance applied by the examiner. |
| Pectoralis major | Upper arm adduction, flexion and internal rotation | Maintain the upper arm in horizontal abduction and the examiner pushes the patient's arm outward. |
| Deltoid | Upper arm abduction | Maintain the upper arm in horizontal abduction and the examiner pushes the patient's elbow down. |
| Biceps brachii | Forearm flexion and external rotation | Maintain elbow flexion and forearm external rotation, and the examiner straightens it out. |

(continued)

**Table 3.3** (continued)

| Muscle | Function | Test Methods |
|---|---|---|
| Triceps brachii | Forearm extension | Maintain elbow extension, and the examiner flexes it. |
| Pronator teres | Forearm internal rotation | Elbow semi-flexion, forearm internal rotation, with resistance by the examiner |
| Wrist extensor | Wrist extension | Maintain wrist dorsiflexion, and the examiner presses down from the back of the hand. |
| Extensor digitorum communis | Extension of metacarpophalangeal joint from index finger to little finger | Extend the distal knuckle of the thumb with resistance applied by the examiner. |
| Extensor pollicis longus | Extension of the distal knuckle of the thumb | Extend the distal knuckle of the thumb with resistance applied by the examiner. |
| Extensor pollicis brevis | Direct extension of the proximal thumb | Extend the distal knuckle of the thumb with resistance applied by the examiner. |
| Abductor pollicis longus | Thumb abduction | Extend the distal knuckle of the thumb with resistance applied by the examiner. |
| Abductor pollicis brevis | Thumb spread in a direction perpendicular to the palm | The patient does this, and the examiner applies resistance at the first metacarpal. |
| Flexor carpi radialis | Wrist flexion and abduction | Maintain wrist flexion, and the examiner applies resistance at the radial palm. |
| Flexor carpi ulnaris | Wrist flexion and adduction | Maintain wrist flexion, and the examiner applies resistance at the ulnar palm. |
| Flexor digitorum superficialis | Flexion of proximal interphalangeal joints from index finger to little finger | Flexion of middle knuckle with resistance applied by the examiner. |
| Flexor disitorum profundus | Flexion of distal interphalangeal joints | Flexion of distal knuckle with resistance applied by the examiner. |
| Flexor pollicis longus | Flexion of distal knuckle of thumb. | Flexion of distal knuckle of thumb with resistance applied by the examiner. |
| Flexor pollicis brevis | Flexion of proximal knuckle of thumb | Flexion of proximal knuckle of thumb with resistance applied by the examiner. |
| Opponens pollicis | Rotation of the first metacarpal forward to the palm | Each interphalangeal joint extends, distal knuckles of thumb and ring finger are close to each other at the palmar side, and the examiner separates them. |
| Lumbricales | Extension of interphalangeal joints | Extension of proximal interphalangeal joints without the application of pressure by the examiner. |
| Dorsal interosseous muscles of hand | Fingers separate (except thumb and little finger) | Separate the extended fingers, and the examiner gathers the middle three fingers together. |
| Palmar interosseous muscles of hand | Fingers gather (except thumb) | Clamp the paper strip with the extended fingers, and the examiner tries to pull it out. |
| Abductor digiti minimi | Little finger abduction | Abduct the extended little finger with resistance applied by the examiner. |
| Iliopsoas | Hip flexion | Supine, knee flexion, maintain hip flexion, and the examiner pushes the thigh toward the foot |
| Quadriceps | Knee extension | Supine, knee extension, and the examiner flexes it. |
| Adductor muscle group | Femoral adduction | Supine, lower limbs extension, two knees together, and the examiner will separate them. |
| Gluteus medius, gluteus minimus | External femoral abduction and internal rotation | Supine, lower limbs extension, two knees apart, and the examiner will make them together. |
| Gluteus maximus | Hip extension. | Prone, lower limbs extension, raising lower limbs, with resistance applied by the examiner |
| Tibialis anterior muscle | Foot dorsiflexion | Maintain foot dorsiflexion, and the examiner presses the dorsum of foot down. |
| Extensor hallucis longus | Toe extension and dorsum of foot | Fix the foot in the middle position, toe extension, with resistance applied by the examiner |
| Extensor digitorum longus | Toe extension and foot dorsiflexion | Fix the foot in the middle position, toe extension, with resistance applied by the examiner |
| Gastrocnemius, soleus | Foot plantar flexion | Knee extension, foot plantar flexion, with resistance applied by the examiner |
| Flexor hallicus longus | Toe plantar flexion | Fix the foot in the middle position, toe plantar flexion, with resistance applied by the examiner at toe distal phalange |
| Flexor digitorum longus | Toe plantar flexion | Fix the foot in the middle position, toe plantar extension, with resistance applied by the examiner |

**Table 3.3** (continued)

| Muscle | Function | Test Methods |
|---|---|---|
| Posterior tibial muscle | Foot varus | Foot plantar flexion, foot internal rotation, with resistance applied by the examiner at the inner margin of foot |
| Peroneal muscles | Foot eversion | Foot plantar flexion, foot external rotation, with resistance applied by the examiner at the outer margin of foot |
| Biceps femoris | Knee flexion | Prone, maintain knee flexion, and the examiner pushes the shank toward the foot |

## 1.4 Coordination

### 1.4.1 Definition

The completion of any action of the body depends on the coordinated movement of a certain group of muscle groups. This coordination mainly depends on the cerebellar function, but the vestibular nerve, optic nerve, deep sensation and extrapyramidal system are all involved.

### 1.4.2 Examination of Coordination

Observe the accuracy of the patient's actions such as dressing, buttoning, fetching, writing and gait, and whether the patient's speech is fluent. In addition, the following tests can also be performed for testing:

#### 1.4.2.1 Finger-to-Nose Test

Instruct the patient to extend one upper limb, touch own nose with the tip of index finger, and open eyes first and then closed, repeat the same action. Note the comparison of bilateral upper limb movements. In the case of cerebellar hemisphere lesions, the finger-to-nose is inaccurate on the affected side, and when approaching the nose, the movements become slower, and action tremors may occur, no significant difference between opening and closing eyes. The finger-to-nose inaccuracy caused by sensory ataxia is very different when opening and closing eyes, that is, the movements are steady and accurate when opening eyes, and difficult to complete when closing eyes.

#### 1.4.2.2 Barany Test

The patient extends his/her upper limb forward, and touches the fixed finger of the examiner with the palm surface of the index finger, then keeps the upper limb straight and raised, leaves the index finger from the examiner's finger to a vertical position of a certain height, and then descends it to the examiner's finger again. Open eyes first and then close eyes, repeat the same action. Note actions of opening and closing eyes and the comparison of the accuracy of bilateral actions. The patients with vestibular ataxia deviate the index fingers to the lesion side when bilateral upper limbs fall; those with cerebellar lesion deviate the affected upper limb to the lateral side; those with deep sensory disturbance cannot touch the target while closing eyes.

#### 1.4.2.3 Alternating Movement

Observe the accuracy and coordination of the patient's rapid, reciprocating motion: (1) Pronation and supination of the forearm, instruct the patient to quickly and alternately contact the bed or table top with the palm and the back of hand. (2) Finger extension and fist-clenching, quickly alternating. Cerebellar ataxia patients' movements are slow, irregular, and clumsy.

#### 1.4.2.4 Heel-Knee-Tibia Test

Instruct the patient to lie in the supine position, raise one lower limb, flex the knee, place the heel on the opposite knee, and then move it down to the ankle along the tibia. Patients with cerebellar ataxia have large and inaccurate movements when lifting the leg and touching the knee, and unstable shaking when moving the heel down along the tibia. It is difficult for patients with sensory ataxia to touch the knee accurately, and the contact with the tibia cannot be maintained when moving down.

#### 1.4.2.5 Rebound Test

Instruct the patient to flex the elbow with force, and the examiner holds his/her wrist to exert force in the opposite direction, and then suddenly releases his/her hand. Normal people terminate the flexion of forearm immediately due to antagonism of antagonist muscles. In the case of cerebellar lesions, due to the lack of antagonism of antagonist muscles, the force of elbow flexion makes the forearm or palm hit the body.

#### 1.4.2.6 Balanced Ataxia Test

1. Romberg's sign: Instruct the patient to stand upright with feet together, extend horizontally his/her hands forward, observe his/fer posture while opening eyes first and then closing eyes. Patients with sensory ataxia can maintain a stable standing posture while opening eyes, but stand unsteadily after closing eyes, which is called positive Romberg's sign. Patients with cerebellar ataxia stand unsteadily with eyes open or closed. Patients with unilateral cerebellar lesion or vestibular lesion fall to the affected side, and those with cerebellar vermis lesion fall to the posterior side.

2. Supine to sit test: Instruct the subject to sit up from the supine position without the aid of hands. Normal people press the lower limbs down while flexing the trunk, while patients with cerebellar ataxia also flex the hips while flexing the trunk, with both lower limbs lifting off the bed, which is called combined flexure sign.

## 1.5 Involuntary Movement

### 1.5.1 Definition

Involuntary movement refers to the uncontrollable and aimless abnormal movement of skeletal muscle when patients are in a state of consciousness.

### 1.5.2 Examination of Involuntary Movement

Abnormal involuntary movements are of various types, ranging from tremor to chorea, from muscle fibrillation to myoclonus. Their only common feature is that they are spontaneous and not controlled by will. Involuntary movements may be regular or random, transient or persistent, predictable or unpredictable, isolated or accompanied by other neurological signs. Common types include tremor, chorea, athetosis, hemiballism, dystonia, spasms, and dyskinesia. It is a certain difficult to evaluate muscle tone. Observe and inquire about the form, location, severity, regularity and process of involuntary movement, as well as its relationship with rest, activity, mood, sleep and temperature, etc., and pay attention to inquiring about family history.

## 1.6 Posture and Gait

### 1.6.1 Definition

Posture and condition of patients in lying, sitting, standing and walking.

### 1.6.2 Examination of Posture and Gait

Observe the patients' starting, foot lifting, foot dropping, stride length, step base, direction, rhythm, stopping and coordinated movements when walking, turning and stopping as instructed. Common gait abnormalities include the following: spastic hemiplegic gait, spastic scissor gait, staggering gait, festinating gait, myopathic gait, and steppage gait.

# 2 Clinical Evaluation of Movement Disorders

Movement Disorders is not due to muscle weakness, but due to the reduction of normal voluntary movements or the increase of abnormal and involuntary motor abnormalities, whereas the former is often accompanied by increased muscle tone. The complexity of movement disorders lies in the fact that the injuries in separate sites can present the same manifestations, while the same site involved in different causes can also have different symptoms, and its clinical diagnosis is mainly based on comprehensive judgment of detailed medical history, comprehensive and meticulous physical examination and necessary auxiliary examinations, etc. The clinical scales for movement disorders are the keys to objective and standardized assessment of disease status, which is particularly important for further clinical diagnosis and development of interventions.

## 2.1 Rigidity

Rigidity, defined as increased resistance to passive stretching, is one of main features of Parkinson's disease. Its evaluation is an integral part of clinical diagnosis and efficacy evaluation. Rigidity usually appears first on one wrist or elbow, so it is often clinically examined at the wrist, especially in the early stages of Parkinson's disease. The Rigidity Item 3.3 in the Movement Disorder Society-sponsored revision of the Unified Parkinson's Disease Rating Scale (MDS-UPDRS) is usually adopted for the clinical evaluation of rigidity, which is graded as follows: 0 = Normal, No rigidity; 1 = Slight, Rigidity only detected with activation maneuver; 2 = Mild, Rigidity detected without the activation maneuver, but full range of motion is easily achieved. 3 = Moderate, Rigidity detected without the activation maneuver; full range of motion is achieved with effort; 4 = Severe, Rigidity detected without the activation maneuver and full range of motion not achieved. Rigidity is judged separately on neck and limbs with patient relaxed in sitting position (Goetz et al. 2007).

KEN, Kuehn and other investigators proposed that the half-point method could be used to improve its sensitivity, which is graded as follows: 0 = Absent; 0.5 = Slightly detectable when activated by mirror or other movements, but only occasionally/intermittently; 1 = Slightly continuously detectable when activated by mirror or other movements; 1.5 = Continuously detectable but very slight; 2 = Mild to moderate; 2.5 = Moderate to marked; 3 = marked, but full range of motion easily achieved; 3.5 = full range of motion achieved with slight difficulty; 4 = Severe, full range of motion achieved with difficulty; 4.5 = Very severe, limited range of motion; however, due to the subjectivity of UPDRS, the scores of the same patient may vary greatly among different examiners, so it can only be used as a qualitative evaluation.

In recent decades, investigators have focused on quantifying rigidity in Parkinson's disease through exercise physiology and biomechanics methods. Some investigators proposed to extract the index from surface electromyography (EMG), but the results showed that surface EMG was greatly affected by electrode and soft tissue status, with a relatively low correlation with rigidity score of UPDRS. Some investigators proposed biomechanical measurements from the perspective of moment and joint, including the mechanical indicators such as viscoelastic properties (VEPs), work, standardized work, peak torque, elastic coefficient, mechanical impedance, which had a higher correlation with rigidity score than EMG measurement. Among them, VEPs had the highest significance and were not affected by speed.

## 2.2 Bradykinesia

Bradykinesia, one of cardinal motor symptoms of Parkinson's disease, is defined as the slowness of voluntary movements and also refers to low-amplitude movements (hypokinesia) and a gradual decrease in speed and amplitude during repetition of movements (sequence effects). MDS-UPDRS III is mainly adopted for clinical evaluation, in which the bradykinesia of limbs is directly assessed in items 4–8, and the akinesia and hypokinesia of trunk and limbs are assessed in items 9–11 and 14. However, because MDS-UPDRS is a discontinuous and coarse-grained scale, the resolution for minor changes in disease progress is limited, and the evaluation results also have strong experience dependence.

For the objective quantitative evaluation of bradykinesia, the timed percussion test using mechanical percussion device was recommended in the Core Assessment Program for Surgical Interventional Therapies in Parkinson's Disease (CAPSIT-PD) in the 1990s. With the development of information sensing technology, a variety of intelligent evaluation methods, such as ex vivo pressure sensor, noctovisor or laser sensor, have been applied to the evaluation of bradykinesia in recent years. In addition, wearable devices enrich the evaluation means of bradykinesia. The commonly used sensors include acceleration sensor, gyroscope, pressure sensor and magnetic motion sensor. The parameters used include amplitude, frequency, speed and frequency band analysis, but the data sample size, unified parameters and classification still need to be expanded (Teshuva et al. 2019).

## 2.3 Tremor

Tremor is an involuntary, rhythmic, oscillatory movement of a part of the body. It is caused by antagonistic muscle alternating or synchronous contraction. The evaluation of patients with tremor includes medical history inquiry, physical examination and specific laboratory examination.

### 2.3.1 Medical History

The history of tremor onset age and tremor development is usually straightforward because tremor is a readily detectable positive symptom for patients and their families. Checking the patients' past handwriting samples may help to determine the exact time of onset. Secondly, it is necessary to guide the patients to tell the inducing, aggravating or alleviating factors of tremor, such as caffeine, alcohol, drugs, exercise, fatigue or stress, and all medications of patients should be completely listed to exclude the possibility of enhanced physiological tremor.

### 2.3.2 Physical Examinations

A detailed neurological examination is performed to identify inducing factors and specific characteristics of tremor (including tremor frequency, pattern and distribution), as well as other possible neurological changes.

#### 2.3.2.1 Evaluation of Inducing Factors

Arm tremor should be observed under three conditions (see Table 3.4): keep the patient's affected upper limbs fully support on their thighs or armrests, raise the upper limbs against gravity, and perform target-directed movements. Most resting tremors subside while changing the arm posture, but re-emerge while maintaining posture, especially maintaining the anti-gravity extension, which is called "re-emergent tremor". Tremor in Parkinson's syndrome often occurs during repetitive movements of the contralateral hand, during walking, and during distraction.

**Table 3.4** Evaluate induced tremors in different states

| Category | Inducing factors | Method of evaluation |
|---|---|---|
| Resting tremor | Tremor occurs when the body part is completely supported and relaxed. | The patient lies flat on a bed or sits on a sofa. Tremors are usually enhanced by cognitive tasks or motor tasks in other parts of the body and can often be temporarily suppressed by spontaneous muscle contractions. |
| Action tremor | Simple kinetic tremor | Tremors are roughly the same throughout voluntary movements. |
| | Intention tremor | Tremor enhances gradually as the affected body part approaches the target. |
| | Task-specific tremor | Occurs when performing a specific task, such as handwriting. |
| | Postural tremor | Occurs when maintaining a particular posture or position, such as holding arms stretched or standing. |
| | Isometric tremor | Occurs when a muscle contracts to overcome a stationary object, such as fist-clenching or squeezing an object. |

Postural and motor tremors are most easily triggered by the following ways: arm extension; shoulder abduction, elbow flexion, index finger maintaining at 1 inch from the face; finger-to-nose test; drinking or pouring water from a paper cup. Writing and drawing may reveal patients with essential tremor who draw curved circles and micrographia in patients with Parkinson's disease. The sensitivity of the test can increase by keeping the patient's hand suspended above the table top and drawing a spiral.

Lower limb tremor should be evaluated in the resting state of the lower limb, in the heel-knee-tibial test, and when standing and walking, respectively. Lower limb tremor is more common in Parkinson's disease.

The gait of patients with essential tremor is almost normal, while the gait of patients with Parkinson's disease is characterized by a shuffling gait with narrow step width, and that of patients with cerebellar lesions is characterized by a wide-based gait with ataxia. The gait of patients with psychogenic tremor may have the nature of performance.

#### 2.3.2.2 Frequency of Tremor

The frequency of tremor is described by the number of oscillations per second, and most pathological tremors have a frequency of about 4–8 Hz. It can be subdivided into <4, 4–8, 8–12 and >12 Hz (see Table 3.5). It usually depends on clinical evaluation, but it can also be accurately measured using motion sensors or electromyography. At the same time, the tremor frequency must be evaluated in the resting state, posture maintenance state and movement process, respectively. The frequency of tremor varies with the etiology.

In addition, the frequency and amplitude of psychogenic tremor are often variable at different times. When the patient is distracted by complex repetitive movements with the contralateral limb, this tremor can become more irregular or disappear completely.

#### 2.3.2.3 Anatomical Distribution of Tremor

Horizontal or vertical tremors of the head are often accompanied by tremors in other parts of the body. But for isolated head tremor, the possibility of cervical dystonia or cerebellar midline syndrome should be considered. Localized tremors of the face, jaw, and lip are more common manifestations of Parkinson's disease. Idiopathic voice tremor is easily recognized, and maintaining long-syllable pronunciation may further enhance voice tremor (see Table 3.6) (Bhatia et al. 2018).

**Table 3.5** Classification of tremors at different frequencies

| Frequency of tremor | Type of tremor |
|---|---|
| High frequency | Enhanced physiological tremor (10–12 Hz), orthostatic tremor (14–20 Hz) |
| Medium to high frequency | Essential tremor (6–12 Hz) |
| Low to medium frequency | Resting tremor in Parkinson's disease (4–6 Hz) |
| Low frequency | Rubral tremor (3–5 Hz) |

**Table 3.6** Classification according to the location of tremor

| Category | Affected part |
|---|---|
| Focal | Only one body part is affected, such as voice, head, jaw or one limb |
| Segmental | 2 or more adjacent body parts in the upper or lower body are affected, such as the head and arms |
| Hemitremor | One side of the body is affected |
| Generalized | Both upper and lower body are affected. |

### 2.3.3 Common Evaluation Scales

Parkinson's disease tremor can be assessed by items 15 to 18 of MDS-UPDRS III for the resting tremor of head, upper limbs and lower limbs and the action or postural tremor of both hands; the common scale for essential tremor is the Fahn-Tolosa-Marin Tremor Rating Scale (FTM-TRS), which contains 21 items with a total score of 84 points.

## 2.4 Dystonia

Dystonia is a kind of movement disorder, and the definition of dystonia was updated by the International Dyskinesia Expert Consensus Committee in 2013, as follows: Dystonia is a movement disorder characterized by sustained or intermittent muscle contractions causing abnormal, often repetitive movements, postures, or both; dystonic movements are typically patterned and twisting, and may be tremulous. Dystonia is often initiated or worsened by voluntary action and associated with overflow muscle activation (Albanese et al. 2013).

### 2.4.1 Classification of Dystonia by Medical History and Clinical Characteristics

The current classification of dystonia is mainly based on clinical manifestations, including age of onset, body distribution, temporal pattern, and other accompanying features. Doctors can guide their diagnosis and treatment based on classification. (Albanese et al. 2013).

According to the onset age of dystonia, it can be divided into 5 types. Infantile dystonia, with patients younger than 2 years old. Childhood dystonia, with patients aged 3–12 years. Adolescent dystonia, with patients aged 13–20 years old. Adulthood dystonia occurs when the patient is over 20 years old. The 20–40 onset is defined as early adulthood dystonia, and after the age of 40, it is defined as late adulthood dystonia.

According to the location and scope of dystonia, it can be classified as localized, segmental, multifocal, generalized, and hemiplegic. Localized dystonia refers to the involvement of a certain part of the body by dystonia. For example, blepharospasm, oromandibular muscle tone disorder, spasmodic dysarthria, spasmodic torticollis, writing spasms, etc. Segmental dystonia refers to the condition that affects two or more connected body parts, such as Meige syndrome, brachial dystonia, etc. When two or more disconnected body parts are simultaneously affected, it is called multifocal dystonia, such as blepharospasm combined with writing spasms. Generalized dystonia refers to the involvement of three or more muscle groups in the head and neck, trunk, or limbs, such as torsion spasms. Hemitonic dystonia refers to the condition where the upper and lower limb muscle groups on the same side are affected, which is relatively rare and usually results from perinatal damage or secondary to trauma or stroke.

Classified by temporal pattern, including disease progression and variability. According to the progression of the disease, it can be divided into static and progressive types. According to variability, it can be divided into four types. Persistent type refers to dystonia persist to almost the same extent, such as spasmodic torticollis. Action specific disorders refer to dystonia that only occur in specific activities or tasks, often related to professions, such as writing spasms. The diurnal type has fluctuating changes throughout the day, such as dopa responsive muscle tone disorder, which manifests as mild symptoms in the morning and worsening in the evening. Paroxysmal type usually induced by certain factors and then spontaneously relieved. According to the different triggering factors, it can be divided into three main forms: (1) Paroxysmal kinesigenic dyskinesia induced by sudden movements; (2) Paroxysmal exercise-induced dyskinesia induced by continuous exercise such as running and swimming; (3) Paroxysmal non-kinesigenic dyskinesia induced by drinking alcohol, tea, and coffee, hunger, fatigue, and emotional fluctuations, etc.

According to the classification of accompanying symptoms, it can be divided into three types. Isolated type refers to dystonia as the only movement symptom, including dystonic tremors. Combined type refers to the dystonia with other movement phenotype, such as myoclonus or Parkinson's syndrome. Complex type refers to the dystonic manifestation combined with other neurological or systemic diseases.

Early-onset dystonia usually begins in a limb, often as inversion of the foot. Initially, the dystonia might be triggered only by vigorous physical activity such as running. Over time, the posturing becomes susceptible to triggering by minimal physical activity, such as walking or standing. Subsequently, the dystonia may be present even at rest.

Dystonia with focal onset in childhood often will progress to generalized dystonia. With early-onset focal dystonia, spread from one leg to other body areas, including the other leg, trunk, arms, and upper body, occurs in approximately 50 to 90 percent of patients, usually within 5 years of onset. However, when early-onset dystonia begins in the arm in late childhood and adolescence, it tends to have a lower likelihood of subsequent spread.

Compared with early-onset dystonia, adult-onset isolated focal or segmental dystonia usually involve the upper body, usually the arm, neck, face or throat, and begin after the age of 30 years. Although symptoms may worsen in the area of involvement or spread to contiguous body regions, adult-onset isolated dystonia rarely become generalized. Late-onset isolated focal or segmental dystonia may involve different body parts. Common and/or important types of focal dystonia include: cervical dystonia (dystonia of neck and shoulders), blepharospasm (dystonia of periocular muscles), oromandibular, lingual, or facial dystonia (dystonia of jaw, oral muscles, tongue, or facial muscles), laryngeal dystonia (dystonia of laryngeal muscles), limb dystonia (dystonia of arm or leg), task-specific or occupational dystonia (dystonia that occurs only with certain activities, such as writer's cramp, musicians focal dystonia or embouchure dystonia).

### 2.4.2 Common Evaluation Scales

A number of rating scales are used in the evaluation of dystonia-related disorders, as well as scales developed for the evaluation of specific disorders. Reliability, validity and sensitivity of the scale are critical in selecting the appropriate scale for a specific disease state (see Table 3.7) (Tarakad 2020).

## 2.5 Tic Disorders

Tic disorder is a rapid, non-rhythmic, repetitive movement, more common in children than in adults, which seems purposeful, but occurs relatively involuntarily, involving a whole muscle or a group of muscles with transient contractions, and usually accompanied by the movement of the affected limbs. Patients can usually temporarily inhibit this movement by focusing their attention, but when their attention is distracted, the movement re-emerges quickly. Active inhibition can lead to an intolerable sense of pressure rise, and there is an urgent need to exercise to focus attention and temporarily inhibit the tic attack. Tic attacks worsen during tension and disappear during sleep. According to clinical characteristics and duration of disease, Fifth Edition of Diagnostic and Statistical Manual of Mental Disorders (DSM-5) classifies tic disorders into three types (see Table 3.8): temporary (provisional) tic disorder, persistent (chronic) motor or vocal tic disorder and Tourette syndrome (TS) (Swedo 2013):

For the evaluation of tic disorder, common clinical scales include Tic Disorder Score (TIS), Hopkins Motor and Vocal Tic Scale (HMVTS) and Yale Global Tic Severity Scale (YGTSS) (Zhong et al. 2006). The most common assessment scale is the Yale Global Tic Severity Scale (YGTSS), which evaluates motor tic and vocal tic, respectively, and evaluates each category of tics in five aspects: the number, frequency, intensity, complexity and interference of tic, with six levels for each item (range 0–5 points). And independently evaluates the injury caused by tic disorder, adds it to the total tic score, and finally gets the total score of the scale. Interpretation of results: the severity of tic in children is judged as mild for <25 points, moderate for 25–50 points and severe for >50 points. At the same time, the scale can also be used to judge the efficacy: excellent response for the reduction rate of >60%; improvement for the reduction rate of 30%–59%; non-response for the reduction rate of <30%.

**Table 3.7** Common assessment scales for dystonia

| Scales | Details and scores |
|---|---|
| Burke Fahn Marsden Dystonia Rating Scale (BFMDRS) | The motion score can be obtained from the monitoring video, which allows the observer to evaluate without knowing. 9 body regions are scored. The scores of each region are derived from the severity and the inducing factors, with the lowest score of 0 and the highest score of 4. Because the eyes, mouth and neck are relatively unimportant in the motor function damage of systemic dystonia, the scores of these areas must also be multiplied by 0.5. The highest score is 120. |
| Unified Dystonia Rating Scale (UDRS) | Assessment of 14 body parts, including upper face, lower face, jaw and tongue, throat, neck, trunk, left and right proximal and distal arms/hands and legs/feet. The upper limb and lower limb use their own scoring criteria, without distinguishing between the proximal and distal limbs. The severity score of each body part ranges from 0 (no dystonia) to 4 (severe dystonia), with a maximum total score of 112. |
| Toronto Western Spasmodic Torticollis Rating Scale (TWSTR) | The scale consists of three parts, including symptom severity, disability and pain, with a score range from 0 to 85. |
| The Blepharospasm Disability Index (BSDI) | Patient self-assessment scale, used to measure the damage of specific activities of daily life caused by blepharospasm.It includes six items, namely driving, reading, watching TV, shopping, walking and doing daily activities. |

## 2.6 Ataxia

Ataxia refers to clumsiness or incoordination of movement, but it is not caused by muscle weakness. If the muscle strength impairment is obvious (grade 4 or less), checking for ataxia is pointless. According to the topical diagnosis of ataxia, ataxia can be divided into sensory, vestibular, cerebellar and cerebral. Although different parts have different causes, hereditary ataxia is the main cause. Symptoms of ataxia include eye movement, speech (dysarthria), limbs, trunk, posture, and gait (see Table 3.9).

The International Cooperative Ataxia Rating Scale (ICARS) was proposed by the Committee of the World Federation of Neurology in 1997, which is the only widely used scale for the evaluation of ataxia in the world and is widely used to evaluate the severity and efficacy of cerebellar ataxia (Trouillas et al. 1997).The scale involves postural and gait disturbances, limb ataxia, dysarthria, and eye movement disorders, with good reliability, but there is repetition between scoring items, and the scoring time is long. In 2006, Schmitz-Hubsch et al. proposed the Scale for the Assessment and Rating of Ataxia (SARA), which contains 8 items that

**Table 3.8** DSM-5 includes three types of tic disorders

| Category | Name | Manifestation |
|---|---|---|
| I | Provisional Tic Disorder (PTD) | Have one or more motor or vocal tics, manifested as recurrent, non-rhythmic, more fixed forms of tic or vocal, that occur many times per day, consistently for a period of at least 4 weeks but less than 1 year, occur before that age of 18 years, and patients caused by certain drugs or medical diseases (such as Huntington's chorea and viral encephalitis) are excluded. |
| II | Chronic Tic Disorder (CTD) | Have one or more motor or (and) vocal tics, in the form described above, the tic and vocalization do not necessarily occur at the same time, but may appear one after the other during the disease, and symptoms manifest for more than 1 year, during which there may be intermittent periods of improvement, but the duration is not more than 3 months. Other conditions are the same as TTD. |
| III | Tourette Syndrome (TS) | Have multiple motor and one or more vocal tics, which occur in the course of the disease, and are sudden, recurrent, non-rhythmic, and relatively fixed forms of tics or vocalizations that affect learning, social interaction, or employment. The medical history is more than 1 year, during which there may be intermittent improvement period, but not more than 3 months. Other conditions are the same as TTD and CTD. |
| IV | Tic disorder Not Otherwise Specified (NOS) | Including tic disorders that do not meet the above three categories of diagnostic indicators, such as those with onset for less than 4 weeks or after 18 years of age. |

**Table 3.9** The origin and clinical symptoms of ataxia

| Location of the lesion | Clinical symptoms |
|---|---|
| On deep sensory pathway | Deep sensory disturbances, gait instability, difficulty walking in the dark |
| Cerebellum | Drunken-type gait, dysmetria, dysdiadochokinesia, tremor, dysarthria, oculomotor dyskinesia, hypotonia |
| Vestibular system | Unsteady standing, postural balance disorder, vertigo, or Dandy syndrome |
| Prefrontal lobes | Postural balance disorder, gait instability, mental disorder, grasp reflex |
| Parietal lobe | Deep sensory disturbances, gait instability, urination and defecation disorder |
| Temporal lobe | The symptoms are mild, transient and difficult to find in the early stage. |

are related to gait (0–8 points), stance (0–6 points), sitting (0–4 points), speech (0–6 points), finger-chase test (0–4 points), nose-finger test (0–4 points), fast alternating movements (0–4 points) and heel-shin test (0–4 points) (Schmitz-Hübsch et al. 2006). SARA has a good correlation with ICARS. Despite the lack of eyeball movement assessment, SARA has more concise language and shorter assessment time, which can be used for daily assessment and follow-up of ataxia.

### 2.7 Chorea

Chorea is a hyperkinetic movement disorder characterized by rapid, random, and irregular movements. The movements are involuntary and nonpatterned with variable speed, timing, and direction, flowing from one body part to another. The unpredictability of chorea is an important feature that distinguishes it from tremor and dystonia. It mainly affects the distal limbs, but can also affect the face and trunk. One patient often has chorea, athetosis (slow flexural writhing), and ballism (large amplitude involuntary movements of the proximal parts of the limbs), which are now considered to belong to the same spectrum of chorea-like disorders. According to the etiology, it is divided into primary (idiopathic, hereditary) and secondary (acquired). Huntington's disease is the most common cause of hereditary chorea. The common cause of acquired chorea in children is group A streptococcal infection, while the common causes of acquired chorea in adults are drug-induced and vascular factors.

Huntington's disease (HD) can exhibit motor, cognitive and psychiatric symptoms similar to Parkinson's disease (Ross et al. 2014). The Unified Huntington's Disease Rating Scale (UHDRS) is the most commonly used for the evaluation in clinical practice and study. This scale evaluates cognition, motor, emotion, behavior and function. The standardized scores of oculomotor function, chorea, dystonia, gait and postural stability are used for the motor part of the scale. The motor score of the evaluation system has been validated and shows good reliability and sensitivity to changes. The occurrence of motor symptoms is a more convincing feature consistent with the disease. The cognitive assessment in the UHDRS scale is performed through the Verbal Fluency Test, Symbol Digit Modalities Test (SDMT) and the Stroop Interference Test, which cover multiple cognitive domains, including visual attention, working memory, symbol coding, psychomotor speed, cognitive flexibility, response inhibition, selective attention, language and executive function. The cognitive subscale has been validated and shows good reliability, but its results in terms of sensitivity to detect change need to be improved.

## 3 Clinical Evaluation of Swallowing Function

Swallowing refers to the process that food or water ingested by the body from outside the body gradually enters the stomach through the mouth, pharynx and esophagus. Normal swallowing is a fluent, coordinated process that can be roughly divided into the oropharyngeal and esophageal phases. Objectively, dysphagia is defined as the abnormal delay in the transport of food or water in oropharyngeal or esophageal phase. In addition, dysphagia can also be subjectively manifested as the sensation of swallowing delay, such as functional dysphagia. Organic diseases of swallowing structures such as the mouth, jaw, throat and esophagus, or functional abnormalities of nerves and muscles involved in swallowing function, will lead to dysphagia. The clinical evaluation of swallowing function can be roughly divided into three parts: physical examination, evaluation scale and auxiliary examination evaluations (Jalil et al. 2015). The following will be classified and summarized.

### 3.1 Physical Examinations

#### 3.1.1 General Conditions

It mainly includes the evaluation on wakefulness degree, spiritual outlook and nutritional status. Patients with poor consciousness such as coma have dysphagia and cannot eat orally, and must be fed by nasogastric tube or given intravenous nutrition. The possibility of malignant tumors or chronic diseases should be warned for dysphagia patients with poor mental and nutritional status.

#### 3.1.2 Higher Cortical Function Examinations

Impaired higher cortical function can also affect normal swallowing function. Higher cortical function examinations mainly include the determination of orientation, comprehension, memory and capacity of calculation. In addition, affective disorders can also cause subjective disturbance of ingestion.

### 3.1.3 Cranial Nerve Examination

Cranial nerve examination is an important link to evaluate dysphagia. It mainly involves the examination of cranial nerves IX, X and VII. The hypoglossal nerve innervates the ipsilateral genioglossus muscle, controlling the movement of the tongue. When evaluating swallowing function, it is necessary to pay attention to whether the patient has tongue deviation, tongue muscle atrophy and tremor, and whether the soft palate is lifted forcefully (Wilkinson et al. 2021). When the unilateral hypoglossal nucleus or hypoglossal nerve is damaged, the tongue deviates to the affected side, with tongue atrophy on the affected side, mostly accompanied by fibrillation of tongue muscle. In the case of supranuclear palsy of the hypoglossal nerve, the tongue deviates to the opposite side, and there is no fibrillation and atrophy of the tongue muscle, which are common in cerebrovascular diseases and brain neoplasms. The palatine muscle is innervated by the vagus and glossopharyngeal nerves. The damage to the glossopharyngeal nerve causes paralysis of unilateral soft palate. Patients with bilateral paralysis of the soft palate often have obvious symptoms of bulbar palsy, such as dysphagia, coughing when drinking water, and hoarseness of voice.

### 3.1.4 Others

A comprehensive and systematic physical examination, such as direct visual examination for facial deformities, oral examination for foreign bodies and auscultation of breath sounds, can help identify causes and assess the severity of dysphagia.

## 3.2 Scale Evaluation

### 3.2.1 Water Swallowing Test

The Water Swallowing Test (WST) was proposed by Japanese researchers Kubota et al. in 1982 to evaluate dysphagia, which is one of the most common and convenient methods for neurologists to screen dysphagia in clinical practice and can be used for the early diagnosis of acute cerebrovascular diseases. In addition, the WST can be used to select the patients with treatment indications and serve as the screening criteria for swallowing angiography.

The implementation of the test requires that patients should be in a state of consciousness and be able to cooperate subjectively to complete the test. Generally, the experimental observer first instructs the patient to drink a few teaspoons of water, and then instructs the patient to drink 30 mL of clear water or liquid at a time if no swallowing problem appears, adjusts the specific amount of water to drink according to the actual situation, observes and records the drinking time, whether there is coughing and choking at drinking water, etc. Determine whether the patient has dysphagia and the degree of dysphagia according to the swallowing speed and reaction of patients.

### 3.2.2 Other Scales

The scale for evaluating swallowing function usually asks the patient about the performance, frequency, severity, chronic degree and accompanying symptoms of dysphagia for further evaluation of the severity and etiological diagnosis of the disease, such as the Mayo Dysphagia Questionnaire (MDQ), and Esophageal Symptoms Questionnaire (ESQ).

## 3.3 Auxiliary Examinations

### 3.3.1 Electronic Fiberoptic Endoscopy

Electronic fiberoptic endoscopy (such as electronic fiberoptic laryngoscopy, esophageal endoscopy) is a common examination for the diagnosis of dysphagia in clinical practice. Through nasal or oral invasive endoscopy, organic hyperplasia or stenosis in the throat and esophagus of patients can be directly observed under the microscope. At the same time, the real-time dynamic observation can also be realized under the electronic endoscope to evaluate the transport of food or water, the contraction of esophageal sphincter and esophageal peristalsis during swallowing, which provides more valuable information for the diagnosis and differential diagnosis of dysphagia. However, it's different from swallowing radiography that electronic fiber endoscopy can only observe local lesions, and cannot realize overall observation.

## 3.4 Videofluoroscopic Swallowing Study

Video-fluoroscopic swallowing study (VFSS) is the "gold standard" for clinical diagnosis of dysphagia. In the process of radiography, dynamic changes of oral, pharyngeal, esophageal and other organs in the whole swallowing process can be observed intuitively and efficiently by giving patients with different textures and volumes of contrast agent food, so as to further clarify the anatomical and physiological mechanism of swallowing function of patients and determine whether patients have dysphagia, etc. In case of organic hyperplasia or stenosis, it can directly observe the imaging abnormalities such as filling defect and niche, as well as the opening of sphincter and esophageal peristalsis, but it is not conducive to the observation of local fine structures. The combination of electronic fiber endoscopy and VFSS is an important method for clinical diagnosis of dysphagia.

# 4 Clinical Language Assessment

## 4.1 Classification of Dysphasia

Dysphasia is divided to two major types: dysarthria or dysphonia, and aphasia (Orchardson 2012). Dysarthria or dysphonia, which results in abnormal changes of tone, pitch and volume in patients and affects their daily interpersonal communication and exchange. Dysarthria or dysphonia may be caused by the organic hyperplasia or stenotic lesions in pharynx, the abnormal movement of musculus vocalis due to lesions in nerves or muscles controlling vocal cords, and the uncoordinated movement of the lips and tongue (Stachler et al. 2018).

Aphasia is common in post-stroke patients. It is an impairment in the expression, use or comprehension of words and sentences due to damage to the language center of the dominant hemisphere and does not involve lesions of vocal organs. Aphasia can be further classified into motor aphasia, sensory aphasia and other types (Tippett 2015). The common ones are as follows:

### 4.1.1 Motor Aphasia (Broca's Aphasia)

It is associated with damage to the Broca's area in the posterior gyrus frontal inferior of the dominant hemisphere. It is generally by prominent expressive language disorder, loss of verbal expression, utterance of single words only, labored speech, and inability to repeat the words of others, usually accompanied by dysgraphia in patients, while they can understand what others say.

### 4.1.2 Sensory Aphasia (Wernicke's Aphasia)

It is associated with damage to the Wernicke's area in the posterior gyrus temporalis superior of the dominant hemisphere. It is manifested by prominent auditory comprehension impairment, inability to understand the words of others, and loss of reading ability. The patients often express fluently but are not aware of what they are saying and cannot repeat what others say.

### 4.1.3 Amnestic Aphasia

It is mostly associated with damage to the posterior gyrus temporalis medius or temporo-parieto-occipital junction of the dominant hemisphere of the brain. It is manifested by prominent anomia. Patients often express fluently with words, have basically normal verbal comprehension, and are able to repeat the words of others but unable to naming an object.

## 4.2 Examination of Dysphasia

The clinical assessment of dysphasia is performed mainly by checking clinical histories and communicating with the patient. The examination of dysphasia involves the following:

### 4.2.1 Auditory Comprehension

When checking the patient's auditory comprehension, give a certain action command to the patient and observe whether the patient responds correctly. For example, ask the patient to touch the right ear with the left hand to judge whether the patient's auditory comprehension is impaired; or ask the patient to explain a certain idiom or repeat a certain passage. Gradually increase the difficulty to assess the patient's ability in understanding short and long sentences and distinguishing synonyms, etc.

### 4.2.2 Verbal Expression

When examining the patient ability of verbal expression, it is necessary to observe whether the patient shows labored speech, difficulty in finding words, stereotyped speech, repetitive speech, disorder in naming objects, etc. The patient ability to narrate a passage may be tested as appropriate. The examiner may have a conversation with the patient on a topic to determine whether the patient has verbal expression disorder.

### 4.2.3 Reading Comprehension

During the examination, ask the patient to explain the meaning of a word, short sentence or long sentence in a book, distinguish English letters or Arabic numerals, or ask the patient to read aloud and retell a passage. Observe the patient's performance in the meanwhile.

### 4.2.4 Dysgraphia

During the examination, ask the patient to write down his/her name on a piece of paper or write randomly, or ask him/her to copy or dictate a passage. Observe the patient's performance in the meanwhile.

## 4.3 Severity of Dysphasia

The classification criterion of Boston Diagnostic Aphasia Examination (BDAE) is now adopted internationally to classify aphasia into the following 6 levels of severity (Goodglass et al. 2011):

Level 0: The words of the patient does not make any sense or the patients loses auditory comprehension;

Level 1: There are very few fragments of words and discontinuous expressions in the patient's verbal expression requiring the listener to speculate and guess. There are limited topics on which the patient can be communicated with, making it difficult for the listener to communicate;

Level 2: With the help of the listener, the patient can communicate about topics that are relatively familiar to him/her, but cannot communicate about unfamiliar topics. Both parties find it difficult to communicate.

Level 3: The patient can discuss daily topics without help or with only a little help but is weak in expression and comprehension, and finds it difficult to communicate about things unfamiliar to his/her daily life;

Level 4: The patient is fluent in speech and have comprehension difficulties, but can basically express their thoughts and words;

Level 5: Dysphasia is minimal. Only the patient himself/herself feels difficulty in communication, while the listener is unaware of it.

## 5 Clinical Pain Assessment

Pain is generally divided into acute pain and chronic pain. Acute pain is often a symptom of certain diseases and lasts for a short period of time. Chronic pain lasts longer, usually for more than 1 month, including more than 40 types such as joint pain, frozen shoulder, migraine, protrusion of intervertebral disc, trigeminal neuralgia, post-herpetic neuralgia, cancer pain, tennis elbow, and vasculitis. Chronic pain brings great misery to patients and seriously affects their quality of life. Various assessment tools assess pain usually from the aspects of pain intensity and pain quality.

### 5.1 Pain Intensity Assessment

#### 5.1.1 Unidimensional Pain Assessment Scales

The first applied clinically was the unidimensional pain assessment, which is simple to understand and applicable for a wide range of people, especially for the elderly, children, and people with low-level literacy (Connelly and Neville 2010). For example, there is the numeric rating scale (NRS), the faces pain scale (FPS), the visual analogue scale (VAS), and the verbal rating scale (VRS), also known as the verbal descriptor scale (VDS).

In NRS (see Fig. 3.1), numbers from 0 to 10 are used to represent pain intensity. In VRS and VDS, the pain intensity from "no pain" to "most painful" are distinguished by "mild, moderate, severe and extreme". It is applicable to simple clinical assessments. In VAS (see Fig. 3.2), a 10 cm-long straight line is used to represent the pain level, with "0 to 10" corresponding to "no pain to most painful". In FPS, 6 different facial expressions (from "smile" to "cry") are used to represent pain severity, which is more visual and easier to understand, and is applicable to all ages. In general, unidimensional clinical assessment scales are simple, easy to use, applicable to a wide range of population, and relatively accurate for pain assessment and clinical efficacy analysis, and hence are widely applied in clinical practice (Chien et al. 2013).

#### 5.1.2 Ancillary Tests for Pain Intensity Assessment

Although the assessment with unidimensional pain scales above is simple and easy to perform, there are problems such as lack of objectivity and difficulty in implementation for patients not cooperating. Therefore, ancillary tests may also be conducted for objective clinical assessment of pain intensity, such as functional magnetic resonance imaging (fMRI), quantitative sensory testing (QST), pain threshold measurement and neuromodulation technique.

### 5.2 Pain Quality Assessment

At present, the common clinical assessment scales for neuropathic pain include LANSS (Leeds assessment of neuropathic symptoms and signs), ID pain, NPQ (neuropathic pain questionnaire), and DN4 (Douleur neuropathiqueen 4) questionnaire. It is easy to operate and implement, and has been gradually promoted in the clinical application.

#### 5.2.1 ID Pain Scale

The questionnaire was completed by the patient (see Table 3.10). There are 6 questions asking about pain sensation and pain location, with a total score of 5 (Portenoy 2006). It is often used for the initial screening and assessment of patients with neuropathic pain.

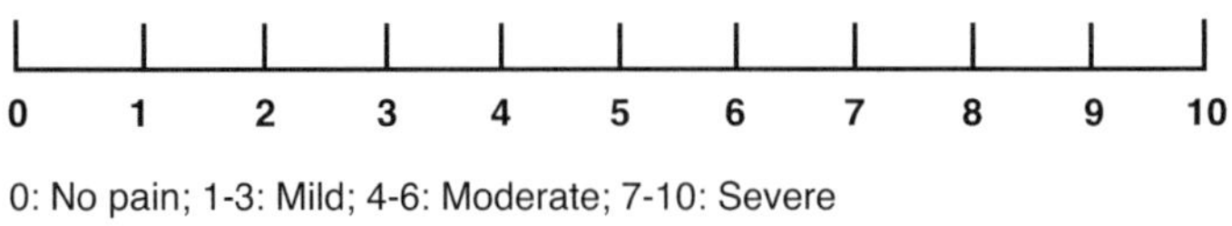


**Fig. 3.1** Numerical rating scale (NRS)

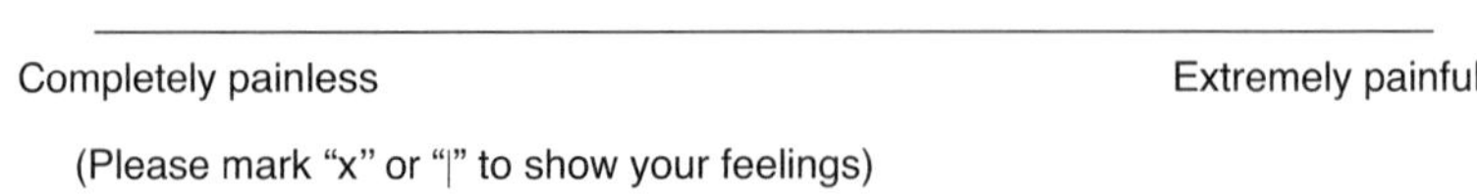


**Fig. 3.2** Visual analogue scale (VAS)

**Table 3.10** ID pain scale

| Question | Score Yes | Score No |
|---|---|---|
| 1. Did the pain feel like pins and needles? | 1 | 1. Did the pain feel like pins and needles? |
| 2. Did the pain feel hot/ burning? | 1 | 2. Did the pain feel hot/ burning? |
| 3. Did the pain feel numb? | 1 | 3. Did the pain feel numb? |
| 4. Did the pain feel like electrical shocks? | 1 | 4. Did the pain feel like electrical shocks? |
| 5. Is the pain made worse with the touch of clothing or bed sheets? | 1 | 5. Is the pain made worse with the touch of clothing or bed sheets? |
| 6. Is the pain limited to your joints? | −1 | 6. Is the pain limited to your joints? |
| Total score: Highest score = 5 lowest score = −1 | | Total score: Highest score = 5 lowest score = −1 |

### 5.2.2 LANSS Scale

It is completed by a physician. It can provide direct clinical information, including 5 items of symptoms and 2 items of clinical signs, with a total score of 24 (Bennett 2001).

Part A is pain questionnaire, including 5 questions: 1) Does your pain feel like strange, unpleasant sensations in your skin? 2) Does your pain make the skin in the painful area look different from normal? 3) Does your pain make the affected skin abnormally sensitive to touch? 4) Does your pain come on suddenly and in bursts for no apparent reason when you're still. 5) Does your pain feel as if the skin temperature in the painful area has changed abnormally? Part B is sensory testing including 1) allodynia; 2) altered pin-prick threshold. The development of the LANSS Pain Scale enabled clarification of the relative contributions of neuropathic symptoms to the diagnostic process. And it attempts to estimate the probability that neuropathic mechanisms contribute to the chronic pain experience in a given patient. Assessors are therefore instructed that "if score is less than 12, then neuropathic mechanisms are unlikely to contribute to the patient's pain" and "if score is 12 or more, then neuropathic mechanisms are likely to contribute to the patient's pain".

### 5.2.3 NPQ Scale

The NPQ scale includes 10 items of perception and 2 items of sensation, whereas the simplified NPQ scale contains only 3 items of numbness, pricking pain, and allodynia.

### 5.2.4 DN4 Scale

DN4 scale is developed by the French Neuropathic Pain Group consisting of both sensory descriptors and signs related to bedside sensory examination (Bouhassira et al. 2005). This provisional questionnaire was not intended to be exhaustive and was deliberately restricted to a minimum of simple and presumably discriminant items requiring yes or no responses. Two questions (I and II) were based on the interview of the patient and two questions (III and IV) were based on a standardized clinical examination.

The DN4 questionnaire is a clinician-administered questionnaire which has been deliberately reduced to a minimum number of items and simplified in its scoring. And it has been proved to be a diagnostic tool which could be easily used by pain specialists or non-specialists in daily clinical practice as a screening tool to better detect neuropathic pain. In addition, the good discriminant properties of the 7 items based on the interview alone suggest that this questionnaire could be used in a very large range of clinical studies including, for example, telephonic surveys and make it suitable for epidemiological studies.

## 5.3 Comprehensive Assessment of Pain Characteristic and Intensity

The above scales for assessing pain intensity and characteristic, although easy to use, are not comprehensive. Sometimes a comprehensive assessment of pain intensity and characteristic is still needed. The McGill pain questionnaire (MPQ) is currently the most widely used in clinical practice to assess the nature and characteristic of pain (Melzack 1975). Dworkin et al. developed the Short-form MPQ-2 (SF-MPQ-2, see Table 3.11) based on the simplified MPQ scale (Dworkina et al. 2009). SF-MPQ-2 contains 22 items (shown in the table below), including 16 questions for nociceptive pain and 6 questions for neuropathic pain. The simplified MPQ-2 scale also refines the classification of pain and clearly defines the method and criteria for measuring hyperpathia, allowing a more comprehensive and accurate assessment of the characteristic and intensity of pain.

| | | | | | | | | | | | | | |
|---|---|---|---|---|---|---|---|---|---|---|---|---|---|
| 1 Throbbing pain | None | 0 | 1 | 2 | 3 | 4 | 5 | 6 | 7 | 8 | 9 | 10 | Worst possible |
| 2 Shooting pain | None | 0 | 1 | 2 | 3 | 4 | 5 | 6 | 7 | 8 | 9 | 10 | Worst possible |
| 3 Stabbing pain | None | 0 | 1 | 2 | 3 | 4 | 5 | 6 | 7 | 8 | 9 | 10 | Worst possible |
| 4 Sharp pain | None | 0 | 1 | 2 | 3 | 4 | 5 | 6 | 7 | 8 | 9 | 10 | Worst possible |
| 5 Cramping pain | None | 0 | 1 | 2 | 3 | 4 | 5 | 6 | 7 | 8 | 9 | 10 | Worst possible |
| 6 Gnawing pain | None | 0 | 1 | 2 | 3 | 4 | 5 | 6 | 7 | 8 | 9 | 10 | Worst possible |
| 7 Hot-burning pain | None | 0 | 1 | 2 | 3 | 4 | 5 | 6 | 7 | 8 | 9 | 10 | Worst possible |
| 8 Aching pain | None | 0 | 1 | 2 | 3 | 4 | 5 | 6 | 7 | 8 | 9 | 10 | Worst possible |
| 9 Heavy pain | None | 0 | 1 | 2 | 3 | 4 | 5 | 6 | 7 | 8 | 9 | 10 | Worst possible |
| 10 Tender | None | 0 | 1 | 2 | 3 | 4 | 5 | 6 | 7 | 8 | 9 | 10 | Worst possible |
| 11 Splitting pain | None | 0 | 1 | 2 | 3 | 4 | 5 | 6 | 7 | 8 | 9 | 10 | Worst possible |
| 12 Tiring-exhausting | None | 0 | 1 | 2 | 3 | 4 | 5 | 6 | 7 | 8 | 9 | 10 | Worst possible |
| 13 Sickening | None | 0 | 1 | 2 | 3 | 4 | 5 | 6 | 7 | 8 | 9 | 10 | Worst possible |
| 14 Fearful | None | 0 | 1 | 2 | 3 | 4 | 5 | 6 | 7 | 8 | 9 | 10 | Worst possible |
| 15 Punishing-cruel | None | 0 | 1 | 2 | 3 | 4 | 5 | 6 | 7 | 8 | 9 | 10 | Worst possible |
| 16 Electric-shock pain | None | 0 | 1 | 2 | 3 | 4 | 5 | 6 | 7 | 8 | 9 | 10 | Worst possible |
| 17 Cold-freezing pain | None | 0 | 1 | 2 | 3 | 4 | 5 | 6 | 7 | 8 | 9 | 10 | Worst possible |
| 18 Piercing | None | 0 | 1 | 2 | 3 | 4 | 5 | 6 | 7 | 8 | 9 | 10 | Worst possible |
| 19 Pain caused by light touch | None | 0 | 1 | 2 | 3 | 4 | 5 | 6 | 7 | 8 | 9 | 10 | Worst possible |
| 20 Itching | None | 0 | 1 | 2 | 3 | 4 | 5 | 6 | 7 | 8 | 9 | 10 | Worst possible |
| 21 Tingling or "pins and needles" | None | 0 | 1 | 2 | 3 | 4 | 5 | 6 | 7 | 8 | 9 | 10 | Worst possible |
| 22 Numbness | None | 0 | 1 | 2 | 3 | 4 | 5 | 6 | 7 | 8 | 9 | 10 | Worst possible |

**Table 3.11** Simplified MPQ-2 Scale (Please put an X through the numbers that best describe the intensity of each of the pain and related symptoms you felt during the past week. Use 0 if the word does not describe your pain or related symptoms.)

# 6 Clinical Cognitive Assessment

It is during the cognitive process that humans understand objective things and acquire knowledge. Cognition includes attention, memory, perception, language, learning, and thinking. The main purposes of cognitive function assessment are to assist in the classification of cognitive impairment, the evaluation of the cognitive impairment degree, and the evaluation of treatment effects. Among them, the cognitive assessment scale tool is commonly used in clinical practice, such as the assessment scale for activity of daily living, the overall cognitive functional assessment scale, the psychobehavioral assessment scale, the single-item cognitive function assessment, and the dementia rating scale.

## 6.1 Overall Cognitive Assessment Scale

### 6.1.1 Wechsler Adult Intelligence Scale

Wechsler adult intelligence scale (WAIS) is used to test the overall intelligence level. The Wechsler Adult Intelligence Scale—Fourth Edition (WAIS-IV) is the most recent edition of the WAIS. It is used to assess intellectual ability in adults and adolescents ages 16–90 and provides information about specific cognitive abilities across various domains. It is frequently used with other instruments in comprehensive evaluations. (Drozdick et al. 2018). It is used to test the overall intelligence level. At present, the "Wechsler Adult Intelligence Scale-Revised by China" (WAIS-RC) is commonly used in China. It includes 11 subtests, divided into a textual and a non-textual part. The textual part is the verbal function test, including 6 subtests of knowledge, comprehension, arithmetic, similarity, digit span and vocabulary. The non-textual part is the operational ability test, including 5 subtests of digit symbols, painting, wooden block building, picture arrangement, and picture piecing. WAIS-RC is accurate for the overall assessment of intelligence, but takes a long time to complete the tests of the whole scale and is often used in scientific research.

### 6.1.2 Mini-Mental State Examination Scale

Mini-Mental State Examination (MMSE), simple to operate and with good reliability, is widely used in clinical practice and often used for dementia screening (Tierney et al. 1997). See the attached Table 3.12. MMSE is mainly performed from the following 7 aspects, namely temporal orientation, place orientation, immediate recall, attention and calculation, delayed recall, language, and visual space. The MMSE scale consists of 30 items, with 1 point for each correct

**Table 3.12** Mini Mental State Examination (MMSE)

| Name______Gender______Age______Education______Diagnosis Department______Inpatient ID______Bed No.______ | | |
|---|---|---|
| **Orientation**<br>What is the (year) (season) (date) (day) (month)? | Score | Maximum score |
| Where are we: (state) (county) (town) (hospital) (floor)? | () | 5 |
| **Registration**<br>Name 3 objects: 1 s to say each. Then ask the patient all 3 after you have said them.<br>Give 1 point for each correct answer. Then repeat them until he learns all 3. Count trials and record. | () | 3 |
| **Attention and calculation**<br>Serial 7's. 1 point for each correct. Stop after 5 answers. Alternatively spell "world" backwards. | () | 5 |
| **Recall**<br>Ask for the 3 objects repeated above. Give 1 point for each correct. | () | 3 |
| **Language** | | |
| Name a pencil, and watch (2 points)<br>Repeat the following "no ifs, ands or buts." (1 point)<br>Follow a 3-stage command:<br>"Take a paper in your right hand, fold it in half, and put it on the floor" (3 points)<br>Read and obey the following:<br>CLOSE YOUR EYES (1 point)<br>Write a sentence (1 point)<br>Copy design (1 point) | () | 9 |
| **Score of 27–30: Normal** **Total score ()**<br>**Score <27: Cognitive impairment**<br>**Rehabilitation diagnosis:** _______________<br>**Time of test using the scale:** **Signature:** | | |

INSTRUCTIONS FOR ADMINISTRATION OF MINI-MENTAL STATE EXAMINATION

ORIENTATION

1. Ask for the date. Then ask specifically for parts omitted, e.g., "Can you also tell me what season it is?" One point for each correct.
2. Ask in turn "Can you tell me the name of this hospital?" (town, county, etc.). One point for each correct.

REGISTRATION

Ask the patient if you may test his memory. Then say the names of 3 unrelated objects, clearly and slowly, about 1 s for each. After you have said all 3, ask him to repeat them. This first repetition determines.

his score (0–3) but keep saying them until he can repeat all 3, up to 6 trials. If he does not eventually learn all 3, recall cannot be meaningfully tested.

ATTENTION AND CALCULATION

Ask the patient to begin with 100 and count backwards by 7. Stop after 5 subtractions (93, 86,79,72,65). Score the total number of correct answers. If the patient cannot or will not perform this task, ask him to spell the word "world" backwards. The score is the number of letters in correct order. E.g., dlrow = 5, dlorw = 3.

RECALL

Ask the patient if he can recall the 3 words you previously asked him to remember. Score 0–3.

LANGUAGE

Naming: Show the patient a wrist watch and ask him what it is. Repeat for pencil. Score 0–2.

Repetition: Ask the patient to repeat the sentence after you. Allow only one trial. Score 0 or 1.

3-Stage command: Give the patient a piece of plain blank paper and repeat the command. Score 1 point for each part correctly executed.

Reading: On a blank piece of paper print the sentence "Close your eyes", in letters large enough for the patient to see clearly. Ask him to read it and do what it says. Score 1 point only if he actually closes his eyes.

Writing: Give the patient a blank piece of paper and ask him to write a sentence for you. Do not dictate a sentence, it is to be written spontaneously. It must contain a subject and verb and be sensible. Correct grammar and punctuation are not necessary.

Copying: On a clean piece of paper, draw intersecting pentagons, each side about 1 in., and ask him to copy it exactly as it is. All 10 angles must be present and 2 must intersect to score 1 point. Tremor and rotation are ignored.

Estimate the patient's level of sensorium along a continuum, from alert on the left to coma on the right.

answer, and 0 points for wrong answers or "don't know", with a total score of 0–30. The normal threshold is classified according to the level of education: illiteracy >17, primary school level >20, junior high school level and above >24. It should be noted that as the evaluation of this scale is limited by the level of education, it may be insensitive to patients with high literacy and may be determined as a "false positive" for patients with low literacy. In addition, the MMSE is mainly performed from the aspect of language functions and is not sensitive enough for the evaluation of non-language functions.

### 6.1.3 Alzhermer's Disease Assessment Scale

The Alzhermer's disease Assessment Scale (ADAS) allows a more comprehensive assessment of common symptoms in AD patients and includes both cognitive and non-cognitive functions. The cognitive subscale (ADAS-cog) contains 12 items for the assessment of cognitive deficits in AD patients, with a total score ranging from 0 (no errors or no impairment) to 75 (severely impaired cognitive function). ADAS can be used both for the early diagnosis of AD and to assess disease progression. Currently, it is mostly used in longitudinal observational follow-up studies and the evaluations of clinical drug efficacy in trials.

### 6.1.4 Severe Impairment Battery

For patients with severe cognitive impairment, many cognitive assessment scales are limited by the patient's inability to properly complete the test. The Severe Impairment Battery (SIB), on the other hand, is aimed at the group with severe cognitive impairment, with tests easy to understand, making up for the deficiencies of other scales. It is divided into several areas such as social communication, memory, orientation, linguistic competence, application skills, and visuospatial skills.

## 6.2 Activities of Daily Living Scale

Activities of Daily Living (ADL) is an assessment of somatic function. It is not only applicable to special groups such as elderly patients, but also to AD patients. The assessment of somatic function and living ability may also reflect cognitive function. The ADL can be used to reflect the severity and progression of the disease, especially for patients with severe cognitive impairment who are unable to complete some scales. The assessment with ADL is simple to conduct, and the information required can be provided by the patient's family or caregivers. Hence, it is suitable for patients with disorders of consciousness and severe cognitive impairment, and unable to cooperate with examinations. There are various types of ADL subscales with different rating criteria, which can be selected as the case may be. The common scales for activities of daily living include the physical-self maintenance scale (PSMS) and the instrumental ADL (IADL).

## 6.3 Psychobehavioral Assessment Scales

### 6.3.1 Neuropsychiatric Inventory

Neuropsychiatric Inventory (NPI) is mainly used for the psychiatric-psychological assessment of patients with abnormal brain functions, and also for the measurement of behavioral abnormalities in patients with dementia. It involves the following 12 behavioral domains: delusions, hallucinations, agitation/aggression, depression/dysthymia, anxiety, hyperthymia/euphoria, emotional indifference/apathy, depression, irritability, abnormal motor behavior, sleep/nocturnal behavior, appetite and eating disorders. Information required by the NPI assessment is primarily provided by the patient's caregiver. The scale hence is not applicable for patients without caregivers.

## 6.4 Dementia Rating Scales

### 6.4.1 Global Deterioration Scale of AD Grading System

Global Deterioration Scale (GDS) classifies the cognitive impairment into 7 grades ranging from "normal" (without cognitive disorder) to "severe cognitive impairment" based on the assessment of the following aspects (Reisberg et al. 1988): memory, orientation (orientation to time, place and person), instrumental activities of daily living (IADL), personality and mood, and activities of daily living (ADL). GDS is non-objective as it is used to make evaluation by asking questions of the patient himself/herself and the caregiver. See the attached Table 3.13.

### 6.4.2 Clinical Dementia Rating

Clinical Dementia Rating (CDR) is performed by a physician based on his/her overall impression of the patient and the patient's family after questioning (Morris 1993). The following 6 domains are evaluated: memory, orientation, judgment and problem-solving abilities, social affairs (working, shopping, group activities, etc.), family and hobbies, and the ability to live alone. The final results are divided into 5 levels of normal, suspicious, mild, moderate, and severe impairment depending on the score. See the attached Table 3.14.

**Table 3.13** Global deterioration scale

| |
|---|
| **Stage 1: No cognitive decline** |
| • No subjective complaints of memory deficit |
| • No memory deficit evident on clinical interviews |
| Yes/No |
| **Stage 2: Very mild cognitive decline** |
| • Subjective complaints of memory deficit, usually manifested as the following: |
| 1. Forgetting where one has placed familiar objects |
| 2. Forgetting the names of familiar persons, with no objective evidence of memory deficit on clinical interviews |
| • No objective deficits in employment or social situations, with appropriate concern regarding symptoms |
| Yes/No |
| **Stage 3: Mild cognitive decline** |
| • Earliest and clear-cut cognitive deficits, with two or more of the following manifestations: |
| 1. The patient may have gotten lost when traveling to an unfamiliar location; |
| 2. Co-workers become aware of the patient's relatively low performance; |
| 3. Word and name finding deficit becomes evident to intimates; |
| 4. The patient remembers little after reading an article or a book; |
| 5. The patient may demonstrate decreased capacity in remembering names upon introduction to new people. |
| 6. The patient may have lost or misplaced an object of value. |
| 7. Attention deficit evident on clinical interviews. |
| • Denial in the psychological defense mechanism begins to manifest in the patient, with mild to moderate symptoms of anxiety. |
| Yes/No |
| **Stage 4: Moderate cognitive decline** |
| • Clear-cut cognitive deficits, as manifested by the following: |
| 1. Decreased knowledge of current and recent events; |
| 2. Deficit in memory of one's personal history; |
| 3. Concentration deficit elicited on successive subtractions; |
| 4. Decreased ability to travel, handle finances, etc. |
| • Generally no deficit in the following areas: |
| 1. Orientation to time and person; |
| 2. Recognition of familiar persons and faces; |
| 3. Ability to travel to familiar locations. |
| • Inability to perform complex tasks; denial is dominant in the psychological defense mechanism; flattening of affect and withdrawal from challenging situations occur. |
| Yes/No |
| **Stage 5: Moderately severe cognitive decline** |
| • At this stage, the patient retains some knowledge of relevant events related to him/herself or others: |
| 1. The patient needs care in activities of daily living but not in eating and urinating/defecating. Many patients have trouble selecting appropriate clothes. |
| 2. During a clinical interview, the patient is unable to recall relevant pieces of his/her lives, e.g., an address or telephone number used for years, names of family members (such as grandchildren), name of high school or college from which the patient graduated, and disorientation to places. |
| 3. An educated patient even has difficulty in successive subtractions (successively subtracting 40 by 4 or 20 by 2). |
| 4. The patient knows his/her own name and usually the names of his/her spouse and only child as well. |
| Yes/No |
| **Stage 6: Severe cognitive decline** |
| 1. The patient forgets his/her spouse's name, most recent experiences and most of the events. |
| 2. The patient can almost always remember his/her own name. |
| 3. The patient can often distinguish between acquaintances and strangers around them. |
| 4. The patient usually cannot recognize surroundings, does not know the year, seasons, etc., with disturbed circadian rhythm. |
| 5. The patient retains some but very little previous experiences and knowledge. |
| 6. The patient may have difficulty adding and subtracting within 10. |

(continued)

**Table 3.13** (continued)

| |
|---|
| 7. The patient needs care in daily life, shows signs of incontinence, needs help to go out, and occasionally goes to familiar places. |
| • There are erratic changes in personality and mood, including: |
| 1. There are delusional behaviors, such as blaming one's spouse for being a liar, talking to imaginary figures, and may talk to one's self in the mirror; |
| 2. There are obsessive-compulsive symptoms, such as repeating the simple behaviors of cleaning; |
| 3. There are symptoms of anxiety, agitation, and even violent behavior never seen before; |
| 4. Cognitive will weakens, such as unable to maintain one thought long enough to decide on behaviors, leading to the loss of willpower. |
| Yes/No |
| **Stage 7: Very severe cognitive decline** |
| 1. The patient loses speech, and is often unable to speak but grunts. |
| 2. The patient needs help with dieting and urinating/defecating. |
| 3. The patient loses basic psychomotor skills, for example, he/she is unable to walk, and his/her brain no longer seems to be able to command the body. A wide range of cortical neurological symptoms and signs are often found from clinical reviews. |
| Yes/No |

**Table 3.14** Clinical dementia rating

<table>
<tr><th>Item</th><th>None<br>0</th><th>Questionable<br>0.5</th><th>Mild<br>1.0</th><th>Moderate<br>2.0</th><th>Severe<br>3.0</th></tr>
<tr><td>Memory</td><td>No memory loss or slight inconstant forgetfulness</td><td>Consistent slight forgetfulness, partial recollection of events, "benign" forgetfulness</td><td>Moderate memory loss; more marked for recent events; defect interferes with everyday activities</td><td>Severe memory loss; only highly learned material retained: New material rapidly lost</td><td>Severe memory loss; only fragments remain</td></tr>
<tr><td>Orientation</td><td>Fully oriented</td><td>Fully oriented except for slight difficulty with time relationships</td><td>Moderate difficulty with time relationships; oriented for place at examination; may have geographic disorientation elsewhere</td><td>Severe difficulty with time relationships; usually disoriented to time, often to place</td><td>Orientation to person only</td></tr>
<tr><td>Judgment and problem solving</td><td>Solves everyday problems and handles business and financial affairs well; judgment good in relation to past performance</td><td>Slight impairment in solving problems, similarities, and differences</td><td>Moderate difficulty in handling problems, similarities, and differences: Social judgment usually maintained</td><td>Severely impaired in handling problems, similarities, and differences; social judgment usually impaired</td><td>Unable to make judgments or solve problems</td></tr>
<tr><td rowspan="2">Community affairs</td><td rowspan="2">Independent function at usual level in job, shopping, and volunteer and social groups</td><td rowspan="2">Slight impairment in these activities</td><td rowspan="2">Unable to function independently at these activities although may still be engaged in some; appears normal to casual inspection</td><td colspan="2">No pretense of independent function outside home</td></tr>
<tr><td>Appears well enough to be taken to functions outside a family home</td><td>Appears too ill to be taken to functions outside a family home</td></tr>
<tr><td>Home and hobbies</td><td>Life at home, hobbies, and intellectual interests well maintained</td><td>Life at home, hobbies, and intellectual interests slightly impaired</td><td>Mild but definite impairment of function at home; more difficult chores abandoned; more complicated hobbies and interests abandoned</td><td>Only simple chores preserved; very restricted interests, poorly maintained</td><td>No significant function in home</td></tr>
<tr><td>Personal care</td><td colspan="2">Fully capable of self-care</td><td>Needs prompting</td><td>Requires assistance in dressing, hygiene, keeping of personal effects</td><td>Requires much help with personal cares, frequent incontinence</td></tr>
</table>

## 6.5 Dementia Diagnosis and Identification Scales

### 6.5.1 Hachinski Ischemic Score

Hachinski Ischemic Score (HIS) was developed by Hachinski in 1975 as a simple examination and identification scale for vascular dementia. It consists of 13 items, mainly for the evaluation of clinical features, and is completed by a physician in a simple way within a short time. There are two scoring methods, Hachinski and Rosen. The Hachinshi scale is scored out of 18 points. Those with a score of 4 or less are considered to have Alzheimer's disease, while those with a score of 7 or above are considered to have vascular dementia.

### 6.5.2 Self-Rating Depression Scale

Self-Rating Depression Scale (SDS) is a self-rating scale for patients with depression and will be scored by the patients themselves. It allows to observe the change and severity of depressive symptoms and also to differentiate dementia from depression. There are 20 questions, and the cut-off standard score of SDS is 53 points. 53–62 points represents mild depression; 63–72 points represents moderate depression; 72 points or above represents severe depression, and score below 53 points represents normal.

## 6.6 Single-Item Cognitive Assessment

### 6.6.1 Vocabulary Learning

It is used to evaluate the patient's immediate and delayed recall. During the test, choose 10 commonly used words and ask the patient to read each word aloud (show 1 word each time). Ask the patient to immediately recall the words they have seen after showing him/her all the words. Score based on the number of the correct words they remember. Repeat this test 3 times and the average thereof is the score of immediate recall of the patient. Two min later, ask the patient to recall the words again, and the number of correct words they remember will be the score of delayed recall. The ratio of the delayed recall score to the immediate recall score is the rate of delayed recall retention.

### 6.6.2 Verbal Fluency Test

It reflects the patient's executive function and semantic memory. Patients are asked to name certain things (e.g., animals, plants, vegetables, fruits, etc.) during the test. It is scored based on the number of things correctly named within 1 min. The sum of the scores of 3 tests is the total score of the test. Positive results are classified according to different educational levels: illiteracy $\leq$15 points, primary school level $\leq$20 points, secondary school level or above $\leq$25 points.

### 6.6.3 Block Building

It is a sub-item of WAIS, aiming to test the patient's visuospatial ability and comprehensive analysis ability. It is often used alone in dementia screening and generates satisfactory evaluation results. The threshold for dementia is classified according to literacy level: illiteracy <10 points, primary school level <15 points, secondary school level or above <20 points.

### 6.6.4 Clock Drawing Test

It is a common clinical test easy to perform and highly sensitive (Royall et al. 1998). It is less affected by factors such as literacy and economic status. The assessment can be performed from various aspects, but the 4-point method is commonly used clinically, that is, the total score is 4 points, with 1 point for drawing a closed circle, 1 point for correct digit position, 1 point for 12 correct digits on the dial, and 1 point for correct pointer position. A total score >2 is normal.

### 6.6.5 Trail Making Test

It is to assess attention and movement speed, simple and easy to operate. There are usually 2 types of tests. In Type A, 25 circles are printed on the paper, labeled with the numbers 1–25. The subject is asked to quickly connect the numbers in sequence. In Type B, there are 25 circles printed on the paper, 13 of which are labeled with numbers 1–13 and the other 12 with letters A–L. The subject is asked to quickly connect the numbers and letters in sequence, i.e., 1-A-2-B-3-C … 12-L-13. Scores will be made based on the time of completion. Generally, it is recognized that Type A reflects the function of the right cerebral hemisphere and shows the primitive perceptual motion rate, whereas Type B reflects the function of the left cerebral hemisphere, including more complex functions such as perceptual motion rate and the conversion of concepts and attention. The trail making test is more sensitive to diffuse and unilateral hemispheric functional impairment and also allows for the screening and evaluation of frontal lobe dysfunction.

In general, each scale has its own emphasis, different reliability and validity. Therefore, they need to be used in combination with each other.

Use all information and make the best judgment. Score each category as independently as possible. Mark in only one box for each category, rating impairment as decline from the person's usual level due to cognitive loss alone, not impairment due to other factors such as physical handicap or depression. Occasionally, the evidence is ambiguous and the clinician's best judgment is that a category could be rated in either one of two adjacent boxes, such as mild (1) or moderate (2) impairment. In that situation, the standard procedure is to check the box of greater impairment.

Aphasia is taken into account by assessing both language and nonlanguage function in each cognitive category. If aphasia is present to a greater degree than the general dementia, the subject is rated according to the general dementia. Supply evidence of nonlanguage cognitive function.

The global CDR is derived from the scores in each of the six categories ("box scores") as follows: Memory (M) is considered the primary category and all others are secondary. CDR = M if at least three secondary categories are given the same score as memory. Whenever three or more secondary categories are given a score greater or less than the memory score, CDR = score of majorities of secondary categories on whichever side of M has the greater number of secondary categories. However, when three secondary categories are scored on one side of M and two secondary categories are scored on the other side of M, CDR = M.

When M = 0.5, CDR = 1 if at least three of the other categories are scored 1 or greater. If M = 0.5, CDR cannot be 0; it can only be 0.5 or 1. If M = 0, CDR = 0, unless there is impairment (0.5 or greater) in two or more secondary categories, in which case CDR = 0.5.

Although applicable to most AD situations, these rules do not cover all possible scoring combinations. Unusual circumstances that occasionally occur in AD and may be expected in non-Alzheimer's dementia as well are scored as follows:

1. With ties in the secondary categories on one side of M, choose the tied scores closest to M for CDR (eg., M and another secondary category = 3, two secondary categories = 2, and two secondary categories = 1; CDR = 2).
2. When only one or two secondary categories are given the same score as M, CUR = M as long as no more than two secondary categories are on either side of M.
3. When M = 1 or greater, CDR cannot be 0; in this circumstance, CDR = 0.5 when the majority of secondary categories are 0.

# 7 Clinical Assessment of Sleep Disorder

A prerequisite for the treatment of sleep disorders is assessment. This chapter will briefly present the clinical assessment of sleep disorder.

## 7.1 Objective Assessment Methods for Sleep Disorder

### 7.1.1 Polysomnography

In addition to clinical signs and symptoms, objective examinations of sleep disorders can help diagnose and identify different types and degrees of sleep disorder. Polysomnography (PSG) is currently the most common test method and has become the gold standard for the assessment of sleep disorder.

The modern PSG is a comprehensive sleep assessment method that allows the continuous collection of physiological parameters during sleep throughout the night and the analysis of sleep amount and structure, electroencephalogram, respiratory function, cardiovascular function and limb muscle movement during sleep, which helps to describe the complex characteristics of sleep and potential sleep disorder.

In 2007, the American Academy of Sleep Medicine (AASM) published the manual of PSG, which is now used as a standardized system to be followed by all AASM-accredited centers and laboratories for sleep disorder. The AASM manual is revised every few years, and the latest version is 2.6. PSG is typically limited to studies in sleep examination rooms and applications in clinical settings.

#### 7.1.1.1 Test Methods

In a PSG test, 2–4 leads are usually used to record EEG, with same specifications and placement of electrodes on scalp as that of routine EEG tests (International 10/20 EEG system). Two leads are applied for recording EOG. There is one lead recording ECG. There are 1–2 leads recording EMG, connected at 1.5 cm on each side from the mandibular midline to record the electromyographic activity of the jaw muscles. The electrodes are placed on the lateral anterior tibia (the distance between the 2 electrodes should exceed 2 cm) to record the electromyographic activity of the anterior tibialis muscle. Respiratory monitoring involves oral and nasal airflow, thoracic or abdominal respiratory motility, $CO_2$ analysis device, thermistor, blood oxygen saturation and snore detector. The effect of body posture on sleep can be determined by recording the change in body posture during sleep. Other tests may be added as needed.

Patients are required to take a PSG test overnight in the sleep laboratory. Generally, each patient will sleep for 2 consecutive nights in the sleep examination room. The first night is for training the patient to adapt to the test environment. The recording time usually starts from the time of natural sleep to the time of spontaneous awakening the next morning.

In order to ensure accurate and reliable test results, the basic requirements for the sleep examination room are keeping it simple and quiet with appropriate temperature and air circulation, and making it as similar as possible to the home environment so that the subject can relax and be comfortable.

#### 7.1.1.2 Indicators of PSG Analysis

With PSG information such as the total bed rest time, total sleep time, sleep efficiency, sleep latency, REM sleep latency, wake time after sleep onset, number of wakes after sleep onset, apnea–hypopnea index, periodic leg movement of sleep index, proportion of each sleep phase and muscle movement can be obtained. The analysis of the above information can help diagnose various sleep disorders.

### 7.1.2 Multiple Sleep Latency Test

Hypersomnolence is one of the complaints most commonly seen in the evaluations of sleep disorder centers. The Multiple Sleep Latency Test (MSLT) is the most common objective test for the assessment of somnolence and is specifically used to diagnose three disorders: narcolepsy with cataplexy (Type 1), narcolepsy without cataplexy (Type 2) and idiopathic somnolence.

#### 7.1.2.1 Test Methods

Before the analysis of MSLT, the PSG test performed through the preceding night must be reviewed to rule out other factors contributing to daytime somnolence.

MSLT tests are usually scheduled 1–3 h after the end of the overnight PSG test. The test needs to be performed in a dark and quiet single room. The entire test involves 5 naps, each lasting 30 min and with an interval of 2 h. It usually starts at 8 a.m., followed by 10 a.m., 12 p.m., 2 p.m. and 4 p.m. All activities during the naps and any factors that may affect the subject's sleep are recorded in detail. Before each test, it is necessary to explicitly ask the subject to start closing eyes and sleeping now. The subject must get up at the end of each 30-min nap and remain awake until the next nap. If the subject is unable to sleep at all, the test should end 30 min after the start of the nap test.

#### 7.1.2.2 Indicators of MSLT Analysis

Mean sleep latency (MSL) $\leq$8 min is considered to be the criterion for pathological somnolence. MSL $\geq$10 is considered normal.

Narcolepsy can be diagnosed with MSL $\leq$8 min and 2 naps (i.e., SOREM) or MSL $\leq$8 min and 1 SOREM and 1 SOREM during overnight PSG (Arand and Bonnet 2019).

### 7.1.3 Actigraphy and Wearable Sleep Tracking Technology

#### 7.1.3.1 Actigraphy

In environments outside the examination room, an actigraphy can capture sleep-wake information in the patient's natural sleep for up to 2 weeks. An actigraphy is a wristwatch-sized device that is small, comfortable and waterproof and is wearable for 24 h a day. It is built in with an accelerometer to measure movement, which works on the principle that movement is high in frequency and amplitude when the wearer is awake, and low in frequency and amplitude when the wearer is asleep.

It can obtain the information of bed rest time, sleep latency, sleep duration and sleep efficiency when the wearer attempts to sleep. In addition, when analyzing sleep modes, it can take into account bedtime, wake-up time and weekday/weekend, and generate different results based thereupon (Meltzer 2018).

An actigraphy is applicable for patients with narcolepsy, delayed sleep-wake phase disorder, sleep deprivation syndrome, and insomnia. However, it is inherently flawed in their specificity for detecting true wakefulness due to its inability to identify wakefulness at rest. In patients with insomnia, an actigraphy can provide an accurate sleep-wake pattern, which may help identify the causes of insomnia, such as prolonged and irregular naps, and detect irregular sleep schedules, for example, sleeping late on weekends (Meltzer 2018). However, it may also overestimate the total sleep time, especially in those who are at rest when awake. Besides, the actigraphy does not allow for staging studies of sleep. Subtle muscle movements, such as periodic limb movements during sleep may not be accurately captured by the actigraphy. This device is also not suitable for patients with obstructive sleep apnea syndrome. The performance differences of individual actigraphies can also affect the data (Smith et al. 2018).

#### 7.1.3.2 Other Wearable Sleep Trackers

Wearable sleep trackers (e.g., wristbands, armbands, smart watches, headbands, rings, sensor clips) can measure a variety of biological signals such as heart rate and its variability, skin conductance and temperature in addition to recording movement, from which behavioral information, including sleep, can be extracted.

In the field of sleep, there is a growing awareness of the potential benefits of using wearable sleep trackers. However, there are inadequate evidence proving the validity, accuracy and reliability of many devices and related systems in assessing various sleep parameters and other metrics. These devices should therefore be used with caution and their results carefully analyzed to avoid generating large inaccurate data sets that can lead to potentially misleading scientific conclusions (de Zambotti et al. 2019; Danzig et al. 2019).

## 7.2 Subjective Assessment Methods for Sleep Disorder

### 7.2.1 Screening of Sleep History and Sleep Disorder of Patients

Asking the patient some simple questions or inquiring about his/her sleep history helps the assessment of sleep disorder. For example, ask the patients if they have had high quality sleep in the past month? Do they feel tired or sleepy during the day? As to the sleep history, ask the patient in detail about the onset, duration, severity and progression of the sleep disorder, including a description of the circadian sleep-wake cycle.

### 7.2.2 Assessment with Questionnaires

#### 7.2.2.1 Sleep Quality Assessment: Pittsburgh Sleep Quality Index

Pittsburgh Sleep Quality Index (PSQI) is one of the most widely used self-report methods for assessing sleep quality. Patients are asked to answer questions about their sleep in the past month. Patients are asked to rate 19 items classified into 7 aspects, namely subjective sleep quality, sleep latency, sleep duration, sleep efficiency, sleep disturbance, use of hypnotics, and daytime functioning. Each item is scored from 0 to 3. The scores of the 7 parts are summed to obtain a total score for overall sleep quality. The total score ranges from 0 to 21. A score larger than 5 suggests a sleep disorder (Kurtis et al. 2018). The disadvantage of PSQI is that the rating is slightly complicated (Frohnhofen et al. 2019). This scale helps the retrospective investigation of sleep quality and sleep disorders while not allowing for a definitive diagnosis of sleep disorders.

#### 7.2.2.2 Assessment of Daytime Somnolence: Epworth Sleepiness Scale

Epworth Sleepiness Scale (ESS) is a subjective scale used for excessive daytime somnolence. In this scale, the subjects are asked to rate their possibility of dozing or falling asleep in 8 different settings during the 1 month prior to the assessment. Each question is scored from 0 to 3. The overall score is the sum of individual scores. The cut-off score is set as 11, with a score above 11 indicating excessive daytime somnolence. ESS shows a weaker correlation than objective tests of excessive daytime somnolence such as MSLT (Kurtis et al. 2018; Frohnhofen et al. 2019).

#### 7.2.2.3 Assessment of Sleep-Disordered Breathing: Berlin Questionnaire

Berlin Questionnaire (BQ) is used to clinically assess the risk of obstructive sleep apnea. The questions of BQ consist of 10 items divided into three sections: snoring, somnolence or fatigue, obesity and hypertension. The score of each part is determined as negative or positive. There is a high risk of sleep apnea if the scores of more than 2 (included) sections are positive, and the risk is low if the scores of only one section or no section are positive. This scale is adequately sensitive but inadequately specific (Kurtis et al. 2018; Frohnhofen et al. 2019).

#### 7.2.2.4 International Restless Leg Scale

International Restless Leg Scale (IRLS) is used to evaluate the severity of restless legs syndrome in the past 1 week. There are 10 questions about the symptoms, frequency, intensity, and causes. Each question is scored from 0–4 as appropriate, and the total score is from 0–40. It is now the most widely used tool to assess the severity and impact of RLS on quality of life. This scale does not have a uniform cut-off score, but allows for group comparisons based on the scores, or for comparison of efficacy before and after treatment (Kurtis et al. 2018).

#### 7.2.2.5 REM Sleep Behavior Disorder Related Scales

Rapid eye movement (REM) sleep behavior disorder screening questionnaire (RBDSQ) is a self-assessment scale used to screen REM sleep behavior disorders. It includes 10 questions about the dream content, the relationship between dream and behavior, injury and neurological disorders. Subjects are asked to choose between "Yes" and "No". The total score is from 0 to 13, with a score larger than 5 considered abnormal. It is less sensitive in patients with neurological disorders or other sleep disorders (Marelli et al. 2016). The RBDSQ in Chinese has been shown to have high sensitivity, specificity and reliability (Li et al. 2017).

The Hong Kong Questionnaire (RBD-HK) has also been gradually and widely used in recent years. The scale consists of 13 subjective questions for patients. Questions 1–5 and 13 correspond to dream-related factors and questions 6–12 correspond to motion-related factors, with a weighted total score of 100.

It is noteworthy that the diagnostic procedure for sleep disorders should take into account a detailed sleep history, sleep-wake patterns and third-party observations, as well as objective measurements if feasible. Although questionnaires facilitate clinical work and have some diagnostic values, sleep disorders should not be diagnosed based on the questionnaires alone.

## 8 Clinical Assessment of Anxiety

Anxiety disorder is a common psychiatric disorder worldwide, characterized by the experience of anxious emotions, with a lifetime prevalence of 9.0%. The diagnosis thereof is primarily based on clinical features. Generalized anxiety disorder is the most common type of anxiety, with a lifetime prevalence of about 4.1%–6.6% and an annual prevalence of about 1.9%–5.1% in the general population. Among patients with mental disorders, at least 1/3 has anxiety disorders in certain forms. As social competition becomes increasingly fierce and stress factors in life build up, there are more anxiety reactions and higher incidence.

At present, the two major diagnostic systems for anxiety disorders are the *Diagnostic and Statistical Manual of Mental Disorders (fifth edition)* (DSM-5) and the *International Classification of Diseases (eleventh edition)*

**Table 3.15** 4 Questions within 90 s

| | | |
|---|---|---|
| Question 1: | Do you think you are a person easily getting anxious or nervous? | Yes (to find out if the patient had an anxious personality or trait) |
| Question 2: | Have you felt more anxious or apprehensive than usual recently? | Yes (to find out if the patient has generalized anxiety) |
| Question 3: | Are there any special occasions or situations that make you more likely to be nervous or anxious? | Yes (to find out if you have any fears) |
| Question 4: | Have you ever had a panic attack, i.e., a sudden onset of intense discomfort or panic, dizziness, feeling breathlessness or having difficulty in breathing, etc.? | Yes (to find out if you have panic attacks) |

If more than 2 of the 4 questions are positive, a further clinical evaluation is required

(ICD-11). The early identification of anxiety disorders is not enough and patients repeatedly visit multidisciplinary clinics for various somatization symptoms but fail to have effective treatments. Hence, the early identification and intervention are particularly important for prognosis. Current scales for the assessment of anxiety disorders include scales for assessments by others and self-assessments. The *Expert Consensus on the Diagnosis and Treatment of Anxiety, Depression and Somatization Symptoms in General Hospitals* (2016) recommends the easy method of "4 questions within 90 s" for a rapid initial screening of anxiety (see Table 3.15). If 2 or more of the 4 questions turn out to be positive, a further clinical assessment is required. The generalized anxiety disorder-7 (GAD-7) is suitable for rapid assessment of generalized anxiety, while self-assessment questionnaires such as the self-anxiety scale (SAS), the symptom checklist-90 (SCL-90) and the hospital anxiety and depression scale (HADS) are suitable for rapid assessment of various types of anxiety. Hospitals with evaluators and appropriate conditions may also choose scales for assessments by others such as Hamilton anxiety scale (HAMA) (Wang et al. 2016). The GAD-7 and the patient health questionnaire-9 (PHQ-9) have been recently recommended by the World Health Organization for their simplicity and their specificity and reliability in terms of diagnosis of anxiety or depression. They are hence widely used in general hospitals and primary institutes of medical care at home and abroad for the diagnosis of anxiety or depression and the efficacy assessment (Kroenke et al. 2010). Proper use of the scales, an essential clinical examination tool, can help clinical staff to identify and assess patient's emotional state as soon as possible. If the patient is as moderate or above according to the scale, a further disease diagnosis is recommended to determine whether it is anxiety disorder. The scales not only help the physician understand the severity of the patient's emotional disorder, but also help the patients aware of their possible emotional problems, which can lay the foundation for them to subsequently receive the diagnosis of emotional disorder and the treatment with anti-anxiety and depression drugs.

## 8.1 4 Questions within 90 s

## 8.2 Generalized Anxiety Disorder-7

The GAD-7 is easy to use, highly sensitive and reliable in specificity, and has been widely used by general hospitals and primary institutes of medical care at home and abroad. It is suitable for the rapid assessment of generalized anxiety (Spitzer et al. 2006). The scale contains 7 items on a 4-point scale of 0–3, and the period under test is the past 2 weeks before the assessment. The scoring criteria are: 0: not at all; 1: those feelings exist some days; 2: those feelings exist more than half of the time; 3: basically, every day. The total scores are summed, with 0–5 representing mild anxiety, 6–10 representing moderate anxiety, and 11–15 representing severe anxiety (see Table 3.16).

If you checked off *any* problems, how *difficult* have these problems made it for you to do your work, take care of things at home, or get along with other people?

| Not difficult at all | Somewhat difficult | Very difficult | Extremely difficult |
|---|---|---|---|
| ☐ | ☐ | ☐ | ☐ |

**Table 3.16** Generalized Anxiety Disorder-7 (GAD-7)

| Over the *last 2 weeks*, how often have you been bothered by the following problems? | Not at all | Several days | More than half the days | Nearly every day |
|---|---|---|---|---|
| 1. Feeling nervous, anxious, or on edge | 0 | 1 | 2 | 3 |
| 2. Not being able to stop or control worrying | 0 | 1 | 2 | 3 |
| 3. Worrying too much about different things | 0 | 1 | 2 | 3 |
| 4. Trouble relaxing | 0 | 1 | 2 | 3 |
| 5. Being so restless that it is hard to sit still | 0 | 1 | 2 | 3 |
| 6. Becoming easily annoyed or irritable | 0 | 1 | 2 | 3 |
| 7. Feeling afraid as if something awful might happen | 0 | 1 | 2 | 3 |
| Total Score = | Add Columns | + | + | |

## 8.3 Self-Rating Anxiety Scale

The self-rating anxiety scale (SAS) was developed by Zung, a professor of Chinese origin, in 1971. It is a simple clinical tool for the analysis of the subjective symptoms of patients, and is very similar to the self-rating depression scale (SDS) from the scale structure to the specific assessment methods. It is suitable for adults with anxiety symptoms. With a 4-point scale, SAS is mainly used to assess the frequency of symptoms. As to the scoring criteria, 1: no or rarely, 2: sometimes, 3: most of the time, and 4: most or all of the time. 15 of the 20 items are stated with negative words and scored in the order of 1–4 above. The remaining 5 items (No. 5, 9, 13, 17, 19) noted with * are stated in positive words and scored in reverse order of 4–1. The main statistical indicator of SAS is the total score. The raw score is obtained by adding the scores of each of the 20 items; the standard score is obtained by multiplying the raw score by 1.25 and then taking the integer. According to the results of Chinese normal, the cut-off value of SAS standard score is 50, with 50 to 59 representing mild anxiety, 60 to 69 representing moderate anxiety, and 70 or more representing severe anxiety (see Table 3.17).

## 8.4 Beck Anxiety Inventory

Beck anxiety inventory (BAI) was developed in 1985 by Aaron T. Beck, USA (Beck et al. 1988). BAI is mainly applicable to adults with anxiety symptoms and can accurately reflect the subjective level of anxiety. The self-experiences of the patient should be assessed within the time frame of "present" or "last week". BAI contains 21 items and is rated on a 4-point scale. As to the scoring criteria, 0: none; 1: mild, without much bother; 2: moderate, feeling uncomfortable but tolerable; 3: severe, barely tolerable. The total score is the sum of the scores of the 21 items, with 0–21 representing low anxiety, 22–35 representing moderate anxiety, and 36 or more representing severe anxiety (see Table 3.18).

## 8.5 Hospital Anxiety and Depression Scale

The hospital anxiety and depression scale (HADS) was created by Zigmond et al. in 1983 and is primarily used to screen for anxiety and depression among patients in general hospitals (Zigmond and Snaith 1983). The scale contains 14 items, 7 of which are used to rate anxiety [A] and 7 of which to rate depression [D]. The scale is scored on a 4-point scale of 0–3, with each question being scored from 0 to 3. There are 6 items of reverse questions. The score ranges of both anxiety and depression are 0 to 21. The threshold varies across studies. According to the criteria set by the original author, the subscales of anxiety and depression are so scored that 0–7 indicates no symptoms, 8–10 indicates suspicious symptoms, and 11–21 indicates definite symptoms. Some studies take the score of 9 as the threshold for anxiety or depression (see Table 3.19).

**Table 3.17** Self-rating Anxiety Scale (SAS)

| | None or a little of the time | Some of the times | Good part of the time | Most or all of the time |
|---|---|---|---|---|
| 1. I feel more nervous and anxious than usual. | 1 | 2 | 3 | 4 |
| 2. I feel afraid for no reason at all. | 1 | 2 | 3 | 4 |
| 3. I get upset easily or feel panicky. | 1 | 2 | 3 | 4 |
| 4. I feel like I'm falling apart and going to pieces. | 1 | 2 | 3 | 4 |
| 5*. I feel that everything is all right and nothing bad will happen. | 4 | 3 | 2 | 1 |
| 6. My arms and legs shake and tremble. | 1 | 2 | 3 | 4 |
| 7. I am bothered by headaches, neck and back pains. | 1 | 2 | 3 | 4 |
| 8. I feel weak and get tired easily. | 1 | 2 | 3 | 4 |
| 9*. I feel calm and can sit still easily. | 4 | 3 | 2 | 1 |
| 10. I can feel my heart beating fast. | 1 | 2 | 3 | 4 |
| 11. I am bothered by dizzy spells. | 1 | 2 | 3 | 4 |
| 12. I have fainting spells or feel like it. | 1 | 2 | 3 | 4 |
| 13*. I can breathe in and out easily. | 4 | 3 | 2 | 1 |
| 14. I get fee lings of numbness and tingling in my fingers, toes. | 1 | 2 | 3 | 4 |
| 15. I am bothered by stomachaches or indigestion. | 1 | 2 | 3 | 4 |
| 16. I have to empty my bladder often. | 1 | 2 | 3 | 4 |
| 17*. My hands are usually dry and warm. | 4 | 3 | 2 | 1 |
| 18. My face gets hot and blushes. | 1 | 2 | 3 | 4 |
| 19*. I fall asleep easily and get a good night's rest. | 4 | 3 | 2 | 1 |
| 20. I have nightmares. | 1 | 2 | 3 | 4 |

**Table 3.18** Beck anxiety inventory

| | Not at all | Mildly | Moderately | Severely |
|---|---|---|---|---|
| 1. Numbness or tingling | 0 | 1 | 2 | 3 |
| 2. Feeling hot | 0 | 1 | 2 | 3 |
| 3. Wobbliness in legs | 0 | 1 | 2 | 3 |
| 4. Unable to relax | 0 | 1 | 2 | 3 |
| 5. Fear of the worst happening | 0 | 1 | 2 | 3 |
| 6. Dizzy or lightheaded | 0 | 1 | 2 | 3 |
| 7. Heart pounding or racing | 0 | 1 | 2 | 3 |
| 8. Unsteady | 0 | 1 | 2 | 3 |
| 9. Terrified | 0 | 1 | 2 | 3 |
| 10. Nervous | 0 | 1 | 2 | 3 |
| 11. Feelings of choking | 0 | 1 | 2 | 3 |
| 12. Hands trembling | 0 | 1 | 2 | 3 |
| 13. Shaky/unsteady | 0 | 1 | 2 | 3 |
| 14. Fear of losing control | 0 | 1 | 2 | 3 |
| 15. Difficulty in breathing | 0 | 1 | 2 | 3 |
| 16. Fear of dying | 0 | 1 | 2 | 3 |
| 17. Scared | 0 | 1 | 2 | 3 |
| 18. Indigestion or discomfort in abdomen | 0 | 1 | 2 | 3 |
| 19. Faint/ lightheaded | 0 | 1 | 2 | 3 |
| 20. Face flushed | 0 | 1 | 2 | 3 |
| 21. Hot/ cold sweats | 0 | 1 | 2 | 3 |
| Total score | | | | |

**Table 3.19** Hospital Anxiety and Depression Scale (HADS). Tick the box beside the reply that is closest to how you have been feeling in the past week. Don't take too long over you replies: your immediate answer is the best

| | | |
|---|---|---|
| | A | I feel tense or 'wound up': |
| | 3 | Most of the time |
| | 2 | A lot of the time |
| | 1 | From time to time, occasionally |
| | 0 | Not at all |
| D | | I still enjoy the things I used to enjoy: |
| 0 | | Definitely as mush |
| 1 | | Not quite so much |
| 2 | | Only a little |
| 3 | | Hardly at all |
| | A | I get a sort of frightened feeling as if something awful is about to happen: |
| | 3 | Very definitely and quite badly |
| | 2 | Yes, but not too badly |
| | 1 | A little, but it doesn't worry me |
| | 0 | Not at all |
| D | | I can laugh and see the funny side of things: |
| 0 | | As much as I always could |
| 1 | | Not quite so much now |
| 2 | | Definitely not so much now |
| 3 | | Not at all |
| | A | Worrying thoughts go through my mind: |
| | 3 | A great deal of the time |
| | 2 | A lot of the time |
| | 1 | From time to time but not too often |
| | 0 | Only occasionally |
| D | | I feel cheerful: |
| 3 | | Not at all |
| 2 | | Not often |
| 1 | | Sometimes |
| 0 | | Most of the time |
| | A | I can sit at ease and feel relaxed: |
| | 0 | Definitely |
| | 1 | Usually |
| | 2 | Not often |
| | 3 | Not at all |
| D | | I feel as if I am slowed down: |
| 3 | | Nearly all the time |
| 2 | | Very often |
| 1 | | Sometimes |
| 0 | | Not at all |
| | A | I get a sort of frightened feeling like 'butterflies' in the stomach: |
| | 0 | Not at all |
| | 1 | Occasionally |
| | 2 | Quite often |
| | 3 | Very often |
| D | | I have lost interest in my appearance: |
| 3 | | Definitely |
| 2 | | I don't take so much care as I should |
| 1 | | I may not take quite as much care |
| 0 | | I take just as much care as ever |
| | A | I feel restless as if I have to be on the move: |
| | 3 | Very much indeed |
| | 2 | Quite a lot |
| | 1 | Not very much |
| | 0 | Not at all |

(continued)

**Table 3.19** (continued)

| | | |
|---|---|---|
| D | | I look forward with enjoyment to things: |
| 0 | | As much as ever I did |
| 1 | | Rather less than I used to |
| 2 | | Definitely less than I used to |
| 3 | | Hardly at all |
| | A | I get sudden feelings of panic: |
| | 3 | Very often indeed |
| | 2 | Quite often |
| | 1 | Not very often |
| | 0 | Not at all |
| D | | I can enjoy a good book or radio or TV programme: |
| 0 | | Often |
| 1 | | Sometimes |
| 2 | | Not often |
| 3 | | Very seldom |

Total score: Depression (D)__________Anxiety (A)____________.

## 8.6 Symptom Check List 90

90-item Symptom Checklist, also known as the symptom check list 90 (SCL-90), was developed by L.R. Derogatis in 1975. The scale contains 90 items involving a wide range of psychosis-like symptoms, ranging from feelings, emotions, thinking, consciousness, and behavior to living habits, relationships, and eating and sleeping. 10 factors are used to reflect 10 aspects of psychological symptoms, namely somatization, obsessive-compulsive symptoms, interpersonal sensitivity, depression, anxiety, hostility, terror, paranoia, psychoticism, and other. The assessment is based on the actual feeling "now" or "during the last week". A 5-point scale is used. To be specific, 0: no, no perception of the problem; 1: mild, perception of the symptom, with no real impact or mild impact on the subject; 2: moderate, perception of the symptom, with some impact on the subject; 3: severe, perception of the symptom, with a considerable impact on the subject; 4: severe, perception of the frequency and intensity of the symptom, with a serious impact on the subject. There are two major statistical criteria of SCL-90, namely, total score and factor score. The total score is the sum of the scores of 90 items, which reflects the severity of the subject's condition. The grand average score is total score/90, which indicates where the subject's self-perception is on a scale of 1 to 5 from the general view. Number of positive items: the number of items with a single score ≥2 out of 90 questions, which indicates the number of items in which the subject has "symptoms"; number of negative items: the number of items with a single score = 1 out of 90 questions, or the number of 90 minus the number of positive items, which indicates the number of "asymptomatic" items. The mean score of positive symptoms: (total score − number of negative items)/number of positive items. It indicates the mean score of "symptomatic" items and indicates the range of severity of the items in which the participant does not feel good about himself/herself. The factor score is the average of each factor score. As each factor reflects a certain aspect of the subject's condition, the factor score reflects the distribution of symptoms of the subject and allows for an analysis with a skeleton map (see Table 3.20).

## 8.7 Hamilton Anxiety Scale

The Hamilton Anxiety Scale (HAMA) was developed by Hamilton in 1959 (Hamilton 1959). It consists of 14 items and is one of the scales commonly used in psychiatric clinics at the very beginning. It is mainly used to assess the severity of anxiety symptoms. The *CCMD-3 Chinese Classification of Mental Disorders Version 3* lists it as an important diagnostic tool for Anxiety disorder. It is often used in clinical practice as a basis for the diagnosis and classification of anxiety disorders. HAMA should be jointly used in an examination by 2 trained evaluators, usually by chatting and observing. The 2 evaluators will score independently at the end of the examination. HAMA is scored on a 5-point scale from 0 to 4. As to the scoring criteria, 0: no symptoms; 1: mild; 2: moderate; 3: severe; 4: extremely severe. A total score range of 0–56, where <17 indicates mild severity, 18–24 mild to moderate severity and 25–30 moderate to severe. HAMA classifies anxiety factors into two categories: somatic and psychogenic. Somatic anxiety: high scores in items from No. 7 to No. 13. Psychogenic anxiety: high scores in items from No. 1 to No. 6 and No.14 (Table 3.21).

**Table 3.20** 90 Items symptom checklist (symptom check list 90, SCL-90). For the PAST WEEK, how much were you bothered by:

| | | Not at all | A little bit | Moderately | Quite a bit | Extremely |
|---|---|---|---|---|---|---|
| 1. | Headaches | 0 | 1 | 2 | 3 | 4 |
| 2. | Nervousness or shakiness inside | 0 | 1 | 2 | 3 | 4 |
| 3. | Unwanted thoughts, words, or ideas that won't leave your mind | 0 | 1 | 2 | 3 | 4 |
| 4. | Faintness or dizziness | 0 | 1 | 2 | 3 | 4 |
| 5. | Loss of sexual interest or pleasure | 0 | 1 | 2 | 3 | 4 |
| 6. | Feeling critical of others | 0 | 1 | 2 | 3 | 4 |
| 7. | The idea that someone else can control your thoughts | 0 | 1 | 2 | 3 | 4 |
| 8. | Feeling others are to blame for most of your troubles | 0 | 1 | 2 | 3 | 4 |
| 9. | Trouble remembering things | 0 | 1 | 2 | 3 | 4 |
| 10. | Worried about sloppiness or carelessness | 0 | 1 | 2 | 3 | 4 |
| 11. | Feeling easily annoyed or irritated | 0 | 1 | 2 | 3 | 4 |
| 12. | Pains in heart or chest | 0 | 1 | 2 | 3 | 4 |
| 13. | Feeling afraid in open spaces or on the streets | 0 | 1 | 2 | 3 | 4 |
| 14. | Feeling low in energy or slowed down | 0 | 1 | 2 | 3 | 4 |
| 15. | Thoughts of ending your life | 0 | 1 | 2 | 3 | 4 |
| 16. | Hearing words that others do not hear | 0 | 1 | 2 | 3 | 4 |
| 17. | Trembling | 0 | 1 | 2 | 3 | 4 |
| 18. | Feeling that most people cannot be trusted | 0 | 1 | 2 | 3 | 4 |
| 19. | Poor appetite | 0 | 1 | 2 | 3 | 4 |
| 20. | Crying easily | 0 | 1 | 2 | 3 | 4 |
| 21. | Feeling shy or uneasy with the opposite sex | 0 | 1 | 2 | 3 | 4 |
| 22. | Feeling of being trapped or caught | 0 | 1 | 2 | 3 | 4 |
| 23. | Suddenly scared for no reason | 0 | 1 | 2 | 3 | 4 |
| 24. | Temper outbursts that you could not control | 0 | 1 | 2 | 3 | 4 |
| 25. | Feeling afraid to go out of your house alone | 0 | 1 | 2 | 3 | 4 |
| 26 | Blaming yourself for things | 0 | 1 | 2 | 3 | 4 |
| 27. | Pains in lower back | 0 | 1 | 2 | 3 | 4 |
| 28. | Feeling blocked in getting things done | 0 | 1 | 2 | 3 | 4 |
| 29. | Feeling lonely | 0 | 1 | 2 | 3 | 4 |
| 30. | Feeling blue | 0 | 1 | 2 | 3 | 4 |
| 31. | Worrying too much about things | 0 | 1 | 2 | 3 | 4 |
| 32. | Feeling no interest in things | 0 | 1 | 2 | 3 | 4 |
| 33. | Feeling fearful | 0 | 1 | 2 | 3 | 4 |
| 34. | Your feelings being easily hurt | 0 | 1 | 2 | 3 | 4 |
| 35. | Other people being aware of your private thoughts | 0 | 1 | 2 | 3 | 4 |
| 36. | Feeling others do not understand you or are unsympathetic | 0 | 1 | 2 | 3 | 4 |

(continued)

**Table 3.20** (continued)

| | | Not at all | A little bit | Moderately | Quite a bit | Extremely |
|---|---|---|---|---|---|---|
| 37. | Feeling that people are unfriendly or dislike you | 0 | 1 | 2 | 3 | 4 |
| 38. | Having to do things very slowly to insure correctness | 0 | 1 | 2 | 3 | 4 |
| 39. | Heart pounding or racing | 0 | 1 | 2 | 3 | 4 |
| 40. | Nausea or upset stomach | 0 | 1 | 2 | 3 | 4 |
| 41. | Feeling inferior to others | 0 | 1 | 2 | 3 | 4 |
| 42. | Soreness of your muscles | 0 | 1 | 2 | 3 | 4 |
| 43. | Feeling that you are watched or talked about by others | 0 | 1 | 2 | 3 | 4 |
| 44. | Trouble falling asleep | 0 | 1 | 2 | 3 | 4 |
| 45. | Having to check and double-check what you do | 0 | 1 | 2 | 3 | 4 |
| 46. | Difficulty making decisions | 0 | 1 | 2 | 3 | 4 |
| 47. | Feeling afraid to travel on buses, subways, or trains | 0 | 1 | 2 | 3 | 4 |
| 48. | Trouble getting your breath | 0 | 1 | 2 | 3 | 4 |
| 49. | Hot or cold spells | 0 | 1 | 2 | 3 | 4 |
| 50. | Having to avoid certain things, places, or activities because they frighten you | 0 | 1 | 2 | 3 | 4 |
| 51. | Your mind going blank | 0 | 1 | 2 | 3 | 4 |
| 52. | Numbness or tingling in parts of your body | 0 | 1 | 2 | 3 | 4 |
| 53. | A lump in your throat | 0 | 1 | 2 | 3 | 4 |
| 54. | Feeling hopeless about the future | 0 | 1 | 2 | 3 | 4 |
| 55. | Trouble concentrating | 0 | 1 | 2 | 3 | 4 |
| 56. | Feeling weak in parts of your body | 0 | 1 | 2 | 3 | 4 |
| 57. | Feeling tense or keyed up | 0 | 1 | 2 | 3 | 4 |
| 58. | Heavy feelings in your arms or legs | 0 | 1 | 2 | 3 | 4 |
| 59. | Thoughts of death or dying | 0 | 1 | 2 | 3 | 4 |
| 60. | Overeating | 0 | 1 | 2 | 3 | 4 |
| 61. | Feeling uneasy when people are watching or talking about you | 0 | 1 | 2 | 3 | 4 |
| 62. | Having thoughts that are not your own | 0 | 1 | 2 | 3 | 4 |
| 63. | Having urges to beat, injure, or harm someone | 0 | 1 | 2 | 3 | 4 |
| 64. | Awakening in the early morning | 0 | 1 | 2 | 3 | 4 |
| 65. | Having to repeat the same actions such as touching, counting, washing | 0 | 1 | 2 | 3 | 4 |
| 66. | Sleep that is restless or disturbed | 0 | 1 | 2 | 3 | 4 |
| 67. | Having urges to break or smash things | 0 | 1 | 2 | 3 | 4 |
| 68. | Having ideas or beliefs that others do not share | 0 | 1 | 2 | 3 | 4 |
| 69. | Feeling very self-conscious with others | 0 | 1 | 2 | 3 | 4 |
| 70. | Feeling uneasy in crowds, such as shopping or at a movie | 0 | 1 | 2 | 3 | 4 |

**Table 3.20** (continued)

| | | Not at all | A little bit | Moderately | Quite a bit | Extremely |
|---|---|---|---|---|---|---|
| 71. | Feeling everything is an effort | 0 | 1 | 2 | 3 | 4 |
| 72. | Spells of terror or panic | 0 | 1 | 2 | 3 | 4 |
| 73. | Feeling uncomfortable about eating or drinking in public | 0 | 1 | 2 | 3 | 4 |
| 74. | Getting into frequent arguments | 0 | 1 | 2 | 3 | 4 |
| 75. | Feeling nervous when you are left alone | 0 | 1 | 2 | 3 | 4 |
| 76. | Others not giving you proper credit for your achievements | 0 | 1 | 2 | 3 | 4 |
| 77. | Feeling lonely even when you are with people | 0 | 1 | 2 | 3 | 4 |
| 78. | Feeling so restless you couldn't sit still | 0 | 1 | 2 | 3 | 4 |
| 79. | Feelings of worthlessness | 0 | 1 | 2 | 3 | 4 |
| 80. | Feeling that familiar things are strange or unreal | 0 | 1 | 2 | 3 | 4 |
| 81. | Shouting or throwing things | 0 | 1 | 2 | 3 | 4 |
| 82. | Feeling afraid you will faint in public | 0 | 1 | 2 | 3 | 4 |
| 83. | Feeling that people will take advantage of you if you let them | 0 | 1 | 2 | 3 | 4 |
| 84. | Having thoughts about sex that bother you a lot | 0 | 1 | 2 | 3 | 4 |
| 85. | The idea that you should be punished for your sins | 0 | 1 | 2 | 3 | 4 |
| 86. | Feeling pushed to get things done | 0 | 1 | 2 | 3 | 4 |
| 87. | The idea that something serious is wrong with your body | 0 | 1 | 2 | 3 | 4 |
| 88. | Never feeling close to another person | 0 | 1 | 2 | 3 | 4 |
| 89. | Feelings of guilt | 0 | 1 | 2 | 3 | 4 |
| 90. | The idea that something is wrong with your mind | 0 | 1 | 2 | 3 | 4 |

**Table 3.21** Hamilton Anxiety Scale (HAMA). Select one of the five responses for each of the 14 questions. 0 = Not present, 1 = Mild, 2 = Moderate, 3 = Severe, 4 = Very severe

| | | | | | |
|---|---|---|---|---|---|
| 1. Anxious mood: Worries, anticipation of the worst, fearful anticipation, irritability. | 0 | 1 | 2 | 3 | 4 |
| 2. Tension: Feelings of tension, fatigability, startle response, moved to tears easily, trembling, feelings of restlessness, inability to relax. | 0 | 1 | 2 | 3 | 4 |
| 3. Fear: Of dark, of strangers, of being left alone, of animals, of traffic, of crowds. | 0 | 1 | 2 | 3 | 4 |
| 4. Insomnia: Difficulty in falling asleep, broken sleep, unsatisfying sleep and fatigue on waking, dreams, nightmares, night terrors. | 0 | 1 | 2 | 3 | 4 |
| 5. Intellectual: Difficulty in concentration, poor memory. | 0 | 1 | 2 | 3 | 4 |
| 6. Depressive mood: Loss of interest, lack of pleasure in hobbies, depression, early waking, diurnal swing. | 0 | 1 | 2 | 3 | 4 |
| 7. Somatic (muscular): Pains and aches, twitching, stiffness, myoclonic jerks, grinding of teeth, unsteady voice, increased muscular tone. | 0 | 1 | 2 | 3 | 4 |

(continued)

**Table 3.21** (continued)

| | | | | | |
|---|---|---|---|---|---|
| 8. Somatic (sensory): Tinnitus, blurring of vision, hot and cold flushes, feelings of weakness, pricking sensation. | 0 | 1 | 2 | 3 | 4 |
| 9. Cardiovascular symptoms: Tachycardia, palpitations, pain in chest, throbbing of vessels, fainting feelings, missing beat. | 0 | 1 | 2 | 3 | 4 |
| 10. Respiratory symptoms: Pressure or constriction in chest, choking feelings, sighing, dyspnea. | 0 | 1 | 2 | 3 | 4 |
| 11. Gastrointestinal symptoms: Difficulty in swallowing, wind abdominal pain, burning sensations, abdominal fullness, nausea, vomiting, borborygmi, looseness of bowels, loss of weight, constipation. | 0 | 1 | 2 | 3 | 4 |
| 12. Genitourinary symptoms: Frequency of micturition, urgency of micturition, amenorrhea, menorrhagia, development of frigidity, premature ejaculation, loss of libido, impotence. | 0 | 1 | 2 | 3 | 4 |
| 13. Autonomic symptoms: Dry mouth, flushing, pallor, tendency to sweat, giddiness, tension headache, raising of hair. | 0 | 1 | 2 | 3 | 4 |
| 14. Behavioral at interview: Fidgeting, restlessness or pacing, tremor of hands, furrowed brow, strained face, sighing or rapid respiration, facial pallor, swallowing, etc. | 0 | 1 | 2 | 3 | 4 |

## 9 Clinical Assessment of Depression

Depressive disorder is a psychiatric disorder with high prevalence and recurrence rates but a low treatment rate. Depressive disorder affects about 350 million people worldwide, and imposes a significant public health burden. Its etiology and pathogenesis have not been fully elucidated. The diagnosis of depressive disorder is mainly based on diagnostic criteria. Currently, there are two diagnostic systems, namely the *Diagnostic and Statistical Manual of Mental Disorders (fifth edition)* (DSM-5) and the *International Classification of Diseases (eleventh edition)* (ICD-11). It is important to raise the public's attention of depressive disorder and, in particular, realize the early diagnosis and effective treatment of this disease. A comprehensive assessment of patients with depressive disorder is the key to proper diagnosis and effective treatment. The clinical assessment scales for depressive disorder used nowadays will be briefly introduced below.

### 9.1 Self-Assessment Scales

Compared with the assessment performed by psychiatrists, self-assessment scales are short and can be frequently repeated by the patients themselves, allowing the patients to aware of their own conditions and recovery status, which not only facilitate the patients to manage their own diseases, but also reduce the deviations caused by the assessor. The value and application of self-assessment scales are greater in clinical research.

The common standardized self-assessment scales for patients include: (1) Brief Patient Health Questionnaire (PHQ-9); (2) Self-rating Depression Scale (SDS); (3) Beck Depression Inventory (BDI); and (4) Quick Inventory of Depressive Symptomatology Self-Rated (QIDS-SR), etc.

#### 9.1.1 Brief Patient Health Questionnaire

The Patient Health Questionnaire (PHQ-9) is a simple and efficient depression self-assessment tool developed on the basis of DSM-IV that does not need a professional assessor to perform the assessment. It has shown good psychological properties when applied in the general population and in different populations with the disease, and is superior to other scales in terms of sensitivity and specificity (Zuithoff et al. 2010). It consists of 9 items, each with 4 options, namely "not at all", "several days", "more than half of the days", and "almost Every day", corresponding to scores of 0, 1, 2 and 3. The total score is 27. The higher the total score, the great the possibility of depression. Scoring criteria: 1–4 indicates normal; 5–9 indicates mild depression; 10–14 indicates moderate depression; 15–19 indicates moderate to severe depression, and 20–27 indicates severe depression (Table 3.22).

#### 9.1.2 Self-Rating Depression Scale

SDS, developed by Zung in 1965, is easy to use and provides a fairly visual representation of the subjective feelings of patients with depression and their changes in treatment (Zung et al. 1965). It is mainly applicable to adults with depressive symptoms, including outpatients and inpatients. However, it is difficult to assess depression with severe retardation symptoms with this scale. In addition, SDS is not effective for people with low literacy or intelligence levels. The scale contains 20 items reflecting the subjective feelings of depression. Each item is divided into 4 levels according to the frequency of symptoms, namely "no or seldom", "occasional", "quite often", and "most or all of the time". 10 of the items are rated positively, with scores of 1, 2, 3, and 4 in that order, and another 10 items are rated negatively, with scores of 4, 3, 2, and 1 in that order. The sum of total scores is total raw score (X), and the standard score (Y) will be obtained by

**Table 3.22** Brief patient health questionnaire (PHQ-9)

| Over the last 2 weeks, how often have you been bothered by any of the following problems? (use "√" to indicate your answer) | Not at all | Several days | More than half the days | Nearly every day |
|---|---|---|---|---|
| 1. Little interest or pleasure in doing things | 0 | 1 | 2 | 3 |
| 2. Feeling down, depressed, or hopeless | 0 | 1 | 2 | 3 |
| 3. Trouble falling or staying asleep, or sleeping too much | 0 | 1 | 2 | 3 |
| 4. Feeling tired or having little energy | 0 | 1 | 2 | 3 |
| 5. Poor appetite or overeating | 0 | 1 | 2 | 3 |
| 6. Feeling bad about yourself or that you are a failure or have let yourself or your family down. | 0 | 1 | 2 | 3 |
| 7. Trouble concentrating on things, such as reading the newspaper or watching television. | 0 | 1 | 2 | 3 |
| 8. Moving or speaking so slowly that other people could have noticed. Or the opposite being so fidgeting or restless that you have been moving around a lot more than usual. | 0 | 1 | 2 | 3 |
| 9. Thoughts that you would be better off dead, or of hurting yourself. | 0 | 1 | 2 | 3 |

taking the integer of the result of multiplying the raw score by 1.25.The upper limit of normal of the total raw score of SDS is 41. The lower this score is, the better the conditions will be. The standard score is the integer of the result obtained by multiplying the total raw score by 1.25. In China, patients with an SDS standard score ≥50 is considered as having depressive symptoms. Depression severity = cumulative score of each item/80. Results: 0.5 or below indicates no depression; 0.5–0.59 indicates minor to mild depression; 0.6–0.69 indicates moderate to severe depression; 0.7 or more indicates severe depression (Table 3.23).

**Table 3.23** Self-rating Depression Scale (SDS)

| Place check mark (√) in correct column. | A little of the time | Some of the time | Good part of the time | Most of the time |
|---|---|---|---|---|
| 1. I feel down-hearted and blue. | 1 | 2 | 3 | 4 |
| 2. Morning is when I feel the best. | 4 | 3 | 2 | 1 |
| 3. I have crying spells or feel like it. | 1 | 2 | 3 | 4 |
| 4. I have trouble sleeping at night. | 1 | 2 | 3 | 4 |
| 5. I eat as much as I used to. | 4 | 3 | 2 | 1 |
| 6. I still enjoy sex. | 4 | 3 | 2 | 1 |
| 7. I notice that I am losing weight. | 1 | 2 | 3 | 4 |
| 8. I have trouble with constipation. | 1 | 2 | 3 | 4 |
| 9. My heart beats faster than usual. | 1 | 2 | 3 | 4 |
| 10. I get tired for no reason. | 1 | 2 | 3 | 4 |
| 11. My mind is as clear as it used to be. | 4 | 3 | 2 | 1 |
| 12. I find it easy to do the things I used to. | 4 | 3 | 2 | 1 |
| 13. I am restless and can't keep still. | 1 | 2 | 3 | 4 |
| 14. I feel hopeful about the future. | 4 | 3 | 2 | 1 |
| 15. I am more irritable than usual. | 1 | 2 | 3 | 4 |
| 16. I find it easy to make decisions | 4 | 3 | 2 | 1 |
| 17. I feel that I am useful and needed. | 4 | 3 | 2 | 1 |
| 18. My life is pretty full. | 4 | 3 | 2 | 1 |
| 19. I feel that others would be better off if I were dead. | 1 | 2 | 3 | 4 |
| 20. I still enjoy the things I used to do. | 4 | 3 | 2 | 1 |

### 9.1.3 Beck Depression Inventory (BDI)

The Beck Depression Inventory, also known as the BDI, was developed by the famous American psychologist A.T. Beck in 1961, and is one of the earliest self assessment scales of depression in the United States, which aims to evaluate the severity of depression. **The early version of the BDI included 21 items, and then a version with only 13 items was introduced in 1972**. The symptoms of the BDI include: depression, pessimism, feelings of failure, lack of satisfaction, guilt feelings, feeling disappointment in oneself, negative tendencies, social withdrawal, indecision, altered self-image, work difficulties, fatigue, and loss of appetite. The duration of the test is recommended to be 5–10 min, and the description of each item is divided into 4 levels, ranked according to the severity of the symptoms in different stages, namely "without this symptom", "mild", "moderate", and "severe", corresponding to 0–3 points, respectively. The total score ranges from 0–39 points, with ≤4 points for no depression or minimal; 5–7 points for mild depression; 8–15 points for moderate depression; and 16 points or higher for severe depression (Table 3.24).

### 9.1.4 Quick Inventory of Depressive Symptomatology Self-Rated (QIDS-SR)

The Quick Inventory of Depressive Symptomatology Self-rated (also known as the QIDS-SR16) was developed by Rush et al. in 2003 based on the DSM-V diagnostic items, and is one of the 3 widely used scales for clinical treatment of depression (Rush et al. 2003). The QIDS-SR is a self-rated

**Table 3.24** Beck depression inventory (BDI)

| | |
|---|---|
| A. (Sadness) | |
| 3 | I am so sad or unhappy that I can't stand it. |
| 2 | I am blue or sad all the time and I can't snap out of it. |
| 1 | I feel sad or blue. |
| 0 | I do not feel sad. |
| B. (Pessimism) | |
| 3 | I feel that the future is hopeless and that things cannot improve. |
| 2 | I feel 1 have nothing to look forward to. |
| 1 | I feel discouraged about the future. |
| 0 | I am not particularly pessimistic or discouraged about the future. |
| C. (Sense of failure) | |
| 3 | feel I am a complete failure as a person (parent, husband, wife). |
| 2 | As I look back on my fife, all I can see is a lot of failures |
| 1 | I feel I have failed more than the average person. |
| 0 | I do not feel like a failure. |
| D. (Dissatisfaction) | |
| 3 | I am dissatisfied with everything. |
| 2 | I don't get satisfaction out of anything anymore. |
| 1 | I don't enjoy things the way I used to. |
| 0 | I am not particularly dissatisfied. |
| E. (Guilt)) | |
| 3 | I feel as though I am very bad or worthless. |
| 2 | I feel quite guilty. |
| 1 | I feel bad or unworthy a good part of the time. |
| 0 | I don't feel particularly guilty. |
| F. (Self-dislike)) | |
| 3 | I hate myself. |
| 2 | I am disgusted with myself. |
| 1 | I am disappointed in myself. |
| 0 | I don't feel disappointed in myself. |
| G. (Self-harm) | |
| 3 | I would kill myself if I had the chance. |
| 2 | I have definite plans about committing suicide. |
| 1 | I feel I would be better off dead. |
| 0 | I don't have any thoughts of harming myself. |
| H. (Social withdrawal) | |
| 3 | I have lost all of my interest in other people and don't care about them at all. |
| 2 | I have lost most of my interest in other people and have little feeling for them. |
| 1 | I am less interested in other people than I used to be. |
| 0 | I have not lost interest in other people. |
| I. (Indecisiveness) | |
| 3 | I can't make any decisions at all anymore. |
| 2 | I have great difficulty in making decisions. |
| 1 | I try to put off making decisions. |
| 0 | I make decisions about as well as ever. |
| J. (Self-image change) | |
| 3 | I feel that I am ugly or repulsive-looking. |
| 2 | I feel that there are permanent changes in my appearance and they make me look unattractive. |
| 1 | I am worried that I am looking old or unattractive. |
| 0 | I don't feel that I look any worse than I used to. |
| K. (Work difficulty) | |
| 3 | I can't do any work at all. |
| 2 | I have to push myself very hard to do anything. |
| 1 | It takes extra effort to get started at doing something. |
| 0 | I can work about as well as before. |
| L. (Fatigability) | |
| 3 | I get too tired to do anything. |
| 2 | I get tired from doing anything. |
| 1 | I get tired more easily than I used to. |
| 0 | I don't get any more tired than usual. |
| M. (Anorexia) | |
| 3 | I have no appetite at all anymore. |
| 2 | My appetite is much worse now. |
| 1 | My appetite is not as good as it used to be. |
| 0 | My appetite is no worse than usual. |

scale for symptom severity that instructs the patients to assess their conditions from sleep status, appetite/weight, speed of movement/sense of calm, mood, self-perception, perception of suicide, sense of worth, and decisiveness. Only 9 items in the scale will be scored, each on a 4-point scale of 0–3. The total score ranges from 0 to 27. The items to be scored are as follows: (1) select the highest score from the 4 items of sleep (items 1–4); (2) select the score of item 5; (3) select the highest score from the 4 items of body weight (items 6–9); (4) select the total score of items 10–14; (5) select the highest score from the 2 items of psychomotor (items 15–16). (6) Sum the scores mentioned above. Judgment of severity based on QIDS-SR16 total score: 6–10 for mild depression; 11–15 for moderate depression; 16–20 for severe depression; 21–27 for extremely severe depression (Table 3.25).

## 9.2 Other Assessment Scales

*Chinese Guidelines for the Prevention and Treatment of Depressive Disorders (Second Edition)* states that specialists may comprehensively and accurately assess the depressive symptoms of patients and check against the self-assessment scales of patients using the Hamilton depression scale (HAMD) and the Montgomery depression rating scale (MADRS) in clinical practice.

### 9.2.1 Hamilton Depression Scale for Depression

HAMD, developed by Hamilton in 1960, is the most common scale for clinical assessment of depression (Hamilton 1960). This scale has 3 versions, namely 17-item, 21-item, and 24-item. The examination of a patient with HAMD is jointly performed by 2 assessors, usually by talking and observing.

**Table 3.25** Quick inventory of depressive symptomatology self-rated (QIDS-SR)

| |
|---|
| Please circle the one response to each item that best describes you for the past 7 days. |
| **1. Falling Asleep:** |
| 0 I never take longer than 30 min to fall asleep. |
| 1 I take at least 30 min to fall asleep, less than half the time. |
| 2 I take at least 30 min to fall asleep, more than half the time. |
| 3 I take more than 60 min to fall asleep, more than half the time. |
| **2. Sleep During the Night:** |
| 0 I do not wake up at night. |
| 1 I have a restless, light sleep with a few brief awakenings each night. |
| 2 I wake up at least once a night, but I go back to sleep easily. |
| 3 I awaken more than once a night and stay awake for 20 min or more, more than half the time. |
| **3. Waking Up Too Early:** |
| 0 Most of the time, I awaken no more than 30 min before I need to get up. |
| 1 More than half the time, I awaken more than 30 min before I need to get up. |
| 2 I almost always awaken at least 1 h or so before I need to, but I go back to sleep eventually. |
| 3 I awaken at least 1 h before I need to, and can't go back to sleep. |
| **4. Sleeping Too Much:** |
| 0 I sleep no longer than 7–8 h/night, without napping during the day. |
| 1 I sleep no longer than 10 h in a 24-h period including naps. |
| 2 I sleep no longer than 12 h in a 24-h period including naps. |
| 3 I sleep longer than 12 h in a 24-h period including naps. |
| **Enter the highest score on any 1 of the 4 sleep items (1–4 above)** ____ |
| **5. Feeling Sad:** |
| 0 I do not feel sad |
| 1 I feel sad less than half the time. |
| 2 I feel sad more than half the time. |
| 3 I feel sad nearly all of the time. |
| **6. Decreased Appetite:** |
| 0 There is no change in my usual appetite. |
| 1 I eat somewhat less often or lesser amounts of food than usual. |
| 2 I eat much less than usual and only with personal effort. |
| 3 I rarely eat within a 24-h period, and only with extreme personal effort or when others persuade me to eat. |
| **7. Increased Appetite:** |
| 0 There is no change from my usual appetite. |
| 1 I feel a need to eat more frequently than usual. |
| 2 I regularly eat more often and/or greater amounts of food than usual. |
| 3 I feel driven to overeat both at mealtime and between meals. |
| **8. Decreased Weight (Within the Last Two Weeks):** |
| 0 I have not had a change in my weight. |
| 1 I feel as if I've had a slight weight loss. |
| 2 I have lost 2 pounds or more. |
| 3 I have lost 5 pounds or more. |
| **9. Increased Weight (Within the Last Two Weeks):** |
| 0 I have not had a change in my weight. |
| 1 I feel as if I've had a slight weight gain. |
| 2 I have gained 2 pounds or more. |
| 3 I have gained 5 pounds or more. |
| **Enter the highest score on any 1 of the 4 appetite/weight change items (6–9 above)** |
| **10. Concentration/Decision Making:** |
| 0 There is no change in my usual capacity to concentrate or make decisions. |
| 1 I occasionally feel indecisive or find that my attention wanders. |
| 2 Most of the time, I struggle to focus my attention or to make decisions. |
| 3 I cannot concentrate well enough to read or cannot make even minor decisions. |
| **11. View of Myself:** |
| 0 I see myself as equally worthwhile and deserving as other people. |
| 1 I am more self-blaming than usual. |
| 2 I largely believe that I cause problems for others. |
| 3 I think almost constantly about major and minor defects in myself. |
| **12. Thoughts of Death or Suicide:** |
| 0 I do not think of suicide or death. |
| 1 I feel that life is empty or wonder if it's worth living. |
| 2 I think of suicide or death several times a week for several minutes. |
| 3 I think of suicide or death several times a day in some detail, or I have made specific plans for suicide or have actually tried to take my life. |
| **13. General Interest:** |
| 0 There is no change from usual in how interested I am in other people or activities. |
| 1 I notice that I am less interested in people or activities. |
| 2 I find I have interest in only one or two of my formerly pursued activities. |
| 3 I have virtually no interest in formerly pursued activities. |
| **14. Energy Level:** |
| 0 There is no change in my usual level of energy. |
| 1 I get tired more easily than usual. |
| 2 I have to make a big effort to start or finish my usual daily activities (for example, shopping, homework, cooking or going to work). |
| 3 I really cannot carry out most of my usual daily activities because I just don't have the energy. |
| **15. Feeling Slowed Down:** |
| 0 I think, speak, and move at my usual rate of speed. |
| 1 I find that my thinking is slowed down or my voice sounds dull or flat. |
| 2 It takes me several seconds to respond to most questions and I'm sure my thinking is slowed. |
| 3 I am often unable to respond to questions without extreme effort. |
| **16. Feeling Restless:** |
| 0 I do not feel restless. |
| 1 I'm often fidgety, wringing my hands, or need to shift how I am sitting. |
| 2 I have impulses to move about and am quite restless. |
| 3 At times, I am unable to stay seated and need to pace around. |
| **Enter the highest score on either of the 2 psychomotor items (15 or 16 above)** ____ |
| **Total Score:____ (Range 0–27)** |

The 2 assessors will independently rate at the end of the examination. The scale is scored before and after treatment to evaluate the severity of the condition and the treatment effect. Most items of HAMD are scored on a 5-point scale ranging from 0 to 4. The criteria for each level are: (0) for none; (1) for mild; (2) for moderate; (3) for severe; and (4) for extremely severe. A few items are graded on a 3-point scale ranging from 0 to 2. The criteria are: (0) for none; (1) for mild to moderate; and (2) for severe. The judgment criteria for the results of HAMD-17 items are as follows. Total score <7: normal; total score of 7–17: possible depression; total score of 17–24: definite depression; total score >24: severe depression (Table 3.26).

**Table 3.26** Hamilton Depression Rating Scale (HAMD-17)

| |
|---|
| 1. Depressed Mood (sadness, hopeless, helpless, worthless) |
| 0-Absent |
| 1-These feeling states indicated only on questioning. |
| 2-These feeling states spontaneously reported verbally. |
| 3-Communicates feeling states nonverbally (i.e. through facial expression, posture, voice and tendency to weep). |
| 4-Subject reports virtually only these feeling states in his/her spontaneous verbal and non-verbal communication. |
| 2. Feelings of guilt |
| 0-Absent. |
| 1-Self-reproach, feels they have let people down. |
| 2-Ideas of guilt or rumination over past errors or sinful deeds. |
| 3-Present illness is a punishment. Delusions of guilt. |
| 4-Hears accusatory or denunciatory voices and/or experiences threatening visual hallucinations. |
| 3. Suicide |
| 0-Absent |
| 1-Feels life is not worth living. |
| 2-Wishes he were dead or any thoughts of possible death to self. |
| 3-Suicide ideas or gestures. |
| 4-Attempts at suicide (any serious attempt rate 4). |
| 4. Insomnia: early |
| 0-No difficulty falling asleep. |
| 1-Complains of occasional difficulty falling asleep (eg, more than 30 minutes). |
| 2-Complains of nightly difficulty falling asleep. |
| 5. Insomnia: middle |
| 0-No difficulty. |
| 1-Subject complains of being restless and disturbed during the night. |
| 2-Waking during the night-any getting out of bed rates 2 (except for purposes of voiding). |
| 6. Insomnia: late |
| 0-No difficulty. |
| 1-Waking in early hours of the morning but goes back to sleep. |
| 2-Unable to fall asleep again if he/she gets out of bed. |
| 7. Work and activities |
| 0-No difficulty. |
| 1-Thoughts and feelings of incapacity, fatigue or weakness related to activities, work or hobbies. |
| 2-Loss of interest in activity, hobbies or work - either directly reported by the subject or indirect in listlessness, indecision and vacillation (feels has to push self to work or activities). |
| 3-Decrease in actual time spent in activities or decrease in productivity. In hospital, rate 3 if subject does not spend at least 3 h a day in activities (hospital job or hobbies) exclusive of ward chores. |
| 4-Stopped working because of present illness. In hospital, rate if subject engages in no activities except ward chores, or if subject fails to perform ward chores unassisted. |
| 8. Retardation-psychomotor |
| 0-Normal speech and thought. |
| 1-Slight retardation at interview. |
| 2-Obvious retardation at interview. |
| 3-Interview difficult. |
| 4-Complete stupor. |
| 9. Agitation |
| 0-None. |
| 1-Fidgetiness. |
| 2-Playing with hands, hair, etc. |
| 3-Moving about, can't sit still. |
| 4-Hand wringing, nail biting, hair-pulling, biting of lips. |
| 10. Anxiety-psychic |
| 0-No difficulty. |
| 1-Subjective tension and irritability. |
| 2-Worrying about minor matters. |
| 3-Apprehensive attitude apparent in face or speech. |
| 4-Fears expressed without questioning |
| 11. Anxiety-somatic |
| 0-Absent. |
| 1-Mild. |
| 2-Moderate. |
| 3-Severe. |
| 4-Incapacitating |
| 12. Somatic symptoms gasrto-intestinal |
| 0-None. |
| 1-Loss of appetite but eating without staff encouragement. Heavy feelings in abdomen. |
| 2-Difficulty eating without staff urging. Requests or requires laxatives or medication for bowels or medication for gastro-intestinal symptoms. |
| 13. Somatic symptoms-general |
| 0-None. |
| 1-Heaviness in limbs, back or head. Backaches, headaches, muscle aches. Loss of energy and fatigability. |
| 2-Any clear-cut symptom rates 2. |
| 14. Genital symptoms |
| 0-Absent. |
| 1-Mild. |
| 2-Severe. |
| 15. Hypochondriasis |
| 0-Not present. |
| 1-Self-absorption (bodily). |
| 2-Preoccupation with health. |
| 3-Frequent complaints, requests for help, etc. |
| 4-Hypochondriacal delusions. |

**Table 3.26** (continued)

| |
|---|
| 16. Loss of weight rated by history |
| 0-No weight loss. |
| 1-Probable weight loss associated with present illness. |
| 2-Definite (according to subject) weight loss. |
| 17. Insight |
| 0-Acknowledges being depressed and ill. |
| 1-Acknowledges illness but attributes cause to bad food, climate, overwork, virus, need for rest, etc. |
| 2-Denies being ill at all. |

### 9.2.2 Montgomery Depression Rating Scale (MADRS)

MADRS is a scale jointly developed by Montgomery at Guy's Hospital in London and Asberg at the Karolinska Institute in Stockholm (Montgomery and Asberg 1979). This scale is simpler than the Hamilton depression rating scale and is used to reflect the effects of antidepressant treatment and to monitor changes in patient conditions. The scale consists of 10 items, each with a score of 0–6, and a total score of 0–60. The higher the score, the severer the depression. Scoring criteria: 0–6 for normal/no symptoms; 7–19 for mild depression; 20–34 for moderate depression; >34 for severe depression (Table 3.27).

As for other assessment tools for depressive disorder, *Chinese Guidelines for the Prevention and Treatment of Depressive Disorders (Second Edition)* also introduces the suicide risk rating, the risk of mania conversion rating, the life quality and social functioning rating, the treatment emergent symptom scale, the Arizona sexual experience scale (ASEX), and the medication adherence rating scale (MARS) and so on. There are also scales for specific populations, such as the Edinburg postnatal depression scale (EPDS) and the Postpartum depression screen scale (PDSS). These scales facilitate a comprehensive assessment of the patient's risk of depressive disorder, the presence of manic symptoms, the disease impact on social functioning, and the adverse reactions and adherence to medications.

**Table 3.27** Montgomery depression rating scale (MADRS)

| |
|---|
| 1. **Apparent Sadness**—Representing despondency, gloom and despair, (more than just ordinary transient low spirits) reflected in speech, facial expression, and posture. Rate by depth and inability to brighten up. |
| 0-No sadness. |
| 1 |
| 2-Looks dispirited but does brighten up without difficulty. |
| 3 |
| 4-Appears sad and unhappy most of the time. |
| 5 |
| 6-Looks miserable all the time. Extremely despondent. |
| 2. **Reported sadness**—Representing reports of depressed mood, regardless of whether it is reflected in appearance or not. Includes low spirits, despondency or the feeling of being beyond help and without hope. Rate according to intensity, duration and the extent to which the mood is reported to be influenced by events. |
| 0-Occasional sadness in keeping with the circumstances. |
| 1 |
| 2-Sad or low but brightens up without difficulty. |
| 3 |
| 4-Pervasive feelings of sadness or gloominess. The mood is still influenced by external circumstances. |
| 5 |
| 6-Continuous or unvarying sadness, misery or despondency. |
| 3. **Inner tension**—Representing feelings of ill-defined discomfort, edginess, inner turmoil, mental tension mounting to either panic, dread or anguish. Rate according to intensity, frequency, duration and the extent of reassurance called for. |
| 0-Placid. Only fleeting inner tension. |
| 1 |
| 2-Occasional feelings of edginess and ill defined discomfort. |
| 3 |
| 4-Continuous feelings of inner tension or intermittent panic which the patient can only master with some difficulty. |
| 5 |
| 6-Unrelenting dread or anguish. Overwhelming panic. |
| 4. **Reduced sleep**—Representing the experience of reduced duration or depth of sleep compared to the subject's own normal pattern when well. |
| 0-Sleeps as usual. |
| 1 |
| 2-Slight difficulty dropping off to sleep or slightly reduced, light or fitful sleep. |
| 3 |
| 4-Sleep reduced or broken by at least 2 h. |
| 5 |
| 6-Less than 2 or 3 h sleep. |
| 5. **Reduced appetite**—Representing the feeling of a loss of appetite compared with when well. Rate by loss of desire for food or the need to force oneself to eat. |
| 0-Normal or increased appetite. |
| 1 |
| 2-Slightly reduced appetite. |
| 3 |
| 4-No appetite. Food is tasteless. |
| 5 |
| 6-Needs persuasion to eat at all. |

(continued)

**Table 3.27** (continued)

| |
|---|
| 6. **Concentration difficulties** - Representing difficulties in collecting one's thoughts mounting to incapacitating lack of concentration. Rate according to intensity, frequency, and degree of incapacity produced. |
| 0-No difficulties in concentrating. |
| 1 |
| 2-Occasional difficulties in collecting one's thoughts. |
| 3 |
| 4-Difficulties in concentrating and sustaining thought which reduces ability to read or hold a conversation. |
| 5 |
| 6 Unable to read or converse without great difficulty. |
| 7. **Lassitude**—Representing a difficulty getting started or slowness initiating and performing everyday activities. |
| 0-Hardly any difficulty in getting started. No sluggishness. |
| 1 |
| 2-Difficulties in starting activities. |
| 3 |
| 4-Difficulties in starting simple routine activities which are carried out with effort. |
| 5 |
| 6-Complete lassitude. Unable to do anything without help. |
| 8. **Inability to feel**—Representing the subjective experience of reduced interest in the surroundings, or activities that norm ally give pleasure. The ability to react with adequate emotion to circumstances or people is reduced. |
| 0-Normal interest in the surroundings and in other people. |
| 1 |
| 2-Reduced ability to enjoy usual interests. |
| 3 |
| 4-Loss of interest in the surroundings. Loss of feelings for friends and acquaintances. |
| 5 |
| 6-The experience of being emotionally paralyzed, inability to feel anger, grief or pleasure and a complete or even painful failure to feel for close relatives and friends. |
| 9. **Pessimistic thoughts**—Representing thoughts of guilt, inferiority, self reproach, sinfulness, remorse and ruin. |
| 0-No pessimistic thoughts. |
| 1 |
| 2-Fluctuating ideas of failure, self-reproach or self depreciation. |
| 3 |
| 4-Persistent self-accusations, or definite but still rational ideas of guilt or sin. Increasingly pessimistic about the future. |
| 5 |
| 6-Delusions of ruin, remorse or unredeemable sin. Self-accusations which are absurd and unshakable. |
| 10. **Suicidal thoughts**—Representing the feeling that life is not worth living, that a natural death would be welcome, suicidal thoughts, and preparations for suicide. Suicidal attempts should not in themselves influence the rating. |
| 0-Enjoys life or takes it as it comes. |
| 1 |
| 2-Weary of life. Only fleeting suicidal thoughts. |
| 3 |
| 4-Probably better off dead. Suicidal thoughts are common, and suicide is considered as a possible solution, but without specific plans or intention. |
| 5 |
| 6-Explicit plans for suicide when there is an opportunity. Active preparations for suicide. |

## 10 Clinical Assessment of Epilepsy

Epilepsy is a chronic disorder characterized by long-standing and recurrent seizures which may interfere with daily activities and quality of life in patients. Scales allow for a comprehensive and rapid assessment of health-related quality of life in adults with epilepsy and an evaluation of each patient's response before and after treatment.

### 10.1 Quality of Life in Epilepsy Inventory-31 (QOLIE-13)

The quality of Life in epilepsy inventory, consisting of 31 items, covers the main issues related to quality of daily life In general, what do you think your quality of life is? Please circle a number between 10 (best quality of life) and 0 (worst quality of life).

The following questions are about how you have been feeling and how you have been in the last month. Please choose the answer that can best describe your feelings. Circle a number between 1 (always) and 6 (never).

1. Do you feel energetic?
2. Are you a nervous person?
3. Do you feel down and nothing can brighten you up?
4. Do you feel at peace?
5. Do you have a lot of energy?
6. Do you feel particularly depressed?
7. Do you feel exhausted?
8. Are you a happy person?
9. Do you feel tired?
10. Are you worried about having another attack?
11. Do you have difficulty in thinking and solving problems (e.g., making plans, making decisions, learning new things, etc.)?
12. Does your health condition limit your social activities (e.g., visiting friends and family)?
13. How has your quality of life been in the last month? (i.e., how are you doing, choosing a number on a scale from "1 = very good, couldn't be better to 5 = very bad, couldn't be worse").

The following questions are about your memory.

14. Have you had difficulty in remembering in the last month? (Choose a number between 1 and 4, 1 = yes, a lot, 4 = no, not at all)

Choose a number between 1 (always) and 6 (never) to indicate how many times you have had difficulty in remembering in the last month, or whether memory difficulties have often interfered with your normal work and life.

15. Do you have trouble remembering things that people have told you?

The following 2 questions are about the attention disorder you may have. Circle a number from 1 (always) to 6 (never) to indicate how many times you have had difficulty in concentrating in the last month, or how many times these difficulties have interfered with your normal work and life.

16. Do you have trouble concentrating while reading?
17. Do you have trouble concentrating on one thing at a time?

The following 2 questions are about the troubles you may have had with certain activities. Circle a number from 1 (very much) to 5 (not at all) to indicate the extent to which your illness or antiepileptic drug has caused troubles during the following periods in the last month.

18. It causes troubles in your spare time (e.g., hobbies, going out).
19. It causes troubles while you are driving, cycling, or motorcycling.

The following questions are about how you feel about seizures.

20. Are you afraid that you will have a seizure in the next month? Choose a number from 1 (very scared) to 4 (not scared at all) to indicate your level of worry/anxiety.
21. Are you worried that you might get hurt during a seizure? Choose a number from 1 (often worried) to 3 (not worried) to indicate your level of worry.
22. Are you worried about embarrassment and other social problems caused by a seizure in the next month? Choose a number from 1 (very worried) to 4 (not worried at all) to indicate your level of worry.
23. Are you worried that taking medication for a long time may cause harm to you? Choose a number from 1 (very worried) to 4 (not worried at all) to indicate your level of concern.

For the following aspects, circle a number from 1 (not bothered at all) to 5 (extremely bothered) to indicate how much they bother you.

24. Seizures.
25. Memory difficulties.
26. Work limitations.
27. Social limitations.
28. Physical side effects of antiepileptic drugs.
29. Psychological side effects of antiepileptic drugs.

How do the following 2 aspects affect your seizures? Please circle a number from 1 (very much) to 3 (no effect) to indicate how they affect your seizures.

30. How do you feel about your health? 100 means very good health, and 0 means very poor health. Please circle a number from 100 (very good) to 0 (very poor) to indicate how you feel about your health. Take epilepsy into account when answering this question.

The initial score of each aspect is equal to the sum of the scores of each question divided by the number of items for that aspect: the initial score of each aspect is multiplied by the weighting score, and the total score is the sum of the scores of each aspect. The higher the score, the better the quality of life.

## 10.2 Quality of Life in Childhood Epilepsy Questionnaire - 16 Items

The following questions are about how your children have been feeling and how your children have been in the last month. Please choose the answer that can best describe your feelings. Circle a number between 0 (always) and 4 (never).

1. Does the child have difficulty in recognizing directions?
2. Does the child have difficulty in playing with complex toys?
3. Does the child have difficulty in playing with simple toys?
4. Does the child have difficulty in remembering what people tell him/her?
5. Does the child feel that no one understands him/her?
6. Does the child feel overwhelmed?
7. Does the child feel frustrated?
8. Does the child have confidence?
9. Is your child's social activity limited compared to other children of his/her age?
10. Is his/her social interaction at school or at work affected?
11. Is he/she isolated?
12. Does he/she have difficulty in making friends?
13. Can he/she play freely in the house like other children of his/her age?
14. Can he/she play sports like other children of his/her age?
15. Can he/she play outside freely like other children of his/her age?
16. Does he/she need more supervision than other children of his/her age?

The total score is the sum of the scores of each aspect. The higher the score, the better the quality of life.

## 10.3 Liverpool Seizure Severity Scale 2.0 (LSSS)

The Liverpool seizure severity scale (LSSS) is used to determine the severity of epileptic seizures by the patient's assessment of their own feelings.

Patients who have had an epileptic seizure in the last 4 weeks answer the following 12 questions:

1. I feel that my severest seizures are mostly:
   3 = Very severe, 2 = Severe, 1 = Mild, 0 = Very mild
2. The duration of my loss of vitality /unconsciousness is often:
   4 = More than 5 min, 3 = 3–5 min, 2 = 1–2 min, 1 = Less than 1 min, 0 = Never
3. I smack my lips, become irritable or behave abnormally when a severe seizure attacks:
   3 = Always, 2 = Often, 1 = Occasionally, 0 = Never
4. After the severest seizure:
   3 = I feel very panicked, 2 = I feel rather panicked.
   1 = I feel a little panicked, 0 = I don't feel panicked at all
5. After the severest attack, the panic feeling lasts for:
   5 = More than 2 h, 4 = 1–2 h, 3 = 6 min to 1 h, 2 = 1–5 min,
   1 = Less than 1 min, 0 = Never feeling panicked
6. When the worse seizure attacks:
   3 = I always fall to the ground, 2 = I often fall to the ground
   1 = I sometimes fall to the ground, 0 = I never fall to the ground
7. After the severest seizure:
   3 = I always have a headache, 2 = I often have a headache
   1 = I sometimes have a headache, 0 = I never have a headache
8. After the severest seizure:
   3 = I always feel lethargy, 2 = I often feel lethargy
   1 = I sometimes feel lethargy, 0 = I never feel lethargy
9. After the severest seizure:
   3 = I always find wetting myself, 2 = I often find wetting myself
   1 = I sometimes find wetting myself, 0 = I never find wetting myself
10. After the severest seizure:
    3 = I always find myself biting my tongue, 2 = I often find myself biting my tongue
    1 = I sometimes find myself biting my tongue, 0 = I never find myself biting my tongue
11. After the severest seizure:
    3 = I always find that I have injured myself (other than biting my tongue)
    2 = I often find that I have injured myself (other than biting my tongue)
    1 = I sometimes find that I have injured myself (other than biting my tongue)
    0 = I never find that I have injured myself (other than biting my tongue)
12. After the severest seizure attacks, I usually can get back to what I was doing after:
    4 = More than 2 h, 3 = 1–2 h, 2 = 6 min to 1 h, 1 = 1–5 min,0 = Less than 1 min

Raw LSSS score = sum of scores of all 12 questions

ICTAL score = (raw LSSS score)/40 × 100

1. Minimum raw score: 0; maximum raw score: 40.
2. Lowest ICTAL score (corrected): 0; highest ICTAL score (corrected): 100.
3. The higher the ICTAL score, the severer the epileptic seizure.
4. If no epileptic seizures have attacked in the last 4 weeks, the ICTAL score is 0.
5. The ICTAL score is "missing" if there are no answers to 4 or more questions.
6. If 1–3 questions are not answered, the average score of the answered questions needs to be calculated and multiplied by 12 to obtain the raw LSSS score.

## 10.4 National Hospital Seizure Severity Scale (NHS3)

The national hospital seizure severity scale (NHS3, where S3 refers to the seizure severity scale) can be used to determine the severity of seizures in patients with epilepsy and to evaluate the efficacy of antiepileptic drugs in clinical trials.

The following indicators are recorded for each seizure type (3 types) of the patient based on the seizures since the last visit:

Indicators

Score of Clinical manifestations in each type of seizure:

1. Generalized seizures
   4 = Existing    0 = None
2. Falling to the ground (including falling from bed).
   4 = Always    4 = Almost always    3 = Often    2 = Rarely    0 = Never
3. Injury
   4 = Burns, scalds, cuts, fractures    3 = Biting the tongue or severe headache
   2 = Mild injury or mild headache    0 = None
4. Gatism
   4 = Always    4 = Almost always    3 = Often    2 = Rarely    0 = Never

5. Loss of awareness with aura
   2 = No aura 1 = Sometimes with aura 0 = Almost always with aura 0 = Always with aura
   0 = No loss of awareness 0 = Seizures only during sleep
6. Time required to recover normal functions
   4 = >3 h 3 = 1–3 h 2 = 11–60 min 1 = 1–10 mi n 0 = <1 min
7. Automatism
   4 = Severe splitting 2 = Mild or topical clonus 0 = None

1. If the frequency of the above is less than 25%, it will be described as occasional.
2. The frequency between 25 and 50% is described as often.
3. If there is only one seizure attack and the above conditions occur during that seizure, it will be described as almost always or always.
4. The degree of injury is scored based on the severest injury

NHS3 score = sum of scores of all indicators +1
Evaluation:

1. Lowest score: 1.
2. Highest score: 27
3. The higher the score, the severer the seizure.

## 10.5 Comorbid Depression in Epilepsy Scale

Patients with epilepsy often suffer from depressive disorder at the same time, which not only worsens seizures, but may also cause increased adverse drug reactions and affect the epilepsy treatment effect and the quality of life of patients. In 2006, American researchers developed the neurological disorders depression inventory for epilepsy (NDDI-E), which can quickly identify patients with co-morbid depressive disorders.

Neurological disorders depression inventory for epilepsy (NDDI-E in Chinese).

Circle the appropriate answer according to your situations in the past 2 weeks, Circle a number 4 (Always or often), 3 (Sometimes), 2 (Remote) and 1 (No).

1. Entangled with everything
2. All is wrong
3. Feeling guilty
4. Better off dead
5. Feeling things are going wrong
6. Hard to cheer up

The higher the score, the higher the likelihood of comorbid depressive disorder.

A score ≥14 indicates that the patient with epilepsy has a comorbid depressive disorder.

## 10.6 Severity Evaluation of Status Epilepticus

1. Status epilepticus severity score (STESS)
   (a) Consciousness: awake, 0 for somnolence or confusion, 1 for coma;
   (b) Worst seizure type: 0 point for simple partial seizures, complex partial seizures, petit mal, myoclonic seizures; 1 point for grand mal, and 2 points for non-convulsive SE;
   (c) Age: 0 point for age <65 years, 2 points for age ≥65 years;
   (d) Seizure history: 0 point for yes, 1 point for no or unknown.

   The total score is the sum of the 4 parts, out of 6 points. A score of STESS ≤2 indicates a good prognosis and a low risk of death.

2. Epidemiology based mortality score in SE (EMSE): EMSE is a score designed to predict mortality during hospitalization based on epidemiological data. The EMSE is scored with reference to independent risk factors associated with the mortality outcome of status epilepticus reported in literatures from the evaluation items of etiology, age, comorbidities and EEG characteristics. Each item includes 4–15 detailed indicators with varying scores. The validation analysis of the EMSE score has shown that the negative and positive predicted values are satisfactory, allowing for an accurate prediction of both survival and death outcomes in status epilepticus, and also allowing for the classification of the severity of patient conditions.
   (a) Etiology: Choose one from the following
      (i) Abnormalities of central nervous system: Score 2
      (ii) Low dose, resistance and poor adherence of antiepileptic drugs: Score 2
      (iii) Multiple sclerosis: Score 5
      (iv) Old cerebrovascular events, craniocerebral injury: Score 7
      (v) Hydrocephalus: Score 8
      (vi) Alcohol abuse: Score 10
      (vii) Drug abuse: Score 11
      (viii) Craniocerebral injury: Score 12
      (ix) Reasons unknown: Score 12
      (x) Craniocerebral tumor: Score 16
      (xi) Metabolism: water-sodium imbalance: Score 17
      (xii) Metabolic disorders: Score 22
      (xiii) Acute cerebrovascular events: Score 26

(xiv) Infection of central nervous system: Score 33
(xv) Ischemic-hypoxic encephalopathy: Score 65

(b) Comorbidities: Select one or more diseases. Calculate the total score.
   (i) Myocardial infarction, congestive heart failure, peripheral vascular disorder, cerebrovascular disease, dementia, chronic lung disease, connective tissue disease, ulcers, mild hepatopathy, diabetes: Score 10
   (ii) Hemiplegia, moderate to severe nephrosis, diabetic target-organ damage, tumors (including leukemia and lymphoma): Score 20
   (iii) Moderate-to-severe hepatopathy: Score 30
   (iv) Metastatic archiblastoma, AIDS: Score 60
(c) Age: Choose one from the rows:
   (i) 21–30: Score 1
   (ii) 31–40: Score 2
   (iii) 41–50: Score 3
   (iv) 51–60: Score 5
   (v) 61–70: Score 7
   (vi) 71–80: Score 8
   (vii) >80: Score 10
(d) EEG features: Choose the severest EEG manifestation
   (i) Burst suppression (spontaneous): Score 60
   (ii) Post-status epilepticus paroxysmal epileptiform discharges (ASIDs lasting seconds to minutes): Score 40
   (iii) Lateral periodic discharges (LPDs): Score 40
   (iv) Generalized periodic discharges (GPDs): Score 40
   (v) No ASIDs, GFDs, ASIDs: Score 0

EMSE = Item a (Etiology) + Item b (Comorbidities) + Item c (Age) + Item d (EEG features)

3. END-IT score [whether including encephalitis, whether there is comorbid nonconvulsive status epilepticus (NCSE), diazepam resistance, image features, and whether tracheal intubation has been performed]: END-IT score is currently the only score that predicts the recovery of limb function 3 months after discharge in patients with status epilepticus. Compared with STESS and EMSE scores, the END-IT score is more predictive

Indicators
Score of Category:

1. Encephalitis:
   1 = Yes  0 = None
2. Comorbid non convulsive epilepsy Status epilepticus:
   1 = Yes  0 = No
3. Diazepam resistance:
   1 = Yes  0 = None
4. Image features:
   2 = Bilateral responsible foci or generalized cerebral edema
   1 = Single responsible focus
   0 = No responsible focus
5. Tracheal intubation:
   1 = Yes  0 = None

Note: END-IT is scored based on whether encephalitis is included, whether there is nonconvulsive status epilepticus (NCSE), diazepam resistance and image features, and whether tracheal intubation has been performed.

## 11 Clinical Assessment of Disorders of Consciousness

Disorders of consciousness (DoC) refer to various states of loss of consciousness caused by severe brain injury, including coma, vegetative state (VS)/unresponsive wakefulness syndrome (UWS), and minimally conscious state (MCS).

### 11.1 Diagnosis of Disorders of Consciousness

Coma is a severe arousal disorder in which the patient is not aware of himself or his surroundings and cannot spontaneously open eyes or be aroused by strong sensory stimuli (Posner et al. 2007). The lack of awareness and also wakefulness distinguishes it from VS/UWS. Patients in coma usually recover in the acute phase (within 4 weeks) or develop prolonged DoC (pDoC) at the end of the acute phase, i.e., more than 28 days of consciousness impairment after various severe brain injuries. Irreversible coma in the absence of brain stem reflexes and spontaneous respirations often predicts brain death.

Under VS/UWS, the basic brain stem reflexes and sleep-wake cycle are retained, with spontaneous or stimulated eye opening but no consciousness and no awareness of himself/herself or surroundings (or with only simple reflex motions unrelated to commands) (Jennett and Plum 1972; Laureys et al. 2010). **Diagnostic criteria for VS/UWS** (Multi-Society Task Force on 1994): (1) Patients lose awareness of themselves and surroundings and are unable to functionally communicate with others; (2) A sleep-wake cycle exists; (3) Autonomic functions of hypothalamus and brain stem are completely or partially preserved; (4) There are no sustained, repetitive, purposeful, or random behavioral responses to visual, auditory, tactile, or noxious stimuli; (5) Patients are unable to understand or express speech; (6) Bowel and bladder in continence; (7) Brain stem reflexes and spinal reflexes are preserved to varying degrees.

In MCS, consciousness is partially preserved, and there are unsustained but clearly identifiable conscious behaviors toward self or environment (e.g., execution of simple commands, visual following, etc.) (Giacino et al. 2002) **Diagnostic criteria for MCS:** Patients have clearly identifiable perceptions of themselves or environment, i.e., showing one or more of the following repeated and sustained behaviors: (1) executing simple commands; (2) expressing yes or no (whether correct or not) through posture or language; (3) expressing comprehensible speech; (4) showing purposeful response rather than reflexive actions to environmental stimuli. **Purposeful behaviors specifically** include: A. Responding to emotionally meaningful auditory or visual stimuli with an appropriate smile or cry. B. Responding directly to questions through gestures or words. C. Displaying a clear relationship between object location and route when finding objects. D. Touching and holding objects in a manner appropriate to their size and shape. E. Eyes tracking or gazing at moving or jumping objects. There is heterogeneity in MCS. It is hence divided into MCS- and MCS + according to the degree of behavior response. MCS- refers to the clinical presence of visual object tracking, orientation of noxious stimuli, and directed voluntary movements, but the inability to complete the activities as required. MCS+ is a state in which patients can move eyes, open close eyes and stably move limbs as required while fail in functional communication with others or intentionally using objects (Bruno et al. 2011). MCS may be a temporary state and considered ended if meaningful functional interactions occur during the recovery period (Giacino et al. 2002). Functional interactive behaviors include: (1) Functional communication (FC): represented by telling or writing signals of yes/no, etc. Enable to accurately answer yes/no to 6 most basic orientation questions in 2 consecutive evaluations. (2) Functional object use (FO): requiring the patient to demonstrate the difference between two objects by motions; or accurately use 2 different objects in 2 consecutive evaluations.

## 11.2 Clinical Assessment of Disorders of Consciousness

As the levels of daily consciousness and waking states of patients with DoC fluctuation, systematic and careful examinations and multiple repeated evaluations are required. It is important to rule out the effects of sedative-hypnotics, antipsychotics, general anesthetics, and muscle relaxants on consciousness prior to evaluation. In addition, anesthesia, dyskinesia, aphasia, and depression can limit the patient's response to the examination and hence should be identified. Currently, disorders of consciousness are mainly assessed by observing the patient's spontaneous and post-stimulus behaviors, i.e., observing whether the patient opens eye to determine "arousal", and observing the execution of commands or other unconditioned reflex behaviors to determine whether the patient has "awareness". In addition, for patients with DoC who cannot open their eyes spontaneously or after stimulation, a passive eye-opening examination after arousal should not be omitted from the daily clinical assessment, along with vertical and horizontal eye movement examinations, so as to identify patients with locked-in syndrome, MCS, and recovery of consciousness. Resistance to passive eye opening is often a sign of preserved consciousness. In patients who do not cooperate with eye movement examinations, a visual tracking examination (looking in the mirror) may be performed. Spontaneous activities should be recorded, including acts such as extubation, scratching nose, grasping sheets, lifting leg, and limb positioning, which all reflect a high level of residual consciousness (Pincherle et al. 2019).

## 11.3 Behavioral Assessment Scales for Disorders of Consciousness

The rate of misdiagnosis of disorders of consciousness can be as high as 10–41% if relying on clinical examinations alone (Schnakers et al. 2009). Standardized neurobehavioral assessment scales may compensate for items that may be missed during clinical examinations and assist in the determination of DoC. As early as 1974, Teasdale and Jennett first applied the Glasgow coma scale (GCS) to the assessment of patients with disorders of consciousness due to craniocerebral injury, and it has since been widely used clinically for its simple quantitative indicators. However, with the development of intensive care medicine, the limitations of GCS in the assessment of patients in coma have gradually shown up: (1) speech assessment is limited in patients after tracheal intubation or tracheotomy; (2) assessment is limited when brain injury involves anatomical sites related to eye opening, speech or movement; (3) GCS does not include the assessment of brain stem functions. Since then, numerous researchers have designed different assessment scales based on clinical experience. They are mainly divided into two categories:

1. Rating scales for the degree of disorders of consciousness: (1) Glasgow-Pittsburgh coma scale (GCS-P) (Edgren et al. 1994): Proposed by Edgren et al. in 1994, GCS-P was developed based on the GCS with the addition of 4 assessment items including pupillary light reflex, brain stem reflex, convulsions and spontaneous breathing. (2) Full outline of unresponsiveness scale (FOUR) (Wijdicks et al. 2005): Designed by Wijdicks et al. in 2005, it includes two additional items of brain stem reflex and respiration on the basis of GCS, and is now consid-

ered applicable to the disorders of consciousness in ICU and superior to GCS (Kondziella et al. 2020). (3) Coma recovery scale—revised (CRS-R) (Giacino et al. 2004). Proposed by Giacino et al., CRS-R contains 6 subscales involving hearing, vision, motor, language, communication, and arousal, with a total of 23 items involving brain stem, subcortical and cortical functions, and is currently a reliable tool for differentiating VS from MCS. It is often used in clinical work as an effective tool for assessing the degree of disorders of consciousness in patients with brain injury (Sattin et al. 2015). (4) Wessex head injury matrix (WHIM) (Shiel et al. 2000): Proposed in Cambridge, UK, in 2000, WHIM consists of 5 aspects of arousal, social behavior, communication, cognition, and memory, with a total of 58 items, and can be used to assess different degrees of disorders of consciousness from coma to consciousness recovery. (5) Other scales, such as the disorders of consciousness scale (DoCS) proposed by Pape et al.; the Chinese vegetative state scale (CVSS) proposed by Zhang Guojin et al.; the Innsbruck coma scale (ICS) proposed by Benzer et al.

2. Outcome rating scales for disorders of consciousness: The Glasgow outcome scale (GOS) and its extension (GOS-E) proposed by Jennett et al. are the main evaluation tool for prognosis, but they still cannot distinguish between VS and MCS. The disability rating scale (DRS) can be extensively applied to the assessment from the onset of injury to the recovery of normal life, but requires the cooperation of other assessment methods. Other scales include the functional independence measure (FIM), Craig handicap assessment and reporting technique (CHART), and the community integration questionnaire (CIQ), etc., for the post-discharge and follow-up assessments of patients.

In clinical practice, appropriate scales should be used according to the stage of disease in which the patients are. FOUR or GCS is recommended for the acute phase, while CRS-R for the recovery phase (Kondziella et al. 2020). Given that the improvement from VS to MCS- or MCS+ in DoC patients is important for prognosis and clinical intervention and judgment, CRS-R is recommended as the preferred tool and GOS-E as an auxiliary scale for prognostic assessment.

## 12 Summary

In order to objectively and accurately assess patients with disorders of consciousness caused by craniocerebral injury as far as possible, the following should be noted: (1) making a comprehensive assessment based on clinical observation indicators; (2) choosing rational assessment methods and prognostic measures at different periods of craniocerebral injury; (3) dynamically and continuously making a comprehensive assessment of patients throughout the period from injury to recovery to instruct strategies of clinical treatment.

## References


Albanese A, Bhatia K, Bressman SB et al (2013) Phenomenology and classification of dystonia: a consensus update. Mov Disord 28(7):863–873

Arand DL, Bonnet MH (2019) The multiple sleep latency test. In: Levin KH, Chauvel P (eds) Handbook of clinical neurology, vol 160. Elsevier, Holland, pp 393–403

Beck AT, Epstein N, Brown G et al (1988) An inventory for measuring clinical anxiety: psychometric properties. J Consult Clin Psychol 56:893–897

Bennett M (2001) The LANSS Pain Scale: the Leeds assessment of neuropathic symptoms and signs. Pain 92:147–157

Bhatia KP, Bain P, Bajaj et al. (2018) Consensus Statement on the classification of tremors. From the task force on tremor of the International Parkinson and Movement Disorder Society. Mov Disord 33(1):75–87.

Bouhassira D, Attal N, Alchaar H et al (2005) Comparison of pain syndromes associated with nervous or somatic lesions and development of a new Neuropathic pain diagnostic questionnaire (DN4). Pain 114:29–36

Bruno MA, Vanhaudenhuyse A, Thibaut A, Moonen G et al (2011) From unresponsive wakefulness to minimally conscious PLUS and functional locked-in syndromes: recent advances in our understanding of disorders of consciousness. J Neurol 258(7): 1373–1384

Chien CW, Bagraith KS, Khan A et al (2013) Comparative responsiveness of verbal and numerical rating scales to measure pain intensity in patients with chronic pain. J Pain 14:1653–1662

Connelly M, Neville K (2010) Comparative prospective evaluation of the responsiveness of single-item pediatric pain-intensity self-report scales and their uniqueness from negative affect in a hospital setting. Pain 11:1451–1460

Danzig R, Wang M, Shah A et al (2019) The wrist is not the brain: estimation of sleep by clinical and consumer wearable actigraphy devices is impacted by multiple patient- and device-specific factors. J Sleep Res 29(1):e12926

Drozdick LW, Raiford SE, Wahlstrom D, Weiss LG (2018) The Wechsler adult intelligence scale—fourth edition and the Wechsler memory scale—fourth edition. In: Flanagan DP, McDonough EM (eds) Contemporary intellectual assessment: theories, tests, and issues. The Guilford Press, pp 486–511

Dworkina RH, Turkb DC, Revicki DA et al (2009) Development and initial validation of an expanded and revised version of the Short-form McGill Pain Questionnaire (SF-MPQ-2). Pain 144:35–42

Edgren E, Hedstrand U, Kelsey S et al (1994) Assessment of neurological prognosis in comatose survivors of cardiac arrest. BRCT I Study Group. Lancet 343(8905):1055–1059

Frohnhofen H, Popp R, Stieglitz S et al (2019) Assessment of sleep and sleep disorders in geriatric patients. Z Gerontol Geriatr 53(2):100–104

Giacino JT, Ashwal S, Childs N et al (2002) The minimally conscious state: definition and diagnostic criteria. Neurology 58(3):349–353

Giacino JT, Kalmar K, Whyte J (2004) The JFK coma recovery scale-revised: measurement characteristics and diagnostic utility. Arch Phys Med Rehabil 85(12):2020–2029

Goetz CG, Fahn S, Martinez-Martin P et al (2007) Movement Disorder Society-sponsored revision of the Unified Parkinson's Disease Rating Scale (MDS-UPDRS): process, format, and clinimetric testing plan. Mov Disord 22(1):41–47

Goodglass H, Kaplan E, Barresi B (2011) Boston diagnostic aphasia examination, 3rd edn. Lippincott Williams & Wilkins, Philadelphia

Hamilton M (1959) The assessment of anxiety states by rating. Br J Med Psychol 32:50–55

Hamilton M (1960) A rating scale for depression. J Neurol Neurosurg Psychiatry 23:56–62

Jalil AA, Katzka DA, Castell DO (2015) Approach to the patient with dysphagia. Am J Med 128(10):1138

Jennett B, Plum F (1972) Persistent vegetative state after brain damage. A syndrome in search of a name. Lancet 1(7753):734–737

Kondziella D, Bender A, Diserens K et al (2020) European Academy of Neurology guideline on the diagnosis of coma and other disorders of consciousness. Eur J Neurol 27(5):741–756

Kroenke K, Spitzer RL, Williams JB et al (2010) The Patient Health Questionnaire Somatic, Anxiety, and Depressive Symptom Scales: a systematic review. Gen Hosp Psychiatry 32(4):345–359

Kurtis MM, Balestrino R, Rodriguez-Blazquez C et al (2018) A review of scales to evaluate sleep disturbances in movement disorders. Front Neurol 29(9):369

Laureys S, Celesia GG, Cohadon F et al (2010) European task force on disorders of unresponsive wakefulness syndrome: a new name for the vegetative state or apallic syndrome. BMC Med 8:68

Li K, Li SH, Su W et al (2017) Diagnostic accuracy of REM sleep behaviour disorder screening questionnaire: a meta-analysis. Neurol Sci 38:1039–1046

Marelli S, Rancoita PM, Giarrusso F et al (2016) National validation and proposed revision of REM sleep behavior disorder screening questionnaire (RBDSQ). J Neurol 263:2470–2475

Meltzer LJ (2018) When is actigraphy useful for the diagnosis and treatment of sleep problems? Paediatr Respir Rev 28:41–46

Melzack R (1975) The McGill Pain Questionnaire: major properties and scoring methods. Pain 1:277–299

Miller A, Heckert KD, Davis BA (2009) The 3-minute musculoskeletal & peripheral nerve exam. Demos Medical Publishing, New York

Montgomery SA, Asberg M (1979) A new depression scale designed to be sensitive to change. Br J Psychiatry 134(4):382–389

Morris JC (1993) The Clinical Dementia Rating (CDR): current version and scoring rules. Neurology 43:2412–2414

Multi-Society Task Force on, P. V. S (1994) Medical aspects of the persistent vegetative state (1). N Engl J Med 330(21):1499–1508

Orchardson R (2012) Aphasia—the hidden disability. Dent Update 39(3):168–174

Pincherle A, Johr J, Chatelle C et al (2019) Motor behavior unmasks residual cognition in disorders of consciousness. Ann Neurol 85(3):443–447

Portenoy R (2006) Development and testing of a neuropathic pain screening questionnaire: ID Pain. Curr Med Res Opin 22:1555–1565

Posner JBSC, Schiff ND, Plum F (2007) The diagnosis of stupor and coma, 4th edn. Oxford University Press, New York

Reisberg B, Ferris SH, de Leon MJ et al (1988) Global Deterioration Scale (GDS). Psychopharmacol Bull 24:661–663

Ross CA, Aylward EH, Wild EJ et al (2014) Huntington disease: natural history, biomarkers and prospects for therapeutics. Nat Rev Neurol 10(4):204–216

Royall DR, Cordes JA, Polk M (1998) Clox: an executive clock drawing task. J Neurol Neurosurg Psychiatry 64:588–594

Rush AJ, Trivedi MH, Ibrahim HM et al (2003) The 16-Item quick inventory of depressive symptomatology (QIDS), clinician rating (QIDS-C), and self-report (QIDS-SR): a psychometric evaluation in patients with chronic major depression. Biol Psychiatry 54:573–583

Sattin D, Minati L, Rossi D et al (2015) The Coma Recovery Scale modified score: a new scoring system for the Coma Recovery Scale-revised for assessment of patients with disorders of consciousness. Int J Rehabil Res 38(4):350–356

Schmitz-Hübsch T, du Montcel ST, Baliko L et al (2006) Scale for the assessment and rating of ataxia: development of a new clinical scale. Neurology 66(11):1717–1720

Schnakers C, Vanhaudenhuyse A, Giacino J et al (2009) Diagnostic accuracy of the vegetative and minimally conscious state: clinical consensus versus standardized neurobehavioral assessment. BMC Neurol 9:35

Shiel A, Horn SA, Wilson BA et al (2000) The Wessex Head Injury Matrix (WHIM) main scale: a preliminary report on a scale to assess and monitor patient recovery after severe head injury. Clin Rehabil 14(4):408–416

Smith MT, McCrae CS, Cheung J et al (2018) Use of actigraphy for the evaluation of sleep disorders and circadian rhythm sleep-wake disorders: an American Academy of Sleep Medicine Systematic Review, Meta-Analysis, and GRADE Assessment. J Clin Sleep Med 14:1209–1230

Spitzer RL, Kroenke K, Williams JBW et al (2006) A brief measure for assessing generalized anxiety disorder: The GAD-7. Arch Intern Med 166(10):1092–1097

Stachler RJ, Francis DO, Schwartz SR et al (2018) Clinical practice guideline: hoarseness (dysphonia) (update). Otolaryngol Head Neck Surg 158(1_suppl):S1–S42

Swedo SE (2013) Diagnostic and statistical manual of mental disorders. In: Kupfer DA (ed) Neurodevelopmental disorders, 5th edn. American Psychiatric Publishing, Washington, DC

Tarakad A (2020) Clinical rating scales and quantitative assessments of movement disorders. Neurol Clin 38(2):231–254

Teshuva I, Hillel I, Gazit E, Giladi N et al (2019) Using wearables to assess bradykinesia and rigidity in patients with Parkinson's disease: a focused, narrative review of the literature. J Neural Transm (Vienna) 126(6):699–710

Tierney MC, Szalai JP, Snow WG et al (1997) Domain specificity of the subtests of the mini-mental state examination. Arch Neurol 54:713–716

Tippett DC (2015) Update in aphasia research. Curr Neurol Neurosci Rep 15(8):49

Trouillas P, Takayanagi T, Hallett M et al (1997) International Cooperative Ataxia Rating Scale for pharmacological assessment of the cerebellar syndrome. The Ataxia Neuropharmacology Committee of the World Federation of Neurology. J Neurol Sci 145(2):205–211

Wang K, Zhu CHY, Chen HB et al (2016) Expert consensus on diagnosis and treatment of anxiety, depression and somatization symptoms in general hospitals. Chin J Neurol 12:908–917

Wijdicks EF, Bamlet WR, Maramattom BV et al (2005) Validation of a new coma scale: the FOUR score. Ann Neurol 58(4):585–593

Wilkinson JM, Codipilly DC, Wilfahrt RP (2021) Dysphagia: evaluation and collaborative management. Am Fam Physician 103(2):97–106

William WC (2013) DeJong's The neurologic examination. Williams & Wilkins

de Zambotti M, Cellini N, Goldstone A et al (2019) Wearable sleep technology in clinical and research settings. Med Sci Sports Exerc 51(7):1538–1557

Zhong Y, Wu J, Xie X et al (2006) Clinical evaluation of children with tic disorder by Yale global severity scale. Chin J Pract Pediatr 21(3):214–216

Zigmond AS, Snaith RP (1983) The hospital anxiety and depression scale. Acta Psychiatr Scand 67(6):361–370

Zuithoff NP, Vergouwe Y, King M et al (2010) The Patient Health Questionnaire-9 for detection of major depressive disorder in primary care: consequences of current thresholds in a cross-sectional study. BMC Fam Pract 11:98

Zung WW, Richards CB, Short MJ (1965) Self-rating depression scale in an outpatient clinic. Further validation of the SDZ. Arch Gen Psychiatry 13(6):508–515

# Noninvasive Electrical Stimulation Basics and Devices

4

Cuiping Xu, Changming Wang, and Runze Chen

Noninvasive electrical neurostimulation (also known as transcranial electrical stimulation, TES) is a noninvasive functional brain modulation technique that applies stimulation generated by electrodes. It modulates synaptic plasticity and neuronal excitability by delivering weak electrical currents to the cerebral cortex, thereby improving pathological conditions or clinical symptoms of patients (Antal et al. 2017; Ritsch et al. 2010). TES mainly includes transcranial direct current stimulation (tDCS), transcranial alternating current stimulation (tACS), transcranial pulsed current stimulation (tPCS), and transcranial random noise stimulation (tRNS).

## 1 Direct Current Stimulation

Direct current is a type of current where electricity always flows in one direction. tDCS is a noninvasive brain stimulation (NIBS) technique that applies a constant current of weak intensity to the scalp to induce changes in the excitability of the cerebral cortex, resulting in physiological and functional effects (Kim et al. 2014). The direction of changes in cortical excitability depends on the polarity of stimulation. In most cases, anode stimulation increases cortical excitability, while cathode stimulation decreases cortical excitability (Zheng et al. 2011).

C. Xu (✉)
Department of Functional Neurosurgery, Beijing Institute of Functional Neurosurgery, Xuanwu Hospital, Capital Medical University, Beijing, China

C. Wang
Department of Neurosurgery, Xuanwu Hospital, Capital Medical University, Beijing, China

R. Chen
Department of Pediatric, Xuanwu Hospital, Capital Medical University, Beijing, China

### 1.1 Waveform Characteristics

Direct current can be categorized as steady and pulsating direct current (Fig. 4.1). Steady direct current is characterized by the constant direction and magnitude, and pulsating direct current is featured with varying magnitude like periodic pulses.

### 1.2 Current Characteristics

The research proves that the electrical stimulation with a current density less than 25 mA/cm$^2$ will not damage the brain tissue (Matsumoto et al. 2017). For example, when the size of the electrode patch is 35 cm$^2$, the current is set to 1 mA, and the measured current density at the target (12 mm below the center of the stimulation electrode) is 0.02857 mA/cm$^2$, the charge per unit area shall not exceed 216 C/cm$^2$. For example, when the electrode of 35 cm$^2$ passes one mA current and the stimulation duration is 15 min, the accumulated electricity is 0.025 C/cm$^2$.

When the stimulation starts, the current slowly rises from 0 mA to the set value, which is called the rising period. When the stimulation ends, the current slowly decreases from the set value to 0 mA, which is called the falling period (Fig. 4.2).

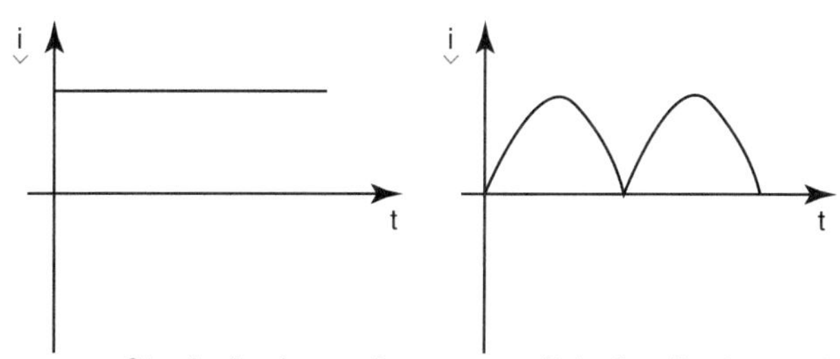


**Fig. 4.1** Waveform of direct current stimulation



Y. Wang (ed.), *Therapeutics of Neural Stimulation for Neurological Disorders*, https://doi.org/10.1007/978-981-99-4538-2_4

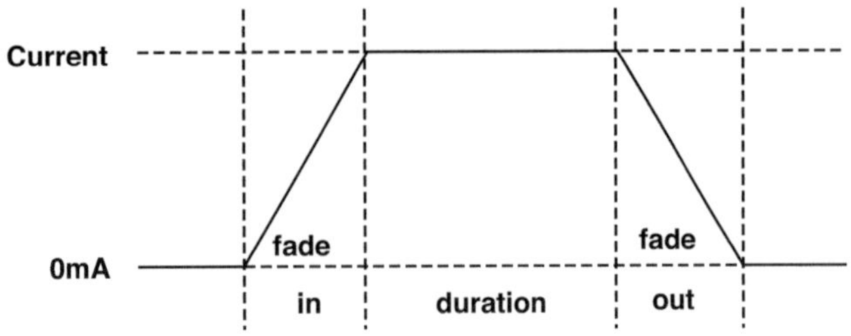


**Fig. 4.2** Direct current pattern diagram

## 2 Alternating Current Stimulation

Alternating current is a type of current where electricity changes its magnitude and direction periodically. Usually, the waveform is a sine wave, but there are actually some non-sinusoidal waveforms such as square, triangle, sawtooth, or ramp wave. What these waveforms have in common is periodic changes in direction and amplitude, so "period" is a very important characteristic of AC waveforms. In a frequency-specific manner, tACS modulates the excitation and inhibition thresholds of target neurons through variable-frequency currents to change the oscillatory rhythm of cortical neurons. A certain frequency of tACS can modulate the firing frequency of neurons, causing a resonance effect. After applying tACS in the alpha band (8–13 Hz) in the occipital region, the energy in the alpha band increased. When 10 Hz tACS was applied, the 10 Hz EEG band had a higher energy peak (Vossen et al. 2015). Studies on the after-effects of tACS have not yet reached clear conclusions.

### 2.1 Waveform Characteristics

#### 2.1.1 Waveform

##### 2.1.1.1 Sine Wave

The peak rises gradually and drops gradually. The duration of rise and fall are equal (Fig. 4.3).

##### 2.1.1.2 Square Wave

In digital electronic devices, square waves are mainly used to represent electrical signals such as voltage, current, and digital output. A square wave waveform has positive and negative swings over time, but the duration of each swing is equal, resulting in a symmetrical waveform. Unlike a sine wave, the wave is flat top at the peak amplitude level with vertical rising and falling edges (Fig. 4.4).

##### 2.1.1.3 Triangle Wave

The triangular wave has a sharper peak. Like the sine wave, the peak rises gradually and has the same periodicity, that is, the duration of rise and fall are equal (Fig. 4.5).

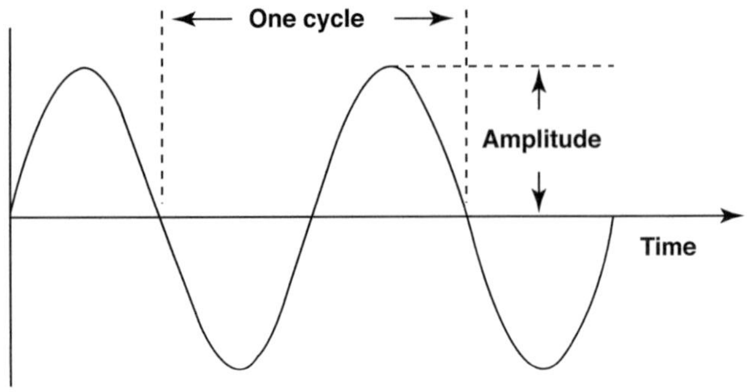


**Fig. 4.3** Sine wave

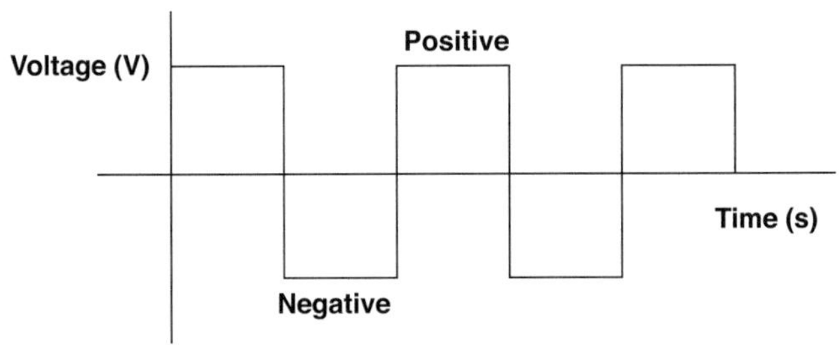


**Fig. 4.4** Square wave

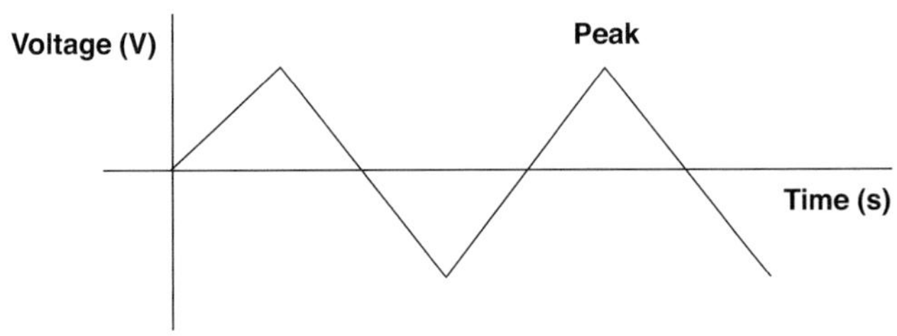


**Fig. 4.5** Triangular wave

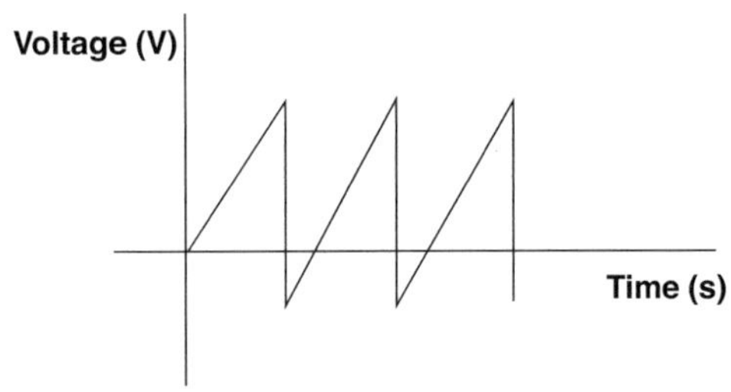


**Fig. 4.6** Positive slope sawtooth wave

##### 2.1.1.4 Sawtooth Wave

As its name implies, the waveform of sawtooth wave looks like sawtooth. According to the speed of rise and fall, the sawtooth wave can be divided into two types. One is a positive slope sawtooth wave; the peak rises slowly and falls rapidly (Fig. 4.6). The other is a negative slope sawtooth wave, the peak rises rapidly and falls slowly (Fig. 4.7).

### 2.1.2 Number of Waveform Phases

Phase is the position of a wave in its cycle at a particular moment. Phase describes the measurement of signal waveform, usually in degrees (angles), also known as phase angle. When the signal waveform changes periodically, the waveform is 360° after one cycle. According to the number of waveform phases, it can be divided into monophasic wave, biphasic wave, triphasic wave, and polyphasic wave (Fig. 4.8). Monophasic wave means that the waveform has only one direction. Biphasic wave means that the waveform has positive and negative directions. Triphasic wave means that the waveform has positive-negative-positive or negative-positive-negative directions. Polyphasic wave means that the waveform has positive-negative-positive-negative-positive--negative directions.

### 2.1.3 Symmetry of Biphasic Waveforms

Biphasic waveforms are categorized as symmetry and asymmetry. Generally, the biphasic waveforms are symmetric (Fig. 4.9).

### 2.1.4 Biphasic Waveforms Charge

Charge balance means that the total concentration of unit positive charge charged by various charged bodies must be equal to the total concentration of unit negative charges. If the charges in the positive-negative direction of the biphasic waveforms are equal, it is called charge balance. If the charges in the positive-negative direction of the biphasic waveforms are not equal, it is called charge imbalance (Fig. 4.10).

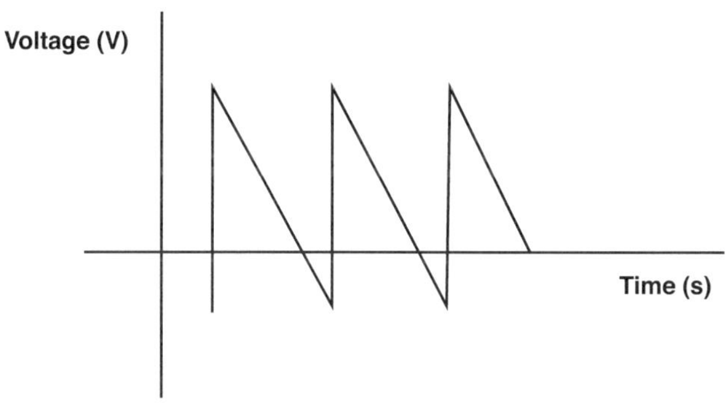


**Fig. 4.7** Negative slope sawtooth wave

## 2.2 Current Characteristics

### 2.2.1 Characterization of Mono-Pulse

A stimulus only sends out one pulse, which is called mono-pulse such as a sine wave, a square wave, a triangular wave, a sawtooth wave, etc.

### 2.2.2 Characterization of Continuous Pulses

Multiple pulses are stimulated at one time to form continuous pulses. There is no time interval between each pulse such as continuous sine wave pulse (Fig. 4.11), continuous square wave pulse (Fig. 4.12), etc.

## 2.3 Current Modulation

### 2.3.1 Amplitude Modulation

The amplitude of AC waveform is the intensity of AV waveform at any particular moment. The amplitude of the AC current varies with the sine of the angle produced by the circular motion of the magnet or coil relative to the zero. When the angle is 90°, the amplitude is one, which is called peak current. When the angle is 270°, the amplitude is −1, which is called the maximum reverse current. The modulation method that makes the amplitude of the modulated wave change according to the changing law of the transmitted signal is amplitude modulation.

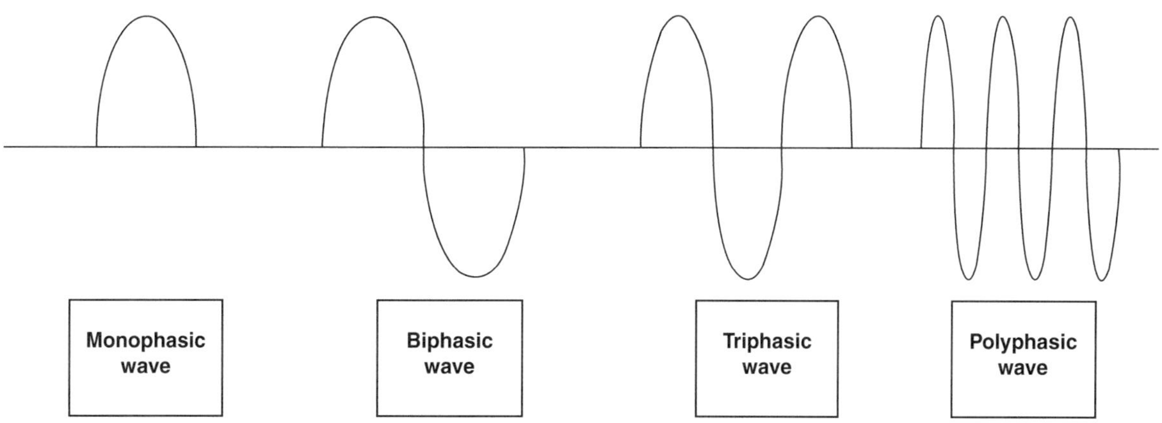


**Fig. 4.8** Classification of phase waves

**Fig. 4.9** Symmetry of biphasic waves

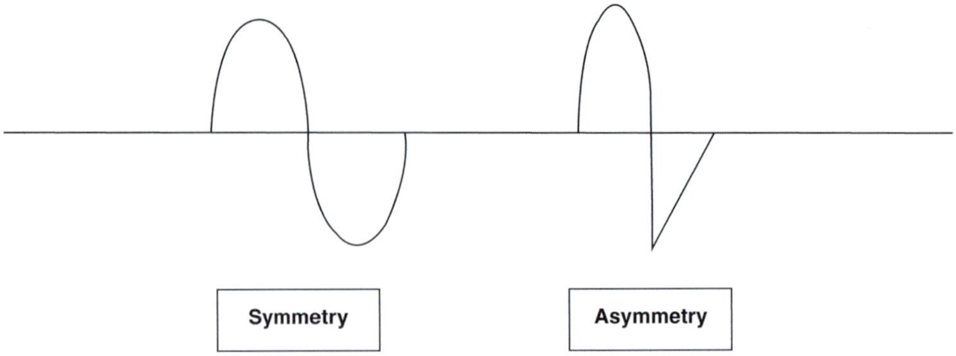


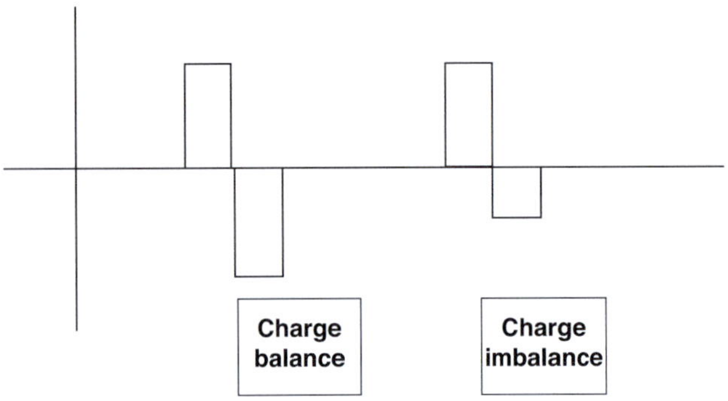


**Fig. 4.10** Charge balance of biphasic wave

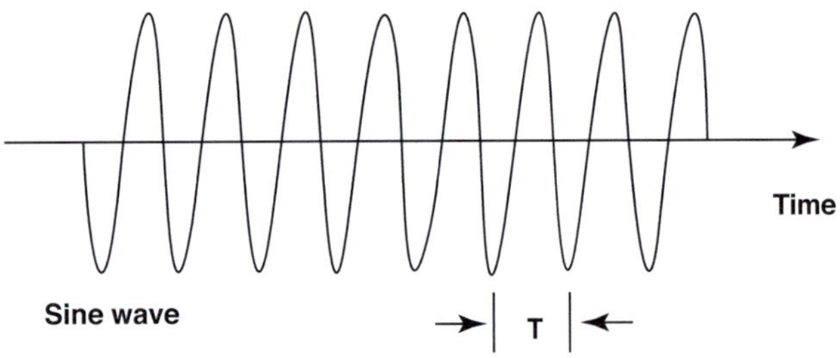


**Fig. 4.11** Continuous sine wave pulse

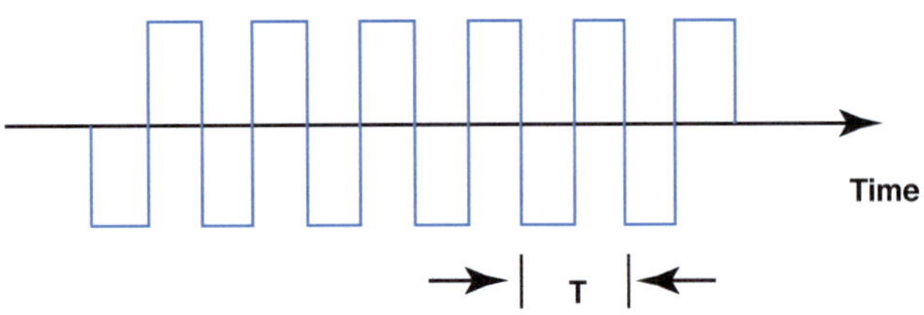


**Fig. 4.12** Continuous square wave pulse

### 2.3.2 Duration Modulation

The pulse duration is known as pulse width, usually in ms. The modulation method that makes the duration of the modulated wave change according to a certain law is duration modulation. If the sonication duration is long, the pulse width is long. If the duration is short, the pulse width is short.

### 2.3.3 Frequency Modulation

The frequency of alternating current refers to the number of periodic changes in unit time. The unit of frequency is Hertz (Hz), which represents the number of cycles experienced by the waveform within 1 s and is inversely related to the cycle. The calculation formula is frequency = 1/cycle. The modulation method that makes the frequency of the modulated wave change with the amplitude of the input signal is frequency modulation.

### 2.3.4 Temporal Modulation

Select specific brain regions, specific stimulation patterns, and appropriate stimulation duration (e.g., 5, 15, 20 min, etc.) for stimulation based on the type of disease so as to achieve the therapeutic effect. The modulation mode, which can obtain different therapeutic effects by changing the stimulation duration, is called temporal modulation.

### 2.3.5 Stimulus Train Modulation

When continuous pulses are used for stimulation, the stimulation of one continuous pulse is called a stimulus train. The frequency of the train can be set as required to achieve the therapeutic effect. The modulation method that makes the modulated stimulus train change according to a certain law is called stimulus train modulation.

## 3 Pulse Electrical Stimulation

Pulse usually refers to a short pulse-like fluctuation of voltage or current, as opposed to a continuous signal that occurs briefly throughout the signal period. The pulse signal is a discrete signal with various waveforms. In contrast to a sine wave, the waveform is discontinuous but can have a certain periodicity. The pulse signal either appears in the same direction or appears in alternating directions of positive and negative.

## 3.1 Waveform Characteristics

### 3.1.1 3Waveforms

Common pulse waveforms include rectangular wave, square wave, spike wave, sawtooth wave, triangular wave, step wave (Fig. 4.13), etc.

### 3.1.2 Number of Waveform Directions

According to the number of waveform directions, it can be divided into unidirectional pulse and bidirectional pulse (Fig. 4.14). The unidirectional pulse signal appears in the same direction, and the bidirectional pulse signal appears in alternating directions of positive and negative.

### 3.1.3 Symmetry of Bidirectional Waveform

Positive pulse and negative pulse appear alternately, and the positive-negative pulse waveform is antisymmetric to the ordinate coordinate, which is called odd symmetric signal, or odd signal.

### 3.1.4 Charge Balance of Bidirectional Waveform

In general, the positive and negative pulse charges of the bidirectional pulses are equal, so it is called charge balance.

## 3.2 Current Characteristics

### 3.2.1 Characterization of Single Pulse

A single pulse signal lasts for a long time before another pulse signal (Fig. 4.15). The pulse delivering time is uncertain. A pulse signal is delivered only when conditions are met such as a rectangular wave pulse, a spike wave pulse, a sawtooth wave pulse, or a triangular wave pulse.

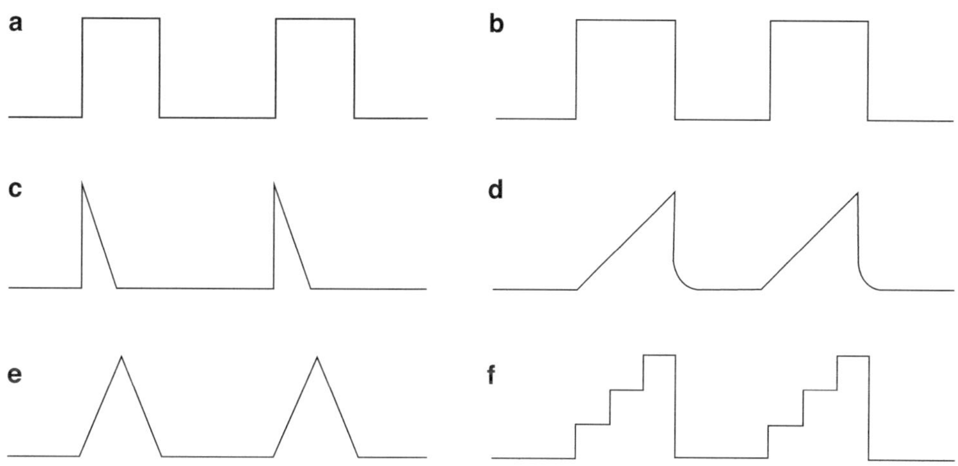


**Fig. 4.13** Common pulse waveforms. (**a**) Rectangular wave; (**b**) square wave; (**c**) spike wave; (**d**) sawtooth wave; (**e**) triangular wave; (**f**) step wave

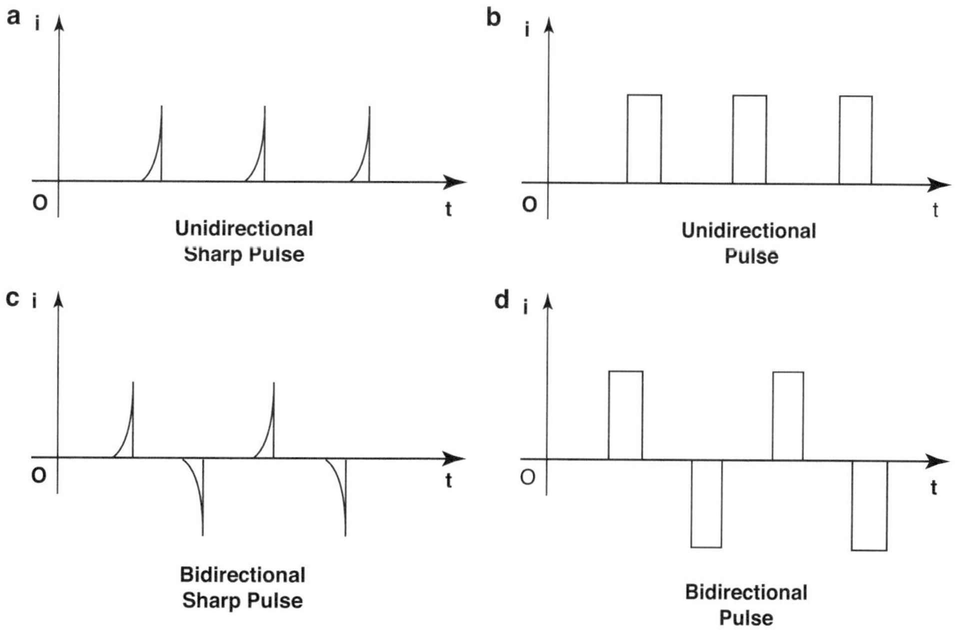


**Fig. 4.14** Unidirectional pulses (upper panel) and bidirectional pulses (bottom panel)

**Fig. 4.15** Single pulse mode diagram

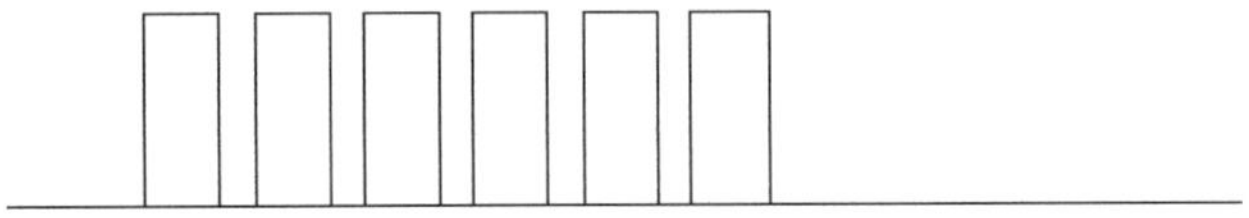

**Fig. 4.16** Continuous pulse mode diagram

### 3.2.2 Characterization of Continuous Pulses

Continuous pulse signal is a continuous and predictable periodic pulse signal. There is a certain time interval between each pulse (Fig. 4.16).

## 3.3 Current Modulation

### 3.3.1 Amplitude Modulation

The pulse amplitude changes with the change law of the signal, which is called pulse amplitude modulation.

### 3.3.2 Duration Modulation

The duration of a pulse is called pulse duration or pulse width. The modulation method that makes the duration of the modulated wave change according to a certain law is duration modulation, also known as pulse width modulation, which is usually calculated in milliseconds (ms).

### 3.3.3 Frequency Modulation

The time interval between pulse signals is called the period, and the number of pulses generated in a unit time (e.g., 1 s) is called the frequency. The measurement unit of frequency is Hertz (Hz). Fast frequency means that there are many pulse signals within 1 s, which is called dense wave (high frequency), generally 50–100 times/s. Slow frequency means that there are few pulse signals within 1 s, which is called sparse wave (low frequency), generally 2–5 times/s. The modulation method that makes the frequency of the modulated wave change with the amplitude of the input signal is frequency modulation.

### 3.3.4 Temporal Modulation

Select specific brain regions, specific stimulation patterns, and appropriate stimulation duration (e.g., 5, 15, 20 min, etc.) for stimulation based on the type of disease so as to achieve the therapeutic effect. The modulation mode, which can obtain different therapeutic effects by changing the stimulation duration, is called temporal modulation.

### 3.3.5 Stimulus String Modulation

When continuous pulses are used for stimulation, the stimulation of one continuous pulse is called a stimulus string. The frequency of the string can be set as required to achieve the therapeutic effect. The modulation method that makes the modulated stimulus string change according to a certain law is called stimulus string modulation.

# 4 Random Noise Stimulation

tRNS is a technique that uses random electrical noise to stimulate specific brain areas through electrodes placed on the surface of the scalp. It is a special type of tACS that can combine various frequencies at will. Similar to the mechanism of tACS, tRNS may interfere with the oscillatory activity of neurons, resulting in increased cortical excitability (Terney et al. 2008). Another possible mechanism is that high-frequency tRNS activates sodium channels, influx of sodium ions, and depolarization of membrane potential.

## 4.1 Waveform Characteristics

The waveform changes according to the stimulation frequency. Usually, the frequency setting range is 100–640 Hz (Fig. 4.17).

## 4.2 Current Characteristics

The tRNS current changes randomly with time, and the current setting range is from −600 μA to 600 μA (Fig. 4.18).

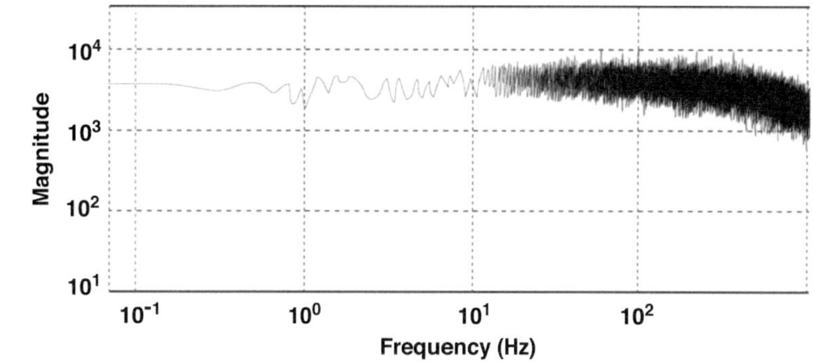


**Fig. 4.17** tRNS waveform characteristics

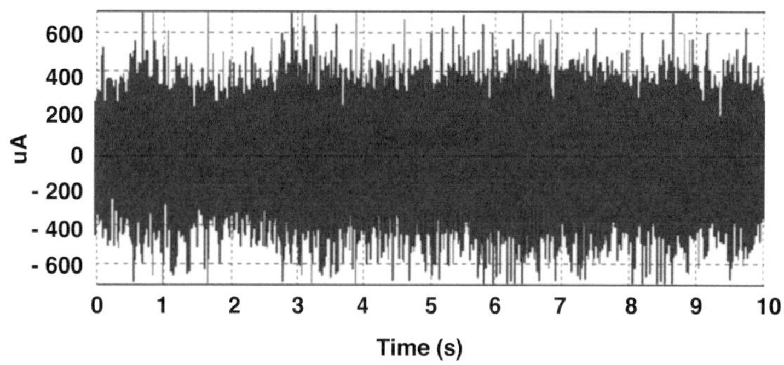


**Fig. 4.18** Current characteristics of tRNS

## 5 Basic Structure of Electrical Stimulator

The noninvasive electrical neurostimulation device consists of stimulation module, control module, impedance detection module, current protection and monitoring module, and battery management module (Fig. 4.19). The patient's scalp impedance directly affects the current intensity acting on the specific cortical region in the brain, which is very important for the treatment effect and safety. Therefore, the impedance detection module is an important part of the device. Currently, the noninvasive electrical stimulation devices adopt the impedance test principle similar to the scalp EEG impedance test and usually cannot test the impedance change in real time.

The working voltage of noninvasive electrical stimulation devices is between 9 and 40 V, the working current is 0.5–5 mA, and some devices reach 15 mA (Antal et al. 2017). The current generated by the equipment starts from the anode, passes through the head tissue, and back to the cathode. To ensure the current output accuracy, the current source output method is usually used. The signal is usually a periodic current signal. According to different current waveforms, it can be divided into sine type, DC type, pulse type, and random noise type.

With the improvement of circuit design and power supply level, a variety of portable electrical stimulation devices have emerged, which greatly improves the possibility of using related technologies in primary medical units and families. Currently, there are also non-medical products that are powered by mobile phone batteries to achieve stimulation. However, the performance of the current and voltage control modules of such products cannot yet be compared with medical-grade products, but related technologies are also constantly developing.

### 5.1 Stimulation Module

The stimulation module includes DA conversion, power amplification, signal conditioning, electrode drive, and other parts to generate waveforms with specific precision and current intensity.

### 5.2 Control Module

The control module is the control center of the entire stimulation system, mainly to control each functional module, such as impedance detection, battery management, current

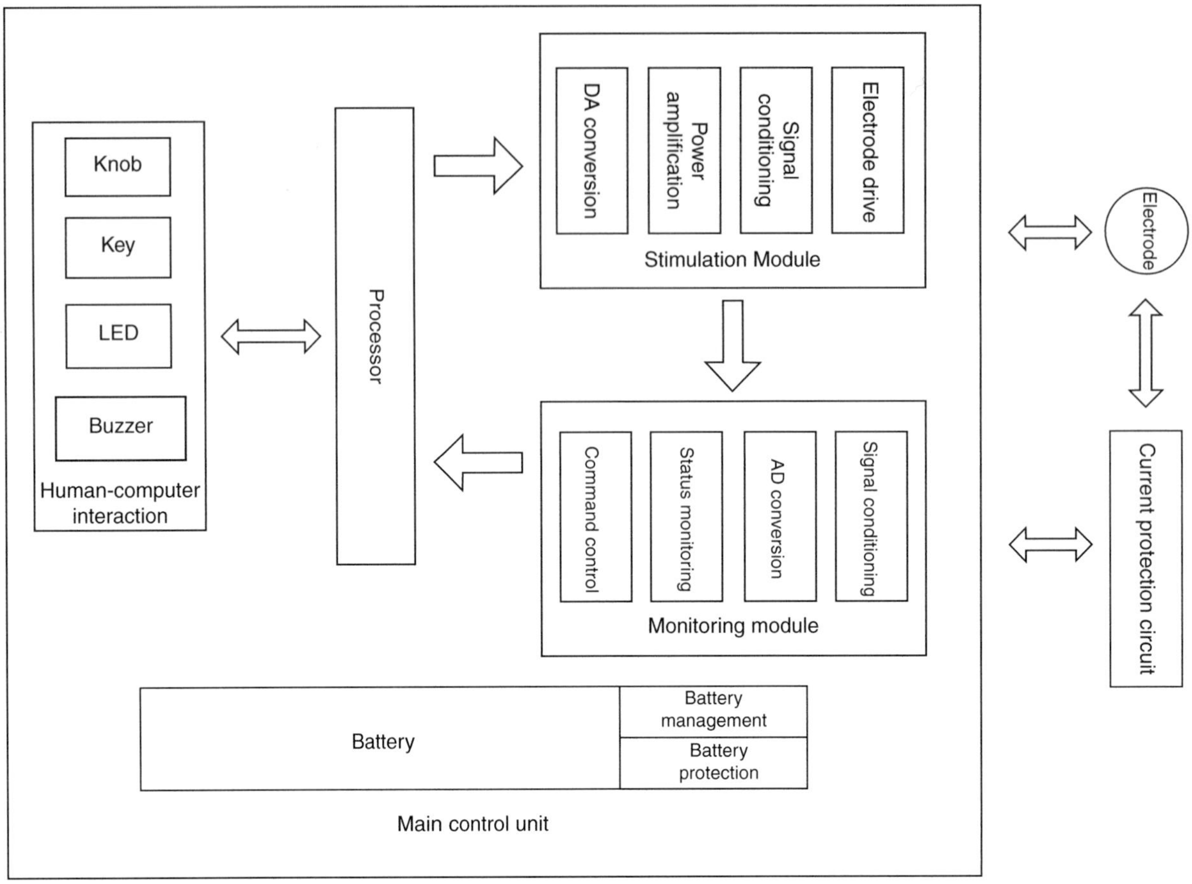


**Fig. 4.19** Schematic diagram of each module of the noninvasive current stimulation device

protection, and other modules. ADC is mainly used to obtain the signal of key regions to be measured, which is controlled by the signal conditioning circuit. The signal conditioning circuit adjusts the amplitude, level, and impedance of the signal to meet the input requirements of the analog-to-digital converter. The AD converter converts it into a digital quantity. The processor calculates the parameters to be measured and determines whether the output needs to be cut off in combination with the current output state.

## 5.3 Impedance Monitoring Module

Impedance detection module is an important part of designing the whole system structure, including impedance detection before stimulation and real-time impedance detection during stimulation. The system measures the ratio of the voltage differences between the electrode and the scalp generated by the current flowing through the human body and uses Ohm's law to calculate the contact resistance.

## 5.4 Current Protection and Monitoring Module

The main function of the current protection module is that when the current is abnormal before or during the stimulation, the system will cut off the output for security. To ensure that the system will not cause harm to human body during stimulation, we need to detect the current in real time. When abnormal current is detected, the professional electrical stimulation device will have a buzzer alarm and cut off the output, which improves the safety and reliability of the equipment. The current protection circuit is designed to prevent abnormal current output. The current self-locking circuit is composed of two power switches and a load resistor.

## 5.5 Batter Management Module

Portable electrical stimulation devices often use lithium battery for power supply, so there is a battery management module, which mainly completes the offline charging of equipment battery and the monitoring of equipment batter power, mainly including USB charging circuit and battery voltage detection circuit, which can also be connected to the host computer through serial port to display the status of current and voltage.

### 5.6 Human–Computer Interaction Module

The human–computer interaction module mainly includes switches, adjustment knob or button, LED display and passive buzzer, etc. The switch is mainly used for the user to start, close, and judge whether the stimulation process starts. The adjustment knob or button is mainly used by the user to adjust the time and intensity of stimulation. LED is mainly used for real-time display of stimulation current intensity. The buzzer will produce an alarm sound when the impedance detection fails. In addition, it will also produce an alarm when the circuit detects abnormal current. Some interactive modules of intelligent stimulation devices are displayed and controlled through mobile app and other software.

## 6 Control of Electrical Stimulation Features and Modes

### 6.1 Waveform Selection

The waveforms of electrical stimulation include single current stimulation, sinusoidal stimulation, pulse stimulation, and noise stimulation. Stimulation modes include single stimulation, paired stimulation, train stimulation, and continuous stimulation. Single current stimulation means that the current value remains constant throughout the stimulation process. Before and after the current stimulation, to avoid sudden current jump, a certain slow increase and slow decrease mode (corresponding to wave rise and wave fall parameters in advanced settings) is usually set to avoid pain caused by current change. The sinusoidal stimulation current waveform is a sine wave, and the current value changes periodically with a specific repetition frequency over time. Similarly, there are systems that use a triangular wave as the stimulation waveform, which increases or decreases linearly with time. Pulse-type current stimulation is usually alternating positive and negative constant current. Some systems design pulses by controlling the presence or absence of current or even complex step type signal combination. For comparison in clinical trials, noise signals with constant energy is often used as the stimulation currents, which is generated by a specific noise generator.

Because the current of single current stimulation is constant, its cathode and anode are fixed. The cathode or anode is placed at one site to improve the inhibition or excitation of the corresponding brain area. The other electrode is only used to form the current circuit. This is also called unipolar mode or DC stimulation mode. Sinusoidal mode is alternating positive and negative polarity, also known as bipolar mode or AC stimulation mode.

### 6.2 Intensity Control

The current intensity determines the therapeutic effect, and excessive current is prone to side effects and even danger. Therefore, it is very important to select the appropriate current intensity. The adjustable current of professional scientific research-grade current stimulation devices can reach up to 3.0 mA, and the minimum current adjustment step can reach 25 μA. However, the current intensity control of portable devices is usually not continuous, and several gears are often set for manual selection, such as 0.5, 1, 1.5, 2, 2.5 mA, etc. Each gear corresponds to an indicator light, which is convenient for the operator to choose flexibly. Adjust the stimulation parameters to visually and clearly see the indicated state. In addition, the device can display the current intensity in real time through the LED bar, which is convenient for users to know the current stimulation state.

### 6.3 Pulse Width and Phase Control

The current of pulse-type stimulation is not continuously presented. The stimulation cycled on/off and the pulse width and interval can also be adjusted. The pulse width (time of periodic change) is less than the interstimulus interval (ISI). The stimulation interval is usually 0.3–2 s. It is gradually adjusted by step, and the increment is usually 0.1 s. Within the range of the stimulation interval, the pulse width can be adjusted by a certain step. The ratio of the pulse width to the stimulation interval (period) is also called duty cycle. The most typical pulse stimulus is a series of stimuli, that is, the parameters of each pulse remain the same and appear continuously. It can also present pulses of specific width in pairs and maintain a certain ISI. In addition, there is a single pulse stimulation. During serial pulse stimulation, the pulse width should not be greater than 50% of the cycle, otherwise it is easy to cause tissue damage.

The sinusoidal and single current stimulations are presented continuously, and the overall stimulation frequency can be adjusted without the concept of pulse width modulation. However, the initial phase of the sinusoidal stimulation can be manually set and can be adjusted within the range of 0–360°, with a step size of 5°.

## 6.4 Frequency Control

Sinusoidal stimulation waveform is a sine wave with a single frequency. Theoretically, the maximum frequency can reach 250 Hz, and the minimum frequency is close to DC (i.e., 0 Hz, single current stimulation). The adjustable frequency step of scientific research devices is at least 0.01 Hz. The frequency of actual clinical equipment is usually predetermined. The commonly used frequencies are 1, 5, 10, 40, and 77.5 Hz. Different stimulation frequencies correspond to different physiological mechanisms, which need to be combined with various diseases, guidelines, and consensus.

The frequency of pulse-type stimulus is different from that of sine type. Its frequency is the reciprocal of the cycle, which refers to the repetition rate of alternating pulse waveform. The time interval between adjacent pulse stimuli determines the frequency. In addition to stimulus with a specific frequency, noise can also be used as stimulus in some control tests. The frequency spectrum can be a normal distribution waveform centered on a certain frequency, or it can be white noise and other types. The stimulation mode of a specific waveform belongs to low-frequency stimulation, and the relevant performance requirements of the device are not high. There are some high-frequency components in the random noise stimulation waveform. When the human body impedance detection is required, a higher bandwidth is necessary, and the sampling frequency of the device needs to be higher. When the noise stimulates impedance detection, the voltage and current change rapidly, and the acquisition speed also requires higher standard.

## 6.5 Power-on and Power-off Time Control

The power-on time of pulse-type stimulation can be set according to the integer multiple of the pulse, usually 1–500 in the software. The total stimulation time can also be set directly. Single current stimulation and sinusoidal stimulation are continuous stimulation, and the total power-on time is used to control the power-on and power-off. The time parameters of the actual stimulation device are divided into five grades, and the stimulation duration is usually 5, 10, 15, 20, and 25 min. The electrical stimulation device for clinical use usually does not exceed 30 min.

The advanced electrical stimulation device also has a timing function. The treatment time reminder can be set through the device or smart terminal to ensure the continuity of treatment and follow the daily treatment situation and total treatment time to form a treatment log.

## 6.6 Output Channel Selection

According to different treatment purposes, different output sites and channels can be selected. The control system of the device controls the system output parameters and modes through the host computer, which can achieve simultaneous multi-channel stimulation, and even two or more simultaneous interventions.

## 6.7 Stimulation Electrode Position Selection

The location of stimulation electrode can follow the EEG 10–20 system (Fig. 4.20) to apply current to specific brain areas. That is, according to the frontal area (denoted by F, the same below), central area (C), parietal area (P), occipital area (O), and temporal area (T), the electrode positions are designed at certain distances to cover all scalp parts. The "z" after the position indicates the midline electrodes, with odd-numbered electrodes to the left of the midline and even-numbered electrodes to the right of the midline. In this way, the electrode position has a certain corresponding relationship with the typical parts of the scalp, which is convenient for pinpointing the position in practice. For example, some studies focus on the memory encoding effect of vocabulary learning in language, and the anode is placed near the left

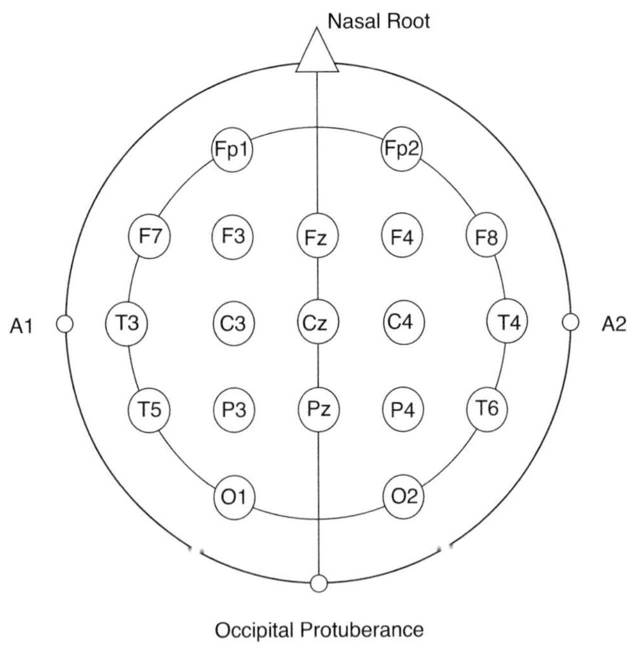


**Fig. 4.20** Schematic diagram of electrode position of international 10–20 system

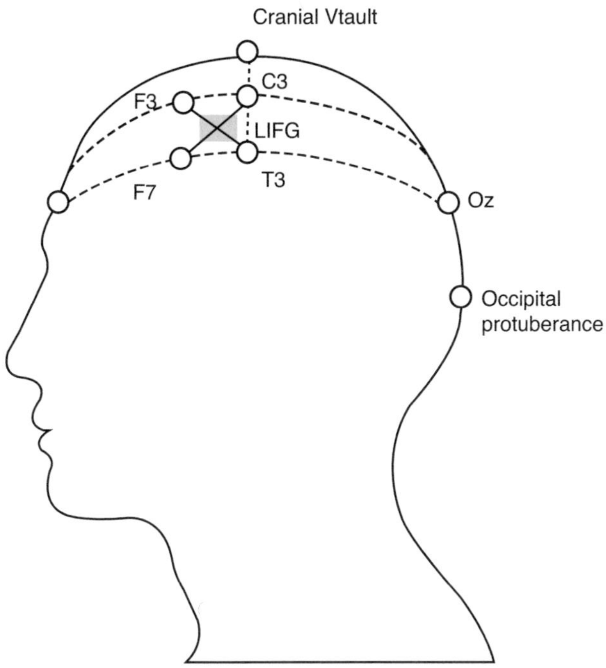


**Fig. 4.21** Schematic diagram of electrode placement site, scalp positioning, and intended action position

Broca area (Fig. 4.21). A convenient way to locate it is the intersection of the lines between F3 and T3, and C3 and F7.

It is worth noting that the channel design of noninvasive electrical stimulation devices is becoming more and more flexible. The traditional micro-current stimulation and electric shock devices usually have one cathode and one anode, respectively. Current with the waveform of DC or AC mode travels in the brain between cathode and anode. Modern electrical stimulation devices not only greatly increase the number of channels, but also have different parameters such as the stimulation frequency of each channel. The device is designed with 32 channels, and the ratio and position of anode and cathode can be arranged according to clinical needs, forming a circuit to act on different brain regions. The selection of channels can be configured through application programs, and the possible action sites and effects can be simulated in advance through current positioning software.

In addition, the waveform, frequency, and phase of the stimulation current of each channel may be different for different systems and different treatment purposes. In recent years, American researchers have designed a bipolar differential frequency electrical stimulation device to produce a differential frequency effect at the position where the current fields of two frequencies overlap. The overlapping region is often a deeper brain region in order to stimulate a specific frequency on a deeper level. These new stimulation schemes need further systematic research and clinical verification.

## 7 Constant Current and Constant Voltage Stimulator

The constant voltage device ensures that the output voltage of the system is of constant value. It is usually used in circuits such as USB power supply. It requires that the internal resistance of the stimulator circuit is very low. No matter the impedance increases or decreases, the output voltage remains constant. It is mainly used in occasions where the quantitative requirements for stimulation are not high or the resistance changes little. Its disadvantage is that the change of impedance will cause the change of current, and the impedance of human scalp is easily affected by sweating and contacts. In addition, constant voltage stimulation is also prone to unsteady current in a single pulse and irregular current curve. The charge may be fast or slow with different effects, resulting in different therapeutic effects in different patients with the same voltage parameters.

In contrast, the constant current stimulator requires a high internal impedance, which is much greater than the skin impedance and can ensure that the output voltage is constant. In conclusion, relative to the external load of the electrical stimulation device, the lower the internal resistance, the better the constant voltage effect, and the higher the internal resistance, the better the constant current effect.

Since the nerve membrane potential is sensitive to electrical stimulation, accumulation of sufficient electricity can change the imbalance between neural excitation and inhibition, and long-term use can improve synaptic plasticity. As a neuromodulation device, the therapeutic effect is related to the total amount of electricity acting on the brain tissue, rather than simply related to the voltage level. Therefore, under the condition of limited treatment time, it is more practical to ensure the output of constant current. Thus, including the popular deep brain stimulation, noninvasive electrical stimulation and other devices generally use constant current stimulators to ensure a constant output current.

When the constant current stimulator is working, the current density is indirectly reflected by the impedance of the head. Due to sweating during the stimulation process and electrode loosening caused by head movement, the impedance will change, and the current density may also change accordingly. Moreover, when the stimulation device is working, the electrodes are in contact with the scalp for a long time. Due to the polarization effect of the electrodes, the skin is prone to itching and pain. The long-term action of high-density current may also cause skin burns. Therefore, it is necessary to control the output of the device. The current is stable, and the skin impedance needs to be monitored to avoid excessive current.

It is worth noting that brain tissue cannot simply be regarded as resistance. The impedance of the nervous system that hinders the flow of charge includes resistance, capacitive reactance, and inductive reactance. Among them, capacitive reactance and inductive reactance are very different. It is largely determined by the speed of the current fluctuation. In a DC circuit, the current is constant. They are basically negligible, and the resistance constitutes the main part of the impedance. However, in the case of pulsed and sinusoidal stimuli, the latter two forces account for a significant portion of the impedance.

In order to ensure constant current, early electrical stimulation devices often used a high-voltage pulse source and a large internal resistance (much larger than the skin resistance) to achieve constant current. This device consumes a lot of power and requires external mains power, which is obviously not suitable for daily use. Currently, the new generation of battery powered portable stimulator achieves equivalent high internal resistance through circuit optimization design, which greatly improves the application of electric stimulator.

## 8 Surface Electrode

Surface electrodes are fundamental parts of a noninvasive direct current stimulation system. A common direct current stimulation system consists of metal or conductive rubber electrode, electrode sponge, electrolyte-based contact medium (e.g., saline, gel, or conductive cream), and shaping materials, such as electrode strap, plastic shell, etc. (Woods et al. 2016; Bikson et al. 2019). Typical direct current stimulation systems include conventional tDCS and high-definition tDCS (HD-tDCS).

### 8.1 Conventional tDCS

#### 8.1.1 Surface Electrode Types

Pad electrodes, as the most representative surface electrodes, are generally rectangular or elliptical. A conductive electrode is usually composed of non-polar metal or conductive rubber with good electrical conductivity and wrapped with a plastic shell. Surface electrodes range in size from 1 to 100 cm$^2$, with common sizes including 4 cm× 4 cm (16 cm$^2$), 5 cm× 5 cm (25 cm$^2$), 5 cm× 7 cm (35 cm$^2$), etc. The size of sponge electrodes is generally variable, depending on specific manufacturer and research needs. Moreover, the size of the electrodes has a significant effect on the distribution of current delivered to the scalp and brain. At a constant current strength level, an increase in electrode size or a difference in the shape of the electrode may result in a different current distribution over the surface area of the scalp and cause a different current distribution in the entire brain (Minhas et al. 2011; Kronberg and Bikson, 2012; Cano et al. 2013; Datta et al. 2008). Therefore, in practice and research, the shape, size and current density of electrode (mA/cm$^2$) should be recorded.

#### 8.1.2 Buffer Substance

Electrochemical reactions occur on the surface of conductive electrodes, and direct contact with the skin should be avoided. Buffer substances are required in the skin and conductive electrodes to prevent chemicals formed on the electrode from entering the skin. They are usually electrolyte-based contact substances, including electrode sponges, gels, or conductive creams impregnated with saline. Among them, the most commonly used buffer substances are electrode sponges and saline (Minhas et al. 2011). In practical applications, the sponge should not be dipped in too much saline. When the electrode sponge is saturated, saline will drain from the sponge and cover the scalp area outside the surface electrode range, which will lead to the expansion of the current delivery area during the treatment, thus affecting the treatment effect.

### 8.2 High-Definition tDCS (HD-tDCS)

Surface electrodes for HD-tDCS are usually 4 × 1 ring electrodes (unlike conventional pad electrodes), in which the center electrode is surrounded by four electrodes of opposite polarity. Common surface electrodes for HD-tDCS include Ag electrodes, Ag AgCl electrodes, rubber particle electrodes, Ag/AgCl ring electrodes, and Ag/AgCl disk electrodes. The Ag/AgCl ring electrode is the most commonly used one, which is a circular electrode with a central hole, and the diameter of each electrode is usually within 12 mm, mostly about 8 mm. The ring electrode is fixed on a special plastic cap with two interconnected layers, and the electrode is placed above and the gel for buffering is injected below. In the HD-tDCS system, each of the electrodes around the "ring" conducts about 1/4 of the current, while the central electrode conducts the entire current. The dominance of these electrodes is that the modulation effect of the cerebral cortex is limited to the radius of the ring electrodes, enhancing the focal length and potentially inducing a larger, longer effect (Kuo et al. 2013; Edwards et al. 2013). In addition, increasing the number of electrodes could support a multi-target stimulation system. Studies have shown that it is well tolerated, and may lead to common adverse reactions similar to conventional tDCS, such as localized skin itching, pricking pain, and burning (Borckardt et al. 2012; Gbadeyan et al. 2016).

## References

Antal A, Alekseichuk I, Bikson M et al (2017) Low intensity transcranial electric stimulation: safety, ethical, legal regulatory and application guidelines. Clin Neurophysiol 128(9):1774–1809

Bikson M, Esmaeilpour Z, Adair D et al (2019) Transcranial electrical stimulation nomenclature. Brain Stimul 12(6):1349–1366

Borckardt JJ, Bikson M, Frohman H et al (2012) A pilot study of the tolerability and effects of high-definition transcranial direct current stimulation (HD-tDCS) on pain perception. J Pain 13(2):112–120

Cano T, Morales-Quezada JL, Bikson M et al (2013) Methods to focalize noninvasive electrical brain stimulation: principles and future clinical development for the treatment of pain. Expert Rev Neurother 13(5):465–467

Datta A, Elwassif M, Battaglia F et al (2008) Transcranial current stimulation focality using disc and ring electrode configurations: FEM analysis. J Neural Eng 5(2):163–174

Edwards D, Cortes M, Datta A et al (2013) Physiological and modeling evidence for focal transcranial electrical brain stimulation in humans: a basis for high-definition tDCS. NeuroImage 74:266–275

Gbadeyan O, Steinhauser M, Mcmahon K et al (2016) Safety, tolerability, blinding efficacy and behavioural effects of a novel MRI-compatible, high-definition tDCS set-up. Brain Stimul 9(4):545–552

Kim S, Stephenson MC, Morris PG et al (2014) tDCS-induced alterations in GABA concentration within primary motor cortex predict motor learning and motor memory: a 7T magnetic resonance spectroscopy study. NeuroImage 99:237–243

Kronberg G, Bikson M (2012) Electrode assembly design for transcranial direct current stimulation: a FEM modeling study. Conf Proc IEEE Eng Med Biol Soc, pp 891–895.

Kuo HI, Bikson M, Datta A et al (2013) Comparing cortical plasticity induced by conventional and high-definition 4 × 1 ring tDCS: a neurophysiological study. Brain Stimul 6(4):644–648

Matsumoto R, Kunieda T, Nair D. (2017) Single pulse electrical stimulation to probe functional and pathological connectivity in epilepsy. Seizure 44:27–36

Minhas P, Datta A, Bikson M (2011) Cutaneous perception during tDCS: role of electrode shape and sponge salinity. Clin Neurophysiol 122(4):637–638

Ritsch B, Reis J, Martinowich K et al (2010) Direct current stimulation promotes BDNF-dependent synaptic plasticity: potential implications for motor learning. Neuron 66:198–204

Terney D, Chaieb L, Moliadze V et al (2008) Increasing human brain excitability by transcranial high-frequency random noise stimulation. J Neurosci 28:14147–14155

Vossen A, Gross J, Thut G (2015) Alpha power increase after transcranial alternating current stimulation at alpha frequency (α-tACS) reflects plastic changes rather than entrainment. Brain Stimul 8:499–508

Woods AJ, Antal A, Bikson M et al (2016) A technical guide to tDCS, and related non-invasive brain stimulation tools. Clin Neurophysiol 127(2):1031–1048

Zheng X, Alsop DC, Schlaug G (2011) Effects of transcranial direct current stimulation (tDCS) on human regional cerebral blood flow. NeuroImage 58:26–33

# Noninvasive Peripheral Nerve and Spinal Cord Stimulation

# 5

Zhaoyang Huang and Yingxue Yang

## 1 Transcutaneous Electrical Nerve Stimulation

Transcutaneous electrical nerve stimulation (TENS) is a noninvasive peripheral stimulation technique that uses electrical current to stimulate nerves for therapeutic purposes. In the application of TENS, pulsed current is delivered through the intact skin surface to activate the underlying nerves. A typical TENS device is capable of modulating pulse width, frequency, and stimulation intensity. TENS may be applied in a high-frequency mode (>50 Hz) with an intensity below the motor threshold (sensory intensity); or in a low-frequency mode (<10 Hz) with an intensity of the motor threshold.

Clinical studies have proved that TENS is highly effective in treating acute and chronic pain, epilepsy, depression, and disorders of consciousness.

### 1.1 Transcutaneous Vagus Nerve Stimulation

Implantable vagus nerve stimulation (VNS) was first approved by the Food and Drug Administration (FDA) for the treatment of refractory epilepsy in 1997. Over the years, VNS has proven to be a promising therapy for a variety of diseases, such as Alzheimer's disease (AD), traumatic brain injury, stroke, post-traumatic stress disorder, rheumatoid arthritis, Crohn's disease, tinnitus, fibromyalgia, epilepsy, depression, and various primary headache disorders.

The clinical use of implantable VNS is limited by high cost and the need for surgical implantation of a pulse generator and electrodes that wraps around the cervical branch of the vagus nerve. In recent years, the novel noninvasive transcutaneous Vagus Nerve Stimulation (t-VNS) has been gradually applied in clinical research. In the application, the t-VNS stimulator delivers stimulation to the branch of the vagus nerve in the neck or external auricle. t-VNS eliminates the need for surgical implantation of a stimulator and has low cost and less side effects, making it a promising therapy in clinical practice. t-VNS has proven effectiveness in the treatment of disorders such as migraine, cluster headache, epilepsy, and depression, but it is essential to conduct clinical studies with large sample size to confirm its efficacy.

#### 1.1.1 Transcutaneous Cervical Vagus Nerve Stimulation

In 2017, the US FDA first approved tcVNS (gammaCore®) for the acute treatment of cluster headache, which is the world's first transcutaneous Cervical Vagus Nerve Stimulation (tcVNS) device. The approval is based on two tcVNS clinical studies for the treatment of acute cluster headache, which evaluated the safety and efficacy of gammaCore for acute cluster headache. Both trials (Trial 1 and trial 2) were prospective, double-blind, placebo-controlled, randomized studies. Trial 1 evaluated a subgroup of 85 patients with cluster headache from a larger (133) cluster headache patient population and found that 34.2% of the patients in the treatment group had pain relief at 15 min. Trial 2 assessed 182 episodes in 27 patients with cluster headache and found that the percentage of patients who received gammaCore treatment that stopped attacks (painless) at 15 min (47.5%) was significantly higher than that in the placebo group. Both trials found that gammaCore was safe and well tolerated, with most adverse events reported as mild and transient that occurred during active treatment.

tcVNS places electrodes at the posterior border of the sternocleidomastoid muscle below the mastoid. tcVNS uses AC consisting of five 5000 Hz pulses that are repeated at a rate of 25 Hz. The waveform of the tcVNS (gammaCore) pulse is approximately a sine wave, which makes the current flow more than 15 times greater than the current used in implantable devices and only causes minimal nociceptive

Z. Huang (✉) · Y. Yang
Department of Neurology, Xuanwu Hospital, Capital Medical University, Beijing, China


Y. Wang (ed.), *Therapeutics of Neural Stimulation for Neurological Disorders*, https://doi.org/10.1007/978-981-99-4538-2_5

pain. Due to its proximity to the sternocleidomastoid (SCM), there is only mild skin sensation and muscle contraction.

In addition to cluster headache, many studies have also investigated the efficacy of tcVNS in the treatment of migraine. In these studies, the gammaCore device was attached to the cervical branch of the vagus nerve, and two stainless steel electrodes produce burst stimulation at 25 Hz. The total stimulation time varied from study to study, but most of them delivered 90 s of stimulation at a time. These findings suggest that tcVNS reduces not only the frequency but also the severity of migraine attacks and the disability resulting from the migraine.

Animal experiments have demonstrated the stimulatory effect of tcVNS on vagus nerve. Chen et al. used animal models to study cortical spreading depression (CSD) and found that acute treatment with VNS, either implanted vagal stimulation or tcVNS, suppressed CSD frequency and increased the electrical threshold for CSD (Chen et al. 2016). Ay et al. compared the effects of iVNS and tcVNS in a rat middle cerebral artery occlusion ischemic stroke model and showed a similar reduction in ischemic area and functional deficits (Ay et al. 2011, 2016).

Functional magnetic resonance imaging (fMRI) directly demonstrated the stimulatory effect of tcVNS on the vagus nerve. Most vagal afferent fibers enter the brain through the jugular foramen and synapse to the nucleus tractus solitaries (the first central relay point of the vagus nerve), where they project directly or indirectly to various structures in the brain (e.g., locus coeruleus, dorsal raphe nucleus, and periaqueductal gray matter). Frangos et al. compared the effects of tcVNS and SCM stimulation and found that tcVNS activated the nucleus solitarius and parabrachial region, primary sensory cortex, basal ganglia, frontal cortex, and insula, while the activities of hippocampus, visual cortex, and spinal trigeminal nucleus decreased (Frangos et al. 2015). These changes are similar to the fMRI studies of iVNS in epilepsy patients and studies of transcutaneous auricular vagus nerve stimulation (taVNS). Results from several animal studies suggest that nVNS produces long-term inhibitory effects on pain-specific neural firing and excitatory neurotransmission activity in the nucleus tractus solitaries and its related structures. The fMRI results provide evidence for the clinical application of human t-VNS.

Evoked potential studies showed that both taVNS and iVNS could elicit vagus somatosensory evoked potentials (VSEP) originated from brainstem. Polak et al. reported specific far-field electroencephalogram (EEG) potentials 2–6 min after stimulation of the auricular vagus nerve. This potential consists of a positive waveform (called P1N1) (Polak et al. 2009). Usami et al. conducted similar studies using iVNS in patients undergoing epilepsy surgery and reported similar P1N1 potentials (Usami et al. 2013). Nonis et al. conducted a similar study using tcVNS, P1N1VESPs were observed in 11 of the 12 subjects with latencies similar to those previously described. Dose-response analysis showed that tcVNS could elicit significant VSEP in more than 80% of subjects at 15 volts (Nonis et al. 2017). These findings suggest that tcVNS can activate vagal afferent fibers.

### 1.1.2 Transcutaneous Auricular Vagus Nerve Stimulation

In 2000, Canadian scholars proposed the technique of percutaneous electrical stimulation of the vagus nerve distribution in the ear, namely transcutaneous auricular vagus nerve stimulation (taVNS). It is a noninvasive regulation technique with the advantage of safety and low price.

The auricular branch of the vagus nerve is the only branch of the vagus nerve on the body surface. It is distributed in the cavity and cymba on the outer side of the auricle. It is a mixed nerve of the facial nerve, glossopharyngeal nerve, and vagus nerve and contains general somatic and visceral sensory. Experimental animal studies found that when the marker was injected into the central end of the auditory branch of the vagus nerve in cats, the markers could be found in the superior vagus nerve, the ipsilateral nucleus tractus solitaries, and the contralateral primary trigeminal sensory nucleus. Stimulating the skin of distribution area of the vagus auricular branch can activate the vagus auricular branch, thus producing a parasympathetic excitatory effect similar to that of stimulating the cervical vagus nerve trunk, which may lead to upward transmission to afferent nucleus tractus solitaries and projection to the locus coeruleus.

In recent years, taVNS has gradually become a research hotspot, and many studies have discussed the efficacy and possibility of taVNS in different diseases.

#### 1.1.2.1 Depression

Hein et al. illustrated the antidepressant effect of 2-week t-VNS treatment using an additive treatment study design. The results showed significant improvement in Beck Depression Scale (BDI: 27.0–14.0), but no significant change in Hamilton Depression Scale (HAMD). There is little information about the stimulus parameters. A 1.5 Hz unipolar square wave was used in this study, and the current intensity was adjusted to the maximum depending on the individual but not distressing (0–600 μA) (Hein et al. 2013).

Fang et al. conducted a single-blind clinical trial to investigate the antidepressant effect of t-VNS alone. Their findings showed significant improvement not only in HAMD (28.5–15.0), but also in Self-Rated Anxiety Scale (SAS: 56.56–42.83) and Self-Rated Depression Scale (SDS: 66.33–50.56). In this study, the stimulation frequency was 20 Hz, the wave width was <1 ms, and the stimulation intensity was adjusted according to the patient's tolerance (4–6 mA) (Fang et al. 2016).

#### 1.1.2.2 Epilepsy

Stefan et al. conducted a pilot study investigating the antiepileptic effect of taVNS. In this study, the effect of taVNS in ten patients with refractory epilepsy who had at least four seizures per month was observed after transcutaneous stimulation of the auricular branch of the vagus nerve via the tragus of the left ear. Stimulation parameters were set to a frequency of 10 Hz, and a pulse width of 0.3 ms. The stimulation intensity was set to the individual's tolerance threshold. For 9 months, participants received 1-h taVNS three times a day. Participants kept seizure diaries to report the frequency of seizures before and during taVNS treatment. Of the seven patients who completed the study, five patients had reduced seizure frequency, indicating that taVNS has antiepileptic effect (Stefan et al. 2012).

He et al. (2013) conducted another pilot study to investigate the therapeutic effect of taVNS on pediatric epilepsy. The stimulation site was the left concha, and the stimulation frequency was 20 Hz, 30 min each time, and three times a day, lasting for 6 months. They found that after 6 months of taVNS treatment, epileptic seizure frequency decreased by 54% (He et al. 2013). Recently, Liu et al. found that 16 of 17 epilepsy patients had an average seizure reduction of 64.4% after 6 months of taVNS treatment. The stimulation frequency was 10 Hz, 20 min each time, and three times a day, lasting for 6 months (Liu et al. 2018).

#### 1.1.2.3 Tinnitus

Lehtimäki et al. designed a pilot study to investigate the efficacy of t-VNS in patients with chronic tinnitus. Ten patients with chronic tinnitus received taVNS stimulation on the left tragus at a frequency of 25 Hz, 45–60 min each time, and a total of seven times. At the end of the study, all participants reported improved mood and reduced tinnitus severity (Lehtimäki et al. 2013).

#### 1.1.2.4 Migraine

Straube et al. studied the therapeutic effect of taVNS on the tragus for migraine. They treated 46 migraine patients with taVNS. The stimulation frequency was 25 Hz, 4 h each time for 3 months. The stimulation frequency in the control group was 1 Hz. They found that compared with 25 Hz active stimulation, the frequency of headache was significantly reduced in 1 Hz stimulation group (Straube et al. 2015).

#### 1.1.2.5 Pain

Johnson et al. were the first to study the effect of transcutaneous electrical stimulation of the ear on the pain threshold. They gave 18 participants low-frequency taVNS at 2.3 Hz for 15 min at three different positions of the auricles. Their results showed that 10 of the 18 participants had increased pain thresholds. Among them, three participants still had this effect for a long time after the stimulation device was turned off (Johnson et al. 1991). Laqua et al. conducted a study to examine the effect of taVNS on pain threshold. When the stimulation frequency changes from 2 to 100 Hz, the stimulation duration is 30 min. Of the 21 participants, pain thresholds increased during taVNS in 15 participants, while pain thresholds decreased during stimulation in 6 participants (Laqua et al. 2014). Busch et al. conducted another study to investigate the effect of taVNS on pain thresholds of different sensory systems. There were 48 participants in the study, and the stimulation frequency was 25 Hz. Different tests were designed to measure different pain thresholds, such as thermal, mechanical, and stress-related pain thresholds. The results showed that mechanical, thermal, and pressure pain sensitivity was inhibited after continuous taVNS for 1 h. There was no significant change in detection thresholds for thermal or mechanical input. These results suggest that t-VNS can affect pain processing and can inhibit different pain patterns (Busch et al. 2013).

### 1.1.3 Safety and Tolerability of t-VNS

t-VNS is generally well tolerated, but there are many different mild side effects. Common side effects included tingling or pain around the stimulation site, and some participants reported itching or redness. Other less common side effects observed in less than one percent of the study population included gastrointestinal problems such as nausea or vomiting, headache, palpitations, facial ptosis, dizziness, hoarseness, and nasopharyngitis. Currently, there is no study that correlates stimulation parameters or doses with the incidence of side effects.

## 1.2 External Trigeminal Nerve Stimulation

External trigeminal nerve stimulation (eTNS) is an emerging noninvasive nerve regulation technique in recent years. This technique acts as a neuromodulator by delivering electrical stimulation from a stimulator connected to electrodes placed in the trigeminal nerve distribution area.

The trigeminal nerve, the thickest of the cranial nerves, is a pair of mixed cranial nerves, containing general somatic sensory fiber and special visceral motor fiber. The trigeminal nerve has three branches: ophthalmic nerve, maxillary nerve, and mandibular nerve. Trigeminal sensory afferents from the spinal trigeminal nucleus have efferent fibers that terminate in the nucleus tractus solitaries and the nucleus locus coeruleus. Studies have shown that trigeminal nerve stimulation can activate a wide range of brain regions including the frontotemporal cortex and hippocampus through the thalamus. Therefore, the antiepileptic effect of eTNS may be through the nucleus tractus solitaries, locus coeruleus, and their related brain regions. Antidepressants may play a role by activating frontotemporal cortex.

Currently, percutaneous trigeminal nerve stimulators are mostly used in clinical practice, and the stimulators send electrical signals to the trigeminal nerve innervation area. The most commonly used stimulation site is one third of 1 cm above the bilateral orbits, stimulating the distribution area of the ophthalmic branch of the trigeminal nerve. The recommended stimulation parameters are: frequency: greater than 100 Hz; current intensity: 10–20 mA, no discomfort; pulse width: 0.5 ms; stimulate for 30 s per min for 12 h every day.

Current clinical studies have shown that eTNS has a positive effect on refractory epilepsy and depression. In 2012, it obtained CE Mark Approval in Europe, which can be used for adjuvant treatment of epilepsy in adults and children over 9 years old. DeGiorgio et al. conducted a series of studies to confirm the efficacy of eTNS in the treatment of refractory epilepsy: In 2003, they reported that two patients with refractory complex partial epilepsy were treated with eTNS for 4 weeks and followed up for 6 months, of whom one patient had a 39% reduction in seizures, and the other patient had a 76% reduction in seizures (DeGiorgio et al. 2003); in 2006, they reported that 7 patients with refractory epilepsy were treated with eTNS, of whom 57% had a greater than 50% reduction in seizures (DeGiorgio et al. 2006). In the feasibility study published in 2009, 14 patients with refractory epilepsy were selected and followed up for 12 months, leading to a 59% reduction in the average number of seizures; and long-term high-frequency stimulation was found to be more effective than low-frequency stimulation (DeGiorgio et al. 2009). In January 2013, a randomized, double-blind, controlled study of eTNS in patients with drug-resistant epilepsy showed that eTNS reduced the frequency of epilepsy (DeGiorgio et al. 2013). In a pilot experiment reported by Zare et al. in 2014, the average number of seizures in eight patients who completed the experiment decreased by 47.9% (Zare et al. 2014). In 2015, Soss et al. conducted a long-term observational study on 50 patients with refractory epilepsy in a prospective open study to observe the efficacy and safety of eTNS on refractory epilepsy. The eTNS stimulation parameters were: 30 s on, 30 s off, duration 250 ms, stimulation frequency 120 Hz. 35 of the 50 subjects completed the study. In the eTNS treatment group, the median monthly reduction in epileptic seizure frequency was 2.39 (27.4%) at 6 months, and 3.03 (34.8%) at 12 months (Soss et al. 2015). In 2020, Gil-López et al. conducted a randomized controlled study to observe the long-term efficacy and tolerance of eTNS in the treatment of focal drug-resistant epilepsy in 40 patients with refractory epilepsy in frontal or temporal lobe who were not suitable for surgery. Their results suggest that at 12 months, 50% of patients in the eTNS treatment group responded to treatment, compared with 0% in the control group. Epileptic seizure frequency in eTNS group decreased by 43.5% from baseline. The efficacy of the temporal lobe epilepsy subgroup was better than that of frontal lobe epilepsy group (55.56% and 45.45%, respectively). The median stimulus intensity was 6.2 mA. eTNS improved quality of life but did not improve anxiety and depression. No related adverse events were observed (Gil-López et al. 2020).

So far, there are few studies on the treatment of depression with eTNS. Shraeder et al. treated five patients (60% female, mean age 49.6 years old), and followed up for 8 weeks. Subjects had electrodes placed on their forehead to stimulate the V1 branches of the trigeminal nerves on both sides for about 8 h each night (8 weeks; approximately 55 nights). The current intensity is adjusted to maintain a comfortable but perceptible stimulation level. During the 2-month follow-up, the remission rate of depressive symptoms reached 70% (Schrader et al. 2011). Shiozawa et al. evaluated 11 patients and showed a mean reduction of 5.72 points (27.6%) on the Hamilton Depression Rating Scale (HDRS-17) after treatment. All patients had a score reduction of at least 50%. Only one patient did not achieve a score indicative of remission (defined as HDRS-17 below eight). In the other two open label trials, the authors reported the efficacy of 11 and five adult patients respectively in an 8-week open outpatient trial. Depressive symptoms were significantly improved without adverse reactions (Shiozawa et al. 2014).

In general, patients have good tolerance to TNS and the most common adverse reaction was facial skin irritation. By changing the electrode position, reduce the stimulation duration or applying one percent hydrocortisone cream to skin (e.g., 12 h per day), discomfort can be relieved. Tingling and headache have also been reported, which can be relieved by reducing the amount of electrical current.

## 2 Transcutaneous Spinal Direct Current Stimulation

Transcutaneous spinal direct current stimulation (tsDCS) places the surface electrode directly on the skin surface of the corresponding spinal cord segment to give local stimulation. It affects the ascending and descending spinal cord pathways and the excitability of spinal cord reflexes. The polarity-specific effects of stimulation produced by tsDCS are opposite to those of tDCS: anodal tsDCS mainly inhibits spinal cord pathway conduction and facilitation reflexes, while cathodal stimulation can improve conduction in the corticospinal tract, spinothalamic tract, and spinal lemniscus, and inhibit reflexes. At the spinal cord level, anodal stimulation acts directly on axons and does not affect postsynaptic motor neuron excitability, whereas cathodal stimulation acts more on interneuron networks. In addition to directly acting on the spinal cord, tsDCS also causes indirect functional changes in the brain.

## 2.1 Fundamentals of tsDCS

tsDCS mainly functions by regulating the activity of neural network. At the neuronal level, the basic mechanism of tsDCS modulation of spinal cord excitability is to cause changes in resting membrane potential excitability. The change of membrane potential excitability is the main mechanism of immediate effect of tsDCS after stimulation. In addition to the immediate effect, tsDCS also has a post-stimulation effect. Changes in neural excitability persist after stimulation ends if stimulation lasts long enough. Further studies confirmed that in addition to changing the polarity of membrane potential, tsDCS can also modulate the synaptic microenvironment, such as changing the activity of NMDA receptor or GABA, thereby regulating synaptic plasticity. tsDCS can also modulate the excitability of spinal cord conduction tracts, including posterior funiculus and spinothalamic tract. In addition, tsDCS can influence the spinal cord by directly activating interneurons at the level of spinal cord segments or by modulating information transmission between the posterior funiculus and the posterior horn, which affect the reflex activity at the level of spinal cord and the electrical activity of the spinal cord circuit, and exert inhibitory effects on spinal cord conduction pathway. The aftereffect mechanism of tsDCS is similar to long-term facilitation of synapses. In human and animal experiments, tsDCS technique has been proved to be effective in modulating spinal cord excitability.

## 2.2 Common Parameters of tsDCS

When applying tsDCS, some commonly used stimulus locations and parameters are available for reference. For example, when regulating the lumbar spinal cord, the stimulation electrode (mostly 5 × 7 cm) is usually placed on the spinous process of the tenth thoracic vertebra, and the reference electrode is placed on the right shoulder. When regulating the cervical spinal cord, the stimulation electrode was placed on the seventh cervical vertebra and the reference electrode was placed on the front of the neck. Whether the placement of the reference electrode on the upper limb or other locations has an effect on the stimulation effect requires further studies. Moreover, the stimulation intensity and duration of most studies were 2–2.5 mA and 15 or 20 min, respectively (Priori et al. 2014). tsDCS usually produces skin tingling for a short time at the beginning and end of stimulation, and sometimes redness of the skin under the electrodes.

## 2.3 Study on the Mechanism of tsDCS in Healthy Population

This part mainly focuses on the effect of tsDCS on the spinal cord ascending and descending tracts and spinal cord reflexes.

Cogiamanian and his colleagues used the changes of somatosensory-evoked potential before and after tsDCS to evaluate the effect of anodal and cathodal stimulation on the ascending spinal cord sensory pathway. They found that anodal stimulation could significantly reduce the P30 amplitude generated by tibial nerve stimulation, and the effect lasted for at least 20 min, while cathodal stimulation had no effect on evoked potential (Cogiamanian et al. 2008). Truini et al. used laser-evoked potentials (LEPs) to evaluate the modulation of anodal and cathodal tsDCS stimulation on the spinal cord. Peripheral nerve photostimulation induces activation of peripheral Aδ fibers, which transmit pain information to the brain through the spinothalamic tract. Studies indicate that anodal tsDCS reduced LEP amplitudes produced by photostimulation on foot, but not cathodal stimulation (Truini et al. 2011). Therefore, the above two studies show that tsDCS can modulate at least two ascending sensory pathways in the spinal cord: the spinothalamic pathway and the spinothalamic tract. tsDCS can modulate sensory conduction pathways. Can tsDCS regulate the corticospinal tract? Lim et al. studied the changes in upper limb motor-evoked potentials before and after cervical cord tsDCS and found that both anodal and cathodal stimulation increased corticospinal excitability but had no effect on H-reflexes (Lim and Shin 2011).

tsDCS can also regulate reflex activities that affect spinal cord level. Reflex activity at the spinal cord level was mainly assessed by H-reflexes (latency, threshold, $H_{max}/M_{max}$) of Ia-monosynaptic connections. Winkler et al. found that the excitability of H-reflex ($H_{max}/M_{max}$) did not change after tsDCS, anodal spinal tDCS induced a decrease in homosynaptic depression whereas cathodal spinal tDCS increased it. The researchers speculate that tsDCS may have no effect on the excitability of alpha motor neurons (no change in $H_{max}/M_{max}$), but rather affects the connection between Ia fibers and alpha motor neurons (Winkler et al. 2010). Regardless of the mechanism, tsDCS may have an ameliorating effect on spinal cord-induced spasticity, considering the decreased inhibition of H-reflex activation in spastic patients. Several studies have since confirmed that tsDCS does not affect the basic properties of the H-reflex (amplitude, latency, and $H_{max}/M_{max}$). The study on the effect of tsDCS on spinal cord

H-reflex showed that tsDCS did not affect the excitability of spinal cord alpha motor neurons. It is speculated that tsDCS may inhibit neural reflexes by adjusting the excitability of the conduction pathway of monosynaptic reflexes.

## 2.4 Clinical Application of tsDCS

In terms of clinical application, previous studies have explored the therapeutic effect of tsDCS on spasticity, stroke, spinal cord injury, and restless legs syndrome.

Bocci et al. conducted a randomized double-blind controlled study of tsDCS in 11 spastic paraplegia patients with a current strength of 2 mA, stimulation duration of 20 min, and a 5-day course of treatment with positive electrodes or sham electrodes placed at the thoracic spinal cord (T10-T12) (Bocci et al. 2019). Ashworth scale scores in the treatment group were significantly improved after treatment and at 2-month follow-up, while there were no significant differences in the 5-min walk test and spastic paraplegia scores between the two groups. In terms of electrophysiological indexes, there was no change in H-reflex, F wave, and motor evoked potentials.

Early clinical studies found that tDCS did not improve the effect of robot-assisted gait training in the sequelae of stroke, which may be related to cortical and spinal cord control. Picelli et al. investigated the improvement of walking ability in stroke patients by combining tDCS, tsDCS, and robot-assisted gait training (Picelli et al. 2015). The anode of tDCS was placed in the primary motor cortex of the injured side, and the current intensity of 2 mA, and each stimulation duration was 20 min, with a total of ten treatments. The study found that tDCS combined with tsDCS could significantly improve the walking ability of stroke patients, and the effect lasted for 2 weeks.

Marangolo et al. explored the therapeutic effect of tsDCS on post-stroke aphasia patients (Marangolo et al. 2017). The stimulation electrodes were placed on the tenth thoracic vertebra and divided into three groups according to different stimulation electrodes. One group was anodal stimulation, one group was cathodal stimulation, and the last one was sham stimulation. In the real stimulation group, the stimulation intensity was 2 mA, each stimulation was 20 min, and the treatment was performed five times a week for a total of 3 weeks. The results showed that the verb naming in the anodal stimulation group was significantly improved compared with the cathodal stimulation group and the sham stimulation group, while noun naming was not significantly different among the three groups.

Hubli et al. performed anodal stimulation of the T11-T12 spinous processes in patients with spinal cord injury at a stimulation intensity of 2.5 mA (Hubli et al. 2013). Stimulating for 20 min improved the spinal cord reflex amplitude by 84%, while the spinal cord reflex amplitude did not change in cathodal stimulation, sham stimulation, and exercise training. They speculated that continuous stimulation may improve neural pathway after spinal cord injury.

Currently, a small number of studies have used tsDCS in the treatment of restless legs syndrome to observe its efficacy and explore its mechanism. Heide et al. placed the anode electrode at the T11 level of the back and the cathode electrode at the right shoulder (Heide et al. 2014). The stimulation current was 2.5 mA, and the stimulation duration was 15 min. The stimulation was performed only once. Results indicate that the symptoms of restless legs syndrome patients with anodal stimulation were significantly relieved, and the VAS score was significantly reduced after treatment. The study on the changes of lower limb H-reflex before and after treatment showed that tsDCS may inhibit the spinal cord pathway of spinal dorsal column and spinothalamic tract in patients with restless legs syndrome. In addition, tsDCS may inhibit the activity of spinal dorsal horn and continuously reduce the activity of two ascending sensory tracts, thereby improving the clinical symptoms of restless legs syndrome in a short term. This study supports the pathophysiological mechanism of spinal cord hyperexcitability in patients with restless legs syndrome and provides the basis for tsDCS as a novel non-drug therapy.

Wang et al. performed anodal stimulation on the T10 spinous process of 30 patients with restless legs syndrome, with a stimulation intensity of 2 mA, each stimulation for 20 min, and continuous treatment for 14 days (Wang et al. 2020). Patient scores on the Restless Legs Scale and Pittsburgh Sleep Scale decreased significantly, the treatment effect lasted for at least 2 weeks, and the patient's sensorimotor network and visual information processing network changed after treatment. We still need to explore the best parameters for tsDCS as a treatment of restless legs syndrome, including stimulation intensity, single stimulation duration, daily stimulation duration, and number of days, in order to maintain long-term curative effect and even completely reverse the state of spinal cord hyperexcitability in patients with restless legs syndrome so as to achieve a complete cure.

## 2.5 Effect and Safety of tsDCS

The follow-up effect of tsDCS can only last for a few hours or even minutes. How to design repeated stimulation parameters to prolong the follow-up effect on the basis of safety and feasibility requires further exploration.

Safety: Although tsDCS is a noninvasive neuromodulation technique, its safety issues cannot be underestimated.

The safety of tsDCS mainly depends on the current density and the total amount of electricity received per unit area. Literature describes that the current density <25 mA/cm$^2$ and total electrical charge <16 C/cm$^2$ will not cause tissue damage. The stimulation parameters used in the current study are far below this standard and only cause numbness in the area covered by the electrode pads at the start of the stimulation and a short time after the end of the stimulation. Studies have shown that there is no significant difference in serum neuronal nonspecific enolase (a sensitive and specific indicator reflecting neuronal damage) before and immediately after tsDCS stimulation, suggesting that direct current stimulation does not cause nerve damage.

## References


Ay I, Sorensen AG, Ay H (2011) Vagus nerve stimulation reduces infarct size in rat focal cerebral ischemia: an unlikely role for cerebral blood flow. Brain Res 1392:110–115

Ay I, Nasser R, Simon B et al (2016) Transcutaneous cervical vagus nerve stimulation ameliorates acute ischemic injury in rats. Brain Stimul 9(2):166–173

Bocci T, Ardolino G, Nigro M et al (2019) Spinal direct current stimulation (tsDCS) in hereditary spastic paraplegias (HSP): a sham-controlled crossover study. Clin Neurophysiol 130(1):e17

Busch V, Zeman F, Heckel A et al (2013) The effect of transcutaneous vagus nerve stimulation on pain perception—an experimental study. Brain Stimul 6(2):202–209

Chen SP, Ay I, Lopes de Morais A et al (2016) Vagus nerve stimulation inhibits cortical spreading depression. Pain 157(4):797–805

Cogiamanian F, Vergari M, Pulecchi F et al (2008) Effect of spinal transcutaneous direct current stimulation on somatosensory evoked potentials in humans. Clin Neurophysiol 119:2636–2640

DeGiorgio CM, Shewmon DA, Whitehurst T (2003) Trigeminal nerve stimulation for epilepsy. Neurology 61(3):421–422

DeGiorgio CM, Shewmon A, Murray D et al (2006) Pilot study of trigeminal nerve stimulation (TNS) for epilepsy: a proof-of-concept trial. Epilepsia 47(7):1213–1215

DeGiorgio CM, Murray D, Markovic D et al (2009) Trigeminal nerve stimulation for epilepsy: long-term feasibility and efficacy. Neurology 72(10):936–938

DeGiorgio CM, Soss J, Cook IA et al (2013) Randomized controlled trial of trigeminal nerve stimulation for drug-resistant epilepsy. Neurology 80(9):786–791

Fang J, Rong P, Hong Y et al (2016) Transcutaneous vagus nerve stimulation modulates default mode network in major depressive disorder. Biol Psychiatry 79(4):266–273

Frangos E, Ellrich J, Komisaruk BR (2015) Non-invasive access to the vagus nerve central projections via electrical stimulation of the external ear: fMRI evidence in humans. Brain Stimul 8(3):624–636

Gil-López F, Boget T, Manzanares I et al (2020) External trigeminal nerve stimulation for drug resistant epilepsy: a randomized controlled trial. Brain Stimul 13(5):1245–1253

He W, Jing X, Wang X et al (2013) Transcutaneous auricular vagus nerve stimulation as a complementary therapy for pediatric epilepsy: a pilot trial. Epilepsy Behav 28(3):343–346

Heide AC, Winkler T, Helms HJ, Nitsche MA et al (2014) Effects of transcutaneous spinal direct current stimulation in idiopathic restless legs patients. Brain Stimul 7(5):636–642

Hein E, Nowak M, Kiess O et al (2013) Auricular transcutaneous electrical nerve stimulation in depressed patients: a randomized controlled pilot study. J Neural Transm (Vienna) 120(5):821–827

Hubli M, Dietz V, Schrafl-Altermatt M, Bolliger M (2013) Modulation of spinal neuronal excitability by spinal direct currents and locomotion after spinal cord injury. Clin Neurophysiol 124(6):1187–1195

Johnson MI, Hajela VK, Ashton CH et al (1991) The effects of auricular transcutaneous electrical nerve stimulation (TENS) on experimental pain threshold and autonomic function in healthy subjects. Pain 46(3):337–342

Laqua R, Leutzow B, Wendt M et al (2014) Transcutaneous vagal nerve stimulation may elicit anti- and pro-nociceptive effects under experimentally-induced pain—a crossover placebo-controlled investigation. Auton Neurosci 185:120–122

Lehtimäki J, Hyvärinen P, Ylikoski M et al (2013) Transcutaneous vagus nerve stimulation in tinnitus: a pilot study. Acta Otolaryngol 133(4):378–382

Lim CY, Shin HI (2011) Noninvasive DC stimulation on neck changes MEP. Neuroreport 22:819–823

Liu A, Rong P, Gong L et al (2018) Efficacy and safety of treatment with transcutaneous Vagus nerve stimulation in 17 patients with refractory epilepsy evaluated by electroencephalogram, seizure frequency, and quality of life. Med Sci Monit 24:8439–8448

Marangolo P, Fiori V, Shofany J, Gili T, Caltagirone C, Cucuzza G et al (2017) Moving beyond the brain: transcutaneous spinal direct current stimulation in post-stroke aphasia. Front Neurol 8:400

Nonis R, D'Ostilio K, Schoenen J et al (2017) Evidence of activation of vagal afferents by non-invasive vagus nerve stimulation: an electrophysiological study in healthy volunteers. Cephalalgia 37(13):1285–1293

Picelli A, Chemello E, Castellazzi P et al (2015) Combined effects of transcranial direct current stimulation (tDCS) and transcutaneous spinal direct current stimulation (tsDCS) on robot-assisted gait training in patients with chronic stroke: a pilot, double blind, randomized controlled trial. Restor Neurol Neuros 33(3):357–368

Polak T, Markulin F, Ehlis AC et al (2009) Far field potentials from brain stem after transcutaneous vagus nerve stimulation: optimization of stimulation and recording parameters. J Neural Transm (Vienna) 116(10):1237–1242

Priori A, Ciocca M, Parazzini M et al (2014) Transcranial cerebellar direct current stimulation and transcutaneous spinal cord direct current stimulation as innovative tools for neuroscientists. J Physiol 592(16):3345–3369

Schrader LM, Cook IA, Miller PR et al (2011) Trigeminal nerve stimulation in major depressive disorder: first proof of concept in an open pilot trial. Epilepsy Behav 22(3):475–478

Shiozawa P, Duailibi MS, da Silva ME et al (2014) Trigeminal nerve stimulation (TNS) protocol for treating major depression: an open-label proof-of-concept trial. Epilepsy Behav 39:6–9

Soss J, Heck C, Murray D et al (2015) A prospective long-term study of external trigeminal nerve stimulation for drug-resistant epilepsy. Epilepsy Behav 42:44–47

Stefan H, Kreiselmeyer G, Kerling F et al (2012) Transcutaneous vagus nerve stimulation (t-VNS) in pharmacoresistant epilepsies: a proof of concept trial. Epilepsia 53(7):e115–e118

Straube A, Ellrich J, Eren O et al (2015) Treatment of chronic migraine with transcutaneous stimulation of the auricular branch of the vagal nerve (auricular t-VNS): a randomized, monocentric clinical trial. J Headache Pain 16:543

Truini A, Vergari M, Biasiotta A, La Cesa S, Gabriele M, Di Stefano G, Cambieri C, Cruccu G, Inghilleri M, Priori A (2011) Transcutaneous spinal direct current stimulation inhibits nociceptive spinal pathway conduction and increases pain tolerance in humans. Eur J Pain 15:1023–1027

Usami K, Kawai K, Sonoo M et al (2013) Scalp-recorded evoked potentials as a marker for afferent nerve impulse in clinical vagus nerve stimulation. Brain Stimul 6(4):615–623

Wang L, Liu C, Hou Y et al (2020) Altered cortical gray matter volume and functional connectivity after transcutaneous spinal cord direct current stimulation in idiopathic restless legs syndrome. Sleep Med 74:254–261

Winkler T, Hering P, Straube A (2010) Spinal DC stimulation in humans modulates post-activation depression of the H-reflex depending on current polarity. Clin Neurophysiol 121:957–961

Zare M, Salehi M, Mahvari J et al (2014) Trigeminal nerve stimulation: a new way of treatment of refractory seizures. Adv Biomed Res 3:81

# Transcranial Electrical Stimulation

# 6

Jing Wang, Sitong Liu, Qihui Zhou, Xiaona Dai, and Jialin Du

## 1 Transcranial Direct Current Stimulation

Electrical current has been used in medicine since the eleventh century. In the 1960s and 1970s, attempts were made to exert electrophysiological effects by transcranially applying direct current to the brains of healthy subjects and patients with psychiatric disorders. However, these studies were suspended due to the unclear mechanism of the technique and lack of scientific evaluation. It was not until 20 years ago that scientists rediscovered transcranial direct current stimulation (tDCS) as a tool to modulate human brain activity and systematically explored its physiological effects. This technique has been widely recognized and applied due to its advantages of high safety, low device cost, ease of operation, and reliable double-blind control.

tDCS is a noninvasive electrical stimulation technique which works through a weak direct current (1–2 mA). The direct current is generated by a special battery that provides a continuous and stable power supply and delivered through a pair of conductive rubber electrode pads. The conductive rubber electrode pads attached to the scalp of the skull consist of an anodic electrode and a cathodic electrode that form a current loop to apply direct current stimulation. Traditionally, a sponge soaked in saline is placed between the conductive rubber and skin to increase the conductivity. There are two common stimulation modes, namely anodal tDCS (atDCS) and cathodal tDCS (ctDCS). As to atDCS, the anodal stimulation electrode is placed in the area of interest and a positive current is applied. As to ctDCS, the cathodal stimulation electrode is placed in the area of interest and a negative current is applied to the target. For conventional tDCS, the size of the two electrode pads (i.e., "anode" and "cathode") usually varies between 25 and 35 $cm^2$, with an "active" electrode above the target area and a reference electrode at another position on the scalp or elsewhere on the body.

### 1.1 Principle and Mechanism

Altering the excitability of neurons is considered to be the main mechanism by which tDCS alters brain activity and thus affects behavioral performance. As research progresses, there are increasing evidence proving that tDCS not only modulates neural activity, but also directly or indirectly modulates non-neural functional activity.

#### 1.1.1 Alteration of Membrane Potential and Cortical Excitability

One of the main mechanisms of tDCS is that it could modulate neuronal excitability by shifting the resting potential of neurons to the depolarization or hyperpolarization threshold. Results from studies in animals also indicate that anodal stimulation increases firing frequency while cathodal stimulation does the opposite. Besides, results from studies in human show that anodal stimulation increases the amplitude of TMS-induced motor-evoked potential, while cathodal stimulation does the opposite, suggesting that anodal stimu-

J. Wang (✉)
Department of Neurobiology, School of Basic Medical Sciences, Capital Medical University, Beijing, China
e-mail: jingw@ccmu.edu.cn

S. Liu
Department of Neurology, Beijing Friendship Hospital, Capital Medical University, Beijing, China

Q. Zhou
Department of Neurology, Beijing Chaoyang Hospital, Capital Medical University, Beijing, China

X. Dai
Department of Neurology, Xuanwu Hospital, Capital Medical University, Beijing, China

J. Du
Department of Pharmacy Phase I Clinical Trial Center, Xuanwu Hospital, Capital Medical University, Beijing, China



Y. Wang (ed.), *Therapeutics of Neural Stimulation for Neurological Disorders*, https://doi.org/10.1007/978-981-99-4538-2_6

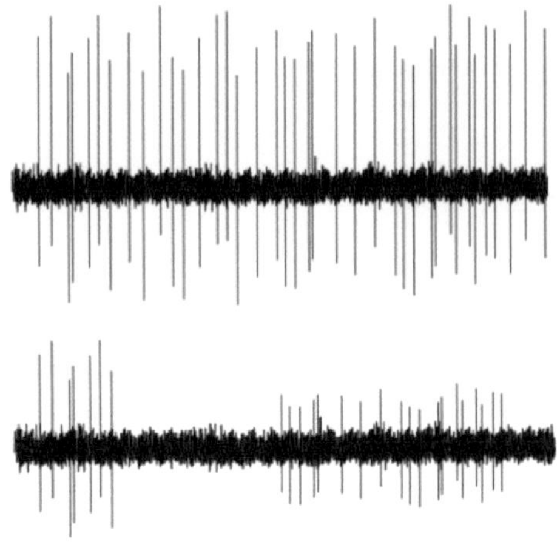


**Fig. 6.1** The effects of tDCS on neuronal excitability

lation increases the excitability of cortical neurons while cathodal stimulation decreases their excitability (Fig. 6.1). It is worth noting that this modulation is not sufficient to induce action potentials, but only to change the response threshold of neurons, which is the reason for its high safety. Moreover, the changes in cortical excitability caused by tDCS are not entirely determined by stimulus polarity, but also by the direction of current flow relative to the direction of the axon.

AtDCS increase the neuronal excitability while ctDCS decrease the neuronal excitability.

#### 1.1.2 Altering Synaptic Plasticity and Cortical Functional Connectivity

The effects of tDCS are not only visible at the time of stimulation, but also persist for a while after the stimulation, known as the aftereffect of tDCS. And the aftereffect of tDCS is related to its induction of synaptic plasticity. The calcium-dependent synaptic plasticity in glutamatergic neurons is believed to play a key role in the persistent aftereffect of tDCS. Moreover, the aftereffect of tDCS was attenuated after antagonizing *N*-methyl-D-aspartate (NMDA) receptors. In addition, neuronal synaptic plasticity corresponds to functional connectivity at the network level. So tDCS may also affect the functional connectivity, synchronization and oscillatory activity in various cortical and subcortical networks. For tDCS acting on the primary motor cortex (M1) (Polanía et al. 2011a, b, 2012), the prefrontal cortex (Keeser et al. 2011), or what happens during slow wave sleep has demonstrated it (Marshall et al. 2004).

#### 1.1.3 Non-neuronal Mechanisms

The neuronal activity is not the only target of tDCS, as almost all tissues and cells are sensitive to electric fields, such as neuroglia cells, endothelial cells, and lymphocytes (Ruohonen and Karhu 2012). Thus, tDCS may regulate the non-neuronal tissues of the brain, the inflammatory responses, the neuroendocrine activity and neurotrophic factors to affect the course of the disease. For example, the conformation of β-amyloid and other pathological proteins will change in case of exposure to appropriate electric fields (Toschi et al. 2009), which may alter their susceptibility to degradation. In addition, direct-current electric field could enhance axonal regeneration and neurite growth (Fehlings and Tator 1992; Pelletier and Cicchetti 2014; Wood and Willits 2006), which may be associated with the improved recovery of neurological functions. Although the relationship between non-neuronal functions and the therapeutic effects of tDCS has not been systematically explored to date, it is one of the non-negligible mechanisms.

### 1.2 Effect Factors

The effect size of tDCS is related to many factors, including various parameters of the stimulus itself such as polarity, intensity, and duration, as well as the brain activities of subjects. Therefore, the effect of tDCS does not always vary linearly along with a single parameter, but in a nonlinear manner under the comprehensive influence of various factors.

#### 1.2.1 Effect of Stimulus Polarity

The classical effects of tDCS polarity on cortical excitability (atDCS increases excitability and ctDCS inhibits excitability) is primarily studied via the stimulation on the motor cortex. However, this effect is not considered to be a constant and universal rule because some factors can turn facilitative effect into inhibitory effect and vice versa. Specifically, the effects of direct current electric fields on local cells in brain are complex, depending on the distance and direction of axon or dendritic axis relative to the electric field, as well as the size and morphology of neurons (Bikson et al. 2004; Gluckman et al. 1996; Purpura and Mcmurtry 1965).

The distance from the stimulating electrode change the particular polarity in the target area. For example, in the cerebral cortex of anesthetized rodents, atDCS increases the spontaneous firing of neurons close to the electrode (depth <500 μm), while ctDCS decreases them (Stagg and Nitsche 2011). However, neurons in the deep layer of cortex are inhibited by atDCS and activated by ctDCS (Stagg and Nitsche 2011). The polarity reversal at a specific depth may be caused by the differences in the lateral connectivity of neurons or the distribution of cortical current source density (CSD), rather than a decrease in current intensity (Rappelsberger et al. 1982). And CSD depth profiles of the human neocortex show a maximal source in layer I (outward current) and a maximal sink in layer II/III (inward current) during oscillations (Csercsa et al. 2010). Thus,

The reduced activities of neurons in the deep layer may be due to the dipole formation enhanced by atDCS.

The direction of neurons is also an important factor in altering the actual effect of tDCS polarity. Pyramidal neurons (the direction from dendrite to axon) parallel to the current field are activated by atDCS and inhibited by ctDCS (Rahman et al. 2013). Similarly, in the cerebellar cortex, the largest modulation is neurons with dendrite-axon orientation parallel to the current vector among Purkinje cells and stellate interneurons (Chan and Nicholson 1986). During atDCS, the apical dendrites of Purkinje cells parallel to the current field are depolarized, while the rest of the dendrites and soma are hyperpolarized (Chan et al. 1988). In contrast, ctDCS depolarizes soma and hyperpolarizes dendrites. Furthermore, Kabakov et al. have revealed that the field excitatory postsynaptic potential (fEPSP) on hippocampal slices will be inhibited to the utmost when action potential propagation vector is in the same direction as DCS vector and fEPSP will be facilitated or maintained when action potential propagation is opposite to DCS vector (Kabakov et al. 2012). In general, axonal direction could determine whether the direct current field is excitatory or inhibitory, whereas dendritic direction could affect the magnitude of the amplitude without the direction of the direct current. In addition, the morphology (size and structure) of neurons could also affect the polarization of tDCS. Some modeling studies have shown that local cathodic stimulation acts on or preferentially activates horizontal fibers that are parallel to electrode, whereas anodic stimulation activates fibers that are perpendicular to electrode (Manola and Holsheimer 2007; Manola et al. 2005, 2007). Besides, the polarity-specific modulation of pyramidal cells is higher than that of non-pyramidal cells (Purpura and Mcmurtry 1965). And in experiments on the brain slices of rodents, neuronal somas of the largest cortical neurons (pyramidal cells in layer V) are most modulated by tDCS depolarization (Radman et al. 2009), which suggests that soma volume could also impact tDCS effects. Obviously, the response magnitude is also affected by dendritic structure. For example, the tDCS polarization is most prominent at the base of CA1 neuron and at the tip of the apical dendrite (Bikson et al. 2004), which may reflect the effect of passive cable's properties on tDCS. Studies in humans have also revealed that the effect of atDCS on cortical excitability depends on the location, size, and direction of the electrode (Ho et al. 2016; Opitz et al. 2015). Thus, the standardization of stimulus should require higher attention (Rampersad et al. 2014).

In conclusion, the polarity-specific effects of tDCS depend on the distance from the stimulus electrode, the current gradient, presynaptic axonal direction, and postsynaptic neuronal morphology. And it is important to note that even if a direct current field induces depolarizing or hyperpolarizing effects on neuronal population through fiber direction and electrode polarity, the physiological effects of stimulation also depend on whether the stimulated network is inhibitory or excitatory.

### 1.2.2 Effects of Current Intensity

A critical parameter of tDCS is current intensity. Many studies have shown that current intensity is an important factor determining whether tDCS produces biological effects and the extent thereof.

The relationship between current intensity on the scalp and intracranial electric field is the first question needs to be clarified. Although the current (in mA) is the stimulus parameter used to regulate, the local electric field (in V/m) ultimately reflects the actual intensity of the stimulus for each brain region. Modeling studies analyze the relationship between the dose applied to the scalp (including current intensity) and the electric field generated in the brain. It is generally recognized that the current intensity (electric field) in the brain will increase linearly with applied current, which is determined by physical properties of tDCS (Bikson et al. 2015).

Whether the brain linearly responds to the increase of electric field is then studied. In animal models, particularly in vitro brain slices (Bikson et al. 2004), the electric field strength can be strictly controlled, so that the response to the electric field be directly tested. Inactive neurons (could not fire action potential spontaneously and it is at an abnormal state due to neurons in the body are active) can be recorded in brain slices. After applying electric fields to the inactive neurons recorded in brain slices, it is found that there is a linear relationship between induced membrane polarization and the polarity and strength of electric field. That is, the greater the external electric field is, the stronger the neuronal polarization will be. However, this relationship will be held only with intensities higher than those in human studies (>10 V/m). For example, Bikson et al. evaluated the effects of uniform direct current electric fields (ranging from 10 to 100 V/m) on the neuronal excitability of hippocampal slices in rats, and reported that membrane polarization is linear generally except field strength exceeding 80 V/m (equivalent to tens of times the tDCS) will cause a nonlinear excitation reaction (Bikson et al. 2004). However, there is no significant and comprehensive evidence for a linear dose–response relationship in inactive or active neurons in animal brain when the electric field strength is below 1 V/m.

In many human studies exploring the tDCS dose–response relationships, they depend heavily on the response of transcranial magnetic stimulation motor-evoked potential (TMS-MEP). In conventional experiments, the linear relationship of tDCS intensity has been investigated by comparing TMS-MEP thresholds and amplitudes before, during, and/or after tDCS. Early normative studies in healthy subjects which use low doses of tDCS (1 mA) find a positive relationship between tDCS intensity and TMS-MEP magnitude. Notably, TMS-MEP reflects a combination of multiple physiological factors including the neuronal excitability at cortical and spinal level and cannot be simply mapped to behavioral changes. And TMS-MEP is usually measured after tDCS (offline) and it may not reflect an online effect on tDCS.

Subsequent studies replicate the findings of low doses tDCS (Strube et al. 2016), but there are also more complicated dose responses. Increasing stimulus intensity and duration (more than 10 min) might alter the extent and direction of excitability change measured by TMS-MEP to alter the dose–response further (Jamil et al. 2017). For example, the activation of motor cortex during atDCS could invert the direction of TMS-MEP modulation, indicating that the direction depends on the state. And the increased intensity of cathode tDCS to 2 mA result in enhancement of TMS-MEP (Batsikadze et al. 2013). Children also show a non-monotonic dose–response relationship even if at varying intensity range. In contrast to adults (Moliadze et al. 2015), ctDCS induce excitatory effects at only 1 mA. Clearly, neurophysiological findings based on human TMS-MEP suggest that the effects of tDCS are not necessarily linear or even monotonic as tDCS intensities increase (even within 1–2 mA). Moreover, the property of modulation is influenced by changes in brain state.

In terms of cognitive and behavioral change in healthy subjects and treatment in patients, the current evidence does not show a definite dose linear relationship between the tDCS intensity and the output of behavior. Previous studies have shown that the effect of tDCS in human is directly related to current intensity (Batsikadze et al. 2013). And the effect of current intensity, like 1 mA and 2 mA, on working memory tasks in healthy people show a non-monotonic dose response to current intensity (Hoy et al. 2013). However, another study illustrates that these two current intensities have no significant effect on working memory tasks (Teo et al. 2011).

tDCS is widely investigated as a potential treatment to improve cognitive activities in neuropsychiatric disorders (Cappon et al. 2016; Hampstead et al. 2017), and only a few studies have involved dose–response relationships. In a meta-analysis for tDCS in major depressive disorders, Brunoni et al. are unable to determine a correlativity between current intensity and tDCS effects (Brunoni et al. 2016). But a cross-over design study show that the remission of tinnitus is positively correlated with current intensity of high density-tDCS (Murray et al. 2015). And TMS-MEP modulation is significantly enhanced by 2 mA tDCS rather than 1 mA in patients with chronic spinal cord injury (Murray et al. 2015). The effects of tDCS stimulating site (DLPFC and M1) and intensity (1 mA and 2 mA) on Parkinson's disease (PD) indicate that only 2 mA anodic tDCS in the dlPFC could improve the accuracy of working memory task (Boggio et al. 2006). Besides, the optimized stimulation parameters (0.1 mA and 5 min) were selected for the treatment of PD in primates, indicating that the total charge (current intensity × stimulus duration) is correlated with the treatment effect (Li et al. 2015).

It is worth noting that the electric field intensity generated by tDCS is associated with enhanced clinical symptoms using current models, such as pain (Mendonca et al. 2011) or TMS-MEP (Bikson et al. 2010). In addition, the relationship is investigated between the changes in electric field strength generated by tDCS on the DLPFC and behavior in a working memory task, and it is found that participants with improved performance on working memory have higher electric field strength in the DLPFC compared to other participants, suggesting the behavior differences in tDCS may be due to individual anatomical differences, which consistent with monotonic dose responses (Kim et al. 2014).

### 1.2.3 Effects of Duration

A persistent effect after stimulation is critical for tDCS to effectively exert intervention. At present, most stimulations last from 20 to 40 min. In order to achieve an aftereffect, stimulation must be sustained at an intensity of 0.6 mA for at least 3 min. Studies have shown that an aftereffect lasting up to an hour after the stimulation, including firing frequency for action potential (Bindman et al. 1964), excitatory post-synaptic potentials (Fritsch et al. 2010) and gamma oscillations (Reato et al. 2010) in rodent cortex. Similarly, tDCS also has an offline effect on MEP modulation in humans (Horvath et al. 2015). However, increasing stimulus duration seem not to lead an improving effect of tDCS (Batsikadze et al. 2013; Nitsche et al. 2015), such as prolonging atDCS from 13 to 26 min leads to a decrease in motor cortex excitability. But studies also report that repetitive stimulation for several days can improve the effect of tDCS.

Therefore, in therapeutic studies, stimulations are often repeated over consecutive days to enhance clinical effects, mostly for five or more days, and the intervention effects of which can last for more than 1 month thereafter (Fregni et al. 2006; Loo et al. 2012; Reis et al. 2009).

## 2 Transcranial Alternating Current Stimulation

Transcranial alternating current stimulation (tACS) is another noninvasive brain stimulation technique where low-intensity biphase currents are applied to specific brain areas through electrodes at specific frequencies to synchronize neural oscillations with the frequency of stimulation applied to the stimulated cortex, which induces neuroplasticity changes in order to modulate neural activity in the brain and produce lasting effects.

tACS applies a sinusoidal alternating current between two electrodes on the scalp. During half a cycle of neural oscillation, one electrode will act as the anode and the other as the cathode. The current intensity will vary with the half-sine wave. During the other half of this cycle, the polarity will be reversed. The polarity functioning as anode previously must now be considered as the cathode, vice versa. In other words, the phase curve of the tACS current, unlike the unidirectional current of the tDCS, alternates periodically between positive and negative voltages. tACS has not been used for a long time and did not become the mainstream until 2008 (about 10 years after the emergence of tDCS), but it has shown to regulate a variety of activities and behaviors like cognition, including learning, memory, creativity. In recent years, studies on tACS have increased gradually, and the technique has been used increasingly.

## 2.1 Principle and Mechanism

The best way to study the action mechanism of tACS and demonstrate its effectiveness is to observe its effects during stimulation. However, it is difficult to figure out its effect because the huge current disturbances produce artifact several orders of magnitude larger than the electrophysiological signals of brain. Therefore, many studies report only the aftereffects of tACS, i.e., the measurement differences before and after stimulation. The mechanisms of tACS are currently recognized to include mainly oscillatory entrainment and neuroplasticity. Among them, it is believed that the immediate effect during stimulation can be explained by neural entrainment (Pikovsky and Maĭstrenko 2003; Thut et al. 2011), while the aftereffect is achieved by neuroplasticity.

### 2.1.1 Oscillatory Entrainment

The term "entrainment" is a phenomenon that occurs in many different natural systems (Pikovsky and Maĭstrenko 2003). In general, entrainment refers to the mutual adaptation or synchronization of rhythmic activity in two systems. One of these is neural oscillator in the brain, whose oscillatory characteristics are variable. In case of neural entrainment, the other system is an external rhythmic driver, such as alternating current, magnetic pulses or flashing lights, sounds, etc.

At the neuronal level, tACS is generally thought to alter the duration of firing without affecting the total firing rate. In a model study with neural networks, the authors shows that only the firing time of action potentials is regulated at low stimulus intensities, while the firing frequency of action potentials does not (Loo et al. 2012). In this case, the action potential will align with some phase of the externally applied sine wave. For example, an electric field strength of 0.7–5 V/m in vitro allows the animal potential to phase lock to follow an oscillating (1–30 Hz) external field (Anastassiou et al. 2011). In successive cycles, this altercation may lead to a cumulative effect on the firing time (Huang et al. 2017). However, it has been shown that under low-frequency stimulation, tACS may induce periodic tDCS-like effects at each half-cycle firing frequency (Chan et al. 1988; Chan and Nicholson 1986; Reato et al. 2010). The sensitivity of neurons to membrane polarization (electric field per mV/mm) decreased with increasing frequency. It is associated with dendritic polarization during alternating current stimulation. This is significantly different from transcranial direct current stimulation.

In humans, physiological changes in brain activity are mostly measured noninvasively by using EEG or MEG. tACS at specific frequencies can shift the brain's rhythmic activity to a pattern that matches the stimulation frequency ["frequency tuning" (Veniero et al. 2015)]. It is currently believed that entrainment has two effects on group activity. Firstly, the amplitude (or power) of neural oscillations increases as more and more neurons are in phase with the entrained stimulation; secondly, the phase of neuronal activity is matched with the stimulation phase. By definition, any entrainment to tACS could only occur during the period of stimulation because internal oscillations need to be coupled to an external oscillator. Although online analysis can capture true neural entrainment, such as the peak alpha activity has been reported in individuals moving toward 10 Hz with tACS (Al et al. 2022), these effects cannot rule out the possibility of artifacts (Noury et al. 2016), especially the responses to very low current intensities, like 0.05 mA (Ruhnau et al. 2016). Many studies have restricted the analysis to post-stimulation to avoid the necessity to deal with the interference of online artifacts. The entrainment is not always observed within a few cycles after the stimulation (Vossen et al. 2015), which is different from persistent aftereffects of tACS (Veniero et al. 2015). Of course, these persistent effects may reflect plasticity-related network changes rather than just "following" effects. Thus, these more persistent effects may reflect changes in the network associated with plasticity, rather than entrainment by itself (Vossen et al. 2015; Wischnewski and Schutter 2017; Zaehle et al. 2010).

In addition to local modulation of brain oscillations, phase coupling between distal cortical regions is important for understanding the effects of tACS on behavioral performance. Indeed, it has been indirectly supported by some results that phase coupling in distal regions occurs through neural synchronization and is implicated in perception and cognitive performance. Therefore, it is necessary to consider a higher order interaction between tACS and neural oscillations in the brain. If we want to modulate brain oscillations through tACS, it will not be a linear process and its effect will not be limited to the frequency of stimulation. First, a linear increase in stimulus intensity may have nonlinear effects on the involved neural tissue. Secondly, the following oscillations will not only affect the oscillations at stimulus frequency, but also the harmonic multiples and subharmonics.

### 2.1.2 Neuroplasticity

It is currently believed that the aftereffects of tACS are related to its effects on synaptic transmission between neurons, as changes in synaptic plasticity. The most prominent model of neuroplasticity is long-term potentiation (LTP). During the stimulating period of tACS, alternating electric fields repeatedly make the cytomembrane depolarize and hyperpolarize, possibly leading to the change of spike timing-dependent plasticity (STDP). Besides, LTP enhances neuronal suprathreshold firing activity as well as neuronal connectivity and signaling transduction, suggesting that frequency and phase information are essential parameters of neural function.

A study in humans finds that changes in motion perception were attributed to modulation of tACS-induced neuroplasticity (Kar and Krekelberg 2014). More precisely, the stimulation led to a change in the intensity of motion adaptation. When visual stimulation is moved in a coherent direction for a relatively long period, the visual system will adapt to that motion and perceives motion in the opposite direction when the visual stimulation stops moving. The strength of this effect is weakened with the application of tACS. A study in macaques apply tACS directly on their scalps with 2 mA and 10 Hz and recorded the local field potential (LFP) from the visual area in the middle temporal cortex while allowing the monkeys to observe the moving visual stimulation (Kar et al. 2017). As same as human data, tACS triggers attenuation of motor adaptation and directly demonstrated changes in the firing pattern of motor sensitive neurons, suggesting the modulation of neuroplasticity, which is thought to underlie the aftereffects.

Notably, neural entrainment and plasticity are not mutually exclusive and may be interdependent (Veniero et al. 2015). The aftereffects of tACS-induced plasticity changes may be related to the different parameters of the oscillations during stimulation. It is demonstrated that the intensity of alpha oscillation after stimulation is positively correlated with the power during alpha-tACS (Helfrich et al. 2014a); and there is a correlation between the change of gamma coherence during and after gamma-tACS stimulation in inter-hemispheric, that is, the stronger the online modulation intensity is, the stronger the aftereffects will be (Helfrich et al. 2014b). Therefore, these findings suggest a relationship between neural entrainment and plasticity, with stronger entrainment predicting stronger aftereffects.

### 2.1.3 Peripheral Nervous Mechanism

During electrical stimulation, the electric field strength observed on the skin may be 20–100 times stronger than in the cortex, easily exceeding the 4–6 V/m threshold for peripheral nerve stimulation (Asamoah et al. 2019). Previous research has shown that tACS could induce pathological tremor (Brittain et al. 2015) and physiological tremor (Mehta et al. 2015) in healthy participants, such as enhanced coherence between hand electromyographic signal and tACS waveform. A series of experiments have confirmed that tACS is accompanied by pathological (~3 Hz) and physiological (~10 Hz) tremor (Asamoah et al. 2019), but this effect does not result from direct cortical stimulation. Besides, local anesthetics were used to reduce the transcutaneous mechanism of tACS. The results showed that the entrainment effect of tremor was attenuated (physiological tremor) or completely disappeared (pathological tremor). In contrast, another experiment shows that tremor is still observed when tACS is applied to the contralateral arm to block its transcranial mechanism.

### 2.1.4 Retinal Stimulation

Phosphene is induced during tACS with 0.25–1 mA, and it will be maximally excited in the low beta and alpha bands of tACS (Kanai et al. 2008). Since endogenous alpha oscillations naturally increase in darkness, the shift toward light-body perception during alpha-tACS is recognized to reflect a direct interaction of endogenous oscillations in the visual cortex. The intensity of flickering light and visual cortical alpha increases when the electrodes of a-tACS are close to the eye (Schutter and Hortensius 2010); the low frequency of phosphine evoked in the dark reflects the increased sensitivity of retinal ganglion cells (Kar and Krekelberg 2012); and even when tACS is applied to the visual cortex, field strengths sufficient to induce visual flickering have also been observed in orbit (Laakso and Hirata 2013). However, it is believed by the current studies that the tACS-induced visual flashing is from retinal rather than cortex (Schwiedrzik 2009), and visual flicker may cause oscillations in visual cortex (such as steady-state visual-evoked potentials), which will affect perceptual and cognitive performance (Schutter 2016; Spaak et al. 2014). In this way, tACS can induce frequency-specific effects on endogenous oscillations in the visual cortex, for example, alpha oscillation entrainment, to indirectly act through sensory stimulation.

In conclusion, the mechanisms underlying tACS are still under investigation. To understand the online effects of tACS, it is relevant to explore the online effects of alternating current stimulation on cortical neurons and neural oscillations. Moreover, the mechanisms underlying the long-term effects still need to be thoroughly investigated. If neuroplasticity is proved to be the mechanism responsible for the after-effect of tACS, the plasticity could be used to establish the long-term effect of tACS and provide evidence to establish tACS as a therapeutic tool.

## 2.2 Effect Factors

### 2.2.1 Cerebral Intrinsic Rhythms of Neural Oscillations

The cerebral intrinsic rhythms of neural oscillations are a very important factor influencing the neural following effect of tACS, which has been confirmed in an experiment with eyes open and closed. Specifically, in the state of high endogenous alpha oscillation (eyes close), a weak entrainment effect is detected, while in the state of low alpha oscillation (eyes open), stronger amplitude is detected and the internal frequency shifts toward the stimulus frequency (Alagapan et al. 2016). At present, it is reported that exogenous tACS cannot lead to the following of intrinsic neural rhythms if nerve tissues do not exhibit rhythmic activity at the corresponding frequency prior to stimulation or if such activity is physiologically impossible. The concept may be extended to the (sub) harmonics of stimulation frequency. Another phenomenon is that the greater the difference between the internal and external frequencies is, the stronger the force required for the external rhythm to entrain the internal vibration will be (Herrmann et al. 2016; Thut et al. 2017). This phenomenon is known as "Arnold Tongue" (Pikovsky et al. 2001) (Fig. 6.2). It has been suggested that the presence of the Arnold Tongue in brain stimulation data can be used as an indicator of the entrainment effect of EEG data in humans (Antal and Herrmann 2016; Notbohm et al. 2016). Therefore, if the Arnold Tongue is detected in a stimulation experiment, it is a powerful indicator suggesting entrainment effect.

A recent meta-study found that selecting stimulation frequency based on individual EEG spectrum is more likely to enhance performance. Some studies indicated that the pre-stimulation alpha amplitude is an important parameter for actual aftereffects (Neuling et al. 2013). Studies on primates also confirmed that externally applied alternating current electric field successfully altered the firing pattern of monoplasts in the neocortex and hippocampus. The stronger the electric field is, the more cells adapted to its firing pattern will be. In awake rats, the same stimulation resulted in a greatly diminished effect when the animals did not show slow endogenous oscillations (Ozen et al. 2010). Therefore, when designing a tACS study, it is important to consider factors that can affect the level of endogenous neural oscillations (like ambient lighting) (Stecher et al. 2017).

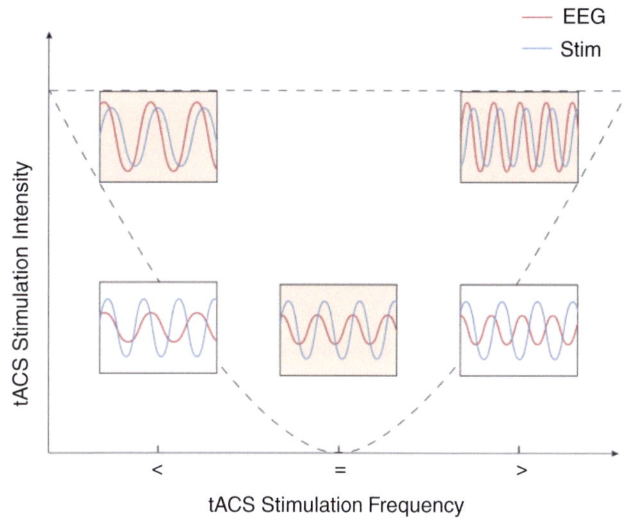


**Fig. 6.2** Arnold Tongue. The horizontal axis represents stimulation frequencies that are less (<), equal (=) or larger (>) than the endogenous frequency. The vertical axis represents the stimulation intensity. The embedded pictures describe the relationship between the endogenous (e.g. EEG, red) and stimulation signals (blue). During stimulation, endogenous oscillations are coupled to the stimulation signal (entrainment) in phase and frequency if within the range of Arnold Tongue (above this curve) and remain unmodulated outside the range of Arnold Tongue (below this curve). When the stimulation frequency is the same as the endogenous frequency, even stimulations with low intensity will lead to the coupling of the two oscillations (the picture in the lower middle). When the stimulation frequency does not match the endogenous frequency (the pictures in lower left and right), the stimulation of same intensity does not produce a following effect. However, if the intensity increases, these mismatched frequencies do produce a following effect (the pictures in upper left and right)

### 2.2.2 Stimulation Modes and Parameters

In addition to the influence of the intrinsic properties of the tissue stimulated, the magnitude of the tACS effect is closely related to various stimulation parameters such as the location, frequency, and phase of stimulation.

**Stimulation Frequency** tACS functions by tuning the intrinsic frequency of the brain using its specific stimulation frequency to obtain the target oscillations, resulting in enhanced activity. This phenomenon is expected given the entrainment and mediating effects of tACS (Thut et al. 2017). But in the few studies assessing the aftereffects of tACS on MEG/EEG, evidences for frequency-specific effects are found to be inconsistent and mixed. Although some of these findings may reflect true cross-frequency interactions, for example, gamma-tACS may inhibit endogenous alpha activity (Helfrich et al. 2016), non-frequency-specific aftereffects may also reflect the influence through other mechanisms or misinformation that are expected to be evenly distributed over the entire frequency band. Similarly, even if there is an evidence of frequency specificity, it may also be due to the bias of the analyzed spectral bands.

**Stimulation Intensity** It is currently recognized that stimulation with a typical intensity of 1 mA can generate an electric field within 0.1–0.3 V/m (Huang et al. 2017; Opitz et al. 2016). And another study tests the intracranial field strength when applying 2 mA tDCS to deep electrodes is 0.1 V/m, indicating a strong electric field in the cortical area, which may correspond to the values measured under tACS (Chhatbar et al. 2018). Thus, these articles validate the previous models of electric fields that were based on brain models derived from MR images (Datta et al. 2013; Neuling et al. 2012a). The finite element models predict an electric field strength of 0.3 V/m in the target area of the cortex, and this intensity is sufficient to affect neuronal firing patterns in animal experiments (Antal and Herrmann 2016).

**Stimulation Duration** Although tACS can modulate brain oscillations and cognitive processes, it is also found that short-term tACS did not achieve the expected effect. Studies in humans have demonstrated that the effects of tACS is obvious when stimulation durations are in the range of minutes; however, the effects are absent with shorter stimulation in animals, for instance, alpha tACS intermittently for 1 s will not produce aftereffects on amplitude or phase of the EEG signal (Struber et al. 2015). Therefore, different researches have focused on the stimulation duration to produce the aftereffects by tACS. In addition, the times of tACS will also affect its effects, for example, long-term treatment groups (21 and 28 days) increase spontaneous gamma band power and finally improved short-term memory function, while the short-term treatment group (7 days) do not (Wu et al. 2022).

**Phase Relationship** Applying opposed-phase gamma-tACS with a phase difference of 180° to the motor-sensitive cortex will reduce the proportion of perceived horizontal motion, while in-phase stimulation (with a phase difference of 0°) will not affect perception (Strüber et al. 2014). By comparing the inter-hemispheric coherence in gamma band in the EEG data recorded before and after tACS, it can be seen that the inter-hemispheric gamma coherence increases during the opposed-phase tACS (Strüber et al. 2014); however, they also point out that their coherence measure only represents the relative stability of gamma oscillations in both hemispheres. The same behavior appearing during the opposed-phase electrical stimulation was found in a follow-up study on focal high-density tACS, but the electrophysiological results show that the inter-hemispheric gamma coherence decreased with reversed tACS (Helfrich et al. 2014a).

## 3 Oscillatory Transcranial Direct Current Stimulation

Oscillatory transcranial direct current stimulation (otDCS) is also a rhythmic transcranial electrical stimulation. otDCS involves a current that has either an anodal or cathodal polarity with respect to the target electrode, i.e., the direction of current flow is either directed inward (toward the electrode) or outward (away from that electrode), respectively. The current intensity oscillates at a specific frequency either between zero and maximum amplitude (Marshall et al. 2006), or with an additional DC offset (Neuling et al. 2012b). Waveform shapes include trapezoids and sinusoids.

### 3.1 Mechanism and Effect Influencing Factors

It is currently believed that otDCS has a similar mechanism to tACS. At present, otDCS is mostly used during sleep regulation to modulate slow waves. An externally applied slow oscillating current produces an electric field intensity as low as 0.5–1 V/m, which is sufficient to acutely evoke oscillatory brain activity in rats (Binder et al. 2014; Ozen et al. 2010). Besides, application of slow oscillatory transcranial direct current stimulation (so-tDCS; frequency <1 Hz) during sleep provides a noninvasive method to enhance slow wave activity. Although this stimulation has also been found to lead to increased spindle activity, but it is unclear whether it alters the cross-frequency coupling between slow and fast spindles.

so-tDCS modulates dominant oscillations with eyes closed during wakefulness, highlighting the influence of brain state on the overall response to so-tDCS. In addition, memory improvements caused by so-tDCS have been repeatedly demonstrated (Ladenbauer et al. 2016; Marshall et al. 2006; Westerberg et al. 2018), while some studies do not find any or opposite effect (Bueno-Lopez et al. 2013; Paßmann et al. 2016).

The effectiveness of otDCS largely depends on the sensitivity of endogenous neural oscillations and the underlying network. Alternatively, plastic changes in learning or "task-evoked" plastic changes reflect network susceptibility. For example, it has been found that when inter-individual differences are assessed by memory quotient (MQ), with subjects divided into high and low score group, and when so-tDCS is performed during early NREM sleep at night, the result showed that only in subjects with high MQ, memory ability on the graph-paired association task is significantly increased after so-tDCS (Koo et al. 2018).

In addition, the use of different electrode sizes and applied current intensities has an impact on the efficacy of otDCS. The different effects of otDCS with different electrode sizes on EEG and behavior have been experimentally studied. The effects of so-tDCS on finger sequence tapping tasks (FSTT) and brain electrical activity were compared when so-tDCS was applied in small (0.50 cm$^2$, maximum current per electrode pair: 0.26 mA) and large (35 cm$^2$; 0.35 mA) electrodes at F3 and F4 electrodes during periods of quiet waking, respectively. And the results show that the performance of the finger sequence tapping task at retest after stimulation decrease when so-tDCS is applied in small electrodes. When applied in small electrodes, so-tDCS suppress alpha standard power in the parietal cortex occurring for 8–12 Hz band, whereas in large electrodes, power at individual alpha frequency (IAF) is suppressed (Koo-Poeggel et al. 2019). These results indicate the different effects of stimulation parameters on brain electrical activity.

In conclusion, the specific mechanism and efficacy of otDCS compared with other electrical stimuli remain unclear. Available findings suggest that its efficacy depends on the state of the brain at the time of stimulation (subject- and/or task-related) as well as stimulation parameters (such as electrode placement and applied current). Therefore, for future research on weak current stimulation, consistent measures of behavior and brain activity must be taken as an important step toward understanding variability and reproducibility in this field.

## 4 Random Noise Stimulation

Transcranial random noise stimulation (tRNS) is a special form of tACS. During tRNS, a low-intensity alternating current is applied, with the current intensity and frequency varying randomly. As with tACS, various forms of noise may be applied based on the frequency range. In most studies on tRNS, a spectrum ranges from 0.1 to 640 Hz (full spectrum) or 101 to 640 Hz (high-frequency stimulation) was used (Saiote et al. 2013). The probability function of noise current stimulation fits with a Gaussian curve or bell-shaped curve, with both mean and variance being zero. Ninety-nine percentage of the generated currents are within ±1 mA. In the frequency domain, all coefficients of the random sequence are similar in magnitude ("white noise").

In the current, the physiological mechanisms of tRNS have not been fully elucidated. Nor are there any animal studies on tRNS elucidating the role of this technique. Although the results of some studies have shown that high frequencies (like 140 Hz) can modulate brain activity, the high frequencies applied by tRNS are believed to have little effect on the degree of neuronal polarization because the neuronal membrane can act as a low-pass filter. In addition, experiments with alternating current electrical stimulation have measured neuronal polarization and estimated the coupling constant between electric field and polarization (Deans et al. 2007). As with tRNS, random fluctuations in external voltage may be sufficient to facilitate the subthreshold neural oscillations to exceed the trigger threshold. Hence, tRNS may just amplify neural activity that already exists prior to stimulation.

One potential online effect of tRNS may be related to the repeated opening of $Na^+$ channel (Terney et al. 2008). The $Na^+$ channel blocker Carbamazepine shows a tendency to inhibit MEP 5–60 min after tRNS (Chaieb et al. 2015). Interestingly, however, the NMDA receptor antagonist Dextromethorphan can block the effects of tDCS but has no significant effect on the increase in excitability induced by tRNS (Chaieb et al. 2015).

Moreover, the effect of tRNS may depend on other mechanisms, such as stochastic resonance (Pavan et al. 2019). Stochastic resonance is a process where brain oscillations below subthreshold are enhanced by the addition of noise and thus become suprathreshold. In short, it is a phenomenon in which, for example, noise is added to amplify signals too weak to exceed the threshold when cerebral neural oscillations are below the threshold. These subthreshold activities at the synaptic level are driven by oscillatory inputs received by neurons from other brain areas that are not strong enough to induce the generation of action potentials. If random noise is added, the sum of the two signals will exceed the threshold at some time. The frequency of the suprathreshold signal is determined by the existing subthreshold neural oscillations. It has been suggested that tRNS could enhance the synchrony of neural firing through the amplification of subthreshold oscillatory activity, thereby reducing the amount of endogenous noise (van der Groen et al. 2018; van der Groen and Wenderoth 2016). However, it is not clear how this process induces long-term changes in human brain activity.

tRNS applied to M1 can increase cortical excitability. For example, tRNS applied at 1 mA for 10 min on M1 induce aftereffects lasting up to 1–1.5 h and improved performance on implicit motor learning tasks (Terney et al. 2008). It has also been reported that high-frequency subdivision of the entire spectrum between 100 and 640 Hz is the functional cause of M1 excitability changes, superior to low-frequency (0.1–100 Hz) stimulation (Terney et al. 2008). In another study, M1 stimulation using tRNS enhanced motor skill learning compared to sham group (Prichard et al. 2014). tRNS played a more progressive role than the time course of anodic tDCS, and tDCS produced a large skill boost immediately after stimulus onset (Prichard et al. 2014).

Interestingly, the aftereffect of tRNS is intensity dependent. Lower intensity stimulation at 0.4 mA tRNS resulted in post-inhibitory effects comparable to those observed with 1 mA or with 140 Hz tACS at 0.4 mA for cathodic tDCS (Moliadze et al. 2012). This suggests that inhibitory neurons in M1 may have a lower threshold, at least for this stimulus. At the behavioral level, several studies have also demonstrated the role of it. Compared with anodal and cathodal tDCS as well as sham stimulation, high-frequency tRNS resulted in post-inhibitory effects while 1 mA tRNS increased MEP compared to sham stimulation and 0.2, 0.6 and 0.8 mA stimulation (Moliadze et al. 2012). This suggests that inhibitory neurons in M1 may have a lower threshold, at least for this stimulus. At the behavioral level, several studies have also demonstrated the role of it. Compared with anodal and ctDCS as well as sham stimulation, high-frequency tRNS, when applied to the visual cortex, significantly improve the abilities of healthy subjects in visual perceptual learning abilities (Fertonani et al. 2011). Interestingly, the effect of improvement will be more prominent if atDCS is applied prior to task performance and tRNS is applied during task performance (Pirulli et al. 2013). In contrast, applying tRNS to the right DLPFC in a classification learning task impairs classification learning ability (Ambrus et al. 2011). These results suggest that tRNS can induce long-term positive and negative changes in cognition and brain function.

Currently, there are few studies on both the mechanisms and effects of tRNS. Whether tRNS can be used as a promising neuromodulation approach remains to be further studied in detail.

## 5 Multichannel Transcranial Electrical Stimulation

Transcranial electrical stimulation is typically designed to be single channel, i.e., a stimulation electrode with a reference electrode. Multichannel stimulation is a novel setup with advantages in terms of stimulation focus, depth, and intensity.

### 5.1 Simultaneous Multichannel Stimulation

Since transcranial electrical stimulation is applied transcutaneously, a large fraction (~75%) of the current applied to the scalp is attenuated by the soft tissues and skull (Vöröslakos et al. 2018). This percentage is even higher when the stimulation electrodes are very close together (short distance from anode to cathode), as the current is more likely to be shunted on the scalp (Faria et al. 2011; Miranda et al. 2006). Multichannel high-density stimulation has been used to optimize the focus of stimulation and enhance spatial accuracy by increasing the number of stimulation electrodes (Dmochowski et al. 2011). With this approach, the stimulation area is more focused than that of large (7 × 5 cm) conventional conductive rubber electrodes, increasing the number of stimulating electrodes to improve spatial accuracy (Dmochowski et al. 2011). The most common simultaneous multichannel high-density stimulation mode is the "4 + 1," where the central electrode is surrounded by four reference electrodes. This setup limits the area of stimulation to that within the four surrounding electrodes, with current flowing between the central electrode connected to one "pole" of the stimulator and the surrounding electrodes connected to the other "pole" of the stimulator. They define high density-TES (HDTES) as any method that delivers a given waveform through multiple electrodes simultaneously, independent of the waveform used or the location of the electrodes.

Currently, the stimulation mode that best fits the target can be determined based on the stimulation target, and the best method of electrode placement can be found using effective methods of numerical optimization (Dmochowski et al. 2011). Stimulation may be so set that it reaches the highest intensity at the stimulation target or it is limited within a restricted space. For example, the optimal electrode fixation montage which is achieved to focused stimulation has 6 electrodes 1 anode and 5 surrounding cathodes. While under settings which is intended to maximize the target intensity, the optimal electrode montage is produced with only two electrodes (Huang and Parra 2019). Clearly, there is a fundamental trade-off between focusing and intensity, and in general, increasing the distance between the anode and cathode electrodes will (a) reduce the proportion of shunt on the scalp, (b) increase the depth and magnitude of the intracranial magnetic field, but (c) cause the electric field in the brain to be more dispersed (non-focal).

### 5.2 Nonsimultaneous Multichannel Stimulation

Another latest way to use multiple electrodes for TES is using interferential AC waveforms (Grossman et al. 2017). This technique is commonly used in the clinical practice of physical therapy, known as "interferential current therapy" (Goats 1990). In the context of brain stimulation, this idea was first proposed half a century ago. Interferential stimulation adopts two different stimulation channels, applying two slightly different frequencies that are beyond the physiological domain. The difference between these two frequencies is the target frequency and this method is interferential cross-stimulation (Grossman et al. 2017). The basic idea of interferential stimulation is that two sinusoidal waveforms with similar frequencies, once superimposed, will produce a sinusoidal waveform modulated by a slower "beat" frequency. For example, a sine wave of 2000 Hz superimposed to a sine wave of 2010 Hz will be modulated with an amplitude of 10 Hz. The strength of this modulation ("modulation depth") is defined as the difference between the maximal and minimal amplitudes of the fast sine waveform.

This is a method which believes that the deep brain can be stimulated in a noninvasive manner (at least for mice) (Grossman et al. 2017). It applies high-frequency (≥1000 Hz) oscillatory electric field at multiple locations on the scalp. As all neurons (even in non-primates) seem to apply a low-pass filter to electrical signals from the neural membrane (i.e., neurons do not discharge from stimulations of extremely high frequency), high-frequency stimulation does not directly cause neurons to discharge. Boyden's team creatively referred to this principle and applied two interfering electric fields on each side of the mouse head, with each electric field differing by only 10 Hz (sinusoidal currents of 2010 and 2000 Hz). They found that the effective stimulation frequency was the difference between the two high frequencies (10 Hz) at the junction of the electric fields. They demonstrated that this method could stimulate deep brain areas, such as the hippocampus, in mice in a noninvasive manner. Nevertheless, there are no studies investigating the use of TI in human or animal disease models so far.

Another nonsimultaneous multichannel stimulation has recently been proposed, in which currents are distributed in space, known as "intersectional short pulse" (ISP). The basic concept is to rapidly switch the current between different electrode pairs. That is, the current passing through multiple electrodes is distributed by temporal multichannel multiplexing (i.e., rapid switching of current between multiple electrode pairs). In this way, currents can be distributed on the scalp to minimize the skin sensation. This kind of increase in fields will produce stronger fields wherever the individual field distributions of each electrode pair overlap or "intersect." For example, suppose current of 6 mA is to be applied through 6 pairs of electrodes (6 + 6 = 12 electrodes in total). At any time point, current is only applied to a single electrode pair. The time consumed by each electrode pair conducting current is 1/6. Thus, when averaged over time, only an average of 1 mA of current is applied to each pair. In one case, the current was switched every 60 μs (at a rate of 16.6 kHz). The basic argument is that the field generated by each pair will polarize the cytomembrane of neurons in the brain. However, as the time constant of the passive membrane is slow (about 30 ms), the polarization will be effectively low-pass filtered. The cytomembrane will blur the time of fast pulse and thus effectively show its response to a continuous current stimulation of 1 mA, the direct current component of a single-phase 6 mA pulse.

The amplitude of the pulses can be modulated at a low rate. The plus/minus characteristics may even be changed. In this way, any desired waveform (slower than 1/30 ms = 33 Hz) can be achieved with ISP stimulation, such as the stimulation that has been demonstrated in a sinusoidal stimulation of 10 Hz, equivalent to 10 Hz tACS.

The intensity was increased by twofold by applying the same maximum current of 1 mA to each electrode. If the limiting factor for sensation is the current at a single electrode, then we have increased the intensity without impacting sensation. Clearly, the electrodes must be far enough apart to avoid cumulative effects on sensation near the electrodes. It has been proven by empirical evidences that the fast pulses, equivalent to a continuous current of 1 mA, may result in lower skin sensation (Paneri et al. 2016). Thus, ISP stimulation contributes to higher overall stimulation intensities by combining two different mechanisms. One of them is to distribute currents in space (equivalent to using a large sponge during conventional TES or distributing the currents across many HD electrodes); and another is to reduce skin sensation by applying high-frequency pulse current.

## 6 Electroconvulsive Stimulation

Electroconvulsive stimulation is an essential method in contemporary psychiatry for treating many patients with severe emotional and mental disorders. It has been applied for more than 80 years with the development of narcotics and muscle relaxants, the side effect of electroconvulsive therapy (ECT) had been reduced and the safety and tolerability increased. At present, ECT has been widely applied to the treatment of neuropsychiatric disorders. In particular, electroconvulsive stimulation is the treatment of choice for patients with severe mental disorders (e.g., severe depression, major mania, and major schizophrenia). Despite the development of new techniques of brain stimulation and new pharmacological agents, there is not yet a therapy that can achieve the efficacy of electroconvulsive stimulation in patients with severe mental disorders.

Due to the lack of knowledge and understanding of modern techniques of ECT, the side effects of it such as memory loss, cognitive impairment, headache, etc., which induced the controversy of ECT. Many previous studies have shown that most patients treated with ECT experienced cognitive impairment and memory loss of varying degrees, mainly manifested as retrograde amnesia and anterograde amnesia of recent events. However, in recent years, it has been gradually found and confirmed that these side effects are mostly transient as the memory can be gradually restored and the headache can disappear soon after treatment. However, due to insufficient evidence on the efficacy and safety of ECT in pregnant and adolescent patients, the choice made based on the benefit-risk ratio varies among the individuals, which requires a rigorous assessment by the attending physicians and adequate weighing by the patients and their families. ECT may also be an alternative option especially for pregnant women and adolescents with no response to drug treatment, serious diseases, high disability rate, and low quality of life. Recent data on the safety and tolerability of electroconvulsive stimulation also strongly supports its widespread application in appropriately selected patients (Kellner et al. 2020). For its extremely low mortality rate (2.1/100,000), electroconvulsive stimulation becomes one of the safest procedures performed under general anesthesia. As to the treatment standard, electroconvulsive stimulation is mainly used to treat severe and treatment-resistant depression, or other indications such as schizophrenia, mania, and catatonia.

## 6.1 Depression

In most European and North American countries, major depressive episodes, either unipolar or bipolar disorders, remain the primary indication for the therapy of ECT stimulation. The efficacy and remission rates in clinical trials have been consistently high, with remission rates typically above 60% and efficacy rates above 70%. Most depression patients treated with ECT stimulation have tried multiple medications [e.g., selective serotonin reuptake inhibitors (SSRIs)] and psychotherapy, which all turn out to be ineffective. The clinical urgency, often marked by strong suicidal tendencies and drive, usually compels the recommendation of ECT stimulation therapy. It has been shown that ECT improves depressive symptoms mainly by inducing the regeneration of neuronal cells, increasing hippocampal volume, and improving the connectivity between the default mode network and hippocampus. A recent health economic study conducted in the United States have shown that it is cost-effective to consider the use of ECT therapy after only two failed trials of antidepressants (Pagnin et al. 2004). Therefore, ECT should be considered as early as possible during the treatment of serious emotional or psychiatric disorders.

## 6.2 Schizophrenia

Schizophrenia is one of the most common indications for ECT in the world. When adopted to treat schizophrenic patients with acute exacerbations, ECT stimulation is mainly intended for positive symptoms while having small effects on negative symptoms. In China, ECT is used more aggressively for patients with schizophrenia, but in other European and American countries, including the United States, ECT is usually reserved for patients with significant affective symptoms, or with well responded to ECT, or with neuroleptic malignant syndrome. Recent literatures have focused on ECT as an add-on treatment for patients with treatment-resistant schizophrenia, including those who are resistant to clozapine. Although a large number of clinical trials, case series and case reports support that ECT is safe and effective in patients with schizophrenia (patients between 30–36 years). Evidences are insufficient to convince the US FDA that ECT should be a standard treatment for schizophrenia. ECT is positively effective in the intermediate clinical response of patients with treatment-resistant schizophrenia. However, for other outcomes, there are no clear and convincing advantages or disadvantages to adding ECT to the standard of care. The early application of electroconvulsive stimulation for schizophrenia, even in first episode of psychosis, is controversial. A recent study on the treatment of treatment-resistant schizophrenia by a combination of clozapine with right-sided monopolar ECT reveals improvements in psychiatric symptoms and cognitive function, suggesting that ECT is a safe and reliable option for treating patients with treatment-resistant schizophrenia. The idea that the application of ECT early in the course of the disease may benefit the subsequent long-term course is controversial.

## 6.3 Mania

ECT has mood stabilizing effects, which shows good effects in treating acute mania and treatment-resistant mania. It is usually adopted for patients with mania whose symptoms are not adequately controlled by multiple medications. It has been found that lithium carbonate combined with ECT can rapidly control acute episodes of manic symptoms and can be timely applied to control acute manic episodes with impulsive behavior before the drugs take effect when there are no contraindications. Whether the efficacy of lithium carbonate combined with ECT for the control of manic episodes is superior to that of lithium carbonate alone is still controversial, requiring further studies. As most manic episodes can be successfully controlled with drug treatment, it is uncommon to apply ECT to manic patients. However, for particularly severe and potentially fatal delirious mania, ECT may be considered as an indication for emergency treatment.

## 6.4 Catatonia

Regardless of etiology or psychiatric comorbidity, patients with catatonia generally respond well to ECT. Patients with catatonia mostly present with complete motion and behavioral syndromes, with some symptoms of catatonia. After no response to benzodiazepines and other drugs, ECT may be preferred as a first-line therapy. Given the severity of many patients with catatonia, stronger forms of electroconvulsive stimulation (e.g., placement of moderate to high doses of bilateral electrodes) may also be preferred. The US FDA lists catatonia as one of the two diseases which can be treated with ECT.

## 7 Transcranial Electrical Stimulation Motor-Evoked Potential

### 7.1 Definition

Motor-evoked potentials (MEPs) are electrical signals recorded in nerve tissues or muscles after stimulation of central motor pathways, the emergence of which has refined clinical electrophysiological techniques for assessing the nervous system, especially for intraoperative neurophysiologic monitoring (Legatt et al. 2016). The transcranial electrical stimulation motor-evoked potential (tESMEP) is used to determine the function of the corticospinal tract and its downstream pathways by the transcranial electrical stimulation (tES) applied to the brain and the recording of neural or myogenic electrical activity. It is commonly used in clinical practice for intraoperative monitoring of spinal cord surgery to determine whether the function of spinal cord is impaired.

The electrical activity recorded when activation of corticospinal tract is induced by the electrical stimulation of the cerebral cortex includes D waves and I waves. The D waves reflect the direct activation of the axons of the pyramidal cells in the stimulated cerebral cortex, while the I waves reflect the indirect activation of the pyramidal cells through the synaptic transmission between neurons in the activated cortex. There can be multiple I waves with approximately equal intervals, reflecting the number of synapses between cortical neurons activated after initial stimulation and the pyramidal cells originating from corticospinal tracts (MacDonald et al. 2013). As I waves are mediated by synaptic activity, they are significantly inhibited by the level of anesthesia during surgery. However, D waves are persistent and can be recorded along the corticospinal tract, so that when used for intraoperative neurophysiological monitoring, they can be recorded from the caudal end of spinal cord to areas with possible risks during surgery. The amplitude depends on the number of rapidly conducting corticospinal tracts in the spinal cord that have been recorded. Because of the uneven distribution of corticospinal tracts in the spinal cord, D waves are recorded with greater amplitude in the cervical cord than in the thoracic cord. Due to the small number of remaining fibers below the tenth vertebra, D wave can't usually be detected (Legatt et al. 2016).

When the cerebral cortex is stimulated to cause muscle contraction, compound muscle action potentials (or myogenic MEPs) can be recorded in multiple muscles. Due to the synapses at the anterior horn cell, myogenic motor-evoked potentials are sensitive to anesthesia, especially for inhalation anesthetics. To this end, anesthetic drugs will be limited when myogenic MEPs are used for intraoperative monitoring. In addition, a cascade of stimuli is required to ensure that myogenic MEPs can be recorded under anesthesia because the postsynaptic potentials of the anterior horn cell can only trigger the threshold potential for muscle contraction when being added up (MacDonald et al. 2013).

Both D waves and myogenic MEPs can be used for intraoperative neurophysiological monitoring. tES produces D-waves under the stimulation site and myogenic MEPs contralateral to the stimulation site.

### 7.2 Parameter Settings

To monitor MEPs in the upper extremities, the stimulating electrodes can be placed at either C1/C2 or C3/C4 of the scalp. Since C3/C4 is closer to the facial motor cortex, mandible muscles and trigeminal nerve stimulation applied here will produce stronger chewing movement. C1/C2 is hence more appropriate than C3/C4. Stimulation electrodes can be placed at CZ/FZ on the scalp to monitor the MEPs in the lower extremities (Legatt et al. 2016). However, the optimal placement site still depends on different patients and surgical conditions. Therefore, it is necessary to test different sites and select the appropriate stimulation site for patients.

As to the monitoring of myogenic MEPs, tES with multiple pulses is required because a single D wave is insufficient for anterior horn cell to reach the threshold potential under surgical anesthesia. Therefore, pulse interval is a factor to be considered. If the pulse interval is too long, the postsynaptic potentials cannot overlap, resulting in the failure to produce temporal summation effect; if the pulse interval is too short, the anterior horn cells are still in a refractory phase when the next stimulus arrives. The most appropriate pulse interval for monitoring myogenic MEPs is 2–4 ms (i.e., the repetition frequency within the pulse is 250–500 Hz) (Deletis 2005).

Stimulators with either constant current or constant voltage can be used for tES. Different machines have different ranges of pulse width and stimulation intensity. Constant voltage stimulators are more commonly used. A short pulse width of 0.05 ms can be used, and the current can reach 1500 mA. Or a long pulse width of 0.5 ms can be used, and the current can be 240 mA. Each sequence generally has 3–9 pulses, and the pulse interval is often 1–5 ms (David and MacDonald 2006). Intraoperative neurophysiological monitoring usually starts with the establishment of a set of standard stimulation parameters (pulse width, stimulation intensity, number of pulses per sequence, and pulse interval) under anesthesia. Parameters are then adjusted according to different patient conditions and surgical needs.

D waves can be recorded by paired electrodes placed near the epidural or subdural spinal cord. The electrodes can be placed by a percutaneous puncture needle or directly by the surgeon in the surgical area. Myogenic MEPs can be recorded by electrodes placed on the muscle of both limbs (David and MacDonald 2006).

### 7.3 Alarm Range

The measurements of wave amplitude can be used to evaluate D waves and myogenic MEPs. The amplitude of both D waves and myogenic MEPs are heights between peaks and troughs (see Fig. 6.3). Similar to that of the sensory-evoked potentials for intraoperative neurophysiological monitoring, the most common alarm criterion of D waves is a 50% decrease in amplitude, which is an inappropriate alarm criterion for myogenic MEPs when they are used for monitoring the spinal cord due to the inherent variability even in the absence of spinal cord injury. Otherwise, it might result in a false-positive result. Currently, there is no consensus on the alarm criteria for monitoring myogenic MEPs. One alarm criterion widely used in the surgery of spinal cord tumor is the disappearance of myogenic MEPs recorded from the muscles of lowest threshold. Nevertheless, it is preferable to inform the surgical team members when the percentage decrease in amplitude exceeds 50% (e.g., 75%, 80%, 90%) rather than waiting until the amplitude disappears completely because a significant decrease in amplitude can occur before the complete disappearance of the electrical records. However, the choice of an alarm criterion of 80% still produces many false-positive results in spinal surgery (Legatt et al. 2016).

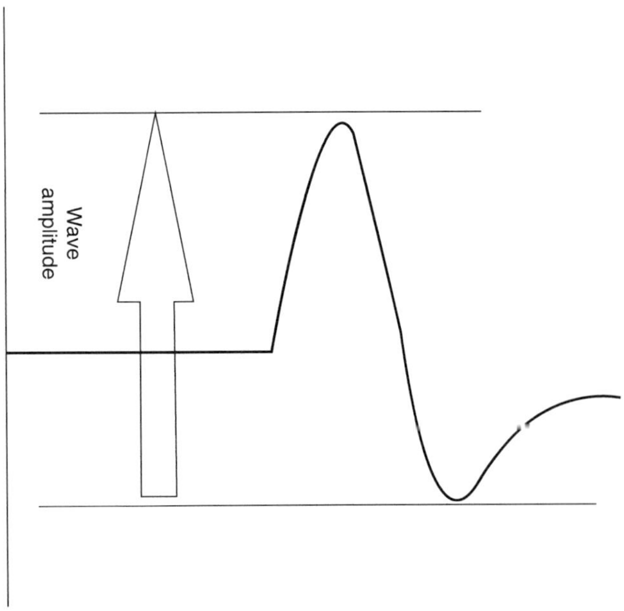


**Fig. 6.3** Schematic diagram of wave amplitude

### 7.4 Safety

Although the current density of tES is well below safe levels, high extracranial current density can cause temporalis contraction and forced jaw closure, which in turn can cause oral injuries that can be prevented by placing a soft bite block in the mouth. Muscle contraction in the axial direction and extremities caused by electrical stimulation may also cause damage to patients. tES, especially prolonged repetitive pulsed stimulation, can induce seizures, which is the most common side effect of direct cortical stimulation. Arrhythmia is very rare and may be caused by other factors. Current contraindications associated with tESMEP include epilepsy, cortical lesions, convex skull defects, elevated intracranial pressure, cardiac disease, vascular dissection, pacemakers, or other implanted biomedical devices. The presence of inexplicable intraoperative seizures and arrhythmias are indications for terminating the tESMEP. In general, however, experts have come to a consensus that the dominance associated with the monitoring of tESMEP far outweigh the possible risks (MacDonald 2002).

## 8 Direct Cortical Motor-Evoked Potential

Motor-evoked potentials (MEPs) are one of the intraoperative neurophysiological monitoring methods widely applied to surgeries in the cerebral motor cortex. It is intended to understand the neurological functions of the motor conduction system, monitor the integrity of motor conduction pathways, maximize resection of foci, and protect motor function. The technique of intraoperative direct electrical stimulation, not only allowing for the intraoperative locating of cortical functions but also enabling the functional monitoring and tracking of subcortical nerve conduction pathway, is a gold standard for locating functional areas in clinical practice, which can improve the complete resection rate of foci while reducing long-term neural dysfunction.

### 8.1 Stimulation Mode of Direct Cortical Motor-Evoked Potential

The direct cortical motor-evoked potential (dcMEP) has been applied to monitor the functions of the corticospinal tract during craniotomy. Compared to the traditional transcranial motor-evoked potential, dcMEP has many advantages, such as not requiring patient movement, thus allowing for continuous monitoring. In addition, dcMEP allows for more focused superficial stimulation. Individual contacts of strip or grid electrodes directly on the cortex realizes more focused stimulation effect by reducing stimulation intensity, stimulation current, and propagation. There are two common stimulation modes, namely bidirectional square wave pulses with a pulse frequency of 60 Hz and a wave width of 1 ms. In this mode, direct electrical stimulation is applied to motor cortex to elicit the expected motor reaction of the patient's contralateral limb or to record the corresponding positive compound muscle action potential (CMAP). The other mode is high-frequency short-train stimulation, with the interstimulus interval set as 2–4 ms, stimuli interval as 0.05–0.5 ms, 3–8 stimuli per train and current intensity <20 mA. Other studies suggest the relevant stimulation parameters can be 40–120 V, 500–1000 Hz, 5–9 pulses/s, 1–3-ms interstimulus interval, monopolar stimulation, and 50-μs pulse width (Tate et al. 2013). The stimulation intensity is gradually adjusted according to the response of MEP. Taniguchi et al. first described the principle of direct stimulation of the cortex in train by short-duration high-frequency multiple pulses (Taniguchi et al. 1993), and they calculated a total charge of <8 μC per train, compared to 600–5000 μC/train for the conventional Penfield electrical stimulation technique with a frequency of 60 Hz. Compared with the motor functions induced by the traditional cortical electrical stimulation, this technique is safer and less likely to cause intraoperative seizures in patients. The probability of causing seizures by dcMEP in some current clinical studies is almost zero, while the success rate of inducing motor potentials is 97%. Szelényi et al. reported the application of both techniques for measuring MEPs (Szelenyi et al. 2011). The incidence of seizures caused by dcMEP was 1.2%, while that caused by the Penfield electrical stimulation technique was 9.5%, suggesting the higher safety of dcMEP. In most European study sites, this technique starts to replace the Penfield technique. Records are made by inserting a pair of needle electrodes into the corresponding muscle to be monitored.

## 8.2 Application of dcMEP in Surgery

Continuous MEP monitoring is possible during cerebral cortical resection, mainly for monitoring the integrity of motor pathways in real time during the tumor surgery involving motor cortices and motor conduction pathways. The stimulation waveform is a biphasic square wave pulse with the same stimulation parameters as that of traditional transcranial motor-evoked potential. The current starts from 10 mA and increases in an increment of 5 mA to elicit a stable CMAP of the patient's contralateral limb muscles as the baseline. Transcortical electrical stimulation at a frequency of 1 time/min is continuously applied during tumor resection, and the obtained CMAP is compared with the baseline. The stimulation frequency can be increased to 5–10 times/min during surgical operations in deep intracranial areas adjacent to the brainstem and the pyramidal tract. The parameters adopted by different sites vary slightly, so do the stimulation parameters during adult and pediatric surgeries.

In resections where the tumor is located in the motor cortex, the challenge is how to monitor the integrity of the patient's motor pathway and the extent of its damage intraoperatively, which requires the real-time monitoring of motor conduction by intraoperative dcMEP (Deletis 2005). Yamamoto et al. applied MEP to 32 cases of glioma in motor cortex and demonstrated that it could satisfactorily monitor the intraoperative motor pathway conduction (Yamamoto et al. 2004). Moiyadi et al. successfully detected MEPs in 26 of 31 patients with motor cortical tumors and found that dcMEP was better than tESMEP in predicting poor neurological prognostic outcomes (Moiyadi et al. 2018). Some studies show that the combined application of MEPs and SEP monitoring during aneurysms resection can distinguish between cortical and subcortical ischemia, thus allowing for the prevention of pure motor hemiparesis due to clamping of lenticulostriate arteries, in which the SEP remains unchanged while the MEP disappears (Maruta et al. 2012). Usually, patients can be protected from hemiparesis by the repositioning of aneurysm clips. Since the motor cortex innervating the lower extremities is in the lobulus paracentralis, deep in the medial face of longitudinal fissure and far from the scalp, it is difficult to induce MEPs in the lower extremities intraoperatively by transcranial stimulation in the state of cortical inhibition by general anesthesia. The conditions of lower extremities can only be reflected indirectly by the MEPs of the upper extremities. In contrast, direct cortical stimulation allows for the insertion of the sheet electrode for stimulation into the lobulus paracentralis along the longitudinal fissure to directly elicit the MEPs in the lower extremities, which provides a reliable basis for judging the patient's postoperative motor functions of low extremities.

In conclusion, for lesions in the motor cortex, it is important to protect the primary motor cortex and corticospinal tract to avoid permanent motor function deficits while maximizing lesion resection under the premise of ensuring the postoperative motor functions of patients, which (Liu Wen et al. 2006) is the goal pursued by surgeons. dcMEPs should be applied to identify the motor cortex. Continuous dcMEPs monitoring should be performed to ensure the integrity of cortical functional areas and corticospinal tract, which can facilitate the surgical resection of lesions in motor cortices.

## References


Al QW, Abubaker M, Kvasnak E (2022) Working memory and transcranial-alternating current stimulation-state of the art: findings, missing, and challenges. Front Psychol 13:822545

Alagapan S, Schmidt SL, Lefebvre J et al (2016) Modulation of cortical oscillations by low-frequency direct cortical stimulation is state-dependent. PLoS Biol 14(3):e1002424

Ambrus GG, Zimmer M, Kincses ZT et al (2011) The enhancement of cortical excitability over the DLPFC before and during training impairs categorization in the prototype distortion task. Neuropsychologia 49(7):1974–1980

Anastassiou CA, Perin R, Markram H (2011) Ephaptic coupling of cortical neurons. Nat Neurosci 14(2):217–223

Antal A, Herrmann CS (2016) Transcranial alternating current and random noise stimulation: possible mechanisms. Neural Plast 2016:3616807

Asamoah B, Khatoun A, Mc Laughlin M et al (2019) tACS motor system effects can be caused by transcutaneous stimulation of peripheral nerves. Nat Commun 10(1):266

Batsikadze G, Moliadze V, Paulus W et al (2013) Partially non-linear stimulation intensity-dependent effects of direct current stimulation on motor cortex excitability in humans. J Physiol 591(7):1987–2000

Bikson M, Inoue M, Akiyama H et al (2004) Effects of uniform extracellular DC electric fields on excitability in rat hippocampal slices in vitro. J Physiol 557(Pt 1):175–190

Bikson M, Datta A, Rahman A et al (2010) Electrode montages for tDCS and weak transcranial electrical stimulation: role of "return" electrode's position and size. Clin Neurophysiol 121(12):1976–1978

Bikson M, Truong DQ, Mourdoukoutas AP et al (2015) Modeling sequence and quasi-uniform assumption in computational neurostimulation. Prog Brain Res 222:1–23

Binder S, Berg K, Gasca F et al (2014) Transcranial slow oscillation stimulation during sleep enhances memory consolidation in rats. Brain Stimul 7(4):508–515

Bindman LJ, Lippold OC, Redfearn JW (1964) The action of brief polarizing currents on the cerebral cortex of the rat (1) during current flow and (2) in the production of long-lasting after-effects. J Physiol 172:369–382

Boggio PS, Ferrucci R, Rigonatti SP (2006) Effects of transcranial direct current stimulation on working memory in patients with Parkinson's disease. J Neurol Sci 249(1):31–38

Brittain JS, Cagnan H, Mehta AR et al (2015) Distinguishing the central drive to tremor in Parkinson's disease and essential tremor. J Neurosci 35(2):795–806

Brunoni AR, Moffa AH, Fregni F et al (2016) Transcranial direct current stimulation for acute major depressive episodes: meta-analysis of individual patient data. Br J Psychiatry 208(6):522–531

Bueno-Lopez A, Eggert T, Dorn H et al (2013) No effects of slow oscillatory transcranial direct current stimulation (tDCS) on sleep-dependent memory consolidation in healthy elderly subjects. Brain Stimul 6(6):938–945

Cappon D, Jahanshahi M, Bisiacchi P (2016) Value and efficacy of transcranial direct current stimulation in the cognitive rehabilitation: a critical review since 2000. Front Neurosci 10:157

Chaieb L, Antal A, Paulus W (2015) Transcranial random noise stimulation-induced plasticity is NMDA-receptor independent but sodium-channel blocker and benzodiazepines sensitive. Front Neurosci 9:125

Chan CY, Nicholson C (1986) Modulation by applied electric fields of Purkinje and stellate cell activity in the isolated turtle cerebellum. J Physiol 371:89–114

Chan CY, Hounsgaard J, Nicholson C (1988) Effects of electric fields on transmembrane potential and excitability of turtle cerebellar Purkinje cells in vitro. J Physiol 402:751–771

Chhatbar PY, Kautz SA, Takacs I et al (2018) Evidence of transcranial direct current stimulation-generated electric fields at subthalamic level in human brain in vivo. Brain Stimul 11(4):727–733

Csercsa R, Dombovári B, Fabó D et al (2010) Laminar analysis of slow wave activity in humans. Brain 133(9):2814–2829

Datta A, Zhou X, Su Y et al (2013) Validation of finite element model of transcranial electrical stimulation using scalp potentials: implications for clinical dose. J Neural Eng 10(3):036018

David B, MacDonald MD (2006) Intraoperative motor evoked potential monitoring: overview and update. J Clin Monit Comput 20(5):347–377

Deans JK, Powell AD, Jefferys JG (2007) Sensitivity of coherent oscillations in rat hippocampus to AC electric fields. J Physiol 583(Pt 2):555–565

Deletis V (2005) What does intraoperative monitoring of motor evoked potentials bring to the neurosurgeon? Acta Neurochir 147(10):1015–1017

Dmochowski JP, Datta A, Bikson M et al (2011) Optimized multi-electrode stimulation increases focality and intensity at target. J Neural Eng 8(4):046011

Faria P, Hallett M, Miranda PC (2011) A finite element analysis of the effect of electrode area and inter-electrode distance on the spatial distribution of the current density in tDCS. J Neural Eng 8(6):066017

Fehlings MG, Tator CH (1992) The effect of direct current field polarity on recovery after acute experimental spinal cord injury. Brain Res 579(1):32–42

Fertonani A, Pirulli C, Miniussi C (2011) Random noise stimulation improves neuroplasticity in perceptual learning. J Neurosci 31(43):15416–15423

Fregni F, Gimenes R, Valle AC et al (2006) A randomized, sham-controlled, proof of principle study of transcranial direct current stimulation for the treatment of pain in fibromyalgia. Arthritis Rheum 54(12):3988–3998

Fritsch B, Reis J, Martinowich K et al (2010) Direct current stimulation promotes BDNF-dependent synaptic plasticity: potential implications for motor learning. Neuron 66(2):198–204

Gluckman BJ, Neel EJ, Netoff TI et al (1996) Electric field suppression of epileptiform activity in hippocampal slices. J Neurophysiol 76(6):4202–4205

Goats GC (1990) Interferential current therapy. Br J Sports Med 24(2):87–92

Grossman N, Bono D, Dedic N et al (2017) Noninvasive deep brain stimulation via temporally interfering electric fields. Cell 169(6):1029–1041.e16

Hampstead BM, Sathian K, Bikson M et al (2017) Combined mnemonic strategy training and high-definition transcranial direct current stimulation for memory deficits in mild cognitive impairment. Alzheimers Dement (N Y) 3(3):459–470

Helfrich RF, Schneider TR, Rach S et al (2014a) Entrainment of brain oscillations by transcranial alternating current stimulation. Curr Biol 24(3):333–339

Helfrich RF, Knepper H, Nolte G et al (2014b) Selective modulation of interhemispheric functional connectivity by HD-tACS shapes perception. PLoS Biol 12(12):e1002031

Helfrich RF, Herrmann CS, Engel AK et al (2016) Different coupling modes mediate cortical cross-frequency interactions. NeuroImage 140:76–82

Herrmann B, Henry MJ, Haegens S et al (2016) Temporal expectations and neural amplitude fluctuations in auditory cortex interactively influence perception. NeuroImage 124(Pt A):487–497

Ho KA, Taylor JL, Chew T et al (2016) The effect of transcranial direct current stimulation (tDCS) electrode size and current intensity on motor cortical excitability: evidence from single and repeated sessions. Brain Stimul 9(1):1–7

Horvath JC, Forte JD, Carter O (2015) Evidence that transcranial direct current stimulation (tDCS) generates little-to-no reliable neurophysiologic effect beyond MEP amplitude modulation in healthy human subjects: a systematic review. Neuropsychologia 66:213–236

Hoy KE, Emonson MR, Arnold SL et al (2013) Testing the limits: investigating the effect of tDCS dose on working memory enhancement in healthy controls. Neuropsychologia 51(9):1777–1784

Huang Y, Parra LC (2019) Can transcranial electric stimulation with multiple electrodes reach deep targets? Brain Stimul 12(1):30–40

Huang Y, Liu AA, Lafon B et al (2017) Measurements and models of electric fields in the in vivo human brain during transcranial electric stimulation. elife 6:e18834

Jamil A, Batsikadze G, Kuo HI et al (2017) Systematic evaluation of the impact of stimulation intensity on neuroplastic after-effects induced by transcranial direct current stimulation. J Physiol 595(4):1273–1288

Kabakov AY, Muller PA, Pascual-Leone A et al (2012) Contribution of axonal orientation to pathway-dependent modulation of excitatory transmission by direct current stimulation in isolated rat hippocampus. J Neurophysiol 107(7):1881–1889

Kanai R, Chaieb L, Antal A et al (2008) Frequency-dependent electrical stimulation of the visual cortex. Curr Biol 18(23):1839–1843

Kar K, Krekelberg B (2012) Transcranial electrical stimulation over visual cortex evokes phosphenes with a retinal origin. J Neurophysiol 108(8):2173–2178

Kar K, Krekelberg B (2014) Transcranial alternating current stimulation attenuates visual motion adaptation. J Neurosci 34(21):7334–7340

Kar K, Duijnhouwer J, Krekelberg B (2017) Transcranial alternating current stimulation attenuates neuronal adaptation. J Neurosci 37(9):2325–2335

Keeser D, Meindl T, Bor J et al (2011) Prefrontal transcranial direct current stimulation changes connectivity of resting-state networks during fMRI. J Neurosci 31(43):15284–15293

Kellner CH, Obbels J, Sienaert P (2020) When to consider electroconvulsive therapy. Acta Psychiatr Scand 141:304–315

Kim JH, Kim DW, Chang WH et al (2014) Inconsistent outcomes of transcranial direct current stimulation may originate from anatomical differences among individuals: electric field simulation using individual MRI data. Neurosci Lett 564:6–10

Koo PC, Mölle M, Marshall L (2018) Efficacy of slow oscillatory-transcranial direct current stimulation on EEG and memory—contribution of an inter-individual factor. Eur J Neurosci 47(7):812–823

Koo-Poeggel P, Böttger V, Marshall L (2019) Distinct montages of slow oscillatory transcranial direct current stimulation (so-tDCS) constitute different mechanisms during quiet wakefulness. Brain Sci 9(11):324

Laakso I, Hirata A (2013) Computational analysis shows why transcranial alternating current stimulation induces retinal phosphenes. J Neural Eng 10(4):046009

Ladenbauer J, Külzow N, Passmann S et al (2016) Brain stimulation during an afternoon nap boosts slow oscillatory activity and memory consolidation in older adults. NeuroImage 142:311–323

Legatt AD, Emerson RG, Epstein CM et al (2016) ACNS guideline: transcranial electrical stimulation motor evoked potential monitoring. J Clin Neurophysiol 33(1):42–50

Li H, Lei X, Yan T et al (2015) The temporary and accumulated effects of transcranial direct current stimulation for the treatment of advanced Parkinson's disease monkeys. Sci Rep 5:12178

Loo CK, Alonzo A, Martin D et al (2012) Transcranial direct current stimulation for depression: 3-week, randomised, sham-controlled trial. Br J Psychiatry 200(1):52–59

MacDonald MD (2002) Safety of intraoperative transcranial electrical stimulation motor evoked potential monitoring. J Clin Neurophysiol 19(5):416–429

MacDonald MD, Skinner S, Shils J et al (2013) Intraoperative motor evoked potential monitoring—a position statement by the American Society of Neurophysiological Monitoring. Clin Neurophysiol 124:2291–2316

Manola L, Holsheimer J (2007) Motor cortex stimulation: role of computer modeling. Acta Neurochir Suppl 97(Pt 2):497–503

Manola L, Roelofsen BH, Holsheimer J et al (2005) Modelling motor cortex stimulation for chronic pain control: electrical potential field, activating functions and responses of simple nerve fibre models. Med Biol Eng Comput 43(3):335–343

Manola L, Holsheimer J, Veltink P (2007) Anodal vs cathodal stimulation of motor cortex: a modeling study. Clin Neurophysiol 118(2):464–474

Marshall L, Mölle M, Hallschmid M et al (2004) Transcranial direct current stimulation during sleep improves declarative memory. J Neurosci 24(44):9985–9992

Marshall L, Helgadóttir H, Mölle M et al (2006) Boosting slow oscillations during sleep potentiates memory. Nature 444(7119):610–613

Maruta Y, Fujii M, Imoto H et al (2012) Intra-operative monitoring of lower extremity motor-evoked potentials by direct cortical stimulation. Clin Neurophysiol 123(6):1248–1254

Mehta AR, Pogosyan A, Brown P et al (2015) Montage matters: the influence of transcranial alternating current stimulation on human physiological tremor. Brain Stimul 8(2):260–268

Mendonca ME, Santana MB, Baptista AF et al (2011) Transcranial DC stimulation in fibromyalgia: optimized cortical target supported by high-resolution computational models. J Pain 12(5):610–617

Miranda PC, Lomarev M, Hallett M. (2006). Modeling the current distribution during transcranial direct current stimulation. Clin Neurophysiol. 117(7):1623–1629. https://doi.org/10.1016/j.clinph.2006.04.009

Moiyadi A, Velayutham P, Shetty P et al (2018) Combined motor evoked potential monitoring and subcortical dynamic mapping in motor eloquent tumors allows safer and extended resections. World Neurosurg 120:e259–e268

Moliadze V, Atalay D, Antal A et al (2012) Close to threshold transcranial electrical stimulation preferentially activates inhibitory networks before switching to excitation with higher intensities. Brain Stimul 5(4):505–511

Moliadze V, Schmanke T, Andreas S et al (2015) Stimulation intensities of transcranial direct current stimulation have to be adjusted in children and adolescents. Clin Neurophysiol 126(7):1392–1399

Murray LM, Edwards DJ, Ruffini G et al (2015) Intensity dependent effects of transcranial direct current stimulation on corticospinal excitability in chronic spinal cord injury. Arch Phys Med Rehabil 96(4 Suppl):S114–S121

Neuling T, Wagner S, Wolters CH et al (2012a) Finite-element model predicts current density distribution for clinical applications of tDCS and tACS. Front Psych 3:83

Neuling T, Rach S, Wagner S et al (2012b) Good vibrations: oscillatory phase shapes perception. NeuroImage 63(2):771–778

Neuling T, Rach S, Herrmann CS (2013) Orchestrating neuronal networks: sustained after-effects of transcranial alternating current stimulation depend upon brain states. Front Hum Neurosci 7:161

Nitsche MA, Polania R, Kuo MF (2015) Transcranial direct current stimulation: modulation of brain pathways and potential clinical applications, Wiley: Hoboken. 233–254

Notbohm A, Kurths J, Herrmann CS (2016) Modification of brain oscillations via rhythmic light stimulation provides evidence for entrainment but not for superposition of event-related responses. Front Hum Neurosci 10:10

Noury N, Hipp JF, Siegel M (2016) Physiological processes non-linearly affect electrophysiological recordings during transcranial electric stimulation. NeuroImage 140:99–109

Opitz A, Paulus W, Will S et al (2015) Determinants of the electric field during transcranial direct current stimulation. NeuroImage 109:140–150

Opitz A, Falchier A, Yan CG et al (2016) Spatiotemporal structure of intracranial electric fields induced by transcranial electric stimulation in humans and nonhuman primates. Sci Rep 6:31236

Ozen S, Sirota A, Belluscio MA et al (2010) Transcranial electric stimulation entrains cortical neuronal populations in rats. J Neurosci 30(34):11476–11485

Pagnin D, de Queiroz V, Pini S, Cassano GB (2004) Efficacy of ECT in depression: a meta-analytic review. J ECT 20:13–20

Paneri B, Adair D, Thomas C et al (2016) Tolerability of repeated application of transcranial electrical stimulation with limited outputs to healthy subjects. Brain Stimul 9(5):740–754

Paßmann S, Külzow N, Ladenbauer J et al (2016) Boosting slow oscillatory activity using tDCS during early nocturnal slow wave sleep does not improve memory consolidation in healthy older adults. Brain Stimul 9(5):730–739

Pavan A, Ghin F, Contillo A et al (2019) Modulatory mechanisms underlying high-frequency transcranial random noise stimulation (hf-tRNS): a combined stochastic resonance and equivalent noise approach. Brain Stimul 12(4):967–977

Pelletier SJ, Cicchetti F (2014) Cellular and molecular mechanisms of action of transcranial direct current stimulation: evidence from in vitro and in vivo models. Int J Neuropsychopharmacol 18(2):pyu047

Pikovsky A, Maĭstrenko IL (2003) Synchronization: theory and application. NATO science series. Kluwer Academic Publishers, Dordrecht, p 258

Pikovsky A, Rosenblum M, Kurths J (2001) Synchronization: a universal concept in nonlinear sciences, vol 12. Cambridge University Press, Cambridge

Pirulli C, Fertonani A, Miniussi C (2013) The role of timing in the induction of neuromodulation in perceptual learning by transcranial electric stimulation. Brain Stimul 6(4):683–689

Polanía R, Paulus W, Antal A et al (2011a) Introducing graph theory to track for neuroplastic alterations in the resting human brain: a transcranial direct current stimulation study. NeuroImage 54(3):2287–2296

Polanía R, Nitsche MA, Paulus W (2011b) Modulating functional connectivity patterns and topological functional organization of the human brain with transcranial direct current stimulation. Hum Brain Mapp 32(8):1236–1249

Polanía R, Paulus W, Nitsche MA (2012) Modulating cortico-striatal and thalamo-cortical functional connectivity with transcranial direct current stimulation. Hum Brain Mapp 33(10):2499–2508

Prichard G, Weiller C, Fritsch B et al (2014) Effects of different electrical brain stimulation protocols on subcomponents of motor skill learning. Brain Stimul 7(4):532–540

Purpura DP, Mcmurtry JG (1965) Intracellular activities and evoked potential changes during polarization of motor cortex. J Neurophysiol 28:166–185

Radman T, Ramos RL, Brumberg JC et al (2009) Role of cortical cell type and morphology in subthreshold and suprathreshold uniform electric field stimulation in vitro. Brain Stimul 2(4):215–228

Rahman A, Reato D, Arlotti M et al (2013) Cellular effects of acute direct current stimulation: somatic and synaptic terminal effects. J Physiol 591(10):2563–2578

Rampersad SM, Janssen AM, Lucka F et al (2014) Simulating transcranial direct current stimulation with a detailed anisotropic human head model. IEEE Trans Neural Syst Rehabil Eng 22(3):441–452

Rappelsberger P, Pockberger H, Petsche H (1982) The contribution of the cortical layers to the generation of the EEG: field potential and current source density analyses in the rabbit's visual cortex. Electroencephalogr Clin Neurophysiol 53(3):254–269

Reato D, Rahman A, Bikson M et al (2010) Low-intensity electrical stimulation affects network dynamics by modulating population rate and spike timing. J Neurosci 30(45):15067–15079

Reis J, Schambra HM, Cohen LG et al (2009) Noninvasive cortical stimulation enhances motor skill acquisition over multiple days through an effect on consolidation. Proc Natl Acad Sci PNAS 106(5):1590–1595

Ruhnau P, Neuling T, Fuscá M et al (2016) Eyes wide shut: transcranial alternating current stimulation drives alpha rhythm in a state dependent manner. Sci Rep 6:27138

Ruohonen J, Karhu J (2012) tDCS possibly stimulates glial cells. Clin Neurophysiol 123(10):2006–2009

Saiote C, Polanía R, Rosenberger K et al (2013) High-frequency TRNS reduces BOLD activity during visuomotor learning. PLoS One 8(3):e59669

Schutter DJ (2016) Cutaneous retinal activation and neural entrainment in transcranial alternating current stimulation: a systematic review. NeuroImage 140:83–88

Schutter DJ, Hortensius R (2010) Retinal origin of phosphenes to transcranial alternating current stimulation. Clin Neurophysiol 121(7):1080–1084

Schwiedrzik CM (2009) Retina or visual cortex? The site of phosphene induction by transcranial alternating current stimulation. Front Integr Neurosci 3:6

Spaak E, de Lange FP, Jensen O (2014) Local entrainment of alpha oscillations by visual stimuli causes cyclic modulation of perception. J Neurosci 34(10):3536–3544

Stagg CJ, Nitsche MA (2011) Physiological basis of transcranial direct current stimulation. Neuroscientist 17(1):37–53

Stecher HI, Pollok TM, Strüber D et al (2017) Ten minutes of alpha-tACS and ambient illumination independently modulate EEG alpha-power. Front Hum Neurosci 11:257

Strube W, Bunse T, Nitsche MA et al (2016) Bidirectional variability in motor cortex excitability modulation following 1 mA transcranial direct current stimulation in healthy participants. Physiol Rep 4(15):e12884

Strüber D, Rach S, Trautmann-Lengsfeld SA et al (2014) Antiphasic 40 Hz oscillatory current stimulation affects bistable motion perception. Brain Topogr 27(1):158–171

Struber D, Rach S, Neuling T et al (2015) On the possible role of stimulation duration for after-effects of transcranial alternating current stimulation. Front Cell Neurosci 9:311

Szelenyi A, Senft C, Jardan M et al (2011) Intra-operative subcortical electrical stimulation: a comparison of two methods. Clin Neurophysiol 122(7):1470–1475

Taniguchi M, Cedzich C, Schramm J (1993) Modification of cortical stimulation for motor evoked potentials under general anesthesia: technical description. Neurosurgery 32(2):219–226

Tate MC, Guo L, McEvoy J, Chang EF (2013) Safety and efficacy of motor mapping utilizing short pulse train direct cortical stimulation. Stereotact Funct Neurosurg 91(6):379–385

Teo F, Hoy KE, Daskalakis ZJ et al (2011) Investigating the role of current strength in tDCS modulation of working memory performance in healthy controls. Front Psych 2:45

Terney D, Chaieb L, Moliadze V, et al. (2008). Increasing human brain excitability by transcranial high-frequency random noise stimulation. J Neurosci; 28(52):14147–14155. https://doi.org/10.1523/JNEUROSCI.4248-08.2008

Thut G, Schyns PG, Gross J (2011) Entrainment of perceptually relevant brain oscillations by non-invasive rhythmic stimulation of the human brain. Front Psychol 2:170

Thut G, Bergmann TO, Fröhlich F et al (2017) Guiding transcranial brain stimulation by EEG/MEG to interact with ongoing brain activity and associated functions: a position paper. Clin Neurophysiol 128(5):843–857

Toschi F, Lugli F, Biscarini F et al (2009) Effects of electric field stress on a beta-amyloid peptide. J Phys Chem B 113(1):369–376 van der Groen O, Wenderoth N (2016) Transcranial random noise stimulation of visual cortex: stochastic resonance enhances central mechanisms of perception. J Neurosci 36(19):5289–5298

van der Groen O, Tang MF, Wenderoth N et al (2018) Stochastic resonance enhances the rate of evidence accumulation during combined brain stimulation and perceptual decision-making. PLoS Comput Biol 14(7):e1006301

Veniero D, Vossen A, Gross J et al (2015) Lasting EEG/MEG after-effects of rhythmic transcranial brain stimulation: level of control over oscillatory network activity. Front Cell Neurosci 9:477

Vöröslakos M, Takeuchi Y, Brinyiczki K et al (2018) Direct effects of transcranial electric stimulation on brain circuits in rats and humans. Nat Commun 9(1):483

Vossen A, Gross J, Thut G (2015) Alpha power increase after transcranial alternating current stimulation at alpha frequency (alpha-tACS) reflects plastic changes rather than entrainment. Brain Stimul 8(3):499–508

Wen L, Hui Q, Shuling L et al (2006) Protective effect of motor evoked potential on motor pathway by stimulating the motor cortex during resection of glioma located in the motor cortex area. Chin J Minim Invasive Neurosurg 11(11):481–483

Westerberg CE, Florczak SM, Weintraub S et al (2018) Non-invasive brain stimulation: a paradigm shift in understanding brain oscillations. Front Hum Neurosci 12:211

Wischnewski M, Schutter D (2017) After-effects of transcranial alternating current stimulation on evoked delta and theta power. Clin Neurophysiol 128(11):2227–2232

Wood M, Willits RK (2006) Short-duration, DC electrical stimulation increases chick embryo DRG neurite outgrowth. Bioelectromagnetics 27(4):328–331

Wu L, Cao T, Li S et al (2022) Long-term gamma transcranial alternating current stimulation improves the memory function of mice with Alzheimer's disease. Front Aging Neurosci 14:980636

Yamamoto T, Katayama Y, Nagaoka T et al (2004) Intraoperative monitoring of the corticospinal motor evoked potential (D-wave): clinical index for postoperative motor function and functional recovery. Neurol Med Chir 44(4):170–182

Zaehle T, Rach S, Herrmann CS (2010) Transcranial alternating current stimulation enhances individual alpha activity in human EEG. PLoS One 5(11):e13766

# 7 Electrical Stimulation Techniques by Implantable Devices

Yiran Duan, Li Wang, and Jia Chen

## 1 Vagus Nerve Stimulation

### 1.1 Principle

The vagus nerve (cranial nerve X) is the longest cranial nerve containing sensory, motor, and parasympathetic nerve fibers that mainly innervates the thoracic and abdominal organs. It is involved in the regulation of the autonomic nervous system, cardiovascular system, respiratory system, gastrointestinal system, immune system, and endocrine system.

Vagal afferent allows to sense a wide variety of internal stimulations, including pressure, pain, stretch, temperature, chemicals, osmotic pressure, and inflammation. Sensory information converges in the vagus nerve nucleus, which transmits information to multiple areas of the brain and transmits modulation information via the vagal descending efferent device. The vagus nerve and its central connections (vagal system) act as an "unconscious internal brain" that integrates the body's "senses" and balances various organs. The vagus nerve regulates heart rate, blood pressure, vascular resistance, airway diameter, breathing, and feeding. Intraluminal nutrients trigger not only the vago-vagal reflex to initiate digestion and peristalsis, but also release endocrine mediators that interact with the vagus nerve (gut–brain signaling pathway). Vagus nerve stimulation (VNS) has demonstrated the role of the vagus nerve system in the regulation of the neuroendocrine-immune axis, mood, pain, and memory. The astrocytes surrounding the vagus nerve also play a regulatory role through neuroactive mediators and neurometabolic coupling. In addition to acute responses, the long-term neuromodulation effects may create a therapeutic approach for certain diseases. Numerous studies have demonstrated the potential of vagus nerve stimulation. There are also studies investigating its potential in treating headache, arthritis, asthma, pain, fibromyalgia, bipolar disorder, and dementia in addition to the Food and Drug Administration (FDA) approved indications: treatment-resistant epilepsy and depression (Yuan and Silberstein 2016).

The sensory afferent cell body of the vagus nerve is located in the ganglia and transmits information to the Nucleus Tractus Solitaries (NTS), which in turn transmits afferent sensory information to other parts of the brain in three main ways: (1) autonomic feedback circuits; (2) direct projection to the medullary reticular formation; (3) projection upward to the forebrain through the parabrachial nuclei (PBN) and the locus coeruleus (LC). The PBN is adjacent to the LC (one of the major areas of the brain containing norepinephrine). The fact that damage to the LC in rats eliminates the ability of the vagus nerve to inhibit seizures suggests that this connection is important for the antiepileptic effects of the vagus nerve (Krahl et al. 1998). PBN/LC directly connects various areas of the forebrain, including the hypothalamus and several thalamic areas that control the insula, orbitofrontal, and prefrontal cortices. PBN/LC is directly associated with the amygdala and caudate nucleus, which may be important for emotion regulation. These anatomical connections of the brainstem and limbic system are functionally associated. The C-fos is a general marker of cell activity. C-fos studies in rats during VNS showed increased activity in the amygdala, cingulum gyrus, LC, and hypothalamus (Naritoku et al. 1995). Recently, Walker et al. (1999) outlined a possible role of NTS in seizure mitigation by the vagal system. By microinjecting an aminobutyric acid agonist or glutamate antagonist into the NTS, they found that the increase in aminobutyric acid or the decrease in glutamate suppressed NTS-induced seizures. These findings suggest that vagus nerve stimulation may alter the expression of NTS, aminobutyric acid, and glutamate, accompanied by secondary changes in the function of the specific limbic structures mentioned above (Walker et al. 1999).

Y. Duan
Department of Neurology, Beijing Friendship Hospital, Capital Medical University, Beijing, China

L. Wang (✉) · J. Chen
Department of Neurology, Xuanwu Hospital, Capital Medical University, Beijing, China



Y. Wang (ed.), *Therapeutics of Neural Stimulation for Neurological Disorders*, https://doi.org/10.1007/978-981-99-4538-2_7

In conclusion, the sensory afferent connections of the vagus nerve produce direct projections to many brain areas associated with neuropsychiatric diseases. These connections reveal that vagus nerve stimulation may create a gateway to brainstem and cerebral connective areas. These circuits may explain the neuropsychiatric effects of vagus nerve stimulation and provide an additional theoretical basis for potential researches and clinical applications.

Vagus nerve stimulation has been used for the treatment of epilepsy without limitations to age or seizure type in Europe (since 1994), the United States (since 1997), and in China (since 1997). As of June 2018, more than 100,000 patients with epilepsy have been treated with vagus nerve stimulation worldwide. Vagus nerve stimulation is an option prior to brain surgery. Many studies have demonstrated the efficacy of VNS for the treatment of intractable epilepsy in children and adults. A large number of uncontrolled studies have shown improved quality of life by VNS treatment. Due to people's understanding of its action mechanism, VNS has also been applied to many other brain functional disorders such as treatment-resistant depression and obesity (Rush et al. 2000; George et al. 2000).

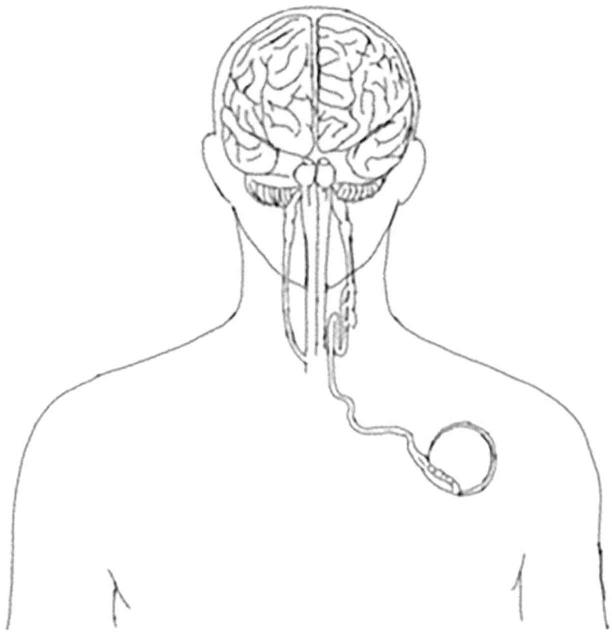

**Fig. 7.1** How to use the VNS system to stimulate vagus nerve. A pocket-size generator is implanted in the left chest wall. The bipolar electrode is wrapped around the left vagus nerve above the neck through a separate incision and then tunneled under the skin into the generator. The physician can control the intensity and frequency of the generator by lifting a telemetry rod connected to a portable computer (not shown in the figure) over the chest wall (and the generator)

## 1.2 Device Description

The vagus nerve stimulator delivering pulses to vagal nerve (Fig. 7.1) is quite similar to the common practice of implanted pacemakers. In both cases, a subcutaneous generator sends electrical signals to the organ through implanted electrodes (Amar et al. 1998). But for sure, the two methods differ in the site of stimulation. Vagus nerve stimulation is performed through an implantable, multi-programmable bipolar pulse generator (pocket-size) that is implanted in the left chest wall and transmits electrical signals to the left vagus nerve through bipolar leads. The pulse generator sends intermittent electrical signals at low frequencies to stimulate the cervical vagus nerve, which then projects through the nucleus tractus solitaries to the limbic lobe, amygdala, hypothalamus, locus coeruleus, raphe nuclei, and related areas of the cerebral cortex. In the vagus nerve stimulation system, the electrodes are wrapped around the cervical vagus nerve near the carotid artery through a separate incision and connected subcutaneously to the generator. Although the implantation of vagus nerve system is always initially performed on inpatients given general anesthesia by a neurosurgeon in most cases, more recently some epilepsy patients have had the device implanted by a vascular surgeon or otolaryngologist through an outpatient operation under local anesthesia. The programmer and software of the stimulator, together with a personal computer, have telemetric communication with the pulse generator, thus enabling noninvasive programming, functional evaluation (device diagnosis and inquiry), and data retrieval. The stimulator system includes electromechanical safety features that can minimize the potential for tissue damage from high-frequency stimulation. In addition, there is a magnet in each patient that turns off stimulation when it is above the pulse generator. When the magnet is removed, the normal stimulation procedure resumes (George et al. 2000).

## 1.3 Precautions for Application

A re-evaluation by the American Academy of Neurology concluded that vagus nerve stimulation is "effective and safe" for epilepsy. Major adverse effects of VNS include infection, bradycardia, cardiac arrest, dyspnea, vocal cord paralysis, voice hoarseness, sore throat, and cough (Handforth et al. 1998). Ninety-nine percentage of adverse events have been judged to be mild or moderate. In conclusion, there are few short-term adverse effects associated with surgery, which are usually resolved. Incision infection is rare and usually responds to antibiotic therapy. Fluid accumulation at the site of infection or without infection occurs in 1–2% of patients and can be cured by inhalation therapy and antibiotics. For rare refractory infections, the generator needs to be removed. About 1% of the unilateral vocal cord paralysis associated with implants may be due to excessive manipulation of the vagus nerve and subsequent deep injury to the vagus artery and its small arteries. In most cases, it will resolve completely within a few weeks. Choosing leads with correct size may reduce this complication, as the chance of this complication doubles when leads with a coil diameter of 2 mm (rather than 3 mm) are used in patients older than 18 years. Post-trauma lead breakage is rare. Stimulation-related adverse reactions (i.e., those that occur only when the vagus nerve is stimulated) can be reduced by decreasing the current level. They can also be terminated by the patient placing a hand-held magnet on the generator to turn off the device until the physician reduces the stimulation intensity. Stimulation-related adverse events include coughing, dyspnea, voice changes, abnormal sensation, headache, and localized pain. The frequency of these adverse events gradually decreases as the treatment continues. Many patients experience hoarseness or changes in voice quality and tingling sensations in the left neck when electronic pulses are delivered. Patients may experience subjective dyspnea or a sensation of neck muscle tightening, but no change in pulmonary function as tested. In case of coughing or sore throat, the current needs to be reduced during stimulation delivery sometimes. Children with a history of dysphagia may develop swallowing disorders during the treatment with vagus nerve stimulation. It may be helpful to adjust the device settings or turn off the stimulator with a magnet during mealtime. These adverse effects rarely result in dysregulation of vagal system treatment.

In patients with epilepsy, the long-term efficacy of vagus nerve stimulation has been maintained or improved. Moreover, the frequency of adverse effects generally decreases as patients adapt to the stimulation. The extensive experience of patients with epilepsy provides important safety data for the continued use of vagus nerve stimulation for 1 year or longer. Side effects diminish as patients adapt to the influences of stimulation. Two years after the initial procedure of implantation, more than 80% (142/172) of patients with epilepsy continue to receive vagus nerve stimulation, indicating that it is well tolerated in the long term (George et al. 2000).

Currently, vagus nerve stimulation cannot be used as a substitute for anticonvulsants. Vagus nerve stimulation is used in combination with drugs in most cases. With a better understanding of the neural basis of the vagus system, it is expected to improve clinical effects by improving the vagus nerve stimulator, thus to save the costs of concomitant medications.

## 2 Implantable Sacral Nerve Stimulation

Sacral nerve stimulation (SNS) is a neuromodulation technique in which electrodes are implanted into the spinal sacral canal to deliver continuous low-frequency electrical pulses to stimulate the corresponding sacral nerves, which produces an excitatory or inhibitory effect on the nerve pathways and thereby modulates the function of the bladder, anal sphincter and pelvic floor muscles. SNS was first applied in animal experiments in the 1970s, soon followed by clinical studies. In the 1990s, SNS was approved for the clinical treatment of urination disorder, urge urinary incontinence, and urinary retention. In 2011, it was approved for defecation dysfunction. Over the past decades, the device and application techniques of SNS have been evolving and improving, including the application of intraoperative fluoroscopic imaging techniques, barbed electrodes, and microstimulators, which makes the permanent implantation of SNS simpler and safer, and also extends its clinical applications. At present, it is not only applied to the treatment of refractory urge urinary incontinence, idiopathic urinary retention, refractory urgency-frequency syndrome, and other refractory lower urinary tract dysfunction as well as defecation dysfunction, but also increasingly applied to the treatment of neurogenic bladder, pelvic pain syndrome, and interstitial cystitis.

### 2.1 Mechanism of Action

The application of sacral nerve stimulation for the treatment of lower urinary tract dysfunction and bowel dysfunction has been studied for decades, while the specific action mechanism of this treatment approach is so complex that it remains unclear. Nevertheless, possible mechanisms include the following:

1. Mechanisms of SNS acting on lower urinary tract dysfunction: (1) SNS activates the somatic afferent component by stimulating the sacral nerve, thereby inhibiting abnormal afferent signals to the bladder and blocking abnormal sensation transmission to the central nervous system. SNS can also inhibit sensory transmission from interneurons to the pontine micturition center: directly inhibiting sacral parasympathetic preganglionic neurons on the efferent pathway. These mechanisms prevent involuntary micturition (micturition reflex) but do not inhibit voluntary micturition, which can treat overactive bladder (OAB) (Chancellor and Chartier-Kastler 2000). (2) SNS helps to relax the pelvic floor muscles, initiate micturition, and restore the function of pelvic floor muscles in patients with urinary retention. Besides, it can attenuate the overprotective reflex, inhibit the excitatory effect of the urethra, and promote bladder emptying (Shaker and Hassouna 1998). (3) In patients with neurogenic bladder, SNS can inhibit the bladder parasympathetic and pelvic efferents, thus regulating bladder and sphincter functions (Wang and Hassouna 2000).

2. Mechanisms of SNS acting on bowel function: (1) SNS can modulate somato-visceral reflexes. Studies have shown that SNS activates the sympathetic efferents via sacral nerve roots, which can inhibit colonic peristalsis, enhance the activity of internal anal sphincter, and improve symptoms of fecal incontinence (Michelsen et al. 2010). However, some other studies suggest that SNS can facilitate colonic peristalsis by activating the parasympathetic efferents. (2) SNS can modulate the afferent of abnormal sensation signals. Similar to the "gate control theory" of pain, SNS can inhibit C-fiber signaling and attenuate rectal hyperactivity by stimulating somatosensory afferent fibers and activating spinal interneurons. It can also inhibit the afferent sensory nerves of the defecation reflex and abnormal excitation of the pontine defecation center, thus attenuating defecation reflex (Hamdy et al. 1998). (3) SNS to the pudendal nerve can enhance the responsiveness of the motor cortex associated with anal activity and the activity of the external anal sphincter.

## 2.2 Operation Process

A rigorous screening of patients is required before performing SNS treatment. Patients who have failed drug and conservative treatment need to be identified should receive detailed urodynamic examination, anorectal dynamics examination, and neurophysiological measurements. In addition, it is necessary to consider the financial situation and general health status of patients, as well as the impact of in vivo implant on other medical examination to weigh the benefits for the patient.

SNS treatment is usually divided into two phases: the first phase involves experience assessment and the second phase involves the permanent implantation of the stimulator. The operation is performed under local anesthesia. Patients do not have to make any special preparation.

### 2.2.1 Treatment Phase of Experience Assessment

The patient is in a prone position with the lower abdomen elevated so that the sacral area is on the horizontal level. The locations of the bilateral S3 sacral habenula perforata are marked on the body surface. After local anesthesia, a puncture needle is inserted to S3 sacral habenula perforata perpendicular to the surface of the sacrum. Thereafter, the external end of the electrode lead is connected to a temporary stimulator to test the motor response and sensory response of the patients and adjust the position of the electrodes. After the position is determined, the barbed electrode is inserted in such a way that the 4 electrode contacts at the head end are as close as possible to the S3 sacral nerve. The electrode depth is then adjusted under intraoperative X-ray fluoroscopy. The response of each contact is tested with an external stimulator. The more contacts showing responses, the better the electrode position. After fixing the electrodes, the external stimulator is connected for experience test. The duration of in-vitro experiential stimulation is 7–14 days, no more than 30 days. The clinical effects are judged by clinical indicators, such as the average degree of urinary urgency, the average number of urinations per day (24 h), the average number of nocturia per day, and the score of life quality. It will be decided whether to proceed to the second phase depending on the improvement degree of the patient symptoms.

### 2.2.2 Phase of Permanent Implantation

During the treatment phase of assessment, if the patient has ≥50% improvement in all clinical symptoms, the treatment will be considered effective, and the second phase may be initiated. The permanent sacral nerve stimulator is attached to the external end of the electrode lead and implanted subcutaneously. A medical programmer is connected for impedance measurement and other related tests. The incision needs to be closed. The stimulation parameters of the stimulator are set according to the effective mode determined during the treatment phase of experience assessment. Patients will make a return visit 2 weeks, 3 months, and 6 months post operation, respectively, and every 6 months thereafter. Each visit involves physical examination, micturition diary, X-ray picture of sacrococcygeal region, stimulator parameter inspection, and the stimulation parameters are adjusted according to symptom improvement. As long as the electrical stimulation is effective, stimulation with voltage as low as possible should be selected to prolong the battery life.

## 2.3 Clinical Application

The typical indications of SNS include various refractory lower urinary tract dysfunctions, such as (1) refractory urge urinary incontinence, (2) refractory urinary frequency, (3) urinary urgency syndrome, (4) idiopathic urinary retention, and defecation dysfunction. Approximately 10,000 patients with urinary incontinence are treated with sacral nerve stimulation each year worldwide. Although the specific treatment mechanism is unknown, several multi-center and large sample studies have demonstrated the definitive and durable efficacy of SNS for refractory urge urinary incontinence and refractory urgency-frequency syndrome (Siegel et al. 2018). A sub-study of a multi-center, retrospective, cross-sectional cohort study in China shows that SNS has good efficacy in patients with symptoms of varying severity (Liu et al. 2019). The conversion rate from Phase 1 to Phase 2 in this study is 79.1%, with low incidence of adverse reactions. It fully indicates that SNS is a safe and effective treatment mode for refractory urge urinary incontinence and refractory urgency-frequency syndrome with definite efficacy and few complications. As to the stimulation frequency, some studies have concluded that there is no significant correlation between frequency and efficacy. A frequency of 14 Hz is usually recommended for optimal efficacy.

In recent years, SNS has also been used in studies on the treatment of neurogenic bladder, pelvic pain syndrome, and interstitial cystitis. A meta-analysis on SNS for neurogenic lower urinary tract dysfunction showed an efficiency of 68% during the treatment phase of experience assessment. The effective rate was 92%, and the incidence of adverse effects was 24% during the 26 months of mean follow-up period after permanent implantation (Kessler et al. 2010). A retrospective analysis of a small sample size demonstrated a synergistic therapeutic effect of SNS on pelvic floor dysfunctions, including defecation and urination disorders and erectile dysfunction in patients with incomplete spinal cord injury. It was reported by a study that the alleviation rate of constipation symptoms could reach 68.8%. Sacral nerve stimulation is generally not recommended for patients with complete spinal cord injury. In 2013, SNS was approved by the US FDA for the treatment of defecation dysfunction. In clinical practices, patients with nerve injury often have multiple manifestations of neurological dysfunctions. SNS may have a therapeutic effect on only some of these dysfunctions with symptom improvement of more than 50%. However, if SNS can improve the quality of life, it may also be recommended for therapy.

Powell et al. summarized the clinical efficacy of 39 cases of interstitial cystitis-bladder pain syndrome treated with SNS (Powell and Kreder 2010). Among the 5-year follow-up, 86% of patients (19 cases) showed a pain relief rate of greater than 50%, and 64% (14 cases) claimed a complete relief of pelvic pain and dysuria. Gajewski et al. reported the efficacy of SNS treatment in 78 patients with pelvic pain syndrome, 46 of whom had long-term SNS implantation. The mean postoperative follow-up period was 5 years, with an overall efficiency of 72% in patients receiving long-term implantation (Gajewski and Al-Zahrani 2011).

Currently, as there is no definite method to predict the efficacy of SNS before treatment initiation, external assessments are necessary. Further studies will suggest some conditions for identification to determine which types of neurological diseases are more suitable for SNS treatment.

## 2.4 Safety

SNS is an invasive neuromodulation technique. In spite of the decades of development of stimulating devices and techniques, its safety deserves our concern. According to some studies, there are some adverse effects in 56% of patients despite 90% of patients being satisfied with the treatment efficacy. The common adverse effect is pain at the stimulation site, with no serious fatal or irreversible adverse events. The incidence of pain at the electrode implantation site is about 4.5–19.1%; the incidence of pain at the stimulator implantation site is about 3.3–19.8%; the incidence of electrode displacement is about 2.2–8.6%; the incidence of infection is about 2.2–14.3%; the incidence of electric shock response is about 5.5–7.9%; and the risk of re-operation is 6.2–39.5% (Gormley et al. 2012).

# 3 Spinal Cord Stimulation

Spinal cord stimulation (SCS), a neuromodulation technique in which stimulation electrodes are directly placed outside the spinal dura of the corresponding segment of the spinal cord, is adjacent to the posterior column, to deliver electrical stimulation to the spinal cord through a pulse simulator. SCS has been developed based on the "gate control" theory of pain signaling. SCS serves as an effective treatment for chronic pain that has been used clinically for 50 years in Europe and the United States. Numerous in-depth studies have been conducted on the action mechanism, indications, expected efficacy, and possible complications of SCS, making it an important technique in the field of clinical treatment of pain today. The indications increase along with the continuous development and improvement of the technique. So far, it has been applied to the treatment of angina, multiple sclerosis, resting pain caused by peripheral vascular disorders, pain caused by peripheral nerve injury, arousal from vegetative state, etc.

## 3.1 Action Mechanism of SCS

### 3.1.1 Gate-Control Mechanism

The gate-control theory of pain signaling suggests that inhibitory interneurons can be activated by stimulating Aβ fibers in the posterior column of the spinal cord, thereby closing the afferent gate of pain signals (Melzack and Wall 1965). Specifically, the segmental neural network of afferent pain signals includes primary afferent Aδ and C fibers, spinal cord dorsal horn projection neurons, and inhibitory interneurons, among which the inhibitory interneurons function as a gate. The afferent impulses from C-fiber activate the spinal cord dorsal horn projection neurons to open the gate, allowing pain signals to be transmitted to the center and thus producing the sense of pain. In contrast, the afferent impulses from the large-diameter myelinated nerve fibers (Aβ fibers) excite the inhibitory interneurons, which inhibits the spinal cord dorsal horn projection neurons, thus closing the gate and preventing the transmission of pain signals.

### 3.1.2 Neurochemical Mechanisms

It has been found that SCS can promote GABA release, increase GABA concentration in the spinal cord dorsal horn and periaqueductal gray matter, and induce delayed GABA release in the spinal cord dorsal horn (Oosten et al. 2012). The animal experiments also show that SCS can promote Ach release in the spinal cord dorsal horn. SCS can also increase the release of neurotransmitters such as 5-hydroxytryptamine, capsaicin, and substance P, causing vasodilation and promoting the release of inhibitory neurotransmitters and thus exerting analgesic effects. Clinical observations show that the analgesic effect can last for a period of time after the cessation of treatment with spinal cord stimulation. The possible reason is that SCS results in the prolonged changes of neuronal excitability and continuously inhibits the afferent pain signals by regulating the release of neurotransmitters.

### 3.1.3 Activation Mechanisms of Sympathetic Nerve

In both animal experiments and clinical studies, SCS has been observed to cause vasodilatation-like phenomena, but the exact mechanism is unclear. Some scholars think that SCS may act directly on the sympathetic nerves of spinal cord segments, inhibiting sympathetic efferent fibers and thus causing a decrease in peripheral vasoconstriction (Jensen et al. 1986). SCS can also increase regional tissue blood flow by increasing the release of vasoactive substances, such as vasoactive peptides and substance P.

## 3.2 Device Description

The spinal cord stimulation system is composed of a stimulator that delivers electrical pulses, a lead connecting the electrodes to the stimulator, electrodes delivering the pulses to the spinal cord, and an external programmable device. After the electrodes are implanted in the spinal epidural cavity, the electrical pulses generated from the stimulator reach the spinal surface through electrode leading to stimulate the spinal cord nerves for therapy. A standard pulse stimulator adopts single-phase DC stimulation signals and can deliver electrical stimuli at relatively low frequencies, such as around 50 Hz. The novel high-frequency stimulator adopts bi-directional stimulus waveforms with a pulse width of approximately 30 μs at a frequency of 10 kHz. The high-frequency spinal cord stimulator does not cause sensory abnormalities and does not require patient feedback for localization. Therefore, the experience assessment and implantation of SCS can be both performed under general anesthesia. However, there is a risk of intraoperative injury not recognized to nerves or spinal cord. Therefore, electrode implantation under general anesthesia should be performed with caution. Electrodes are divided into percutaneous puncture electrodes and surgical electrodes. The percutaneous puncture electrodes are needle-shaped and cylindrical, while the surgical electrodes are flat, with multiple contacts at the end of the electrodes arranged in sequence. Electrodes can also be implanted by percutaneous puncture and surgical operation. Although implantation by surgical operation is more invasive, the electrodes by this method are more securely fixed, the incidence of electrode dislodgement is low, the current transmission is concentrated, and the stimulation range is wider. The open-loop stimulation protocol is adopted for conventional spinal cord stimulator. To be specific, the stimulation parameters can only be set manually by the physician or the patient through an external device. In addition, the stimulator does not automatically adjust the stimulation parameters when the patient's pain signal fluctuates or when the distance between the electrode and the target nerve changes due to a change in the patient's positioning. The new closed-loop spinal cord stimulator has been successfully developed and gradually applied to clinical practice. It is compatible with MRI and allows wireless charging of the stimulator, which greatly promotes the clinical application of this technology.

## 3.3 Operation Process

The procedure of spinal cord stimulation is divided into two phases: experiential treatment and long-term treatment.

### 3.3.1 Experiential Treatment

1. The physician will perform a minimally invasive procedure for the patient in the operating room to place electrodes in the area of the nerve roots on the dorsal side of the spinal cord in the epidural space and primarily confirm the location of the electrodes through intraoperative imaging.
2. The external end of the electrode lead is connected to an external temporary stimulator.
3. The patient remains awake throughout the procedure. The surgeon adjusts the stimulation parameters of the stimulator to produce a tingling sensation in the patient to suppress the sensation of pain. The electrodes are positioned properly as long as the feeling of numbness produced by the stimulated spinal cord segment covers the area of pain. Electrodes are then fixed.
4. Experiential treatment. During the test, the physician adjusts the stimulation parameters (polarity, pulse frequency, intensity, and width) according to the patient's experience, observes the treatment and adverse effects, and instructs the patient to use the stimulator.

### 3.3.2 Long-Term Treatment

1. During the experiential treatment period, when symptoms are relieved by more than 50%, the treatment is considered effective, and the permanent stimulator can be implanted under the skin in the lower abdomen for replacement. The implantation procedure is also performed in the operating room.
2. At discharge, the physician will set the original stimulation parameters based on the data from the testing period. Patients can turn on the stimulator by themselves and perform regular treatment.
3. Patients will return regularly for the assessment of effects and the adjustment of treatment parameters.

## 3.4 Clinical Application

### 3.4.1 Spinal Cord Stimulation for Chronic Intractable Pain

SCS has been used in the treatment of chronic pain for more than 40 years. Due to its characteristics of minimal invasion, treatment reversibility, parameter individualization, and high safety profile, SCS becomes an important technique in the field of clinical treatment of pain today. It is currently used to treat complex regional pain syndromes, failed back surgery syndromes, post-amputation phantom limb pain, ischemic limb pain, and peripheral nerve injury-induced pain. The action mechanism of SCS for the treatment of chronic pain involves neurophysiological and neurotransmitter mechanisms by stimulating the posterior column of the spinal cord and activating interneurons, which inhibits neurons that are sensitive to injury and increases the release of inhibitory neurotransmitters. This technique is now quite well established in the treatment of chronic pain, with more than 50,000 patients treated with spinal cord stimulation each year and an overall efficiency of about 80%.

### 3.4.2 Spinal Cord Stimulation for Intractable Angina

Spinal cord stimulation for the treatment of intractable angina pectoris is a safe and effective adjunctive therapy that exerts analgesic and anti-ischemic effects. Spinal cord stimulation stabilizes the electrical activity of intracardiac neurons, increases inhibitory neuropeptides, and decreases the release of excitatory amino acids. It may reduce pain, thereby decreasing myocardial contractility and myocardial oxygen consumption, with favorable results for revascularization. The American Heart Association's guidelines for the management of angina also include SCS as a IIb recommendation for the treatment of intractable angina.

### 3.4.3 Spinal Cord Stimulation for Limb Ischemic Pain

Spinal cord stimulation was first reported for the treatment of ischemic disorders in 1976. SCS may inhibit central sympathetic nerve activity and attenuate sympathetic vasoconstriction. SCS may also increase the release of vasorelaxation factors from nerve endings. These effects may reduce vascular resistance, increase local blood flow, and promote revascularization (Naoum and Arbid 2013), thus preventing the progression of ulcers, reducing patient pain, delaying amputation, and ultimately improving the quality of life in patients.

### 3.4.4 Spinal Cord Stimulation for Prevention and Treatment of Cerebral Ischemia

Isono et al. found that cerebral blood flow increased to 140% of the original level after cervical spinal cord stimulation in animal tests, which lasted until 15 min after the end of spinal cord stimulation (Isono et al. 1995). The effect of SCS-induced increase in cerebral blood flow may be related to both sympathetic and parasympathetic pathways that innervate the cerebral vessels, which subsequently cause an increase in the release of vasoactive factors. Electrical stimulation to cervical cord can enhance micro-circulatory cerebral blood flow with little damage to spinal cord tissues. It provides neurology with new therapeutic means in the treatment of cerebral ischemia, cerebral vasospasm, and vegetative state.

### 3.4.5 Other Applications of Spinal Cord Stimulation

With the in-depth development of studies on the action mechanism, pathophysiology, and safety of SCS techniques, spinal cord stimulation has been increasingly used. Electrical stimulation to thoracic spinal cord activates respiratory muscles, increases airway pressure, and improves respiratory function. SCS is also used in the treatment of cancer pain, dyskinesia, heart failure, multiple sclerosis, and restless legs syndrome.

## 3.5 Safety

Spinal cord stimulation is characterized with mature techniques and high safety, which has been widely applied in the field of pain treatment. However, there are some possible complications: (1) Cerebrospinal fluid leakage: the electrode penetrating the dura mater often results in cerebrospinal fluid extravasation, thus triggering intracranial hypotension headache. (2) Displacement of electrodes: usually occurring within a few days after implantation. Strenuous physical activity should be avoided in the early stage of electrode implantation. (3) Local infection: prophylactic antimicrobial agents are usually used after SCS surgery to reduce the incidence of postoperative infection. (4) Spinal dural hematoma and local injury: it mostly occurs after electrode implantation by surgery (spinal laminectomy), which is a serious but rare complication of SCS. Therefore, electrodes should be implanted by percutaneous puncture whenever possible.

In conclusion, the technique of SCS has been repeatedly adopted in basic research and clinical applications as it is minimally invasive, safe, effective, and adjustable. Spinal cord stimulation has successfully solved some complex pain problems in clinical practices. In recent years, SCS therapy has been widely used in various fields such as pain treatment, neurosurgery, neurology, rehabilitation, cardiology, etc. SCS will become a safe and effective adjunctive therapy.

## 4 Deep Brain Stimulation

Deep brain stimulation (DBS) is a new type of neuromodulation therapy in recent years, which relies on the stereotactic surgery to precisely implant stimulation electrodes into the target cerebral nuclei and adjust the stimulation parameters to change the excitability of the nuclei and thus control the disease symptoms. In recent years, with the development of stereotactic neurosurgery techniques and neuroimaging, DBS has replaced traditional surgical approaches for the treatment of dyskinesia associated with essential tremor and Parkinson's disease (PD). In addition, its application has been extended to other disorders such as dystonia, Tourette syndrome, epilepsy, chronic pain, and some psychiatric disorders. DBS was approved by FDA in 1997 for essential tremor, and subsequently approved for PD, primary dystonia, and obsessive-compulsive disorder (OCD), respectively, and was then approved for epilepsy in 2019 (Pycroft et al. 2018; Lozano et al. 2019). These indications are also approved in Europe under Conformité Européenne (CE) Mark including a recent approval for refractory epilepsy in 2010 (Aum and Tierney 2018). In China, it was approved as therapy for PD, tremor, dystonia, and epilepsy.

As a clinical tool, DBS has several advantages over other surgical approaches in neuromodulation. For example, DBS, with relatively less invasive nature, allows clinicians to gradually increase stimulation parameters to maximize benefit and minimize side effects and makes it possible to interact directly with symptomatic targets. DBS can also be used as a scientific research tool to study the pathological mechanisms of brain dysfunction. In addition, DBS, as a focused intervention means with anatomical resolution typically at the millimeter level, provides support for network theories of brain dysfunction by demonstrating that focal dysfunction and interventions can profoundly affect the whole-brain networks (Lozano et al. 2019).

So far, researchers have developed over 40 stimulation targets for DBS, among which the common ones include the globus pallidus internus (GPI), subthalamic nucleus (STN), dorsal STN, ventral intermediate nucleus (VIM), centromedian-parafascicular nuclei complex (CM-pf), and pedunculopontine nucleus. DBS applied to a specific target modulates the brain function associated with that target or affects multiple neural circuits through that target. In contrast, a symptom (or syndrome) can be affected by stimulation of multiple targets involved in that function or symptom.

### 4.1 Principle

At present, the efficacy of DBS in the treatment of essential tremor, PD, and dystonia is well recognized, while its mechanisms are still being explored. An in-depth understanding of the mechanisms of DBS will not only help to further improve the efficacy of DBS and reduce side effects, but will also contribute to the study of the pathophysiological mechanisms of the diseases.

It is often believed that the basic purpose of DBS is to regulate pathological neural activity in targeted circuits in the brain. Due to the successful application of DBS for essential tremor and PD, many studies on the mechanism mainly investigate the basal ganglia-thalamocortical circuitry (BGTC). It has been found by some studies that there are connections between multiple cerebral nuclei in this circuit, and there is also a direct connection between the cerebral cortex and the STN pathway. DBS can increase the release of excitatory and inhibitory neurotransmitters in the BGTC circuit. The release of neurotransmitters at specific nuclei can in turn contribute to the restoration of the firing pattern of the entire circuit or to the desynchronization of abnormal oscillations (hypothesis of neural interference or modulation). In this process, astrocytes may play a role in promoting neurotransmitter release and inhibiting side effects (Udupa and Chen 2015).

Researchers have proposed various hypotheses for the action mechanism of DBS. For example, the electrode stimulation of the nuclei affects the surrounding neuron cell body, axons, and nerve fibers, producing synaptic inhibition. High-frequency stimulation depletes neurotransmitters, thus affecting synaptic information transmission around the electrodes. Some researchers have also found that through current, chemical, or other regulatory pathways, the objects controlled by DBS are not limited to structures such as local neuron cell body, axons, monosynapses, and polysynapses and can directly modulate pathological neural networks

(Jakobs et al. 2019). The therapeutic effects of DBS depend on a variety of factors, including electrode stimulation parameters, physiological characteristics of the target site, electrode structure, surrounding tissue characteristics, and disease stages. This suggests that DBS may not have a single mechanism of action, but rather a synergistic work of multiple mechanisms, which is poorly understood and needs to be further investigated. In addition, most of the previous studies on DBS have focused on movement disorders, and studies on its action mechanisms on mood and cognitive functions are still in their infancy and need to be further investigated in the future.

## 4.2 Device Description

DBS consists of the implantation part, physician operation part, and patient operation part. The specific components include intracerebral stimulation electrodes, subcutaneous leads, neurostimulators, and extension leads.

### 4.2.1 Implantation Part

**Neurostimulator** It is an instrument that generates high-frequency pulses and is the core part of the system, including a set of microcircuit system and a battery. The stimulator is implanted subcutaneously near the clavicle and transmits electrical pulses to target nuclei in the deep brain via extension leads. The battery capacity of the neurostimulator is fixed and usually can be used for 6–8 years. It needs to be replaced or charged when the battery is running out.

**Extension Lead** It is an insulated lead placed subcutaneously connecting the electrodes of deep brain stimulation to the subcutaneous neurostimulator at the clavicle.

**Electrode** The intracerebral stimulation electrode is 1.27 mm in diameter and has 4 stimulation contacts at the end, with 0.5 mm or 1.5 mm interval. The operator implants the stimulation electrodes into specific cerebral nuclei by stereotactic technique, and electrical pulses are used to stimulate and suppress firing to achieve the therapeutic effect.

### 4.2.2 Patient Operation Part

**Patient Controller** It is also known as simple remote control. The patient attaches the remote control to the skin surface at the site of the implanted neurostimulator and controls the neurostimulator through buttons. Using the unilateral controller, the patient can switch the machine on and off and check the system on/off status to check the battery status. With the bilateral controller, in addition to the functions of the unilateral controller, the patient can also adjust the stimulation parameters appropriately under the guidance of the physician.

### 4.2.3 Physician Operation Part

**Stimulator for Neurological Test** It is used by clinicians to test the patient's response to DBS during the procedure of electrode implantation.

**Medical Programmer** Clinicians can use this instrument to check the operating status of the neurostimulator, adjust the stimulation parameters according to the state of illness and treatment response, and check the battery usage.

Traditionally, DBS is implanted for awake patients using frame-based stereotactic surgery to realize precise positioning. In recent years, traditional stereotactic frames have been gradually replaced by new frame-free positioning devices, supported by intraoperative imaging for confirmation of electrode placement.

## 4.3 Precautions for Application

The key to the success of DBS surgery lies in the choice of indication (symptoms of diseases), stimulation target (anatomical function), and stimulation parameters (functional modulation). The first step is to make a correct diagnosis to ensure that the patient is fully compatible with the indications for the procedure and then to develop a surgical plan for DBS implantation. This process requires multidisciplinary collaboration: neurologists are responsible for clarifying the diagnosis and assisting surgeons in selecting patients with complying indications, while radiologists and electrophysiologists assist in target identification and the validation of electrical stimulation to targets (Hartmann et al. 2019). Patients with PD or tremor usually undergo motor symptom tests before and after drug treatment to assess the disease severity. Patients with epilepsy need to have an EEG examination. Some patients may receive a neuropsychological evaluation during the test. Patients with obsessive-compulsive disorder must complete the obsessive-compulsive scale test.

DBS is generally safe and effective in patients with appropriate indications, but there are still some risks and possible side effects, which are usually mild and reversible. Risks of DBS surgery include: (1) 1% chance of cerebral hemorrhage and stroke; (2) infection; (3) equipment failure; (4) lack of therapeutic benefits in some symptoms; (5) headache; (6) worsening mental or emotional state. The DBS neurostimulation device must be set up according to each patient's needs and tolerability. The current, frequency, pulse width, voltage, and on/off schedule of the device need to be adjusted by an experienced neurophysiologist, usually a neurologist or neurosurgeon. During stimulation, possible side effects include (1) temporary tingling sensation in the face or extremities; (2) muscle tension; (3) verbal or visual problems; (4) balance disorders.

## 5 Responsive Brain Stimulation

Responsive brain stimulation (RBS), also known as responsive neurostimulation (RNS), is a neuromodulation means for drug-resistant epilepsy, which functions based on the principle of monitoring focal abnormal firing in the cerebral cortex and suppressing pathological electrical activity by applying responsive neurostimulation to terminate epileptic seizures.

### 5.1 Principle

Responsive neurostimulation therapy is based on the phenomenon that researchers have observed in intracranial EEG monitoring, i.e., electrical stimulation can stop epileptic afterdischarge. In intracranial EEG monitoring, electrical stimulation is a common means of brain localization. Partial electrical stimulation can lead to afterdischarge, which may evolve into clinical seizures. On the other hand, subsequent electrical stimulation can block the aforementioned afterdischarge and prevent seizures (Hartshorn and Jobst 2018).

### 5.2 Device Description

The RBS instrument mainly consists of cortical electrodes, deep electrodes, and nerve stimulator. The cortical electrodes are surgically placed on the cerebral surface around the epileptogenic focus, while the deep electrodes are inserted deeper into the epileptogenic focus, both of which are used to detect EEG, transmit it to a processor to predict seizures by specific algorithms, and then apply stimulation to terminate the seizures. These electrodes are connected to a neurostimulator inside the skull. The contacts on the electrodes can be set as anode or cathode, and the stimulator itself can act as a cathode. The neurostimulator can analyze the EEG signals, automatically identify the characteristic EEG signals before seizure, and release electrical stimulation in real time. After the device is implanted, a system transmits EEG information to the patient's online data management system for the clinician's reference, allowing the clinician to identify cortical EEG patterns that can lead to epileptic seizures. Continuously fine tuning the parameters can increase the seizure recognition and control rate, which contributes to the complete termination of seizures. The seizure pattern detected by the implanted electrodes can trigger up to 5 electrical stimuli (0.5–12 mA), each lasting 100–200 ms. This kind of stimulation can be applied thousands of times per day.

### 5.3 Clinical Application

RBS has been applied in clinical practice for a relatively short time, and the currently available device is the RNS stimulator system from NeuroPace, which was approved by the FDA in 2013 for the treatment of intractable epilepsy. The initial clinical trial of RBS was a multi-center, randomized, double-blind controlled trial that included 191 adult patients randomized to stimulation and non-stimulation groups (Morrell and Group RNSSiES 2011). At the 5th month, the seizure rate was reduced by 41.5% in the stimulation group and by 9.4% in the non-stimulation group, with a statistically significant difference. In addition, the efficacy of RBS gradually increased as the treatment course prolonged.

A study published on Neurology in 2015 reported the results of a multi-center and prospective clinical trial on RBS for the treatment of intractable focal epilepsy (Bergey et al. 2015). The study enrolled 256 patients and showed that the patients' seizure reduction rates after 1, 2, 3, and 6 years after implantation were 44%, 53%, 60%, and 66%, respectively, with more than 60% of patients experiencing a 50% or more reduction rate in epileptic seizure frequency within 3 months. Two-point-four percent of patients experienced postoperative adverse events, with the most common being soft tissue infections. The results of this study confirmed that the implanted RBS system could effectively and securely treat adult patients with intractable focal epilepsy by responsive stimulation.

Geller et al. reported 111 patients with medial temporal lobe epilepsy treated with RBS (Geller et al. 2017). These patients were followed up for 6 years on average. Among them, the duration of disabling seizures was reduced by 66.5% and response rates (>50%) continued to improve, reaching 64.6% at the sixth year. The seizure-free interval was 3 months or longer in 45% of patients and >6 months in 29%. Another study on 126 patients with neocortical epilepsy showed a mean reduction rate in epileptic seizure frequency of 51–70%. These results demonstrated the long-term effectiveness of RBS.

Several studies have investigated the quality of life and mood of patients using RBS (Hartshorn and Jobst 2018). Meador et al. found that 2 years after RBS implantation, 44% of patients reported an improvement in quality of life while 16% reported a decrease in quality of life. Some improvements were also seen in the assessment scores of depression scale. Other researchers measured neurocognitive function in patients using RBS. After 2 years of RBS treatment, patients showed improvements in naming (23.5% patients having high scores in the Boston Naming Test) but no changes in verbal learning. The improvement in naming was more pronounced in patients with temporal neocortical epilepsy compared to those with medial temporal lobe epilepsy.

One advantage of RBS is the function allowing patients to remotely upload seizure data, which can be used to track seizure trends and patterns over time and help physicians individualize the medication or neuromodulation therapy for patients. As seizure data accumulates over time, RBS may help us better explore the underlying mechanisms of epilepsy.

### 5.4 Applicable Population

Responsive brain stimulation may be considered for patients unable to receive surgical resection but with lesions that can be localized. Patients suitable for this treatment include those with multiple foci, such as dual-focus epilepsy, or those with seizure focus located in functional areas, such as focus in motor cortex or language area. Intracranial EEG is often used to identify the foci of seizures and to determine that patients are not suitable for surgical treatment.

## References

Amar AP, Heck CN, Levy ML et al (1998) An institutional experience with cervical vagus nerve trunk stimulation for medically refractory epilepsy: rationale, technique, and outcome. Neurosurgery 43(6):1265–1276

Aum DJ, Tierney TS (2018) Deep brain stimulation: foundations and future trends. Front Biosci (Landmark Ed) 23(1):162–182

Bergey GK, Morrell MJ, Mizrahi EM et al (2015) Long-term treatment with responsive brain stimulation in adults with refractory partial seizures. Neurology 84(8):810–817

Chancellor MB, Chartier-Kastler EJ (2000) Principles of sacral nerve stimulation (SNS) for the treatment of bladder and urethral sphincter dysfunctions. Neuromodulation 3(1):16–26

Gajewski JB, Al-Zahrani AA (2011) The long-term efficacy of sacral neuromodulation in the management of intractable cases of bladder pain syndrome: 14 years of experience in one centre. BJU Int 107(8):1258–1264

Geller EB, Skarpaas TL, Gross RE et al (2017) Brain-responsive neurostimulation in patients with medically intractable mesial temporal lobe epilepsy. Epilepsia 58(6):994–1004

George MS, Sackeim HA, Rush AJ et al (2000) Vagus nerve stimulation: a new tool for brain research and therapy. Biol Psychiatry 47(4):287–295

Gormley EA, Lightner DJ, Burgio KL et al (2012) Diagnosis and treatment of overactive bladder (non-neurogenic) in adults: AUA/SUFU guideline. J Urol 188(6 Suppl):2455–2463

Hamdy S, Enck P, Aziz Q et al (1998) Spinal and pudendal nerve modulation of human corticoanal motor pathways. Am J Phys 274(2):G419–G423

Handforth A, DeGiorgio CM, Schachter SC et al (1998) Vagus nerve stimulation therapy for partial-onset seizures: a randomized active-control trial. Neurology 51(1):48–55

Hartmann CJ, Fliegen S, Groiss SJ et al (2019) An update on best practice of deep brain stimulation in Parkinson's disease. Ther Adv Neurol Disord 12:1756286419838096

Hartshorn A, Jobst B (2018) Responsive brain stimulation in epilepsy. Ther Adv Chronic Dis 9(7):135–142

Isono M, Kaga A, Fujiki M et al (1995) Effect of spinal cord stimulation on cerebral blood flow in cats. Stereotact Funct Neurosurg 64(1):40–46

Jakobs M, Fomenko A, Lozano AM et al (2019) Cellular, molecular, and clinical mechanisms of action of deep brain stimulation—a systematic review on established indications and outlook on future developments. EMBO Mol Med 11(4):e9575

Jensen MP, Karoly P, Braver S (1986) The measurement of clinical pain intensity: a comparison of six methods. Pain 27(1):117–126

Kessler TM, La Framboise D, Trelle S et al (2010) Sacral neuromodulation for neurogenic lower urinary tract dysfunction: systematic review and meta-analysis. Eur Urol 58(6):865–874

Krahl SE, Clark KB, Smith DC et al (1998) Locus coeruleus lesions suppress the seizure-attenuating effects of vagus nerve stimulation. Epilepsia 39(7):709–714

Liu X, Meng L, Zhang W et al (2019) A multi-center study of sacral neuromodulation in the treatment of overactive bladder. J Mod Urol 24(11):897–901

Lozano AM, Lipsman N, Bergman H et al (2019) Deep brain stimulation: current challenges and future directions. Nat Rev Neurol 15(3):148–160

Melzack R, Wall PD (1965) Pain mechanisms: a new theory. Science 150(3699):971–979

Michelsen HB, Worsøe J, Krogh K et al (2010) Rectal motility after sacral nerve stimulation for faecal incontinence. Neurogastroenterol Motil 22(1):36–41

Morrell MJ, Group RNSSiES (2011) Responsive cortical stimulation for the treatment of medically intractable partial epilepsy. Neurology 77(13):1295–1304

Naoum JJ, Arbid EJ (2013) Spinal cord stimulation for chronic limb ischemia. Methodist Debakey Cardiovasc J 9(2):99–102

Naritoku DK, Terry WJ, Helfert RH et al (1995) Regional induction of fos immunoreactivity in the brain by anticonvulsant stimulation of the vagus nerve. Epilepsy Res 22(1):53–62

Oosten E, Janssen S, Gerrard S et al (2012) The pain-relieving effect of SCS and the induction of the release of spinal dorsal horn intracellular GABA. J Pain Pract 12(S1):15

Powell CR, Kreder KJ (2010) Long-term outcomes of urgency-frequency syndrome due to painful bladder syndrome treated with sacral neuromodulation and analysis of failures. J Urol 183(1):173–176

Pycroft L, Stein J, Aziz T (2018) Deep brain stimulation: an overview of history, methods, and future developments. Brain Neurosci Adv 2:2398212818816017

Rush AJ, George MS, Sackeim HA et al (2000) Vagus nerve stimulation (VNS) for treatment-resistant depressions: a multicenter study. Biol Psychiatry 47(4):276–286

Shaker HS, Hassouna M (1998) Sacral root neuromodulation in idiopathic nonobstructive chronic urinary retention. J Urol 159(5):1476–1478

Siegel S, Noblett K, Mangel J et al (2018) Five-year follow-up results of a prospective, multicenter study of patients with overactive bladder treated with sacral neuromodulation. J Urol 199(1):229–236

Udupa K, Chen R (2015) The mechanisms of action of deep brain stimulation and ideas for the future development. Prog Neurobiol 133:27–49

Walker BR, Easton A, Gale K (1999) Regulation of limbic motor seizures by GABA and glutamate transmission in nucleus tractus solitarius. Epilepsia 40(8):1051–1057

Wang Y, Hassouna MM (2000) Neuromodulation reduces c-fos gene expression in spinalized rats: a double-blind randomized study. J Urol 163(6):1966–1970

Yuan H, Silberstein SD (2016) Vagus nerve and vagus nerve stimulation, a comprehensive review: part I. Headache 56(1):71–78

# Fundamentals of Magnetic Stimulation Devices

# 8

Weimin Wang and Yicong Lin

## 1 Fundamentals of Electromagnetism

This section is intended to help readers deepen their understanding of the electromagnetic principles applied in magnetic stimulation devices by introducing basic electromagnetic terminology at first, then briefly summarizing the development of electromagnetism from the perspectives of electromagnetic phenomena and basic laws, where Faraday's law of electromagnetic induction that links electricity and magnetism in a real sense will be presented, and finally introducing the influence of conductance interface on the distribution of induced electric field with Faraday's law of electromagnetic induction as the core and Maxwell's equations as the supplement (Chen 2000).

### 1.1 Introduction to the Electromagnetic Principles of Magnetic Stimulation Devices

A magnetic stimulation device is related to electromagnetism in two aspects, namely the RLC (resistance inductance capacitance) circuit of magnetic stimulation used to generate the stimulation waveform and the electromagnetic field distribution of magnetic stimulation coils (Ueno et al. 1988).

The RLC circuit of magnetic stimulation mainly consists of the energy-storage capacitor $C$, circuit equivalent resistor $R$, and coil inductor $L$ (i.e., the magnetic stimulation coil), which is used to generate electromagnetic oscillation. Current changes caused by the coil inductor $L$ produce a varying magnetic field, which further generate a vortex current in human brain and cause an electrophysiological response (Fig. 8.1).

Therefore, the electromagnetic principle of magnetic stimulation devices is mainly related to Faraday's law of electromagnetic induction in electromagnetism. In addition, the specific discussion requires the application of Maxwell's equations and their related solutions in electromagnetic theory (Chen et al. 2004).

### 1.2 Electromagnetic Terminology

#### 1.2.1 Electric Charge

There are only two kinds of electric charges in nature, namely positive charge, which is generated by rubbing a glass rod with silk, and negative charge, which is generated by rubbing a hard rubber rod with fur. The positive and negative charges are relative. The above nomenclature was first proposed by Franklin, which is still in use today. The same kind of charges repel each other, and different kinds of charges attract each other. The amount of charge carried by an object is known as the charge quantity of the object.

#### 1.2.2 Electric Field (Electrostatic Field)

Any charge excites an electric field in the space around itself. The electric field exerts a force on any other charge in it, i.e., the electric field force. The magnitude of the electric field force can be expressed by multiplying the point charge quantity of the object by the component of the electric field intensity in the direction of the force. Charges interact with each other through the electric field, and the specific quantity of this interaction can be expressed by Coulomb's law.

#### 1.2.3 Magnetic Poles

Take a bar magnet as an example. When a bar magnet is immersed in a heap of iron filings and pulled out, it can be observed that there is a large amount of iron fillings attracted at both ends, indicating that the magnetism is particularly

W. Wang
School of Electronics, Peking University, Beijing, China
e-mail: wmw@pku.edu.cn

Y. Lin (✉)
Department of Neurology, Xuanwu Hospital, Capital Medical University, Beijing, China



Y. Wang (ed.), *Therapeutics of Neural Stimulation for Neurological Disorders*, https://doi.org/10.1007/978-981-99-4538-2_8

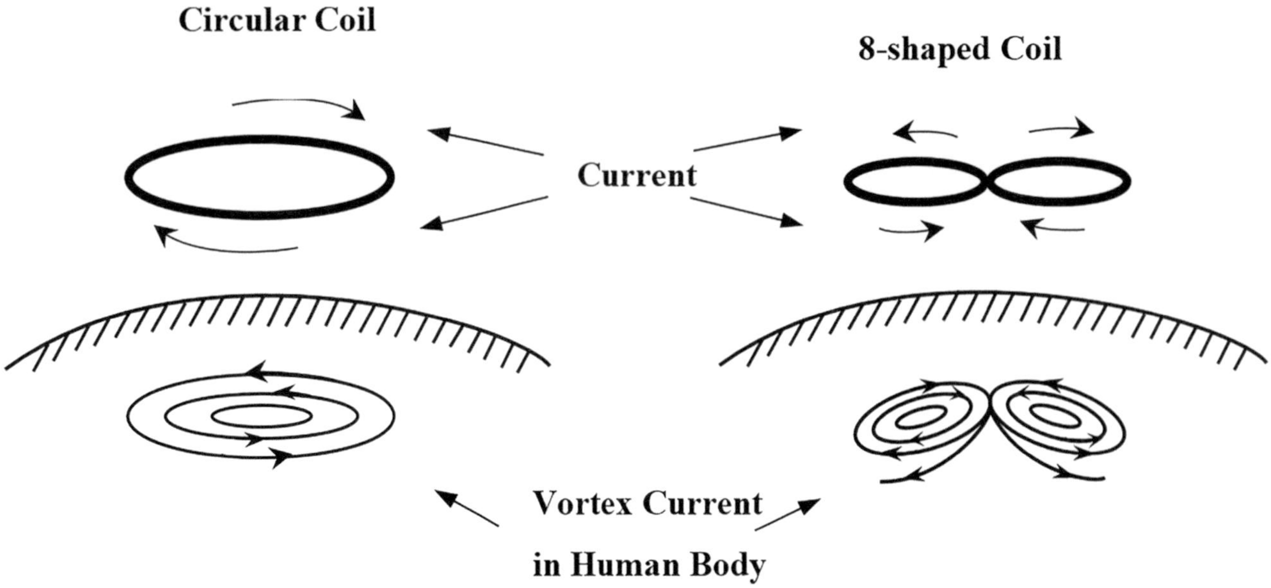


**Fig. 8.1** Model of induced electric field of a magnetic stimulation coil

strong. The area with particularly strong magnetic strength is called the magnetic pole, while the middle area without magnetism is called the neutral zone. When the middle of the bar magnet or narrow magnetic needle is supported or suspended so that it can rotate freely in the horizontal plane, the two magnetic poles always point in the north-south direction (deviating from the geographic north-south direction strictly). Hence, the end pointing to north is called the north pole (indicated by N), and the end pointing to south is called the south pole (indicated by S). Opposite poles are attracted to each other, while the same poles repel each other (Zhao and Chen 2011).

### 1.2.4 Magnetic Field (Magnetostatic Field)

Magnetic poles or currents produce a magnetic field in the space around themselves. In fact, since the basic source of the magnetism of a matter is the motion of electrons (electron spin, electron rotation around the nucleus, etc.) within microscopic particles such as atoms and molecules that make up the matter, it can be concluded that a moving charge (i.e., an electric current) generates a magnetic field in the space around itself; the magnetic field exerts a force on any other pole or current in it, which is essentially the interaction between the moving charges (i.e., currents) that is transmitted through the magnetic field and quantified by Ampere's law.

Note: The magnetic interaction between charges is different from the Coulomb interaction. The latter exists between charges whether they are at rest or in motion, while the former exists only between charges in motion.

### 1.2.5 Electromagnetic Field

Electric and magnetic fields are collectively called the electromagnetic field. Depending on the variation of electromagnetic fields over time, electromagnetic fields can be divided into the following:

#### 1.2.5.1 Static Fields (Electrostatic Fields and Magnetostatic Fields)

Neither electric fields nor magnetic fields change with time, but both have different values at different spatial locations.

#### 1.2.5.2 Time-Harmonic Electromagnetic Fields

Time-varying changes of electromagnetic fields can be expressed in harmonic function, but can have different amplitudes and phases in different spatial locations, which usually can be expressed in complex numbers.

#### 1.2.5.3 Time-Dependent Electromagnetic Fields

Time-varying changes of electromagnetic fields at a spatial point can be expressed in ordinary time function. If transformed to the frequency domain, its spectrum contains a variety of frequency components.

The electromagnetic field is a form of matter, which can exist independently of electric charge and current and has its own laws of motion and the same properties of energy and momentum as material objects (i.e., matters composed of atoms, molecules, etc.). In other words, it has materiality. The materiality of electromagnetic field is obvious only when the electromagnetic field changes rapidly (i.e., in electromagnetic waves).

### 1.2.6 Time-Varying Electromagnetic Fields

A time-varying electromagnetic field, time-varying field for short, refers to the unified electromagnetic field formed by magnetic field excited by time-varying electric field and the electric field excited by time-varying magnetic field when both electric and magnetic fields change over time. Time-varying electromagnetic fields contain time-harmonic electromagnetic fields and time-dependent electromagnetic fields. Depending on the different frequencies of electromagnetic fields changing over time, time-varying electromagnetic fields (Xia and Wang 2013) can be divided into quasi-static electromagnetic fields and rapidly varying electromagnetic fields, in which the quasi-static electromagnetic field changes at a smaller frequency, the field-like radiation effect and delay effect can be ignored, so the method of dealing with steady electromagnetic field (steady state field) to approximately deal with relevant problems can be used. For rapidly varying electromagnetic fields, as the field changes at a high frequency, the radiation effect and delay effect cannot be ignored. As the electromagnetic field generated by the magnetic stimulation coil in the magnetic stimulation device changes at a low frequency, it can be regarded as a quasi-static electromagnetic field, so that the electric and magnetic fields can be calculated and analyzed separately using the method of analyzing static fields. The physical laws of time-varying electromagnetic fields are in line with Maxwell's equations.

### 1.2.7 Magnetic Flux Density

MFD is the abbreviation of magnetic flux density and also an alias of magnetic induction intensity. It represents the number of magnetic lines of force perpendicularly passing through to a unit area and quantitatively reflects the density of magnetic lines of force. The strength and size of a magnetic field are usually expressed by the magnetic induction intensity $\vec{B}$. To be specific, the stronger the magnetic field is, the larger the value of the magnetic induction intensity $\vec{B}$ and the denser the magnetic lines of force will be; the weaker the magnetic field is, the smaller the value of the magnetic induction intensity $\vec{B}$ and the sparser the magnetic lines of force will be.

### 1.2.8 Electromagnetic Waves

Electromagnetic waves are propagated by electromagnetic oscillations in space (vacuum or medium), which depends on the excitation of the vortex electric field by a changing magnetic field (see the second equation of Maxwell's equations) and the excitation of the vortex magnetic field by a changing electric field (displacement current) (see the fourth equation of Maxwell's equations). The mutual excitation between varying magnetic fields and varying electric fields and the corresponding propagation are shown in Figs. 8.2 and 8.3, respectively.

## 1.3 Electromagnetic Phenomena and Fundamental Laws

### 1.3.1 Electrical Phenomena

The electricity development in history is shown in Table 8.1. In 1785, Coulomb introduced electricity into the field of quantitative scientific research.

### 1.3.2 Coulomb's Law, Gauss's Theorem, and Electric Field

Coulomb's law suggests that "the force between two charged points is inversely proportional to the square of the distance between them," as shown in the following equation:

$$\vec{F}_{12} = k\frac{q_1 q_2}{r_{12}^2}\hat{r}_{12} = \vec{E}_{12}q_2, \tag{8.1}$$

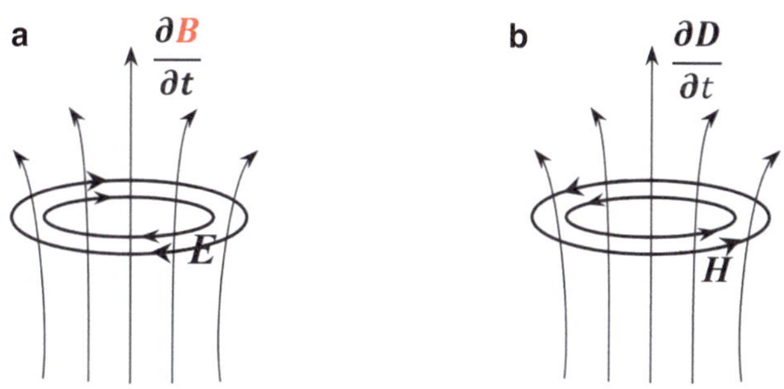


**Fig. 8.2** Mutual excitation of varying magnetic fields and varying electric fields. (**a**) The excitation of the vortex electric field by a changing magnetic field; (**b**) the excitation of the vortex magnetic field by a changing electric field

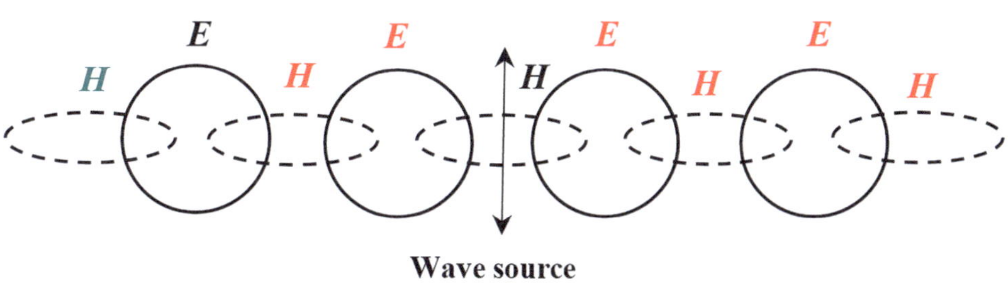


**Fig. 8.3** Example of propagation of electromagnetic oscillations in space

**Table 8.1** Overview of electricity development in history

| Time | Discovered by | Progress of electrical phenomena |
|---|---|---|
| B.C. | | Frictional electricity, lightning |
| 1731 | Gray | Difference between a conductor and an insulator |
| 1745 | Leyden | Leyden jar (capacitor) |
| 1752 | Franklin | Electrical properties (positive and negative electricities) |
| 1758 | Verkerk | Polarization of dielectrics |
| 1775 | Volta | Induced electrification |
| **1785** | **Coulomb** | **Coulomb's law** |

where $\vec{F}_{12}$ represents the force between two charged points, $q_1$ and $q_2$ represent the charge quantity of two point charges, respectively; $r_{12}$ refers to the distance between two point charges; $\hat{r}_{12}$ refers to the unit vector from charge $q_1$ pointing to charge $q_2$. $\vec{E}_{12}$ refers to the electric field intensity generated by the charge $q_1$ at the position of the charge $q_2$. The proportionality coefficient $k$ is determined by experiments and the unit system. It is determined in the International System of Units (SI) as:

$$k = 9\times10^9 = \frac{1}{4\pi\varepsilon_0}, \quad \varepsilon_0 \cong 8.854\times10^{-12}\left(\frac{\mathrm{F}}{\mathrm{m}}\right).$$

The electric field intensity is defined as the force per unit positive charge. In combination with Eq. (8.1), the electric field intensity produced by a point charge $Q$ at a distance of $r$ can be expressed as:

$$\vec{E} = \frac{1}{4\pi\varepsilon_0}\frac{Q}{r^2}\hat{r}. \tag{8.2}$$

It is therefore easy to prove that

$$\nabla\times\vec{E} = 0, \tag{8.3}$$

the electrostatic field is irrotational.

Gauss's theorem suggests that the electric flux passing through any closed surface is equal to the total electric charge in the closed loop divided by $\varepsilon_0$, as shown in Eq. (8.4).

$$\oint_S \vec{E}\cdot \mathrm{d}\vec{S} = \frac{Q}{\varepsilon_0}. \tag{8.4}$$

Substitution of Eqs. (8.2) into (8.4) yields the expression of Coulomb's law, which can prove that Gauss's theorem is a consequence of Coulomb's law.

### 1.3.3 Magnetic Phenomena

The magnetism development in history is shown in Table. 8.2. In 1820, Ampere and Biot Savart introduced magnetism into the field of quantitative scientific research.

In 1820, Ampere and Biot Savart introduced magnetism into the field of quantitative scientific research.

**Table 8.2** Overview of magnetism development in history

| Time | Discovered by | Progress of magnetic phenomena |
|---|---|---|
| 200 B.C. | | Sinan spoon (compass) |
| Eleventh century | Shen Kuo | Compasses were recorded |
| Twelfth century | | Compasses were used for navigation |
| Early fifteenth century | Zheng He | Geomagnetic navigation marine map |
| Late sixteenth century | Gilbert | Integration of many magnetic phenomena |
| **1820** | Oersted<br>**Ampere**<br>**Biot**<br>**Savart** | The magnetic effect of electric current<br>**Ampere's law**<br>**Biot–Savart law** |

### 1.3.4 Ampere's Law, Biot–Savart Law, Ampere's Circuit Law, and Magnetic Field

Ampere's law suggests that "the force between two current elements is inversely proportional to the square of the distance between them," as shown in the following equation:

$$\vec{F}_{12} = k_m \frac{\left(I_2\mathrm{d}\vec{l}_2\right)\times\left(I_1\mathrm{d}\vec{l}_1\times\hat{r}_{12}\right)}{r_{12}^2} = \left(I_2\mathrm{d}\vec{l}_2\right)\times\mathrm{d}\vec{B}_{12}, \tag{8.5}$$

where $\vec{F}_{12}$ represents the force between two current elements; $\left(I_2\mathrm{d}\vec{l}_2\right)$ and $\left(I_1\mathrm{d}\vec{l}_1\right)$ represent two current elements, respectively; $r_{12}$ refers to the distance between two current elements; $\hat{r}_{12}$ refers to the unit vector from current element $\left(I_1\mathrm{d}\vec{l}_1\right)$ pointing to current element $\left(I_2\mathrm{d}\vec{l}_2\right)$; $\mathrm{d}\vec{B}_{12}$ refers to the magnetic induction intensity generated by the current element $\left(I_1\mathrm{d}\vec{l}_1\right)$ at the position of the current element $\left(I_2\mathrm{d}\vec{l}_2\right)$, i.e., the magnetic field. The proportionality coefficient $k$ is determined by experiments and the unit system. It is determined in the International System of Units (SI) as:

$$k_m = 10^{-9} = \frac{\mu_0}{4\pi}, \mu_0 = 4\pi\times10^{-7}\left(\frac{\mathrm{H}}{\mathrm{m}}\right).$$

In combination with Eq. (8.4), the Biot–Savart law, i.e., the magnetic induction intensity produced by a current source $I\mathrm{d}\vec{l}$ at a distance of $r$ can be expressed as:

$$\mathrm{d}\vec{B} = \frac{\mu_0}{4\pi}\frac{I\mathrm{d}\vec{l} \times r\text{Æ}}{r^2}, \vec{B} = \bullet_{C}\, \mathrm{d}B. \tag{8.6}$$

It is therefore easy to prove that

$$\nabla \cdot \vec{B} = 0, \tag{8.7}$$

the magnetostatic field is solenoidal.

Ampere's circuit law suggests that "the line integral of magnetic induction intensity along any closed circuit C is equal to the algebraic sum of all currents passing through the circuit multiplied by $\mu_0$," as shown in Eq. (8.8).

$$\oint_C \vec{B} \cdot \mathrm{d}\vec{l} = \mu_0 I. \tag{8.8}$$

Substitution of Eqs. (8.6) into (8.8) yields the expression of Ampere's law, which can prove that Ampere's circuit law is a result of Ampere's law.

### 1.3.5 Faraday's Law of Electromagnetic Induction and Lenz's Law

Since the discovery of the magnetic effect of electric current by Oersted in 1820, people started to study the opposite of "electricity generating magnetism": "magnetism generating electricity." It was not until 1831 that Faraday discovered that when the magnetic flux changes through a conductor circuit, there will be an induced current and an induced electromotive force in the circuit, and the induced electromotive force is closely related to the change rate of the magnetic flux, thus discovering the famous Faraday's law of electromagnetic induction. This law reveals the experimental fact that "magnetism generates electricity"—a changing magnetic field can generate an electric field, which lays the theoretical foundation for the birth of the electric generator (Lin et al. 2000).

Faraday's law of electromagnetic induction, abbreviated as Faraday's law or the law of electromagnetic induction, can be expressed as "the induced electromotive force $\varepsilon$ through any closed circuit is equal to the negative of change rate of the magnetic flux through the circuit," as shown in the following equation

$$\varepsilon = \oint_C \vec{E} \cdot \mathrm{d}\vec{l} = -\frac{d\Phi}{\mathrm{d}t}, \Phi = \int_S \vec{B} \cdot \mathrm{d}\vec{S}, \tag{8.9}$$

where $\Phi$ represents the magnetic flux through the circuit; the first-order derivative of $\Phi$ with respect to time (i.e., $\frac{d\Phi}{\mathrm{d}t}$) refers to the change rate of the magnetic flux through the circuit; the minus sign before the change rate indicates that the induced electromotive force $\varepsilon$ is always the opposite of the rate of change of the magnetic flux $\frac{d\Phi}{\mathrm{d}t}$ in terms of plus and minus. This plus–minus relationship between the induced electromotive force and the change rate of magnetic flux can be used to determine the direction of the vortex current produced in human brain by a magnetic stimulation device.

In 1834, Lenz proposed a method to directly determine the direction of the induced current so that the direction of the induced electric potential could be represented by the direction of the induced current. This method is known as the Lenz's law, which is so expressed that "the direction of the induced current in a closed circuit is always acted as that the magnetic field it excites prevents the change (increase or decrease) of the magnetic flux causing the induced current." To discuss with examples, the following right-hand rule is adopted by convention: the four bent fingers of the right hand, except for the thumb, represent the winding direction of the selected circuit, and the straightened thumb represents the normal direction of the circuit (perpendicular to the plane of the winding circuit).

Take the direction of vortex current generated by a magnetic stimulation device in human brain for example. Assuming that the magnetic stimulation coil in the device is on the horizontal level and the current through the coil keeps increasing in intensity and remains clockwise in direction for a certain period of time, the magnetic field generated by the energized coil is also increasing in intensity. In addition, according to the right-hand rule, it can be determined that the magnetic field generated by the coil points vertically downward and the magnetic induction lines pass vertically through the human brain. As the magnetic flux through a certain area of the closed circuit of human brain continuously increases, a vortex current is generated in the human brain. In combination with the Lenz's law, the direction of the induced current (vortex current) should be such that the magnetic field it excites prevents the change in magnetic flux that causes the induced current (an increase in this case). Therefore, the direction of the magnetic field excited by the vortex current should be vertically upward to prevent the increase in magnetic flux in the circuit. According to the right-hand rule, the direction of the vortex current in human brain should be counterclockwise here (Parazzini et al. 2017).

In practice, the direction of the vortex current excited by the changing magnetic field will change accordingly because the direction of the current through the magnetic stimulation coil will alternate and thus lead to a change in the direction of the magnetic field.

## 1.4 Influence of the Conductance Interface on the Distribution of Induced Electric Fields

### 1.4.1 Vortex Electric Field and Vortex Current

Experiments have showed that the induced electromotive force excited by a changing magnetic field is completely independent of the type and nature of the conductor, which indicates that the induced electromotive force is caused by the changing magnetic field itself. After analyzing some phenomena of electromagnetic induction, Maxwell keenly perceived that the phenomenon of induced electromotive force foreshadowed new effects of electromagnetic fields. He believed that even if there was no conductor circuit, the changing magnetic field would excite an electric field around it, known as the induced electric field or vortex electric field. This kind of electric field is in common with the electrostatic field in terms of the force acting on the charge and differs from the electrostatic field in two aspects. First, the vortex electric field is not excited by the charge, but by the changing magnetic field; second, the electric field lines expressing the vortex electric field are closed. Therefore, it is not a conservative field (or potential field), expressed by the following equation:

$$\varepsilon = \oint_C \vec{E} \cdot \mathrm{d}\vec{l} = -\frac{d\Phi}{\mathrm{d}t} \neq 0. \tag{8.10}$$

Here $\vec{E}$ represents the vortex electric field indeed. Unlike the electrostatic field, the vortex field is rotational. Under the action of a vortex electric field, a closed circuit of conductor (e.g., in the human brain) can produce a vortex current.

### 1.4.2 Maxwell's Equations

The phenomena and basic laws of electromagnetism reflect the development of people's understanding from the mutual independence of electricity and magnetism to the interconnection between both. In the middle of the nineteenth century, Maxwell summarized all the previous knowledge about electricity and magnetism using the method of mathematical analysis, so that electricity and magnetism could be expressed in a concise and complete mathematical form, which was summarized in four equations:

$$\begin{cases} \oint_C \vec{E}\cdot \mathrm{d}\vec{l} = -\dfrac{d\Phi}{\mathrm{d}t} \text{ (Faraday's law of electromagnetic induction)} \\ \oint_C \vec{B}\cdot \mathrm{d}\vec{l} = \mu_0 I \text{ (Ampere's circuit law)} \\ \oint_S \vec{E}\cdot \mathrm{d}\vec{S} = \dfrac{Q}{\varepsilon_0} \text{ (Gauss's theorem)} \\ \oint_S \vec{B}\cdot \mathrm{d}\vec{S} = 0 \text{ ("Gauss's theorem" for magnetic fields)} \end{cases} . \tag{8.11}$$

These four equations were obtained under the conditions of vacuum and static fields.

#### 1.4.2.1 Integral Form

Later on, Maxwell's equations were further applied to dielectrics and time-varying fields, making Maxwell's equations applicable to all cases. The resulting Maxwell's equations in integral form is shown in the following equation.

$$\begin{cases} \oint_C \vec{E}\cdot \mathrm{d}\vec{l} = -\dfrac{\partial}{\partial t}\int_S \vec{B}\cdot \mathrm{d}\vec{S} \text{ (Faraday's law of electromagnetic induction)} \\ \oint_C \vec{B}\cdot \mathrm{d}\vec{l} = I_f + \dfrac{\partial}{\partial t}\int_S \vec{D}\cdot \mathrm{d}\vec{S} \text{ (Ampere's circuit law)} \\ \oint_S \vec{E}\cdot \mathrm{d}\vec{S} = Q_f \text{ (Gauss's theorem)} \\ \oint_S \vec{B}\cdot \mathrm{d}\vec{S} = 0 \text{ ("Gauss's theorem" for magnetic fields)} \end{cases} . \tag{8.12}$$

The charge conservation law (or current continuity equation) can be derived from the above equations:

$$\nabla \cdot \vec{J}_f + \frac{\partial \rho_f}{\partial t} = 0. \tag{8.13}$$

#### 1.4.2.2 Differential Form

Using Stoke's theorem and Divergence theorem, the Maxwell's equations in integral form, i.e., Eq. (8.11), can be rewritten in differential form:

$$\begin{cases} \nabla \times \vec{E} = -\frac{\partial \vec{B}}{\partial t} \\ \nabla \times \vec{H} = \vec{J} + \frac{\partial \vec{D}}{\partial t} \\ \nabla \cdot \vec{D} = \rho \\ \nabla \cdot \vec{B} = 0 \end{cases}, \tag{8.14}$$

where $\vec{E}$ is the electric field intensity (in V/m); $\vec{H}$ is the magnetic field intensity (in A/m); $\vec{D}$ is the electric displacement vector (flux density) (in C/m$^2$); $\vec{B}$ is the magnetic induction intensity (flux density) (in Wb/m$^2$ or T); $\vec{J}$ is the volume current density (in A/m$^2$); and $\rho$ is the volume charge density (in C/m$^3$).

#### 1.4.2.3 Boundary Conditions

The unique solution of Maxwell's equations can be determined only if the boundary conditions are known. The boundary conditions refer to the relationships that the electromagnetic field vectors on the interface of different media should satisfy. Maxwell's equations can be applied to the interior of any continuous medium. On the interface of two different media (such as air and human body), however, the electromagnetic field may change abruptly, invalidating the Maxwell's equations in differential form, while the integral form is still applicable. The following boundary conditions of the electromagnetic field can be inferred as a result:

$$\begin{cases} \hat{n} \times \left( \vec{E}_1 - \vec{E}_2 \right) = -\vec{J}_{ms} \\ \hat{n} \cdot \left( \vec{D}_1 - \vec{D}_2 \right) = \rho_s \\ \hat{n} \times \left( \vec{H}_1 - \vec{H}_2 \right) = \vec{J}_s \\ \hat{n} \cdot \left( \vec{B}_1 - \vec{B}_2 \right) = \rho_{ms} \end{cases}, \tag{8.15}$$

where $\hat{n}$ is the unit normal vector pointing from medium 2 to medium 1. The values on the right of the equation, from top to bottom, represent the surface magnetic current density $\vec{J}_{ms}$, the surface charge density $\rho_s$, the surface current density $\vec{J}_s$, and the surface magnetic charge density $\rho_{ms}$.

As magnetic charge and magnetic current mathematically assumed do not exist in the real world, the values on the right of the first and fourth equations in Eq. (8.15) are always zero for any real physical interface but can be non-zero for a virtual equivalent interface.

If both media are not an ideal electric or magnetic conductor, there will be not free charge accumulation on the surface, and both surface charge and surface current are zero in this case.

If medium 2 is a perfect electric conductor (PEC), Eq. (8.15) can be rewritten as the boundary condition of the ideal conductor:

$$\begin{cases} \hat{n} \times \vec{E}_1 = 0 \\ \hat{n} \cdot \vec{D}_1 = \rho_s \\ \hat{n} \times \vec{H}_1 = \vec{J}_s \\ \hat{n} \cdot \vec{B}_1 = 0 \end{cases}. \tag{8.16}$$

When solving practical problems, not all boundary conditions are necessary as they are not completely independent of each other. For example, the first equation in Eq. (8.16) is sufficient to find out the conductor boundary, while the second and third equations can be used to calculate the surface charge and current distribution, respectively.

## 2 Magnetic Stimulation Modes

Transcranial magnetic stimulation can be applied in multiple stimulation modes, such as single-pulse transcranial magnetic stimulation (sTMS), paired-pulse transcranial magnetic stimulation (ppTMS), and repetitive transcranial magnetic stimulation (rTMS), as well as the patterned rTMS that has been developed on the basis of rTMS in recent years, including theta burst stimulation (TBS) and quadri-pulse stimulation (QPS) (Chen et al. 2008; Yiftach and Abraham 2012).

### 2.1 Single-Pulse Transcranial Magnetic Stimulation (sTMS)

In the 1980s, single-pulse transcranial magnetic stimulation (sTMS) was used to study the motor system by applying single-pulse transcranial magnetic stimulation to the cerebral motor cortex at a certain intensity while recording myoelectric activity in the periphery, i.e., the motor-evoked response (MEP). Together with electrophysiological and/or neuroimaging techniques, it enabled the assessment of brain dynamics (see Fig. 8.1).

sTMS, usually applied by a hand-held device, only outputs one pulse at a time. When one pulse is delivered, the energy stored in the capacitor is released, resulting in a time-varying current pulse at a certain intensity passing through the coil. The time-varying current in the coil induces a short time-varying pulsed magnetic field, with the pulse perpendicular to the plane of the coil and the surface of the brain tissue. This time-varying magnetic field in turn induces an electric field in the brain that is parallel to the coil plane. This electric field generates a current in the tissue and causes depolarization of the cytomembrane and hence induces an action potential. Thus, the electrical energy stored in the capacitor of the transcranial magnetic stimulation device is eventually converted into an induced current in brain without being significantly attenuated by tissues (e.g., skull, dura mater) with high electrical impedance (Chen et al. 2008).

sTMS can be used to (1) assess the presence and severity of neurological dysfunction, (2) study the pathophysiology of disease, (3) monitor disease progression, and (4) evaluate the action mechanism of various therapies. A variety of sTMS-related electrophysiological examinations are available now to study the neurophysiological functions of the motor system, such as motor threshold (MT), input–output curve, central motor conduction time (CMCT), and cortical silent period (CSP). To accurately make a clinical assessment, a combination of multiple TMS protocols and neuroimaging measures is usually required.

A single TMS pulse of sufficient intensity delivered to the primary motor cortex activates corticospinal (or corticobulbar) neurons and produces multiple descending volleys, which are referred to as indirect waves. These volleys depolarize spinal (or bulbar) motor neurons. The peripheral volleys produce a relatively synchronous response in the target muscle, the motor-evoked potential (MEP), which can be recorded through surface electromyographic electrodes.

#### 2.1.1 Motor Threshold

MT is the minimum stimulus strength that produces MEP (usually 50–100 μV) in the target muscle, in at least half of 10 consecutive trials (Rossini et al. 1994). It reflects the excitability and the local density of a central core of excitatory interneurons and corticospinal neurons in the muscle representation in the primary motor cortex, and the excitability of the target spinal (or brainstem) motor neurons (Hallett 2000). MT can be evaluated at rest (resting MT, RMT) or during slight isometric activation of the target muscle (active MT, AMT). AMT is lower than RMT. This facilitatory effect of tonic muscle contraction mainly depends on enhanced excitability of the motor neuron pool, although corticospinal neuron excitability is also slightly increased (Mazzocchio et al. 1994).

#### 2.1.2 Input–Output Curve

The input–output curve is obtained by measuring MEP amplitudes across a wide range of intensities as sTMS pulses are applied over the cortical hot spot of the target muscle at rest or during tonic activation. In the resting muscle, MEP amplitude shows a sigmoid increment with increasing stimulus strength up to a plateau level. The curve is related to the number of corticospinal neurons activated at the different stimulus intensities (Ridding and Rothwell 1997) and depends on the strength of the corticospinal projections relative to the target muscle as well (Abbruzzese et al. 1999). It represents a sensitive but relatively nonspecific measure of the excitability of the corticospinal system.

#### 2.1.3 Central Motor Conduction Time

CMCT is the conduction time between motor cortex and spinal (or bulbar) motor neurons and is calculated by subtracting the peripheral motor conduction time (PMCT) from the MEP latency. PMCT can be estimated by the F-wave method or by spinal root stimulation (Rossini et al. 1994).

#### 2.1.4 Cortical Silent Period

CSP indicates a transient silence in isometric voluntary electromyographic activity of the target muscle produced by sTMS of the contralateral primary motor cortex. CSP duration increases with the stimulus intensity. Various cortical and subcortical areas projecting to the primary motor cortex modulate these inhibitory phenomena (von Giesen et al. 1994).

## 2.2 Paired-Pulse Transcranial Magnetic Stimulation

### 2.2.1 Paired-Pulse Transcranial Magnetic Stimulation Between Cerebral Cortices

Paired-pulse transcranial magnetic stimulation (ppTMS) outputs a pair of pulses each time at a certain interval. The two pulses can be output to the same magnetic stimulation coil to stimulate the same site in pair, or to two separate magnetic stimulation coils to stimulate different sites in succession. Between them, the first one is the conditioning stimulus (CS) and the second one is the test stimulus (TS).

For paired magnetic stimulation, two sets of charge/discharge system, two sets of silicon-controlled switch, and two energy-storage capacitors are required to charge the capacitors separately and apply paired stimulations in turn according to the set time. The two sets of devices are connected by special modules. Each set allows to set the stimulation intensity independently. The time interval between stimulations can be set by changing the special module or through an external hardware and software interface.

ppTMS is used to detect the excitability and inhibition of the cerebral cortex, intercortical conduction, and functional integrity. Commonly used electrophysiological parameters include short interval intracortical inhibition, long interval intracortical inhibition, and intracortical facilitation. Although the mechanisms of inhibition and facilitation are not completely clear, these techniques have been increasingly and widely applied due to the satisfactory reproducibility thereof.

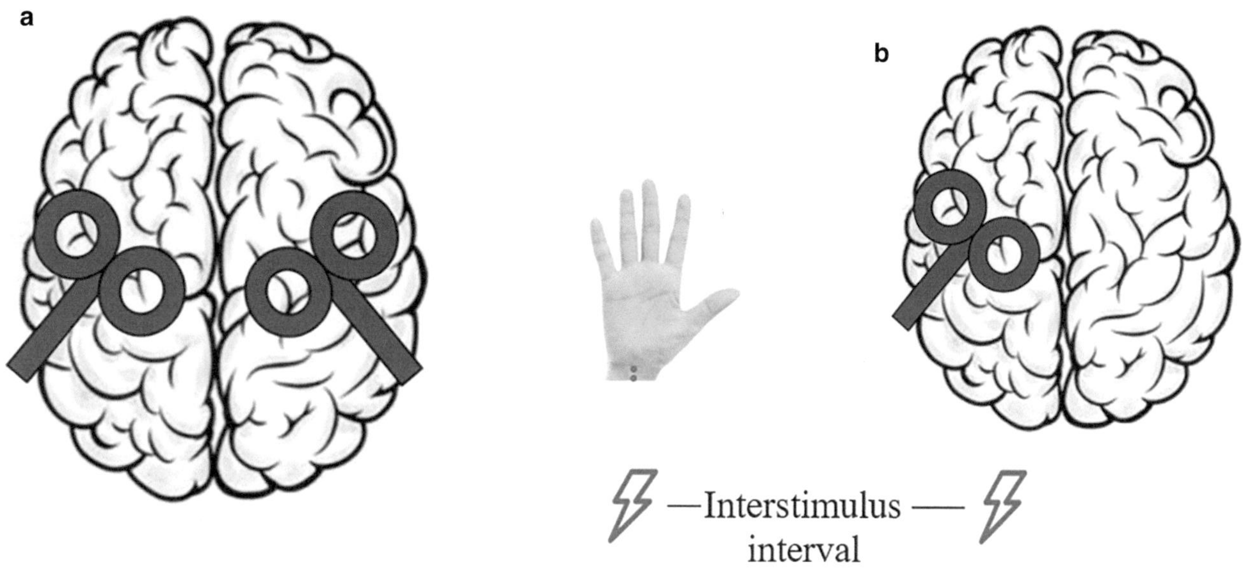


**Fig. 8.4** ppTMS. (**a**) Shows the inter-hemisphere; ppTMS. (**b**) Shows PAS

### 2.2.2 Paired Associative Stimulation

Paired associative stimulation (PAS) refers to a stimulation mode in which the paired stimulations exceed the cerebral range, with one electrical pulse stimulates the peripheral nerve and the other magnetic pulse stimulates the cerebral cortex at a certain interval. This stimulation mode can induce long-term depression or long-term potentiation in the cerebral cortex.

PAS can be applied by a general TMS device, which can be made either by outputting the pulse of electrical stimulation to delay magnetic stimulation, or by applying magnetic stimulation to delay electrical stimulation (Fig. 8.4).

The application of repetitive paired magnetic stimulation can result in long-term modulation of cortical excitability. The property of modulation is mainly determined by the time interval between peripheral electrical stimulation and transcranial magnetic stimulation.

## 2.3 Repetitive Transcranial Magnetic Stimulation

Repetitive transcranial magnetic stimulation (rTMS) outputs more than two regularly repeated magnetic stimuli at a time. Low-frequency rTMS is applied at frequencies equal to and less than 1 Hz, while high-frequency rTMS is applied at frequencies equal to and higher than 5 Hz. Both have different physiological effects and risks. In general, low-frequency stimulation inhibits cortices with low risks, while high-frequency stimulation above 10 Hz is prone to increase the cortical excitability with relatively higher risks.

### 2.3.1 Conventional Repetitive Transcranial Magnetic Stimulation

For conventional stimulation with frequencies greater than 1 Hz, the pulses are usually delivered intermittently for the safety of the subject and equipment. The duration of each series of stimulus is called the train duration or train length, measured in seconds. The duration between each series without output is called series interval. The number of stimulating pulses in each series is equal to the stimulation frequency multiplied by the series length. Stimulating pulses can be output continuously when the conventional stimulation frequency is equal to and less than 1 Hz. The higher the stimulation frequency is, the shorter the train length and the longer the train interval will be.

### 2.3.2 Patterned Repetitive Transcranial Magnetic Stimulation

Patterned rTMS is an improvement based on the conventional stimulation pattern by adding parameters such as intra-cluster frequency, intra-cluster pulse number, inter-cluster pulse number, and inter-cluster frequency. As to theta burst stimulation (TBS), each cluster has 3 pulses; the intra-cluster frequency is 50 Hz, and inter-cluster frequency is 5 Hz. Due to the limitations of the charging time and power of magnetic stimulation devices, the parameters of TBS currently used are mostly 50 Hz, with 3 pulses per cluster. TBS with a continuously repetitive frequency of 5 Hz is known as continuous theta burst stimulation (cTBS), which is capable of producing inhibitory effects. The intermittent TBS (iTBS) is often applied by using cTBS stimulation for 2 s, with an 8 s pause, which is capable of producing excitatory effects.

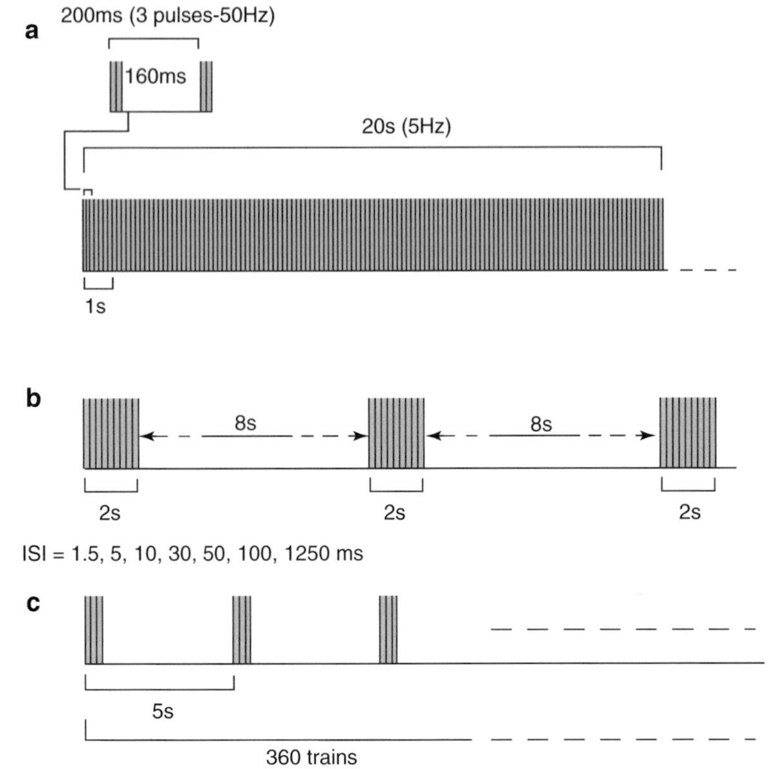


**Fig. 8.5** Patterned repetitive transcranial magnetic stimulation. (**a**) Shows continuous theta burst stimulation (cTBS); (**b**) shows intermittent theta burst stimulation (iTBS), and (**c**) shows quadri-pulse stimulation (QPS)

QPS is another patterned rTMS, which requires the connection of four different stimulators to the same coil through a specially designed combination module, so as to combine the interstimulus interval and inter-cluster interval of four independent single-phase pulse stimulations in various ways (Hamada et al. 2008). Studies have showed that QPS with interstimulus interval of less than 10 ms can increase cortical excitability, while interstimulus interval of 30–100 ms can inhibit cortical excitability (Fig. 8.5).

# 3 Basic Structure of Magnetic Stimulator

## 3.1 Multi-Channel High-Frequency and High-Voltage Power Supply System for Transcranial Magnetic Stimulation

The high-voltage pulse power generator is the core of the transcranial magnetic stimulation device, mainly including high-voltage pulse power supply, energy-storage capacitor, silicon-controlled switch, and backward diode. It is designed to accept the control command from the computer, charge the energy-storage capacitor, and control the discharge of the energy-storage capacitor.

As to the charging of capacitors, the charging techniques widely adopted for capacitors include the Direct Current (DC) charging with current-limiting resistor, power-frequency Induction Capacitor (LC) resonance charging, and high-frequency switching charging. The method of DC charging with current-limiting resistor has a simple circuit structure, but the charging speed gradually decreases and the overall efficiency is relatively low, which is suitable for low voltage charging application. Power-frequency LC control is relatively simple, with a fast-charging speed as a whole. The single power-frequency transformer is large in size but low in charging accuracy, making it suitable for the situation where there is no special requirement for charging accuracy and device size. The charging accuracy of the high-frequency switching charging is higher than that of the power-frequency LC resonant charging technique and facilitates the modularity and miniaturization of the overall device, which is suitable for high-power fast charging.

The specific structure of a multi-channel high-voltage pulse power generator for transcranial magnetic stimulation is shown in the figure (Fig. 8.6) below:

The multi-channel high-voltage pulse power generator for transcranial magnetic stimulation reduces the complexity and costs of the magnetic stimulator system and improves the power supply efficiency of the system from the perspective of power supply design by integrating the individual circuit modules into a single power generator. The specific principles and components are described as follows:

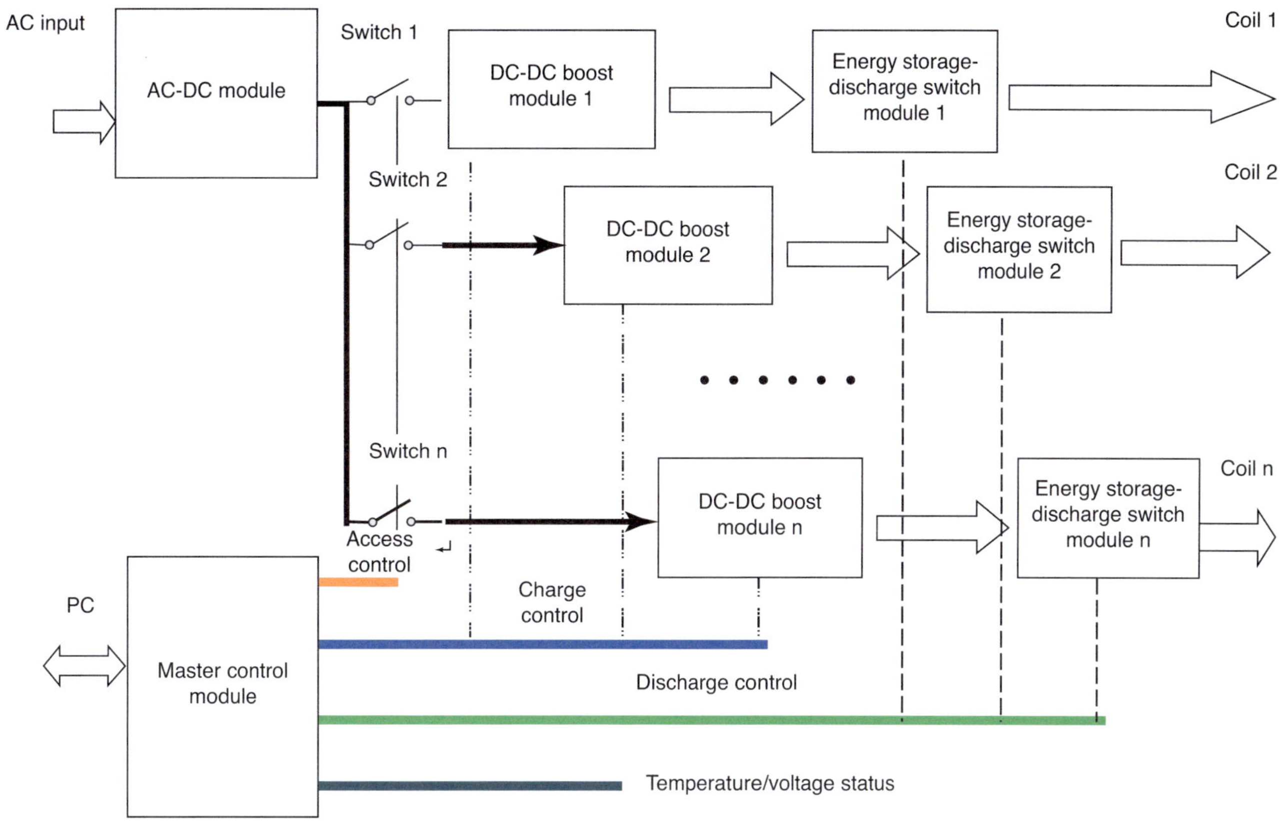


**Fig. 8.6** Block diagram of multi-channel high-voltage power supply for TMS

The multi-channel high-voltage pulse power generator for transcranial magnetic stimulation consists of four parts: Alternating Current to Direct Current (AC–DC) module, n sets of Direct Current to Direct Current (DC–DC) boost modules ($n \geq 2$), n sets of energy storage—discharge switch modules ($n \geq 2$), and master control module. Among them, AC–DC module converts one channel of input AC voltage into n channels of output DC voltage, which are connected in parallel to the inputs of n sets of DC–DC booster modules (Corthout et al. 2001).

N sets of DC–DC booster modules boot the DC voltage output from the AC–DC module into n channels of DC high voltage, supplying power to n energy-storage capacitors in n sets of energy storage-discharge switch modules, which is characterized by the fact that the access of the DC–DC booster module is controlled by the control module, the output voltage range is adjustable from 0 to 2000 V, and the value of output voltage is controlled by the DC voltage signal of 0–10 V from the control board.

Among n sets of energy storage-discharge switch modules, each set includes an energy-storage capacitor, a silicon-controlled discharge switch, and a protection diode. The master control module controls the on/off status of each silicon-controlled switch to excite to the external stimulation coil.

The master control module includes an interface for data acquisition card, an opto-coupler isolation circuit, and an acquisition and control circuit. The acquisition card interface is connected to the computer. Three kinds of control signals are sent out by the computer, after being isolated by the opto-coupler and processed by the signal acquisition and control circuit. The master control module collects the temperature/voltage status of the power supply and inputs it to the computer for processing through the acquisition card interface.

The master control module sends out three kinds of control signals. To be specific, n channels of high/low electrical level signals control the access of n sets of DC–DC modules; n channels of 0–10 V voltage signals control the charging voltage of n sets of DC–DC modules within 0–2000 V; n channels of clock pulse signals when outputting n channels control the on/off status of silicon-controlled switch in n sets of energy storage-discharge switch module, with the discharge frequency from 0 to 100 Hz.

The input of n sets of DC–DC booster module is from the same AC–DC module, which improves the overall efficiency of the power supply compared to the way of using multiple high-voltage power supplies.

The charging power, energy-storage capacitor, discharge switch, and stimulation coil between each channel of the multi-channel high-voltage pulse power generator are completely independent, which can meet various needs of magnetic stimulation therapy (Fang et al. 2018). In the first application scenario, n coils are used for n patients, so that one device can treat multiple patients at the same time. In the second application scenario, n coils are used to stimulate n sites simultaneously for the same patient, extending the application field of treatment. In the third application scenario, n energy-storage capacitors are used for the same coil, and the stimulation frequency of a single coil can be increased by changing the discharge clock of each channel to superimpose the stimulation frequency.

### 3.2 Energy-Storage Devices

The selection of an energy-storage capacitor is influenced by the capacitance withstand voltage and the maximum peak current, generally, with a certain margin. The capacitance loss factor should also be as small as possible in order to reduce the AC loss.

The larger the capacitance of the capacitor is, the larger the peak current generated by the circuit will be. However, excessive capacitance will increase the period of the discharge waveform. In addition, the slow charging speed of a large capacitor will reduce the stimulation frequency.

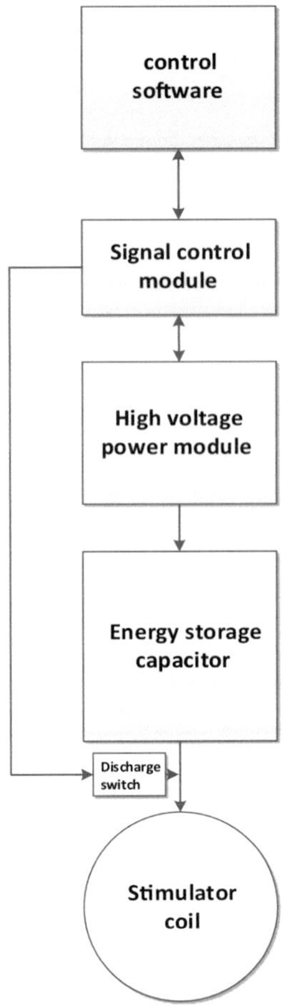


**Fig. 8.7** Overall structure of magnetic stimulator

### 3.3 Stimulator Unit

The main principle of the magnetic stimulator mostly used at present is to charge the energy-storage capacitor and then instantaneously discharge to the RLC circuit consisting of the stimulation coil, the energy-storage capacitor, and the stimulation switch to generate a pulsed magnetic field for stimulating biological tissues.

The specific structure of the magnetic stimulator is shown in the figure (Fig. 8.7) below, which mainly consists of functional control software, control modules of stimulation and charging signal, high-voltage power supply module, energy-storage capacitor, discharge switch, stimulation coil, and other modules. When the magnetic stimulator works, firstly, the host computer's functional software will set the key parameters such as the stimulation intensity, frequency, and time, and then the signal control module will generate the corresponding charging and discharging pulse signals according to the software settings to control the output of the high-voltage power supply module, the charging of the energy-storage capacitor, and the discharge switch, and finally, when the charging voltage of the energy-storage capacitor matches the stimulation intensity set by the software, the discharge switch will turn on to instantaneously discharge to the RLC circuit composed of the energy-storage capacitor and the stimulation coil to generate the desired pulsed magnetic field (Peterchev et al. 2011).

In addition, in order to obtain the maximum pulsed magnetic field under the same conditions, it is necessary to study in detail the parameters of the various parts that make up the RLC circuit and ensure that it operates under under-damping conditions. Of course, conditions should be considered, such as the capacitance value of the energy-storage capacitor, withstand voltage parameters, resistance and inductance of the stimulation coil, and the pulse width of the stimulation pulse.

### 3.4 Magnetic Pulse Generation

In order to obtain the maximum magnetic field strength, i.e., the maximum change rate of current with respect to time, the discharge circuit of the entire magnetic field needs to be in an under-damping state. The under-damping waveform is shown in the following figure (Fig. 8.8):

The magnetic stimulators mostly seen in the market are mainly divided into single-phase magnetic stimulators and dual-phase magnetic stimulators.

### 3.4.1 Single-Phase Stimulators

The pulse generation circuit of typical single-phase magnetic stimulator is shown below (Fig. 8.9).

Assuming that $t_m$ is the moment when the pulse current reaches its maximum value; when $t \leq t_m$, the current in the circuit satisfies the differential equation for dual-phase stimulation; when $t > t_m$, the current in the circuit satisfies the differential equation for single-phase stimulation, according to the circuit, we can get the following equation:

$$L\frac{\mathrm{d}i}{\mathrm{d}t} + R_1 i = -R_2 i_r = V_c$$

$$i = i_r + i_c$$

$$C\frac{\mathrm{d}V_C}{\mathrm{d}t} = -i_c. \tag{8.17}$$

The equation of the current $I(s)$ obtained after Laplace transformation is as follows:

$$I(s) = \frac{\left(s + \frac{1}{R_2 C}\right) i(t = t_m)}{s^2 + \left(\frac{1}{R_2 C} + \frac{R_1}{L}\right) s + \frac{\frac{R_1}{R_2} + 1}{LC}}. \tag{8.18}$$

The single-phase pulse (Fig. 8.10) rises rapidly from zero to peak and then drops gradually to zero. The single-phase stimulator mainly consists of an energy-storage capacitor bank and a discharge control circuit. It is operated by the host computer software, firstly, the set voltage is output from the power supply module to charge the energy-storage capacitor and then the discharge circuit will generate single-phase current pulse to turn on the discharge circuit switch to instantaneously discharge to the discharge circuit composed of the coil, energy-storage capacitor, and other devices, thus generating instantaneous high-intensity magnetic field pulse (Goetz et al. 2013).

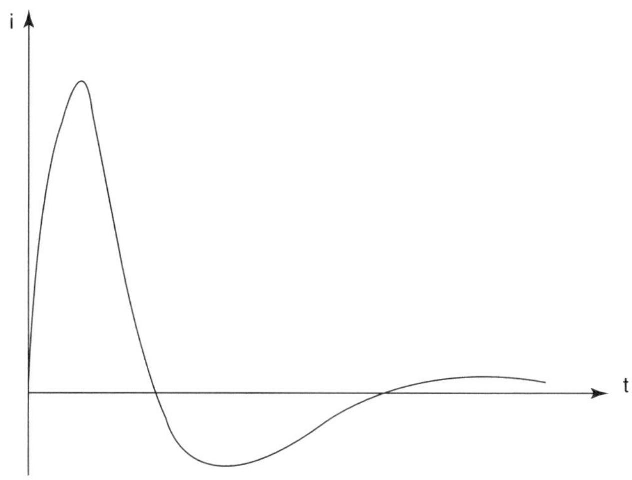


**Fig. 8.8** Waveform of discharge circuit in under-damping state

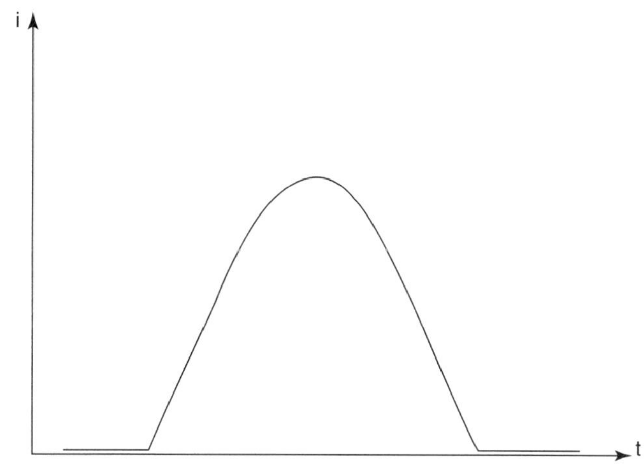


**Fig. 8.10** Single-phase pulse waveform

**Fig. 8.9** Pulse generation circuit of typical single-phase magnetic stimulator

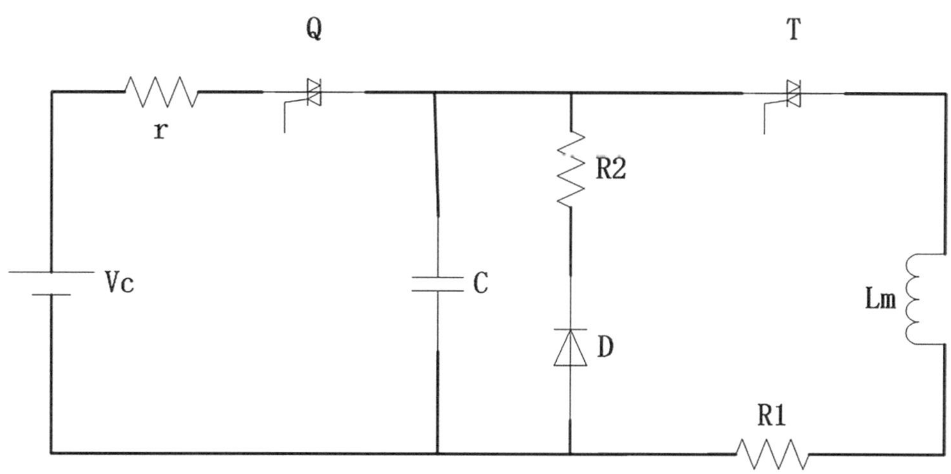

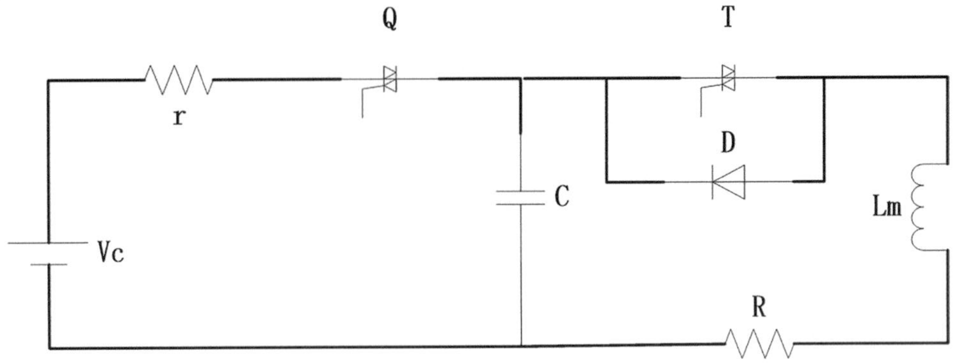


**Fig. 8.11** Pulse generation circuit of typical dual-phase stimulator

### 3.4.2 Dual-Phase Stimulator

The pulse generation circuit of typical dual-phase magnetic stimulator is shown below (Fig. 8.11):

The differential equation of the circuit is as follows:

$$L\frac{\mathrm{d}^2 i}{\mathrm{d}t^2}+R\frac{\mathrm{d}i}{\mathrm{d}t}+\frac{i}{C}=0. \tag{8.19}$$

$L$, $R$, and $C$ are the inductance, resistance, and capacitance of the circuit. The initial conditions of the pulse current $i(t)$ are $\frac{\mathrm{d}i}{\mathrm{d}t}(t=0)=\frac{V_C}{L}$ and $i(t=0)=0$; $V_C$ represents the set voltage of the energy-storage capacitor. The current $I(s)$ equation can be obtained after the Laplace transformation of the time domain equation in the formula above:

$$I(s)=\frac{V_C}{Ls^2+Rs+\frac{1}{C}}. \tag{8.20}$$

The expression for the coil current can also be obtained by an inverse transformation:

$$i(t)=\frac{V_c}{\omega L}e^{\frac{R}{2L}t}\sin(\omega t). \tag{8.21}$$

In the above equation, $R$ represents the series equivalent resistance of the capacitor, coil, and switch; $\omega$ represents the resonant frequency. When the parasitic resistance $R > 0$, the resonant frequency of the circuit is as follows:

$$\omega=\frac{1}{\sqrt{LC}}. \tag{8.22}$$

The dual-phase current (Fig. 8.12) pulse is a periodically damped sinusoidal pulse, which is an optimized design based on the original single-phase pulse. The continuous development of the magnetic stimulator poses an increasingly high requirement for its frequency, while it is difficult to further improve the frequency of single-phase pulse. In this case, the design advantages of dual- and multi-pulse come into play. The hardware composition of a dual-phase stimulator is similar to that of a single-phase stimulator, except that a dual-phase current pulse is used in driving the discharge circuit to work. Currently, the common rTMS devices mainly adopt dual-phase current pulses.

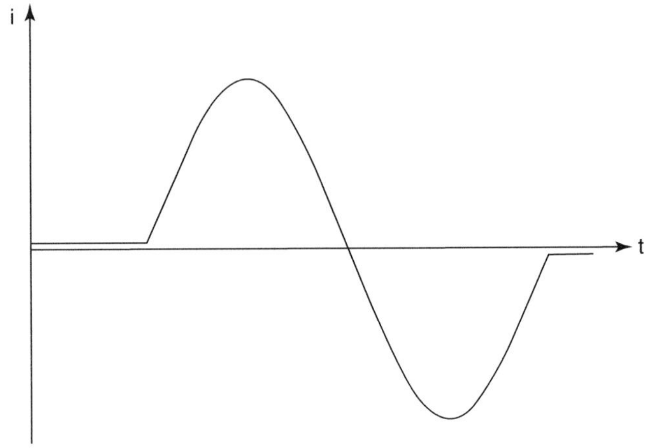


**Fig. 8.12** Dual-phase pulse waveform

This circuit structure shown in Fig. 8.11 is called dual-phase pulses structure as the circuit generates pulsed currents in two opposite directions, and the magnetic pulses generated in the space near the coil are also in two opposite directions. The circuit is capable of recovering energy, thus significantly improving the energy utilization of the stimulation device.

When other conditions are the same, the efficiency of the traditional single-phase pulse wave acting on the membrane potential is only lower than that of the traditional dual-phase pulse; however, its development is limited as its tail current leads to a serious reduction of circuit efficiency. Nevertheless, the stimulation effect of single-phase pulse is better than that of dual-phase pulse in practical applications and researches as it can provide higher selectivity of stimulation and better neuromodulation effects, which places a demand on the development of high-frequency pulse devices that can effectively generate single-phase pulses or single-phase electric fields.

### 3.5 Magnetic Pulse Sequence

In a single magnetic stimulation, the pulsed magnetic field, excited by instantaneous discharge to a coil, is manifested in the following figure (Fig. 8.13), with the horizontal axis representing the time axis and the vertical axis representing the amplitude of stimulation pulse.

**Fig. 8.13** Magnetic pulse sequence in single stimulation

This form of pulse is commonly used in medical therapy to collect the patient's threshold, that is, to trigger a single magnetic stimulation within a certain time to complete the acquisition of the threshold.

The form of the magnetic pulse sequence of rTMS is as follows:

The magnetic pulse sequence of repetitive stimulation is shown in Fig. 8.14. The stimulation frequency, interval, duration, and other key parameters can be set. Currently, it is the working mode of magnetic stimulator commonly used on the market.

In addition to the conventional magnetic pulse sequences of single stimulation and repetitive stimulation, the sequences in stimulation modes can also be customized according to the customer demands. For example, the TMS device of Tianjin Timus Medical Technology Co., Ltd. adopts the customized magnetic pulse sequence, as shown in Fig. 8.15.

In addition to individually setting stimulation frequency, intensity, and duration, various modes may be customized and combined according to the specific demands, which represents the development direction of magnetic stimulator technology in the future.

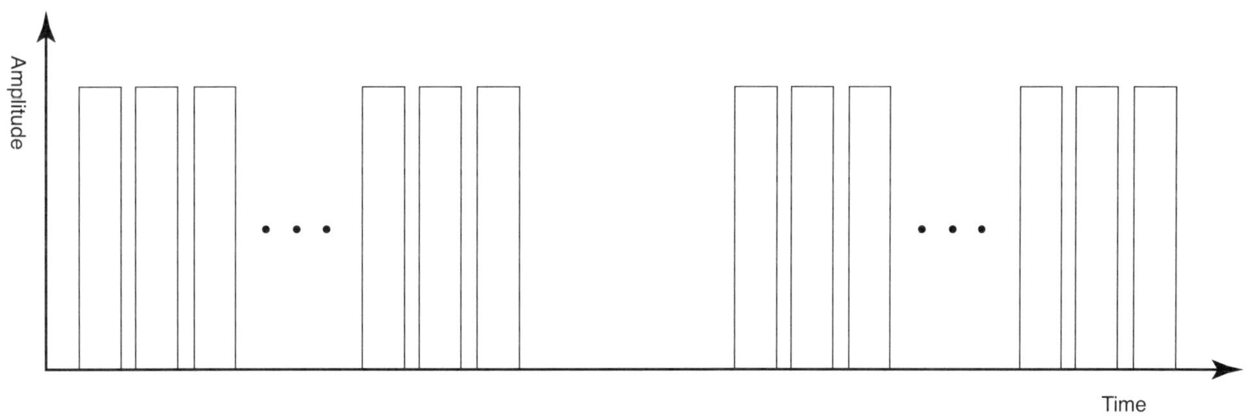


**Fig. 8.14** Magnetic pulse sequence of repetitive stimulation

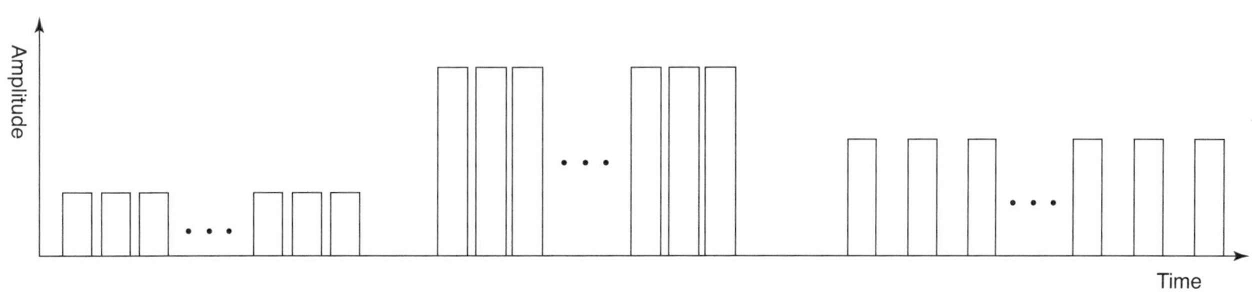


**Fig. 8.15** Customized magnetic pulse sequence. The amplitude, interval, duration, and frequency of the stimulation pulses can be setting arbitrarily

### 3.6 Characteristics of Magnetic Pulse

The parameters of magnetic pulses mainly include the stimulation depth, focus degree, stimulation intensity, etc., which are affected by factors such as the high-voltage power supply and stimulation coil.

The stimulation depth is mainly related to the output of the high-voltage power supply, the distance of the stimulation coil from the target on the patient's scalp, the shape design of the coil. One way to stimulate deeper target site is to directly increase the excitation of stimulation coil, that is, to increase the output of high-voltage power supply. Although it is easy to operate, with the deepening of the stimulation depth, the attenuation of magnetic field accelerates. In order to achieve the stimulation effect, the output of high voltage needs to be further enhanced, which causes severe stimulation of the superficial cerebral and facial nerves, inducing a feeling of sharp pain in the face and scalp of the subject and even resulting in some other side effects. Reducing the distance from the stimulation coil to the target on the patient's scalp by preparing the coil as thin as possible on the side that fits the subject and keeping it as close as possible to the subject's target cortex during stimulation can lessen the attenuation of the original magnetic field. In addition, the coil can be designed as special shapes such as "H" shape, cone shape, etc., so that the stimulation can reach the deep site based on the principle of vector superposition of the induced electric field in the target area (Pascual-Leone et al. 2002).

The degree of focus is mainly targeted for coils with focal points for stimulation, such as 8-shaped coils. The focus area of the coil increases, and the stimulation depth of the coil deepens with the increase of the coil inner diameter. In general, the focus degree of the coil is inversely proportional to the stimulation depth.

Stimulation intensity is not only related to the output of high-voltage power supply, but also related to the parameters of the stimulation coil. When the output intensity of the magnetic stimulator reaches a certain degree, the voltage excitation required by the stimulation coil increases with the number of turns and the inner diameter of the coil. In the meanwhile, optimizing the inductance of the stimulation coil during design can reduce energy consumption and increase efficiency.

## 4 Magnetic Stimulation Coils

Magnetic stimulation coils are the key component of transcranial magnetic stimulation, which determine the intensity, depth, range, and focus degree of the magnetic stimulation (Roth et al. 2007; Farzan 2014). The standard coils of transcranial magnetic stimulation in the market are usually round and figure-of-8 shaped. There are also some special coils that can be customized, such as H-coil and double coil. Coils can be classified into free cooling, air cooling, external liquid cooling, and internal liquid cooling depending on the cooling method. The various types of coils are shown in Fig. 8.16. Commonly, the stimulation depth of coil is 1.5–3.0 cm. In case of stimulation at 120% resting motor threshold, its action depth is generally no more than 2 cm. The commonly used figure-of-8 coil has a stimulation depth ranging from 2 to 2.5 cm. The stimulation range and depth increase as the stimulation intensity builds up (Zhi et al. 2013). Here some common magnetic stimulation coils are presented.

### 4.1 Round Coil

Round coil is the simplest type of magnetic stimulation coil and is the first kind of coil used in history. The outer diameter of the coil is usually 8–15 cm. The varying current within the coil induces a reverse ring current in the cerebral cortex. The magnetic field is strongest at the center of the coil, while the current induced in the cerebral cortex is strongest at the rim of the coil (Roth et al. 2007). This is because the magnetic lines in the middle of the coil cannot pass through the skull. Therefore, it is important to note when using the coil that the site where the magnetic stimulation functions is not the center of the round coil. Stimulation depth is related to the coil diameter. When two round coils have the same inductance

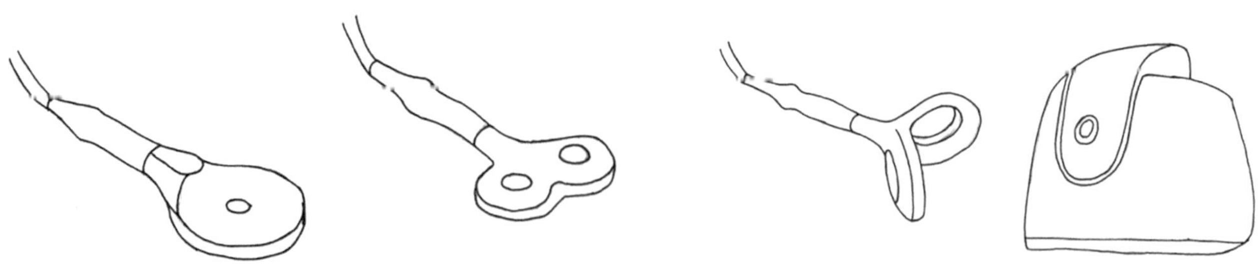

**Fig. 8.16** Various types of magnetic stimulation coils. From left to right: round coil, figure-of-8 coil, double coil, and H-coil

and device output, the round coil with a smaller diameter has higher surface magnetic induction intensity but shallower stimulation effect, while that with a larger diameter is just the opposite, with lower surface magnetic induction intensity but deeper stimulation effect.

The greatest disadvantage of round coils is poor focus. On the one hand, the coil covers a large area; on the other hand, it fails to precisely determine the radius of the strongest field strength. If the coil is slightly skewed to touch the scalp on one side, the stimulation range narrows, but the stimulation effect is also significantly weakened.

## 4.2 Figure-of-8 Coil

If two toroidal coils are placed side by side, with the current direction of one coil clockwise and of the other counter-clockwise, the current direction at the middle junction of the two coils is the same, and the induced electric field will be superimposed, reaching the highest at the junction, with the smallest stimulation area, which produces the effect of focused stimulation. This is known as a figure-of-8 coil. This coil is capable of realizing focused stimulation in a defined and limited area. The range and depth of stimulation are related to the stimulation intensity. Due to its outstanding performance of focused stimulation, the figure-of-8 coil has been widely used in clinical practice and scientific researches. As a figure-of-8 coil consists of two small toroidal coils with a smaller diameter than that of a standard toroidal coil, the figure-of-8 coil has a shallower stimulation depth than the toroidal coil (Zhi et al. 2013).

## 4.3 Double Coil

A double coil consists of two toroidal coils, with opposite current directions, forming an angle of approximately 90°. The superimposed magnetic field strength at the middle point is greater than that of a planar figure-of-8 coil. The biconical coil has higher local stimulation intensity and deeper stimulation depth while is less focused than the figure of 8 coil, so it is suitable for the motor cortex that innervates the muscles of the lower limbs and pelvic floor.

## 4.4 H-Coil

The H-coil, also known as the "deep coil," is formed by bending a number of parallel leads fitting the curvature of the skull at right angles in a three-dimensional direction, so that the magnetic field vectors around the leads overlap and add up in the target area of stimulation. The magnetic field around each lead is not strong enough to cause scalp pain. As the magnetic field vectors of many spherical leads add up and focus, the stimulation depth can reach 6 cm (Harel et al. 2014; Roth et al. 2007).

## 4.5 Sham Coils

In clinical double-blind trials, a sham coil that is the similar to the real stimulation coil is required to serve as a control. A perfect sham stimulation coil has the same shape, weight, stimulation sound and mechanical action of hitting the scalp, and complicated feelings during scalp muscle contraction as the real stimulation coil. In practice, such a sham stimulation coil does not exist.

One way is to tilt the stimulation coil at an angle and keep it out of the subject's view. Depending on the tilt angle and the type of coil, this method can reduce the field strength by 24–73%, but the stimulation sensation at the contact point on the scalp is significantly different from that of the real stimulus. Besides, the operator is not blinded. More importantly, the residual stimulation field strength still produces a biological effect. Another method is to place two coils with the same current direction side by side, with the current direction in the middle stimulation site being opposite, which can significantly reduce the resulting magnetic field intensity. However, it still produces a weak magnetic field around the stimulation site (Lisanby et al. 2001). There is another kind of sham coil which installs an electromagnetic shielding layer made of materials with low magnetic permeability on the surface of the stimulation coil to prevent the magnetic field spreading in space from stimulating the target area, so as to shield the magnetic field of the stimulation coil.

## 4.6 Coils with Cooling Function

Conventional coils cool naturally. Although the coils are relatively thin and light and can deliver stimulation on both sides, they cannot continuously deliver magnetic stimulation pulses. At 100 Hz, for example, stimulation can last for only a few seconds. When the duration of continuous application of transcranial magnetic stimulation (TMS) is extended, or when the frequency and intensity of stimulation increase, the coil will heat, limiting the use of the transcranial magnetic stimulation. As a result, stimulation coils with various cooling methods have emerged, mainly including air-cooled coils and water-cooled coils. Air-cooled coils are relatively large, heavy, and can only allow for single-sided stimulation, making them less convenient to operate. Water-cooled coils are thicker and heavier and also cannot be turned over for use. Besides, the risk of high-voltage leakage is increased.

# 5 Safety Issues of Magnetic Stimulation Devices and Coils

Repetitive transcranial magnetic stimulation (rTMS) is a noninvasive stimulation method for human brain, which can provide repetitive pulsed magnetic fields of sufficient intensity and density to activate neurons and modulate cortical excitability. This section is mainly intended to describe the safety and adverse effects of rTMS devices and instructions for clinical use.

## 5.1 Safety Issues

### 5.1.1 Heating

Single-pulse transcranial magnetic stimulation (sTMS) itself causes a very small temperature rise in brain tissues, estimated to be less than 0.1 °C. In contrast, the direct heating around deep brain stimulation electrodes increases temperature up to 0.8 °C. Vortex currents induced in conductive surface electrodes and implants cause them to heat up. The temperature rise depends on the shape, size, orientation, conductivity, and nature of the surrounding tissue of the electrode or implant, as well as the type, location, and stimulation parameters of the TMS coil. Silver and gold electrodes are highly conductive and can result in overheating and resulting burn skin. Using low conductive plastic electrodes can reduce heat generation. Radial incisions of the electrodes and skull plates can also mitigate heat generation by cutting off the vortex current path. Skull plates made of titanium tend to have a lower potential for heating due to the low conductivity of titanium and the radial incisions. Brain implants such as aneurysm clips and stimulation electrodes can also heat up. Brain tissues heated up to a temperature above 43 °C can lead to irreversible damages (U. S. Food and Drug Administration 2011).

### 5.1.2 Force and Magnetization

TMS may exert forces on some brain implants, which may cause them to displace. The forces acting on ferromagnetic objects tend to be greater than those acting on non-ferromagnetic conductors. Titanium skull plates are non-ferromagnetic and low conductive and may have radial incisions that reduce induction forces. Some titanium skull plates may be safe for TMS. The net energy of a stainless-steel aneurysm clip is typically less than 1010 J, which corresponds to a vertical movement of less than 0.3 μm for the aneurysm clip and is unlikely to cause clinical problems (Loo et al. 2011). A cochlear implant consists of a loop antenna, a permanent magnet, an electronic chip implanted under the scalp, and electrodes implanted in the cochlea, which may be moved or demagnetized by TMS pulses. Therefore, TMS should not be applied to subjects who have received cochlear implants. Jewelry, glasses, watches, and other items on the head that may be conductive or magnetic should be removed during cranial magnetic stimulation to prevent interaction with the magnetic field.

### 5.1.3 Induced Voltage

Strong magnetic field pulses from the TMS coil can generate high voltage in nearby leads and electronic devices. The leads connected to electrodes on the scalp should be kept free of and should be twisted together to reduce magnetically induced voltage. TMS can generate voltage on the electrode leads, which can result in unexpected stimulation of the brain, whether the implant is turned on or off. Such materials include brain implants, such as deep brain stimulation (DBS) system, epidural electrode arrays for cortical stimulation and cochlear implants, contain intracranial electrodes connected to subcutaneous leads in the scalp. A significant limitation of the in vitro safety studies on induced voltage is that only the induced voltage between paired contacts on the electrode leads is tested, while the induced voltage between the electrode contacts and the contacts formed by the implantable pulse generator (IPG) housing is not measured. Therefore, according to the existing in vitro safety studies, the induced voltages and currents may significantly underestimate the magnitude of induction in human body (Rossi et al. 2009).

In addition to the voltages and currents induced in the stimulation leads, the electromagnetic pulses generated by TMS can cause failure or even damage of the internal circuit of the electronic implant near the TMS coil.

### 5.1.4 TMS in Patients with Implanted Electrodes

A large number of studies on TMS have been conducted in patients with implanted electrodes in the central and peripheral nervous system. Most of them adopt sTMS; some adopt paired-pulse transcranial magnetic stimulation (ppTMS), and a few adopt repetitive transcranial magnetic stimulation (rTMS). These studies are mainly intended to (a) assess the effects of TMS on the activities of central nervous system by recording TMS-induced responses or evaluating changes in sustained spontaneous electrophysiological activity after TMS in subjects through implanted electrodes and (b) assess the stimulation effects of implanted electrodes on nervous system structures revealed by TMS induced responses. Three major types of electrodes have been used in studies on patients with implanted electrodes: (1) epidural electrodes (implanted in the cerebral cortex or spinal cord); (2) deep brain electrodes; or (3) peripheral or cranial nerve stimulation electrodes (e.g., vagus nerve electrodes).

According to studies, TMS coils appear to be safe for use in patients implanted with stimulator in central and peripheral nervous systems when they are not close to the IPG system. However, we lack detailed information about what a safe distance between the TMS coil and the implanted stimulator is and how the shape and angle of the coil, etc. affect this relationship. Therefore, TMS should only be applied to patients with implanted stimulator if there is a scientifically or medically compelling reason to do so. The TMS procedures should be performed in strict accordance with predefined experimental protocols and settings and under the appropriate supervision of an institutional review board or ethics committee.

TMS is considered safe for individuals with VNS systems, pacemakers, and spinal cord stimulator, as long as the TMS coil does not activate components located in the neck or chest. If the TMS coil discharges close to the implanted lead connecting the electrode to the IPG, it may generate potentially significant voltages and currents between the electrode lead and the IPG, which may result in unexpected neural stimulation and pose a safety risk. This may occur during DBS and cortical stimulation with epidural electrodes. Additional safety studies should be performed to evaluate the magnitude of induced voltages and currents in the implanted stimulation system. In the end, TMS should not be applied to subjects with cochlear implants due to the multiple potentially unsafe interactions between TMS pulses and the cochlear implant (Brüel 1980).

### 5.1.5 Magnetic Field Exposure

The total duration is so short that there are no significant risks of magnetic field exposure to a single TMS or rTMS. However, the typical rTMS therapy for psychiatric applications (e.g., 10 Hz, 20 pulses, 5 s, for 20 courses) will yield a period of exposure lasting about 5 s in total. Theoretically, this exposure will fall into the radio frequency (RF) range supposing that each pulse of continuous stimulation lasts about 250 μs. However, it is not clear whether high-intensity pulse stimulation of TMS has the same long-term effects as continuous, low-intensity occupational exposure. What is even less clear is whether the effects of long-term exposure to rTMS are altered by concomitant medication. Nevertheless, it is noteworthy that long-term exposure to electromagnetic field at levels higher than those possible with TMS appears to be safe (Gates et al. 2010).

## 5.2 Adverse Reactions

### 5.2.1 Epilepsy

Seizures are caused by hypersynchronous firing of neuronal populations in the gray matter, mainly due to an imbalance between inhibitory and excitatory synaptic activities. Seizures can be induced by rTMS when the pulse frequency and stimulation intensity are relatively high and the interval between stimulation sequences is short. Theoretically, rTMS may cause seizures at two different periods associated with stimulation: (1) during or after rTMS training and (2) recurrent seizure during sequelae due to the modulation of cortical excitability (i.e., the kindling effect). The risk of rTMS-induced seizures is of course very low.

Kindling effect: In addition to immediate seizures, rTMS may have a lasting effect on seizure thresholds. The kindling effect is a process that occurs in animals, where the original subconvulsive stimulation is repeatedly applied, resulting in a gradual enhancement of induced electrical activities of nerves and eventually leading to seizures. The classical kindling effect is achieved by applying stimulations at a regular interval, with a specific combination of stimulation intensity, frequency, and stimulation duration. Currently, there are no reports on seizures or recurrent spontaneous seizures after TMS in human subjects or in clinical practice.

To avoid seizures, the maximum stimulation duration is limited. The risk of seizures will increase if such limit is exceeded. Considering the intensity and frequency, the maximum safe stimulation n duration for healthy adults is shown in Table 8.3. It is noteworthy that what in discussion is the safety limit of rTMS used as a stand-alone therapy.

### 5.2.2 Syncope

Vasodepressor (neurocardiogenic) syncope is a common response to anxiety and physical and psychological discomfort and may be more likely to occur than seizures during rTMS. Common symptoms include visceral discomfort, nausea, dizziness, pallor, sweating, tonic stiffness, convulsion, incontinence, and hallucination. Typically, the main difference between syncope and seizures in clinic is whether consciousness is rapidly recovered within seconds rather than min.

The initial measures for a suspected seizure and syncope are same. TMS should be terminated immediately. The sub-

**Table 8.3** Maximum safe stimulation duration for seizure avoidance (in s)

| | Intensity (% of motor threshold) | | | | | | | | | | | | | |
|---|---|---|---|---|---|---|---|---|---|---|---|---|---|---|
| Frequency (Hz) | 80–100 | 100 | 110 | 120 | 130 | 140 | 150 | 160 | 170 | 180 | 190 | 200 | 210 | 220 |
| 1 | >1800 | >1800 | >1800 | 360 | >50 | >50 | >50 | >50 | 27 | 11 | 11 | 8 | 7 | 6 |
| 5 | >10 | >10 | >10 | >10 | >10 | 7.6 | 5.2 | 3.6 | 2.6 | 2.4 | 1.6 | 1.4 | 1.6 | 1.2 |
| 10 | >5 | >5 | >5 | 4.2 | 2.9 | 1.3 | 0.8 | 0.9 | 0.8 | 0.5 | 0.6 | 0.4 | 0.3 | 0.3 |
| 20 | 2.05 | 2.05 | 1.6 | 1.0 | 0.55 | 0.35 | 0.25 | 0.25 | 0.15 | 0.2 | 0.25 | 0.2 | 0.1 | 0.1 |
| 25 | 1.28 | 1.28 | 0.84 | 0.4 | 0.24 | 0.2 | 0.24 | 0.2 | 0.12 | 0.08 | 0.12 | 0.12 | 0.08 | 0.08 |

ject should be assisted to lean properly without experiencing any impact, and his/her airway breathing and circulation should be evaluated. Unless there are tonic-clonic seizures, the subject should be turned to one side to help clear the airway and avoid inadvertent aspiration. Subjects with convulsion should be turned to one side immediately after the convulsion stops and remain in this position until their consciousness recovered. Delays in the recovery of consciousness of more than 30 s after a seizure require further medical evaluation.

### 5.2.3 Neuropsychological and Motor Effects

The possibility of rTMS disrupting neuropsychological function during actual application has been well recognized. It has been reported that rTMS is used as an investigation probe that in essence produces a temporary functional impairment to localize cortices involved in a specific task. However, the safety issue of rTMS as a therapy is primarily concerned with the presence of any lasting effects after rTMS, rather than during the stimulation. Many studies have tested the neuropsychological functions of patients with depression before and after treatment with rTMS in the left prefrontal cortex. It has been proved by most studies that results of cognitive function tests improve over time. So far, studies have tested the functions in a wide range of neuropsychological domains, the results of which suggest that rTMS is unlikely to cause significant neuropsychological impairment.

### 5.2.4 Hearing Impairment

When energized, the stimulation coil can produce a strong broadband noise artifact due to rapid mechanical deformation, along with a ticking sound. This noise sounds mild, but the intensity of the brief sound is often underestimated, which can exceed a sound pressure level of 140 dB, above the safety level that the auditory system can tolerate. High-power levels, sound pressure levels, and stimulation frequencies can induce severe hearing loss. After exposure to TMS, a small percentage of adults experience a transient increase in auditory threshold. Up to now, there is one case of permanent threshold shift in an individual receiving transcranial stimulation with H-coils and while not using earplugs. In most cases, transcranial stimulation does not result in hearing changes when hearing protection is taken. A report on hearing safety of pediatric cases showed that there was no change in hearing in 18 children without hearing protection. However, the sample size was too small to ensure hearing safety in children. In addition, toddlers are of particular concern because their ear canal resonance is different from that of adults. As their heads are smaller, TMS coils are closer to the ear. Besides, there is no hearing protection suitable for toddlers.

Therefore, it is recommended that (1) all subjects receiving rTMS take hearing protections (ear plugs or ear muffs); (2) individuals who experience hearing loss, tinnitus, or ear congestion after completing TMS be promptly evaluated for hearing; (3) patients with known pre-existing noise-induced hearing loss or receiving concurrent treatment with ototoxic medications (aminoglycosides, cisplatin) receive TMS only when the risk-benefit profile is satisfactory, for example, receiving rTMS for tinnitus; and (4) cochlear implant recipients should not receive TMS.

### 5.2.5 Localized Pain and Headache

sTMS is generally well tolerated as most participants feel painless. When triple stimulation technique is used, some discomfort may be caused due to the associated threshold peripheral stimulation. Sometimes, when TMS is applied, especially rTMS, the muscles and nerves near the stimulation coil are activated by TMS, and the muscle contractions and changes in cerebral blood flow due to the stimulation can cause a sensation of pain in some subjects. Indeed, this is the most common side effect of TMS. The intensity of pain experienced by different subjects varies, depending on the individual's susceptibility, position on scalp, coil design, and the intensity and frequency of stimulation. Patients and subjects should be warned that TMS may not be comfortable and may cause pain. However, this is negligible in terms of safety. So far, only a small proportion of patients in clinical trials of TMS discontinue treatment due to pain (2%).

The muscle contractions induced by TMS may be relevant, especially those away from the vertex. Neck pain may be related to the forced posture and head immobilization during rTMS or may be possibly related to cerebellar stimulation. Most subjects/patients experience localized pain, including toothache, during TMS, the effects of which disappear rapidly. However, headaches occasionally persist after TMS. In such cases, oral administration of common analgesics may be helpful.

### 5.2.6 Fever or Skin Burns

During TMS, vortex currents can be generated by any conducting material within the magnetic field. This brings about potential problems such as heat generation or movement of metals (e.g., leads and electrodes) or failure of electrical equipment. During rTMS, vortex currents generated by EEG electrodes on metal surfaces near the stimulation coil can generate heat and burn skin. Heat generation is related to electrode size and conductivity, as well as stimulation parameters. The radial incision of the electrode can reduce its tendency to heat up by cutting off the current path. Induced currents in the leads can also be reduced by moving the leads away from the circuit near the excitation coil. It is recommended that TMS be applied with extreme caution to any patient with implanted electronic devices (e.g., pacemakers, intracardiac catheters, drug pumps) and only if the benefits outweigh the risks. Intracranial metal implants are also noteworthy as they may heat or oscillate in a rapidly changing magnetic field.

### 5.2.7 Histopathological/Structural Changes

The histopathological changes that may be generated by TMS are mainly due to the hyperexcitability of massive neurons and are directly related to the stimulation effects. An animal experiment where electrical stimulations were repeatedly applied to brain showed that there were always histopathological damages under different combinations of charge and charge density at each stage during the 7-h stimulation at 50 Hz. The charge transfer during rTMS is very small compared to direct electrical stimulation. In addition, rat experiments have confirmed the relative safety of rTMS, with most not showing any histopathological changes. In most studies, histopathological changes are not found after rTMS treatment in humans. Therefore, rTMS is relatively safe.

### 5.2.8 Effects of Electromagnetic Field

The peak field strength of a conventional TMS coil is 2 T (Tesla) or less, which decays rapidly as it gets far away from the coil surface. Therefore, the head is only exposed to a strong and rapidly changing magnetic field. Although the local field strength is strong, the duration of exposure to rTMS is extremely short compared to the exposure to common environmental sources, even it is repeatedly applied. Long-term exposure to electromagnetic fields has not been proven to result in any health risk. However, it is necessary to be vigilant to the long-term effects of TMS on patients and researchers.

### 5.2.9 Psychiatric Complications

Although most researchers are interested in using TMS for psychiatric disorders, particularly depression, there have been some reports on the psychiatric side effects in patients with depression treated with TMS. In all cases, the psychiatric side effects caused by TMS are transient and resolved with the discontinuation of TMS or a rapid response to medication. However, subjects receiving rTMS should be informed of the risk of possible psychiatric complications, especially mania or hypomania.

### 5.2.10 Effects on Pregnant Patients

The risk of TMS on fetal development remains unknown. Although one prospective study suggests that exposure to magnetic fields ≥16 milligauss (unrelated to TMS) in early pregnancy may increase the risk of abortion. Given the limited knowledge in this area, rTMS should be applied to women of childbearing age with caution and applied to pregnant women only after careful consideration of the potential risk-benefit profile.

## References


Abbruzzese G, Assini A, Buccolieri A et al (1999) Comparison of intracortical inhibition and facilitation in distal and proximal arm muscles in humans. J Physiol 514:895–903

Brüel PV (1980) The influence of high crest factor noise on hearing damage. Scand Audiol Suppl Suppl 12):25–32

Chen AC, Zhang WT, Han JS. Transcranial magnetic stimulation (TMS): physiology, psychology, brain mapping and clinical applications. Sheng Li Ke Xue Jin Zhan. 2004;35(2):102–6.

Chen R (2000) Studies of human motor physiology with transcranial magnetic stimulation. Muscle Nerve Suppl 9:S26–S32

Chen R, Cros D, Curra A et al (2008) The clinical diagnostic utility of transcranial magnetic stimulation: report of an IFCN committee. Clin Neurophysiol 119:504–532

Corthout E, Barker AT, Cowey A (2001) Transcranial magnetic stimulation. Which part of the current waveform causes the stimulation? Exp Brain Res 141:128–132

Fang X, Ding HF, Huang YH et al (2018) Improved intracranial induced electrical field in transcranial magnetic stimulation with semiellipse coil pair. IEEE Trans Appl Supercond 28:4901306

Farzan F (2014) Single-pulse transcranial magnetic stimulation (TMS) protocols and outcome measures. In: Rotenberg A, Horvath JC, Pascual-Leone A (eds) Transcranial magnetic stimulation, 1st edn. Human Press, New York, pp 69–116

Gates JR, Dhuna A, Pascual-Leone A (2010) Lack of pathologic changes in human temporal lobes after transcranial magnetic stimulation. Epilepsia 33:504–508

Goetz SM, Nam TC, Gerhofer MG et al (2013) Analysis and optimization of pulse dynamics for magnetic stimulation. PLoS One 8:1–12

Hallett M (2000) Transcranial magnetic stimulation and the human brain. Nature 406:147–150

Hamada M, Terao Y, Hanajima R et al (2008) Bidirectional long-term motor cortical plasticity and metaplasticity induced by quadripulse transcranial magnetic stimulation. J Physiol 586:3927–3947

Harel EV, Rabany L, Deutsch L et al (2014) H-coil repetitive transcranial magnetic stimulation for treatment resistant major depressive disorder: an 18-week continuation safety and feasibility study. World J Biol Psychiatry 15:298–306

Lin VW, Hsiao IN, Dhaka V (2000) Magnetic coil design considerations for functional magnetic stimulation. IEEE Trans Biomed Eng 47:600–610

Lisanby SH, Gutman D, Luber B et al (2001) Sham TMS: intracerebral measurement of the induced electrical field and the induction of motor-evoked potentials. Biol Psychiatry 49:460–463

Loo CK, Mcfarquhar TF, Mitchell PB (2011) A review of the safety of repetitive transcranial magnetic stimulation as a clinical treatment for depression. Int J Neuropsychopharmacol 1:131–147

Mazzocchio R, Rothwell JC, Day BL et al (1994) Effect of tonic voluntary activity on the excitability of human motor cortex. J Physiol 474:261–267

Parazzini M, Fiocchi S, Chiaramello E (2017) Electric field estimation of deep transcranial magnetic stimulation clinically used for the treatment of neuropsychiatric disorders in anatomical head models. Med Eng Phys 43:30–38

Pascual-Leone A, Davey M, Wassermann EM, Rothwell J, Puri B (eds) (2002) Handbook of transcranial magnetic stimulation. Edward Arnold, London

Peterchev AV, Murphy DL, Lisanby SH (2011) Repetitive transcranial magnetic stimulator with controllable pulse parameters. J Neural Eng 8:036016-1–036016-13

Ridding MC, Rothwell JC (1997) Stimulus/response curves as a method of measuring motor cortical excitability in man. Electroencephalogr Clin Neurophysiol 105:340–344

Rossi S, Hallett M, Rossini PM et al (2009) Safety, ethical considerations, and application guidelines for the use of transcranial magnetic stimulation in clinical practice and research. Clin Neurophysiol 120:2008–2039

Rossini PM, Barker AT, Berardelli A et al (1994) Non-invasive electrical and magnetic stimulation of the brain, spinal cord and roots: basic principles and procedures for routine clinical application. Report of an IFCN committee. Electroencephalogr Clin Neurophysiol 91:79–92

Roth Y, Amir A, Levkovitz Y et al (2007) Three-dimensional distribution of the electric field induced in the brain by transcranial magnetic stimulation using figure-8 and deep H-coils. J Clin Neurophysiol 24:31–38

U. S. Food and Drug Administration (2011) Repetitive transcranial magnetic stimulation (rTMS) systems-Class II Special Controls Guidance for Industry and FDA Staff [EB/OL]. https://www.fda.gov/medical-devices/guidance-documents-medical-devices-and-radiation-emitting-products/repetitive-transcranial-magnetic-stimulation-rtms-systems-class-ii-special-controls-guidance

Ueno S, Tashiro T, Harada K (1988) Localized stimulation of neural tissues in the brain by means of a paired configuration of time-varying magnetic fields. J Appl Phys 64:5862–5864

von Giesen HJ, Roick H, Benecke R (1994) Inhibitory actions of motor cortex following unilateral brain lesions as studied by magnetic brain stimulation. Exp Brain Res 99:84–96

Xia M, Wang J (2013) Introduction to electromagnetic field theory and calculation methods. Peking University Press, Beijing, pp 3–15

Yiftach R, Abraham Z (2012) Basic principles and methodological aspects of transcranial magnetic stimulation. In: Miniussi C, Paulus W, Rossini PM (eds) Transcranial magnetic stimulation, 1st edn. CRC Press, Boca Raton, pp 3–34

Zhao K, Chen X (2011) Electromagnetics, 3rd edn. Higher Education Press, Beijing, pp 11–12

Zhi DD, Sarah HL, Angel VP (2013) Electric field depth-focality tradeoff in transcranial magnetic stimulation: simulation comparison of 50 coil designs. Brain Stimul 6:1–13

# 9 Basics of Magnetic Stimulation Technique

Weimin Wang, Yuzhou Guan, Hua Lin, Wensi Hao, Siran Li, Qilin Zhou, Xueli Song, and Yicong Lin

## 1 Magnetic Stimulation of Brain Conductors

### 1.1 Basics of Physiology

The basic unit of brain tissue is the cell. Brain tissue has special functions and is composed of well-organized cells with similar shapes, functions, and properties. The cell is the fundamental unit of life, which consists of membranes, nucleus, and cytoplasm containing a large amount of intracellular fluid. A biological membrane is mainly composed of a lipid bilayer with embedded proteins, as well as a small number of sugars, inorganic ions, and water molecules, its main function is to control the movement of ions and molecules in and out of cells. The "channels" formed by the proteins embedded in the surface allow for ion exchange inside and outside the cell, exhibiting selective permeability. Intra- and extracellular fluids are electrolytes with good electrical conductivity, and the electrical performance of cells depends on the membrane's ionic behavior.

W. Wang (✉)
School of Electronics, Peking University, Beijing, China
e-mail: wmw@pku.edu.cn

Y. Guan
Department of Neurology, Peking Union Medical College Hospital, Beijing, China

H. Lin · W. Hao · S. Li · Y. Lin
Department of Neurology, Xuanwu Hospital, Capital Medical University, Beijing, China

Q. Zhou
Department of Neurology, Fuwai Hospital, National Center for Cardiovascular Diseases, State Key Laboratory of Cardiovascular Disease, Chinese Academy of Medical Sciences and Peking Union Medical College, Beijing, China

X. Song
Department of Neurology, Yuquan Hospital, Tsinghua University (Tsinghua University Hospital of Integrated Traditional Chinese and Western Medicine), Beijing, China

The electrical activations of nerves and muscles are associated with the behavior of passing membranes for sodium, potassium, and other irons. Relevant studies have shown that potassium concentration outside excitable cells is much lower than that inside cells and intracellular sodium and chlorine concentrations are even lower. Differences in intracellular and extracellular concentrations make ions diffuse from areas of high concentration to areas of lower concentrations at the diffusion rate determined by membrane permeability and concentration differences. The potential difference across the membrane arises from ions being charged. This potential difference creates an electrical field in the membrane.

In the resting state, there is an electric potential difference of about −70 mV across the two sides of the membrane, with the inside of the cell negative with respect to the outside. When cells are stimulated, the electric potential difference between the two sides of the membrane changes, accompanied by a weak current. If the intracellular electric potential changes in a positive direction, that is, the normal resting potential decreases, then the stimulus is excitatory. If the intracellular electric potential changes in a negative direction with respect to the extracellular potential, that is, the normal resting potential increases, then the stimulus is inhibitory or hyperpolarized. Neurons, also known as nerve cells, consist of a cell body, dendrites, and an axon. Dendrites resemble a tree-like structure, receive input information from other cells, and transmit them to the soma. An axon is an upper extension leaving nerve, that carries electrical signals from the cell body, and is the main unit of conductivity of neurons. The flow of information between cells is actually the flow of electrical current, and the macroscopic expression of the flow of electrical currents between brain cells is the electrical potential occurred in the brain cortex. When the characteristics of the tissue change, the electrical potential between the two sides of the membrane will change, accordingly, the electrical current will also change.


Y. Wang (ed.), *Therapeutics of Neural Stimulation for Neurological Disorders*, https://doi.org/10.1007/978-981-99-4538-2_9

Electrical conductivity is variable, depending on changes in brain tissue. Pathological changes in the tissue will lead to changes in the characteristics of the tissue, changes in the ion movement state of neurons in the tissue, changes in the discharge state, and consequent changes in the electrical conductivity of the tissue and in the induced magnetic field. For example, when a tumor develops in the brain, the electrical conductivity is half that of normal brain tissue; during a seizure, the conductivity increases by 20%, etc. When the conductivity of a part of tissue changes, the firing pattern of neurons in adjacent tissues and the conductivity at tissue boundaries are also affected. When the conductivity at the tissue boundaries changes, which can promote or inhibit the transmission of neuronal electrical signals, thereby indirectly affecting the induced magnetic field. The induced electric field creates the induced magnetic field in the brain, which in turn affects neuronal firing. In addition, as the highest nerve center regulating human functions, gray matter is composed of various neurons, glial cells, and an abundance of nerve fibers, forming a complex communication system between neurons. When gray matter changes, in which many neuronal firing patterns change, affecting the induced magnetic field more than other tissues.

### 1.2 Basics of Physics

Magnetism is one of the most common phenomena in the material world. As a kind of magnetic medium, the magnetism of biological tissue comes from the biological tissue structure, the physiological process and the influence on the organisms caused by the external magnetic field. The magnetic field of human tissues is roughly divided into the magnetic field generated by the motion of charged particles in human bodies, such as the magnetic field generated by ions motions in the physiological process, the magnetic field of neurons, muscles, and other excitable tissues, etc.; and the induced magnetic field generated by biological magnetic materials, such as the magnetic field produced from the liver, kidney, and other organs under the external magnetic field.

Electric field and magnetic field can be converted to each other. Electric conductor can produce a magnetic field surrounding it, the current in the conductor varies with time, and the resultant magnetic field also varies with time. The time-varying magnetic field will produce an induced magnetic field, in which the closed conductor produces an induced current and the non-closed conductor generates an electrical potential.

Since the amplitude of the electroencephalogram (EEG) signals reflecting the advanced functions of neurons is much lower than the signal amplitude produced by the spontaneous electrical activities, they are often covered by the signals generated by spontaneous electrical activities and are not easy to be observed. We can use the external alternating magnetic fields to produce the stimulus signals to induce certain advanced brain functional changes. The rapid time-varying current in the coil can induce the generation of a pulsed magnetic field, under the influence of the magnetic and electric field, ion channels in the cell membrane open, and ions flow through the cell membrane to generate the action potential. This action potential passes along the nerve fiber and activates large pyramidal neurons, leading to microscale changes in axons and further electrophysiological and functional changes. The magnetic field can penetrate through the scalp, skull, and other structures in a noninvasive manner and without attenuation. When it arrived at the brain cortex, it will generate the corresponding current. Once reaching a certain threshold of intensity, it can promote or inhibit nerve activities. This kind of excitatory or inhibitory effect can be shown as either long-lasting effects or immediate effects.

In all human tissues, the electric conductivity of nerve tissues is relatively high, while the electric conductivities of muscle and bone are relatively low (Wang Liwei 2000). Therefore, magnetic field stimulation can generate a large induced current in the nervous system, neurons of the cerebral cortex are very sensitive to the induced currents induced by magnetic changes. When the magnetic stimulation effect arrives at the motor neurons in the spinal cord through the corticospinal tract, MEP and compound muscle action potential on the target muscle can be recorded. The generation of evoked potential is induced by magnetic stimulation, first producing an induced electric field at the neurons, and then the induced electric field acts on the nerves to generate a neural response potential. The evoked potential can spread to other cortical areas, its excited neurons act on surrounding neurons and induce other neurons to generate excitements, forming the propagation of the evoked potential.

## 2 Peripheral Magnetic Stimulation

Almost 190 years ago, Faraday discovered that a current varying with time creates a magnetic field, that can trigger another current in a nearby medium. The physiological foundation of magnetic stimulation is Faraday's Laws of electromagnetic induction, which makes changing magnetic field create a current in the conductor of the human body (Maccabee et al. 1991). Around 60 years ago, Kolin first confirmed that a certain magnetic field can stimulate nerves in animal models. In 1982, scholars from the University of Sheffield first developed a practical magnetic peripheral stimulator and used it to simulate human peripheral nerves. The varying current flow passing to the coil creates a magnetic field around the coil, when the magnetic field passes into the human body, it can create a voltage difference between any two points, thereby creating a current. Peripheral magnetic stimulation (PMS) is also known as transcutaneous magnetic stimulation (Maccabee et al. 1988). When magnetic stimulation is applied to the peripheral nerve, the stimulation coil can be placed in any location on the body surface where peripheral nerves walk by, and compound muscle action potential (CMAP) and distal sensory nerve action potentials (retrograde) are recorded on the effector muscle. Its stimulation effect on nerves is similar to electrical stimulation, inducing motor nerve excitation to facilitate muscle contraction. Since it can stimulate peripheral nerves without attenuation, the stimulation in the brachial plexus, lumbosacral plexus nerve, nerve root, deep nerve radial nerve, and the sciatic nerve all can obtain muscle action potential. Sensory nerve action potentials can be recorded by stimulating local skin (especially the deep sensory receptors). If the stimulation is at the wrist, muscle action potentials can be recorded in the muscles of the palm, and sensory nerve action potentials can be recorded in the fingers or forearm. Unlike electrical stimulation, magnetic stimulation does not need a flow of current through electrodes or mediums such as skin and tissue interface and it does not stimulate the nerve tissues directly. Since magnetic stimulation can pass any medium without attenuation, it can stimulate deep spinal nerve roots or deep muscles, the attenuation of magnetic stimulation is only associated with the distance away from the coil, therefore, magnetic stimulation does not need to touch patients' skin directly, also the patients do not need to undress. Due to the weak excitatory effect on the sensory nerves, magnetic stimulation rarely induces pain.

The mechanism of peripheral magnetic stimulation is to recruit peripheral afferents and induce affect cortical activities. The induced cortical remodeling has been clinically proven to be a treatment method for stroke. Magnetic stimulation regulates muscle excitability by directly exciting sensorimotor fibers and muscle neuromuscular junctions. It has been shown that stimulating nerve roots and muscles can reduce muscle spasms and pain.

The peripheral magnetic stimulation coil can be shared with the transcranial stimulation coil. The peripheral magnetic stimulation coil is generally smaller than the transcranial coil and has a lower maximum output power. The most commonly used head coils (high magnetic field generator) include round and 8-shaped coils. Under the same magnetic field, 8-shaped coils create a stronger current than round coils but decay faster with distance. Round coils are more unfocused but create a deeper magnetic field.

Therefore, the stimulation and recording sites of motor nerve stimulation and sensory nerve stimulation of peripheral magnetic stimulation are the same as those of electrical peripheral nerve stimulation. The intensity of peripheral nerve stimulation is commonly 60% of the maximum output (usually 2.1–3.8 T). By far, the parameter settings of peripheral magnetic stimulation have not reached an agreed standard (Beaulieu and Schneider 2013). The wave latency of muscle CMAP generated by magnetic peripheral nerve stimulation is the same as that of electrical stimulation, and its amplitude is higher than that of electrical stimulation.

Nerve root stimulation: nerve root stimulation can be applied to study proximal nerve function by stimulating nerve roots. Central motor conduction time (CMCT) can be calculated by subtracting phrenic nerve conduction time (PNCT) from cortical motor stimulation latency.

## 2.1 Parameter Settings

(1) Mode: there are continuous "on" mode and "on-off" mode. The "on" mode is to keep stimulating during the treatment or testing period; the "on-off" mode is an interval switch mode, and the "off" period can avoid coil overheating, but it is still unclear how long the "off" period should last. (2) The number of stimulations: the total number of stimulation is a key parameter of the cortex in transcranial magnetic stimulation (TMS), but there is no agreed standard for the number of peripheral nerve stimulation. (3) Stimulation frequency: low-frequency (<1 Hz) transcranial magnetic stimulation has an inhibitory effect on the cortex, while high-frequency (>5 Hz) stimulation has an activation effect, however, the role of stimulation frequency in the peripheral nerve is still unclear. (4) Intensity: the intensity of peripheral nerve stimulation is also expressed by the magnetic field intensity and the percentage of the maximum magnetic field, but the actual magnetic field intensity through the tissue cannot be detected. Factors that affect intensity include the shape of the coil and the distance from the tissue to the coil, the muscle contracting strength reflects the intensity of stimulation (Kanjanapanang et al. 2020).

Maccabee et al. found that the wave amplitude of CAMP was the highest when the round coil runs perpendicular to the nerve, and the latency was the same when the head coil runs perpendicular or parallel to the nerve. And the sensory nerve conduction showed that the wave amplitude of SNAP was the highest when the head coil runs perpendicular to the nerve, and the latency was the shortest. However, the wave amplitude was the same when the 8-shaped coil runs perpendicular or parallel to the nerve.

Repetitive magnetic stimulation at the periphery (rPMS) restores motor function by repeated stimulation to nerve roots, peripheral nerves, and muscles. The rPMS of nerve roots may cause an immediate decrease in the H reflex amplitude in normal people, stimulating muscles cannot affect the excitability of the spinal cord but may cause the delayed transmission of the neuromuscular junction.

### 2.2 Safety

(1) Thermal effects: the magnetic head coil leads to local heating, so when applying magnetic stimulation locally, the application time and intensity should be adjusted based on the tolerance of local tissues. Normally, the heat tolerance temperature of body tissue is 43 °C. The device is usually equipped with the temperature display, it will show the temperature once it exceeds 40 °C. The air-cooling device and water-cooling device should be equipped on the coil. (2) Magnetic field effect: patients with magnetospheric metal implants can refer to the indications for MRI examination. (3) Effects on electronic products: magnetic field can interfere with small electronic products such as deep brain implants or cochlear implants, especially heart pacemakers. So having electronic implants in the body is a contraindication to magnetic stimulation. (4) Pediatric application: the safety of single pulse or paired transcranial magnetic stimulation in children over 2 years of age has created a consensus, but the studies about the safety of peripheral nerve stimulation are limited. (5) Pregnant women: direct stimulation on the lumbar spine of pregnant women should be avoided. It is recommended that the magnetic head coil should be placed at a distance of 70 cm from the pregnant women (Chalfouh et al. 2020).

### 2.3 Clinical Application

Upper limb MEPs are recorded from abductor digiti minimi. Wrists (ulnar nerves), Erb's point (brachial plexus), C7 spinous process [cervical nerve roots (Schmid et al. 1990)], and cortex are stimulated in order. Lower limbs MEP are usually recorded in the tibialis anterior, stimulating points are in the popliteal fossa (peroneal nerve), hip (sciatic nerve), T12 spinous process (lumbar nerve roots), and cortex.

Clinical application: (1) Myofascial pain syndrome: peripheral magnetic stimulation treatment for trapezius myofascial pain is proved to improve the subjective and objective scores after 3 months. (2) Traumatic brachial plexus disease: magnetic stimulation treatment for post-traumatic brachial plexus injury improved the electrophysiological parameters, hand-grip strength, and pain score. (3) Post-traumatic peripheral neuralgia: pain score decreased 60–100% after low-frequency 0.5 Hz of magnetic stimulation. (4) Acute and chronic low back pain: acute low back pain can be immediately improved after magnetic stimulation treatment. It can be seen that the cortical pain area is not activated, and the combination of exercise and training has a positive effect on pain and functional score. (5) Reduce muscle spasms: peripheral nerve stimulation may relieve muscle spasms through the stimulation of the nerve root. (6) Increase muscle strength: the mechanism behind that peripheral nerve stimulation can increase muscle strength may be the proprioceptive effect of peripheral nerve stimulation on the brain. (7) Language obstacles: the study for the cases of stroke patients with language obstacles suggests that language function can be restored after peripheral nerve stimulation.

Studies of rPMS treatment for stroke showed that rPMS do not show a more significant benefit in improving muscle strength and muscle tension when compared with the sham group.

## 3 Magnetic Stimulation of the Spine

When magnetic stimulation coils are placed at the spine, stimulating the spinal cord can simultaneously stimulate the conduction tract and nerve roots to excite muscles to create the action potential, which is called transcutaneous magnetic stimulation of the spine (tsMSS). Round coils or 8-shaped coils can be used for magnetic stimulation of spinal cord nerve roots. Upper limbs MEP are usually recorded in abductor digiti minimi, stimulation points are in order of wrists (ulnar nerves), Erb's point (brachial plexus), C7 spinous process (cervical nerve roots), and cortex, for checking whether the function of peripheral nerve segments is normal. Lower limbs MEP are usually recorded in the tibialis anterior, stimulating points are in the popliteal fossa (peroneal nerve), hip (sciatic nerve), T12 spinous process (lumbar nerve roots), and cortex.

### 3.1 Parameter Settings

(1) Mode: there are continuous "ON" mode and "ON-OFF" discontinuous mode. The continuous "ON" mode can result in high local skin temperature. The differences in stimulation effects of round coils and 8-shaped coils on the muscle and spinal cord have not yet come to a conclusion. The stimulation points of the 8-shaped coils are more focused, while round coils can stimulate a deeper and larger range, thereafter, round coils are recommended for spinal cord stimulation, and EMG monitoring is used in the stimulating and surrounding segment muscles to regulate and avoid stimulation in a larger range. (2) The number of stimulations: there is no unified standard, 400–5000 times; The longer the stimulation last, the stronger the roles in motor-sensory pathways. (3) Stimulation frequency: currently the effects of different frequencies on the spinal cord have not been determined, lower-frequency is usually used for pain stimulation. Spinal cord stimulation commonly selects 10–20 Hz of high-frequency stimulation to activate motor never, higher frequencies need to be verified for safety. (4) Intensity: stimulation intensity is the key parameter to determine the magnetic stimulation effects. The stimulation intensity depends on the depth of the

stimulated parts, a 1.2 T magnetic field is usually selected, and higher intensity needs to be verified for safety. (5) Sham stimulation selection: in randomized double-blind trials, sham stimulators that can induce similar body sensations or muscle contractions are needed for double-blind studies. (6) The difference in the latency between the single transcranial magnetic stimulation (TMS) and magnetic spinal cord stimulation when recording CMAP in the same muscle can estimate the conduction velocity of the pyramidal tract, that is, CV = (cortical latency − spinal cord latency)/the distance between stimulation sites.

### 3.2 Comparison of Magnetic Spinal Cord Stimulation and Electrical Stimulation

Compared with electrical stimulation, magnetic spinal cord stimulation has the benefits of being painless and noninvasive and can penetrate deep tissues without attenuation. It can also be used in children and can stimulate different muscles individually. It has better therapeutic effects for stroke, multiple sclerosis, and spinal cord injury. However, its stimulation points cannot be accurately located, the coils need to be cooled after overheating, and the stimulation range expands after increasing the intensity, the equipment is expensive, and the time of location and placing is long, the coils shift after muscle contraction, has a weaker therapeutic effect on complete spinal cord injury.

Many researchers did clinical studies on the spinal cord or peripheral magnetic stimulation; however, all studies were small-scale observation experiments with dozens of cases. From Nielsen in 1995 to Beaulieu in 2015, studies on multiple sclerosis, stroke, spinal cord injury, and peripheral neuralgia were conducted.

### 3.3 Clinical Application

Beaulieu and Schneider reviewed the literature on rPMS published before March 2015. Among the 24 papers included, 16 studies had administrated rPMS to treat sensorimotor impairments caused by several neurological diseases and 8 studies had used rPMS to reduce pain.

Twenty-two of the 24 papers provided information on the type of coil, with seven studies using round coils (RC) and 12 studies using figure-of-eight coils (Fof8). The remaining three studies used both Fof8 and RC. In more detail, 11 of the 12 studies using Fof8 coils were designed to stimulate superficial muscles or nerves, while all seven studies using figure-of-eight coils were designed to stimulate deeper conductive structures, such as the spinal roots. The reason for this difference is the different focusability and depth of penetration of the two types of coils. First, the RC is less focal with a stimulus area equal to its diameter, whereas the Fof8 produces a magnetic field with a relatively accurate focus. Second, the magnetic field of Fof8 decreases faster with distance than that of RC. Therefore, the authors suggest that RC is more effective for stimulating deep conductive structures (e.g., nerve roots), while Fof8 is suitable for selectively recruiting superficial structures (e.g., muscles and nerves) without co-activating nearby tissues.

Seven studies provided sufficient detail about the orientation of the coil relative to the target structure, but the direction of the current within the coil was not provided by any studies. Four studies used data from basic studies to determine the coil orientations, and another study performed pilot test to determine the optimal coil location/orientation. The authors conclude that while we can select coil orientation to adequately stimulate the nerve trunk or spinal roots based on previous basic studies, there is little evidence regarding other peripheral conduction structures (e.g., muscles, whose fiber orientation varies from person to person). To induce stronger muscle contractions or better-tolerated sensations when stimulating structures like muscles, different combinations of coil orientation and current direction may be tested.

Seventeen studies used intermittent stimulation and provided details of the ON period (active stimulation) and OFF period (pause) for rPMS application, and other studies applied either continuous stimulation or their combination. In the intermittent protocol, the OFF period was about 4 times longer than the ON period. However, the duration of ON period seemed to be arbitrary, while the longer duration of OFF period seemed to arise from the technical limitations of the rPMS stimulators. The authors suggest that future studies should test the after-effects of ON versus ON/OFF protocols and report any side effects.

Twenty-two studies reported the total duration or number of rPMS stimulation. In summary, the duration of stimulation and the number of stimuli were highly variable across studies (13.4 ± 8.8 min and 2914 ± 2552 stimuli). The authors concluded that there were insufficient data to show whether the longer rPMS durations or larger numbers of stimuli better impact the sensorimotor systems.

Eleven studies applied rPMS at frequencies of 20–25 Hz, 6 studies used 15 Hz or lower, 4 studies used theta burst stimulations, only one study on pain conditions used a very low frequency (≤0.5 Hz). There appears to be no evidence to determine whether rPMS frequency affects the outcomes.

Most studies used percentages of motor threshold or spinal motion threshold to show the intensity of stimuli. Seventeen studies applied suprathreshold intensity, 4 used subthreshold intensities, 1 used both suprathreshold and subthreshold protocols. Notably, all studies using subthreshold intensities were intended to treat pain reductions, as evidence from electrical stimulation studies suggested that stimulation which was insufficient to induce muscle contraction is effective in reducing pain. However, some studies have shown that above-threshold intensities could also reduce pain. Therefore, the authors conclude that intensity appears to be a determinant of rPMS after-effects.

(1) Multiple sclerosis (MS): In 1995–1997, Nilsen made a repeated stimulation, 3750–10,000 times, on 60 cases of MS with round head coils at the T8 level, 0.95–1.26 T, and the results showed MS functional scores were improved. (2) Spinal cord injury, (3) stroke, Heldmann and Struppler, Struppler et al. stimulated over 2000 times on forearms or fingers in 70 cases of stroke using 8-shaped coils with an intensity of threshold at 20–25 Hz, there existed controversy over the improvement effect. Beaulieu stimulated the tibialis anterior, the results showed lower limb function was improved. The suprahyoid muscle was stimulated at 20 Hz for 3 s and ceased after 27 s in patients with dysphagia after stroke, 90% of the minimum motor threshold was stimulated for 6 days with the deglutition training, and the deglutition function was significantly improved when compared with that before stimulation. (4) Traumatic brain injury, (5) cerebral palsy, Flamand stimulated the lower limbs of cerebral palsy using 8-shaped coils with an intensity of threshold at 15 Hz for 600 times. (6) Traumatic brachial plexus neuropathy, (7) PD, the studies on Parkinson's freezing of gait with 25 Hz and 90% of the minimum motor threshold showed an immediately visible improvement in gait, (8) Fibromyalgia, (9) Complex regional pain: Pujol, Smania et al. stimulated the trigger point of pain in more than 60 patients with facial muscle pain syndrome using 8-shaped or round coils at 20 Hz below the threshold intensity, and the pain score was decreased. (10) Acute and chronic back pain. Studies of acute low back pain showed that the VAS score of the magnetic stimulation group decreased more significantly than that of the sham stimulation group after 10 days of treatment at 20 Hz for 25 s, 20% maximum stimulation, and 4000 times per day. Stimulations on the L4 spinous process of the spine at 20 Hz, 40% maximum output, 10 s stimulation with 30 s interstimulus interval, 6000 times per day, and 1 week's exercise training, suggested that chronic back pain was improved.

## 4 TMS-Induced Motor Evoked Potentials

Motor evoked potentials (MEP) refers to a technique for detecting compound muscle action potentials by stimulating motor cortex or central nervous system-related motor pathways. In clinical practice, MEP is used to detect the overall synchrony and integrity of motor pathways. A stimulator connected to the EMG machine applies electrical and magnetic stimulation to the subject's motor pathway, while the EMG machine collects the measured muscle action potentials in real time. Electrical stimulation MEP is often accompanied by pain; in contrast, magnetic stimulation MEP is more widely used in clinical practice due to its stability, painlessness, and noninvasiveness.

Transcranial magnetic stimulation (TMS) was introduced to elicit MEP in 1985 (Alisauskiene et al. 2005). According to Faraday's law of electromagnetic induction, a capacitor charged to 4 kV may generate a current of 8000 A in the coil after being discharged rapidly, whereby a magnetic field perpendicular to the coil appears and reaches a maximum value within 160 μs. This high magnetic field penetrates the skull and induces electrical currents in the cerebral cortex. The induced current stimulates brain tissue, leading to changes in nervous system excitability. The potentials induced by muscle contraction in the motor cortex stimulated by the induced current may be acquired by electromyography, such potentials are known as TMS-MEP (Banerjee et al. 1993).

Compared with direct electrical stimulation, the TMS magnetic field induces weaker currents on the surface, resulting in almost no pain. Due to its safety, noninvasiveness, and painlessness, magnetic stimulation MEP is commonly used in neuroelectrophysiology for evaluating the functional integration of corticospinal and corticobulbar pathways.

### 4.1 MEP Recording

Single-pulse TMS stimulates the primary motor cortex and generates descending impulse along corticospinal motor pathway activating spinal motor neurons and their innervated muscles. MEP is recorded when muscle is in contraction.

MEP stimulation sites: when recording the MEP of abductor pollicis brevis, the TMS stimulation target is placed 2 cm before C3 or C4, when recording the MEP of tibialis anterior, the TMS target is located 2 cm before Cz. The stimulation intensity of TMS is 120% of the resting motor threshold. For those who cannot be induced for resting motor threshold, 90% of the maximum output intensity of the machine should be used when recording MEP of upper limbs, and 100% of the maximum output intensity of the machine should be used when recording MEP of lower limbs. The MEP recording electrode is usually an Ag-Cl electrode, the cathode recording is located at the proximal end, and the anode recording is located at the distal end, the distance between the two ends is about 2 cm, and the electrode impedance is less than 5 kΩ, the ground wire is attached to the forearm of one side (see Fig. 9.1).

The distribution of magnetic field and induced electrical field depends on the shape and size of the TMS coils. Generally, the larger the TMS coils, the deeper the energy can reach, but the ability to focus is simultaneously reduced. Special round TMS coils have expanded stimulation areas and high energy. It is mainly used for inducing the MEP of

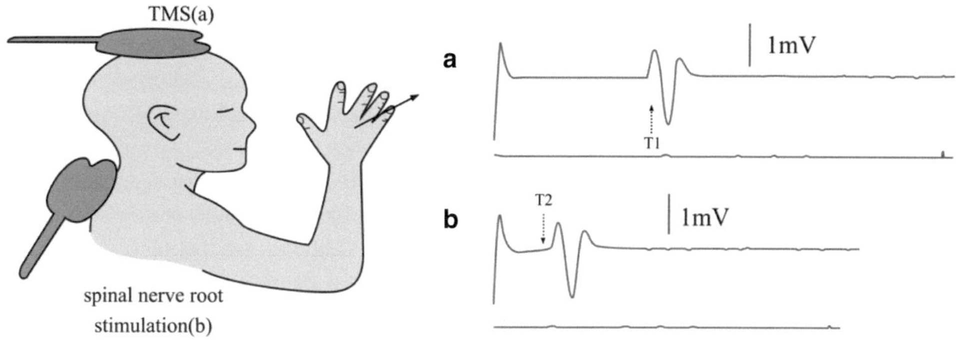


**Fig. 9.1** Diagram of MEP. (**a**) MEP waveform recorded on the innervated hand muscles by stimulating M1 region. (**b**) MEP waveform induced by stimulating cervical spinal nerve roots. T1: MEP latency induced by M1 stimulation. T2: MEP latency induced by stimulation of cervical spinal nerve roots. Central motor conduction time (CMCT) = T1 − T2

lower limbs, but the shortcoming of round coils is its no focusing for accurate stimulation. The stimulation depth of 8-shaped coils is light, but the ability to focus is stronger, and it can stimulate the target cortex precisely, so it has been more widely used in clinical practice.

### 4.2 Facilitation Effect

Studies have shown that the persistent slight active contraction of the target muscle can reduce the resting motor threshold, increase the MEP amplitude, and simultaneously shorten the latency, which is clinically called the facilitation effect (Abbruzzese et al. 1999). The active contraction of the muscle makes the synchronous activity of the descending nerve impulse from the motor cortex increase, strengthen the energy, or promote the re-input of the corresponding segment level.

### 4.3 Influencing Factors

Except for simple muscle active contraction, there are many methods that can facilitate MEP, such as lasting hand movement, imagined lasting hand movement, stimulating afferent sensory nerves, and high-frequency TMS. Isometric contraction of arm muscles can induce the facilitation of post-exercise MEP. If the clinical application is to confirm the delay of central motor nerve conduction, so active contraction of the target muscle is needed to promote response facilitation.

## 5 Cortical Silent Period

Cortical silent period (CSP) is an electromyographic activity suppression lasting approximately 100–200 ms arising from TMS-elicited MEP in the corresponding cortical motor cortex of the target muscle in a state of active, slight, and sustained contraction. The duration of such interruption of electromyographic activity is known as cortical silent period (Alisauskiene et al. 2005) (see Fig. 9.2).

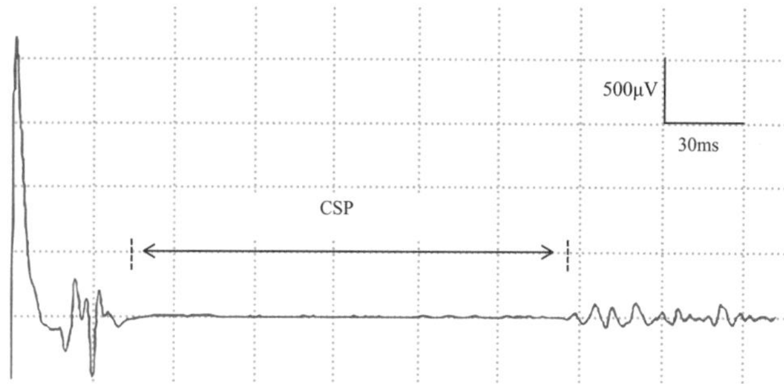


**Fig. 9.2** Diagram of CSP. CSP was characterized by an induced transient inhibition of ongoing motor activity after single-pulse TMS evoked muscle response during target muscle activation

### 5.1 Measurement Method

Stimulus intensity is 120% of the rest motor threshold (RMT), continuous slight contraction of affected abductor pollicis brevis (15–20% of the maximum contraction), stimulating the motor cortex of the patient's brain, and recording CSP. The length of the CSP is significantly prolonged, which cannot be explained by the refractory period of neurons. Typical CSP often occurs after an MEP, which can also be induced by the subthreshold stimulus. The CSP is stable in normal people and extends with the increase of stimulus intensity.

### 5.2 Application Value

CSP originates from the cerebral cortex, mainly produced by inhibitory intermediate neurons. Some studies suggest that CSP may be a result of GABA mediated activation of local inhibitory neural circuits while $GABA_A$ receptor-mediated inhibitory mechanism participates in the first 50–75 ms of the CSP and subsequently is thought to be mediated by cortical $GABA_B$ receptor. CSP is thought to be able to reflect intercortical inhibition. The longer the CSP, suggesting the lower the cortical excitability, and the higher the cortical inhibition. Whether for healthy subjects or patients, the CSP is recorded without the induction of MEP, confirming the excitatory and inhibitory descending cortical pathways are divided.

## 6 Paired-Pulse TMS Evoked Potentials

Paired-pulse transcranial magnetic stimulation (ppTMS) is a new technique for studying the changes of cortical excitability at different magnetic stimulation intervals. The generation of this kind of paired-stimulus pulses needs two stimulus capacitors that can lock the time relationship. The first released pulse is a conditional stimulus, followed by a testing stimulus (see Fig. 9.3). The time difference between the conditional stimulus and the testing stimulus is known as the interstimulus interval (ISI). When detecting different cortical excitability indexes using ppTMS, parameters of conditional stimulus, the intensity of the testing stimulus, and ISI are all adjusted accordingly. Paired-pulse evoked potentials can be used to study the excitatory facilitation effect and inhibitory effect in the ipsilateral cortex or bilateral intercortices (Müller-Dahlhaus et al. 2008).

When ISI was 1–4 ms of paired-pulse stimulus, the amplitude of MEP induced by the testing stimulus was lower than that induced by the single suprathreshold stimulus, suggesting MEP was inhibited, and this phenomenon is called intracortical inhibition (ICI). When ISI was 6–15 ms of paired-pulse stimulus, the amplitude of MEP induced by the testing stimulus was higher than that induced by the single suprathreshold stimulus, this phenomenon is called intracortical facilitation (ICF) (Fisher et al. 2002). Researchers found that ICI reflects the activity of inhibitory interneurons which was mediated by $GABA_A$ type. ICF is the mixed result of the strong facilitation role of glutamatergic neurons mediated by the *N*-methyl-D-aspartic acid (NMDA) receptor and the weak inhibitory role mediated by the $GABA_A$ type receptor (Chen 2004; Florian et al. 2008; Alisauskiene et al. 2005). When ISI was 100–200 ms of paired-pulse stimulus, both the conditional and testing stimuli are defined as the suprathreshold stimulus. Although the testing stimulus appeared after the CSP induced by the conditional stimulus, however, MEP was inhibited and even couldn't be induced during the testing stimulus process, named long-interval intracortical inhibition (LICI). LICI is mediated by $GABA_B$-type receptors, its changes reflect the influence of intracortical inhibitory mechanism on cortical excitability (Chen 2004; Florian et al. 2008; Irlbacher et al. 2007).

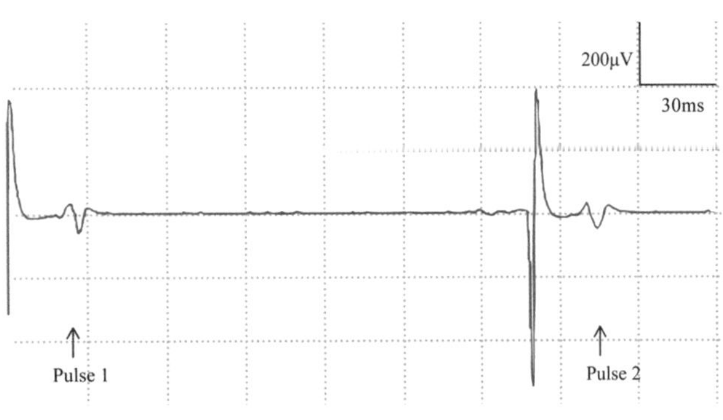


**Fig. 9.3** The diagram of paired-pulse evoked potentials

### 6.1 Measurement Method

First, confirm the mean MEP, that is 120% RMT single-pulse stimuli, 10 times, the mean MEP amplitude was determined as the mean MEP. ISI is paired-pulse stimuli at 2, 5, 10, and 15 ms, the intensity of the conditional stimulus is 80% RMT, and the intensity of the testing stimulus is 130% RMT. With a total of 10 times of stimulation, the mean MEP is the MEP of test response (TR). For the paired-pulse stimuli with the ISI of 150–300 ms, the intensity of both conditional stimuli and testing stimuli is 120% RMT above the threshold for 10 times. TR and conditioning response (CR) of MEP amplitudes are recorded, mean MEP is taken. The time interval between two pairs of paired-pulse stimuli needs to be longer than 15 s. The ratio of the MEP amplitude of paired-pulse stimuli TR at 2, 5, 10, and 15 ms to the average MEP amplitude (TR/MEP) was calculated. The ratio of the MEP amplitude of paired-pulse stimuli TR at 150, 300 ms to the MEP amplitude of CR was calculated.

## 7 Repetitive Transcranial Magnetic Stimulation

Transcranial magnetic stimulation (Jean-Pascal 2019) is a magnetic stimulation technique that uses a time-varying pulse magnetic field to act on the central nervous system (mainly the cerebrum), change the membrane potential of cortical neural cells, make it generate induced current, affect brain metabolism and neural electrical activity, thus causing a series of physiological and biochemical reactions. Repetitive transcranial magnetic stimulation (rTMS) is when two or more clusters of regular magnetic stimulation are output every time, repetitive frequency ≤1 Hz is called low-frequency rTMS, while the repetitive frequency >10 Hz is called high-frequency rTMS. Except for causing a series of physiological and biochemical changes, different patterns of repetitive transcranial magnetic stimulation can affect brain neural function and neural regulation.

## 7.1 Stimulus Frequency Effect

The division of repetitive transcranial magnetic stimulation between high frequency and low frequency is mainly according to the physiological function and risk of different frequencies of stimuli, it is agreed that low frequency below 1 Hz can induce the inhibition of cortical function and have no stimulus risks, but the function of high-frequency stimulus is the contrary, generally, stimulus with a frequency higher than 10 Hz is easy to induce increased cortical excitability, and an increased risk of side effects.

From the perspective of patient safety and tolerance, when the frequency of repetitive transcranial magnetic stimulation is no less than 5 Hz, the stimulation is divided into stimulus trains, and the time of each stimulus train is called train duration or length, the unit is expressed in seconds. There is an output off between each string, the time of output off between trains is called train interval. The number of stimulus impulses in each train is equal to frequency times train length. When the stimulus frequency is no more than 1 Hz, the stimulus impulse can be continuously output. The higher the frequency, the shorter the string length, and the longer the string interval.

## 7.2 Patterned Repetitive Transcranial Magnetic Stimulation (prTMS)

Patterned repetitive transcranial magnetic stimulation (patterned rTMS, prTMS) is a technical term proposed in the International TMS Symposium 2008. The content and meaning of prTMS are obviously different from the stimulation sequences of conventional rTMS, with the addition of various theta burst stimulation (TBS) modes, where each cluster is equivalent to one pulse in conventional rTMS, and the combination of multiple cluster stimuli is equivalent to one stimulus train in conventional rTMS.

TBS simulates the bursting discharge mode of physiological action potentials of the central nervous system, such as the common 5-Hz clustered action potential modes for nerve activity in the hippocampus, and 5 Hz is in the frequency range of theta wave of electroencephalogram. TBS (theta burst stimulation) most commonly used currently is a mixed stimulation mode, i.e., 5-Hz TBS, with multiple pulses in each cluster. At present, the most frequently used pulse frequency within the cluster is high frequency of 50 Hz, with three pulses forming a cluster. TBS with a continuous repetition frequency of 5 Hz is called cTBS, and intermittent TBS is called iTBS. A wide variety of TBS modes can be created by changing several parameters such as intra-cluster frequency, intra-cluster pulse number, and inter-cluster interval.

## 7.3 Effects of Stimulus Mode

Although cTBS mode is the continuous high-frequency stimulus with 50 Hz within 5 Hz, it cannot induce the increased excitability of neural function but can induce the inhibitory function of neural function rapidly. iTBS mode can induce an increase in long-term excitability produced by the neural function. These stimulus modes are different from the traditional stimulus. Compared with the traditional stimulus, the stimulating amount of iTBS only needs 80% of MT, and the stimulation duration declined to 1/10 of the original. The stimulation has obvious modulation effects on the nerve with few side effects and has clinical application value, which can meet the clinical needs of a large number of daily patients.

## 7.4 Multi-Target Repetitive Transcranial Magnetic Stimulation

Amounts of studies demonstrate that the brain dysfunction disorder is caused by abnormal nerve oscillations in different regions of the brain. Therefore, multi-target repetitive transcranial magnetic stimulation is expected to become an effective way to regulate this oscillation. There are studies in which 1 Hz of repetitive transcranial magnetic stimulation, with two targets of the right dorsolateral prefrontal cortex and right intraparietal sulcus, was used to treat patients with generalized anxiety disorder, all subjects were randomly divided into three groups, and the stimulus interval between two targets was 10 ms, 20 ms, and 50 ms, respectively, the results suggested that stimulus interval at 20 ms and 50 ms can improve the clinical symptoms of patients with generalized anxiety disorder (Yicong et al. 2020). In another study in which 1 Hz of repetitive transcranial magnetic stimulation, with two targets of the supplementary motor area and orbitofrontal cortex, was used to treat patients with obsessive-compulsive disorder, the Yale-Brown Obsessive-Compulsive Scale (YBOCS) of patients decreased by 33%, the Montgomery-Asberg Depression Rating Scale (MADRS) of patients improved by 48%, the clinical global impression rating scale (CGI-S) of patient improved by 33% (Richard and Daniel 2017). In other study in which 1 Hz of repetitive transcranial magnetic stimulation, with two targets of the primary motor cortex and dorsal premotor cortex, was used to treat patients with PD in order to improve their motor symptoms, however, the results suggested that this regimen cannot improve the motor symptoms of patients with PD (Christopher et al. 2019). The treatment effect and parameters of multi-target repetitive transcranial magnetic stimulation still lack sufficient evidence, waiting for future further research to clarify.

# 8 Combined Peripheral and Central Stimulation

## 8.1 Basic Concepts

Peripheral and central stimulation, first proposed by Stefan et al. (2000), is called paired associative stimulation (PAS, see Fig. 9.4) (Stefan et al. 2000). PAS is a new type of non-invasive brain stimulation (NIBS) technique, it is a method combining the electrical peripheral nerve stimulation (PNS) and transcranial magnetic stimulation (TMS). PNS excitation afferent sensory pathways are used, and TMS was provided in extracranial target regions after appropriate time intervals, exciting specific postsynaptic neurons. After repetitive stimulations mentioned above, it has a mechanism similar to spike-timing-dependent plasticity (STDP), which makes nerves remodeled, thus resulting in changes in the cerebral cortex excitability.

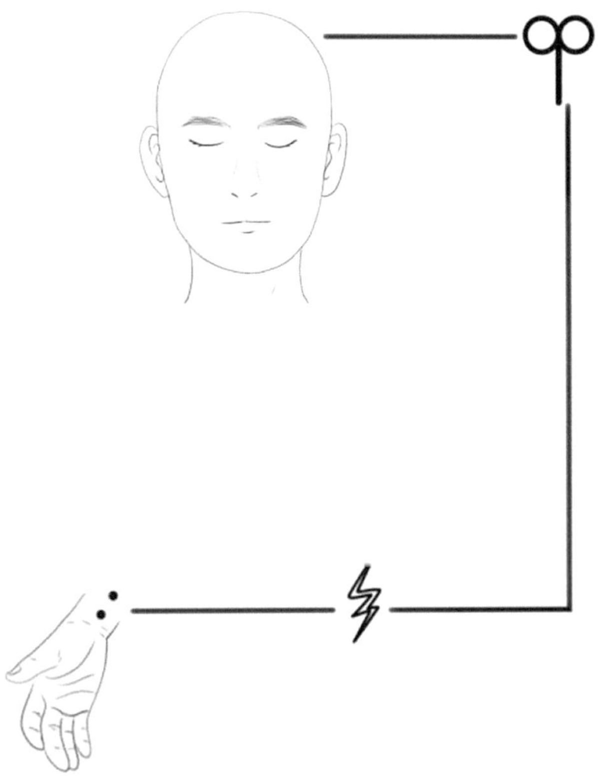

**Fig. 9.4** The diagram of paired associative stimulation

## 8.2 Principle of Action

The generation of the PAS method is based on Hebbian's STDP mechanism. When two excitatory signals, one weak and one strong, are input, the temporary order of presynaptic and postsynaptic stimuli determined the induction of long-term potentiation (LTP) or long-term depression (LTD) effects (Dan and Poo 2004). Experiments showed that changes in synaptic effects induced by associated excitatory signals depend on the order of presynaptic and postsynaptic neural excitation. If the postsynaptic neuron already produced an excitatory postsynaptic potential under the input stimulation, followed by another action potential, an LTP-like effect is induced. On the contrary, if the postsynaptic neuron already produced an action potential, then input stimuli make it produce an excitatory postsynaptic potential, and an LTD-like effect is induced. This important rule is called Hebbian's temporally asymmetric rule. PNS precedes TMS in most cases of PAS application. The induction of LTP-like or LTD-like postsynaptic effects in the application depends on the interstimulus interval (ISI) between PNS and TMS. If the ISI between PNS and TMS is slightly longer than the time for PNS-induced nerve pulse reaching the cerebral cortex, an LTP-like effect is induced, and the excitability of the cerebral cortex increases. On the contrary, the LTD-like effect is induced, and the excitability of the cerebral cortex decreased.

## 8.3 Action Parameters

PAS has two key parameters, ISI and stimulus frequency.

The first parameter, ISI, is the key parameter of PAS, which determines the direction where PAS changes the cerebral cortex's excitability. Currently, the most commonly used ISIs is 10 ms (denote as PAS10) and 25 ms (denote as PAS25). Multiple studies have shown that the excitability of the cerebral cortex increased after PAS25 treatment, and the excitability of the cerebral cortex decreased after PAS10 treatment. Notably, since the mechanism of PAS is similar to STDP, depends on the relative time of presynaptic and postsynaptic neural excitation, however, the association at this time point is not conclusive. Because there are usually multiple synaptic pathways involved in PAS examples, and LTP-like or LTD-like effects induced by PAS are associated with multiple receptors. The effects of PAS10 and PAS25 were blocked when using *N*-methyl-D-aspartate (NMDA) receptor antagonists, the changes induced by γ-aminobutyric (GABA) A receptor antagonists contribute to the enhanced cortical excitability. So, this uncertainty should be considered in the application of PAS.

The second parameter is the frequency of the stimulus. Studies showed that the frequency of PAS has a certain effect on the intensity of cortical excitability changes induced by itself. Wischnewski et al. conducted a meta-analysis of 104 PAS-treated healthy subjects and found that 0.05 Hz and 0.2 Hz of excitatory PAS have the strongest effect (Wischnewski and Schutter 2016). Another study showed that the increase of PNS frequency in PAS can increase the efficacy of PAS.

### 8.4 Action Characteristics

PAS-induced LTP-like or LTD-like effects have their own unique advantages. PAS's effects appear rapidly and are long-lasting, this advantage is superior to single repetitive peripheral nerve stimulation or repetitive transcranial magnetic stimulation. Studies showed that the time of excitability changes in the cerebral cortex induced by PAS was shorter, which can be shortened to 30 min, and the number of stimuli was less. The duration of action can last for at least 60 min after PAS. The direction of excitability changes in the cerebral cortex induced by PAS depends on ISI, rather than the frequency of the stimulus like repetitive transcranial magnetic stimulation. This feature allows PAS to reduce the subjects' headaches and other discomfort caused by high-frequency stimuli and reduce the risks of induced epileptic seizures. The effect of PAS has better region-specificity. Studies showed a specific increase in a small region of event-related potentials in the hand region of the cerebral cortex when PAS acted on the peripheral nerve and cerebral sensory cortex. The same region-specific results also appear when the brain's motor cortex is stimulated, which is different from the wide range of impacts induced by only using TMS, indicating that PAS has a stronger region-specificity. The excitability changes of the cerebral cortex induced by PAS are not just limited by the local cortical neurons in stimulated spots, but also can be transmitted by neural fiber network connections to adjacent or distant cortical regions, including cortical regions of the contralateral hemisphere.

### 8.5 Common Stimulation Methods and Applications

PAS commonly selects median or ulnar nerves for PNS and cerebral primary motor cortex or primary sensory cortex for TMS by measuring the changes in motor evoked potential (MEP) or somatosensory evoked potentials before or after PAS treatment to determine the changes in cerebral cortical excitability. Previously PAS is mainly used for the studies on the cerebral cortical remodeling in healthy people. In recent years, some clinical studies have applied PAS to the pathophysiological mechanism studies and treatment of disease. In the pathophysiological mechanism studies of diseases, the regional specificity of PAS is used to locate cerebral cortical lesions in focal dystonia such as writer's cramp and study the features of cerebral cortical remodeling in patients such as Parkinson's disease and senile depression. In the treatment of diseases, studies of PAS are mainly used for patients with spinal cord injury and stroke. Many studies showed that PAS treatments for stroke patients significantly improved limb dysfunction and swallowing function of patients, leading to a significant decrease in the rate of aspiration. Some of the studies demonstrated that the response rates of PAS for stroke patients are higher than repetitive transcranial magnetic stimulation. In learning and memory, many studies suggested that the motor learning ability is changed after using PAS, which supplied a more promising future for the application of PAS.

In conclusion, PAS, as a new type of, noninvasive and convenient non-frequency-dependent NIBS technique, has shown an important role in cerebral cortical remodeling. In the future, this method may contribute to explaining some pathophysiological mechanisms of diseases and may also become a treatment for diseases associated with abnormal remodeling of the human central nervous system.

## 9 Magnetic Seizure Therapy

Magnetic seizure therapy (MST) is a noninvasive technique that uses a high-frequency strong pulsed magnetic field to continuously stimulate the cerebral cortex for several seconds to induce a convulsive seizure. Modern MST equipment treats with stimulation of 100 Hz, 2 Tesla, and less than 10 s, repetitive implementations can be applied to treat neuropsychiatric disorders.

### 9.1 History of MST

MST was applied to non-human primates in 1998. MST was first applied to humans in 2000. A patient received MST twice during the course of electroconvulsive therapy (ECT). After that, 12 times MST treatments were successfully achieved in another patient. At that time, due to the limited performance of transcranial magnetic stimulation equipment, MST only can perform 40–50 Hz of stimulation which was 100% of the device output intensity. But, 50 Hz cannot induce a convulsive seizure stably, and the curative effect was not obvious. The reason may be related to the less powerful MST equipment and insufficient induction of stronger convulsive seizures. In late 2000, high-dose MST (HD-MST) was able to be developed, which can deliver a stable 100 Hz of continuous stimulation for 10 s and output

1000 pulses. After that, HD-MST was applied to monkeys, contrasted with electroshock, the effects of HD-MST with 400–1000 pulse train on neurophysiology and neuroanatomy were less than those of electroshock, and neurocognition still remained. In 2006, White et al. compared the HD-MST with electroshock in the patients with depression and found that HD-MST can improve the depression symptoms significantly, the consciousness recovery after treatment was faster than electroshock, and the need for muscle relaxants was less.

## 9.2 Implementation of MST

Similar to ECT, one course of MST often includes 8–12 times stimulations, within 3–4 weeks. Patients need to receive the treatment under general anesthesia and with muscle relaxants to avoid muscle twitches. Similar to ECT, a soft occlusal block is usually placed to protect the tooth (although the jaw contraction cannot be seen). Earplugs are recommended to protect the hearing of patients and researchers (Cretaz et al. 2015). Electroencephalogram was recorded during the peri-ictal period as a marker of curative effect (Perera et al. 2004).

## 9.3 Equipment and Parameters of MST

MST technique requires the equipment to have performances like high output intensity, high output frequency, long continuous output time, high power, good heat dissipation, and so on. The first generation of MST device can achieve a train stimulation with 40 Hz, 6.3 s (252 pulses per string) and induce convulsion in non-human primates, it was also the first device to successfully induce magnetic convulsions in humans. The second generation of MST device can supply a continuous stimulation with 50 Hz, 8 s (400 pulses per string) at 100% output strength. The third generation of HD-MST can implement with 100 Hz, and output 1000 pulses in 10 s. The fourth generation of HD-MST devices allows higher frequencies and longer train stimulation, but its stimulation strength declined from 100 to 49% at the highest frequency (250 Hz) of output.

When performing MST, large coils and coils with the poor focusing property are usually selected as stimulation coils (such as round coils) to stimulate the middle part of the cranial top, close to the bilateral motor cortex, which is easy to induce convulsions. The commonly used 8-shaped coils are hard to induce convulsions since their focusing points are too small, and not suitable for MST (Engel and Kayser 2016).

## 9.4 Clinical Application of MST

Since MST is a new technique that is still in the experimental stage, there are great differences in the study schemes of each center, most of the experiments are open research.

Studies showed that MST has a specific anti-depression role. White et al. compared the curative effect of MST with that of electroshock on the treatment of depression and found that the score of the Hamilton depression scale decreased significantly in both groups, but electroshock was more effective (an average reduction of 24 points in the electroshock group versus 18 points in MST group, $P < 0.05$). In 2011, Kayser et al. reported the first random study, 20 cases of patients with major depression and bipolar depression who received MST or electroshock treatment (Kayser et al. 2011). They found that there was no significant difference in symptom improvement between the two groups. Scale scores decreased by 11.9 in the electroconvulsive group and decreased by 12.4 in MST group. In MST group, scores of 6/10 patients decreased by over 50%. In the electroshock group, scores of 4/10 patients decreased by over 50%. After that there have been dozens of clinical data of patients, suggesting that MST had good anti-depression efficacy (Hoy et al. 2013). In Fitzgerald et al. studies, scores of depression decreased from 26.7 to 19.2, and scores of 5/13 patients decreased by more than 50% (Fitzgerald et al. 2013).

Compared with ECT, MST produces a shorter period of post-episode confusion in patients. The side effects of MST on cognitive function were smaller than that of electroshock treatment, in addition, it also played role in improving attention deficit, anterograde and retrograde amnesia (Kayser et al. 2015). Studies found that MST had the tendency of improving cognitive function, which may mean that MST had reversed the cognitive deficits that come from depression.

With therapeutic effects and less side effects, MST is considered a very promising new treatment for major depressive disorder. However, the clinical experience is limited and it cannot be promoted for further application. Further studies on its mechanism and randomized control studies for its efficacy validation are needed in the future.

## References

Abbruzzese G, Assini A, Buccolieri A et al (1999) Comparison of intracortical inhibition and facilitation in distal and proximal arm muscles in humans. J Physiol 514(3):895–903

Alisauskiene M, Truffert A, Vaiciene N et al (2005) Transcranial magnetic stimulation in clinical practice. Medicina (Kaunas) 41(10):813–824

Banerjee TK, Mostofi MS, Us O et al (1993) Magnetic stimulation in the determination of lumbosacral motor radiculopathy. Electroencephalogr Clin Neurophysiol 89(4):221–226
Beaulieu LD, Schneider C (2013) Effects of repetitive peripheral magnetic stimulation on normal or impaired motor control. A review. Clin Neurophysiol 43:251–260
Chalfouh C, Guillou C, Hardouin J et al (2020) The regenerative effect of trans-spinal magnetic stimulation after spinal cord injury: mechanisms and pathways underlying the effect. Neurotherapeutics 17(4):2069–2088
Chen R (2004) Interactions between inhibitory and excitatory circuits in the human motor cortex. Exp Brain Res 154(1):1–10
Christopher F, Charlotte D, Timo BW et al (2019) Dual-site transcranial magnetic stimulation for the treatment of Parkinson's disease. Front Neurol 10:174
Cretaz E, Brunoni AR, Lafer B (2015) Magnetic seizure therapy for unipolar and bipolar depression: a systematic review. Neural Plast 2015:521398
Dan Y, Poo MM (2004) Spike timing-dependent plasticity of neural circuits. Neuron 44:23–30
Engel A, Kayser S (2016) An overview on clinical aspects in magnetic seizure therapy. J Neural Transm 123:1139–1146
Fisher RJ, Nakamura Y, Bestmann S et al (2002) Two phases of intracortical inhibition revealed by transcranial magnetic threshold tracking. Exp Brain Res 143(2):240–248
Fitzgerald PB, Hoy KE, Herring SE et al (2013) Pilot study of the clinical and cognitive effects of high-frequency magnetic seizure therapy in major depressive disorder. Depress Anxiety 30:129–136
Florian J, Müller-Dahlhaus M, Liu Y et al (2008) Inhibitory circuits and the nature of their interactions in the human motor cortex a pharmacological TMS study. J Physiol 586(2):495–514
Hoy KE, Thomson RH, Cherk M et al (2013) Effect of magnetic seizure therapy on regional brain glucose metabolism in major depression. Psychiatry Res 211:169–175
Irlbacher K, Brocke J, Mechow JV et al (2007) Effects of GABA(A) and GABA(B) agonists on interhemispheric inhibition in man. Clin Neurophysiol 118(2):308–316
Jean-Pascal L (2019) Transcranial magnetic stimulation. Handb Clin Neurol 160:559–580
Kanjanapanang N, Munakomi S, Chang K-V (2020) Peripheral magnetic stimulation (transcutaneous magnetic stimulation). In: StatPearls. StatPearls Publishing, Treasure Island (FL)
Kayser S, Bewernick BH, Grubert C et al (2011) Antidepressant effects, of magnetic seizure therapy and electroconvulsive therapy, in treatment-resistant depression. J Psychiatr Res 45:569–576
Kayser S, Bewernick BH, Matusch A et al (2015) Magnetic seizure therapy in treatment-resistant depression: clinical, neuropsychological and metabolic effects. Psychol Med 45:1073–1092
Maccabee PJ, Amassian VE, Cracco RQ et al (1988) An analysis of peripheral motor nerve stimulation in humans using the magnetic coil. Electroencephalogr Clin Neurophysiol 70(6):524–523
Maccabee PJ, Amassian VE, Cracco RQ et al (1991) Stimulation of the human nervous system using the magnetic coil. J Clin Neurophysiol 8(1):38–55
Müller-Dahlhaus JF, Orekhov Y, Liu Y et al (2008) Interindividual variability and age-dependency of motor cortical plasticity induced by paired associative stimulation. Exp Brain Res 187(3):467–475
Perera TD, Luber B, Nobler MS et al (2004) Seizure expression during electroconvulsive therapy: relationships with clinical outcome and cognitive side effects. Neuropsychopharmacology 29:813–825
Richard CH, Daniel PW (2017) Positive clinical response to treatment of obsessive-compulsive disorder using dual-site transcranial magnetic stimulation. Aust N Z J Psychiatry 51(6):642–643
Schmid UD, Walker G, Hess CW et al (1990) Magnetic and electrical stimulation of cervical motor roots: technique, site and mechanisms of excitation. J Neurol Neurosurg Psychiatry 53:770–777
Stefan K, Kunesch E, Cohen LG et al (2000) Induction of plasticity in the human motor cortex by paired associative stimulation. Brain 123:572–584
Wang Liwei X (2000) A new treatment for depression: transcranial magnetic stimulation. Shanghai J Psychiatry 12:41–42
Wischnewski M, Schutter DJLG (2016) Efficacy and time course of paired associative stimulation in cortical plasticity: implications for neuropsychiatry. Clin Neurophysiol 127(1):732–739
Yicong L, Peiqiong C, Kun Y et al (2020) Efficacy of repetitive dual-site paired associative transcranial magnetic stimulation in the treatment of generalized anxiety disorder. Brain Stimul 13(5):1170–1172

# 10 Transcranial Photobiomodulation

Hongfeng Duan and Penghui Song

## 1 Transcranial Photobiomodulation Device

### 1.1 Theoretical Basics of Photobiomodulation

Photobiomodulation (PBM) is a novel interventional therapy that stimulates brain tissue with light at infrared to near-infrared wavelengths, which is applied in the treatment of various neurological and psychological disorders including Alzheimer's disease, Parkinson's disease, depression, stroke, and traumatic brain injury. Studies also suggest that PBM is effective in enhancing cognitive functions such as sustained attention and short-term memory in normal individual. The widely recognized PBM mechanism is that the absorption of light by neurons leads to an increase in mitochondrial activity and also an increase in cerebral blood flow (Barrett and Gonzalez-Lima 2013). Other studies suggest that the sustained effect of light can only be explained by activating signal pathways and transcription factors that cause changes.

Clinical studies have found that light at different wavelengths has variable therapeutic effects. In transcranial PBM, it is challenging to deliver sufficient dose to achieve optimal stimulation due to the exponential attenuation of light penetration in tissue. To overcome this challenge, natural orifice stimulation (e.g., intraoral, intranasal, and intraaural stimulation) has been proposed to improve the stimulation effect on top of traditional transcranial stimulation (Hamblin 2016).

H. Duan
Beijing PsycheArk Science and Technology Development Co., Ltd., Beijing, China
e-mail: duanhongfeng@xlfz.cn

P. Song (✉)
Department of Neurology, Xuanwu Hospital, Capital Medical University, Beijing, China

### 1.2 PBM Modes and Modulation Control

PBM generally has two modes: pulse wave stimulation and continuous wave stimulation. In these two modes, different stimulation signals may be further generated by modulation of pulse wave signal and continuous wave signal. Complex treatment regimens may be developed based on these stimulation regimens in combination with the target site, timing, and sequence of stimulation, so as to meet the need for digital precise therapy. Wavelength, waveform, light intensity, duration, target site, and sequence are the main influences and key regulatory factors of PBM (Freitas and Hamblin 2016). In terms of stimulation modality, in addition to single light stimulation, multimodal combined modes such as photoelectric stimulation, photomagnetic stimulation, photoacoustic stimulation, and photoelectromagnetic stimulation are also available.

#### 1.2.1 Wavelength Selection

Due to the influence of intracranial neurons on the change of absorption coefficient at different wavelengths, the optical window from red light to near-infrared light is usually selected as the range of wavelengths for transcranial PMB, as shown in Fig. 10.1. The wavelength range should be selected based on the photothermal effect as well as the location and depth of action. From an engineering perspective, the availability of appropriate light sources (more cost-effective, from more stable sources) should be considered in wavelength selection. Typical wavelengths include 670 nm, 808/810 nm, and 1064 nm; optionally, broadband light sources with more wavelengths can also be used (Salehpour et al. 2018). More wavelengths may lead to higher flexibility, but also higher cost and technical difficulty. A green or blue laser may be selected as the light source in some special cases, but is not considered here.


Y. Wang (ed.), *Therapeutics of Neural Stimulation for Neurological Disorders*, https://doi.org/10.1007/978-981-99-4538-2_10

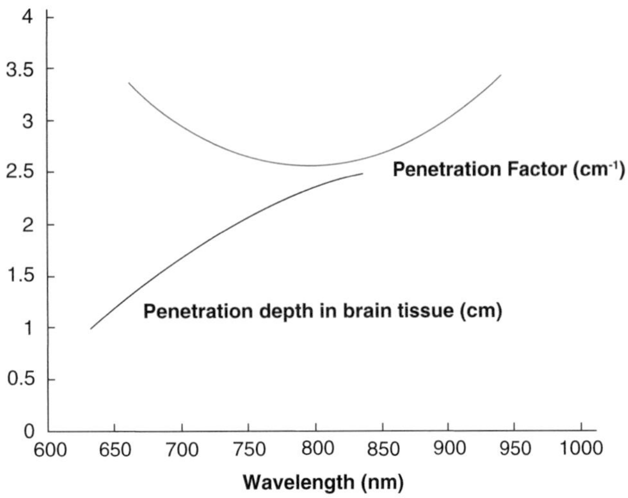


**Fig. 10.1** Optical window

### 1.2.2 Optical Modulation

According to the modulation mode, optical modulation is divided into intensity modulation, phase modulation, polarization modulation, and frequency and wavelength modulation. Since wavelengths are usually preset, optical modulation for transcranial PBM usually involves intensity modulation, phase modulation, and polarization modulation. Optical modulation (regardless of pulse or continuous wave) enables constant and variable frequency modulation on the time axis and constant and variable intensity modulation on the intensity axis. It is possible to output design-specified variable light waves for precise modulation of electronic medicine, just like the components and ratios of drug products. Commercially available products are rarely used in such applications. Existing applications use a pulsed light wave with 10 Hz frequency to resonate with the half-life of the opening and closing of mitochondrial membrane ion channels, so as to achieve better results at lower power (Schiffer et al. 2009).

### 1.2.3 Intensity Control

Optical modulation has actually mentioned phase and polarization modulation, and other methods of modulating light intensity from the perspective of drug components. The intensity control here refers to the luminous flux and irradiance of light. Because luminous flux and light irradiation intensity are not only related to the emission power, but also related to the cumulative time of action, we can control the intensity by adjusting the emission power of the laser and the duty cycle of the pulse signal. From the perspective of digital medicine, intensity control here refers to the amount of active ingredients per unit time. On the other hand, American National Standards Institute (ANSI) standard maximum laser irradiance allowed for human body is 320 mW/cm$^2$. Taking safety issues into account, the maximum laser irradiance is set to 250 mW/cm$^2$ (Scholkmann et al. 2014). As mentioned earlier, the irradiance is a physical quantity related to total action time. So, if we use pulsed light wave with lower action time, we can obtain a higher instant laser power under the same irradiance, which allows the light to penetrate deeper into the brain.

### 1.2.4 Time Control

Time control refers to the time dimension from the broader treatment plan perspective. Time control is about control of the duration of each treatment, for example, a treatment with continuous irradiation for 4 min or with irradiation for 8 min (30 s of irradiation per minute followed by a 30 s pause); for a longer duration, we may take into account factors such as number of sessions per day or number of treatment days (Tian et al. 2016). The luminous fluxes of the above two treatment plans are the same, but the treatment effect may differ. This is another interesting and significant aspect of time control beyond the total dosage control.

### 1.2.5 Location Control

Location control involves stimulation targets; the selection of specific stimulation targets appears to be a purely medical issue. However, considering the registration of functional and structural brain image, if it is a multi-target stimulation, there is a stimulation sequence among different targets and a connection between independent stimulations. Both of these aspects require the transcranial PBM device to have the corresponding location control function. Currently, existing mature solutions such as 3D stereotactic and neuronavigation system can accomplish location control.

### 1.2.6 Programmable Control

Strictly speaking, programmable control is not an independent element. It is an automatic control form and methods different from the elements above. The above five elements can be divided into three categories. The wavelength directly uses the laser's corresponding wavelength. The latter four optical modulation and intensity control are mostly realized through the software and hardware design of the equipment, which is the focus of programmable control. Time and location control are broadly manifested in human–equipment interaction behaviors rather than the equipment itself (Wang et al. 2017). Of course, if the treatment plan is preset from the perspective of medical robot, it also belongs to programmable control. But this is not the focus on the discussion.

The purpose of programmable control design is to achieve remote personalized precision medication through on-site and even remote individualized adjustment according to the guidance of the quantitative signal of patient's brain function.

## 1.3 Structure of Transcranial PBM Device

A transcranial PBM device may be hand-held or head-mounted based on the probe fixation method (as shown in Fig. 10.2) to meet the needs of different medical scenarios. A transcranial PBM device may be wearable, tabletop, and trolley based on the mobility of the host. A transcranial PBM device, regardless of its type, has similar working principle and composition; it consists of the probe (including laser, laser collimation, cooling), power supply system, signal modulation system, and control system. If it is a feedback type, a signal acquisition and a processing feedback system are also required. This part can be either separated from or integrated with the stimulation modulation system. The integrated system has better signal synchronization and feedback guidance, which can accomplish real-time feedback (Song et al. 2020).

### 1.3.1 Luminous Light Source

For the transcranial PBM device, the primary component is the light source. The laser as the light source usually uses semiconductor laser or light emitting diode (LED). Although the laser has high performance, it has disadvantages such as high price, large volume, and large power consumption, LED is the opposite.

If we only look at the decomposition of the performance indicators of the light source device, it has three important indicators. The first is the light intensity, the second is the divergence angle, and the third is the spectral full width at half maximum wavelength. The laser operation needs to pay attention to the threshold current and the working current, and the selection of light source indicators is more based on the device requirements.

Semiconductor laser has a narrow light source wavelength band. Although the precision of semiconductor is within 1 nm, which is true for the spectral full width at half maximum wavelength of a given laser, but the standard nominal value of the light source of general commercial laser is often within a few to 10 nm. For example, if we buy an 810-nm light source, the value we actually get may be 800–810 nm. However, LED's full width at half maximum wavelength is of 10 of nanometers. Second, the luminous power of semiconductor diode is much higher than that of LED. The divergence angle of light diode (LD) laser is only a few tenths of a degree, while that of LED is more than 10° or even 20°. So, it seems that the LD laser with high power, good monochromaticity, and small divergence angle should completely beat LED light source, but this is not the case. Regarding optical power, the upper limit of ANSI safety standard is 320 mW/$cm^2$. If target area of the light source is designed to be one square centimeter, the power of the selected light source would at least reach this threshold. Because of the energy loss due to collimated broadcasting, the optical power reaching the scalp is lower than the light source power. We also would not allow the equipment to work at 100% full power. In engineering design, we usually want the device to operate for a long time at half of its maximum power. Even so, working at a power of 620 mW is easy for both LD laser and LED laser. Therefore, LD laser's high power does not have much advantage. Regarding divergence angle, the distance from the exit surface to the incident surface is very close, usually a little more than 10 mm or at most a few centimeters away. In addition, the skull is a substance with large scattering coefficient. So the light is dispersed in the brain, and the requirement of collimation is low. The remaining factor is monochromaticity, which seems very important, but academia has not reach consensus about its quantification for years. There is also no final conclusion in fields that require higher wavelength precision, including near-infrared functional brain imaging. Therefore, quantitatively speaking, LED laser is more commonly used as the light source for transcranial PBM due to its safety, convenience, low-cost, and portability.

In terms of using the devices, we are also concerned about the switching time of the light source. Faster switching time means that we can get pulsed signals with higher frequency. The steeper the rising and falling edges, the more likely we get ideal rectangular pulses.

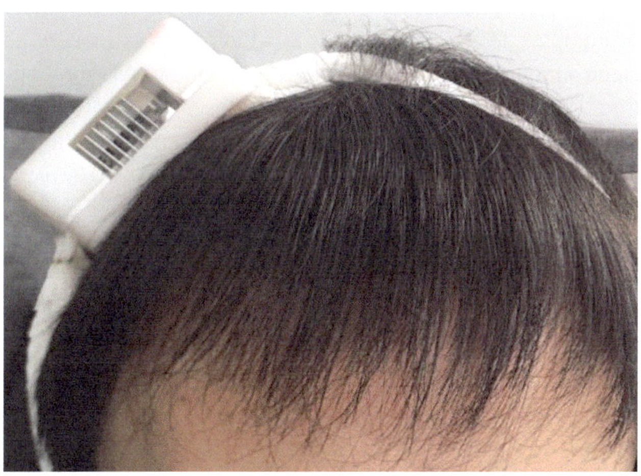

**Fig. 10.2** Transcranial PBM device

### 1.3.2 Power Supply System

If we choose LED as the light source, then the power supply system is not a big problem, especially when we use DC power supply. We can meet the system requirements by designing a common voltage stabilizer. If we choose the laser diode as the light source, then a constant current source is necessary. When using AC power supply, the medical constant voltage is a common choice. If the requirements are higher, an external regulated power supply will be designed as a system accessory and be placed at places with unstable voltage.

### 1.3.3 Optical Modulation System

When it comes to optical modulation, many would think of the "optical model" used for optical modulation and demodulation in optical communication. If we do not consider the monitoring feedback of quantitative signals, the modulation alone is more like the dispensing process of traditional medicines. If we build the system from an ordinary LED light source, what we get at this time is a continuous light wave. Pulse modulation is when we want to turn the continuous light wave into a pulsed light wave of specific frequency. There are many kinds of pulse modulation, from various choppers to direct on-off switching control of LED. For LED with a breaking time of tens of nanoseconds, it is not difficult to produce a pulse of thousands of hertz. With this method, we can basically get the pulse signal we want by controlling the on-off switching of LED through the host computer combined with Field Programmable Gate Array (FPGA). For a continuous light wave, working voltage is usually adjusted and amplified through circuit design, so as to modulate light intensity and generate output power of triangular or sine waves. The current transcranial PBM still use constant continuous wave, which can be greatly improved with further mechanism research. Of course, the modulation is not a technical show. Technically speaking, there are always mature solutions. So the difficulty here lies in the research on the mechanisms of light's effect on the brain and the pathogenesis of diseases, which would provide guidance for optical modulation.

### 1.3.4 Control System

The control system is mainly composed of hardware FPGA engineering programmable control unit, combined with host computer program, to accomplish equipment control and meet the needs of medical protocols. We spent a lot of ink on transcranial PBM equipment. In fact, the telemedicine cloud platform, which is relatively opposite and seamlessly connected to the hardware devices, is not only a key part of the control system, but also the main component of the entire transcranial PBM system. Electronic medical record, which consists of structured inquiry system, quantitative brain signal feedback system, patient treatment, and disease course file system, is the core of telemedicine cloud platform. Through the telemedicine cloud platform, targeted treatment plans can be personalized according to patients' different conditions and personal characteristics. The whole personalized treatment plan should be given by AI intelligent diagnosis and treatment system based on big data, all of which depend on the convolution feedback between treatment parameters and their effectiveness. In particular, remote program regulation is emphasized because personalized precise treatment at home is realized and the strong vitality of transcranial PBM is shown.

## 2 Transcranial PBM Technique

Transcranial near-infrared stimulation (tNIRS) is a novel, noninvasive neural regulation technique that stimulates biological response via near-infrared light at 630–1064 nm. The near-infrared light up-regulates mitochondrial function by regulating the activity of cytochrome oxidase, thereby increasing local neuronal metabolism. Near-infrared light has strong penetrability and can penetrate the scalp and skull into the brain. It is easier to pass through and less dispersed by the brain tissue. It is also easy to control the direction, range, and energy level. The operation is simple and painless. Near-infrared light can penetrate the entire skill and reach deeper brain tissue than red light. Near-infrared light can be used to treat various diseases such as muscle pain, neuropathic pain, headache, and wound treatment (Farzad et al. 2018).

In NIR spectroscopy, the clinical application of near-infrared light can be traced back to mid-1980s when it was used to monitor the brain status of newborns and fetuses. PBM is an application of phototherapy, usually from red to near-infrared lasers or incoherent light sources such as LED. As a novel therapy, PBM therapy improves brain function by stimulating neural activity. Because of its safety, it has been used continuously for wound healing and pain relief for a long time in the past 50 years. In addition, there is considerable evidence that PBM can also change behavioral expressions such as enhancing cognition and improving depressive symptoms. Nowadays, the therapeutic effect of PBM in depression, anxiety, stroke, brain injury, AD, and Parkinson's disease has received increasing attention.

Infrared light enhances the cytochrome C oxidase (CCO or complex IV) in mitochondria, which is a component of electron transport chain and a key compound of ATP production. Vitro experiments show that laser irradiation can increase the respiration level of mitochondria, leading to the amplification of mitochondrial products such as ATP, nicotinamide adenine dinucleotide (NADH), protein, and ribonucleic acid (RNA). tNIRS can increase the cellular respiratory process of neurons by increasing energy and cyclic adenosine monophosphate (cAMP) levels and indirectly regulate neuronal activity. Konstantinovic et al. extended this idea in a study, which emphasized that the changes in intracellular calcium concentration induced by cortical trauma and the regulation of $Na^+$ $K^+$-ATPase activity are related to neuropathology (such as stroke and traumatic brain injury). They hypothesized that tNIRS has membrane-stabilizing effects (increase in $Na^+$ pump activity caused by laser irradiation is the underlying mechanisms), which is an important contributor to positive clinical effects in early acute stroke studies. Besides the potential effect of CCO in tNIRS, the second hypothetical mechanism by which near-infrared light affects neurons is through the dissociation of nitric oxide (NO) and oxygen. NO is an essential cellular signaling molecule and a powerful neurotransmitter in the central nervous system that induces synaptic plasticity. Separating NO from CCO compound through laser inducement, the ongoing cellular respiration rate in mitochondria can continue unhindered even under stressful situations. Since neurons are highly dependent on oxygen, tNIRS operates through photo-bioenergy mechanism and leads to metabolic and hemodynamic changes that promote neural functions. PBM therapy can increase cerebral blood flow (CBF) and enhance brain metabolism and antioxidant defense. It promotes neuroprotection by regulating anti- and pro-apoptotic mediators, inflammatory signaling molecules, and neurotrophic factors (Wang et al. 2017).

## 2.1 Transcranial PBM Modulates Cortical Excitability

tNIRS can penetrate the scalp and skull to reach the superficial layers of the cerebral cortex. Song Penghui et al. found that 820 nm tNIRS can regulate the excitability of primary motor cortex (M1). 12 right-handed healthy adults were given 820 nm tNIRS. The stimulation target was left M1, and the stimulation duration was 4 min. Motor evoked potentials (MEPs) were measured at an interval of 5 min between baseline before stimulation and 30 min after stimulation. Results showed that the cortical excitability of the 12 participants increased by 20–40% after 4 min of 820 nm tNIRS, and the MEP value increased significantly after tNIRS. During sham stimulation, there was no significant difference in MEP amplitude at any time point. Compared with sham stimulation, MEP amplitude in 820 nm tNIRS stimulation group increased significantly. MEP amplitude was the highest 15 min after stimulation, followed by time points of 20 min and 10 min. The results showed that tNIRS at 820 nm induced a transient increase in excitability in the stimulated cortex (Fig. 10.3) (Song et al. 2020).

Chaieb et al. used tNIRS at 810 nm in the hand area of the primary motor cortex for 10 min. Single pulse and dual pulse transcranial magnetic stimulation (TMS) were used to assess cortical excitability levels in corticospinal pathways and intracortical circuits. Serial reaction time task (SRTT) was used to study the effect of tNIRS on learning. By evaluating the mean amplitude of MEP stimulated by single-pulse TMS, they observed that the amplitude decreased significantly 30 min after stimulation compared with baseline. Therefore, applying 810 nm tNIRS to M1 for 10 min can

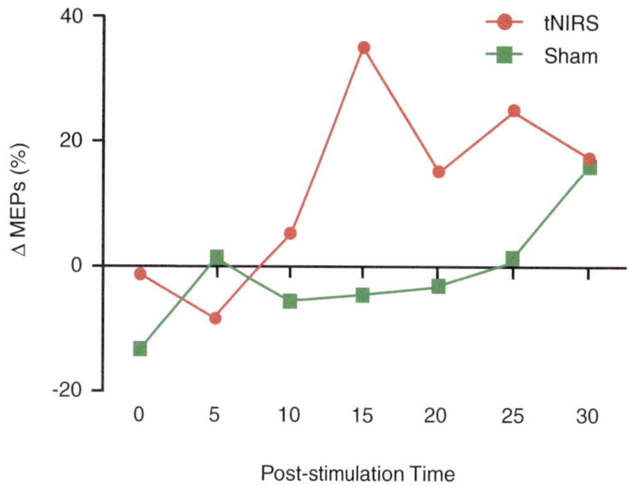


**Fig. 10.3** Rate of change of motor evoked potential in left motor cortex via near-infrared stimulation. The left motor cortex was stimulated by transcranial near-infrared light for 4 min, and the MEP amplitude was measured every 5 min. The rate of change of MEP increased significantly after transcranial near-infrared stimulation, suggesting that near-infrared light can activate cortical function. *MEP* motor evoked potential, *tNIRS* transcranial near-infrared stimulation, *Sham* sham stimulation

inhibit cortical excitability. Moreover, short interval intracortical inhibition (SICI) increased significantly, and intracortical facilitation (ICF) decreased significantly after tNIRS. SICI reflects intracortical inhibition and is mediated by $GABA_A$ receptors, while ICF may be mediated by glutamate system. Therefore, tNIRS may promote the intracortical inhibitory network and/or inhibit the intracortical facilitation of corticospinal motor neurons by increasing GABAergic neurotransmission and/or reducing glutamatergic effect, which results in MEP amplitude inhibition. This experiment further demonstrated that tNIRS can be used to change the excitability of the brain cortex in healthy adults.

## 2.2 Clinical Application of Transcranial PBM

### 2.2.1 Anxiety Disorder

Anxiety refers to irrational feelings of worry or fear due to internal conflict of emotions or mind. Anxiety may arise under specific circumstances. It may also be habitual or become a common feeling. Anxiety comes in many forms, including social anxiety, self-compulsion, post-traumatic stress, and even phobia. Anxiety has a severe negative influence on human body such as shortness of breath, headache, palpitation, and weakness in the limbs.

Existing studies have confirmed the clinical efficacy of tNIRS in the treatment of patients with generalized anxiety and depression. Fredric Schiffer et al. conducted an experiment to treat 10 patients with depression and anxiety and suggested that tNIRS could significantly improve the psychological scale score. One patient with MDD comorbid with anxiety received tNIRS treatment for 31 months, which was well tolerated for 9 months. The symptoms continued to improve. The total anxiety Average Sales Quantity score decreased by 50%, and depressive symptoms were significantly alleviated. Foreign scholars used NIRS measurement [OxyHb] to confirm that cerebral cortex function in patients with anxiety disorder is abnormal. Most results show that the functional activity of prefrontal lobe and its adjacent temporal lobe is low. Literature found that the excitability of left dorsolateral prefrontal cortex in anxiety patients decreased (Shaw et al. 2010). Wang Yuping's research team used 820 nm tNIRS to stimulate the dorsolateral prefrontal cortex of patients with anxiety disorder. The preliminary results showed that transcranial PBM could significantly improve the symptoms of anxiety patients and reduce the score of Hamilton Anxiety Scale.

### 2.2.2 Depressive Disorder

Depression is a state of low mood and aversion to activities, which has impact on person's thoughts, behaviors, feelings, and physical health. Depressed people may experience feelings of depression, anxiety, emptiness, despair, guilt, irritability, and restlessness. They may lose interest in their usual hobbies, lose their appetite or overeat, become less social, and have problems such as memory loss, difficult concentrating, and indecisiveness. Depression is also accompanied by symptoms such as insomnia, lethargy, fatigue, and may even induce suicide.

Development of effective and sustainable methods and modalities for the treatment of major depressive disorder has long been recognized as a worldwide goal. Following the exposure to a series of chronic unpredictable mild stressors, animals develop symptoms similar to those seen in humans with depression, such as loss of interest (loss of the ability to experience pleasure, a core symptom of depression), weight loss or slow weight gain, reduced motor activity, and sleep disorder. Wu et al. studied Wistar rats. After 5 weeks of chronic stress stimulation, tPBM was applied three times a week for 3 weeks (810 nm, 100 Hz with 20% duty cycle, 120 J/cm$^2$). They observed significant improvement in the performance of forced swim test (FST). In a similar study, Salehpour et al. compared two different lasers (630 nm, 89 mW/cm$^2$; 810 nm, 562 mW/cm$^2$, pulsed at 10 Hz, 50% duty cycle). Results showed that the 810 nm group was better than the 630 nm group in FST test, elevated plus maze test, and reduction in blood cortisol level (Wu et al. 2012). The first clinical study on depression and anxiety was published by Schiffer et al. (2009). They placed a fairly small area of LED array (1 W 810 nm) on the forehead of patients with severe depression and anxiety. Results indicated that Hamilton Depression Scale (HAMD) and Hamilton Anxiety Scale (HAMA) improved after a single treatment for 2 weeks. In addition, they found that the regional cerebral blood flow (rCBF) in frontal pole increased during light transmission using near-infrared spectroscopy. Schiffer et al. treated nine patients with comorbid anxiety and post-traumatic stress disorder using tPBM. The study showed that there was significant improvement with anxiety during Week 2 and Week 4.

Wang Yuping's research team found that compared with the normal people, the gray matter volume and cortical thickness of the left frontal pole cortex in depressive patients were significantly reduced. By analyzing the correlation between the gray matter of the left medial frontal pole cortex and the occurrences of depression, the duration of each occurrence, and the score of self-reported depression scale of depressive patients, they found that the gray matter volume of the left medial frontal pole cortex showed a significant negative correlation with the severity of depression. Later, the team treated depressive patients with 820 nm near-infrared stimulation. The preliminary results showed that after 2 weeks of near-infrared treatment, depression symptoms were significantly improved and the Hamilton Depression Scale scores were reduced.

### 2.2.3 Alzheimer's Disease

Alzheimer's disease (AD), commonly known as senile dementia, is a neurodegenerative disease with slow onset which deteriorates over time. The most common early symptom is the loss of short-term memory (difficulty remembering recent events). As the disease progresses, symptoms may gradually appear, including language disorders, disorientation (including easy to get lost), emotional instability, motivation loss, inability to take care of themselves, and many other behavioral problems. When symptoms get worse, patients are often disconnected from the family or society and gradually lose physical function, eventually leading to death.

De Taboada et al. conducted a study on an APP transgenic AD mouse. From the age of 3 months, different doses of tPBM (810 nm) were used three times a week for 6 months. When administered in a dose-dependent manner, Aβ plaques decreased significantly in the brain. tPBM alleviated the behavioral effects caused by late amyloid deposition and reduced the expression of inflammatory markers in transgenic mice (Saltmarche et al. 2017). In clinical research, near-infrared PBM treatment significantly improved sleep quality, emotional state, EEG mode, and cognitive function including memory and attention. In addition, delivering a red laser into the brain through arterial catheter can improve the cerebral blood flow of AD patients. Using a transcranial intranasal PBM device, 19 participants with memory/cognitive impairment were randomized into real and sham stimulation groups for a 12-week study and a 4-week study treatment-free follow-up, which were assessed with MMSE and ADAS-cog scale. Results showed that the real stimulation group with moderate to severe impairment showed significant improvement after 12 weeks (Naeser et al. 2014). ADAS-cog score also improved significantly. The sleep quality of patients was improved, the number of anger outbreaks was reduced, and the anxiety symptoms were reduced. During the 4-week treatment-free follow-up, the patients' condition stabilized, and no adverse symptoms were reported.

### 2.2.4 Stroke

Stroke refers to the death of brain cells caused by cerebral ischemia. Stroke can be categorized into ischemic stroke caused by vascular obstruction; and hemorrhagic stroke caused by bleeding. Common stroke symptoms include inability to move one side of the limb or weakness in one side of the body, inability to understand others' words, inability to speak, dizziness, and loss of vision in half of the visual field. Patients with hemorrhagic stroke may experience sudden and severe headaches. Symptoms usually appear soon after a stroke occurs. If the symptoms disappear within 24 h, it is called "Transient Ischemic Attack" (TIA). Stroke patients who receive treatment as soon as possible can reduce the chance of permanent damage.

Numerous preclinical animal studies have demonstrated that applying near-infrared laser (810 nm) on the head at different times has beneficial effects on neurological performance and lesion reduction after inducing acute stroke. This method can activate the anti-inflammatory and anti-apoptotic in the brain and help the proliferation of neurons. In a series of three clinical trials called "nervous system effectiveness and safety trial," NEST-1 recruited 120 patients between 40 and 85 years old, who were diagnosed with ischemic stroke with measurable neurological deficits. The purpose of the first clinical trial is to demonstrate the safety and effectiveness of laser treatment of stroke within 24 h. Within 18 h after stroke, the application of tPBM can significantly improve the prognosis of stroke patients. NEST-2 recruited 660 patients between 40 and 90 years old. They were randomly assigned to one of the two groups. Significant results were found for patients with moderate to severe (but not for severe) stroke patients who received real stimulation ($P < 0.04$) (Lampl et al. 2007). Compared with sham stimulation, near-infrared had no significant difference in the occurrence of adverse symptoms.

### 2.2.5 Parkinson's Disease

Parkinson's disease (PD) is a chronic neurodegenerative affecting the central nervous system, mainly the motor nervous system. Its symptoms usually appear slowly over time. The most obvious symptoms in the early stage are tremor, limb stiffness, decreased motor function, abnormal gait, and potentially cognitive and behavioral impairments. Dementia is quite common in patients with severe conditions, and more than one-thirds of cases will also develop major depressive disorder and anxiety disorder. Other possible symptoms include perception, sleep, and emotional deficits. Most studies on PD treatment with PBM were carried out in animal models using small molecular substances (MPTP or 1-methyl-4-phenyl-1,2,3,6-tetrahydropyridine) on mice. MPTP is an impurity among illicit stimulants that can cause Parkinson-like symptoms (due to loss of substantia nigra cells) in young people taking the drug. Mice were treated with tPBM (670 nm LED, 40 mW/cm$^2$, 3.6 J/cm$^2$) 15 min after each MPTP injection, which repeated four times in more than 30 h. The number of dopaminergic cells increased significantly in mice treated with tPBM. A subsequent study showed similar results in a chronic mouse model of MPTP-induced Parkinson's disease. In a study of PD patients, motor and cognitive functions improved after 2 weeks of tPBM treatment. A clinical report of PBM in the treatment of Parkinson's disease recruited eight patients with advanced PD aged from 18 to 80 in a non-randomized study with no controls. Participants received tPBM treatment on the head. Laser targeted the brain stem, bilateral occipital lobe, parietal, temporal, frontal lobes, and along the sagittal suture. Visual analogue scale (VAS) was used to record the severity of their balance, gait, physical stiffness, cognitive function, bed rolling movements, and pre-verbal process difficulties at the end of the study, with 10 being the most severe and 0 being asymptomatic. All participants had improved VAS score compared with baseline. There were statistically significant reductions in VAS scores for gait and cognitive function, and significant reductions in physical stiffness and speech difficulty (Peoples et al. 2012).

### 2.2.6 Traumatic Brain Injury

People recovering from moderate or severe Traumatic Brain Injury (TBI) are accompanied by a variety of long-term symptoms including cognitive impairment (e.g., memory loss, impaired executive function, difficulty concentrating), headache, sleep disturbances, and depression. In early TBI studies, transcranial LED therapy (633/870 nm) can improve self-awareness, self-regulated social function, and sleep quality. The application of high-energy near-infrared laser has significant clinical effects on TBI patients. It can significantly reduce headache symptoms, improve sleep quality, and enhance cognitive and emotional states. Meanwhile, 785 nm irradiation can improve the alertness and consciousness of TBI patients with severe disorders of consciousness (Morries et al. 2015).

## 2.3 Summary

As a new and noninvasive brain stimulation technique, tNIRS is widely used in the treatment of depression, anxiety, stroke, traumatic brain injury, Alzheimer's disease, and Parkinson's disease. Additionally, tNIRS-induced cellular effects include changes in intracellular calcium level, regulation of $Na^+$ $K^+$ ATPase activity, increased $Na^+$ pump activity, and membrane stabilization. Thereby, tNIRS is capable of increasing cerebral blood flow, enhancing brain energy metabolism and antioxidant capacity. As a means of treatment to promote nerve enhancement or neuroprotection, tNIRS will become one of the most promising nerve rehabilitation strategies in the coming years.

## References

Barrett DW, Gonzalez-Lima F (2013) Transcranial infrared laser stimulation produces beneficial cognitive and emotional effects in humans. Neuroscience 230:13–23

Farzad S, Javad M, Farzin K et al (2018) Brain photobiomodulation therapy: a narrative review. Mol Neurobiol 55(8):6601–6636

Freitas LF, Hamblin MR (2016) Proposed mechanisms of photobiomodulation or low-level light therapy. IEEE J Sel Top Quantum Electron 22(3):7000417

Hamblin MR (2016) Shining light on the head: photobiomodulation for brain disorders. BBA Clin 6:113–124

Lampl Y, Zivin JA, Fisher M et al (2007) Infrared laser therapy for ischemic stroke: a new treatment strategy: results of the NeuroThera effectiveness and safety Trial-1 (NEST-1). Stroke 38:1843–1849

Morries LD, Cassano P, Henderson TA (2015) Treatments for traumatic brain injury with emphasis on transcranial near infrared laser phototherapy. Neuropsychiatr Dis Treat 11:2159–2175

Naeser MA, Zafonte R, Krengel MH et al (2014) Significant improvements in cognitive performance post-transcranial, red/near-infrared light-emitting diode treatments in chronic, mild traumatic brain injury: open-protocol study. J Neurotrauma 31(11):1008–1017

Peoples C, Spana S, Ashkan K et al (2012) Photobiomodulation enhances nigral dopaminergic cell survival in a chronic MPTP mouse model of Parkinson's disease. Parkinsonism Relat Disord 18:469–476

Salehpour F, Mahmoudi J, Kamari F et al (2018) Brain photobiomodulation therapy: a narrative review. Mol Neurobiol 55(8):6601–6636

Saltmarche AE, Naeser MA, Ho KF et al (2017) Significant improvement in cognition in mild to moderately severe dementia cases treated with transcranial plus intranasal photobiomodulation: case series report. Photomed Laser Surg 35(8):432–441

Schiffer F, Johnston AL, Ravichandran C et al (2009) Psychological benefits 2 and 4 weeks after a single treatment with near infrared light to the forehead: a pilot study of 10 patients with major depression and anxiety. Behav Brain Funct 5:46

Scholkmann F, Kleiser S, Metz AJ et al (2014) A review on continuous wave functional near-infrared spectroscopy and imaging instrumentation and methodology. NeuroImage 85(Pt 1):6–27

Shaw VE, Spana S, Ashkan K et al (2010) Neuroprotection of midbrain dopaminergic cells in MPTP-treated mice after near-infrared light treatment. J Comp Neurol 518:25–40

Song P, Han T, Lin H et al (2020) Transcranial near-infrared stimulation may increase cortical excitability recorded in humans. Brain Res Bull 155:155–158

Tian F, Hase SN, Gonzalez-Lima F et al (2016) Transcranial laser stimulation improves human cerebral oxygenation. Lasers Surg Med 48(4):343–349

Wang X, Tian F, Reddy DD et al (2017) Up-regulation of cerebral cytochrome-c-oxidase and hemodynamics by transcranial infrared laser stimulation: a broadband near-infrared spectroscopy study. J Cereb Blood Flow Metab 37(12):3789–3802

Wu X, Alberico SL, Moges H et al (2012) Pulsed light irradiation improves behavioral outcome in a rat model of chronic mild stress. Lasers Surg Med 44:227–232

# Transcranial Ultrasonic Neurostimulation

**11**

Hairong Zheng, Lili Niu, Chunyan Liu, and Tingting Zhang

**Overview**
Medical ultrasound is a common technology for clinical diagnosis and treatment, with the advantages of noninvasive, non-radioactive, easy and fast to use, and low equipment cost. Ultrasound is a mechanical wave that can enter the body in a noninvasive manner. During the transmission of ultrasound in the human body, echoes are generated due to the different acoustic impedances of different tissues and organs. As the echoes carry information about the difference in acoustic impedance of tissues, the echo information can be obtained to demodulate the corresponding tissue information and obtain structural images of tissues and organs to help diagnose diseases. In addition to image diagnosis, ultrasound technology has been successfully applied to the clinical treatment of tumors and other diseases, such as the elimination of lesions by thermal ablation using high-intensity ultrasound. Recent advances in ultrasound technology have enabled the application of ultrasound wave to open the blood-brain barrier in a noninvasive manner and facilitate the delivery of drugs to areas of brain diseases, which enhances the therapeutic effect of brain tumors and other diseases. In addition, studies have shown that acoustic radiation force can open mechanosensitive ion channels to modulate the electrical activity of neurons, further expanding the prospects for the application of ultrasound in the treatment of brain diseases.

H. Zheng · L. Niu (✉)
Shenzhen Institutes of Advanced Technology, Chinese Academy of Sciences, Shenzhen, China
e-mail: hr.zheng@siat.ac.cn; ll.niu@siat.ac.cn

C. Liu · T. Zhang
Department of Neurology, Xuanwu Hospital, Capital Medical University, Beijing, China

## 1 Biological Effects of Ultrasound

Ultrasound is a mechanical wave with a frequency above the range of human hearing (>20 kHz), generated by the vibration of an object (sound source) and propagated by the compression or expansion of a propagating medium (e.g., air, water, or tissue) (Warwick and Pond 1968). During the interaction between ultrasound and biological tissue, three basic ultrasound biological effects are produced: fluctuating, mechanical, and thermal effects. These effects are the basis for the application of ultrasound to the medical field. As ultrasound propagates in the form of waves, there will be refraction, reflection, scattering, and diffraction in two media with different acoustic impedances. Based on their interference properties, physically dispersed arrays of ultrasound transducers can harness both beneficial and destructive interference effects to focus ultrasound energy on a single point. Focused ultrasound (FUS) can be specifically focused at a specific depth in the target area through the superficial tissue and has excellent tissue penetration and localization (Ter Haar 2001).

### 1.1 Thermal Effects

Numerous studies including ex vivo and in vivo have discussed the thermal effects of ultrasound (Smith et al. 2001; Baker et al. 2001; Chang et al. 2002; Levenback et al. 2012). The temperature induced in the tissue by focused ultrasound depends mainly on the ultrasound intensity and tissue kinetic viscosity. The thermal effect of ultrasound is currently being applied to the treatment of tissues following musculoskeletal injuries (Baker et al. 2001). At higher intensities, ultrasound can generate temperatures sufficient to denature proteins and coagulate tissues (Lee and Park 1999). On the basis of this research, high-intensity focused ultrasound has been used to ablate pathological tissues such as kidney stones, myomas, cancer, or intracranial tumors (Cheng et al. 1997; Wu et al. 2001; Adams et al. 1996).



Y. Wang (ed.), *Therapeutics of Neural Stimulation for Neurological Disorders*, https://doi.org/10.1007/978-981-99-4538-2_11

### 1.2 Mechanical Effects

The non-thermal effects of ultrasound include mechanical, cavitation, and physicochemical effects. Ultrasound opening the blood-brain barrier and ultrasound modulation of cell membranes in the nervous system are mechanical effects (McDannold et al. 2008; van den Bijgaart et al. 2017). Ultrasound is a mechanical wave that produces mechanical effects by moving and changing the cellular volume and contents of biological tissues through radiation forces. The lipid bilayer of cytomembrane is mobile, and there are mechanosensitive ion channels in the membrane, which can recognize mechanical force changes in the cytomembrane and respond rapidly to the changes by converting mechanical signals into electrical or chemical signals. Ultrasound activates mechanosensitive ion channels by mechanical effects, i.e., changing the membrane potential, which in turn affects the activity of voltage-gated ion channels and then changes neuronal excitability (Kubanek et al. 2018).

### 1.3 Cavitation Effects

High-intensity ultrasound primarily produces cavitation, a phenomenon in which microbubbles (cavitation bubbles) in the tissue oscillate, grow, and accumulate in response to the ultrasound field and rupture when the energy reaches a certain threshold. The cavitation effect is divided into inertial (transient) cavitation and non-inertial (steady-state) cavitation. Inertial cavitation refers to the rapid expansion to violent collapse of microbubbles under the action of ultrasound. Non-inertial cavitation, on the other hand, refers to the repeated expansion and compression of microbubbles in the presence of ultrasound. Inertial cavitation is associated with a higher risk of tissue damage than non-inertial cavitation (Miller 2007). So far, studies have been conducted to artificially inject microbubbles into the body and use FUS to open the blood-brain barrier for local targeted drug delivery based on the cavitation effect (Meng et al. 2021).

## 2 Basic Structure of Transcranial Ultrasonic Stimulation System

The *Nature* in 2017 and *Science* in 2020 stated that "ultrasound can be used to modulate brain activities and treat diseases" suggesting that ultrasonic neurostimulation is a new neuromodulation technology for functional brain diseases, characterized by noninvasiveness, deep penetration, and high spatio-temporal resolution. However, before putting it into practice, advanced ultrasonic neurostimulation devices should be developed, which is a prerequisite to solve the basic problems in the field of ultrasonic neuromodulation and promote the rapid development of this field. Transcranial ultrasonic neurostimulation device consists of four core parts: ultrasonic transducers, ultrasonic transcranial electronic control system, the localization system, and monitoring and safety assessment system for ultrasonic stimulation responses. With further development of the research, scale-span ultrasonic neurostimulation devices targeting different levels of cells, small animals and non-human primates have been reported. Precise focusing of transcranial ultrasound and visualization of ultrasonic stimulation process are the key technical obstacles of current transcranial ultrasonic neurostimulation devices to realize accurate and safe ultrasound neurostimulation. Further improvement and refinement of the ultrasonic neurostimulation system will promote the clinical process of ultrasonic neuromodulation technology in the treatment of functional brain diseases such as epilepsy, depression, Parkinson's disease, drug addiction, and sleep disorder.

### 2.1 Structure of Transcranial Ultrasonic Stimulation System

The ultrasonic transducers are the core component of the device, which can be divided into single-array and multi-array ultrasonic transducers according to the number of array elements. The use of ultrasound transcranial electronic control system to give a certain phase delayed excitation signal to different array elements can achieve noninvasive, precise, multi-point, and dynamic transcranial ultrasound stimulation in the deep brain region. For magnetic resonance-guided ultrasound stimulation system, the positions of ultrasound focus are localized by the detection of tiny position shift caused by acoustic radiation force in tissues using the magnetic resonance acoustic radiation force imaging and the detection of blood oxygen changes caused by ultrasonic stimulation using blood oxygenation level-dependent magnetic resonance imaging. The monitoring and safety assessment system for ultrasonic stimulation responses mainly uses magnetic resonance temperature imaging and electroencephalography to assess the modulation effect and safety of ultrasonic stimulation and can be applied to the selection of stimulation target screening and exploration of deep brain nuclei function, etc.

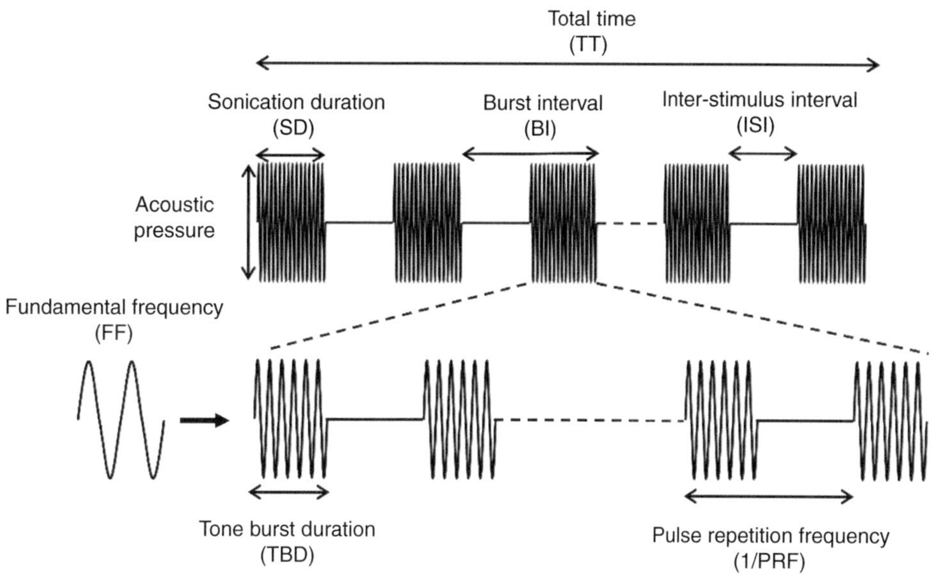


**Fig. 11.1** Ultrasonic stimulation parameters

## 2.2 Transcranial Ultrasonic Stimulation Parameters

Transcranial ultrasonic stimulation is a multi-parameter adjustable neuromodulation technique, as shown in Fig. 11.1, with the parameters mainly including fundamental frequency (FF), acoustic pressure (AP), pulse repetition frequency (PRF), sonication duration (SD), and duty cycle (DC).

The higher the fundamental frequency, the higher its spatial resolution, but as the fundamental frequency increases, the energy attenuation also increases. Previous studies have shown that ultrasound at frequencies of 0.25–0.65 M Hz is most effective in eliciting the activation of intact brain circuit (Tufail et al. 2011).

The evaluation of ultrasound intensity is usually quantified using spatial peak pulse average intensity ($I_{SPPA}$) and spatial peak temporal average intensity ($I_{SPTA}$). The Food and Drug Administration (FDA) requires that $I_{SPTA}$ for ultrasonic imaging be no higher than 720 mW/cm$^2$ and $I_{SPPA}$ no higher than 190 W/cm$^2$.

The pulse repetition frequency describes the number of pulses sent per unit time. The neuronal modulation is related to the pulse repetition frequency, by changing which it is possible to adjust the effect of ultrasound on the modulation of excitatory or inhibitory neural activity of neurons and the modulation duration. Ultrasound stimulation of neurons in mouse cortical brain slice neurons shows a significant increase in the frequency of action potentials using a pulse repetition frequency of 1000 Hz, while ultrasound can suppress the neuronal excitability using a pulse repetition frequency of 100 Hz (Asan et al. 2021).

Tone-burst duration (TBD), also known as pulse length, is defined as the time from the beginning to the end of the pulse, and adjusting the TBD allows for bidirectional modulation of the neurons. It was shown that ultrasonic stimulation with a TBD of 0.5 ms suppressed visual cortical, whereas with a TBD of 50 ms, motor cortex was activated and forelimb activity was observed in rabbits (Yoo et al. 2011a).

Duty cycle is the time percentage of the ultrasound signal in each pulse. Ultrasound can transmit as continuous or impulse waves. However, when the duty cycle increases, the thermal effect produced by ultrasound transmitting through the tissue is enhanced. At present, most experimental studies on ultrasonic neuromodulation choose to use pulsed ultrasound to ensure safety. Finding the optimal duty cycle that minimizes the neural activation threshold is also one of the directions that researchers are now continuously exploring.

Sonication duration (SD) is defined as the total time from the start of the first pulse to the termination of the last pulse. Evidence from preclinical studies suggests that long durations favor the inhibition of cortical neurons, while short durations result in excitation (Fomenko et al. 2020; Xia et al. 2021).

# 3 Applications of Ultrasonic Neuromodulation System

## 3.1 Cellular Level

The underlying mechanism of low-intensity focused ultrasound (LIFUS)-induced neuromodulation is still poorly understood. As mentioned before, three potential physical mechanisms could contribute to the modulation of neuronal excitability induced by ultrasound stimulation, including fluctuation effect, mechanical effect, and thermal effect. The thermal effects of ultrasound manifest as perturbations in the level of neuronal activity, such as ultra-structural synaptic changes, reduction in the number of synaptic vesicles, and expansion of pre- and post-synaptic connections (Shealy and Henneman 1962). Since the pulsed form minimizes the tissue heat-up (<0.1 °C) at lower intensities, the modulation by LIFUS for the thermal effects of neural activity may be negligible (Kim et al. 2014). Non-thermal effects of LIFUS include acoustic cavitation effects, altered ion channel permeability, and indirect modulation of neural activity through membrane deformation. Due to the fundamental frequency of ultrasound, it was difficult to generate acoustic cavitation without the microbubbles, especially for small input power application (Holland and Apfel 1990). Recent observations have confirmed the hypothesis that LIFUS exerts its effects in part through the force of acoustic radiation impinging on local ion channels. When ultrasound wave is applied to a target neuron under steady pressure, the mechanical energy delivered stretches and deforms the cytomembrane; thereby, the gating dynamics, channel conformation, and molecules associated with the cytomembrane are directly altered. These changes ultimately affect the ion flux through the lipid bilayer and thus affect the probability of neuronal firing (Tyler 2012; Kubanek et al. 2018). Based on this, the mechanosensitive channels expressed on the surface of neuronal membrane can be activated by LIFUS (Azadeh et al. 2021; Rabut et al. 2020). Whether ultrasound directly exerts its effect (piezoelectricity) through conformational changes of the tertiary structure or causes local membrane depolarization, thereby in turn activating the voltage-gated channel (flexible electricity), remains to be investigated. The compression and thinning of the surrounding medium during ultrasound stimulation has shown mechanical changes in plasma membrane tension, which may alter the mobility and permeability of the lipid bilayer (Sassaroli and Vykhodtseva 2016). In addition, as momentum is transferred from the resonant lipid bilayer to the surrounding extracellular and intracellular environment, periodic pressure changes may cause the formation of microjet (including turbulent and vortex) (Tyler 2011; Riley 1998). Both mechanisms can contribute to intracellular flow of ions and consequently enhance neuronal firing.

Combining ultrasonic transducers with conventional calcium imaging and electrophysiological techniques, research platforms for the exploration of ultrasound modulation of neuronal electrical activity at the cellular level can be established to study the potential mechanisms of ultrasound neuromodulation in vitro. In 2008, the research group of W.J. Tyler at Arizona State University found that ultrasound excited hippocampal neurons by activating voltage-sensitive sodium and calcium channels (Tyler et al. 2008). In 2016, researchers at the University of Washington including Jianmin Cui pointed out that focused ultrasound could trigger transmembrane currents by acting on mechanosensitive potassium and sodium ion channels (Kubanek et al. 2016). However, conventional ultrasonic transducers are large in size and are difficult to compatible with calcium imaging and electrophysiology. In 2017, the team from Shenzhen Institute of Advanced Technology, Chinese Academy of Sciences, designed neurostimulation chip to generate surface acoustic wave with frequencies ranging from 22 to 40 MHz, by using interdigital transducers. The acoustic wavelength is close to the size of cells, ideal for neuron stimulation on cultured cells, brain slices, or *C. elegans* (Ye et al. 2018; Lin et al. 2018; Zhou et al. 2017). The chip has transparent piezoelectric substrate; thus, it was compatible with fluorescent microscope and electrophysiology system, such as patch clamp. Zheng et al. found that ultrasonic stimulation directly activate the amphid sheath glia (ASH), a polymodal sensory neuron of *C. elegans*, and elicit avoidance behavior using the chip. They revealed for the first time that acoustic radiation force opens the mechanosensitive ion channels (MscL, Piezo1) to modulate the activity of neurons (Ye et al. 2018; Qiu et al. 2019). Then, several groups have proven that ultrasound can precisely modulate neuronal activity by the expression of mechanosensitive ion channels (Fig. 11.2) (Huang et al. 2020; Duque et al. 2022; Azadeh et al. 2021; Rabut et al. 2020).

## 3.2 Level of Small Animals

Over the past five decades, the effects of ultrasound in animal spinal cord, cortex, subcortical structures, and peripheral nerves have been explored (King et al. 2013; Tufail et al. 2010, 2011). It has been found that FUS can result in stable changes in the potentials of the rat cerebral cortex, hippocampus, thalamus, and caudate nucleus, causing conduction inhibition in the cortex as well as in the subcortical structures, leading to corresponding functional block (Kim et al. 2015; Yuan et al. 2016). Ultrasound stimulation of the motor cortex in rats (Kim et al. 2014), mice (King et al. 2013, 2014), and rabbits (Yoo et al. 2011a) could induce the activity of contralateral muscle groups under low levels of anesthesia. Electrophysiological studies revealed that ultrasound may have short- and long-lasting inhibitory effects on neural pathways. Previous studies in rat (Kim et al. 2015), rabbit (Yoo et al. 2011a), and cat (Barnard et al. 1955) have indicated that FUS could suppress visual evoked potentials. Additionally, several studies have investigated the role of FUS in suppressing epileptiform activity in various animal models of epilepsy. FUS at 200 kHz can reduce seizures in a mouse model of mesial temporal lobe epilepsy induced by kainate (Hakimova et al. 2015). A small pilot study on the same seizure rat model showed that MRI-guided FUS of the hippocampus at 690 kHz could terminate seizure activity (Yoo et al. 2010). Similarly, FUS could reduce the duration

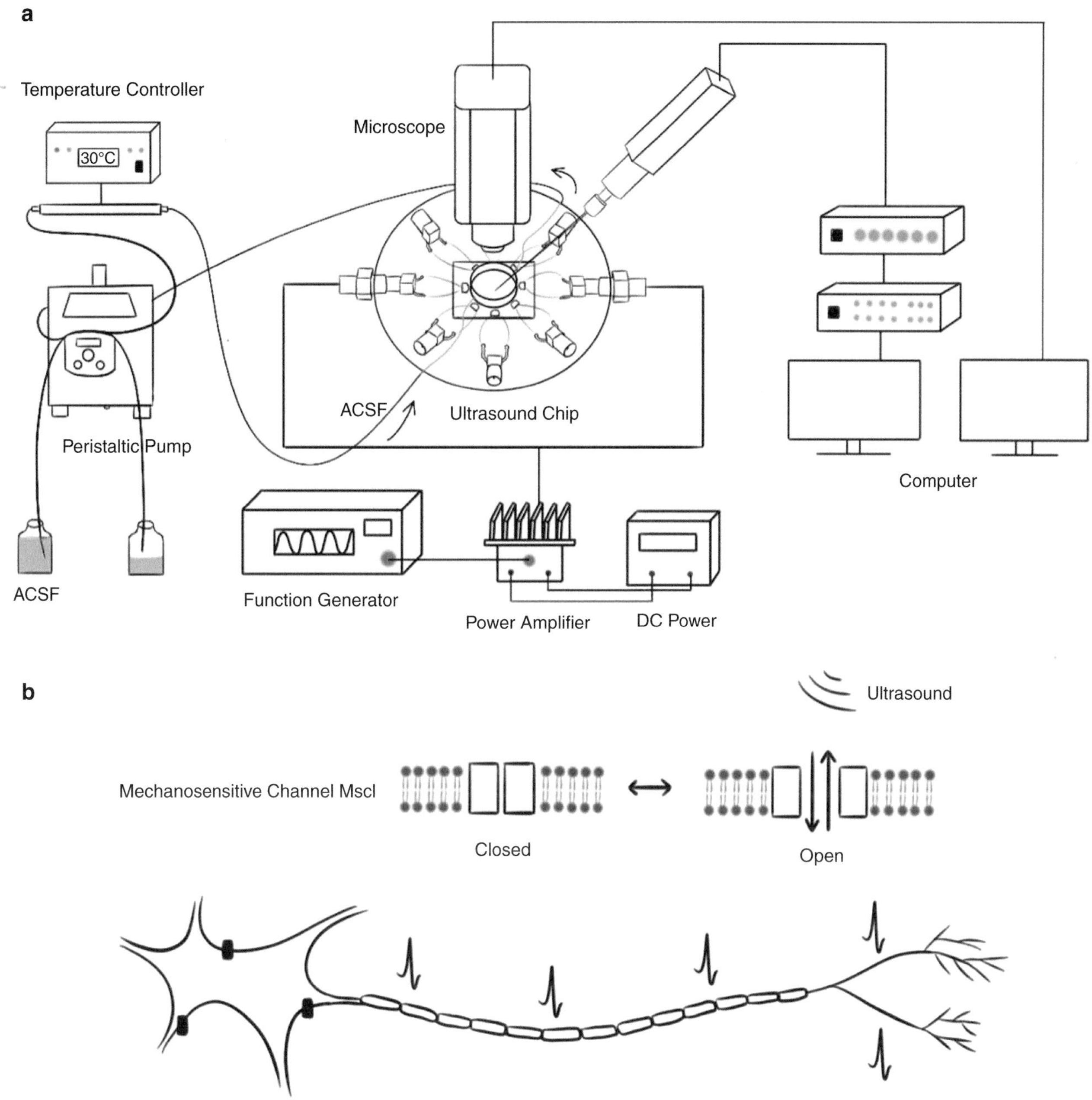


**Fig. 11.2** (**a**) Schematic illustration of the ultrasound device, which generated the acoustic pressure to stimulate neurons in the chamber during the electrophysiological recording or fluorescence measurement. (**b**) Ultrasonic control of neural activity through the activation of the *Escherichia coli* mechanosensitive channel of large conductance (MscL)

of acute seizures induced by pentetrazol in anesthetized rats (Min et al. 2011). A significant limitation of the experiments described above is the application of anesthesia during ultrasound administration to prevent animal movement from interfering with the accuracy of ultrasound localization, while anesthetics may influence judgments of ultrasound effects (King et al. 2014; Yoo et al. 2011b).

In order to achieve ultrasound stimulation treatment while the animal is awake and to ensure its motor dexterity, small-volume, lightweight head-mounted ultrasound transducer is usually customized. An adapted collimator is fixed to the rodent's skull with dental cement, and the head-mounted transducer is then mounted on the collimator. This type of stimulation facilitates real-time observation of behavioral changes in the animal during ultrasound neurostimulation and excludes the interference of anesthetics on the experimental results. In 2018, Li et al. made a focused head-mounted ultrasonic transducer with a weight of about 1 g by attaching a lens with a curve surface radius of 7 mm to a 2 MHz disk-shaped PZT-4 to allow the ultrasound field pass through the lens to focus (Li et al. 2019). In 2019, Zhou et al. made a spherical head-mounted ultrasonic transducer with a diameter of about 10 mm and a weight of about 1 g by pressing the PZT-5H/epoxy 1–3 piezoelectric composite into a spherical component, assembling it with leads, matching layer and backing, and then soldering a special coaxial cable and connector to the assembled component (Zhou et al. 2019). In 2020, Jin et al. made a needle transducer with a significantly reduced size using PZT-5, which allowed rats to wear multiple probes to receive ultrasound stimulations for ultrasound therapy at multiple targets (Jin et al. 2020). In addition to the manufacture transducers by conventional methods, micromachined ultrasound transducer (MUT) was designed based on semiconductor fabrication methods, which can also be applied to ultrasound neuromodulation studies. In 2019, Kim et al. made a 32-array capacitive micromachined ultrasound transducer (cMUT), which was a ring transducer with good magnetic compatibility and weighed about 0.8 g. The head-mounted transducer is a new tool to study diseases with significant behavioral manifestations. In addition, the head-mounted transducer is compatible with electrophysiological techniques and fiber-optic recording technology, facilitating the exploration of the impact of physical mechanisms of ultrasound neurostimulation (Kim et al. 2019). A number of results have been obtained from the ultrasound neuromodulation based on the above two systems in disease model experiments of small animals. In 2015, Guo et al. used ultrasound to stimulate ischemic areas in cerebral artery embolized rats and found that ultrasound stimulation reduced the volume of cortical infarct and improved the motor ability of rats (Guo et al. 2015). In 2018, Zhang et al. found that FUS of the prelimbic cortex elevated the expression of brain-derived neurotrophic factors and improved depressive symptoms in the rats of depression model (Zhang et al. 2018). In 2021, a study by Zhou et al. showed that FUS of the subthalamic nucleus alleviated motor symptoms and mitigated the loss of dopaminergic neurons in mice model of Parkinson's disease (PD), showing a neuroprotective effect (Zhou et al. 2021). In 2020, a study by Bobola et al. showed that FUS of the hippocampus reduced amyloid aggregation on Alzheimer's disease (AD) mice (Bobola et al. 2020). In 2020, a study by Chen et al. found that FUS could attenuate abnormal discharges in the brain areas of epileptic rats (Fig. 11.3) (Chen et al. 2020).

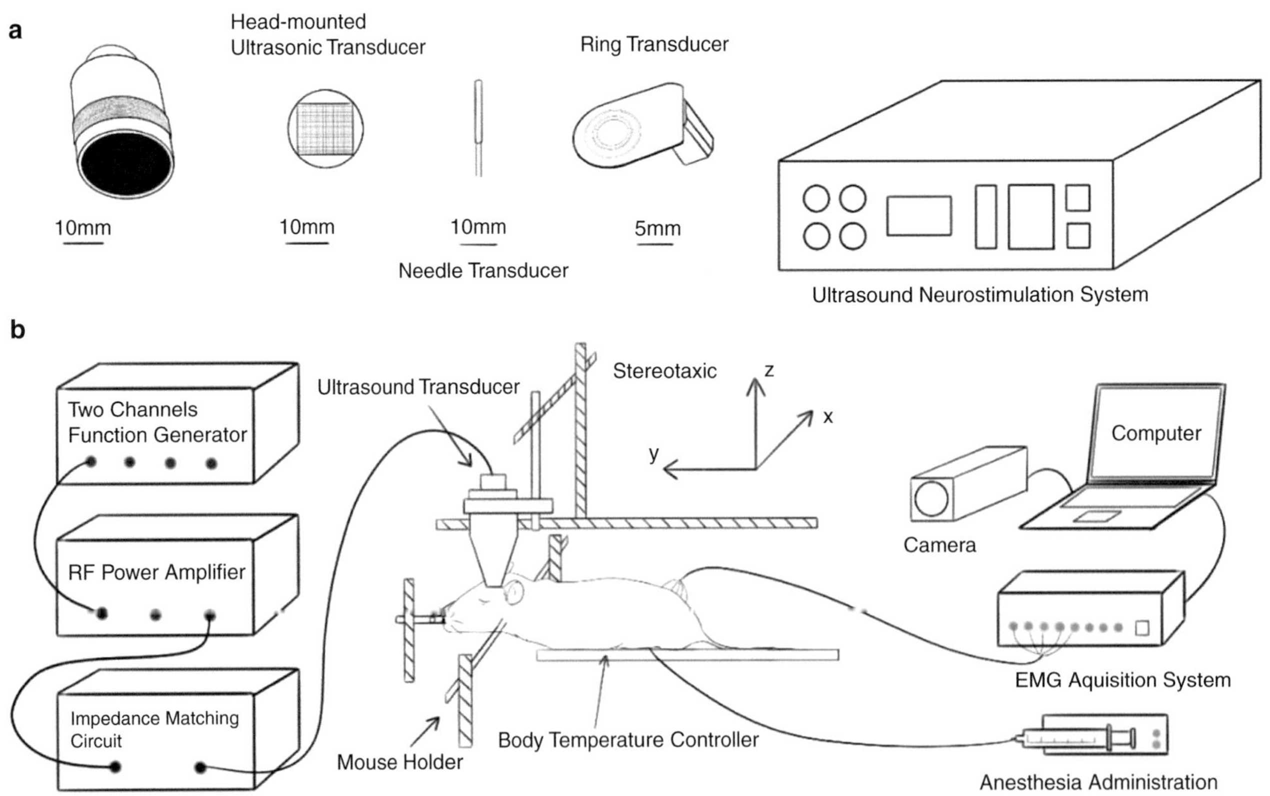


**Fig. 11.3** Ultrasonic neuromodulation device at the level of small animals. (**a**) Ultrasound transducers and system used for neuromodulation. (**b**) Schematic of the ultrasound neuromodulation system for small animals

### 3.3 Level of Large Animal and Human Brain

For non-human primates, ultrasonic stimulation could modulate neural activity (Kubanek et al. 2019). In 2013, Thomas et al. showed that FUS was delivered in the left frontal eye field and changed the direction of eye movements (Deffieux et al. 2013). In 2018, Yang et al. combined MRI guidance technology with ultrasonic stimulation, so that ultrasound can accurately focus on the 3a/3b area of the monkey brain, and found that ultrasound can stimulate the target area and the brain area downstream of the neural circuit under fMRI (Yang et al. 2018). In 2019, Folloni et al. indicated that FUS can be used to alter neural activity transiently and reversibly in subcortical and deep cortical areas with high spatial specificity. Ultrasound stimulation has a significant regulatory effect on brain functional activities (Fouragnan et al. 2019). The effect of ultrasound not only exists during stimulation but also exists after stimulation for a period. Sallet et al. showed that only 40 s of ultrasonic stimulation could induce neural activities lasting up to 1 h (Verhagen et al. 2019). In 2020, Zheng et al. found that ultrasound on the frontal cortex of monkeys can significantly improve the seizure induced by penicillin (Zou et al. 2020; Lin et al. 2020).

Clinically, in 2014, Tyler et al. showed that ultrasound stimulation significantly attenuated the amplitudes of somatosensory evoked potentials elicited by median nerve stimulation and FUS can alter EEG oscillatory dynamics through local mechanical perturbation of discrete cortical circuits (Mueller et al. 2014). At the same time, they indicated that ultrasound focused on the left primary sensory cortex and found that it could reduce the EEG amplitude caused by median nerve stimulation. In 2015, Seung-Schik Yoo et al. show that ultrasound on the hand somatosensory cortex can cause evoked potentials in the cortex, accompanied by transient tactile sensation in the opposite hand area (Lee et al. 2015). In their subsequent study, multiple ultrasonic transducers were simultaneously fixed to the head so that various tactile senses could be produced after simultaneous stimulations in the ipsilateral hemisphere applied by a combination of multiple single-array ultrasound transducers. In 2016, Monti et al. fixed an ultrasound probe on a surgically ground "trough" in the skull of a patient with disorders of consciousness to apply ultrasonic stimulation, which greatly improved the conscious state of the patient (Monti et al. 2016). In 2018, Legon et al. used a single-array ultrasound probe to focus on the thalamus to reduce the amplitude of somatosensory evoked potentials (Legon et al. 2018).

Meanwhile, several studies have reported the effects of FUS on mood improvement and the modulation of brain functions in patients with diseases. In 2013, Hameroff et al. used a clinical ultrasound device at 8 MHz and found that 15 s of FUS of the prefrontal cortex enhanced the mood of chronic pain patients, which lasted up to 40 min (Hameroff et al. 2013). In 2016, Monti et al. applied MRI-navigated FUS to the thalamus of a patient with post-traumatic disorders of consciousness for ten times, which lasted 30 s (Monti et al. 2016). In 2020, Beisteiner et al. applied FUS to the right sensory cortexes of 17 AD patients for 2–4 weeks, which improved functional networks and cognitive abilities for up to 3 months in AD patients (Beisteiner et al. 2020). Meanwhile, Sanguinetti et al. showed that FUS stimulation of the right prefrontal cortex enhanced self-reported mood states and also decreased FC in resting-state networks related to emotion and mood regulation (Sanguinetti et al. 2020).

In general, studies in humans have demonstrated that FUS modulates neural activity in sensory motor and visual cortices to some extent. Functional changes induced by FUS include emotion improvement, phosphenes, and abnormal feelings, depending on the intracerebral target sites and parameters of ultrasonic stimulation. The effect of ultrasound on behavior remains unclear. The neuromodulation effect of FUS on humans leaves us with many questions. For example, the differences in ultrasound intensity and processing parameters used in different experiments make the optimal parameters of ultrasound neuromodulation uncertain, whether different neural circuits respond differently to the same acoustic parameters, whether ultrasound-mediated effects are binary or in a dose-response relationship, and whether FUS causes neuroplasticity or long-term changes in gene expression in the brain. Future neuromodulation studies with the application of a wider range of ultrasound parameters and more precise localization of stimulation targets, as well as corresponding neural mechanism studies, will make it possible to elucidate the exact neuromodulatory effects of FUS and apply it to the treatment of functional brain diseases (Darrow 2019).

**Conclusions**

Deep brain stimulation (DBS), transcranial direct current stimulation (tDCS), and transcranial magnetic stimulation (TMS) are commonly adopted for neuromodulation techniques. Although DBS was approved by the FDA in 1997 for the treatment of primary tremor, it is an invasive treatment method that requires artificial implantation of stimulation electrodes into the brain and regular battery replacement, with a high risk of infection and low patient acceptance. tDCS and TMS are noninvasive neuromodulation methods, while they have low spatial resolution and penetration depth. Ultrasonic neuromodulation uses ultrasound as the energy transmission carrier, and its wave energy can effectively penetrate biological tissues at different depths (including skull). More importantly, the use of multi-array ultrasonic transducer and phased-array technique can precisely regulate the ultrasound transmission path and focus the ultrasound energy on the designated biological tissues. Ultrasound wave is a mechanical wave and, compared with TMS, is technically easier to use with magnetic resonance at the same time, which guarantees the accuracy and safety of treatment. Therefore, compared with traditional electrical or magnetic neurostimulation techniques, ultrasonic neurostimulation has the advantages of deep penetration, controllable targets, and noninvasive and precise treatment. However, the range of ultrasound parameters and brain areas of stimulation remain to be tested, and further prospective studies are needed to refine the stimulation protocols to determine its dose-response effects on specific areas of neural activity. In addition, many cortical and subcortical brain target sites need to be explored to investigate the potential electrophysiological and behavioral responses. Human trials on ultrasound brain stimulation for the treatment of epilepsy, AD, PD, disorders of consciousness, and stroke are now underway. The greatly promising ultrasonic neuromodulation can serve as not only a neuromodulation tool but also a novel technique for noninvasive brain-computer interface.

## References


Adams JB, Moore RG, Anderson JH et al (1996) High-intensity focused ultrasound ablation of rabbit kidney tumors. 10(1):71–75

Asan AS, Kang Q, Oralkan O et al (2021) Entrainment of cerebellar Purkinje cell spiking activity using pulsed ultrasound stimulation. Brain Stimul 14(3):598–606

Azadeh SS, Lordifard P, Soheilifar MH et al (2021) Ultrasound and sonogenetics: A new perspective for controlling cells with sound. Iran J Pharm Res 20(3):151–160

Baker KG, Robertson VJ, Duck FA (2001) A review of therapeutic ultrasound: Biophysical effects. Phys Ther 81(7):1351–1358

Barnard JW, Fry WJ, Fry FJ et al (1955) Effects of high intensity ultrasound on the central nervous system of the cat. J Comp Neurol 103(3):459–484

Beisteiner R, Matt E, Fan C et al (2020) Transcranial pulse stimulation with ultrasound in Alzheimer's disease—a new navigated focal brain therapy. Advanced Science 7(3):1902583 https://doi.org/10.1002/advs.201902583

Bobola MS, Chen L, Ezeokeke CK et al (2020) Transcranial focused ultrasound, pulsed at 40 Hz, activates microglia acutely and reduces Aβ load chronically, as demonstrated in vivo. Brain Stimul 13(4):1014–1023

Chang WHS, Sun JS, Chang SP et al (2002) Study of thermal effects of ultrasound stimulation on fracture healing. Bioelectromagnetics 23(4):256–263

Chen SG, Tsai CH, Lin CJ et al (2020) Transcranial focused ultrasound pulsation suppresses pentylenetetrazol induced epilepsy in vivo. Brain Stimul 13(1):35–46

Cheng SQ, Zhou XD, Tang ZY et al (1997) Iodized oil enhances the thermal effect of high-intensity focused ultrasound on ablating experimental liver cancer. J Cancer Res Clin Oncol 123(11):639–644

Darrow DP (2019) Focused ultrasound for neuromodulation. Neurotherapeutics 16:88–99. https://doi.org/10.1007/s13311-018-00691-3.

Deffieux T, Younan Y, Wattiez N et al (2013) Low-intensity focused ultrasound modulates monkey Visuomotor behavior. Curr Biol 23(23): 2430–2433. https://doi.org/10.1016/j.cub.2013.10.029

Duque M, Lee-Kubli CA, Tufail Y et al (2022) Sonogenetic control of mammalian cells using exogenous transient receptor potential A1 channels. Nat Commun 13(1):600

Fini M, Tyler WJ (2017) Transcranial focused ultrasound: a new tool for non-invasive neuromodulation. Int Rev Psychiatry 29(2):168–177

Fomenko A, Chen KS, Nankoo JF et al (2020) Systematic examination of low-intensity ultrasound parameters on human motor cortex excitability and behavior. Elife 9:e54497

Fomenko A, Neudorfer C, Dollapiazza RF et al (2018) Low-intensity ultrasound neuromodulation: an overview of mechanisms and emerging human applications. Brain Stimul 11(6):1209–1217

Fouragnan EF, Chau BKH, Folloni D et al (2019) The macaque anterior cingulate cortex translates counterfactual choice value into actual behavioral change. Nat Neurosci 22(5): 797–808. https://doi.org/10.1038/s41593-019-0375-6

Guo T, Li H, Lv Y et al (2015) Pulsed transcranial ultrasound stimulation immediately after the ischemic brain injury is neuroprotective. IEEE Trans Biomed Eng 62(10):2352–2357

Hakimova H, Kim S, Chu K et al (2015) Ultrasound stimulation inhibits recurrent seizures and improves behavioral outcome in an experimental model of mesial temporal lobe epilepsy. Epilepsy Behav 49:26–32

Hameroff S, Trakas M, Duffield C et al (2013) Transcranial Ultrasound (TUS) Effects on Mental States: A Pilot Study. Brain Stimulation 6(3): 409-415. https://doi.org/10.1016/j.brs.2012.05.002.

Holland CK, Apfel RE (1990) Thresholds for transient cavitation produced by pulsed ultrasound in a controlled nuclei environment. J Acoust Soc Am 88(5):2059–2069

Huang YS, Fan CH, Hsu N et al (2020) Sonogenetic modulation of cellular activities using an engineered auditory-sensing protein. Nano Lett 20(2):1089–1100

Jin Y, Li Y, Ye Y et al (2020) Development of multi-layer lateral-mode ultrasound needle transducer for brain stimulation in mice. IEEE Trans Biomed Eng 67(7):1982–1988

Kim H, Chiu A, Lee SD et al (2014) Focused ultrasound-mediated non-invasive brain stimulation: Examination of sonication parameters. Brain Stimul 7(5):748–756

Kim H, Park MY, Lee SD et al (2015) Suppression of EEG visual-evoked potentials in rats through neuromodulatory focused ultrasound. Neuroreport 26(4):211–215

Kim H, Kim S, Sim NS et al (2019) Miniature ultrasound ring array transducers for transcranial ultrasound neuromodulation of freely-moving small animals. Brain Stimul 12(2):251–255

King RL, Brown JR, Newsome WT et al (2013) Effective parameters for ultrasound-induced in vivo neurostimulation. Ultrasound Med Biol 39(2):312–331

King RL, Brown JR, Pauly KB (2014) Localization of ultrasound-induced in vivo neurostimulation in the mouse model. Ultrasound Med Biol 40(7):1512–1522

Kubanek J, Shi J, Marsh J et al (2016) Ultrasound modulates ion channel currents. Sci Rep 6(1):24170

Kubanek J, Shukla P, Das A et al (2018) Ultrasound elicits behavioral responses through mechanical effects on neurons and ion channels in a simple nervous system. J Neurosci 38(12):3081–3091

Kubanek J, Brown J, Ye P et al (2019) Transcranial ultrasound selectively biases decision-making in primates. bioRxiv:486134. https://doi.org/10.1101/486134

Lee W, Kim H, Jung Y et al (2015) Image-guided transcranial focused ultrasound stimulates human primary somatosensory cortex. Sci Rep 5(1): 8743. https://doi.org/10.1038/srep08743

Lee SJ, Park KH (1999) Ultrasonic energy in endoscopic surgery. Yonsei Med J 40(6):545–549

Legon W, Ai L, Bansal P (2018) Neuromodulation with single-element transcranial focused ultrasound in human thalamus. Hum Brain Mapp 39(5): 1995–2006. https://doi.org/10.1002/hbm.23981

Levenback BJ, Sehgal CM, Wood AKW (2012) Modeling of thermal effects in antivascular ultrasound therapy. J Acoust Soc Am 131(1):540–549

Li GF, Qiu WB, Zhang ZQ et al (2019) Noninvasive ultrasonic neuromodulation in freely moving mice. IEEE Trans Biomed Eng 66(1):217–224

Lin Z, Meng L, Zou J et al (2020) Non-invasive ultrasonic neuromodulation of neuronal excitability for treatment of epilepsy. Theranostics 10(12):5514–5526. https://doi.org/10.7150/thno.40520

Lin Z, Zhou W, Huang X, et al (2018) On-chip ultrasound modulation of pyramidal neuronal activity in hippocampal slices. Advanced Biosystems 2(8): 1800041. https://doi.org/10.1002/adbi.201800041

McDannold N, Vykhodtseva N, Hynynen K (2008) Blood-brain barrier disruption induced by focused ultrasound and circulating preformed microbubbles appears to be characterized by the mechanical index. Ultrasound Med Biol 34(5):834–840

Meng Y, Hynynen K, Lipsman N (2021) Applications of focused ultrasound in the brain: from thermoablation to drug delivery. Nat Rev Neurol 17(1):7–22

Miller DL (2007) Overview of experimental studies of biological effects of medical ultrasound caused by gas body activation and inertial cavitation. Prog Biophys Mol Biol 93(1–3):314–330

Min BK, Bystritsky A, Jung KI et al (2011) Focused ultrasound-mediated suppression of chemically-induced acute epileptic EEG activity. BMC Neurosci 12:23

Monti MM, Schnakers C, Korb AS et al (2016) Non-invasive ultrasonic thalamic stimulation in disorders of consciousness after severe brain injury: a first-in-man report. Brain Stimulation: Basic, Translational, and Clinical Research in Neuromodulation 9(6): 940–941. https://doi.org/10.1016/j.brs.2016.07.008

Mueller J, Legon W, Opitz A, Sato T F & Tyler W J (2014) Transcranial focused ultrasound modulates intrinsic and evoked EEG dynamics. Brain Stimul 7(6): 900–908. https://doi.org/10.1016/j.brs.2014.08.008

Qiu Z, Guo J, Kala S et al (2019) The mechanosensitive ion channel Piezo1 significantly mediates in vitro ultrasonic stimulation of neurons. Science 21:448–457

Rabut C, Yoo S, Hurt RC et al (2020) Ultrasound technologies for imaging and modulating neural activity. Neuron 108(1):93–110

Riley N (1998) Acoustic streaming. Theor Comput Fluid Dyn 10(1):349–356

Sanguinetti JL, Hameroff S, Smith EE et al (2020) Transcranial Focused Ultrasound to the Right Prefrontal Cortex Improves Mood and Alters Functional Connectivity in Humans 14: 52. https://doi.org/10.3389/fnhum.2020.00052

Sassaroli, E., Vykhodtseva, N (2016) Acoustic neuromodulation from a basic science prospective. J Ther Ultrasound 4:17. https://doi.org/10.1186/s40349-016-0061-z

Shealy CN, Henneman E (1962) Reversible effects of ultrasound on spinal reflexes. Arch Neurol 6:374–386

Smith NB, Temkin JM, Shapiro F et al (2001) Thermal effects of focused ultrasound energy on bone tissue. Ultrasound Med Biol 27(10):1427–1433

Ter Haar GR (2001) High intensity focused ultrasound for the treatment of tumors. Echocardiography 18:317–322. https://doi.org/10.1046/j.1540-8175.2001.00317.x

Tufail Y, Matyushov A, Baldwin N et al (2010) Transcranial pulsed ultrasound stimulates intact brain circuits. Neuron 66(5):681–694

Tufail Y, Yoshihiro A, Pati S et al (2011) Ultrasonic neuromodulation by brain stimulation with transcranial ultrasound. Nat Protoc 6(9): 1453–1470. https://doi.org/10.1038/nprot.2011.371

Tyler WJ (2011) Noninvasive neuromodulation with ultrasound? A continuum mechanics hypothesis. Neuroscientist 17(1):25–36

Tyler WJ (2012) The mechanobiology of brain function. Nat Rev Neurosci 13(12):867–878

Tyler WJ, Tufail Y, Finsterwald M et al (2008) Remote excitation of neuronal circuits using low-intensity, low-frequency ultrasound. PLoS One 3(10):e3511

van den Bijgaart RJE, Eikelenboom DC, Hoogenboom M et al (2017) Thermal and mechanical high-intensity focused ultrasound: perspectives on tumor ablation, immune effects and combination strategies. Cancer Immunol Immunother 66(2):247–258

Verhagen L, Gallea C, Folloni D et al (2019) Offline impact of transcranial focused ultrasound on cortical activation in primates. eLife 8: e40541. https://doi.org/10.7554/eLife.40541

Warwick R, Pond J (1968) Trackless lesions in nervous tissues produced by high intensity focused ultrasound (high-frequency mechanical waves). J Anat 102(Pt 3):387–405

Wu F, Chen WZ, Bai J et al (2001) Pathological changes in human malignant carcinoma treated with high-intensity focused ultrasound. Ultrasound Med Biol 27(8):1099–1106

Xia X, Fomenko A, Nankoo JF et al (2021) Time course of the effects of low-intensity transcranial ultrasound on the excitability of ipsilateral and contralateral human primary motor cortex. NeuroImage 243:118557

Yang PF, Phipps MA, Newton AT et al (2018) Neuromodulation of sensory networks in monkey brain by focused ultrasound with MRI guidance and detection. Sci Rep 8(1):7993. https://doi.org/10.1038/s41598-018-26287-7

Ye J, Tang S, Meng L et al (2018) Ultrasonic control of neural activity through activation of the mechanosensitive channel MscL. Nano Lett 18(7):4148–4155. https://doi.org/10.1021/acs.nanolett.8b00935

Yoo SS, Bystritsky A, Lee JH et al (2011a) Focused ultrasound modulates region-specific brain activity. NeuroImage 56(3):1267–1275

Yoo SS, Kim H, Min BK et al (2011b) Transcranial focused ultrasound to the thalamus alters anesthesia time in rats. Neuroreport 22(15):783–787

Yoo SS, Jung K, Zhang Y et al (2010) Non-invasive suppression of animal-model chronic epilepsy using image-guided focused ultrasound. Proc Int Soc Magn Reson Med 18:105–111

Yuan Y, Yan JQ, Ma ZT et al (2016) Noninvasive focused ultrasound stimulation can modulate phase-amplitude coupling between neuronal oscillations in the rat hippocampus. Front Neurosci 10:348

Zhang DQ, Li HD, Sun JF et al (2018) Antidepressant-like effect of low-intensity transcranial ultrasound stimulation. IEEE Trans Biomed Eng 66(2):411–420

Zhou W, Wang JJ, Wang KY et al (2017) Ultrasound neuro-modulation chip: activation of sensory neurons in Caenorhabditis elegans by surface acoustic waves. Lab Chip 17(10):1725–1731 https://doi.org/10.1039/c7lc00163k

Zhou H, Niu LL, Xia XX et al (2019) Wearable ultrasound improves motor function in an MPTP mouse model of Parkinson's disease. IEEE Trans Biomed Eng 66(11):3006–3013

Zhou H, Meng L, Xia XX et al (2021) Transcranial ultrasound stimulation suppresses neuroinflammation in a chronic mouse model of Parkinson's disease. IEEE Trans Biomed Eng 68(11):3375–3387

Zou J, Meng L, Lin Z, et al (2020) Ultrasound Neuromodulation inhibits seizures in acute epileptic monkeys. iScience 23(5):101066. https://doi.org/10.1016/j.isci.2020.101066

# Simultaneous Brain Stimulation and Acquisition

# 12

Li Wang, Kai Wang, Gongjun Ji, Penghui Song, Di Wang, and Xiating Zhang

Transcranial magnetic stimulation (TMS) is a noninvasive stimulation technique applied to the cerebral cortex for the treatment of functional brain disorders, but also for the research of brain mechanisms. The combination of TMS with functional brain examination techniques such as fMRI, positron emission tomography (PET), and electroencephalography (EEG) makes it possible to observe brain network activities associated with specific functional cortices and elucidate the association between cortical networks and behaviors. It can also reveal neural mechanisms and efficacy predictive markers associated with rTMS treatment. This chapter will describe TMS and several functional brain examination techniques, the combination of TMS with these techniques, and related studies on their applications.

## 1 Simultaneous TMS-PET

### 1.1 TMS and PET

The combination of TMS with PET makes it possible to detect cortical excitability and connectivity, characterize TMS-induced brain network reorganization patterns at the systemical level, and investigate causal connections in brain circuits by measuring TMS-induced effects of cortical areas on activity in other brain areas. TMS-PET is thus considered as a tool for investigating the mechanisms of brain network connectivity. In addition, it helps to understand the biological effects of rTMS. The combination is available in online and offline modes. In online mode, PET is performed simultaneously with TMS; in offline mode, PET is performed not simultaneously with TMS, but before or after TMS. Before getting into the TMS-PET technique, let's take a brief look at common TMS modes and PET tracer imaging.

#### 1.1.1 TMS Modes

TMS may modulate cortical excitability in different stimulation modes, commonly including single-pulse TMS (spTMS), paired-pulse TMS (ppTMS), and rTMS. These modes may induce excitability changes in specific cortices. As the induced changes in cortical excitability are variable in different TMS modes, the combination of TMS and PET is believed to be promising in many applications.

#### 1.1.2 PET Imaging

PET can be used for the detection of cerebral blood flow (CBF), molecular metabolism, and receptors.

**CBF Measurement** Common imaging agents include water, radon, fluoromethane, and butanol labeled with positron-emitting radionuclides, which penetrate the blood-brain barrier into the brain for CBF measurement.

**Metabolism** It is commonly used to investigate glucose metabolism, oxygen metabolism, and protein anabolism. F-FDG is the most common glucose derivative used in studies on glucose metabolism in the brain. Imaging agents commonly used in protein anabolism studies include $^{11}$C-labeled methionine (C-MET), leucine ($^{1}$C-Leu), and F-labeled tyrosine (F-Tyr).

L. Wang (✉)
School of Life Sciences, Beijing Institute of Technology, Beijing, China

K. Wang · G. Ji
The School of Mental Health and Psychological Sciences, Anhui Medical University, Hefei, China

P. Song · X. Zhang
Department of Neurology, Xuanwu Hospital, Capital Medical University, Beijing, China

D. Wang
School of Engineering Medicine, Beihang University, Beijing, China
e-mail: wangdi09@buaa.edu.cn



Y. Wang (ed.), *Therapeutics of Neural Stimulation for Neurological Disorders*, https://doi.org/10.1007/978-981-99-4538-2_12

**Receptor Assay** Many neuropsychiatric disorders are characterized by abnormalities in presynaptic and postsynaptic dopamine, 5-hydroxytryptamine, and acetylcholine receptors; thus, imaging of these neurotransmitters has become a research hotspot in the field of PET. Receptor assay is a significant focus in the research of neuropsychiatric disorders, which allows clinicians to make more accurate evaluations before patients develop significant clinical symptoms, thus providing a reference for the determination of molecular targets for therapeutic development.

## 1.2 Simultaneous TMS-PET Acquisition

### 1.2.1 Introduction of Simultaneous TMS-PET Acquisition

Simultaneous TMS-PET acquisition is defined as the implementation of PET scanning during TMS. This approach allows for observing the excitability changes of the brain network during the application of TMS to specific cortical sites and identifying stimulus-induced changes in brain activity in real time. This approach also provides a basis for exploring the working mechanism of function-related network in the stimulated areas and its association with behavior.

### 1.2.2 Applications of Simultaneous TMS-PET Acquisition

#### 1.2.2.1 Single-Pulse TMS-PET

Over the past decade, functional brain imaging has shifted from depicting cognitive, perceptual, and sensorimotor-related circuits to exploring the interactions between different functional brain circuits by analyzing the cerebral blood flow or blood oxygenation level-dependent signals in subjects during task or resting state. However, the deduction of causal relationships in brain connectivity is limited by the existing imaging methods due to a low temporal resolution.

"Perturbation" has long been used as a pathway for studying brain-behavior relationships. For example, irreversible perturbations (disruption of specific brain areas) suggest that the inferior frontal gyrus is critical for language production and the hippocampus in narrative memory. With single-pulse TMS, "cortical excitability" was assessed by stimulating the primary motor cortex to evoke motor responses. The association between the stimulated area and specific behaviors can also be explored by interfering with ongoing neural activity in the stimulated area (virtual lesions). For example, a study (Mulckhuyse et al. 2017) applied inhibitory rTMS to the right posterior parietal cortex in healthy subjects and found that the stimulation increased the time cost of the subject's response to threatening attentional stimulation.

#### 1.2.2.2 Paired-Pulse TMS-PET

Paired-pulse stimulation is a rapid sequential stimulation of two TMS pulses (pulse interval of 1–30 ms) to cortical sites, which is used for the assessment of intracortical inhibition and intracortical facilitation. Short pulse intervals (1–5 ms) caused intracortical inhibition, mainly mediated by gamma-aminobutyric acid. In contrast, long pulse intervals (8–30 ms) caused intracortical facilitation, mainly mediated by glutamate.

Paired-pulse TMS makes it more likely to study the combination of TMS and PET. Paired magnetic pulses target the activation of cortical neurons with different anatomical and functional connections depending on the stimulation intervals. Strafella and Paus (2001) performed PET examination before TMS stimulation, during single-pulse TMS in the left M1 area, 3-ms paired TMS and 12-ms paired TMS in eight healthy volunteers, and a significant positive correlation was observed between the CBF difference (TMS3-TMSsing) and the amount of suppression of electromyogram (EMG) response, as well as between the CBF difference (TMS12-TMSsing) and the amount of facilitation of EMG response. It suggests that electromyography responses are mediated by different cortical interneurons and that paired pulses with different stimulation intervals could induce different patterns of rCBF changes.

#### 1.2.2.3 rTMS-PET

rTMS can induce persistent neuroplastic changes in cortical function. The cortical excitability changes induced by rTMS can last up to several hours, which makes rTMS an important tool to study brain network plasticity.

In healthy subjects, Paus et al. (1997) first explored the effect of rTMS on rCBF with PET. 10-Hz rTMS with different numbers of pulses (5–30) was applied to the left frontal eye field in six healthy volunteers. It was found that rCBF changes in the parietal and medial parieto-occipital cortices were positively related to the number of magnetic stimulations. In contrast to the stimulation of the frontal eye field, rCBF changes induced by magnetic stimulation of the motor cortex of hands were negatively correlated with the number of stimulations. It suggests that the intrinsic properties of the stimulated cortex could influence the TMS effect. Siebner et al. (1998) performed PET detection in healthy subjects when low-frequency rTMS was applied to the left sensorimotor cortex and during the mimicry of rTMS-induced arm movements and found that the increase in blood volume in the sensorimotor cortex was greater, the affected brain areas were more extensive, and the activation of supplementary motor area was more pronounced during the mimicry of arm movements. By assessing the role of TMS to the medial frontal gyrus in modulating brain activity during the execution of

Stroop task, it was found that the increase in rCBF during TMS was correlated with the decrease in rCBF of the anterior cingulate cortex (Hayward et al. 2007), suggesting that TMS to the medial frontal gyrus could affect the activities of the anterior cingulate cortex. A study (Felipe et al. 2016) applied rTMS to the primary motor cortex at 3 Hz, 5 Hz, 10 Hz, and 15 Hz and performed PET simultaneously. A structural equation model was used to analyze the effective connectivity of the motor network during stimulation. It was found that the motor network had the strongest path coefficients under 5-Hz magnetic stimulation, which may be the best stimulation frequency to study the motor network. The cerebral premotor cortex was preferentially activated under rTMS at 10 Hz, while the supplementary motor area was preferentially activated at 15 Hz. It suggested that specific frequencies could be used to target specific connections in the motor network.

In terms of diseases, PET detection was performed during post-stroke speech tasks in a study on 11 patients with middle cerebral artery occlusion (Winhuisen et al. 2005). 4-Hz rTMS was applied to the area of maximal activation in the inferior frontal gyrus during task performance, and it was found that the inferior frontal gyrus was activated after stimulation (left activation in three patients and bilateral activation in eight patients). The worse the response of the right inferior frontal gyrus to rTMS, the worse the performance in the task. It suggests that the right inferior frontal gyrus is critical for maintaining language function in stroke patients. To investigate the antidepressant mechanism of prefrontal rTMS, one study performed PET detection when rTMS was applied to depressive patients and found that active stimulation induced a more pronounced increase in rCBF in the ipsilateral anterior cingulate cortex, as well as more sustained activation in the medial prefrontal cortex, ventral lateral prefrontal lobe, and ventral striatum. The effect on the ventral striatum may reflect how rTMS modulates the dopaminergic reward system of the limbic midbrain, which may be an important antidepressant mechanism.

## 1.3 Offline TMS-PET Acquisition

Offline means that TMS stimulation and PET examination are not performed simultaneously in temporal and spatial dimension, with PET performed before or after rTMS. Offline TMS-PET acquisition is easier to implement because it does not require hardware compatibility and is suitable for assessing the immediate effects of rTMS and the persistent neuroplastic changes produced by rTMS treatment. It can be used not only for pathological studies of brain diseases but also for exploring the treatment mechanisms and treatment efficacy prediction of rTMS.

### 1.3.1 Applications in Healthy Individuals

One study applied low-frequency rTMS to the primary motor cortex and dorsal premotor cortex and performed PET examination after stimulation. It revealed that stimulation of the dorsal premotor cortex could modulate metabolic activities in several brain areas including the frontal motor cortex and its parietal and prefrontal cortices. In contrast, stimulation of the primary motor cortex affected fewer brain areas. After detecting rCBF during resting-state and free finger movements with PET when low-frequency rTMS was applied (Lee et al. 2003), it was found that activities in the stimulated M1 area increased and more extensive changes in activity in the premotor cortex of the non-stimulated hemisphere were induced. M1-stimulated sites were less responsive to inputs from other motor cortices, indicating persistent disruption at the rTMS stimulation sites. In contrast, the linking between the left motor cortex of hands and the premotor cortex was strengthened. This compensatory mechanism supports the motor system to maintain normal function and provides the neural basis for the motor system to compensate for the adverse effects of focal lesions (e.g., stroke). The detection of rCBF changes in 40 healthy subjects after high-frequency rTMS was applied to the left DLPFC revealed that active rTMS could result in the redistribution of blood flow in the network under default condition, with increasing rCBF in the left medial temporal lobe/hippocampus and decreasing rCBF in the precuneus and cerebellum, whereas pseudo-stimulation did not cause any change in rCBF (Shang et al. 2018). It suggests that high-frequency rTMS could produce a beneficial effect by inducing rCBF redistribution in the brain.

In addition to blood flow and glucose metabolism, PET can be used to assess the effects of rTMS on neurotransmission and elucidate the molecular transmitter basis of brain circuit activity. The effect of high-frequency rTMS (10 Hz) to the left DLPFC on the serotonin synthesis capacity in the brain was observed in ten healthy subjects (Sibon et al. 2007). It was found that stimulations could result in decreased metabolism in the left parahippocampal gyrus and right insula and increased metabolism in the right cingulate gyrus and cuneus. It suggests that magnetic stimulation to the left DLPFC could modulate tryptophan/5-hydroxytryptamine metabolism in the limbic system possibly due to the antidepressant mechanism of rTMS. Dopamine D2 receptor was traced right after 10-Hz rTMS was applied to the right and left DLPFC, respectively, in healthy subjects (Sang et al. 2009). It was found that stimulating the left side could cause a decrease in receptor occupancy in the subgenual and pregenual areas of the ipsilateral cingulate gyrus and in the medial orbitofrontal cortex, whereas stimulating the right side generated no significant effect on dopamine receptor binding, which confirmed that DLPFC can modulate extrastriate dopamine. Strafella et al. (2001) used PET to examine the changes in extracellular dopamine concentrations after high-frequency rTMS was applied to the left DLPFC in healthy subjects and found that stimulation caused a decrease in dopamine receptor binding in the left dorsal caudate nucleus, suggesting that stimulation of the prefrontal lobe could increase the dopamine release of the ipsilateral caudate nucleus, which supported the cortico-striatal model of dopamine release. Continuous theta-burst stimulation (cTBS) of θ pulse was applied to healthy subjects when they were performing a card-sorting task, from which significant hemispheric lateralization of dopamine release post-stimulation was observed (Ji et al. 2008). Stimulation to the left but not right DLPFC could affect the subject performance in sorting cards and activities of the caudate nucleus.

### 1.3.2 Studies on Application to Diseases

Exploring the effects of rTMS on brain function by combining rTMS with PET may reveal treatment-related neural mechanisms. A double-blind, crossover, placebo-controlled study (Speer et al. 2000) including ten patients with depression assessed cerebral blood flow changes in the left prefrontal lobe after 10 days of 1-Hz or 20-Hz rTMS and found that 20-Hz stimulation caused increases in rCBF in the bilateral prefrontal lobe, cingulate gyrus, insula, basal ganglia, hippocampus, thalamus, and left amygdala. In contrast, 1-Hz low-frequency stimulation resulted in decreased rCBF in the right prefrontal lobe, left medial temporal lobe, left basal ganglia, and left amygdala. The two forms of stimulation induced opposite emotional changes, suggesting that high-frequency and low-frequency rTMS have different effects on brain function in depressive patients. The detection of extracellular dopamine concentrations in the putamen after high-frequency rTMS was applied to the bilateral motor cortices in patients with PD revealed that stimulations resulted in reduced dopamine binding in the ipsilateral putamen mass (Antonio et al. 2005), which suggested abnormal cortico-striatal transmission in patients with PD. In a comparative study on patients with primary focal dystonia and healthy controls, Siebner et al. (2003) examined the changes in rCBF produced by low-frequency rTMS to the left dorsal premotor cortex and found that stimulations had a greater effect on neural activity in patients than in healthy subjects, manifested by more pronounced rCBF reduction in the bilateral and medial premotor cortex, putamen, and thalamus.

PET can also be used to predict treatment efficacy of rTMS. Nadeau et al. (2002) adopted SPECT to study eight patients with treatment-resistant depression and examined the changes in rCBF before and after ten sessions of 20-Hz rTMS to the prefrontal lobe. It was found that the blood flow in the left amygdala in those who responded to treatment reduced compared to those who did not and the blood flow in the orbitofrontal or anterior cingulate cortex reduced in those who responded to treatment but did not change in those who did not respond to treatment. Mottaghy et al. (2002) adopted SPECT to assess the effect of ten sessions of high-frequency rTMS (10 Hz) to the left DLPFC on rCBF in depressive patients and the predictability of rCBF for efficacy. It was found that there was significant asymmetry in activity between the bilateral hemispheres before treatment, which was reversed after treatment. rCBF in the limbic system was negatively correlated with efficacy, whereas that in several cortices was positively correlated with efficacy, suggesting that rCBF may be used to predict the antidepressant effect of rTMS. Fifteen patients with treatment-resistant depression were enrolled and given 2 weeks of high-frequency rTMS to the left DLPFC (Baeken et al. 2015). PET scans were performed before treatment and at week 1 and at week 2 of treatment. It was found that higher sgACC metabolism predicted more significant antidepressant efficacy and that treatment response was associated with a significant reduction in sgACC metabolic activity, whereas patients not responsive to treatment did not show any changes in sgACC metabolic activity. It indicates that sgACC function is a biological marker mediating the antidepressant response to rTMS.

The abnormal sites of brain function detected by PET may also provide a reference for the stimulation target selection of rTMS treatment. One study (Plewnia et al. 2007) included nine patients with tinnitus and found increased activity in the left medial temporal gyrus and inferior temporal gyrus, left temporo-parietal associative cortex, and posterior cingulate cortex after PET detection. After applying low-frequency rTMS to one of these cortical areas, it was found stimulations could mitigate tinnitus, which revealed the feasibility and effectiveness of individualized PET-guided rTMS treatment for tinnitus.

## 1.4 Limitations and Outlook

There are some limitations in the combination of TMS and PET, which need to be further resolved. One limitation of applying TMS in brain function studies is the precise positioning of the stimulation sites. At present, the positioning of magnetic stimulation is often done with reference to EEG electrode positions or under the navigation of individual anatomical images, whereas it is more scientific to position sites according to abnormal activation areas analyzed from individual functional imaging. In the future, stimulation site positioning based on individualized functional image can improve stimulation accuracy and efficacy. Another limitation is that the TMS coils currently used cannot act on deeper cortical and subcortical structures. With conventional TMS coils, it is possible to stimulate brain areas about 30–40 mm from the scalp convex. New coils (H-coil) are expected to overcome this deficiency. Nevertheless, the evidence of clinical applications of H-coil still needs to be accumulated. It should be also noticed that when TMS is applied simultaneously during PET imaging, changes in rCBF in brain areas distant from the stimulated site do not necessarily reflect the connectivity between brain areas. For example, noise induced by an electromagnetic coil can cause sustained activation of the auditory system. In addition to auditory stimulation, rapidly changing magnetic fields can also stimulate trigeminal afferent fibers, which in turn activate the somatosensory system. Emotional responses to sensory discomfort induced by rTMS may also lead to the activation of the insula or anterior cingulate cortex.

## 2 Simultaneous TMS-fMRI Acquisition

### 2.1 fMRI Study Paradigm

Since the 1990s, fMRI techniques based on blood oxygenation level-dependent (BOLD) signals have been widely used in various fields of brain science. As shown by searching ["fMRI" and "brain"] in PubMed, nearly 4000+ related papers were published annually (2014–2019). The fMRI technique uses endogenous hemoglobin as a biomarker to indirectly portray neuronal activity by detecting changes in the ratio of deoxygenated and oxygenated hemoglobin in the venule. Current studies on fMRI basically explore brain functions from two perspectives (Raichle 2010). One is to emphasize that brain activity is reactive, producing different responses to different external environmental stimulations, the study paradigm of which is task-based fMRI. The other believes that brain activities are mostly intrinsically spontaneous and related to the acquisition and consolidation of information, the study paradigm of which is resting-state fMRI. During task-related experiments, subjects are usually asked to perform relevant tasks, while the activation features of their brain activities are observed. In this paradigm, there are generally two conditions, namely, target condition and control condition. The neural mechanisms associated with the target task are analyzed by comparing the changes in brain activity under different conditions. This study paradigm requires a certain level of subject involvement. In contrast, resting-state fMRI does not require the subject to do any specific task during the scan. The inspection process is no different from a routine clinical examination for subjects and is particularly suitable for studying brain function in specific populations (e.g., anesthetized and comatose patients). Investigators can characterize spontaneous mode of neural activity according to functional connectivity between different brain areas at low frequencies (0.01–0.1 Hz) and correlate it with behavioral patterns or clinical symptom scores to draw conclusions like those of task-based fMRI.

In general, fMRI is prominently advantageous as it allows noninvasive in vivo study of brain function at high spatial and temporal resolution. Brain images obtained with conventional MRI equipment (1.5 or 3 T) generally have spatial and temporal resolutions at the millimeter and second levels (new sequences developed in recent years can increase the temporal resolution to the millisecond level). Over the past 30 years, fMRI has been extensively used in studies on brain science, which provides abundant research evidences for the understanding of brain functions and disease mechanisms. However, it should be noted that fMRI is a "correlation" study on the brain and behavior by characterizing brain activity in response to specific behaviors, making it difficult to draw conclusions about causal relationship. TMS, as a noninvasive neuromodulation technique, allows investigators to make causal inferences about the function of specific brain areas by interfering with the activities therein. Due to the complementary advantages of TMS and fMRI, the combination of the two plays an extremely important role in neuroscience and clinical studies.

### 2.2 Simultaneous TMS-fMRI Acquisition

In the late 1990s, investigators in the US Medical University of South Carolina carried out a series of explorations of the simultaneous TMS-fMRI acquisition. The simultaneous TMS-fMRI acquisition technique requires the integration of TMS with MRI scanner in order to acquire immediate images of brain functions right after TMS. Since both techniques have specific requirements for the local magnetic field, the simultaneous operation of the two without processing poses a series of safety hazards and data problems. These problems are divided into two categories: static and dynamic. The former is caused by the integration of TMS device, while the latter is caused by conflicts between the operation of TMS and MRI.

Due to the persistence of static high field strengths (1.5 or 3 T) in MRI device, all components of TMS device containing ferromagnetic materials (e.g., coils) need to be changed to MRI-compatible materials in order to exist between magnets for safety reasons. Even so, MRI-compatible TMS coils may still cause distortion of fMR images. In studies, the interference can be reduced by shortening the readout time of EPI sequences or by adopting stronger imaging gradients. TMS components except for coils can be placed outside the scanner room or enclosed in an RF-isolated box. However, when these components are placed away from the scanner, the long cable length from the coil to the host weakens the effective output of the coil, and therefore, the requirement for TMS host power is higher accordingly. The problem with simultaneous TMS and fMRI operation is mainly that fMR image quality can be affected by TMS. The RF current generated during TMS operation will enter the scanner along the stimulation coil, reducing the fMRI signal-to-noise ratio and causing adverse effects such as image distortion and aberration. The transient magnetic field generated during TMS operation can further distort MR images. The degree of its effect is related to several factors, such as the strength of the transient magnetic field, the position and orientation of the stimulation coil, etc. To avoid such adverse effects, MR images should be acquired after a certain time (>100 ms) from the end of TMS (Bohning et al. 1999).

In addition, as the stimulation range of TMS is relatively focused, conventional TMS studies usually utilize real-time navigation techniques to precisely position the target site with the help of neuroimaging. However, studies on simultaneous TMS-fMRI acquisition suggest that the integration of navigation system into magnets may pose a series of safety and interference issues. More importantly, the limited space of the MRI compartment and the presence of MRI and TMS coils during scanning further reduce the available range, which makes it difficult for optical imaging-based navigation systems to locate the relative positions of head and coils in real time. Therefore, target spots are normally marked on the scalp outside the scanner room. Before formal experiments, TMS coils and target spots can be marked by vitamin tablets to confirm the overlapping degree of the two on the image.

## 2.3 Application

### 2.3.1 Neuromechanisms of TMS

After years of research and demonstration, the US Food and Drug Administration (FDA) has approved TMS for the treatment of treatment-resistant depression and obsessive-compulsive disorder. In addition, there are numerous clinical randomized controlled trials demonstrating the therapeutic effects of TMS technique on other neuropsychiatric disorders. However, in contrast to this wide range of clinical applications, we are unclear on the mechanisms by which TMS works. Combining fMRI with TMS provides a noninvasive and repeatedly verifiable approach for the study of TMS mechanisms. Initially, such studies focused on the motor system for two main reasons. Firstly, the peripheral effects produced by stimulating the motor system can be achieved quickly and directly by observing the electrical activity of specific muscle groups; secondly, it is easy to position target spots on primary motor cortex by evoking twitches in the hand muscles.

It was found that short-term stimulation of the primary motor cortex can induce fMRI signal changes in other regions within the motor network (e.g., supplementary motor cortex, premotor cortex, etc.). The degree of effect is related to the stimulation intensity. As the stimulation increases subthreshold intensity to suprathreshold intensity, motor evoked potentials in peripheral muscles increase nonlinearly, and hemodynamic indicators within the brain (e.g., whole-brain activation counts, activation intensity, mean lump size, etc.) show similar nonlinear changes (Fox et al. 2006). Interestingly, these changes in brain function beyond the targets are often more consistently observed than the targets themselves. Activities at targets will significantly change following suprathreshold stimulation, whereas it is not the case for subthreshold stimulation. Therefore, such changes may not be directly induced by TMS. One possibility is that the somatosensory signals generated by peripheral muscle movements due to suprathreshold stimulation activate primary sensory cortex and secondarily cause changes in the activities of the motor cortex. In contrast, even if the target is not activated, areas spatially distant from the target while within the same functional network as the target show changes in activities under multiple sequences of TMS. Such changes at distance can be observed even when the stimulation target is not in the primary motor cortex and peripheral muscle activity does not change. Therefore, it is unlikely to be caused by the activation of peripheral signal return. By applying stimulation of different (supra- and subthreshold) intensities, Hanakawa et al. explored in detail the relationship between stimulation intensity and corresponding degrees of hemodynamics (Hanakawa et al. 2009). They found that the degree of response of various brain areas within the motor system increases linearly with the intensity of subthreshold stimulation. A steep increase in the hemodynamic response of multiple nodes within the network was seen as the stimulation intensity gradually increased from the subthreshold to the suprathreshold, while the functional changes of different nodes within the target network were different. It has been suggested that the response degree of different nodes to TMS may be influenced by the connectivity of the nodes to targets. Hawco et al. (2018) conducted a study on simultaneous fMRI for TMS to the dorsolateral prefrontal cortex (the treatment target of depression). The results showed that although the characteristics of brain functional changes induced by TMS (high intensity vs. low intensity) were significantly positively correlated with the whole-brain connectivity of the target at the group level, the individual variances were large, with some subjects showing a significant negative correlation.

In addition to factors such as stimulation intensity and target position, the state of subjects during TMS, especially the excitability of target circuit, was closely related to the modulation effect of TMS. The higher the excitability at the time of stimulation, the better the modulation effect is likely to be. For example, the excitatory state of peripheral muscles (tense or relaxed) during TMS directly affects the changes of muscle evoked potentials (Fujiwara and Rothwell 2004) and the balance of excitation and inhibition within the cortex (Ortu et al. 2008). The excitability thresholds in the occipital cortex are significantly lower in migraine patients than in healthy controls (Aurora et al.

1998). In the motor system, Bestmann et al. (2008) applied short-term TMS pulse to subjects grasping hands and at rest, respectively. Significant stimulation effects (high intensity vs. low intensity) were observed at the TMS target (left premotor cortex), contralateral premotor cortex, and primary motor cortex, but there were state-dependent changes in specificity. The activity in the contralateral area got enhanced with the presence of stimulation during grasping movement, while it became weak with the presence of stimulation in the resting state. The connectivity between the target and the contralateral area was significantly higher in the grasping state than in the resting state. It suggested that TMS may selectively activate brain areas with dominant connectivity to the target area. This state-dependent TMS effect also exists in the visual system. The frontal eye field (FEF) and the intraparietal sulcus (IPS) are two superior control areas in the visual system. Ruff et al. (2008) modulated the excitability of the primary cortex by applying visual stimulations and applying TMS to FEF or IPS in the meanwhile. The results showed that TMS acting on the FEF did not differ in its functional effects on the visual cortex in the two states, whereas that acting on the IPS had significantly different after-effects in the two states.

### 2.3.2 Applications in Neuroscience

In the field of cognitive neuroscience, TMS has been successfully used to explore the causal relationship between brain function and behavior. Generally, it is believed that if a subject's behavior or cognitive performance changes after TMS intervention, it is related to the function of the stimulation targets. It should be noted that if there is no change in the observed indicators, it does not mean that they are not related to the function of the target area in the brain. However, how do stimulation targets involve in behavioral changes? Is it because the target's specialized functions are modulated or because the synergistic functions with other brain areas or networks are affected? The exact mechanisms cannot be simply explained by behavioral studies. The study paradigm of simultaneous TMS-fMRI acquisition, on the other hand, can well answer this question.

Early behavioral studies suggested that TMS to the right parietal lobe could strengthen the tactile sense of the ipsilateral hand and that this behavioral change could be caused by the interaction between the left and right hemispheres. However, there is a lack of specific data to support this. With the simultaneous TMS-fMRI acquisition, Blankenburg et al. (2008) found that TMS to the right parietal lobe significantly increased the responsiveness of the left primary sensory cortex to external sensory input. It was also found by subsequent neuropsychological experiments that TMS to the right parietal lobe increased the subject's perception of electrical stimulation of the right median nerve, the primary processing brain area of which happened to be the primary sensory cortex. Although it is believed by most of the studies that there is a top-down modulation of the visual cortex by FEF and IPS, the exact mechanism of modulation is not clear. Minimally invasive neuromodulation studies in non-human primates have revealed that stimulation of FEF can modulate the functional activity in the visual cortex, which directly proved the top-down modulation of FEF. Although noninvasive neuroimaging studies in humans have consistently concluded that there is an inter-modulation of activities between the frontoparietal cortex and visual cortex, there is a lack of direct evidence for causal relationship. Ruff et al. (2008) observed the functional modulation effect of stimulation to the frontal eye field on the visual system with the simultaneous TMS-fMRI acquisition. When TMS was applied to FEF, the posterior parietal cortex V1–V4 showed significant changes, the extent of which was related to the stimulation intensity, and showed a certain topological pattern. Increasing the stimulation intensity significantly weakened the activity in the central visual cortex and enhanced the activity in the peripheral visual cortex. This intensity-dependent topographic pattern, which was not sensitive to whether the visual cortex was stimulated by peripheral visual input, also did not appear when stimulation was applied to the cranial parietal target. When investigators adopted the same stimulation paradigm but changed the target to IPS, modulation characteristics different from those of FEF were observed. The modulation of IPS only affected the primary visual cortex that had no peripheral visual input.

In addition to the primary processing cortex (e.g., visual and tactile systems), the simultaneous TMS-fMRI acquisition can also be used to analyze the relationship between superior cognitive networks. Some investigators proposed that the cognitive processing in the human brain is the result of interactions among functionally specialized brain networks. Three of the most important brain networks for cognitive function are the default mode network (DMN), the salience network (SN), and the central executive network (CEN). Imaging studies have suggested a negative modulation effect of SN and CEN on the DMN, which, however, is mainly evidenced by correlation studies. Hence, there are direct evidences supporting it. Chen et al. (2013) stimulated the SN and CEN prefrontal nodes with TMS while acquiring functional MRI data from the whole brain under resting state. Psychophysiological interaction-based analysis revealed that stimulation of the CEN node produced a significant negative relationship between DMN and CEN/SN activity, whereas stimulation of the SN node did not result in this association. On the other hand, 1-Hz inhibitory TMS to CEN nodes resulted in a shift of the DMN signal from the lower- to the higher-frequency band in the spectrum and an increase in excitability. This study used TMS modulation to visualize the negative modulation effect of CEN instead of SN on the activity of DMN.

The increase in hemodynamic response of the target brain area and its functionally related brain areas appearing after TMS is generally considered as an indication of functional upregulation. However, same hemodynamic changes, when occurring in different cognitive processes, may induce different behavioral manifestations, which may be both gaining and weakening. If TMS interferes with information processing in the target brain area during cognitive processing, which in turn leads to behavioral abnormalities, it can also be inferred that the target brain area may be one of the core nodes of this cognitive function. Based on this idea, Sack et al. (2007) explored the role of bilateral parietal lobes in visuospatial tasks. Although fMRI studies suggest that bilateral intraparietal sulci are involved in visuospatial information processing, behavioral studies of TMS suggest that behavioral abnormalities in visuospatial processing are observed only when there is interference with the right parietal lobe, whereas interference with the left does not result in similar effects. It suggests a significant right hemispheric dominance in terms of visuospatial abilities. However, it is unclear whether the abnormalities induced by TMS are due to the abnormal function of the right intraparietal sulcus or to its impaired synergy with other brain areas. In a study on simultaneous TMS-fMRI acquisition, Sack et al. found that TMS not only reduced the activity of the stimulation target (right intraparietal sulcus) but also resulted in reduced function in the ipsilateral medial frontal lobe. Moreover, these imaging changes were significantly correlated with the prolonged response of subjects in the visuospatial task. This phenomenon was not present when the left intraparietal sulcus was stimulated. No correlation between behavior and imaging metrics was seen when the right intraparietal sulcus was stimulated if the task paradigm was changed to a color discrimination task. This study clearly demonstrated that the same stimulation may induce completely different behavioral and imaging changes depending on the target or task paradigm.

### 2.3.3 Application to Clinical Studies

For many neuropsychiatric disorders, significant organic lesions or changes in blood indicators cannot be detected during routine clinical examinations. As a noninvasive method for brain function studies, fMRI has been widely used to explore the neurological mechanisms of clinical disorders. Most studies locate disease or cognitive performance-related pathological circuits and propose corresponding hypotheses on mechanisms by comparing brain functions in resting or task states between patients and healthy controls. However, it is difficult to test theoretical hypotheses from the perspective of causal relationship in fMRI studies alone. Modulation of theoretical pathological circuits with TMS is an important way to verify the relevant hypotheses. Its application to psychiatric disorders such as depression and schizophrenia is described below as an example.

A large-sample, multi-center, randomized controlled study has shown that excitatory rTMS sequences acting on the left dorsolateral prefrontal cortex can effectively alleviate clinical symptoms in patients with depression. Based thereupon, it was approved by the US FDA in 2008 as one of the clinical treatments for treatment-resistant depression. It was believed by neuroimaging studies that the pathogenesis of depression was related to the dysfunction of the cingulate cortex, amygdala, and other nodes of the emotional circuit. However, follow-up studies suggested that the actual effectiveness of this method was about 30%, with large individual variance in efficacy. Vink et al. (2018) investigated the neural mechanisms of this treatment by adopting the simultaneous TMS-fMRI acquisition. According to the theoretical hypothesis of the protocol, TMS to the dorsolateral prefrontal cortex should be able to modulate the function of emotion-related brain networks or nodes, especially the subgenual cingulate gyrus. However, only four of the nine subjects showed activation of the subgenual cingulate gyrus along with the TMS to the dorsolateral prefrontal cortex. It showed that only a small proportion of patients had their emotional networks effectively modulated with this treatment protocol. It also suggested that conduction of an experiment of simultaneous TMS-fMRI acquisition before treatment may be effective in screening patients who are sensitive to treatment. On the other hand, this study positioned stimulation targets based on anatomical markers. Due to the superior functional properties of the frontal cortex, inter-individual functional variability is great. In addition, not all sub-areas uniform in anatomical positioning are functionally connected to the emotional circuit. Therefore, by detecting multiple dorsolateral prefrontal sub-areas with the simultaneous acquisition, positioning allows the effective transmission of TMS to the subgenual cingulate gyrus, which, if taken as individualized treatment target, will improve the efficiency of treatment.

Imageology meta-analyses for schizophrenia have shown that brain volume, activation patterns, and functional connectivity in prefrontal lobes differ between patients and healthy controls. However, since these studies only suggest an association between brain imaging and psychiatric abnormalities, they do not indicate causal relationships, nor do they distinguish inhibitory neural activity from excitatory neural activity. The results of task-based fMRI studies are also influenced by factors such as subject engagement and task complexity. This may also be one of the reasons why the results are not consistent across different schizophrenia samples using similar study paradigms. Changing the cognitive task to a TMS, which does not require active participation of the subject, allows for a more objective comparison of the differences between patients and healthy controls. Webler et al. (2020) applied TMS to the left dorsolateral prefrontal cortex and found based on the fMRI data acquired in real time that patients with schizophrenia had a significantly stronger response in the left prefrontal lobe than healthy controls and an even more significant reduction in bilateral prefrontal functional connectivity. Studies on simultaneous TMS-fMRI directly showed abnormally high excitability of the left prefrontal lobe in schizophrenia patients, suggesting that strategies to reduce excitability should be considered in subsequent treatment studies.

Simultaneous TMS-fMRI can also be used to explore the effects of drugs on brain functions, such as cortical excitability. Lamotrigine is a sodium channel blocker commonly used for anticonvulsant therapy in epilepsy syndromes. Multiple randomized double-blind controlled studies have shown that it is also effective for treating patients with bipolar disorder, while the mechanism is not clear. Some investigators suggested that lamotrigine may function by reducing the excitability of some brain areas in the emotional circuit, without direct research evidences however. Li et al. (2004) observed the characteristics of brain functional changes when the primary motor cortex and prefrontal lobe were stimulated by TMS, respectively, in subjects orally taking lamotrigine or placebo. When stimulation was applied to the motor cortex, it was found the activation of some brain areas of the motor system in the medication group was significantly weaker than in the placebo group. When stimulation was applied to the prefrontal lobe, it was found the activation of some brain areas of the limbic system was significantly stronger than in the placebo group. These findings suggested that the modulation effect of lamotrigine on functional excitability in the brain was specific, demonstrating that the simultaneous TMS-fMRI acquisition could function as an effective method for studies on the mechanism of drug treatment.

Cocaine addiction is one of the difficult types of treatment-resistant substance abuse disorders. Its high relapse rate is associated with many factors, including the enhanced tendency or rigorous demand for drugs as well as the inadequate management of impulsivity. In neuroimaging, these two phenomena correspond to the limbic circuit (medial prefrontal cortex, ventral striatum, etc.) and the executive control circuit (dorsolateral prefrontal lobe, dorsal striatum, etc.), respectively. Hanlon et al. (2013) first verified these two specific circuits using the simultaneous TMS-fMRI acquisition. The activation of the dorsal striatum was higher than that of the ventral striatum when TMS was applied to the dorsolateral prefrontal lobe, whereas the activation of the ventral striatum was stronger than that of the dorsal striatum when the medial prefrontal cortex was stimulated. It has been found by both task-based fMRI and resting-state fMRI studies that the prefrontal activity in cocaine-addicted patients is different from that in healthy controls, while the lesion characteristics suggested by different studies vary. Hanlon et al. (2016), by using the simultaneous TMS-fMRI acquisition, found that the ventral striatal function in the group receiving stimulation to the medial prefrontal cortex was significantly lower than that in the control group, whereas the dorsal striatal activity evoked by the stimulation applied to the dorsolateral prefrontal lobe was not different from that in the control group. It suggested that in cocaine-addicted patients, the ability to mobilize executive control systems is the same as in normal persons, while the function of non-drug information in activating limbic circuit is significantly weaker than in normals. Hence, the latter should be used as a target network when developing novel treatment protocols.

### 2.3.4 Summary and Development Direction

TMS and fMRI are complementary as noninvasive tools in neuromodulation and neuroimaging studies. Not only can fMRI assist TMS with high spatial resolution for real-time precise modulation, but it can also detect the changes in brain function induced by TMS through hemodynamic characteristics. On the other hand, the modulation ability of TMS makes it possible to verify the theoretical hypothesis of brain function in neuroimaging from the perspective of causal relationship. The real-time combination of the two makes them an important research tool in brain science. As mentioned above, (1) fMRI allows investigators to observe the local and brain network mechanisms of TMS in real time and can clarify the characteristics of brain functional changes induced by different stimulation parameters; (2) by interfering with a specific node in the brain functional network with TMS and analyzing the response of other nodes through fMRI data in the meanwhile, investigators can clearly see the driving (excitation or inhibition) relationships between different brain areas within the network; and (3) the simultaneous TMS-fMRI acquisition is an important method to verify hypotheses of disease mechanisms and an important way to bridge the basics and clinical settings.

Because of its safety and noninvasive operation, TMS has been increasingly applied to clinical treatments and studies. However, due to the uncertainty of numerous TMS parameters and the long clinical study period, the progress of related studies remains slow. In contrast, as simultaneous TMS-fMRI acquisition can rapidly display specific brain functional characteristics of the individuals, with the maturing of this technology and the popularization of related hardware and software, it will play an increasingly important role in individualized TMS treatment in the clinical setting:

1. Precise positioning of targets. With the commonly used eight-shaped coil, TMS can generate relatively focused induction currents to stimulate local brain areas. Obviously, the focusing feature can truly play its role when the target is precisely positioned. The current target positioning methods can be roughly divided into three

categories, which are based on cranial markers or EEG systems, structural brain imaging, and functional brain imaging, respectively. The first one is not suitable for individualization at all, while the latter show individualization from the aspect of functional positioning. Although few studies on simultaneous TMS-fMRI acquisition are designed with functionally individualized targets, this is undoubtedly a direction of individualized studies in the future.

2. Brain area-specific stimulation intensity. Due to the safety risk of repetitive high-intensity TMS (e.g., inducing seizures, etc.), the motor threshold of each subject is measured, and the stimulation intensity is controlled to a certain range in most studies, so as to properly avoid the side effects of stimulation. Even when the stimulation target is at the associated cortices, the stimulation threshold of the motor system is adopted during intervention in most studies. Using a unified threshold is obviously unreasonable due to the variabilities of cortical excitability in different brain areas or networks. Simultaneous TMS-fMRI acquisition provides an effective method for detecting the threshold of the associated cortices. In future studies, the minimum intensity that can induce the functional activation of the target brain area or network found after observing different stimulation intensities can be defined as the stimulation threshold of that brain area, so as to apply specific stimulation to specific brains in individuals.
3. Efficacy prediction. Although the immediate after-effects of TMS are in local brain areas, they can still be transmitted to other nodes that are far apart through the connection network of the target site. The therapeutic mechanism of repetitive TMS is also closely related to this network effect of TMS. For example, the therapeutic effect of TMS to the dorsolateral prefrontal lobe on depression is closely associated with functional changes at nodes in the emotion circuit, such as the subgenual area of the cingulate gyrus. Due to the large difference in the individualized efficacy of TMS treatments in clinical settings, investigators have been searching for the cause of this difference, in order to predict individualized efficacy. From the perspective of TMS parameters, this variation may be related to target accuracy and stimulation intensity mentioned above, while the simultaneous TMS-fMRI acquisition is an excellent approach to address these issues. Before formal treatment, conducting an experiment on simultaneous TMS-fMRI acquisition will help the investigators to assess whether the target brain network in subject (e.g., emotional circuit in depression, motor circuit in PD, etc.) has been effectively modulated under different stimulation parameters. It is clear that better clinical efficacy will be obtained from repetitive TMS treatment in subjects showing more prominent modulation effects.

## 3 Simultaneous TMS-EEG

### 3.1 Simultaneous TMS-EEG Acquisition and Analysis

The two main techniques for detecting brain function today are based on hemodynamics (e.g., fMRI) or cerebral metabolism (e.g., PET) and electromagnetism.

Electroencephalogram (EEG) is a brain function detection method that collects real-time electrical activity of brain neurons from the scalp, featured with ultra-high temporal resolution and noninvasive detection. EEG can be divided into spontaneous EEG and evoked EEG: as to spontaneous EEG, changes in the electrical activity of brain neurons spontaneously generated in the absence of specific external stimulation and as to evoked EEG, changes in electrical potential caused by specific stimulation particularly delivered to sensory organs, such as auditory evoked potentials and visual evoked potentials. Event-related potentials are evoked potentials obtained when subjects actively participate in cognitive activities, which reflect changes in not only physiological processes but also, more importantly, the mental processes of subjects when performing cognitive tasks. It is one of the important tools in cognitive neuroscience. However, EEG has some drawbacks as EEG signal analysis can indicate a certain relationship between brain cognitive activity and cognitive behavior, while it cannot characterize the nature underlying the relationship.

#### 3.1.1 Development of Simultaneous TMS and EEG

Multi-modal brain function detection has been increasingly applied to neuroscience studies, such as the combination of functions and anatomical information contained in fMRI with EEG. In addition, TMS combined with EEG (TMS-EEG) (stimulating specific brain areas with TMS and then simultaneously recording EEG changes) allows to stimulate the neural activity of specific cortices and record the transient changes of EEG signals after TMS using EEG, from which the effects of the perturbed areas on the whole brain network connectivity can be inferred and the top-down modulation mechanism of cognitive processes can be known (Cracco et al. 1989). It is of great significance for both basic neuroscience studies and clinical studies.

Cracco was the first to combine TMS with EEG in 1989 to obtain EEG signals on the scalp when TMS was applied. However, the quality of TMS-EEG signals acquired by previous devices is poor, and the major artifact signal resulted from the noise interference generated by the transient discharge of the magnetic stimulation coil. Usually, TMS results in a pulse signal of hundreds of millivolts for several 100 μs at the electrodes. As EEG amplifiers are highly sensitive microvolt instruments, the amplifier will become saturated, and the data will change abruptly. It normally takes time for the amplifier saturation to disappear. It is worse that the filtering circuit of the amplifier and the filtering algorithm of the software fluctuate drastically due to the abrupt data change, which extends the effect to hundreds of milliseconds or even longer. Only very few domestic manufacturers can provide EEG amplifiers for use under TMS conditions, which usually employ the technology of simultaneous switching and use sampling holders at present. Although they can solve the problems well, 10 ms of real signals cannot be acquired theoretically.

The combination of TMS-EEG can instantaneously change the excitability of the cerebral cortex and has the advantages of real-time recording of the electrical activity of the cerebral cortex by EEG, which allows simultaneous recording of the changes of electrical activity of the cerebral cortex induced by magnetic stimulation and thus understanding of the real-time changes in stimulation-induced activity of the human brain. As TMS not only acts on target sites but also indirectly affects other brain areas closely related to the neurological function of the site, TMS-EEG has been increasingly used to study the connectivity between specific stimulation sites and other brain areas. As a promising tool, it can be used to study the neural circuits of human cortical connections, the intrinsic connections of brain networks, and the underlying neurophysiological mechanisms. TMS-EEG has the following main advantages. Firstly, the information on the temporal pattern of responses evoked by TMS allows investigators to study the causal relationships between brain regions. Secondly, TMS-EEG highlights the changes in information over a short time and shows the temporal evolution of connections among brain areas, allowing investigators to successfully identify cognitive processing during the performance of specific tasks or in different states of the brain. Thirdly, EEG allows to directly measure brain electrical activity, thus helping investigators explore the nature underlying the connections between excitatory and inhibitory brain networks (Paus et al. 2001).

Target localization: After performing a head MRI scan for each subject, the images obtained thereof are used to construct a 3D head model using the Brainsight navigation system. Select the stimulation target required for the test, and match the virtual coils in the navigation system with the magnetic stimulation coils by the infrared fluorescent localization function, so as to match the reconstructed 3D head model with the subject's head and precisely find the corresponding stimulation target on the scalp.

Determination of resting motor threshold: Stimulate the left cerebral hemisphere, and record motor evoked potential (MEP) in the right abductor pollicis brevis. Use Ag/AgCl circular electrodes, with positive and negative electrodes on the tendon and muscle belly of the right abductor pollicis brevis, respectively, and the ground electrode on the subject's wrist. Seat the subject quietly, and apply a single-pulse stimulation, with the coil plane tangential to the skull and the tail of the coil facing the occipitalia. Apply TMS at a place 2 cm before and left the Cz point at 100% of the maximum output of the machine to determine the stimulation degree that can evoke the MEP of the right abductor pollicis brevis, then decrease the stimulation intensity at 5% intervals from 100% of the maximum output, and apply ten times of TMS at each stimulation intensity until the MEP cannot be evoked by all ten times of TMS. Then increase the stimulation intensity at a 1% interval. Take the minimum stimulation intensity that can evoke a wave amplitude ≥50 μv MEP in five out of ten times of TMS as the resting motor threshold (RMT).

### 3.1.2 Simultaneous TMS-EEG Acquisition

Seat subjects in a relaxed position, and put an EEG cap with 10–20 system distribution on the head. Electrodes are arranged according to the international 10–20 system, with the magnetic stimulation intensity of 90% RMT and a single-pulse TMS interval of 4 s. Run EEG at the same time. The correct rate and response time during task performance will be recorded automatically by the computer. The data sampling rate of EEG amplifier is 1024 Hz; the low-pass filtering is 50 Hz, and the high-pass filtering is 0.5 Hz. Lead F5 records magnetic stimulation pulse signals and is located between leads F7 and F3 in the left frontal lobe. It allows to precisely mark the stimulation time point on the EEG signal, thus helping the subsequent extraction and processing of feature information after EEG acquisition.

### 3.1.3 Simultaneous TMS-EEG Data Analysis

A brain network consists of interconnections among different brain areas. Brain connections are divided into anatomical connections, functional connections, and effective connections. An anatomical connection is a neuroanatomical structure with direct connections among brain areas. A functional connection is defined as the temporal correlation of neurophysiological events spatially distant. In addition, an effective connection exists when one brain area effectively affects another brain area, which indicates the important information of direction. Granger causality, partial directed coherence, and directional transfer functions are all part of frequency domain causality analysis and are now widely used to study effective connection in brain networks. These methods, based on the hypothesis that connections are stationary and invariant within a selected time window, use multivariate models to calculate connections between network nodes. In fact, the connections of brain areas change all the time. The directed transfer function (DTF) is a common method for frequency domain casualty analysis and has been widely applied to the source localization in scalp EEG, cortical EEG, and magnetoencephalography and to the analysis of information stream transmission. However, DTF is mainly applied to stationary signal analysis. However, during seizures or after perturbation by magnetic stimulation, EEG signals may change drastically within a short time. Hence, DTF is obviously not suitable for non-stationary signal analysis. A new adaptive directed transfer function (ADTF), also known as time-varying DTF, has been proposed for studying the time-varying connectivity among different brain areas within certain frequency bands. ADTF adopts a multivariable adaptive autoregression model and calculates the directed information stream between two nodes at each time point by Kalman filtering algorithm to analyze the effective connectivity of time-varying brain networks. This method reveals the connectivity among different brain areas at different frequencies as a function of time. Therefore, the combination of TMS-EEG with the methods of causality analysis allows to investigate the connectivity of specific cortical cognitive processing and detect the changes in dynamic connection of brain networks over a certain period of time. The analysis of TMS-EEG with ADTF can show the enhanced and attenuated time-varying EEG network connectivity within a certain time after stimulation, which is important for us to analyze the dynamic changes of network connectivity.

TMS-EEG data are mainly processed with the MATLAB software. It usually consists of the following steps: (1) EEG data preprocessing, (2) ADTF processing based on multivariable adaptive autoregression (MVAAR) model, and (3) plotting time-varying EEG network.

## 3.2 TMS-ERP Analysis Technique

Previously, when superior mental activities in the brain needed to be objectively evaluated, such as cognitive processes, it was difficult to attribute them only to changes in specific parts of human brain tissue or neurons. Therefore, studies targeting specific mental activities were very limited. It was after 1960 that Sutton first defined the event-related potential (ERP) and mapped the neurophysiological changes in the brain during the complete cognitive process by the averaging of brain evoked potentials recorded from the head surface. ERP has been widely adopted by psychologists as it reflects cognitive processes in the brain.

ERP is also known to psychologists as cognitive potential, i.e., brain potential recorded from the head surface when the subject processes a study object. Despite its wide application, ERP has its unique limitations. First, ERP can only be detected in specific brain area. Second, the components of each ERP have their own unique characteristics. Third, ERP is fixed in time. Its latency and the application of stimulation are time-locked.

ERP has become one of the main means to study cognitive activity in the brain and reflect dynamic cognitive neural processes because of its high temporal resolution of EEG signals. So far, the commonly adopted ERP waveforms include P100, N100, P200, N200, N270, and P300. ERP components such as the mismatch negativity and N400 are greatly influenced by perceptual stimulation and reflect different cognitive processes. Results of our studies show that when a given outcome does not match the correct outcome, a widely distributed negative wave can be recorded approximately 270 ms after the presentation of the stimulation, with its highest amplitude in the occipital, frontal, and temporal areas on both sides, i.e., the N270 potential. Our experiments reveal that N270 potential will appear in the presence of different digit, color, and shape stimulations. The N270 potential also appears in ERP when the stimuli mismatched spatial locations.

The attentional system regulates various aspects of visual information processing. After the TMS perturbation of a specific part of the attentional system (virtual impairment or stimulation), we observed changes in the whole attentional system, providing the most direct evidence for studying the attentional dynamics processes at the millisecond level. The TMS perturbation was given between S1 and S2 in a dual-feature matching task under the visual stimulation paradigm (Fig. 12.1) for the observation of the effect of functions in the stimulated area on cognitive activity.

### 3.2.1 Data Acquisition of TMS-ERP

TMS targets are also precisely localized by the Brainsight navigation system, with the stimulation intensity of 90% RMT generally and in the mode of single pulse. Theta-burst stimulation (TBS), also known as short-burst rapid pulse transcranial magnetic stimulation, which is often used for virtual injuries, can also be adopted. For TMS-ERP data acquisition, magnetic stimulation needs to be compatible with EEG and cognitive task triggering software (e.g., Stim2, E-prime, etc.), so as to record both response time and correct rate, and the acquisition of EEG information is the same as that of TMS-EEG.

### 3.2.2 Data Analysis of TMS-ERP

The TMS-ERP data are preprocessed and analyzed with MATLAB and the EEGLAB toolboxes. Independent component analysis aims to exclude components including blinks, eye movements, and muscle artifacts, to ensure that each subject performs an average of more than 30 tests under each condition. It has been demonstrated that TMS to the left dorsolateral prefrontal cortex and right temporal lobe could improve the processing speed in the dual-feature matching task. TMS affects the electrical activity of the sensory areas of the brain, which regulate information storage, i.e., the processing of physical features of stimulation, and are responsible for information discrimination, which thus increases the processing capacity of task-relevant stimulations and plays a crucial role in properly accomplishing goal-driven

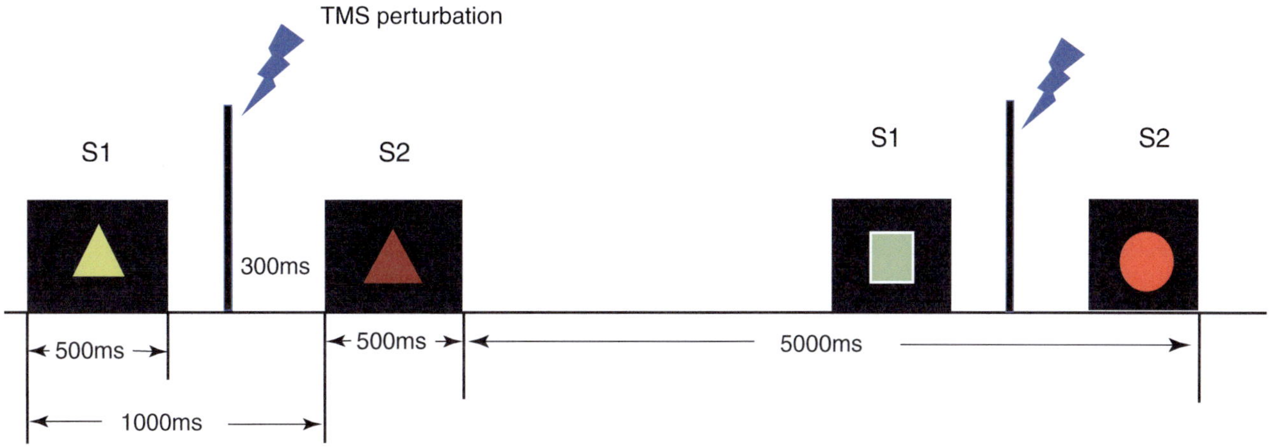


**Fig. 12.1** Schematic diagram of TMS-ERP: S1 and S2 both have the features of color and shape, which are presented successively to form stimulation pairs, 160 stimulation pairs, 4 types with equal probability in a random manner, with TMS perturbation given 300 ms before S2 presentation; subjects are asked to press buttons after S2 presentation according to test requirements

tasks. TMS to the sensorimotor area modulates the signal-to-noise ratio in the posterior cerebral cortex to increase storage capacity, which optimizes the efficiency of monitoring conflict and execution and saves cognitive resources for memory trajectory reconstruction.

### 3.2.3 TMS Evoked Potential

TMS evoked potential (TEP) is a new technique for assessing the spatio-temporal distribution of inter-cortex connections. When a TMS pulse is given to a cortex, the activity conduction mediated by it can be traced by scalp TEP waveforms and topography, thus allowing a direct measurement of brain connectivity. This method allows for the evaluation of conduction velocity or efficiency. TEP has been mostly applied to observe the after-effects of TMS on the brain, changes in network connection between different brain areas after magnetic intervention, and connections between brain areas under different brain states.

Paus et al. used white noise to mask the sound of magnetic stimulations, thus attenuating the interference caused thereby. Paus et al. stimulated the left motor cortex and obtained TEP including P30, N45, and N100. The analysis of TEP showed that there was short-term β (15–30 Hz) simultaneous stimulation activity after TMS, confirming that TMS could induce simultaneous excitability of neurons.

TEP degree was strongly related to TMS intensity, with posterior head stimulation more likely to spread to multiple parts of the brain. TEP was significantly different at 10 ms, 25 ms, and 50 ms, with 10 ms referred to as the fast component of the TEP, when the peak fluctuations were greatest. Studies have also shown that early peaks of TEP (10–30 ms) reflect excitatory neurotransmission and late peaks (40–50 ms, 100–200 ms) reflect inhibitory neurotransmission.

The results of TEP directly reflect cortical excitatory information and are not restricted to motor or non-motor cortex. Studies have shown that there are also significant differences in TEP before and after alcohol consumption, with significant TEP changes in the right frontal and left parietal cortices, especially in the right prefrontal cortex. Investigators believe that alcohol significantly reduces TEP amplitude, confirming that alcohol consumption can reduce excitability in the prefrontal cortex.

## 3.3 Application

TMS-EEG has been widely used in cognitive neuroscience because of its ability to present dynamic changes in brain networks with high temporal resolution.

### 3.3.1 Clarifying Time-Varying Brain Networks in Resting State

Guo Shuting et al. studied the differences in brain network changes induced by different modes of stimulations applied to the right prefrontal dorsolateral single target and the right frontoparietal dual targets with ADTF and found that there were two different brain network changes within 80–2048 ms after stimulation compared with pre-stimulation, i.e., weakening and strengthening patterns. In the early post-stimulation period, each mode of stimulation induced a similar network weakening pattern dominated distributed mainly over the right posterior head. In the middle and late periods, different patterns of brain networks were induced, in which the right prefrontal dorsolateral single-pulse magnetic stimulation and stimulations to other frontoparietal targets induced a weakened network pattern with the left parietal lobe as the core node, while the right parietal single-pulse magnetic stimulation was manifested by a weakened network pattern mainly distributed over the left temporal lobe. In other words, the early magnetic stimulation showed a weakened form in the frontoparietal lobe compared with that stimulation before, while it showed target specificity in the late period and suggested that the stimulation of the right dorsolateral prefrontal lobe (both single-target and double-target stimulation) resulted in a weakening effect on the left parietal activity (an important site of the right frontoparietal network modulation).

In the early post-stimulation period, different types of stimulation evoked different patterns of network enhancement and showed specificity of stimulation targets as well as time intervals. In the middle and late periods, a network enhancement pattern ranging from stimulation targets to the whole brain was presented. It was found that the core nodes of the time-varying enhanced brain network evoked by all stimulation modes all included the right central and right prefrontal cortices, suggesting that when the right frontal or parietal or right frontoparietal cortices received magnetic stimulations, enhanced activities in these two cortices would be observed. It suggests that they are important areas for the right frontoparietal connection. The single-target stimulation resulted in a pattern of brain network changes specifically related to the stimulation target. As to the dual-target stimulation, compared with the single-target simulation, there were a more significant left frontal weakened effect in the early period and a right central frontal enhancing effect in the late period when stimulations were applied to the right dorsolateral prefrontal cortex at intervals of 10 ms and 50 ms; when the stimulation interval was 100 ms, however, the effects of the right dorsolateral prefrontal cortex on the ipsilateral frontal lobe were not as significant as that at other intervals and resembled the combined effect of the right dorsolateral prefrontal cortical and right frontal single-target stimulations, with more significant effect in the right frontal lobe (Song et al. 2019a, b).

### 3.3.2 Brain Network Reorganization During the Cognitive Tasks

Selective attention and sustained attention are mainly controlled by the dorsal and ventral attention networks. The functional link between them may be explained as follows: The dorsal attention network filters the signals driven by the stimulation of the ventral attention network, while stimulation-driven signals from the ventral attention network interrupt the activity of the dorsal attention network and shift the attention to other task-related stimulations. During the top-down cognitive control, information processing is influenced by the allocation of goals and attentional resources, including the emphasis on task-related information and the exclusion of irrelevant information. There is inter-hemispheric asymmetry between attentional networks, leading to the functional lateralization of the medial frontal gyrus.

Single-pulse magnetic stimulations to the middle frontal gyrus induced changes in the time-varying network across attentional modes, including enhanced and diminished connectivity. During visual and selective attention tasks, there are changes presented by enhanced left frontal to posterior connectivity and attenuated posterior to prefrontal connectivity. In contrast, during sustained attention tasks, there are changes presented by enhanced connectivity in the bilateral posterior cortices and diminished connectivity in the prefrontal cortex (Fig. 12.2). It means that the middle frontal gyrus has different connection tendencies in different attentional modes and that the right middle frontal gyrus plays an important role in sustained attention tasks. Previous studies have shown that TMS applied to either side of the frontal lobe affects attentional ability. It also suggests a possible role of spatial bias in the right frontal lobe during attention (Song et al. 2019a, b).

### 3.3.3 Evaluation on Prognosis of Brain Functional Diseases

A study by Tong Han et al. found that low-frequency magnetic stimulation to the right frontal lobe could significantly alleviate anxiety symptoms in patients with generalized anxiety disorder. In addition, investigators used TMS-EEG to observe how low-frequency stimulation to the right frontal lobe modulates the cortical network and found that compared with the pre-treatment period, single-pulse TMS applied to the right frontal lobe in anxious patients after treatment resulted in enhanced connections from the right frontal lobe to the frontal midline and left parietal lobe, followed by increased information output from the left parietal lobe, a relay key node, to the frontal midline and right frontal lobe, and finally the left frontal lobe and left frontal polar continued to deliver information to bilateral temporal and parieto-occipital areas, showing a significant increase in connections between anterior and posterior cortices. Meanwhile, the information output from the right and left temporal lobes was significantly weakened in the generalized anxiety patients after treatment. It suggests that the therapeutic mechanism of low-frequency magnetic stimulation to the right frontal lobe may be accomplished by the enhancement of information output from the left frontal lobe and frontal polar while weakening the information output from the left and right temporal lobes.

The investigators from Wang Yuping's team applied ADTF to analyze TMS-EEG signals and found that there were brain network abnormalities in primary insomnia patients. Compared with normal controls, primary insomnia patients showed enhanced connectivity in the frontal midline and left posterior temporal areas and reduced connectivity in the right central and right temporal areas after receiving treatment of 1-Hz low-frequency magnetic stimulation.

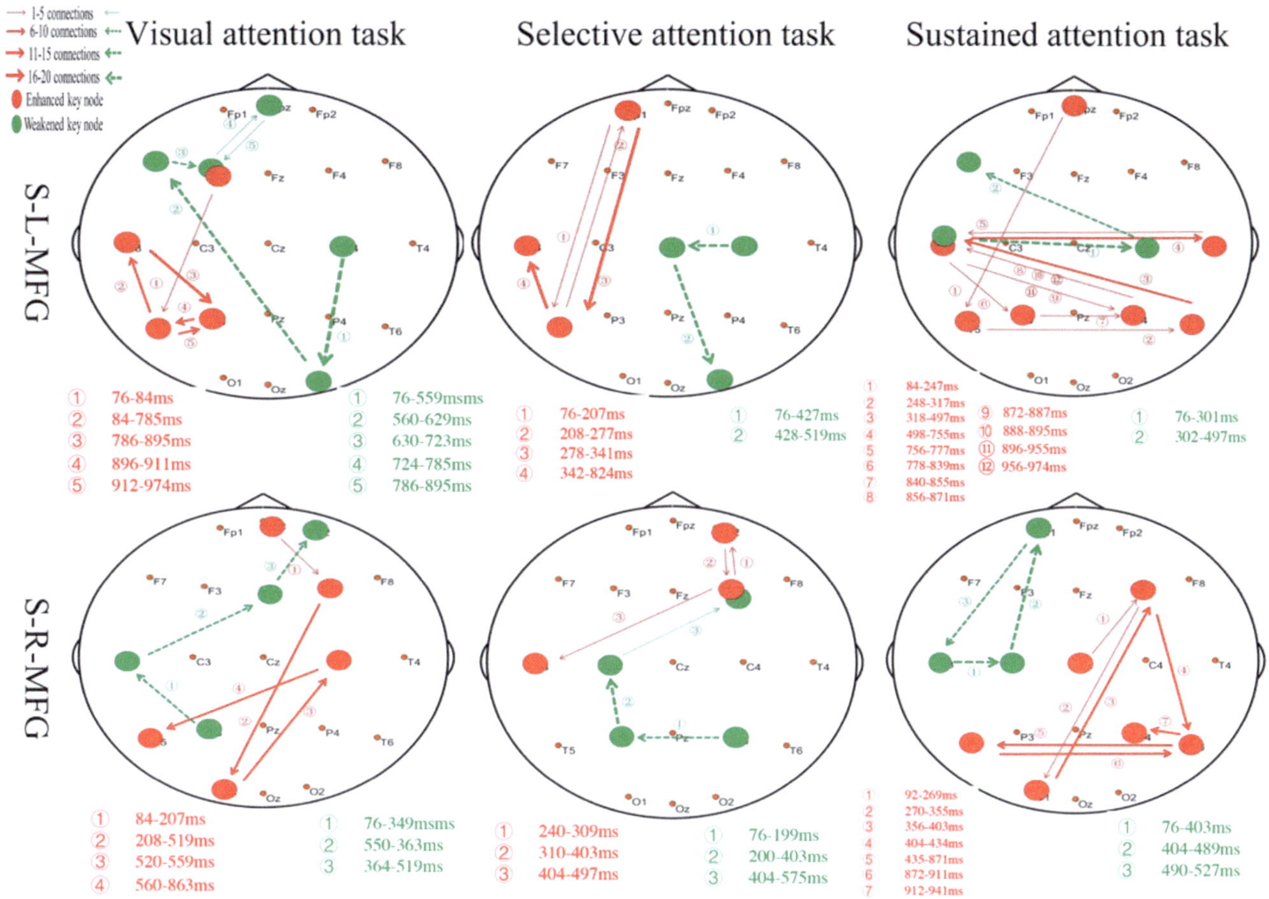


**Fig. 12.2** Changes of effective connections recorded by EEG while transcranial magnetic stimulations are applied to perturb the left and right medial frontal gyrus. Note: L, left; R, right; red, enhanced connections; green, diminished connections. Perturbation of the left gyrus frontalis medius: During the simple visual task alone and the selective attention task, it is mainly manifested by enhanced subnetwork connections in the left anterior-posterior cortices and weakened subnetwork connection in the right posterior cortex. During the sustained attention task, the connections in the left and right posterior cortices were significantly enhanced, while the connections in the left anterior cortex were weakened. Perturbation of the right medial frontal gyrus: During the simple visual task, it is manifested by enhanced subnetwork connections in the right anterior cortex and the left posterior cortex and weakened connections in the left anterior cortex. During the selective attention task, the connections in the right and left anterior cortices were enhanced, while the connections in the left posterior cortex were weakened. During the sustained attention task, the connections in the left and right posterior cortices were significantly enhanced, while the connections in the left anterior cortex were weakened

Studies on functional magnetic resonance revealed the disruption of functional connectivity within the default mode network in patients with primary insomnia. The regional functional connectivity between the medial prefrontal cortex and right medial temporal lobe was significantly reduced in patients with primary insomnia compared to patients with good sleep quality. Primary insomnia appears to alter the functional connectivity between the parietal and frontal lobes. Patients with primary insomnia showed lower fractional anisotropy scores, mainly in the right anterior limb of the internal capsule, the right posterior limb, the right anterior corona radiata, the right superior corona radiata, and the right superior fasciculi longitudinales. After the application of stimulation to the right posterior parietal cortex, the connectivity between the frontal and temporal lobes was reduced, corresponding to the main areas of the DMN, which thus confirmed the abnormalities in the default mode network in patients with primary insomnia.

In a study on the TMS treatment for mild cognitive impairment, Wang Ye et al. found that after a single TMS applied to the right precuneus, healthy patients experienced enhanced anterior cortical information transmission in the early period and shift of transmission nodes to the posterior cortex and enhanced inter-hemispheric connection in the late period. Correspondingly, in healthy patients, the posterior cortical information transmission was weakened in the early period, and the transmission nodes shifted to the anterior cortex in the late period. It suggested that the horizontal connections between the bilateral frontal lobes were enhanced in the early period after stimulation, while connections between the bilateral posterior temporo-occipital cortices were suppressed. It was the opposite in the late period. In contrast, the opposite relationship between frontal-anterior temporal connection and occipital-posterior temporal connection was not observed on patients after stimulation and before receiving the treatment of magnetic stimulation; the enhancement of inter-hemispheric connectivity in all periods was weaker than that in healthy subjects. In addition, reduced longitudinal connections from both frontal polars to the parieto-occipital area were observed in patients before treatment, which were close to the midline and the default mode network, which possibly suggested a decrease in default network connectivity in the patients. Compared with healthy controls, patients showed connectivity deficits in the early period before magnetic stimulation treatment, manifested by reduced occipital to midline and anterior temporal and frontopolar to occipital information stream, which was prominent on the right side. The connectivity deficit in the early period was partially corrected after magnetic stimulation treatment. In contrast, the enhanced connectivity in the pretreatment patients appeared at a slightly later stage than the healthy controls, with increased information output from the central frontal lobe, followed by increased output from the right anterior temporal lobe and right occipital lobe to the prefrontal lobe. After treatment, over-enhancement in the late period was suppressed. After 460 ms, the left temporal output to the central frontal lobe was reduced, followed by reduced left temporal and bitemporal connections to other cortices and then by multiple nodes, including the left anterior temporal lobe, left frontal lobe, frontal polar, and occipital area, showing a cooled state of information exchange.

However, TMS-EEG examination has some drawbacks, such as relatively poor spatial resolution, difficulty in tracing the EEG signal, and amplifiers compatible with magnetic stimulation needing further improvement. Attention should be paid to the contraindications and side effects of TMS-EEG examination. Because of the intervention of magnetic stimulation, TMS-EEG examination is normally not recommended for patients with metal implants such as cerebrovascular stents, cardiac stents, and pacemakers. It is prohibited for patients with skull defects and pregnant women. The side effects of TMS-EEG examination are mainly the same as those of magnetic stimulation. As low-frequency magnetic stimulation is adopted by TMS-EEG in most cases, symptoms such as dizziness and headache are less likely to occur after stimulation, and the effect on hearing is also very small, let alone seizures. Thus, TMS-EEG examination is safe and has few side effects.

TMS-EEG, as an emerging multi-modal test, has the advantages of easy operation, high temporal resolution, and accurate response to dynamic connectivity changes in the brain network. It has played an active role in clinical physiological assessment and evaluation of the efficacy of related treatments. It will possibly become a biological marker for detecting certain disease states someday, with extensive application prospects in clinical and scientific researches.

## 4 Cortico-cortical Evoked Potential

In recent years, our understanding of brain function has some changes. Originally, it was thought that all brain functions were the responsibility of specific areas. With the development of neuroscience, it is gradually realized that many functions of the brain are under the coordinated control of neural networks composed of multiple brain areas. This has led to a slow rise in the study of the functional connectivity between brain areas. Such studies can be traced back to the autopsy studies conducted by the pioneers of anatomy in the eighteenth and nineteenth centuries. Subsequently, some investigators traced nerve pathways in animals using tracers. These techniques provided abundant data to understand the connections between brain areas. However, as they could not be applied to the living human brain, they were unable to reflect the functional relationships between human brain areas. It was not until the advent of MRI that neurologists gained access to imaging techniques such as fMRI and diffusion tensor imaging (DTI) for studies on the living human brain. However, imaging science was only an indirect approach and was clearly not enough. The advent of cortico-cortical evoked potential (CCEP) broke this deadlock. CCEP is an electrophysiological technique based on intracranial EEG recording, i.e., applying direct electrical stimulation to local brain areas and then recording potential responses in the vicinity of the stimulating electrode and/or sites at distance in a time-locked relationship to the electrical stimulation. To make an analogy, the inter-cortical fibers are like telephone lines, and the original techniques examined the presence of these telephone lines or the integrity, while CCEP makes a direct telephone call to see if it can be connected. Therefore, in terms of detecting functional connections between brain areas, this approach is the most direct.

The recording of evoked cortical responses by direct electrical stimulation of the cortex can be dated back to the studies of Adrian in the first half of the twentieth century, who applied single-pulse electrical stimulation (SPES) to the cortex of anesthetized cats or monkeys using electrodes and recorded cortical evoked potentials in the vicinity of the stimulating electrodes (Adrian 1936). These potentiometric responses were known as "direct cortical responses" (DCR) later. In the 1960s, Buser, Bancaud, and their colleagues in Paris made the first attempts to detect connections in the human brain using SPES, where SPES (rectangular pulses of 2–6 mA intensity, 0.1 Hz frequency, and 1 ms duration) was applied intraoperatively using deep electrodes to detect connections between the ipsilateral amygdala and hippocampus (Buser et al. 1969). Later, Wilson et al. used long-term implanted depth electrodes to detect cortical responses in adjacent and distant sites and applied SPES (biphasic rectangular pulses of 0.1 ms duration, 4–5 mA intensity, 0.1 Hz frequency, 30–50 times) to the medial temporal lobe and recorded cortical evoked potentials in ipsilateral and contralateral medial temporal lobes (Wilson et al. 1991). Goldring et al. pioneered the application of SPES to the sensorimotor cortex via subdural electrodes and recorded DCR responses in the immediate vicinity (Goldring et al. 1994). In the early twenty-first century, research groups of the Cleveland Clinic (Matsumoto et al. 2004, 2017), the University of Iowa (Howard et al. 2000), and King's College London (Valentin et al. 2005) independently developed their own methods with SPES to detect effective cortical connectivity or cortical excitability. Since then, SPES has been commonly applied in the field of epilepsy surgery to track functional and epileptic networks and to detect epileptogenicity. SPES is a type of intracranial electrical stimulation, which is characterized by low stimulation frequency and aims to induce DCR or CCEP in the vicinity or distant sites of the stimulated brain areas and thus analyze the functional connectivity between brain areas. The term "CCEP" is widely used because of its clarity and was originally named by the research group of the Cleveland Clinic (Matsumoto, Nair, Lüders, and their colleagues) in 2004. In clinical studies, CCEP is mainly used to study epilepsy-related and functional brain networks. With CCEP, the electrophysiological relationship between the stimulation site and the recording site can be determined for the in vivo tracking of connections between different brain areas and seizure-related networks in humans. In addition, CCEP test is easy and quick to operate, requiring only 1 min to complete a single test using online real-time averaging technology, and can be performed under general anesthesia without the patient's cooperation. Compared with the cortical electrical stimulation in epilepsy surgery, the stimulation energy of CCEP is too small to induce epilepsy. Because of these advantages, CCEP can satisfy the requirements of intraoperative monitoring.

### 4.1 Methodology of CCEP

The recording of CCEP is practical as it does not require the subject to perform any specific task, with the subject simply lying or sitting on the bed. To avoid inducing seizures, CCEP studies are usually performed after recording seizures and administering baseline doses of antiepileptic drugs. In our experience, CCEP-induced seizures are extremely rare. The SPES is performed with the patient in a quiet and comfortable environment. A representative pair of electrodes was selected to form a bipolar circuit. In our institution, the Epilepsy Center of Xuanwu Hospital, we applied square-wave pulses of 0.3 ms duration at a fixed frequency of 1 Hz for 25 s, followed by an exchange of positive and negative electrodes with another 25 s of electrical stimulation. This consisted of a set of stimulations, containing 50 electrical stimulations of square-wave single pulse. The purposes of alternating polarity were (1) to reduce stimulation artifacts, (2) to avoid the accumulation of charge in the cortex (for safety reasons), and (3) to avoid the polarization of platinum electrode. The current intensity was set as 80–100% of the current intensity capable of producing a clinical signal or afterdischarge during the application of standard 50-Hz electrical stimulation. If there were no clinical signs or intracranial afterdischarge during the application of 50-Hz electrical stimulation, the intensity was set as 10–12 mA (subdural electrodes) and 4 mA (deep electrodes or stereotactic electrodes used in stereoelectroencephalography (SEEG)). Other institutional investigators adopted different types of electrical pulses or intervals. The pulse width was usually 0.1–1 ms. The intensity of stimulation depended on the type of electrode (subdural electrode vs. deep electrode/SEEG) and the pulse width. Compared with deep electrodes or SEEG, the extra-parenchymal location and wide surface area of the subdural stimulating electrodes resulted in greater resistance and lower charge density. Duration also affected charge density. For example, with longer pulse widths (e.g., 1 ms), the stimulation intensity should be one-third of the intensity we used (pulse width of 0.3 ms) in order to deliver the same charge density. Intensities of deep electrodes or SEEG ranged from 1–3 mA (pulse width of 1 or 3 ms) to 4–8 mA (pulse width of 1 ms or 0.3 ms). The frequency of stimulation also varied between institutions and required the consideration of analysis window and total acquisition time. Studies on network tracking adopted higher stimulation frequencies, mostly 1 Hz, while some studies used frequencies of 0.5 Hz or 2 Hz. Lower stimulation frequencies such as 0.1 or 0.2 Hz, causing delayed responses, were adopted for mapping the epileptogenic cortex. Different parameters may be a potential confounding factor evoking responses, while this was not a serious problem as one of the most studied pathways, the dorsal language pathway (via the arcuate fasciculus), was especially consistent among the institutions using different parameters.

We performed an average superposition of the original stimulation recordings by a program written based on MATLAB. A set of EEG signals containing 50 stimulations was taken for average superposition with the moment of onset of each stimulation as the zero point and 200 ms before the zero point and 800 ms after the stimulation as the observation window. The purpose of average superimposition was to preserve the EEG activity that had a time-locked relationship with the single-pulse electrical stimulation, while the spontaneous EEG activity would be counteracted during the average superimposition. Typically, CCEP is followed by

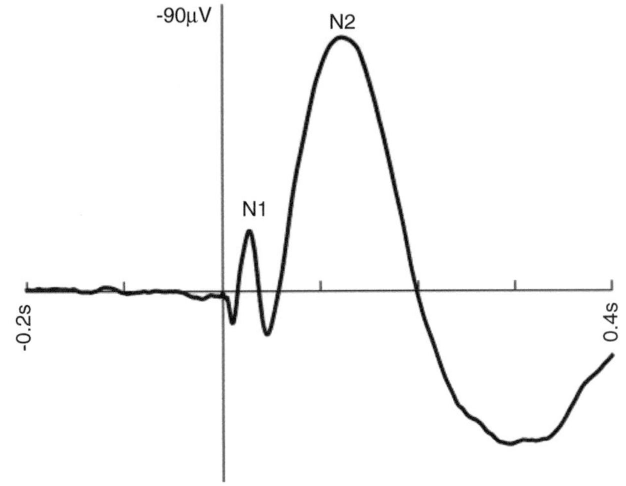


**Fig. 12.3** Cortico-cortical evoked potential

two amplitude deflections (see Fig. 12.3), with the peak of the first amplitude deflection typically occurring 10–50 ms after the stimulation moment (N1) and the second typically occurring 50–300 ms after the stimulation moment (N2). It is recognized by current studies that the first peak deflection represents a direct fiber connection between the brain area where the stimulating electrode is located and the brain area where the recording electrode is located, while the second peak deflection represents an indirect connection between the two brain areas. Since the deep electrode or SEEG may be on the surface or bottom of the cortex and cause a phase deflection of the waveform signal, it is not necessary to distinguish the direction of these two amplitude deflections.

## 4.2 Mechanism of CCEP

The formation of CCEP involves two processes: the physiological changes in the cortex on which the stimulating electrodes are located during stimulation and the process of evoking potentials on the cortex on which the recording electrodes are located. Although the mechanism of CCEP is unclear so far, there is a consensus among investigators that pyramidal cells play an important role in the process. Some zoological studies have shown that when a stimulating current is applied to the local cortical surface, the dendrites of granule cells and pyramidal cells in layers II, III, V, and VI are activated. With the enhancement of excitability, the pyramidal cells in the middle or deeper part of the cortex are finally activated, which project nerve impulses outward along efferent fibers to cortical or subcortical structures connected to them. This is followed by the generation of evoked potentials in the corresponding cortex, where the pyramidal cells still play an important role. A typical CCEP waveform has two negative waves, N1 and N2. The early N1 peak is formed due to the transmission through a direct cortico-cortical connection, whereas the formation of the N2 peak may be associated with a cortico-subcortical structure-cortical circuit, where the amplitude and latency of the N1 wave correlate significantly with the strength of the connection between the two brain areas.

## 4.3 Application of CCEP

### 4.3.1 In Vivo Tracking of Functional Brain Networks

#### 4.3.1.1 Language Network

In an intracranial EEG study, Matsumoto et al. found that CCEP could be recorded in the cortex in Wernicke's area (i.e., the language area in the lateral convexity of the temporal lobe and parietal lobe) and at the bottom of temporal lobe when stimulation is applied to Broca's area (i.e., the language area before the sulcus centralis and above the sylvian). Correspondingly, direct electrical stimulation of Wernicke's area caused CCEP in Broca's area and at the cortical base of the temporal lobe. It revealed that Broca's area and Wernicke's area, the main parts of human language network, consisted of the language system by two-way connection. In addition, Enmsu et al. used the CCEP approach to study the functional reorganization in Wernicke's area (i.e., the damaged language area is compensated by the ipsilateral adjacent site or transferred to the contralateral homologous brain area) and found that the brain area of CCEP response showed two patterns of relationship with the posterior language area (Wernicke's area). To be specific, the posterior language area localized by direct electrical stimulation of the cortex was located wholly or partially in the CCEP distribution area. The results of this study showed that the functional reorganization in the posterior language area (Wernicke's area) may be related to its functional transfer from the end connecting the anterior and posterior language areas (manifested as the recording site of CCEP) to the surrounding cortex and indicated that language areas could be identified beyond the response sites by CCEP.

#### 4.3.1.2 Motor Network

The CCEP reveals a motor-related human cortico-cortical network that connects (1) anatomically homologous areas of medial and lateral motor cortices (such as the area from the supplementary motor area (SMA) to the lateral area of the premotor/primary motor cortex (M1)) with (2) areas of medial and lateral motor cortices localized from the somatic cortex (such as the area from the facial cortex of SMA to the hand-facial cortices of the precentral gyrus). Clinically, these circuits may explain epileptic discharges afferent to or from the SMA, as well as atypical motor responses seldom seen during electrical cortical stimulation.

#### 4.3.1.3 Limbic System Network

The limbic system network has been extensively studied due to its involvement in the seizure network of medial temporal lobe epilepsy and its important brain functions including memory. Since the pioneering work of Buser and the colleagues thereof in the 1960s and Wilson and the colleagues thereof in the 1990s, the connectivity of the limbic system network has been extensively studied. The research group from Cleveland studied the limbic system network by SEEG and found strong connections between the hippocampus and the ipsilateral fornix and posterior cingulate cortex. After increasing the number of subjects (28 patients), they found that areas in the limbic network were closely connected through circuits to specific ipsilateral and contralateral areas, which may reflect different functional roles.

#### 4.3.1.4 Other Cortical and Subcortical Networks

In addition to the abovementioned networks, there are various studies on other functional networks based on CCEP. By means of SEEG, French investigators tried to trace the connectivity of deep structures including the insula and the thalamus and found that the medial pulvinar nucleus of the thalamus was connected with various cortices (e.g., temporal neocortex, temporo-parietal junction, frontal-capital area, and insula). Likewise, there were complex connections between the insula and other brain areas. In addition, the application of CCEP between the two hemispheres demonstrated the existence of functional connections between bilateral motor cortices and bilateral medial subtemporal areas.

### 4.3.2 Detection of Epileptogenicity and Epileptic Network

SPES has been used for two main purposes in epilepsy: detection of cortical excitability, i.e., epileptogenicity, and detection of epileptic networks. Alarcon and the colleagues thereof were the first to use SPES to detect epileptogenicity. By applying SPES at lower frequencies (e.g., 0.1 or 0.2 Hz), they found two sets of cortical responses: early responses and delayed responses. Early responses corresponded to CCEP and could be recorded from both normal and epileptic cortices. Delayed responses, which usually occurred 100 ms to 1.5 s after electrical stimulation, were specific against the epileptogenic area and may be considered as markers of epileptogenicity. Excision of the brain area causing the delayed response allowed good seizure control. The Cleveland School, based on the notion that CCEP reflects cortical excitability at the site of stimulation and/or recording, focused on waveform changes in the CCEP/early response, with seizure patterns characterized by repetitive spike waves showing greater CCEP amplitude than seizures characterized by focal paroxysmal rapid activity. These early or delayed responses could be taken as indicators of epileptogenicity. Other indicators include high-frequency oscillations (HFO) counterparts in early responses and SPES-induced high-frequency activity (HFA) in delayed responses. It has not been conclusively established whether the epileptic network is connected differently than in the normal network. Lacruz et al. found that the ipsilateral and contralateral connections between the normal and epileptic hemispheres were similar. Later graphical analysis of CCEP confirmed the network topology was alike between epileptic seizure areas and the normal cortex. The increased N1 wave amplitude in the epileptic network suggested that the strength of functional connections rather than their distribution would be influenced by epilepsy.

CCEP is a powerful tool in exploring human brain functional networks, which is straightforward, effective, easy, and quick to operate, featured with good spatial resolution and excellent temporal resolution. However, it is also limited by several factors. As an invasive electrophysiological technique, it can currently only be applied to intraoperative monitoring of brain functional network in surgeries for patients with epilepsy and brain tumors. Besides, its results are affected by the lesioned brain tissue and are also limited by the electrode coverage. The distribution of electrodes varies among individuals, making it difficult to summarize the rules. In addition, the generation mechanism of CCEP is complex and has not been clearly investigated. Given that most studies suggest a bidirectional connection between brain areas, investigators also doubt its directionality. Recent studies have shown that CCEP is also influenced by the state of consciousness of patients. Therefore, the physiological significance of the CCEP remains to be further elucidated. Nonetheless, CCEP has greatly contributed to the understanding of brain functional connectivity and to the study of cortical excitability, hence with a high value of clinical application.

## 5 Simultaneous CCEP-MEG Recording

### 5.1 Overview of CCEP-MEG

At present, there are two major approaches for the study of human brain network connectivity, especially effective connectivity, which are invasive and noninvasive. As mentioned earlier, CCEP is a powerful invasive method for studying brain function by implanting electrodes into the brain to record local neural activity directly from the cortical surface and thus assessing the directionality and causality of human cortical networks (Krieg et al. 2017). Scalp EEG is a noninvasive method that has been widely applied in the study of physiological or pathological activity in large brain networks. In recent years, with the development and maturation of electrophysiological techniques, the role of magnetoencephalography (MEG) in the study of brain function has also become increasingly evident.

MEG is a noninvasive brain function examination technique that detects the dynamic changes of magnetic field signals in the brain by a highly sensitive superconducting quantum interference device (SQUID), which can be used to locate abnormal discharge lesions in patients with seizures and to study the dynamic mechanisms of brain function. The MEG device has hundreds of detection channels covering the entire brain for an all-round detection of the brain's magnetic field signals. The recorded magnetic signals are analyzed by mathematical models to calculate their anatomical location, signal intensity, and direction and are fused with cranial MRI. This process is known as magnetic source imaging (MSI), which can be used to localize epileptogenic focus and functional areas. Although scalp EEG and MEG are both neurophysiological techniques, they have their own characteristics. Different brain tissues, such as the peripheral cerebrospinal fluid, skull, and scalp, have different electrical conductivity, which has a significant effect on the spread of EEG activity. Therefore, source localization based on EEG signals may affect the accuracy of localization. On the other hand, the penetration rate of magnetic field in the media of the brain, cerebrospinal fluid, skull, and scalp is hardly affected by different tissues. Source localization using magnetic signals is hence much more accurate than EEG signals. In addition, MEG is featured with both millisecond temporal and spatial resolution, thus enabling more accurate detection of changes in activity in different brain regions than EEG.

Both MEG and stereoelectroencephalography (SEEG) record postsynaptic activity originating from pyramidal cells, but each has its limitations (Kakisaka et al. 2012; Gavaret et al. 2016). SEEG directly records cortical activity, but its spatial coverage is limited as the recording perspective of individual electrodes is narrow. Hence, it needs to focus on areas with excellent signal-to-noise ratios. In contrast, MEG enables the dynamic observation of brain networks across the brain, but is affected by factors such as inverse operation or field diffusion. In addition, as brain activity fluctuates over time, recording using the two methods separately does not ensure capturing brain activity exactly in the same state. If the two modalities can be used to record simultaneously, the average responses to the same neural events in different modalities can be compared. In addition, neuronal activities recorded by MEG need to be summarized from the overall level of neurons and are transmitted within a large range of brain structures with hierarchical organization and connectivity. Therefore, it is unclear to which extent the activities can be reliably extracted from deep sources such as the amygdala, parahippocampal gyrus, thalamus, and basal ganglia. To explain the relationship between distal surface signals and actual brain activity, the best strategy is to implant electrodes into the brain to simultaneously record MEG or EEG (Shigeto et al. 2002; Santiuste et al. 2008).

Currently, there are some technical challenges in the simultaneous recording of MEG and SEEG (Dubarry et al. 2014; Badier et al. 2017). First, the intracranial electrode connecting device is large and sometimes difficult to completely be placed in the MEG helmet, which affects signal detection. In addition, the SEEG device can cause characteristic artifacts in the original MEG signals, which affects the quality of MEG recordings. There have been several studies on improving simultaneous recording techniques, such as selecting appropriate electrodes and fixation techniques to deal with material compatibility and physical limitations of detection devices, using totally electrically insulated EEG signal amplifiers to address electromagnetic perturbations, etc. In addition to improving techniques, careful arrangements are needed during simultaneous recordings to ensure successful recordings and reduce the retention time of patients in the MEG monitoring room, so as to maximize the cooperation effect.

## 5.2 Clinical Application of CCEP-MEG

There have been several studies on CCEP-MEG simultaneous recording. Dubarry et al. once stimulated a patient with treatment-resistant epilepsy implanted with SEEG using visual stimuli. Simultaneously acquired EEG, MEG, and SEEG data were analyzed for testing the feasibility of the simultaneous recording of brain activity by these three modalities, and Dubarry et al. proposed a strategy for analyzing such data (Dubarry et al. 2014). The results revealed that the method of simultaneous recording by the three modalities was feasible and that there were clear and meaningful correlations between certain signal components. However, due to the complexity of the recording process, studies were only conducted with individual cases. It is too early to apply the results to other cases or to the general population. Nevertheless, the study also proposed some strategies that could improve the data quality during CCEP-MEG simultaneous recording. For example, the SEEG connector interfered with the MEG signals to some extent and should be kept as far away as possible from the MEG detector and the area of interest and be prevented from physical contact with the patients as far as possible.

As mentioned above, CCEP is a powerful tool for detecting the functional connectivity of cortical networks and analyzing the function of deep brain structures using electrodes implanted in the brain, while MEG, which covers the whole brain, supplements the spatial drawback of

SEEG. The combination of the two recording methods allows us to gain further insight into the network connectivity between deep brain structures and the whole cortex. It is an emerging trend to combine invasive and noninvasive methods to study brain functions, which will undoubtedly facilitate the development of basic neuroscience and clinical studies.

## References

Adrian ED (1936) The spread of activity in the cerebral cortex. J Physiol 88:127–161

Aurora SK, Ahmad BK, Welch KM et al (1998) Transcranial magnetic stimulation confirms hyperexcitability of occipital cortex in migraine. Neurology 50:1111–1114

Badier JM, Dubarry AS, Gavaret M et al (2017) Technical solutions for simultaneous MEG and SEEG recordings: towards routine clinical use. Physiol Meas. 38:N118-NN27 38:N118

Baeken C, Marinazzo D, Everaert H et al (2015) The impact of accelerated HF-rTMS on the subgenual anterior cingulate cortex in refractory unipolar major depression: insights from 18FDG PET brain imaging. Brain Stimul 8:808–815

Bestmann S, Swayne O, Blankenburg F et al (2008) Dorsal premotor cortex exerts state-dependent causal influences on activity in contralateral primary motor and dorsal premotor cortex. Cereb Cortex 18:1281–1291

Blankenburg F, Ruff CC, Bestmann S et al (2008) Interhemispheric effect of parietal TMS on somatosensory response confirmed directly with concurrent TMS-fMRI. J Neurosci 28:13202–13208

Bohning DE, Shastri A, McConnell KA et al (1999) A combined TMS/fMRI study of intensity-dependent TMS over motor cortex. Biol Psychiatry 45:385–394

Buser P, Bancaud J, Talairach J et al (1969) Amygdalo-hippocampal interconnections in man. Physiological study during stereotaxic explorations[J]. Electroencephalogr Clin Neurophysiol 26(6):637

Chen AC, Oathes DJ, Chang C et al (2013) Causal interactions between fronto-parietal central executive and default-mode networks in humans. Proc Natl Acad Sci U S A 110:19944–19949

Cho SS, Strafella AP (2009) rTMS of the left dorsolateral prefrontal cortex modulates dopamine release in the ipsilateral anterior cingulate cortex and orbitofrontal cortex. PLoS One 4:e 6725

Cracco RQ, Amassian VE, Maccabee PJ et al (1989) Comparison of human transcallosal responses evoked by magnetic coil and electrical stimulation. Electroencephalogr Clin Neurophysiol 74:417–424

Dubarry AS, Badier JM, Trebuchon-Da FA et al (2014) Simultaneous recording of MEG, EEG and intracerebral EEG during visual stimulation: from feasibility to single-trial analysis. NeuroImage 99:548–558

Fox PT, Narayana S, Tandon N et al (2006) Intensity modulation of TMS-induced cortical excitation: primary motor cortex. Hum Brain Mapp 27:478–487

Fujiwara T, Rothwell J (2004) The after effects of motor cortex rTMS depend on the state of contraction when rTMS is applied. Clin Neurophysiol 115:1514–1518

Gavaret M, Dubarry AS, Carron et al (2016) Simultaneous SEEG-MEG-EEG recordings overcome the SEEG limited spatial sampling. Epilepsy Res 128(67):72

Goldring S, Harding GW, Gregorie EM (1994) Distinctive electrophysiological characteristics of functionally discrete brain areas: a tenable approach to functional localization[J]. J Neurosurg 80(4): 701–709

Hanakawa T, Mima T, Matsumoto R et al (2009) Stimulus-response profile during single-pulse transcranial magnetic stimulation to the primary motor cortex. Cereb Cortex 19:2605–2615

Hanlon C, Canterberry M, Taylor J et al (2013) Probing the frontostriatal loops involved in executive and limbic processing via interleaved TMS and functional MRI at two prefrontal locations: a pilot study. PLoS One 8:e67917

Hanlon C, Dowdle L, Moss H et al (2016) Mobilization of medial and lateral frontal-striatal circuits in cocaine users and controls: an interleaved TMS/BOLD functional connectivity study. Neuropsychopharmacology 41:3032–3041

Hawco C, Voineskos A, Steeves J et al (2018) Spread of activity following TMS is related to intrinsic resting connectivity to the salience network: a concurrent TMS-fMRI study. Cortex 108:160–172

Hayward G, Mehta M, Harmer C et al (2007) Exploring the physiological effects of double-cone coil TMS over the medial frontal cortex on the anterior cingulate cortex: an H2(15)O PET study. Eur J Neurosci 25:2224–2233

Howard MA, Volkov IO, Mirsky R et al (2000) Auditory cortex on the human posterior superior temporal gyrus. J Comp Neurol 416:79–92

Kakisaka Y, Kubota Y, Wang ZI et al (2012) Use of simultaneous depth and MEG recording may provide complementary information regarding the epileptogenic region. Epileptic Disord 14:298–303

Ko JH, Monchi O, Ptito A et al (2008) Theta burst stimulation-induced inhibition of dorsolateral prefrontal cortex reveals hemispheric asymmetry in striatal dopamine release during a set-shifting task – a TMS–[11C] raclopride PET study. Eur J Neurosci 28:2147–2155

Krieg J, Koessler L, Jonas J et al (2017) Discrimination of a medial functional module within the temporal lobe using an effective connectivity model: a CCEP study. NeuroImage 161:219–231

Lee L, Siebner HR, Rowe JB et al (2003) Acute remapping within the motor system induced by low-frequency repetitive transcranial magnetic stimulation. J Neurosci 23:5308–5318

Li X, Tenebäck C, Nahas Z et al (2004) Interleaved transcranial magnetic stimulation/functional MRI confirms that lamotrigine inhibits cortical excitability in healthy young men. Neuropsychopharmacology 29:1395–1407

Matsumoto R, Kunieda T, Nair D (2017) Single pulse electrical stimulation to probe functional and pathological connectivity in epilepsy. Seizure 44:27–36

Matsumoto R, Nair DR, LaPresto E et al (2004) Functional connectivity in the human language system: a cortico-cortical evoked potential study. Brain 127:2316–2330

Mottaghy F, Keller C, Gangitano M et al (2002) Correlation of cerebral blood flow and treatment effects of repetitive transcranial magnetic stimulation in depressed patients. Psychiatry Res 115:1–14

Mulckhuyse M, Engelmann JB, Schutter DJLG et al (2017) Right posterior parietal cortex is involved in disengaging from threat: a 1-Hz rTMS study. Soc Cogn Affect Neurosci 12:1814–1822

Nadeau S, McCoy K, Crucian G et al (2002) Cerebral blood flow changes in depressed patients after treatment with repetitive transcranial magnetic stimulation: evidence of individual variability. Neuropsychiatry Neuropsychol Behav Neurol 15:159–175

Ortu E, Deriu F, Suppa A et al (2008) Effects of volitional contraction on intracortical inhibition and facilitation in the human motor cortex. J Physiol 586:5147–5159

Paus T, Jech R, Thompson CJ et al (1997) Transcranial magnetic stimulation during positron emission tomography: a new method for studying connectivity of the human cerebral cortex. J Neurosci 17:3178–3184

Paus T, Sipila PK, Strafella AP (2001) Synchronization of neuronal activity in the human primary motor cortex by transcranial magnetic stimulation: an EEG study. J Neurophysiol 86:1983–1990

Plewnia C, Reimold M, Najib A et al (2007) Dose-dependent attenuation of auditory phantom perception (tinnitus s) by PET-guided repetitive transcranial magnetic stimulation. Hum Brain Mapp 28:238–246

Raichle ME (2010) Two views of brain function. Trends Cogn Sci 14:180–190

Ruff CC, Bestmann S, Blankenburg F et al (2008) Distinct causal influences of parietal versus frontal areas on human visual cortex: evidence from concurrent TMS-fMRI. Cereb Cortex 18:817–827

Sack A, Kohler A, Bestmann S et al (2007) Imaging the brain activity changes underlying impaired visuospatial judgments: simultaneous FMRI, TMS, and behavioral studies. Cereb Cortex 17:2841–2852

Salinas FS, Franklin C, Narayana S et al (2016) Repetitive transcranial magnetic stimulation educes frequency-specific causal relationships in the motor network. Brain Stimul 9:406–414

Santiuste M, Nowak R, Russi A et al (2008) Simultaneous magnetoencephalography and intracranial EEG registration: technical and clinical aspects. J Clin Neurophysiol 25:331–339

Shang Y, Xie J, Peng W et al (2018) Network-wise cerebral blood flow redistribution after 20 Hz rTMS on left dorso-lateral prefrontal cortex. Eur J Radiol 101:144–148

Shigeto H, Morioka T, Hisada K et al (2002) Feasibility and limitations of magnetoencephalographic detection of epileptic discharges: simultaneous recording of magnetic fields and electrocorticography. Neurol Res 24:531–536

Sibon I, Strafella AP, Gravel P et al (2007) Acute prefrontal cortex TMS in healthy volunteers: effects on brain 11C-alphaMtrp trapping. NeuroImage 34:1658–1664

Siebner HR, Filipovic SR, Rowe JB et al (2003) Patients with focal arm dystonia have increased sensitivity to slow-frequency repetitive TMS of the dorsal premotor cortex. Brain 126:2710–2725

Siebner HR, Willoch F, Peller M et al (1998) Imaging brain activation induced by long trains of repetitive transcranial magnetic stimulation. Neuroreport 9:943–948

Song P, Lin H, Li S et al (2019a) Repetitive transcranial magnetic stimulation (rTMS) modulates time-varying electroencephalography (EEG) network in primary insomnia patients: a TMS-EEG study. Sleep Med 56:157–163

Song P, Lin H, Liu C et al (2019b) Transcranial magnetic stimulation to the middle frontal gyrus during attention modes induced dynamic module reconfiguration in brain networks. Front Neuroinform 13:22

Speer AM, Kimbrell TA, Wassermann EM et al (2000) Opposite effects of high and low frequency rTMS on regional brain activity in depressed patients. Biol Psychiatry 48:1133–1141

Strafella AP, Paus T (2001) Cerebral blood-flow changes induced by paired-pulse transcranial magnetic stimulation of the primary motor cortex. J Neurophysiol 85:2624–2629

Strafella AP, Ko JH, Grant J et al (2005) Corticostriatal functional interactions in Parkinson's disease: a rTMS/[11C] raclopride PET study. Eur J Neurosci 22:2946–2952

Strafella AP, Paus T, Barrett J et al (2001) Repetitive transcranial magnetic stimulation of the human prefrontal cortex induces dopamine release in the caudate nucleus. J Neurosci 21:RC157

Valentin A, Alarcon G, Honavar M et al (2005) Single pulse electrical stimulation for identification of structural abnormalities and prediction of seizure outcome after epilepsy surgery: a prospective study. Lancet Neurol 4:718–726

Vink J, Mandija S, Petrov P et al (2018) A novel concurrent TMS-fMRI method to reveal propagation patterns of prefrontal magnetic brain stimulation. Hum Brain Mapp 39:4580–4592

Webler R, Hamady C, Molnar C et al (2020) Decreased interhemispheric connectivity and increased cortical excitability in unmedicated schizophrenia: a prefrontal interleaved TMS fMRI study. Brain Stimul 13:1467–1475

Wilson CL, Isokawa M, Babb TL et al (1991) Functional connections in the human temporal lobe. I. Analysis of limbic system pathways using neuronal responses evoked by electrical stimulation. Exp Brain Res 85(2):279–292

Winhuisen L, Thiel A, Schumacher B et al (2005) Role of the contralateral inferior frontal gyrus in recovery of language function in poststroke aphasia: a combined repetitive transcranial magnetic stimulation and positron emission tomography study. Stroke 36:1759–1763

# 13 Multimodal Brain Stimulation Techniques

Tao Han and Penghui Song

In the past three decades, transcranial magnetic stimulation (TMS) and transcranial direct current stimulation (tDCS), as the most widely used noninvasive neuromodulation techniques, have made great achievements in the treatment of various neuropsychiatric diseases such as depression, generalized anxiety disorder, pain, and AD. Transcranial near-infrared stimulation (tNIRS), as a novel neuromodulation technique, has also attracted the attention of researchers. However, it must be acknowledged that the clinical application of these techniques still faces many challenges. In addition to the uncertainty of stimulation parameters such as stimulation target and stimulation frequency, insufficient modulation efficiency is another important challenge. Studies have confirmed that modulation efficacy is not linearly correlated with stimulation intensity and duration. Therefore, simply increasing the stimulation intensity or prolonging the stimulation duration cannot correspondingly enhance its regulatory efficacy. In recent years, the combination of different neuromodulation techniques has become a hot research topic.

## 1 tDCS-TMS Technique

### 1.1 tDCS-TMS Theory and Device Development

In recent years, in order to improve the modulation efficiency of noninvasive neuromodulation technique, scholars have successively carried out the sequential application research of these two techniques. Relevant research conclusions are inconsistent. The effects of pre-stimulation on test stimuli can be enhanced, weakened, or even reversed. There is no relevant report on the simultaneous application of TMS technique and tDCS at the same target. The main reason is that it is technically difficult to stably and safely output direct current under the complex changing magnetic field environment. Professor Wang Yuping of Xuanwu Hospital of Capital Medical University and Professor Wang Weimin of Peking University jointly developed a tDCS-TMS device, which uses a new type of electrode to solve the problems of stable direct current (DC) output and electrode heating during the tDCS-TMS. The team examined the relationship between the time required to change the pH of the electrode interface and the current density and electrode material of the electrode surface and then developed the stimulation electrode on the multilayer electrode surface. On this basis, an electrode cap is successfully provided for the combined system of TMS and tDCS, which is suitable for complex electromagnetic environment. At the same time, the magnetic stimulation and direct current stimulation of tDCS-TMS device developed by the team combined the software and hardware systems, which achieve the purpose of integrated stimulation.

### 1.2 Physiological Evaluation of Neuromodulation Effect of tDCS-TMS

The clinical application of any neuromodulation technique depends on its clear verification of physiological effect. The research about physiological effects of noninvasive neuromodulation therapy mainly have been done in focuses on the following four areas: (1) TMS motor evoked potential (TMS-MEP), (2) event-related potentials (ERPs), (3) electroencephalographic spectral analyses, and (4) functional magnetic resonance imaging (fMRI). The index that has been consistently recognized is the amplitude of TMS-MEP. Related studies of other indicators did not reach a con-

T. Han
Department of Neurology, Shandong Provincial Hospital Affiliated to Shandong First Medical University, Jinan, Shandong, China

P. Song (✉)
Department of Neurology, Xuanwu Hospital, Capital Medical University, Beijing, China



Y. Wang (ed.), *Therapeutics of Neural Stimulation for Neurological Disorders*, https://doi.org/10.1007/978-981-99-4538-2_13

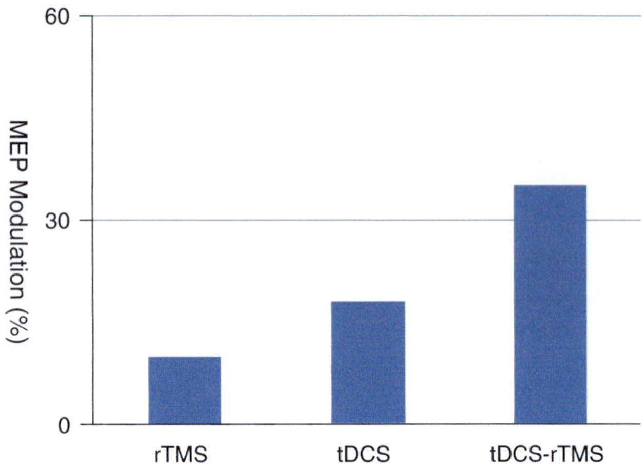


**Fig. 13.1** Subject's motor evoked potential (MEP) modulation at each time point after sham stimulation, tDCS, rTMS, and tDCS-rTMS. *tDCS* transcranial direct current stimulation; *rTMS* repetitive transcranial magnetic stimulation; *tDCS-rTMS* simultaneous transcranial direct current stimulation and repetitive transcranial magnetic stimulation

sistent conclusion. In the study conducted by the team led by Professor Wang Yuping, the dynamic changes of the amplitudes of single-pulse TMS-MEP before and after different stimulations were monitored in healthy subjects, and the effects of different stimulation modes on motor cortex excitability were evaluated.

Relevant studies first confirmed that cathodal tDCS at 1 mA and rTMS at 1 Hz can significantly inhibit the motor cortex excitability. Secondly, the study showed that the simultaneous application of 1 mA cathodal tDCS and 1 Hz rTMS at the same target for 20 min had a stronger inhibitory effect on the excitability of the motor cortex than that of electrical stimulation or magnetic stimulation alone (Fig. 13.1). This inhibition existed immediately after stimulation and remained significant until 30 min after stimulation (Han et al. 2019).

## 1.3 Synergistic Mechanism of tDCS-TMS

The modulation mechanism of tDCS on cortical excitability mainly includes the following three aspects (Rahman et al. 2013). First, it can regulate the resting membrane potential of neurons by modulating the concentration of extracellular ions. That is, the anode causes depolarization, while the cathode causes hyperpolarization. This mechanism is thought to act during stimulation and within 5 min after stimulation. Second is altering the activity of synapses to regulate the cortical excitability. That is, the anode causes a similar long-term potentiation effect, while the cathode causes a similar long-term depression effect. This mechanism is thought to persist during stimulation with a duration over 7 min and within 2 h after the stimulation. Third, some studies have confirmed that the neuromodulatory effects of tDCS also include non-synaptic mechanisms, including changes in neuronal membrane function. The presence of a persistent electric field induces changes not only in local ion concentration but also in transmembrane proteins as well as current-related [H+]. These three aspects work together to achieve the modulation of tDCS on cortical excitability. Transcranial magnetic stimulation is a noninvasive neuromodulation technique that modulates cortical excitability by stimulating cortical neurons. rTMS has been repeatedly shown to alter cortical excitability and cortical inhibition over the past three decades (Fitzgerald et al. 2006).

Insufficient modulation efficiency about either electrical stimulation or magnetic stimulation is still a problem. In two successive stimulation techniques, on condition that similar effects evoked by priming and test protocol, respectively, how the priming protocol influence the effect of test protocol can be interpreted by homeostatic and nonhomeostatic plasticity theory (Hassanzahraee et al. 2018). In short, the homeostatic and nonhomeostatic plasticity functioning are adjusted by the threshold of long-term depression (LTD) and long-term potentiation (LTP); interestingly, this threshold is not fixed but dynamically slid (Hassanzahraee et al. 2018). If the effects of priming protocol were high enough to displace the threshold, homeostatic metaplasticity will be activated, and the expected effect of test protocol will be reversed (Quartarone et al. 2005). Contrarily, when priming protocol activity is not high enough to displace the threshold, nonhomeostatic plasticity will be activated, and it will intensify the expected effect of the test protocol (Hassanzahraee et al. 2018). Although the underlying mechanism of how sliding threshold changes is not clear, the most potential explanation is the existence of "critical time window" in effective duration after the priming protocol (Hassanzahraee et al. 2018). It is suggested that the critical time window for homeostatic plasticity covered the middle third of the effective duration after the priming protocol and nonhomeostatic plasticity may occur in the rest of this effective duration (Hassanzahraee et al. 2018). Based on this theory, the synergistic effect of tDCS-TMS could be interpreted by nonhomeostatic mechanisms. For tDCS-TMS, no matter tDCS or rTMS ahead of the "critical time window" of another protocol, which will ignite nonhomeostatic mechanisms to consolidating inhibitory effects.

Essentially, any neuromodulation technique that modulates motor cortex excitability should be ultimately implemented on the pyramidal neurons. The effect of tDCS on the resting membrane potential of neurons is an important mechanism to achieve its neuromodulatory effect. But in fact, it is not clear what polarity changes will occur in different components of neurons, including cell bodies, dendrites, and axons, and their terminals under direct current stimulation with different parameters. Rahman et al. conducted relevant research and found that when direct current stimulation acts on cortical pyramidal neurons, the depolarization of cell body is often accompanied by the hyperpolarization of apical dendrites, while the hyperpolarization of cell body is often accompanied by the depolarization of apical dendrites. Cell body hyperpolarization and dendritic branch depolarization were involved in the inhibitory effect of 1 mA cathode tDCS. It is well known that the modulation of cortical excitability differs considerably between resting and activated states. These differences are manifested in the balance of firing rate, excitatory synaptic transmission, inhibitory synaptic transmission, frequency spectrum, and amplitude of oscillatory. All these factors can affect the efficacy of neuromodulation. Therefore, these factors must be considered when discussing the synergistic effect of tDCS-TMS (Rahman et al. 2013). Paulus et al. summarized an important mechanism often neglected in transcranial magnetic stimulation, namely, "shunting inhibition". The so-called shunting inhibition refers to the increase of synaptic conductivity without significant changes in membrane potential, leading to the formation of short-circuit currents between adjacent synapses, which in turn weakens the membrane depolarization caused by a given current and its effect on the voltage-gated channel of post-synaptic membrane. According to this theory, rTMS can activate shunting inhibition and increase the conductivity of the soma membrane. If tDCS acts on the same target, the membrane hyperpolarization induced by tDCS mentioned above is inhibited by the increase of mem brane conductivity caused by TMS. Therefore, the author believes that the synergistic effect of the two neuromodulation techniques is because they act on different targets of the cortical circuit and ultimately produce a stronger regulatory effect.

The structure of the human brain is extremely complex with many folds. Extracranial weak current can reach the surface of the cerebral cortex only after passing through the skin, skull, and cerebrospinal fluid. Therefore, it is difficult for previous studies to fully explain the exact electric field distribution of transcranial electrical stimulation. Opitz et al. studied the spatial distribution of the intracranial electric field of transcranial electrical stimulation and the electric field intensity that actually reaches the cortex for effective regulation. Studies have shown that the electric field intensity of transcranial electrical stimulation on the cortical surface is 0.5–1 V/m, which confirms that tDCS is a technique of continuous low-voltage regulation (Opitz et al. 2016). The electric field intensity of TMS is significantly different from that of tDCS. Magnetic stimulation can pass through the skull without attenuation and then generate an induced electric field in the cortex. The intensity of this electric field is much greater than that of transcranial electrical stimulation. Casali et al. applied simultaneous high-density electroencephalography (hd-EEG) to analyze the electric field distribution and intensity of TMS at the same time. The study selected a standard, data-driven approach to analyze the result. Main indicators include significant current density (SCD), phase-locking (PL), and significant current scattering (SCS). SCD is used to analyze the current intensity generated by TMS, PL refers to the effect of TMS on the phase of brain rhythm oscillation, and SCS reflects the cortical area that can be expanded by the TMS. The results suggest that the electric field intensity generated by TMS with threshold intensity is about 50 V/m (Casali et al. 2010). Danner et al. included 50 healthy subjects to study the electric field intensity of threshold intensity navigation monophasic and biphasic TMS. They applied the navigation software to calculate the electric field intensity at a distance of 25 mm from the scalp and the individualized cortical electric field intensity. The study found that the overall range of electric field intensity is 50–100 V/m (Danner et al. 2012). The above three studies suggest that the synergistic effect of tDCS-TMS may be the sum of a rapidly changing high-intensity modulation of intrinsic neural circuit imposed by TMS and a sustained slow modulation of spontaneous neuronal activity by tDCS, in which the effect of the latter on dendrites is four times greater than that of the cell body. Through these two different pathways, the changes of calcium current in pyramidal neurons are finally brought together to achieve the coordinated regulation of cortical excitability.

## 1.4 Clinical Application of tDCS-TMS

After completing the verification of the physiological effects of tDCS-TMS, Professor Wang Yuping's team included patients with disorders of consciousness for the treatment of tDCS-TMS based on individual brain network analysis. The treatment lasted for 2 weeks, which is 14 consecutive days. Bilateral inferior parietal lobules were given simultaneous 1.5 mA anodal tDCS and 5 Hz TMS at the same target. The stimulation duration was 20 min on each side. The study confirmed that the treatment strategy significantly improved the patient's state of consciousness after 2 weeks of treatment and this improvement still existed 1 month after treatment. Applying EEG and fMRI, they also demonstrated this treatment strategy effectively improved the patient's abnormal brain network connections.

Wang's group recruited 84 AD patients to investigate the efficacy and safety of tDCS-TMS over the bilateral angular gyrus. Patients were randomized to receive tDCS-TMS, single-rTMS, single-tDCS, or sham stimulation for 4 weeks. They also confirmed that tDCS-TMS produced greater improvement in Neuropsychiatric Inventory scores than single-tDCS and single-rTMS. Interestingly, rTMS-tDCS-induced Neuropsychiatric Inventory scores' improvement was significantly associated with cognitive function improvement. Their research demonstrated that tDCS-TMS could effectively improve AD-related neuropsychiatric symptoms and sleep quality (Hu et al. 2022).

The tDCS-TMS applied to the same target overcomes the shortcomings of previous single-modal stimulation, and based on the multimodal quantification of brain functional network characteristics, it realizes the targeting of key nodes in the cerebral cortex network. The combination of TMS and tDCS with different mechanisms of action and synergistic effects can comprehensively improve the efficacy of electromagnetic stimulation to promote neural remodeling and ultimately achieve the therapeutic goal of fundamentally improving the prognosis of brain dysfunction diseases. It is of great significance to optimize relevant parameters, include more patients with brain function diseases for multicenter large-sample clinical research, and realize the clinical translation of this technique, which may create a new situation for the neuromodulation treatment of brain function diseases.

## 2 Electro-optical Simultaneous Brain Stimulation Technique

### 2.1 Overview of Electro-optical Simultaneous Stimulation Technique

Transcranial near-infrared stimulation (tNIRS) is a new noninvasive neuromodulation technique, also known as photobiomodulation, which uses near-infrared light with a wavelength of 630–1064 nm to stimulate biological response. Due to its excellent penetrability, near-infrared light penetrates the scalp and skull into the brain and easily passes through the brain barrier with less dispersion in the brain. Unlike other stimulation techniques, photostimulation may change the cortical excitability by regulating intracellular energy metabolism to improve brain function. It has shown great promise in a wide range of neuropsychiatric applications. tNIRS absorbs photons through cytochrome c oxidase (CCO) to generate ATP to increase the mitochondrial function of neurons. Since mitochondria provide energy for nerve and synaptic activities and are closely related to synaptic and neurotransmission, 820 nm tNIRS can alter motor cortex excitability.

Noninvasive brain stimulation is effective in many clinical trials, but it also has a monostimulatory low modulatory effect. Therefore, with the increasing demand for the efficacy of neurostimulation treatment, multimodal stimulation has gradually begun its preclinical application research. As indicated by the research results of the electromagnetic combination technique introduced above, multimodal stimulation will be a major development direction of neurostimulation technique in the future. In promoting neural excitability, tNIRS works by modulating mitochondrial metabolism, while tDCS works by regulating membrane potential. These two kinds of noninvasive stimulation have different effects and may be complementary. Therefore, the combination of tNIRS and tDCS may produce obvious synergistic effects.

Wang et al. evaluated whether the combined application of 820 nm tNIRS and anodal tDCS produces a synergistic effect. The effects of tNIRS and tDCS on motor cortex excitability were evaluated by MEPs. A total of 16 healthy subjects were included in the study. They received electro-optical simultaneous stimulation, tNIRS, tDCS, and sham stimulation. The stimulation duration was 10 min. MEP amplitude was measured at an interval of 5 min within 30 min immediately after stimulation. The change rate between the MEP amplitude and the baseline amplitude was calculated. The results showed that compared with the application of tNIRS or tDCS alone, the simultaneous application of 820 nm tNIRS and 2 mA anodal tDCS induced a synergistic enhancement of cortical excitability in the primary motor cortex. The synergistic effect on MEP amplitude appeared immediately after stimulation and lasted for at least 30 min after stimulation (Fig. 13.2).

Obviously, the above results should prove that electro-optical simultaneous stimulation has a stronger effect than a single stimulation. Since increases in MEP amplitudes are thought to reflect increased synaptic excitability at the motor cortex, we propose that the synergistic effect of electro-optical simultaneous stimulation is associated with increased synaptic plasticity. Anodal tDCS has been reported to alter cortical excitability and synaptic plasticity by upregulating $Na^+$, $K^+$, ATPase, and the actions of N-methyl-D-aspartate (NMDA) and γ-aminobutyric acid (GABA) receptors (Funke 2013). Furthermore, several studies have shown that tNIRS regulates synaptic plasticity by increasing energy production in mitochondria, such as CCO oxidation, and by oxidizing ADP to ATP. One possible reason for the electro-optical synergy is that the $Na^+$, $K^+$, and ATPase required for tDCS are provided by tNIRS to promote mitochondrial metabolism. In addition, tNIRS has been shown to increase ATP synthesis in glial cells, thereby increasing synapse-induced long-term potentiation (LTP), which suggests that it may lead to the rapid influx of $Ca^{2+}$ and promote the release of glutamate acting on NMDA receptors by inducing cell membrane depolarization. Another plausible speculation is that simultaneous stimulation of tDCS and tNIRS promotes synaptic plasticity by doubling transmitter release. Therefore, it seems more likely that tNIRS, which has a synergistic effect on tDCS, promotes the processes of LTP and long-term depression (LTD) at excitatory glutamatergic synapses and increases the presynaptic input to NMDA receptors, resulting in a significant increase in MEP amplitude that can persist for up to 30 min.

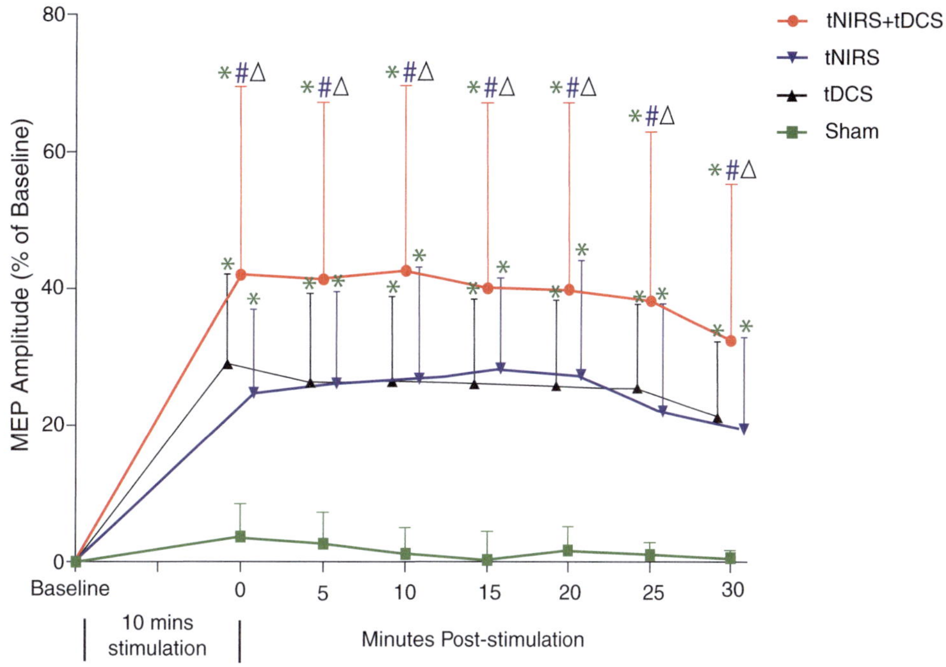


**Fig. 13.2** Amplitude change rate of motor evoked potential at each time point after stimulation under four stimulation modes. *MEP* motor evoked potential; *tNIRS* transcranial near-infrared stimulation; *tDCS* transcranial direct current stimulation; *tNIRS + tDCS* transcranial near-infrared stimulation and transcranial direct current stimulation

## 2.2 Electro-optical Simultaneous Stimulation Parameter Setting

With the rapid development of neuroimaging and the clinical application of PET-MRI, magnetoencephalogram (MEG), and image post-processing fusion technique, more structural and functional changes related to brain functional diseases can be found. For example, focal epilepsy caused by cortical dysplasia or gray matter heterotopia could be well displayed by MRI and PET. MEG can also accurately locate the dipole position of abnormal discharge in the brain through special algorithms, so as to guide clinical treatment. After obtaining the position of the target, it is necessary to carry out accurate navigation. For noninvasive stimulation, the infrared navigation system is mainly used to determine the corresponding scalp stimulation location. First, the image data with intracranial positioning (MRI, MEG, PET) are obtained and marked. The individualized scalp membrane is reconstructed with Brainsight software in three dimensions, and the corresponding scalp stimulation targets are found through infrared optical induction.

tDCS anodal electrode is generally placed at the target, and the cathodal electrode can be placed on the corresponding side or on the forehead or mastoid on the opposite side. The electro-optical simultaneous stimulation device has a small hole in the center of the anodal electrode, which is convenient for the insertion of the near-infrared light device. The diameter of the hole should be consistent with the diameter of the light source range of the near-infrared light device to ensure that all near-infrared light can irradiate the target and the current inflow in the anode is maximized. The intensity of electrical stimulation generally does not exceed 2 mA. A three-layer concentric sphere model can also be established through COMSOL simulation software, and the amount and range of current entering the cortex under a certain stimulation intensity can be obtained through numerical calculation. The current mainstream near-infrared light stimulation devices include light sources (mostly light-emitting diodes), emitters, and heat sink. The wavelengths of near-infrared light stimulation are mostly 810–830 nm and 1064 nm, and the stimulation time ranges from 4 to 15 min (Tsai and Hamblin 2017). Considering the near-infrared photothermal effect, the single stimulation duration and intensity of near-infrared light cannot be too large, and there will be a large loss of near-infrared light through the skull. About 20% of the total energy is actually irradiated to the cerebral cortex (Hamblin 2016). It is worth noting that the hair has a very strong ability to absorb near-infrared light. So, when performing electro-optical simultaneous stimulation, try to cut the hair in the stimulation target area as short as possible to reduce its absorption of near-infrared light, thereby increasing the regulatory effect.

## 2.3 Potential Clinical Application of Electro-optical Simultaneous Stimulation

Currently, there are few clinical studies on electro-optical simultaneous stimulation. According to its physiological effects, joint regulation can enhance neurometabolic activity, which can be used as a means to treat diseases with weakened brain function, such as depression and AD.

### 2.3.1 Electro-optical Synchronous Stimulation in the Treatment of Depressive Disorder

Depressive disorder is a type of mood disorder characterized by significant and persistent depression, which has a significant impact on the patient's social function. It has high recurrence rate and high suicide rate and is accompanied by different degrees of cognitive impairment, sleep disturbance, decreased volitional activity, and somatic symptoms. There are a variety of therapies for depression. Clinically, medication is the main treatment. Since 1995, when repetitive high-frequency TMS was used for the first time to treat people with depression who are resistant to drugs, TMS has been clinically verified for the treatment of depression. However, recent studies have shown that the overall clinical effective rate (25%) of TMS on the left dorsolateral prefrontal lobe did not achieve the expected effect.

Several studies using transcranial near-infrared light to monitor cerebral blood flow in different parts of depressive disorder patients and healthy people have found that the bilateral frontal polar cortex was significantly reduced in depressive disorder patients, both at rest and under tasks. After a period of psychotherapy for some depressive patients who had never received medication, it was also found that there was a significant increase in blood flow in the bilateral frontal polar cortex. Zhou et al. found that tDCS positive excitation of the cerebral cortex may improve the clinical symptoms of patients with major depressive disorder and tDCS may exert its antidepressant effect by regulating the relationship between the brain's emotional processing and executive control network (Nitsche et al. 2004). The synergistic effect can be produced by giving the electro-optical stimulation of the frontal pole. Therefore, the electro-optical stimulation may be a potential new technique for the treatment of depression. The stimulation target can be positioned at the left frontal pole through navigation, and the stimulation parameter is 820 nm near-infrared light combined with 2 mA anodal direct current stimulation. The stimulation is performed twice a day, 20 min each time, with a 20 min interval between the two stimulations. After a 2-week continuous treatment, the effective rate of transcranial synchronous stimulation treatment group was 50%, and that of single-mode stimulation group was 30%.

### 2.3.2 Treatment of Alzheimer's Disease by Electro-optical Simultaneous Stimulation

AD is the most common type of dementia. Its clinical manifestations are memory impairment and other advanced cognitive dysfunction. According to the 2019 World AD Report, there are 50 million AD patients in the world, and it is expected to increase to 152 million by 2050. More efficient, safe, and noninvasive neuromodulation therapy has become a new way of AD treatment. Among them, TMS is a common noninvasive neuromodulation technique. TMS is based on Faraday's law of electromagnetic induction, which generates electrical current through a magnetic field that passes through the skull. Studies have shown that 20 Hz high-frequency TMS stimulation of the precuneus can adjust the default mode network connection and help improve memory impairment in AD patients. However, TMS machines are expensive, noisy, and contraindicated in patients with implanted stents, and high-frequency regulation easily induces epilepsy, thus limited in clinical application.

Previous studies have used 630 nm near-infrared light-emitting diodes to irradiate APP/PS1AD mice. The results show that after near-infrared light irradiation, the synthesis of Aβ in the mice is directly destroyed in vivo, thereby improving the cognitive function of the mice. Another study showed that 1064 nm near-infrared light stimulation of the forehead can enhance people's sustained attention and short-term memory (Buckley et al. 2017). In an exploratory study of nasal near-infrared light stimulation for the treatment of memory impairment, five patients with mild to moderate cognitive impairment were treated with 810 nm near-infrared light stimulation, and the cognitive scale measures were significantly improved. Li Xiaoqin et al. showed that the AD patients had a decrease in cerebral blood flow metabolism. Direct current stimulation was given to AD patients (Voss et al. 2020). The results showed that the average velocity of cerebral blood flow in the electrical stimulation treatment group was significantly improved, the MMSE score was significantly improved, and the ADL score was significantly decreased. The patients' cognitive function and quality of life ability improved. Thus, electrical stimulation can also produce positive results in the treatment of AD, and the mechanism is similar to that of near-infrared light stimulation. Therefore, we believe that electro-optical simultaneous stimulation will also have better curative effect in the treatment of AD in the future.

## 2.4 Safety and Side Effects of Electro-optical Stimulation

tDCS is a safe noninvasive stimulation technique. Nitsche et al. observed no abnormal changes in brain tissue after 30 min and 1 h of tDCS stimulation through MRI T1 sequence and diffusion-weighted imaging sequence. The current is gradually increased from a low current intensity to a predetermined current intensity, and the stimulation current is gradually reduced as needed when the stimulation ends. The stimulation current commonly used in clinical trials is generally below 2 mA, and the mainstream current is 1 mA intensity. Clinical tDCS treatment duration is generally maintained at 20 min each time. Near-infrared light stimulation penetrates the scalp and skull into the brain in the form of light, which is also very safe within a reasonable light intensity and irradiation time. The side effects of electro-optical stimulation are mainly due to the damage caused by excessive illumination or current intensity parameters. For example, when it exceeds 320 Mw/cm$^2$, the thermal damage effect of heat accumulation may occur. At the same time, when performing near-infrared light stimulation, direct exposure to the eyes should be avoided so as not to cause damage to the retina. If the current intensity of direct current stimulation is too high, it will cause burning pain on the skin at the stimulation location and even more serious burns. Therefore, reasonable selection of treatment parameters can effectively reduce or even avoid side effects during electro-optical simultaneous stimulation.

## 2.5 Summary

Electro-optical stimulation can produce more efficient synergistic effects than single stimulation, so it has broad application prospects and scientific research value in diseases with reduced clinical brain function. During electro-optical simultaneous stimulation therapy, therapeutic targets and parameters should be reasonably selected to reduce side effects.

## References

Buckley RF, Schultz AP, Hedden T et al (2017) Functional network integrity presages cognitive decline in preclinical Alzheimer disease. Neurology 89:29–37

Casali AG, Casarotto S, Rosanova M et al (2010) General indices to characterize the electrical response of the cerebral cortex to TMS. NeuroImage 49:1459–1468

Danner N, Kononen M, Saisanen L et al (2012) Measure of cortical excitability. J Neurosci Methods 203:298–304

Fitzgerald PB, Fountain S, Daskalakis ZJ (2006) A comprehensive review of the effects of rTMS on motor cortical excitability and inhibition. Clin Neurophysiol 117:2584–2596

Funke K (2013) Quite simple at first glance complex at a second: modulating neuronal activity by tDCS. J Physiol 591:3809

Hamblin MR (2016) Shining light on the head: photobiomodulation for brain disorders. BBA Clin 6:113–124

Han T, Xu Z, Liu C et al (2019) Simultaneously applying cathodal tDCS with low frequency rTMS at the motor cortex boosts inhibitory aftereffects. J Neurosci Methods 324:108308

Hassanzahraee M, Zoghi M, Jaberzadeh S (2018) How different priming stimulations affect the corticospinal excitability induced by noninvasive brain stimulation techniques: a systematic review and meta-analysis. Rev Neurosci 29:883–899

Hu Y, Jia Y, Sun Y et al (2022) Efficacy and safety of simultaneous rTMS-tDCS over bilateral angular gyrus on neuropsychiatric symptoms in patients with moderate Alzheimer's disease: a prospective, randomized, sham-controlled pilot study. Brain Stimul 15(6):1530–1537

Nitsche MA, Niehaus L, Hoffmann KT et al (2004) MRI study of human brain exposed to weak direct current stimulation of the frontal cortex. Clin Neurophysiol 115:2419–2423

Opitz A, Falchier A, Yan CG et al (2016) Spatiotemporal structure of intracranial electric fields induced by transcranial electric stimulation in humans and nonhuman primates. Sci Rep 6:31236

Quartarone A, Rizzo V, Bagnato S et al (2005) Homeostatic-like plasticity of the primary motor hand area is impaired in focal hand dystonia. Brain 128:1943–1950

Rahman A, Reato D, Arlotti M et al (2013) Cellular effects of acute direct current stimulation: somatic and synaptic terminal effects. J Physiol 591:2563–2578

Tsai SR, Hamblin MR (2017) Biological effects and medical applications of infrared radiation. J Photochem Photobiol B 170:197–207

Voss MW, Weng TB, Narayana-Kumanan K et al (2020) Acute exercise effects predict training change in cognition and connectivity. Med Sci Sports Exerc 52:131–140

# Movement Disorders 14

Mingwei Wang, Qinying Ma, Yuan Geng, Yuqing Zhang, Hua Wei, Chunyan Liu, Xiaofei Jia, and Ying Sun

## 1 Parkinson's Disease

### 1.1 Noninvasive Stimulation Technique

Parkinson's disease (PD) is a common neurodegenerative disease in the middle-aged and elderly people. The pathophysiological changes associated with the disease include gradual loss of dopaminergic neurons in the substantia nigra and decreased dopamine levels in the striatum of the basal ganglia. PD is clinically characterized by motor and non-motor symptoms. The four primary motor symptoms of PD include resting tremor, bradykinesia and slow movement, rigidity (muscle stiffness), and postural instability; non-motor symptoms include depression, anxiety, cognitive impairment, autonomic nerve dysfunction, mental and behavioral abnormalities, etc. At present, the main treatment of PD is levodopa replacement therapy, which can obviously improve the patients' motor symptoms and quality of life in a long period of time from onset. With the increasing loss of dopamine neurons, the effect of drug therapy is getting worse, and patients who take drugs for a long time may also develop motor complications such as dyskinesia and motor fluctuation. In order to find a better treatment, investigators have found that the neuromodulation stimulation technology can play an active role in the regulation of both motor and non-motor functions in PD patients and has gradually been widely used in clinical practice (Wagle Shukla et al. 2016; Lefaucheur et al. 2020; Delgado-Alvarado et al. 2020). This chapter summarizes and evaluates the latest evidences supporting the development of non-drug therapy for PD in recent years.

#### 1.1.1 Repetitive Transcranial Magnetic Stimulation

##### 1.1.1.1 Treatment Principle

Transcranial magnetic stimulation (TMS), based on the electromagnetic induction principle discovered by Faraday in 1838, is a safe and noninvasive method of electrically stimulating human cortical neurons, which can regulate the activity of neurons locally and remotely and emit a series of pulses. The time-varying electromagnetic field can be used to induce an electric current strong enough to stimulate living tissues. The electromagnetic field is generated by the instantaneous current in the coil placed on the head of the patient, which penetrates the scalp and skull to the brain and induces a second current, causing depolarization of the brain cell membrane at the stimulated site. The electromagnetic field can penetrate tissues, including the epidermis, dermis, and subcutaneous tissues, as well as tendons, muscles, and bones. The impact on target organs depends on the magnitude of the magnetic field strength, treatment intensity, and duration. The stimulation modes of TMS mainly include single pulse stimulation, paired pulse stimulation, repetitive transcranial magnetic stimulation (rTMS), and burst mode pulse stimulation.

At present, rTMS is the most widely used in the treatment of PD. The effect of rTMS in changing cortical excitability depends on the duration and mode of stimulation. Since this type of pulse generally does not penetrate more than 2 inches into the brain, it is possible to selectively target specific brain areas. Usually, the patient will feel a pulse-type tap on the head. Generally speaking, rTMS frequency of about 1 Hz can induce the inhibition on cortical excitability, and over

M. Wang · Q. Ma · Y. Geng
Brain Aging and Cognitive Neuroscience Laboratory of Hebei Province; Neuromedical Technology Innovation Center of Hebei Province; Department of Neurology, Hebei Hospital of Xuanwu Hospital, Capital Medical University, the First Hospital of Hebei Medical University, Shijiazhuang, Hebei, China

Y. Zhang (✉) · X. Jia
Department of Functional Neurosurgery, Beijing Institute of Functional Neurosurgery, Xuanwu Hospital, Capital Medical University, Beijing, China
e-mail: yuqzhang@vip.163.com

H. Wei · C. Liu · Y. Sun
Department of Neurology, Xuanwu Hospital, Capital Medical University, Beijing, China



Y. Wang (ed.), *Therapeutics of Neural Stimulation for Neurological Disorders*, https://doi.org/10.1007/978-981-99-4538-2_14

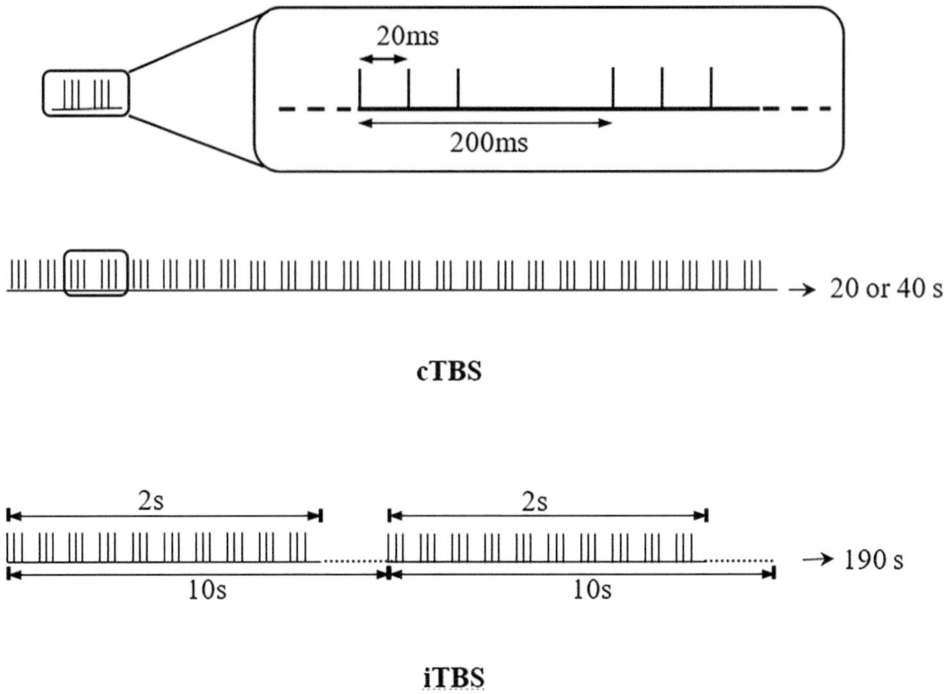


**Fig. 14.1** Different TBS stimulation schemes

5 Hz will increase cortical excitability in a short time. rTMS induces motor evoked potentials (MEPs) of muscles by stimulating limb movement and cortical supplementary motor area (SMA). The degenerative changes of dopaminergic neurons in the substantia nigra of PD patients lead to the dysfunction of basal ganglia circuit and inhibit the initiation of motor action by the cortex, resulting in bradykinesia and movement reduction. rTMS stimulation can promote the release of endogenous dopamine in the brain and inhibit the decomposition of dopamine in the brain, thereby improving the motor symptoms of patients. In addition, it can also inhibit the release of the neurotransmitter GABA and promote neuronal excitability, thereby improving non-motor symptoms of patients.

Theta-burst stimulation (TBS) is a new stimulation method being discussed at present. Some observations show that it can improve the motor symptoms of PD patients, such as rigidity and abnormal posture and gait, effectively shorten the 10-m back and forth exercise time of PD patients, and improve their walking ability. The potential for clinical popularization remains to be further explored.

#### 1.1.1.2 Treatment Regimen

Clinically, different frequencies of rTMS are often used to improve PD symptoms. Motor cortex (MC) and SMA are stimulation areas that are often selected for treating motor symptoms. In the treatment of non-motor symptoms, the MC and dorsolateral prefrontal cortex (DLPFC) are the main stimulation sites, especially the left dorsolateral prefrontal cortex (Randver 2018).

In recent years, TBS developed on the basis of classical rTMS is a novel rTMS stimulation mode. The stimulation scheme is shown in Fig. 14.1: the stimulation frequency of 50 Hz and the interval of 20 ms and three stimulations as a group of burst every 200 ms, i.e., five TBSs per second. TBS is classified as continuous TBS (cTBS) and intermittent TBS (iTBS) depending on the presence or absence of time interval. cTBS refers to stimulation, with no intermission, lasting 20 s or 40 s, with a total stimulation pulse of 300 or 600; iTBS refers to stimulation for 2 s and intermittent for 8 s, with a total stimulation pulse train of 600. Many observational studies have found that cTBS is similar to low-frequency rTMS, which produces inhibitory effect and can slightly reduce intracortical facilitation and reduce the excitability of the motor cortex; the effect of iTBS is similar to that of high-frequency rTMS, which can produce excitatory stimulation and improve the excitability of the motor cortex.

At present, there is still a large international controversy about which treatment parameter plays a decisive role in the treatment of PD, and even some study findings are contradictory, which need to be further confirmed (Chung et al. 2016; Horn et al. 2020; Eggers et al. 2015).

#### 1.1.1.3 Clinical Application

**Low-Frequency rTMS**

Meta-analysis of multiple randomized controlled studies has shown that rTMS can improve the motor ability of PD patients. Previous studies have suggested that low-frequency rTMS can enhance intracortical inhibition, prolong the silent period of the central cortex, and reduce the release of abnormal nerve impulses. Based on the characteristics of increased cortical excitability in PD patients, it is generally believed that the application of low-frequency rTMS can improve the symptoms by reducing cortical excitability. Studies have found that 86% of PD patients reported significant improvement in symptoms after low-frequency rTMS treatment with different schemes and the score of Part III of the Unified Parkinson's Disease Rating Scale (UPDRS) used to track the progress of PD decreased by 5.05%. Some scholars in China applied the stimulation frequency of 1 Hz to patients using three different intensities, respectively. The results showed that the symptoms of PD patients in the high-intensity (25 and 30%) groups improved significantly and the efficacy was long, lasting for nearly 3 months. In addition, some scholars suggest that appropriate treatment regimens should be selected according to the different clinical subtypes of PD patients, because their studies show that low-frequency rTMS has different therapeutic effects on motor symptoms of different subtypes of PD patients.

Cognitive dysfunction, mood, and psychotic symptoms are common non-motor symptoms in PD patients. About 20–50% of PD patients develop cognitive impairment in the early stage of onset, of which about 40% will progress to PD with dementia; depression is also the most common non-motor manifestation of PD with a mean prevalence of about 40%, ranging from 27 to 76%. The application of low-frequency rTMS therapy can improve the common non-motor symptoms of PD patients. Studies have reported that the cognitive function scores of memories, orientation, attention, visual space/executive function, naming, speech, abstract ability, and aspect of early PD patients treated with low-frequency rTMS were significantly improved and basically restored to the normal level, which was significantly higher than those of patients in the control group.

### High-Frequency rTMS

More studies support that high-frequency rTMS can improve the clinical symptoms of PD patients, and the motor part score of UPDRS shows that its efficacy can be stable for a period of time. Some scholars selected ten PD patients without drug treatment and used 5-Hz rTMS to stimulate contralateral M1 area of upper extremities with severe symptoms for PD patients. With the stimulation of the middle frontal lobe as the sham stimulation control, they found that the total score of UPDRS decreased after 1 h of true stimulation at M1 area and the motor symptoms were significantly improved. In a study, 10-Hz rTMS was given to stimulate the left motor cortex in 12 PD patients in the drug effect off-stage, and it was observed that the myotonia and bradykinesia of the right upper extremity were improved. Khedr et al. randomly divided 55 PD patients into 4 groups: The first 2 groups were given 25-Hz rTMS for stimulation in the motor cortex of bilateral upper and lower extremities, respectively; the third group was treated with 10-Hz rTMS stimulation; and the fourth group was treated with stimulation in the occipital lobe as the control group. The results showed that the motor symptoms of patients in the two groups receiving 25-Hz stimulation were significantly improved, followed by the 10-Hz group, which were all superior to the control group, and the observed effect lasted for 1 month.

Similar to low-frequency rTMS, high-frequency rTMS is effective in the improvement of non-motor symptoms such as cognitive impairment, mood disturbances, sleep disorder, and fatigue in PD patients and has a better stimulation effect than low-frequency rTMS. Its efficacy is related to the stimulation site. It has been reported that stimulation of the parietal lobe (rather than the motor cortex) can increase the sleep duration, reduce sleep fragmentation and awakening time of PD patients, and improve sleep quality, but does not significantly improve motor symptoms and moods.

### TBS

TBS mode, which is widely discussed at present, is generally divided into cTBS and iTBS according to whether the stimulation lasts or not, with the characteristics of high frequency and low intensity. cTBS can reduce cortical excitability, with slight intracortical inhibition. When applying cTBS to PD patients and healthy controls, investigators observed a decrease in cortical excitability in M1 area, but the severity of resting tremor in patients failed to be improved. Eggers and other scholars found that cTBS stimulation at SMA of the brain can improve the UPDRS Part III score and slightly improve the upper limb movement symptoms of PD patients in the "off" period of drug treatment. Compared with cTBS, it is generally believed that iTBS can improve cortical excitability but has no significant effect on intracortical facilitation. Suppa et al.'s study of iTBS stimulation in M1 area showed that no LTP-like neuroplasticity changes were observed during the "on" or "off" period of dopaminergic drugs (Suppa et al. 2011). Kishore et al. put forward the opposite result (Kishore et al. 2017). Compared with the normal control, PD patients showed no neuroplasticity changes after cTBS or iTBS stimulation.

In addition, Huang et al. found that the efficacy of TBS was related to treatment with anti-Parkinsonian agents and duration of treatment (Huang et al. 2010). The result showed that PD patients without dopa-responsive dystonia in the whole-dose levodopa group responded to TBS treatment; however, those in the half-dose levodopa did not respond to TBS treatment. In addition, the study found that PD patients with dopa-responsive dystonia responded to TBS treatment when half the usual dose of levodopa was administered. Therefore, it is speculated that the reactivity to TBS treatment is related to the clinical characteristics of PD patients

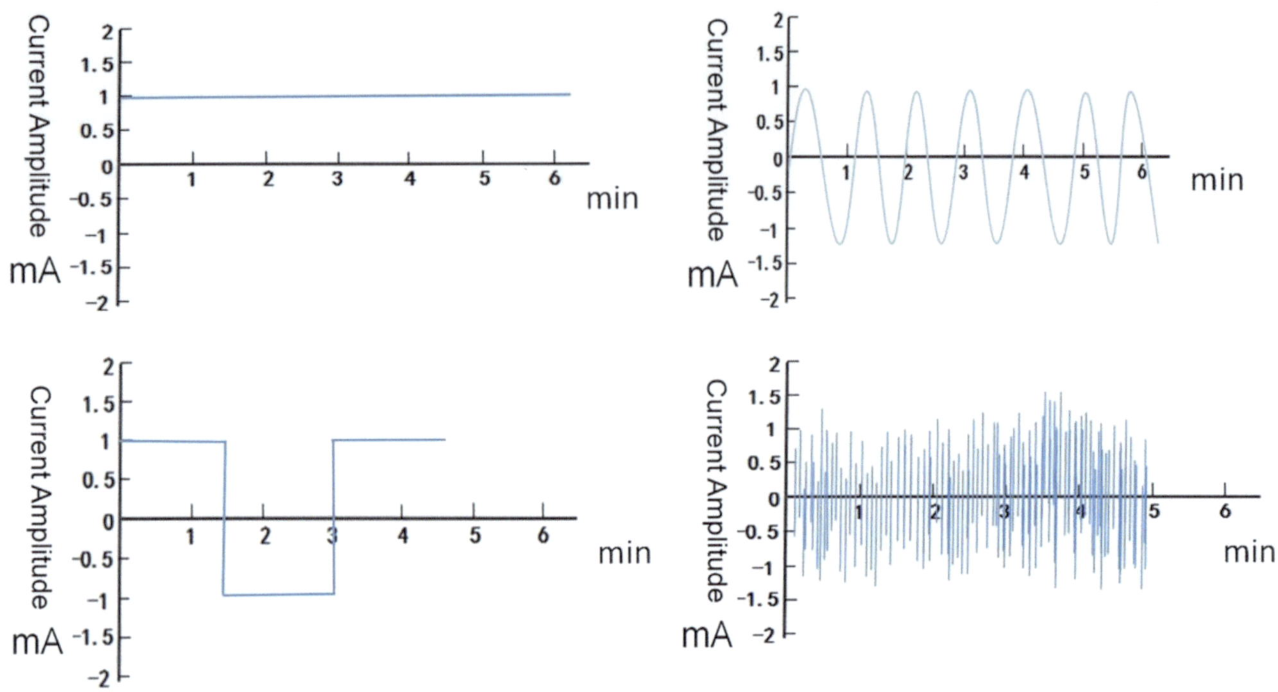


**Fig. 14.2** Noninvasive brain electrical stimulation mode

receiving drug therapy for a long time. Other studies have found that PD patients with stable effect of levodopa drug therapy responded to TBS stimulation, while PD patients with motor fluctuation but no dystonia complications only responded to iTBS but not cTBS treatment, while PD patients with motor fluctuation and dystonia complications did not respond to TBS treatment. Benninger et al. found in their study that iTBS treatment only improved the mental state but did not change the motor symptoms and cortical excitability (Benninger et al. 2011). At present, there are few studies involving the therapeutic effect of TBS mode on PD patients, and there is still lack of strong evidence-based medical evidence to confirm which parameters (stimulation sites, stimulation frequency, and stimulation time) have an exact therapeutic effect on motor or non-motor symptoms.

### 1.1.2 Transcranial Electrical Stimulation

Transcranial electrical stimulation (tES) is another potential treatment option for neurodegenerative diseases, including transcranial direct current stimulation (tDCS), transcranial alternating current stimulation (tACS), transcranial pulsed current stimulation (tPCS), and transcranial random noise stimulation (tRNS). In all of these tES techniques, electrical current is delivered to the scalp through two electrodes and conducted to target areas of the brain, thereby altering local neuronal excitability. tDCS is delivered by a constant current stimulator via two electrodes (anode and cathode); the anode electrode plays a role in reducing the resting membrane potential of cortical neurons and increasing the excitability of neurons, while the cathode electrode plays a role in reducing the excitability of neurons. tACS is delivered by a biphasic sinusoidal alternating current stimulator; tPCS is delivered by a monophasic rectangular pulse stimulator with regular intervals; tRNS is delivered by a random alternating current stimulator with varying amplitude and frequency (Fig. 14.2).

#### 1.1.2.1 Transcranial Direct Current Stimulation

At present, the vast majority of exploration on the treatment of PD is tDCS. Below, we will elaborate on motor symptoms and non-motor symptoms, respectively.

**Gait and Balance Function**

The excitability of the primary motor cortex (M1 area) has changed in patients with idiopathic PD. The low dopaminergic state of the basal ganglia may contribute to the adaptive beneficial increase of cortical excitability to compensate for the globus pallidus-thalamus-cortex hypodynamia. The enhancement of cortical excitability by anodal tDCS may further enhance this compensatory mechanism and improve motor function, and anodal tDCS may also induce dopamine release in the basal ganglia by activating glutamatergic corticostriatal fibers. Multiple studies on the stimulation of the motor cortex (M1, 1–2 mA, 13–30 min) or the dorsal nucleus have shown that most patients have short-term improvement in gait after stimulation (Ganguly et al. 2020).

The role of the prefrontal striatal pathway in freezing of gait (FOG) of PD patients has been confirmed, and motor, cognitive, and limbic circuits converge in the common output nuclei, medial globus pallidus/substantia nigra pars reticulata, to inhibit the movement of the pedunculopontine nucleus. Motor and cognitive processing of corticostriatal parallel circuits and intrastriatal integrated circuits is impaired in PD patients. During walking, PD patients rely more on DLPFC for cognitive control to compensate for their motor deficits, while more challenging walking like dual tasks requires more DLPFC activation. Therefore, stimulation of multiple brain regions, such as the prefrontal cortex and the motor cortex, may provide better efficacy for gait in PD patients. Studies have confirmed that multi-target tDCS is effective in reducing the decoupling between motor and cognitive networks; improving the connectivity among the prefrontal cortex, motor cortex, and subcortical structures; and increasing the release of dopamine outside the striatum. FOG improvement can be achieved with tDCS delivered to the bilateral premotor cortex and M1 (1 mA, 20 min), bilateral premotor cortex and M1 or prefrontal cortex (2 mA, 20 min), bilateral DLPFC (2 mA, 20 min), bilateral premotor cortex alone (2 mA, 20 min), and M1 and left DLPFC (2 mA, 20 min).

Reference parameter settings: anode current of 1–2 mA, 7–20 min/time, 2–5 sessions in M1 area or left DLPFC, which can be combined. Multi-target stimulation is also possible. tDCS can be combined with physical therapy.

In PD patients, upper extremity function is gradually affected, initially asymmetrical, manifested as impaired dexterity, abnormal strength, and poor coordination of the arms. Benninger et al. found that the administration of anode tDCS (2 mA, 20 min/time, eight times) in the bilateral M1/premotor cortex or prefrontal cortex significantly improved the motor function of upper extremities and the effect could last until the third month of follow-up. To the contrary, the administration of stimulation in SMA (2 mA, 13 min) or DLPFC (ten times, 2 mA, 20 min) did not show significant improvement in upper extremity movement.

Reference parameter settings for improving upper extremity motor function: anode placed in the M1/bilateral M1/premotor cortex or prefrontal cortex, current 1–2 mA, lasting 20–25 min, single/multiple treatments.

Recommendation: Motor area/premotor area/SMA tDCS may be effective for the motor function of PD (grade C) patients; anode prefrontal tDCS may be ineffective for the motor function of PD (grade B) patients.

### Improvement in Cognitive Function

Cognitive dysfunction is one of the most common non-motor symptoms in PD patients. Dopaminergic depletion leads to frontostriatal circuit dysfunction involving the DLPFC and dorsal caudate, which is the main cause of PD executive dysfunction. The tDCS in the left DLPFC can improve working memory performance by enhancing local cortical excitability. The beneficial effects were found during the task (online effect, 2 mA, 20 min, left DLPFC) and after the task (offline effect, 2 mA, 20 min, left/right DLPFC). The online effect of tDCS is due to the changes in the polarization of neuron membrane, while the offline effect is related to long-term potentiation, long-term inhibition, and long-term synaptic plasticity (Fregni et al. 2021).

Reference parameter settings: anode placed in the left DLPFC, current 1–2 mA, 20–25 min/time for 10–16 times.

Recommendation: Anodal DLPFC tDCS may be effective for the cognitive function of PD (grade B).

### Improvement in Impulsive and Pathological Gambling Behavior

Pathological gambling is one of the main side effects of dopamine agonists in the treatment of PD patients. Dysfunction of the orbitofrontal-striatal circuit may be responsible for this dangerous and impulsive behavior. In the Iowa gambling task study using cathode tDCS (single sequence, 2 mA, 10 min, starting 2 min ago, covering all tasks, current density 0.057 mA/cm$^2$), the decision-making has been improved. The cathodal tDCS of the right DLPFC may have reduced pathological overload of the prefrontal-striatal pathway.

Reference parameter settings: Cathode placed in the right DLPFC (single stimulation, 2 mA, 10 min, starting 2 min ago, covering all tasks, current density 0.057 mA/cm$^2$). No recommendation.

### Improvement in Verbal Fluency

Patients with PD have phonological and semantic fluency-related deficits due to frontal and temporal lobe dysfunction. In contrast to the left temporoparietal cortex, the modulation of verbal and phonemic fluency can be observed when tDCS (2 mA, 20 min) treatment is performed in the left DLPFC. The tDCS of DLPFC increases the functional connectivity of verbal fluency network including the frontal lobe, parietal lobe, and fusiform area.

Reference parameter settings: anode placed in the left DLPFC (2 mA, 20 min).

### Improving Fatigue

The study of Forogh et al. showed that the tDCS (8 times, 0.06 mA/cm$^2$ current, 20 min/time) of the bilateral DLPFC had a good effect on the fatigue of PD patients. At the same time, the use of left anode and right cathode stimulation combined with occupational therapy had no obvious effect on daytime sleepiness, and the improvement of fatigue might be related to the improvement of mood and depressive symptoms.

Reference parameter setting: the current of the left anode and right cathode of DLPFC is 0.06 mA/cm$^2$, 20 min/time, eight times in total.

#### 1.1.2.2 Transcranial Alternating Current Stimulation

The possible mechanism of tACS is that it interacts with spontaneous brain oscillations at a specific frequency and even triggers spontaneous brain oscillations. Reference stimulation parameters for improving PD patients' symptoms: phase-locked stimulation, 10–20 Hz, adjusted according to the higher power spectrum in EEG; stimulation site, cortical M1 area, 15 min; and stimulation cycle of 2 weeks (Ganguly et al. 2020).

#### 1.1.2.3 Transcranial Pulsed Current Stimulation

The possible mechanism of tPCS is the stimulation of direct current component in the neural network, the changes in the excitability of neurons due to the tension depolarization of resting membrane potential, the phase effect of pulsating current on/off characteristics, etc. At present, there are only studies on gait and balance function of PD patients. Reference stimulation parameters: anode placed in M1 area, single-phase current (unidirectional) waveform, pulse width of 33.3 μs, and pulse interval of 33.3 μs.

#### 1.1.2.4 Transcranial Random Noise Stimulation

The possible action mechanisms of tRNS are as follows: stochastic resonance enhances synaptic signals, induces long-term potentiation by changing the potency of N-methyl-D-aspartate receptor, and activates sodium channels, achieving neural plasticity; it may interfere with the ongoing oscillation effect (Ganguly et al. 2020).

In conclusion, exploring tDCS in PD holds great promise. Even with the best dopaminergic drugs, symptoms such as non-motor symptoms and drug-induced dyskinesia are also difficult to manage. tDCS is affordable, portable, cost-effective, and well tolerated. However, more massive data and more rigorous blinding studies are needed to screen more suitable stimulation sites (M1/DLPFC/cerebellum) and effective stimulation parameters (intensity and duration).

Most studies are high-quality randomized controlled trials on idiopathic PD. Although tDCS may be beneficial to motor function improvement, and tDCS of DLPFC may improve cognition, it seems that a feedback loop does exist between motor and cognitive function. There is no significant difference between the quantitative analysis indicators grouped by the Unified Parkinson's Disease Rating Scale (UPDRS) and speed (bradykinesia) and the cognitive indicators grouped by TMT B in each time slot, and there is a significant mixed motor effect (but no cognitive effect). These are the most commonly used quantifiable results in studies. In 2017, due to the diversity of outcome measures and goals, Lefucheur et al. did not make recommendations on the motor or non-motor/cognitive benefits of tDCS in PD, although they believed that the combination of tDCS and rehabilitation or startup might improve the efficacy. A recent meta-analysis found significant short-term benefits with an amount of 0.36% ($P = 0.001$) but not significant long-term benefits, and multi-regional targets improved outcomes compared to single brain targets. A meta-analysis on FOG in Parkinson's syndrome showed that after treatment with rTMS and tDCS, the Freezing of Gait Questionnaire (FOG-Q) in the PD group alone was improved more significantly, while TMS showed a better therapeutic effect in combination with UPDRS III score. As for the improvement of cognitive function, another meta-analysis of three studies found that tDCS did not significantly improve cognitive function.

Result selection is challenging for motor and cognitive function. Scales used clinically often incorporate multiple aspects of motor function. For example, UPDRS Part III scores speech, facial expression, tremor, rigidity, hand/leg/foot movements, bradykinesia, posture/stability, and gait. Different stimulation goals and parameters (and whether the patient is in the on/off phase of the drug) will affect these motor symptoms in different ways. Outcome measures and study protocols also vary by cognition. A larger PD motor/cognitive RCT is required to determine optimum practices program, minimal clinically important difference (MCID) of different outcomes, and duration of effect (Beretta et al. 2020).

## 1.2 Implantable Device Stimulation Technique

In the early stage of PD, general drug treatment can effectively control the related symptoms. However, with the progression of the disease, patients in the advanced stage often have decreased efficacy and even severe motor complications. Deep brain stimulation (DBS) has become an important treatment for adjuvant drug therapy due to its minimally invasive and adjustable advantages. DBS surgery is a treatment that adopts the brain stereotactic technique and implants deep stimulation electrodes into the nucleus of the deep brain to act on the corresponding neuron population with specific stimulation frequency, pulse width, and voltage, thus improving the symptoms such as PD tremor and rigidity by changing the excitability of neurons.

From the 1950s to 1960s, stereotactic resection of posteroventral pallidum or ventralis intermedius nucleus (Vim) was the main surgical treatment for PD. DBS therapy began in the 1970s and was initially used to treat intractable pain and epilepsy. In the 1980s, Benabid et al. found that DBS stimulation of the ventral intermediate nucleus (VIM) could control the tremor symptoms in PD patients. In 1993, Benabid used the current stimulation of the subthalamic nucleus (STN) to treat the advanced symptoms and the "on-off" period fluctuation in PD patients, with good therapeutic effects. Since then, DBS technology has been widely used in the field of movement disorders such as PD. In 1998, the first operation of deep brain stimulation for PD was carried out in China. In 1999, the NMPA approved DBS treatment of PD. This chapter mainly describes the clinical application and progress of DBS in the treatment of PD.

### 1.2.1 Treatment Principle

With the update and progress of study technology, the specific mechanism of DBS in the treatment of PD has been further understood. DBS treatment of PD involves complex polysynaptic response, thus affecting pathological signal transduction. From the perspective of biochemistry, DBS can also improve the symptoms of the disease by changing the levels of neurotransmitters; however, the clinical study in this respect is still uncertain. In addition, some radiographic studies have found that there are different metabolic patterns in the brain of PD patients, such as PD tremor-related metabolic pattern and PD-related metabolic pattern, which also indicates that the pathophysiological mechanisms of dyskinesia symptoms such as tremor symptoms and muscle rigidity in PD patients are different. DBS can affect these metabolic patterns by stimulating different targets, so as to control the symptoms such as tremor and rigidity. There is no clear conclusion in the current mechanism study, and there is still a long way to go for the exploration on the mechanism and principle of DBS in the treatment of PD.

### 1.2.2 Treatment Regimen

DBS treatment of PD is not only a simple surgical implantation of current stimulation devices, from preoperative diagnosis and clinical evaluation, operation planning, to postoperative program control, drug management, and rehabilitation treatment, involving the mutual cooperation among multiple-discipline fields such as functional neurosurgery, neurology, imaging, and rehabilitation medicine. The standardized and multi-disciplinary cooperative diagnosis and treatment of PD has become a trend. However, PD patients vary significantly from individual to individual in their symptoms and response to treatment, and there is currently no fixed treatment regimen or mode. In clinical treatment, patients' condition should be fully understood, and whether or not patients can undergo the operation, when to operate, and what surgical target to choose should be clarified. While following the treatment principles, individual variances should also be paid attention to, and the treatment expectations of different patients at different disease stages should be considered to determine the appropriate DBS operation timing and operation method, so as to achieve a more satisfactory treatment effect.

#### 1.2.2.1 Indications and Contraindications of DBS Surgery

**Surgical Indications**

1. Primary Parkinson's disease, meeting the *Diagnostic Criteria of Parkinson's Disease in China* issued in 2016, or the diagnostic criteria of the International Parkinson and Movement Disorder Society in 2015, or genetic PD and various genotypes of PD with good response to levodopa.
2. Drug efficacy has significantly decreased, or significant motor complications (dyskinesia, motor fluctuation) occur, affecting the quality of life of patients.
3. Adverse drug reactions occur, affecting the efficacy of drugs.
4. There are tremors that the medication cannot control.

**Surgical Contraindications**

1. Severe cognitive impairment, anxiety, and depression affect the patients' ability of daily living. Because the program control during and after surgery needs frequent interaction and cooperation between doctors and patients, patients with significant dysgnosia or mental disorders have great difficulty in communicating with doctors.
2. Parkinson-plus syndromes: multiple system atrophy (MSA), progressive supranuclear paralysis, the disease develops rapidly, and the efficacy cannot be maintained for a long time.
3. Severe concomitant diseases, or bleeding tendencies, or poor systemic symptoms cannot tolerate surgery.

#### 1.2.2.2 Preoperative Evaluation

Before DBS surgery, the comprehensive evaluation of PD patients' motor symptoms, "dyskinesia" and other complications, non-motor symptoms, cognitive ability, and ability of daily living needs to be made to determine whether the patients meet the surgical indications and whether there are contraindications. The evaluation results can predict the postoperative efficacy as reference information for target selection and also provide the basis for postoperative drug adjustment and determination of programmed parameters. Generally speaking, the maximum improvement rate ≥30% in levodopa drug stress test is indicated for surgery.

#### 1.2.2.3 Operation Time

1. Course of disease: In principle, DBS surgery is recommended for PD patients with a course of disease ≥5 years. For patients with the course of disease <5 years but meeting the criteria for clinical diagnosis of primary PD, the indication of surgery is clear, and it is recommended that the course of disease be broadened to 4 years.
2. Severity of the disease: For PD patients with "on-off" fluctuation, surgical treatment may be considered if the Hoehn and Yahr stage is 2.5–4.0 in the off state.
3. Age: Patients under 75 years old and elders should be in good general condition. Studies have found that early DBS intervention can greatly improve the quality of life of patients. DBS surgery can be well tolerated even in patients with a short course of disease. The combination of DBS and medication can appropriately reduce drug dosage and achieve more satisfactory efficacy. Actually, it

is more important to select the right patient than the time point in the development of the disease. As long as it is a suitable PD patient, under the condition of fully considering the risk/benefit ratio, surgical treatment can have a good efficacy expectation.

#### 1.2.2.4 Target Selection

At present, the most commonly used targets of DBS surgery for PD are the STN and globus pallidus internus (GPi). Randomized controlled studies show that both of them can improve the motor symptoms of PD. The advantages of STN-DBS lie in that it not only has good clinical efficacy for tremor, myotonia, and bradykinesia but also can reduce the dosage of dopaminergic drugs according to the specific situation after surgery, with less power consumption. STN-DBS can indirectly alleviate dyskinesia by reducing the dosage of drugs, but it may cause side effects such as dysphonia, paresthesia, cognitive impairment, and even mental disorder. The overall efficacy of GPi-DBS is similar to that of STN-DBS. GPi-DBS may be superior to STN in improving dyskinesia in the "on" stage and has relatively few side effects. But generally, the dosage cannot be reduced after surgery. STN should be given priority to patients with the goal of drug reduction; patients with cognitive or affective disorders should have preference for GPi. DBS surgery mainly improves motor symptoms and may improve non-motor symptoms to some extent. There is no conclusion yet on the effects of GPi- and STN-DBS on cognition, depression, and anxiety.

Vim is an important target for the treatment of various tremors, including PD tremor. Vim-DBS has no obvious effect on muscle rigidity, bradykinesia, and dyskinesia caused by levodopa in PD patients, but it is especially effective for resting tremor of the upper extremity, and long-term stimulation may generate tolerance, which needs to be overcome by increasing stimulation intensity. Bilateral high-intensity current stimulation may lead to dysarthria. For patients with bilateral refractory tremor, the deep brain stimulation of the posterior subthalamic area PSA-DBS can be tried to control tremor with less stimulation intensity, but the number of cases is limited currently, and it is not routine treatment.

The pedunculopontine nucleus (PPN), a potential target, has gradually attracted the attention of clinicians. PPN-DBS can be considered for primary PD with postural gait disorder. PPN-DBS alone is not ideal for the improvement of core motor symptoms such as tremor, rigidity, and bradykinesia. STN-DBS stimulation can be combined according to the condition. However, there are few PPN-related studies at present, the localization of low-field magnetic resonance is difficult, and the variability of the currently available study results is large, so further large-sample randomized controlled trials are needed to validate its clinical efficacy.

#### 1.2.2.5 Surgical Steps

First, the head MRI scan should be perfected before surgery to obtain radiographic data. On the day of surgery, the head frame (CRW or Leksell head frame) is mounted under local anesthesia, and then the head CT scan is performed. Then, MRI and CT data are fused by using the surgery planning system or the mechanical arm-assisted positioning system. With the help of radiographic data, the coordinates of targets in the standard brain atlas are converted into the target coordinates of the stereotaxic apparatus. With the development of radiograph and image co-registration technology, novel MRI sequences such as susceptibility-weighted imaging, quantitative susceptibility imaging, and diffusion tensor imaging have been gradually applied to preoperative positioning and postoperative evaluation to achieve more accurate target nucleus positioning. However, it has not been widely used yet, and further studies are needed to evaluate its effect. After entering the room, the patient is in the supine position. The needle entry point is selected according to the requirements of the target and needle track, the scalp is cut in an arc or straight cut, and the skull is drilled; then the dura mater is cross-cut, and the plastic base ring is stably installed at the cranial hole.

The accuracy of the location of the electrode implantation is the key to satisfactory surgical results, so the re-confirmation of the intraoperative target is particularly important. The intraoperative target and the depth of electrode implantation can be confirmed according to the results of microelectrode recording (single channel or multichannel), or the threshold of adverse motor and sensory reactions of intraoperative prestimulation, or the results of intraoperative CT scan, O-arm, or MRI scan. The deviation of electrode position may lead to side effects. For patients with electrode implantation under local anesthesia, the target position can be confirmed during surgery by observing the improvement of clinical symptoms of patients and the occurrence of specific side effects. After the electrode position is satisfactory, the trocar sleeve is removed, the electrode position is fixed with the electrode clip, and the bone hole is closed with the electrode cap. The implantation of the electrode extension line and the pulse generator requires general anesthesia, and the wound is sutured layer by layer after confirming that the components of each part are connected correctly and the impedance is within the normal range.

#### 1.2.2.6 Postoperative Program Control

DBS surgery is only the first step in the surgical treatment of PD patients. Postoperative program control is an important link of the long-term front of DBS treatment. The accurate location of surgical targets combined with professional postoperative program-controlled management can play a maximum role. Personalized and standardized postoperative program-controlled management can improve the clinical symptoms of patients to the greatest extent and achieve the ultimate goal of surgical treatment.

The startup for patients generally is on weeks 2–4 after DBS surgery, of course, and can also be on early after surgery according to the situation. Generally, the startup stimulation parameters are 130–160 Hz, the pulse width is 60–90 μs, and the voltage is adjusted according to the improvement degree of the patient. Generally, 3–6 months of startup after surgery, the patient needs the program control, adjustment, and optimization of stimulation parameters and electrode contacts again. For patients programmed after surgery, not only the patient's medical history should be inquired in detail, but also careful physical examination and evaluation should be conducted to understand the current real and objective conditions and severity of patients. Low-frequency, frequency conversion, and crossed electrical pulse stimulation can be considered for patients with advanced disease who develop FOG, etc. The ultimate goal of program control should be to avoid stimulating other structures while stimulating the target neural structure and to obtain the maximum improvement of clinical symptoms with the minimum stimulation intensity.

#### 1.2.2.7 Postoperative Drug Management

There is no clear consensus or clinical guideline for the adjustment of drugs after surgery, which is usually performed according to the patient's condition and the doctor's experience. Some PD patients may develop malignant syndrome after acute drug withdrawal, manifested as fever, disorders of consciousness, and increased serum creatine kinase, which may be fatal in severe cases, so it is not recommended to stop taking drugs of anti-Parkinson's disease after surgery. Generally, patients can start taking drugs of anti-Parkinson's disease on the same day after surgery, and the initial drug therapy after surgery is usually the same as that before surgery. In the long-term postoperative management of patients, the dosage of drugs can be adjusted according to the progress of the disease and the change of stimulation parameters. Due to the synergism of DBS and drugs of anti-Parkinson's disease, it is advisable to adjust the drug after the program control has reached a steady state. To be sure, the effect of the combination management of proper postoperative program control and drugs is greater than that of single therapy, which can effectively improve the quality of life of patients.

### 1.2.3 Clinical Application

#### 1.2.3.1 Refer to the Following Guidelines and Consensus for Diagnosis and Deep Brain Stimulation Surgery-Related Therapies

*Diagnostic Criteria for Parkinson's Disease in China* (2016), Diagnostic Criteria of International Parkinson and Movement Disorder Society (MDS) (2015), Chinese Expert Consensus on Deep Brain Stimulation Therapy for Parkinson's Disease (Second Edition), and Chinese Guidelines for the Treatment of Parkinson's Disease (Fourth Edition)

#### 1.2.3.2 Safety, Complications, and Management of DBS Treatment of PD

The safety of DBS surgery has always been a key concern of patients and doctors. Studies have found that the incidence of complications of DBS surgery in PD patients is significantly lower than that in non-PD patients. The complications of DBS surgery are mainly divided into procedure-related complications, implanted hardware-related complications, and startup stimulation-related complications. Procedure-related complications include incision infection, epilepsy, and intracranial hemorrhage. The proportion of intracranial hemorrhage reported in the literature is about 1–4%. Hypertension, advanced age, poor treatment of cerebral superficial vein, excessive loss of cerebrospinal fluid, too close to the midline of the puncture site, and rough puncture operation are all risk factors of bleeding. Therefore, it is recommended to avoid passing through the sulcus and ventricle as much as possible in the surgical route, use fibrin sealant to prevent excessive loss of cerebrospinal fluid, strictly perform aseptic procedures during surgery, minimize the number of punctures, gentle operate during puncture, and withdraw the tube core slowly. The blood pressure of patients should be strictly controlled during and after surgery. Hardware-related complications mainly include electrode dislocation, infection, stimulator exposure, and fracture of electrodes or leads. For the patients with infection and exposed stimulator, after necessary debridement and anti-infective therapy, it is recommended to remove the intracranial electrode, extension line, or pulse generator as soon as possible and then re-implant the corresponding device after the infection is effectively controlled. Stimulation-related complications are mainly caused by unreasonable setting of stimulation parameters or poor implantation site of stimulation electrodes. Patients may suffer from muscle spasm, convulsion, paresthesia, dyskinesia, diplopia, and dysarthria. If the adjustment of stimulation parameters still fails to improve these problems, it may be necessary to remove the original electrode, change its position, reposition, and implant a new electrode.

In general, DBS is relatively safe. Procedure-related and hardware-related complications can be reduced by strictly grasping the indications of surgery before surgery, skillfully and normatively performing the surgical procedure by the operator during surgery, as well as standardized management and education for patients after surgery. And most of the complications can be effectively solved through early detection, diagnosis, and treatment.

#### 1.2.3.3 Long-Term Efficacy of DBS in the Treatment of PD

Although DBS surgery cannot cure PD, the results of currently available studies all support that DBS can improve the symptoms of PD patients for a long time and significantly improve the quality of life of patients. Some studies have found that the efficacy of surgery can last for up to 10 years. Especially for patients who have tried to control symptoms with conventional drugs for many times but have not responded well, DBS may provide long-term symptom relief. The currently available data support the use of this clinical method to a great extent. Some studies believe that early DBS intervention can obtain better long-term efficacy. Considering the possibility of misdiagnosis for early PD, it is not recommended to blindly and prematurely perform surgery for young patients whose symptoms can be well controlled by traditional drugs. So far, the efficacy of DBS has been confirmed, but the evidences of long-term benefit of DBS still need to be supported and validated by large sample trial data.

# 2 Dystonia

## 2.1 Noninvasive Stimulation Technique

### 2.1.1 Introduction

Characteristics of dystonia are abnormal postures and involuntary twisting and repetitive movements due to excessive muscle contraction. Although dystonic syndromes are a set of heterogeneous disorders, the pathophysiological mechanisms of some forms of dystonia have been well elucidated. Noninvasive brain stimulation can regulate the neural activities of one or more network nodes involved in different dystonia. The future development of noninvasive brain stimulation (NIBS) in the treatment of various dystonia deserves attention (Erro et al. 2017).

At present, it is believed that the occurrence of dystonia symptoms is related to the reduction of inhibition at different levels for the central nervous system, abnormal (excessive) cortical plasticity, and dysregulation of sensorimotor integration. These pathophysiological mechanisms may explain some common clinical phenomena in dystonia, such as hyperkinesis and the presence of sensory symptoms. It has been confirmed that such abnormality occurs at all levels of the central nervous system. Therefore, dystonia is currently considered to be a disease involving abnormalities of the sensorimotor cortex, basal ganglia, and cerebellar network.

The current main symptomatic treatment for dystonia is local injection of botulinum toxin. The efficacy of injection therapy for cervical dystonia and blepharospasm is good, but the efficacy for patients with focal hand dystonia is poor, with great side effects. In addition, botulinum toxin cannot be injected for dystonia in many parts of the body and systemic dystonia. Some dystonia (i.e., DYT1, DYT6, DYT11, DTY28, tardive dystonia, etc.) can be treated with DBS, but DBS is not suitable for all patients with dystonia. Therefore, it is necessary to find other new treatments. NIBS can cause changes in the excitability of one or more nodes in the brain network, becoming a new treatment for dystonia.

### 2.1.2 Treatment Principle

At present, there are two main NIBS techniques: TMS and tES. These techniques act noninvasively on the scalp, avoiding DBS-related complications and side effects of systematic drug therapy. Theoretically, both techniques can act on selected cortical areas and regulate specific cortical/subcortical networks related to clinical manifestations. However, this technique has a tendency to spread from the target area of the brain to the adjacent areas, which may affect the selection and application of this technique.

In addition to the two techniques, the transcutaneous electrical nerve stimulation (TENS) can also regulate motor cortex excitability, but it is traditionally not considered as NIBS. The theoretical basis for the proposal of TENS is that the afferent stimulation causes lasting changes in the original sensorimotor cortex. The effect of TENS on the center depends on the frequency of stimulation. Low frequency (1–4 Hz) increases the excitability of sensorimotor cortex, while high frequency (>50 Hz) decreases the excitability. It is inferred that after TENS, the excitability of peripheral (M wave) and spinal cord (H wave) remains unchanged, which confirms that plasticity changes occur in the central nervous system. The exact mechanism of TENS in regulating sensorimotor excitability is unclear, and some authors speculate that it may cause changes of long-term potentiation in inhibitory synapses.

### 2.1.3 Treatment Regimens and Clinical Application

#### 2.1.3.1 Dystonia in Childhood

So far, the potential therapeutic effect of NIBS in children with dystonia has been evaluated in three studies, all of which adopted tDCS therapy. In 2012, Young et al. (2013) conducted an open-label study of cathode tDCS in ten children with dystonia caused by different etiologies (two of them were idiopathic dystonia and the other eight were secondary dystonia). The contralateral motor cortex of the affected limb was subjected to a single sequence of cathodal stimulation, and no improvement was found in the Barry-Albright Dystonia (BAD) scale. However, it was observed that there was a non-significant decrease in hyperkinesis in the electromyography-related examinations. Therefore, they conducted the repeated trial in 14 children with dystonia using a double-blind controlled design of cathodal tDCS. Similarly, there was no obvious improvement in BAD scale, but there was a significant decrease in hyperkinesis, although the magnitude of impact was very small. Bhanpuri et al. (2015) treated nine patients with dystonia of different causes, trying to explore whether repetitive tDCS would produce cumulative effect. They used a double-blind controlled design, in which five sequences of cathodal or anodal stimulation were applied to the contralateral motor cortex of the affected limb, and the study result showed no improvement in clinical symptoms and electromyography. However, the study suggested that cathodal stimulation of the motor cortex may be effective and anodal tDCS may even exacerbate dyskinesia in some patients. The major deficiencies of the above three studies are that there are different etiologies in the included patients, so it is difficult to draw an exact conclusion. In addition, the sample size is too small to perform subtype comparison for patients with primary and secondary dystonia. Some patients could not tolerate the stimulation intensity of 1–2 mA, so it was reduced to less than 1 mA. All patients except one completed the trial, and no obvious side effects were observed. Therefore, a double-blind controlled study with expanded sample size is needed in similar patients.

#### 2.1.3.2 Dystonia in Adults

##### 2.1.3.2.1 Focal Hand Dystonia

**Repetitive TMS/Theta-Burst Stimulation (TBS)**

For the study of focal palmar dystonia, there are great differences in the design of including patients. Most studies select patients with writer's cramp, with the principle adopting inhibitory stimulation in the contralateral cortical motor area of the affected limb. Siebner et al. (1999) found that when a single 1-Hz rTMS was given in the M1 area, the symptoms of writer's cramp were reduced, which may be due to the normalization of cortical-cortical inhibition and the prolongation of cortical silent period. Murase et al. (2005) used 0.2-Hz rTMS in the premotor cortex (PMC) in a single-blind controlled study, and the symptoms of writer's cramp were also improved. In one study, TMS-PET were observed in seven patients with focal hand dystonia, and dorsal PMC (dPMC) was stimulated by a single sequence of 1-Hz rTMS for 30 min to evaluate the perfusion before and after stimulation. Although cerebral blood flow in the medial and lateral premotor cortex, putamen, and thalamus decreased significantly, neither the symptoms of writer's cramp nor clinical scores improved. On the contrary, Borich et al. (2009) gave PMC five consecutive 1-Hz rTMS, and the writing function was significantly improved for 10 days after stimulation. Studies have shown that no efficacy was observed after a single sequence of stimulation, while repeated sequences would produce cumulative effect. Huang et al. conducted consecutively five sequences of cTBS on PMC using a single-blind controlled design, and there was no any clinical improvement despite a significant increase in cortical inhibition. In a double-blind controlled study, Havrankova et al. evaluated 11 patients with writer's cramp, whose S1 areas were stimulated using 5 consecutive sequences of 1-Hz rTMS. The subjective and objective evaluation was improved for 2 weeks, although the score of the Burke-Fahn-Marsden Dystonia Rating Scale did not change.

**Transcranial Electrical Stimulation**

Buttkus et al. (2010, 2011) reported that in patients with writer's cramp, a single sequence of cathodal and anodal tDCS was conducted in the contralateral M1 area of the affected limb, without effect. Benninger et al. conducted cathodal tDCS in the M1 area for 20 min at a time and three-sequence courses within 1 week in 12 patients with writer's cramps, without significant effect. On the contrary, Furuya et al. (2014) stimulated bilateral M1 areas in patients with writer's cramp, with cathodal electrodes placed on the affected hemisphere, in conjunction with motor function training; the clinical symptoms were significantly improved. This indicated that the bilateral hemispheric tDCS combined with the motor function training of both hands was beneficial to the accuracy of finger motor training. Rosset-Llobet et al. (2015) also obtained similar results by repetitive training (five consecutive times a week for 2 weeks) with bilateral hemispheric tDCS in the sensorimotor cortex area in conjunction with sensorimotor training. The above studies indicated that bilateral hemispheric tDCS stimulation (with the cathode placed in the contralateral hemisphere of the affected limb) could significantly improve motor function. Sadnicka et al. conducted tDCS in ten patients with writer's cramp, and the anode electrode was placed in the ipsilateral cerebellar hemisphere. No improvement in clinical symptoms and cortical excitability was observed.

**Transcutaneous Electrical Nerve Stimulation (TENS)**

In a randomized, double-blind, placebo-controlled study (Tinazzi et al. 2005) of ten patients with writer's cramp, the forearm flexor was electrically stimulated at a frequency of 50 Hz, and the muscle contraction intensity with the stimulation intensity lower than the pain threshold. The electrical stimulation was performed at an interval of 2 s sequence (100 stimuli/sequence), with a pause of 2 s, a total of 20 min, and 10 sequences of stimulation treatment for 2 weeks. It was found that the subjective and objective evaluations of writing were all improved, although the disability score did not change, and the effect lasted for about 6 weeks. In another study, ten patients with writer's cramp were treated with TENS in forearm flexor, which showed improvement in writing function, accompanied by a significant decrease in the motor evoked potential amplitude of forearm flexor and an increase in the motor evoked potential amplitude of forearm extensor.

2.1.3.2.2 Craniocervical Dystonia

**Repetitive TMS/Theta-Burst Stimulation (TBS)**

There are few studies on rTMS or cTBS in the treatment of cervical dystonia, and there is only one study on rTMS in the treatment of blepharospasm. Koch et al. (2014) conducted a double-blind sham-controlled study in 18 patients with cervical dystonia using repeated cTBS in bilateral cerebellar hemispheres and found that the symptoms of the experimental group were significantly relieved, while the results of the control group were negative. However, there was no significant benefit at 2–4 weeks of follow-up. In addition, they also confirmed that there was also significant improvement when cTBS was combined with paired stimulation, indicating that the regulation of cerebellar brain inhibition (CBI) could indeed change some pathophysiological mechanisms of dystonia and improve clinical symptoms. Pirio Richardson et al. (2015) conducted a single-blind, blank-controlled low-frequency rTMS study in patients with cervical dystonia and found that the stimulation in M1 area and dorsal PMC area significantly improved the Toronto Western Spasmodic Torticollis Rating Scale (TWSTRS), but there was no significant difference. There were also studies (Zittel et al. 2015) on low-frequency rTMS in S1/M1 area, which confirmed the normalization of short-interval afferent inhibition (SAI), but there was no change in clinical symptoms. Kranz et al. (2010) conducted a double-blind, sham-controlled study in 12 patients with blepharospasm using a single sequence of low-frequency rTMS. The stimulation target was the anterior cingulate cortex, and the MEP amplitude was highest at the orbicularis oculi muscle. The result showed that both the doctor's assessment and the patient's subjective symptoms were improved and the blink reflex loop showed "normalization."

**Transcranial Electrical Stimulation**

At present, there are only case reports on tES in the treatment of cervical dystonia. One of them used tDCS, and the other used tDCS combined with tACS. In 2013, the study of Angelakis et al. (2013) showed that 15-Hz tACS in C3/C4 area (supplementary motor area as target area) had a better effect than cathode tDCS in C4 area, with clinical symptoms significantly improved, TWSTRS score reduced by about 50%, and TWSTRS pain score reduced by nearly 75%. This efficacy lasted for 1 month, indicating that the multiple-sequence tACS had cumulative effects. On the contrary, Bradnam et al. (2014) used repeated-sequence anode tDCS to stimulate the bilateral cerebellar hemisphere and the contralateral M1 area of the side with severe symptoms and found that TWSTRS score was reduced by about 40%, TWSTRS pain score was reduced by 55%, and other dystonia-related biochemical scale scores were also improved. In this study, the stimulation treatment conducted 1 week after botulinum toxin injection showed that tDCS might enhance the effect of botulinum toxin. The electrophysiological changes of cortical motor excitability also reflected the changes of clinical symptoms.

## 2.2 Implantable Device Stimulation Technique

DBS emerged in the past 30 years, and it plays an increasingly important role in the treatment of dystonia. Many studies have proved its efficacy in improving patients' symptoms and quality of life. In 2003, DBS was successively approved by the European Union and the US FDA for the treatment of dystonia. In 2016, the NMPA approved DBS for the treatment of primary dystonia. Combined with the recent study progress, this part introduces the principle, schemes, and clinical application of DBS in the treatment of dystonia.

### 2.2.1 Treatment Principle

The treatment principle of DBS remains unclear. The globus pallidus internus (GPi) is the main output site of the basal ganglia. Radiographic studies have shown that the primary role of GPi-DBS is to normalize pathologically overactive motor activation responses and patterns in these patients. It is also believed that DBS increases the output of GPi signals and activates the peripheral fiber pathways, resulting in complex regulatory effects on the whole basal ganglia-thalamic cortex network. The study found that recordings of GPi neuronal activity obtained from patients with implanted DYT1 showed an abnormal discharge pattern, that is, excessive synchronization of low-frequency activity. This abnormal oscillation in GPi may be a manifestation of abnormal plasticity distortion network function. Clinical response caused by DBS usually delays for several days or weeks, and some patients may not develop symptoms for a long time after stimulation stops. This fact may support the hypothesis that DBS can cause unknown changes in neuroplasticity. There is clear evidence that DBS may modulate cortical excitability; DBS stimulation interrupts pathological somatosensory activity and is associated with changes in the number and connection of cortical neurons, supporting the hypothesis that DBS causes changes in neuroplasticity (Macerollo et al. 2020).

### 2.2.2 Treatment Regimen

The diagnosis and treatment of dystonia is complex and special, so functional neurosurgeons and neurologists need to cooperate with each other to jointly formulate DBS surgical protocol and perform postoperative management.

#### 2.2.2.1 Indications

1. Isolated (idiopathic or hereditary) systemic dystonia and isolated (idiopathic or hereditary) segmental dystonia (grade A recommendation) with disabling motor symptoms and ability of daily living and severe pain which cannot be effectively improved by non-surgical therapies such as oral medication.
2. Isolated (idiopathic or hereditary) focal dystonia (such as cervical dystonia, oromandibular dystonia, and writer's cramp) (grade B recommendation) with disabling motor symptoms and ability of daily living which cannot be effectively improved by non-surgical therapies such as oral medication and botulinum toxin.
3. DBS surgery can be considered first for a diagnosed DYT1 systemic, segmental dystonia (grade B recommendation).
4. Some moderate and severe acquired dystonia with poor response to non-surgical treatment mainly refers to drug delayed systemic, segmental, and focal dystonia (grade C recommendation).
5. For some neurodegenerative diseases (such as neurodegeneration with brain iron accumulation and acanthocytosis) with poor response to non-surgical drug treatment, with dystonia (systemic, segmental, focal) as the prominent manifestation, with or without symptoms of other movement disorders, DBS can be tried with caution.

#### 2.2.2.2 Patient Selection

**Classification and Severity Assessment**

Prior to DBS surgery, patients with dystonia should first be classified and assessed for severity. At present, the classification of dystonia is mainly based on two main lines, clinical characteristics and etiology, which includes onset age, symptom distribution, time pattern, and accompanying symptoms according to clinical characteristics. According to the etiology, it includes nervous system pathological, hereditary or acquired, and idiopathic. Severity includes dysfunction and social disability. Generally, surgery can be considered when dystonia is the only or most prominent symptom leading to impairment of function and quality of life and when drug therapy fails.

**Age and Course of Discasc**

The surgical age of patients should range from 7 to 75 years; DBS surgery can provide better outcomes in young populations; elderly patients up to 80 years can also be involved after individualized assessment of benefits and risks; for patients <7 years old and >3 years old with severe primary systemic dystonia, age should not be used as a single exclusion indicator. The course of disease should not be used as an independent factor to evaluate whether patients are suitable for DBS surgery, but studies have confirmed that the shorter the duration of the course of disease, the greater the degree of benefit.

**Comorbidities**

Comorbidities include the following:

1. Mild structural abnormalities in the basal ganglia should not be contraindications for DBS.
2. Mild to moderate depression, anxiety, and schizophrenia stabilized by drug treatment should not be contraindications for DBS.
3. DBS is not recommended for the treatment of such dystonia in patients with mental diseases such as severe (refractory) depression, anxiety, and schizophrenia and mental symptoms which cannot be effectively controlled by drugs.
4. DBS is not recommended for the treatment of such dystopia with obvious cognitive dysfunction that is sufficient to affect the patient's ability of daily living.
5. For elderly patients with mild cognitive impairment, specialists should carefully evaluate the risks and benefits and decide whether to operate after full communication with patients and their families.
6. DBS is not recommended for the treatment of such dystonia in patients with obvious medical comorbidities affecting surgery or survival.

**Excluded Diseases**

Excluded diseases include the following:

1. DBS surgery is not recommended for such patients with dopa-responsive dystonia, because levodopa replacement therapy works well in these patients.
2. Oral medication is preferred for paroxysmal dystonia, and DBS treatment is not recommended for such patients.

#### 2.2.2.3 Evaluation and Examination

1. Radiographic examination: Brain MRI scan should be performed before surgery to exclude or confirm some structural abnormalities. If MRI is not applicable, CT can also be used instead.
2. Motor assessment: Motor assessment can quantify the severity of patients' symptoms, establish a clear baseline feature, facilitate doctors to formulate treatment regimens, and provide reference for postoperative efficacy.
3. Cognitive and psychiatric assessment: Severe cognitive impairment (dementia) and mental illness (schizophrenia) are contraindications for DBS surgery, and preoperative cognitive and psychiatric assessment helps to exclude the related contraindications for optimal efficacy.
4. Gene testing: Preoperative examination should include gene testing, because some types of gene mutations respond better to DBS than others. For example, DYT1 has better DBS results than DYT6, and DBS also shows positive therapeutic effects in patients with myoclonus-

dystonia syndrome with double mutations in DYT1 and DYT11. In addition, gene testing is helpful to further clarify the cause (Wu and Li 2018).

#### 2.2.2.4 Target Selection

The original DBS target for the treatment of dystonia was the thalamic Vim, but it has now been almost completely replaced by GPi. In recent years, more studies have confirmed the effectiveness and safety of STN as a target for the treatment of dystonia. In addition, some studies are still focused on the thalamus or cerebellum. It is possible that different subtypes of dystonia respond better to stimulation of different targets, which implies different pathophysiological mechanisms.

**GPi**

GPi is the most commonly selected target for DBS surgery in dystonia patients, which can be used to treat refractory primary systemic (or segmental) dystonia, late-onset dystonia, cervical dystonia, DYT1 dystonia, non-DYT1 primary systemic dystonia, axial dystonia, myoclonus-dystonia, and Meige syndrome. The most common side effects of stimulation at this site are dysarthria and tonic contraction of the face and arms. However, these side effects are generally reversible upon the adjustment of stimulation parameters. Some reports have described bradykinesia, freezing of gait, and Parkinson's symptoms after chronic GPi-DBS, and patients often report worsening activities involving upper and lower extremities that were not previously affected by dystonia.

**STN**

STN-DBS has been reported for the treatment of primary systemic dystonia, primary segmental dystonia, and refractory late-onset dystonia. STN-DBS is also effective for focal dystonia, such as oromandibular dystonia, cervical dystonia, writer's cramp, and Meige syndrome. Recent studies have shown that STN-DBS and GPi-DBS are equally effective in patients with isolated, systemic, or focal (cervical) dystonia, but with STN-DBS, the side effects of bradykinesia and freezing of gait are less frequent, and STN stimulation parameters are lower than GPi, resulting in longer battery life. In addition, current studies support the effectiveness and long-term tolerability of STN-DBS in the treatment of dystonia, but larger-scale comparative studies with GPi are needed.

**Thalamus**

Vim can be used for dystonia tremor. Vim-DBS may be considered for patients with dystonia characterized by tremor as a major disability. GPi-DBS and Vim-DBS can be combined in patients with severe dystonia tremor and dystonia coexisting. DBS in the more anterior part of the ventral thalamus, such as the ventralis oralis posterior nucleus (Vop) and ventralis oralis anterior nucleus (Voa), also shows good results in patients with more phasic or tonic components than tonic tremor.

**Cerebellum**

The cerebellum has become an attractive and promising target in the treatment of neurological diseases because of its several connections to the important cortical and subcortical structures related to motor control. However, there are few studies to evaluate its effect on dystonia, generally focusing on TMS or tES effect. It has been reported that ten patients with spasm and dystonia secondary to cerebral palsy (CP) received DBS in the anterior cerebellar lobe. Dystonia was improved by 25% in five of these patients. With the deepening of the understanding for pathophysiology of dystonia, the cerebellum may provide a new target for DBS (Cury et al. 2018).

#### 2.2.2.5 Postoperative Management

1. Radiographic reexamination after DBS surgery: Postoperative cranial imaging examinations are routinely performed to reconfirm the position of electrode implantation and identify whether there are intracranial hemorrhage, intracranial pneumatosis, and cerebral edema after surgery. There are no special requirements for a cranial CT scan, and a cranial 1.5 T MRI scan can be performed postoperatively, but it is necessary to refer to the specific MRI system and the special parameter setting and limitation of DBS system.
2. Postoperative drug adjustment: According to the improvement of symptoms, the dosage can be gradually reduced. Do not withdraw the medicine quickly, so as not to cause discomfort to patients.
3. Timing of the first startup program control after DBS treatment: Patients are usually in good condition. The startup can be performed after cerebral edema subsides, but for patients with severe dystonia symptoms, the startup may be performed as soon as possible.
4. Setting of stimulation parameters of DBS at startup: Most frequencies are 130 Hz, and the pulse width is 60 μs. The voltage can be adjusted according to the patient's symptom improvement and response, generally not exceeding 3 V.
5. Changes of stimulation parameters for long-term DBS therapy: Within half a year after surgery, two to three times of parameter adjustment is usually required to achieve the optimum efficacy; usually, the pulse width of STN long-term stimulation is 60–120 μs, and GPi is 90–150 μs. After the parameter adjustment with the GPi nucleus as the target, the efficacy appears after several hours to several days, which is slightly longer than that in the STN nucleus. Usually, after long-term stimulation (more than 1 year), the stimulation efficacy is stable, and there is no need to continuously increase stimulation parameters.
6. Other non-surgical treatment: For patients who have undergone DBS surgery for more than half a year, the resistance of DBS system is normal, and the electrode

implantation position is confirmed again. However, patients who still have symptoms of focal and segmental dystonia affecting their quality of life after several parameter adjustments can be treated with non-surgical treatment such as oral medicine, botulinum toxin, and rehabilitation (Wu and Li 2018).

### 2.2.3 Clinical Application

#### 2.2.3.1 Clinical Effects of DBS in the Treatment of Primary Dystonia

Panov et al. conducted a retrospective chart survey (cohort study) on 47 patients with DYT1 in a 10-year follow-up study. These patients were treated by a single surgical team during the 10-year period and followed up for up to 96 months (average 46 months). The results showed that, according to the Burke-Fahn-Marsden Dystonia Rating Scale (BFMDRS), after 2 years of DBS treatment, the severity of symptoms of patients decreased to below 20% of baseline and the disability score decreased to 30% of baseline. Follow-up showed a durable improvement in symptoms. All dystonia-related medications were discontinued in 61% of patients at the last follow-up. Ninety-one percent of patients discontinued at least one type of medication (Panov et al. 2013).

In a multicenter retrospective study, Brüggemann et al. evaluated the GPi-DBS results of 8 patients with DYT6, 9 patients with DYT1, and 38 patients with non-DYT at the early stage (1–16 months) and the late stage (22–92 months) by BFMDRS score. Results showed that at the early follow-up, the mean reduction in dystonia severity was greater in DYT1 (BFMDRS score: 260%) and non-DYT dystonia (BFMDRS score: 252%) patients than in DYT6 dystonia patients (BFMDRS score: 232%; $P = 0.046$). However, in the late follow-up, there was similar improvement in all three groups (DYT6, −42%; DYT1, −44%; non-DYT, −61%) for GPi-DBS (Brüggemann et al. 2015).

A multicenter, sham stimulation-controlled, randomized study in Germany showed that among 40 patients with segmental or systemic dystonia who were treated with bilateral GPi-DBS, the symptoms of dystonia in the treatment group were reduced to a greater extent ($P < 0.001$). After 6 months of follow-up, the symptoms of dystonia decreased by 48% on average. Systemic dystonia subgroup showed continuous improvement of symptoms during 3-year and 5-year follow-up (Kupsch et al. 2006). A recent retrospective study of STN-DBS therapy in 14 patients with idiopathic dystonia showed continuous improvement in the BFMDRS motor score 1 month, 1 year, and 5 years after DBS surgery, which were 59.0%, 85.0%, and 90.8%, respectively. In addition, it is reported that patients' quality of life is constantly improving and there are no significant side effects. Other published papers have also shown the beneficial clinical effects of DBS in the treatment of primary dystonia (Deng et al. 2018).

#### 2.2.3.2 Clinical Effects of DBS in the Treatment of Secondary Dystonia

Gruber et al. conducted a randomized controlled trial in 25 patients with tardive dystonia/dyskinesia to evaluate the safety and effectiveness of globus pallidus DBS in the treatment of tardive dystonia. The patients were randomly divided into 2 groups in a 1:1 ratio: 12 cases in the active stimulation group and 13 cases in the pseudostimulation group. The severity of dystonia/dyskinesia was assessed by blinded video at 3 months (blinded phase) and 6 months (open-label phase) after DBS in the globus pallidus. The results showed a 22.8% improvement in dystonia severity in the stimulation group and a non-significant improvement in the control group (12.0%) compared to their respective baseline severity after 3 months. In the open-label phase of active treatment in both groups, the dystonia severity of patients was improved by 41.5% (Gruber et al. 2018).

Deng et al. conducted a long-term follow-up for ten patients with refractory TD after bilateral STN-DBS treatment, with a mean follow-up time of 65.6 ± 30.4 months (range: 12–105 months). The results showed significant and continuous improvement. At the first follow-up, the BFMDRS motor and disability scores increased by 55.9% ± 28.3% and 62.6% ± 32.0%, respectively, and the Abnormal Involuntary Movement Scale (AIMS) score increased by 53.3 ± 26.7%. At the second follow-up, the BFMDRS motor and disability scores increased further by 87.3% ± 17.0% and 84.3% ± 22.9%, respectively, and the AIMS score increased by 88.4 ± 16.1%. Such benefit persisted at the last follow-up (Deng et al. 2017).

Perinatal hypoxia is manifested as cerebral palsy (CP), which is one of the common causes of dystonia and often has poor response to drug therapy. A prospective study of 13 adult CP patients treated with GPi-DBS showed a 24% improvement from baseline in motor function after 12 months of treatment with GPi-DBS; however, individual changes in BFMDRS showed large differences, ranging from −7.5% (deterioration) to +55% (improvement) (Vidailhet et al. 2009). An analysis of DBS in 68 CP patients showed that the efficacy of GPi-DBS was not obvious. BFMDRS motor and disability scores were improved by 23.6% and 9.2%, respectively, after 12 months (Ramirez-Zamora et al. 2016). However, a recent long-term study of 15 CP patients showed a 49% improvement in motor after a mean follow-up of 4.4 years. For other secondary dystonia, such as post-stroke dystonia and post-traumatic dystonia, there are also reports about the clinical effect after DBS treatment, but the results vary greatly (Macerollo et al. 2020).

In summary, DBS has a reliable clinical effect in the treatment of primary dystonia. For secondary dystonia, except that TD shows a good clinical effect, the clinical effects for other types of secondary dystonia vary greatly, so it should be used with caution.

#### 2.2.4 Recommended Guidelines

Chinese expert consensus on deep brain stimulation therapy for dystonia, Chinese expert consensus on the diagnosis of dystonia, Chinese expert consensus on the treatment of dystonia, and EFNS guidelines on the diagnosis and treatment of primary dystonia

## 3 Tourette Syndrome

### 3.1 Noninvasive Stimulation Technique

Tourette syndrome (TS) is a complex neuropsychiatric syndrome with onset in childhood or adolescence, which is characterized by chronic motor and vocal tics. As TS often co-occurs with obsessive-compulsive disorder (OCD), attention-deficit hyperactivity disorder (ADHD), and other comorbidities, psychotropic drugs are often used to treat TS; however, due to adverse reactions associated with such drugs, clinicians are looking for more effective treatment options. In addition to the emerging comprehensive behavioral interventions in recent years, the application of noninvasive neurostimulation techniques also plays a positive role in the treatment of this disease.

#### 3.1.1 Treatment Principle

With an unestablished pathophysiology, TS is generally considered to be related to the functional cortico-striatal-thalamic-cortical (CSTC) circuit abnormalities (Zhishun Wang et al. 2011). It may be related to increased striatal dopamine release, glutamatergic imbalance, and GABAergic interneuron dysfunction.

##### 3.1.1.1 Transcranial Magnetic Stimulation

Previous electrophysiological studies (Chen et al. 1997) showed that TS patients experienced increased motor cortical excitability and defective motor inhibition, which was manifested by shortened cortical silent period (CSP) and reduced short-interval intracortical inhibition (SICI); reduced SICI in TS patients was associated with the severity of motor tics (Gilbert et al. 2004). Studies (Orth and Rothwell 2009) have found that TS patients with ADHD had increased intracortical facilitation (ICF), another marker of motor cortical inhibition. These findings were consistent with the neuropathological studies of the basal ganglia in TS patients. The distribution of GABAergic interneurons in TS patients changed (Sowell et al. 2008), and the concentration of GABA in the primary sensorimotor cortex decreased. In addition, studies (Orth and Munchau 2013) have shown that short SAI decreased in TS patients, suggesting that such patients had reduced thalamic motor cortical inhibition, thus also leading to hyperexcitability of the motor cortex. Functional imaging studies (Swain et al. 2007) have shown that the activities of SMA and limbic system in TS patients increased before tic; the metabolism of SMA and DLPFC increased. Therefore, increasing inhibition and decreasing excitability of motor cortex in the above areas can theoretically improve the severity of tic symptoms.

rTMS can change the excitability of the cortex through magnetic fields with different stimulus frequencies and intensities. Generally, low-frequency stimulation ($\leq$1 Hz) plays a role in inhibiting local neurons, which can inhibit the excitability of the local cortex and lead to the reduction of local cerebral blood flow and metabolism; high-frequency stimulation (> 5 Hz) plays a role in facilitating local neurons, which can increase local cortical excitability and local cerebral blood flow and metabolism. The abovementioned regulatory effects of rTMS are not only limited to the stimulation to the cortex below the coil but also can exert regulatory effects on the distal cortical and subcortical areas through the cortico-cortical and cortico-subcortical connections. Therefore, many studies have applied the rTMS regimen to TS patients, trying to achieve the therapeutic goal of reducing tics by reducing cortical excitability.

##### 3.1.1.2 Transcranial Direct Current Stimulation

tDCS regulates the excitability of neurons by changing their resting membrane potential. Anodal stimulation plays a role in increasing cortical excitability, while cathodal stimulation plays a role in inhibiting neuronal excitability. The regulation time of tDCS on cortical activity excitability exceeds its direct stimulation time. In recent years, some studies have begun to try tDCS for the treatment of TS patients.

#### 3.1.2 Treatment Regimen

##### 3.1.2.1 rTMS

Due to the reduced inhibition of the motor cortex in TS patients, the previous selected targets of rTMS for the treatment of TS were mostly located in the bilateral SMA, MC, PMC, prefrontal cortex (PFC), etc. Previous meta-analyses (Chih-Wei Hsu et al. 2018) showed that the stimulation to bilateral SMA was superior to that in other sites in the improvement of tic symptoms. Functional imaging studies of TS patients have confirmed that SMA activity of patients increased before tic and was positively correlated with the severity of tic. SMA was widely associated with the areas involving motor control and played an important role in the neural network connecting cortical, thalamic, and basal ganglia pathways. However, when the target was located in MC, PMC, or PFC, the symptoms of patients were not improved significantly.

At present, there is still a lack of consistent protocol for the stimulation parameters of rTMS in the treatment of TS (Table 14.1). Meta-analysis (Chih-Wei Hsu et al. 2018) showed that in the selection of stimulation frequency, based on the pathophysiological mechanism of increased motor cortical

**Table 14.1** Study on rTMS for the treatment of TS

| Study | Study type | No. of patients | Control | Stimulation site | Stimulation frequency | Stimulation intensity | Number of pulses per day × days | Results |
|---|---|---|---|---|---|---|---|---|
| Münchau et al. (2002) | RCT, single, crossover | 16 (complicated with OCD: 7) | Sham coil | L MC/L PMC | 1 Hz | 80% RMT | 1200 × 2 | No significant improvement in symptoms |
| Jeong-Ho Chae et al. (2004) | RCT, crossover | 8 (complicated with OCD, 4; complicated with ADHD, 3) | 45°coil | L MC/L PFC | 1 Hz/15 Hz | 110% RMT | 2400 × 5 | After high-frequency stimulation to PFC, the tic symptoms were improved more significantly, but there was no difference between the stimulation protocols |
| Orth et al. (2005) | RCT, single, crossover | 5 (complicated with ADHD: 2) | Sham coil | Bilateral PMC | 1 Hz | 80% AMT | 1800 × 2 | No significant improvement in symptoms |
| Mantovani et al. (2006) | OLT | 10 (OCD, 5, TS, 3, OCD and TS, 2) | No | Bilateral SMA | 1 Hz | 100% RMT | 1200 × 10 | Significant improvement in tic and OCD symptoms |
| Mantovani et al. (2007) | OLT | 2 (complicated with OCD: 2) | No | Bilateral SMA | 1 Hz | 110% RMT | 1200 × 10 | Significant improvement in tic symptoms, but a recurrence in one case |
| Kwon et al. (2011) | OLT | 10 (complicated with OCD, 1; complicated with ADHD, 3) | No | Bilateral SMA | 1 Hz | 100% RMT | 1200 × 10 | Significant improvement in tic symptoms |
| Le et al. (2013) | OLT | 25 | No | Bilateral SMA | 1 Hz | 110% RMT | 1200 × 20 | Significant improvement in tic symptoms |
| Wu et al. (2014) | RCT, double, parallel | 12 (active, 6, in which complicated with OCD, 4, and complicated with ADHD, 5; sham, 6, in which complicated with OCD, 4, and complicated with ADHD, 3) | NA | Bilateral SMA | 30 Hz (cTBS) | 90% RMT | 2400 × 2 | No significant inter-group difference in symptomatic improvement |
| Landeros-Weisenberger et al. (2015) | RCT, double, parallel | 20 (active, 9, in which complicated with OCD, 3, and complicated with ADHD, 5; sham, 11, in which complicated with OCD, 2, and complicated with ADHD, 3) | Sham coil | Bilateral SMA | 1 Hz | 110% RMT | 1800 × 15 | No significant inter-group difference in symptomatic improvement |
| Bloch et al. (2016) | OLT | 12 (complicated with OCD, 6; complicated with ADHD, 4) | No | Bilateral SMA | 1 Hz (H-coil) | 110% RMT | 1200 × 20 | Significant improvement in tic symptoms of patients with OCD |
| Singh et al. (2018) | OLT | 3 (complicated with OCD: 2) | No | Bilateral SMA | 1 Hz | 110% RMT | 900 × 20 | Significant improvement in tic symptoms of patients with OCD |

*RCT* randomized controlled trial; *OCD* obsessive-compulsive disorder; *L* left; *MC* motor cortex; *PMC* premotor cortex; *RMT* resting motor threshold; *ADHD* attention-deficit hyperactivity disorder; *PFC* prefrontal cortex; *YGTSS* Yale Global Tic Severity Scale; *AMT* active motor threshold; *OLT* open-label trial; *TS* Tourette syndrome; *SMA* supplementary motor area; *cTBS* continuous theta-burst stimulation excitability in TS patients, previous studies mostly used low-frequency rTMS (1 Hz) to reduce cortical excitability in patients. Only one study (Jeong-Ho Chae et al. 2004) in 2004 tried to treat TS patients with low-frequency (1 Hz) and high-frequency rTMS (15 Hz) and found that after high-frequency stimulation to PFC, the score reduction rate of the Yale Global Tic Severity Scale (YGTSS) was obvious, but there was no significant difference in the efficacy of different stimulation frequencies.

#### 3.1.2.2 tDCS

At present, there is still a lack of strict randomized controlled studies on tDCS for the treatment of TS, so there is no recognized treatment regimen.

The selected targets for tDCS in the treatment of TS are mostly located in bilateral SMA and pre-SMA. In the studies by Eapen (Eapen et al. 2017) and Katherine (Katherine et al. 2019), tDCS targets were all SMA, and the tic symptoms of TS patients were significantly improved by treatment.

At present, there is a lack of recognized stimulation parameters for tDCS in the treatment of TS.

### 3.1.3 Clinical Application

#### 3.1.3.1 rTMS

Of the 11 studies on rTMS for the treatment of TS, listed in Table 14.1, except for the stimulation targets of PMC, MC, and PFC in 3 early studies (Münchau et al. 2002; Chae et al. 2004; Orth et al. 2005), the stimulation targets in other studies were all bilateral SMA. There were five randomized, controlled studies in all the studies, and the results were all negative. Among them, three early exploratory studies on targets of PMC, MC, and PFC have proved that the treatment failed. In another study (Wu et al. 2014) using cTBS, only 2400 × 2 days of stimulation quantity was given to bilateral SMA of patients. Mean YGTSS scores decreased in both active and sham groups. However, no significant difference in video-based tic severity rating was detected between the two groups. In the study (Landeros-Weisenberger et al. 2015), the treatment course of low-frequency rTMS was relatively long (3 weeks), mainly for patients with severe tic or self-injurious tic for at least 4 months, but the results were still negative, suggesting that longer or other alternative stimulation protocol may be needed in future studies.

Another six were open-label studies, in four (Mantovani et al. 2006, 2007; Kwon et al. 2011; Le et al. 2013) of which, the tic symptoms of TS patients were significantly improved after stimulation of 1-Hz rTMS in bilateral SMA, and the study of Le (Le et al. 2013) showed that the clinical efficacy could last up to 6 months. However, an important limitation of these studies was the lack of control groups. The subjects included in three (Mantovani et al. 2006; Kwon et al. 2011; Le et al. 2013) of these studies were not confirmed to suffer from refractory TS. In open-label studies, there was another study using deep rTMS (Bloch et al. 2016); although the treatment regimen of 1 Hz stimulation to bilateral SMA was also used, the clinical tic symptoms were not significantly improved. In TS patients with OCD, two studies (Bloch et al. 2016; Singh et al. 2018) showed that after low-frequency stimulation to bilateral SMA, the tic symptoms of TS patients with OCD were improved significantly, further suggesting that the comorbidities of TS patients may affect the efficacy of rTMS. At present, there is no effective randomized controlled study on the treatment of TS with low-frequency rTMS in SMA, and there is no international recommended regimen.

#### 3.1.3.2 tDCS

Of the five studies on tDCS for the treatment of TS, listed in Table 14.2, most of the studies had fewer subjects. In 2019, only ten patients were included in the study of Katherine (Katherine et al. 2019). However, it was still difficult to evalu-

**Table 14.2** Study on tDCS for the treatment of TS

| Study | No. of patients | Control | Stimulation site | Stimulation current | Stimulation time | Electrode area of stimulation electrode | Results |
|---|---|---|---|---|---|---|---|
| Mrakic-Sposta S Mrakic-Sposta et al. 2008 | 2 (complicated with OCD: 1) | Yes | L MC | 2 mA | 15 min * 5 day | Cathode, 21 cm² | Improvement in tic symptoms |
| Carvalho et al. 2015 | 1 | No | Bilateral pre-SMA | 1.425 mA | 30 min * 10 day | Cathode, 25 cm² | Significant improvement in tic symptoms, lasting for 6 months |
| Eapen et al. 2017 | 2 | Yes | Bilateral SMA | 1.4 mA | 20 min * 18 day | Cathode, 25 cm² | Improvement in tic symptoms |
| Behler et al. 2018 | 3 (complicated with OCD: 2) | No | Bilateral pre-SMA | 2 mA | 60 min * 5 day | Cathode, 35 cm² | No significant improvement in tic symptoms |
| Katherine et al. 2019 | 10 (complicated with OCD, 1; complicated with ADHD, 3) | Yes | Bilateral SMA | 1 mA | 20 min * 1 day | Cathode, 35 cm² | No significant improvement in tic symptoms |

*OCD* obsessive-compulsive disorder; *L* left; *MC* motor cortex; *SMA* supplementary motor area; *ADHD* attention-deficit hyperactivity disorder ate the true reliability of its therapeutic effect due to only one course of treatment. In the studies with effective treatment, the stimulation targets involved the left MC, bilateral pre-SMA, and bilateral SMA, among which in the studies with the stimulation targets in the bilateral pre-SMA or bilateral SMA, the effects of treatment regimens varied according to different stimulation parameters, which could be related to different stimulation parameters. In the studies with effective treatment, the current intensities were concentrated at 1.4–2 mA, the stimulation electrodes were all negative, and the electrode areas were 21–25 $cm^2$. Most of case reports involved in studies had a short follow-up period, and only the study of Carvalho (Carvalho et al. 2015) showed that the clinical efficacy could be maintained for 6 months. In the future, the large-scale RCT study on tDCS for the treatment of TS will be needed to explore more reliable treatment parameters.

## 3.2 Implantable Device Stimulation Technique

Vandewalle et al. reported for the first time the application of DBS in refractory TS in 1999. A 42-year-old male patient underwent bilateral thalamic nucleus DBS surgery, with the reduction of tic by 90% after a 1-year follow-up (Vandewalle et al. 1999). Since then, more and more scientific researchers have devoted themselves to the studies on DBS for the treatment of TS and have achieved significant study results. Combined with the recent study progress, this part introduces the treatment principle, schemes, and clinical application of DBS in the treatment of TS.

### 3.2.1 Treatment Principle

DBS has shown to play a role in improving symptoms at multiple targets in TS patients, but its mechanism of action remains to be further elucidated. Supported by the hypothesis of high dopaminergic innervation in TS pathology and the use of dopamine antagonists in TS treatment, some studies have explored the role of DBS in dopaminergic regulation of striatal-thalamic pathway. The first report of DBS regulating dopaminergic transmission was that a 22-year-old patient had an improvement in tics by bilateral thalamic stimulation. Six months later after surgery, the positron emission computed tomography (PET) result showed that the binding potential of dopamine receptor decreased by 16.3% during the on-stimulation period compared with non-stimulation conditions, suggesting that DBS might indirectly lead to the reduction of dopamine release. In addition, according to the published control study, the availability of D2/3 receptor at baseline increased compared with the healthy control group (n = 20). Further studies of 18F-FP-CIT (N-3-fluoropropyl-2-beta-carboxymethoxy-3-beta-(4-iodophenyl)nortropane) PET scans in three patients undergoing DBS in the thalamus showed similar results for dopamine binding during DBS. Although these studies were limited by small sample sizes and potential anesthetic effects, these studies suggested that dopamine regulation could be a partial way for DBS to play a therapeutic role.

Previous studies have found that microelectrode recordings of thalamic nuclei during DBS surgery typically exhibit burst discharge patterns with burst intervals within the low-frequency (δ/θ) range (2.5–8 Hz) from 0.12 s to 0.4 s. Similar results were also obtained in a study showing low-frequency oscillations, suggesting that low-frequency activity and attenuated thalamic beta activity may be related to the pathophysiology of TS. This was supported by clinical phenotypes. The low-frequency oscillation in a patient with mild tic but severe OCD was significantly less than that in patients with severe tic. In addition, Maling et al. demonstrated that DBS-mediated gamma synchronization enhancement was associated with relief of symptoms of TS. Therefore, transferring the oscillation power of the basal ganglia and thalamus from low frequency to high frequency may be one of the effects of DBS on TS patients (Akbarian-Tefaghi et al. 2016).

### 3.2.2 Treatment Regimen

#### 3.2.2.1 Patient Selection

**Inclusion Criteria**

1. DSM-V diagnosis of TS made by clinical experts.
2. Age >18 years old; for patients under 18 years old, the decision should be made after discussion by the ethics committee.
3. Severity of tics: Yale Global Tic Severity Scale (YGTSS) >35/50.
4. Tic is the main cause of disability.
5. Tic does not respond to conservative treatment (failed in treatment with three drugs in trials): (a) one α-adrenergic agonist; (b) two dopamine antagonists (including one typical and one atypical); and (c) at least one additional class of drugs (e.g., clonazepam, topiramate, or tetrabenazine).
6. The complicated medical, neurological, and mental diseases are treated and stabilized for at least 6 months.
7. Neuropsychological characteristics show that patients can accept the requirements of operation and follow-up and possible adverse results.

**Exclusion Criteria**

1. Active suicidal or homicidal ideation within 6 months
2. Active or recent drug abuse
3. Structural brain lesions in MRI
4. Neurological or psychiatric disorders that may increase the risk of surgery or interfere with postoperative management
5. Personality disorder or psychogenic tic

#### 3.2.2.2 Target Selection

So far, more than 100 cases of TS treated with DBS have been reported in the literature, but there is a lack of rigorous study comparing the therapeutic effects in different brain targets, so there are almost no data to guide the target selection of general or specific patients. The target selection is mainly based on the patient's symptom characteristics and the experience of each center (Schrock et al. 2015).

**Thalamus**

The centromedian-parafascicular complex (CM-Pf) of the thalamus is the most commonly used target. Many studies have found that DBS of CM-Pf can not only improve motor tics and vocal tics in TS patients but also improve coexisting psychiatric disorders in TS patients, such as OCD, depression, and anxiety.

**Anteromedial Globus Pallidus Internus (amGPi)**

amGPi-DBS has been successfully applied in the treatment of TS. In 1 study, 15 TS patients were treated with amGPi-DBS. The results showed that the severity of tics was significantly reduced at follow-up (YGTSS total score decreased by 38% on average, the vocal score decreased by 38% on average, and the motor score decreased by 33% on average). At the overall level, amGPi-DBS had no significant effect on comorbid OCD, depression, and anxiety. However, the investigators determined that a group of patients with severe OCD symptoms as defined by the Yale-Brown Obsessive-Compulsive Scale (Y-BOCS) had a 39% improvement after treatment with amGPi-DBS. Although this study did not find an overall effect on depression, other studies have reported improvements in comorbid depression after application of amGPi-DBS.

**Posteroventral Globus Pallidus Internus (pvGPi)**

A retrospective study reported that the pvGPi-DBS played a role in improving motor tic in three adolescents with refractory TS. In addition, pvGPi-DBS also played a role in stabilizing the symptoms of comorbid OCD in one patient. These findings indicate that pvGPi-DBS can be used as a safe and effective intervention for the treatment of adolescent TS patients with tic and OCD symptoms.

**Globus Pallidus Externus (GPe)**

There are few studies selecting GPe as a therapeutic target, and a case report described a refractory TS patient with significant improvement in tic and mental health after bilateral GPe-DBS. When the stimulation was unexpectedly interrupted due to battery exhaustion, some TS symptoms of the patient reappeared. These findings suggest that GPe can also be considered as a potentially effective DBS target for the treatment of severe and refractory TS.

**Nucleus Accumbens (NAc) and Anterior Limb of the Internal Capsule (ALIC)**

The role of DBS of NAc and ALIC in the treatment of TS has been evaluated in some case studies. In a 26-year-old TS patient, tic severity was reduced by 50% following bilateral NAc-DBS. The role of NAc-DBS and ALIC-DBS in reducing the tic severity has also been confirmed in other case reports. However, a case report showed that a 38-year-old TS patient had a 53% reduction in OCD symptoms at the 3-month follow-up after NAc-DBS, which persisted for 36 months, but the patient continued to experience recurrent depressive episode. The observations highlight a treatment precaution that DBS in NAc/ALIC area may cause emotional side effects, including depression and hypomania.

**Subthalamic Nucleus (STN)**

STN is the most common target of DBS for the treatment of PD. There is evidence that this target may also control symptoms in TS patients. It is reported that in a 38-year-old PD patient with TS, the tic frequency was improved by 89% after 6 months of bilateral STN-DBS treatment and 97% after 12 months. In another study, four patients with TS received DBS in bilateral GPI and bilateral STN simultaneously. The investigators obtained the local field potential (LFP) and electromyogram records of patients within 3–5 days after DBS implantation. These results suggested that the stimulation to STN and GPi could improve the symptoms of acute TS by regulating the oscillation of basal ganglia neurons. These studies showed that STN could also be used as the target of DBS in the treatment of TS (Xu et al. 2020).

#### 3.2.2.3 Stimulation Parameters

On the one hand, the stimulation parameters of DBS in TS are based on its positive effects on tics and behavior disorders, and on the other hand, they are based on avoiding the adverse reactions caused by stimulation. It is reported that for three commonly used targets, namely, the thalamus, GPi, and NAc/ALIC, the stimulation frequencies range from 110 to 180 Hz. Besides, except the thalamus, some studies have reported the positive effect of 70 Hz. It is speculated that that the pedunculopontine nucleus (PPN), as a part of the reticular activating system (RAS), has similar electrophysiological properties to the thalamic nucleus targeting DBS in TS. RAS output or arousal can be stabilized during continuous stimulation at 40–60 Hz (Andrade and Visser-Vandewalle 2016).

#### 3.2.2.4 Evaluation of Clinical Efficacy

To evaluate the clinical efficacy of DBS, the effect of DBS on tic, related behavior disorders, and side effects caused by stimulation must be recorded. The most commonly used TIC grading scale is the Yale Global Tic Severity Scale

(YGTSS). For a more objective evaluation, videotaping should also be used to document the patient's condition with and without stimulation. Tic should be evaluated by two independent investigators on these videos, and ideally, neither patients nor investigators were aware of the stimulation. The investigators should perform regularly detailed psychiatric and neuropsychological assessments for patients. The clinical effect of DBS is related to the accuracy of the position of electrode implantation. The cranial imaging examination should be performed after surgery to determine the position of electrode implantation (Ackermans et al. 2013).

### 3.2.3 Clinical Application

#### 3.2.3.1 Randomized Controlled Study

In 2007, Maciunas et al. conducted a randomized double-blind study. In this study, five patients received bilateral CM-Pf-DBS implantation in the thalamus. The randomization period began 1 month after DBS implantation. The patient experienced 1 week in each of the following states: both stimulators were off, the left was on and right off, the right was on and left off, and both stimulators were on. Subjective and objective blinded outcomes were evaluated at the end of each week. Under the bilateral stimulation state, the modified RVBTS score decreased by 4.2 points (tic decreased by 53%), with statistically significant difference. Motor and vocal tic counts and YGTSS and TS symptom list scores were both improved. After 3 months of bilateral stimulation, similar assessment was performed for the results in a non-blind way, and the results showed that patients continued to benefit (Maciunas et al. 2007).

In 2008, Welter et al. conducted a randomized, double-blind, crossover study in three patients with severe TS to explore the effects of CM-Pf and/or GPi high-frequency stimulation. In the crossover design, subjects were randomly assigned to four stimulation conditions: (1) bilateral thalamic stimulation, (2) bilateral GPi stimulation, (3) combined stimulation to the bilateral GPi and thalamus, and (4) no stimulation. Each stimulation condition was maintained for 2 months, and the patients were examined monthly by an unaware clinician. The study showed that bilateral GPi stimulation significantly improved patients' YGTSS scores (tic severity of patients 1, 2, and 3 decreased by 65%, 96%, and 74%, respectively). Bilateral CM-Pf stimulation reduced the tic severity by 64%, 30%, and 40%, respectively. No further decrease in tic severity (60%, 43%, and 76%) was found during the combined stimulation to the bilateral GPi and thalamus, whereas the motor symptoms recurred in patients during the non-stimulation period (Welter et al. 2008).

In 2011, Ackermans et al. conducted a randomized, double-blind, crossover study in six TS patients to evaluate the effectiveness and safety of Cm-Spv-Voi stimulation to the thalamus. Patients were randomly assigned into two groups after surgery: group A, stimulator on at the first 3 months and off at the next 3 months, and group B, the opposite stimulation protocol. The stimulators were switched on in both groups at the following 6 months. The assessment was performed before and 3, 6, and 12 months after surgery. The results showed that the tic severity during the "on-stimulation" period was significantly lower than that during the "off-stimulation" period and the YGTSS score was improved by 37% ($P = 0.046$). There was a sustained stimulation effect at postoperative 1 year, with a 49% improvement in the YGTSS score compared with preoperative assessment ($P = 0.028$) (Ackermans et al. 2011).

In 2015, Kefalopoulou et al. performed GPi-DBS surgery on 15 patients (13 of whom completed the double-blind assessment) in a randomized, double-blind, crossover study, who were then randomized in a 1:1 ratio to receive stimulation or not for the next 3 months and then switched to the opposite state for another 3 months. The results showed that the mean YGTSS total score of patients was 87.9 (SD: 9.2) at baseline, 80.7 (SD: 12.0) at non-stimulation state, and 68.3 (SD: 18.6) at stimulation state. After Bonferroni correction, the pairwise comparisons of YGTSS total scores were significantly lower at the end of the stimulation state than at the non-stimulation state, with a mean improvement of 12.4 points (95% CI: 0.1–24.7), $P = 0.048$, equal to a difference of 15.3% (95% CI: 5.3–25.3) (Kefalopoulou et al. 2015).

In 2017, Welter et al. performed bilateral anterior internal globus pallidus (aGPi)-DBS surgery on 19 patients (16 of whom completed the double-blind evaluation) in a randomized, double-blind, controlled study. They were randomized in a 1:1 ratio 3 months after surgery and received active or pseudostimulation in a double-blind method at the following 3 months. The results showed no significant differences in YGTSS score changes between the active and pseudostimulation groups from the start to the end of the 3-month double-blind period (Welter et al. 2017).

The above five randomized controlled studies, except the last one, all showed the positive effect of DBS in the treatment of TS. More recently, Martinez-Ramirez et al. summarized data of 185 patients with TS who underwent DBS surgery at 31 institutions in 10 countries worldwide from 2012 to 2016. The results showed that 1 year after DBS implantation, the mean YGTSS total score decreased from 75.01 at baseline to 41.19, which almost decreased by 45.1% (Martinez-Ramirez et al. 2018). Generally speaking, the effect of DBS in the treatment of TS is encouraging, and DBS is a promising treatment for some drug-refractory or critically ill patients. In the future, large-scale randomized controlled studies are required to demonstrate the safety and effectiveness of DBS.

#### 3.2.3.2 Adverse Reactions

Adverse reactions can be divided into stimulation-related events and surgery-related events, and some stimulation effects vary according to different targets.

Stimulation to the thalamus is associated with transient blurred vision, dysarthria, recurrent tension headaches, single seizures, oculomotor dyskinesia, mood disturbance, erectile dysfunction, paresthesia, weight gain, and apathy. Stimulation to pvGPi is associated with increased anxiety, depression, and memory impairment. Stimulation to amGPi is associated with higher levels of anxiety, limb dyskinesia, and hypomania. Stimulation to NAc/ALIC is associated with depression and hypomania. Apathy, fatigue, dizziness, and weight change are the most common adverse reactions during the application of DBS. In most cases, these adverse reactions can be reduced or eliminated by the adjustment of stimulation parameters, but the effects should be limited when targets are selected.

Surgery-related adverse reactions mainly focus on hardware failures and infection, and the latter appears to be more severe than other dyskinesia in TS. This is put forward through a retrospective study of 272 patients undergoing DBS surgery, 39 of whom were TS patients. In these patients, the overall infection rate was 3.7%, while that of TS patients was 18%. Higher infection rates in TS patients were associated with forced removal of surgical scars in some patients (Akbarian-Tefaghi et al. 2016).

### 3.2.4 Recommended Guidelines

European clinical guidelines for Tourette syndrome and other tic disorders. Part IV: deep brain stimulation (2011), Tourette syndrome deep brain stimulation: a review and updated recommendations (2014), and Practice guideline recommendations summary: treatment of tics in people with Tourette syndrome and chronic tic disorders (2019)

## References


Murase N, Rothwell JC, Kaji R et al (2005) Subthreshold low-frequency repetitive transcranial magnetic stimulation over the premotor cortex modulates writer's cramp. Brain 128(Pt 1):104–115

Ackermans L, Duits A, van der Linden C et al (2011) Double-blind clinical trial of thalamic stimulation in patients with Tourette syndrome. Brain 134(Pt 3):832–844

Ackermans L, Neuner I, Temel Y et al (2013) Thalamic deep brain stimulation for Tourette syndrome. Behav Neurol 27(1):133–138

Akbarian-Tefaghi L, Zrinzo L, Foltynie T (2016) The use of deep brain stimulation in Tourette syndrome. Brain Sci 6(3):35

Andrade P, Visser-Vandewalle V (2016) DBS in Tourette syndrome: where are we standing now? J Neural Transm (Vienna) 123(7):791–796

Angelakis E, Liouta E, Andreadis N et al (2013) Transcranial alternating current stimulation reduces symptoms in intractable idiopathic cervical dystonia: a case study. Neurosci Lett 533:39–43

Behler N, Leitner B, Mezger E et al (2018) Cathodal tDCS over motor cortex does not improve Tourette syndrome: lessons learned from a case series. Front Behav Neurosci 24(12):194

Benninger DH, Berman BD, Houdayer E et al (2011) Intermittent theta-burst transcranial magnetic stimulation for treatment of Parkinson disease. Neurology 76(7):601–609

Beretta VS, Conceição NR, Nóbrega-Sousa P et al (2020) Transcranial direct current stimulation combined with physical or cognitive training in people with Parkinson's disease: a systematic review. J Neuroeng Rehabil 17(1):74

Bhanpuri NH, Bertucco M, Young SJ et al (2015) Multiday transcranial direct current stimulation causes clinically insignificant changes in childhood dystonia: a pilot study. J Child Neurol 30:1604–1615

Bloch Y, Arad S, Levkovitz Y (2016) Deep TMS add-on treatment for intractable Tourette syndrome: a feasibility study. World J Biol Psychiatry 17(7):557–561

Borich M, Arora S, Kimberley TJ (2009) Lasting effects of repeated rTMS application in focal hand dystonia. Restor Neurol Neurosci 27:55–65

Bradnam LV, Frasca J, Kimberley TJ (2014) Direct current stimulation of primary motor cortex and cerebellum and botulinum toxin A injections in a person with cervical dystonia. Brain Stimul 7:909–911

Brüggemann N, Kühn A, Schneider SA et al (2015) Short- and long-term outcome of chronic pallidal neurostimulation in monogenic isolated dystonia. Neurology 84(9):895–903

Buttkus F, Baur V, Jabusch HC et al (2011) Single-session tDCS-supported retraining does not improve fine motor control in musician's dystonia. Restor Neurol Neurosci 29:85–90

Buttkus F, Weidenmüller M, Schneider S et al (2010) Failure of cathodal direct current stimulation to improve fine motor control in musician's dystonia. Mov Disord 25:389–394

Carvalho S, Goncalves OF, Soares JM et al (2015) Sustained effects of a neural-based intervention in a refractory case of Tourette syndrome. Brain Stimul 8(3):657–659

Chen R, Classen J, Gerloff C et al (1997) Depression of motor cortex excitability by low-frequency transcranial magnetic stimulation. Neurology 48(5):1398–1403

Chung SW, Hill AT, Rogasch NC et al (2016) Use of theta-burst stimulation in changing excitability of motor cortex: a systematic review and meta-analysis. Neurosci Biobehav Rev 63:43–64

Cury RG, Kalia SK, Shah BB et al (2018) Surgical treatment of dystonia. Expert Rev Neurother 18(6):477–492

Delgado-Alvarado M, Marano M, Santurtún A et al (2020) Nonpharmacological, nonsurgical treatments for freezing of gait in Parkinson's disease: a systematic review. Mov Disord 35(2):204–214

Deng Z, Pan Y, Zhang C et al (2018) Subthalamic deep brain stimulation in patients with primary dystonia: a ten-year follow-up study. Parkinsonism Relat Disord 55:103–110

Deng ZD, Li DY, Zhang CC et al (2017) Long-term follow-up of bilateral subthalamic deep brain stimulation for refractory tardive dystonia. Parkinsonism Relat Disord 41:58–65

Eapen V, Baker R, Walter A et al (2017) The role of transcranial direct current stimulation (tDCS) in Tourette syndrome: a review and preliminary findings. Brain Sci 7(12):161

Eggers C, Günther M, Rothwell J et al (2015) Theta burst stimulation over the supplementary motor area in Parkinson's disease. J Neurol 262(2):357–364

Erro R, Tinazzi M, Morgante F et al (2017) Non-invasive brain stimulation for dystonia: therapeutic implications. Eur J Neurol 24:1228–1243

Fregni F, El-Hagrassy MM, Pacheco-Barrios K et al (2021) Evidence-based guidelines and secondary meta-analysis for the use of transcranial direct current stimulation (tDCS) in neurological and psychiatric disorders. Int J Neuropsychopharmacol 24(4):256–313

Furuya S, Nitsche MA, Paulus W et al (2014) Surmounting retraining limits in musicians' dystonia by transcranial stimulation. Ann Neurol 75:700–707
Ganguly J, Murgai A, Sharma S et al (2020) Non-invasive transcranial electrical stimulation in movement disorders. Front Neurosci 14:522
Gilbert DL, Bansal AS, Sethuraman G et al (2004) Association of cortical disinhibition with tic, ADHD, and OCD severity in Tourette syndrome. Mov Disord 19(4):416–425
Gruber D, Südmeyer M, Deuschl G et al (2018) Neurostimulation in tardive dystonia/dyskinesia: a delayed start, sham stimulation-controlled randomized trial. Brain Stimul 11(6):1368–1377
Horn MA, Gulberti A, Gülke E et al (2020) A new stimulation mode for deep brain stimulation in Parkinson's disease: theta burst stimulation. Mov Disord 35(8):1471–1475
Huang YZ, Rothwell JC, Lu CS et al (2010) Reversal of plasticity-like effects in the human motor cortex. J Physiol 588(Pt 19):3683–3693
Hsu CW, Wang LJ, Lin P (2018) Efficacy of repetitive transcranial magnetic stimulation for Tourette syndrome: a systematic review and meta-analysis. Psychiatry Q 89(3):645–665
Chae J-H, Nahas Z, Wassermann E et al (2004) A pilot safety study of repetitive transcranial magnetic stimulation (rTMS) in Tourette's syndrome. Cogn Behav Neurol 17(2):109–117
Katherine D, Georgina MJ, Elena N et al (2019) Effects of single-session cathodal transcranial direct current stimulation on tic symptoms in Tourette's syndrome. Exp Brain Res 237(11):2853–2863
Kefalopoulou Z, Zrinzo L, Jahanshahi M et al (2015) Bilateral globus pallidus stimulation for severe Tourette's syndrome: a double-blind, randomised crossover trial. Lancet Neurol 14(6):595–605
Kishore A, James P, Krishnan S et al (2017) Motor cortex plasticity can indicate vulnerability to motor fluctuation and high L-DOPA need in drug-naïve Parkinson's disease. Parkinsonism Relat Disord 35:55–62
Koch G, Porcacchia P, Ponzo V et al (2014) Effects of two weeks of cerebellar theta burst stimulation in cervical dystonia patients. Brain Stimul 7:564–572
Kranz G, Shamim EA, Lin PT et al (2010) Transcranial magnetic brain stimulation modulates blepharospasm: a randomized controlled study. Neurology 75:1465–1471
Kupsch A, Benecke R, Müller J et al (2006) Pallidal deep-brain stimulation in primary generalized or segmental dystonia. N Engl J Med 355(19):1978–1990
Kwon HJ, Lim WS, Lim MH et al (2011) 1-Hz low frequency repetitive transcranial magnetic stimulation in children with Tourette's syndrome. Neurosci Lett 492(1):1–4
Landeros-Weisenberger A, Mantovani A, Motlagh MG et al (2015) Randomized sham controlled double-blind trial of repetitive transcranial magnetic stimulation for adults with severe Tourette syndrome. Brain Stimul 8(3):574–581
Le K, Liu L, Sun M et al (2013) Transcranial magnetic stimulation at 1 hertz improves clinical symptoms in children with Tourette syndrome for at least 6 months. J Clin Neurosci 20(2):257–262
Lefaucheur JP, Aleman A, Baeken C et al (2020) Evidence-based guidelines on the therapeutic use of repetitive transcranial magnetic stimulation (rTMS): an update (2014-2018). Clin Neurophysiol 131(2):474–528
Macerollo A, Sajin V, Bonello M et al (2020) Deep brain stimulation in dystonia: state of art and future directions. J Neurosci Methods 340:108750
Maciunas RJ, Maddux BN, Riley DE et al (2007) Prospective randomized double-blind trial of bilateral thalamic deep brain stimulation in adults with Tourette syndrome. J Neurosurg 107(5):1004–1014
Mantovani A, Leckman JF, Grantz H et al (2007) Repetitive transcranial magnetic stimulation of the supplementary motor area in the treatment of Tourette syndrome: report of two cases. Clin Neurophysiol 118(10):2314–2315
Mantovani A, Lisanby SH, Pieraccini F et al (2006) Repetitive transcranial magnetic stimulation (rTMS) in the treatment of obsessive-compulsive disorder (OCD) and Tourette's syndrome (TS). Int J Neuropsychopharmacol 9(1):95–100
Martinez-Ramirez D, Jimenez-Shahed J, Leckman JF et al (2018) Efficacy and safety of deep brain stimulation in Tourette syndrome: the international Tourette syndrome deep brain stimulation public database and registry. JAMA Neurol 75(3):353–359
Mrakic-Sposta S, Marceglia S, Mameli F et al (2008) Transcranial direct current stimulation in two patients with Tourette syndrome. Mov Disord 23(15):2259–2261
Münchau A, Bloem BR, Thilo KV et al (2002) Repetitive transcranial magnetic stimulation for Tourette syndrome. Neurology 59(11):1789–1791
Orth M, Kirby R, Richardson MP et al (2005) Subthreshold rTMS over pre-motor cortex has no effect on tics in patients with Gilles de la Tourette syndrome. Clin Neurophysiol 116(4):764–768
Orth M, Munchau A (2013) Transcranial magnetic stimulation studies of sensorimotor networks in Tourette syndrome. Behav Neurol 27(1):57–64
Orth M, Rothwell JC (2009) Motor cortex excitability and comorbidity in Gilles de la Tourette syndrome. J Neurol Neurosurg Psychiatry 80(1):29–34
Panov F, Gologorsky Y, Connors G et al (2013) Deep brain stimulation in DYT1 dystonia: a 10-year experience. Neurosurgery 73(1):86–93
Pirio Richardson S, Tinaz S, Chen R (2015) Repetitive transcranial magnetic stimulation in cervical dystonia: effect of site and repetition in a randomized pilot trial. PLoS One 10:e0124937
Ramirez-Zamora A, Kaszuba B, Gee L et al (2016) Clinical outcome and intraoperative neurophysiology for focal limb dystonic tremor without generalized dystonia treated with deep brain stimulation. Clin Neurol Neurosurg 150:169–176
Randver R (2018) Repetitive transcranial magnetic stimulation of the dorsolateral prefrontal cortex to alleviate depression and cognitive impairment associated with Parkinson's disease: a review and clinical implications. J Neurol Sci 393:88–99
Rosset-Llobet J, Fàbregas-Molas S, Pascual-Leone A (2015) Effect of transcranial direct current stimulation on neurorehabilitation of task-specific dystonia: a double blind, randomized clinical trial. Med Probl Perform Art 30:178–184
Schrock LE, Mink JW, Woods DW et al (2015) Tourette syndrome deep brain stimulation: a review and updated recommendations. Mov Disord 30(4):448–471
Siebner HR, Tormos JM, Ceballos-Baumann AO et al (1999) Low-frequency repetitive transcranial magnetic stimulation of the motor cortex in writer's cramp. Neurology 52:529–537
Singh S, Kumar S, Kumar N et al (2018) Low-frequency repetitive transcranial magnetic stimulation for treatment of Tourette syndrome: a naturalistic study with 3 months of follow-up. Indian J Psychol Med 40(5):482–486
Sowell ER, Kan E, Yoshii J et al (2008) Thinning of sensorimotor cortices in children with Tourette syndrome. Nat Neurosci 11(6):637–639
Suppa A, Marsili L, Belvisi D et al (2011) Lack of LTP-like plasticity in primary motor cortex in Parkinson's disease. Exp Neurol 227(2):296–301
Swain JE, Scahill L, Lombroso PJ et al (2007) Tourette syndrome and tic disorders: a decade of progress. J Am Acad Child Adolesc Psychiatry 46(8):947–968
Tinazzi M, Farina S, Bhatia K et al (2005) TENS for the treatment of writer's cramp dystonia: a randomized, placebo-controlled study. Neurology 64:1946–1948
Vandewalle V, van der Linden C, Groenewegen HJ et al (1999) Stereotactic treatment of Gilles de la Tourette syndrome by high frequency stimulation of thalamus. Lancet 353(9154):724
Vidailhet M, Yelnik J, Lagrange C et al (2009) Bilateral pallidal deep brain stimulation for the treatment of patients with dystonia-

choreoathetosis cerebral palsy: a prospective pilot study. Lancet Neurol 8(8):709–717

Wagle Shukla A, Shuster JJ, Chung JW et al (2016) Repetitive transcranial magnetic stimulation (rTMS) therapy in Parkinson disease: a meta-analysis. PM R 8(4):356–366

Wang Z, Maia TV, Marsh R et al (2011) The neural circuits that generate tics in Tourette's syndrome. Am J Psychiatry 168:1326–1337

Welter ML, Houeto JL, Thobois S et al (2017) Anterior pallidal deep brain stimulation for Tourette's syndrome: a randomised, double-blind, controlled trial. Lancet Neurol 16(8):610–619

Welter ML, Mallet L, Houeto JL et al (2008) Internal pallidal and thalamic stimulation in patients with Tourette syndrome. Arch Neurol 65(7):952–957

Wu SW, Maloney T, Gilbert DL et al (2014) Functional MRI-navigated repetitive transcranial magnetic stimulation over supplementary motor area in chronic tic disorders. Brain Stimul 7(2):212–218

Wu YW, Li DY (2018) Chinese expert consensus on deep brain stimulation therapy for dystonia. Chin J Neurosurg 34(06):541–545

Xu W, Zhang C, Deeb W et al (2020) Deep brain stimulation for Tourette's syndrome. Transl Neurodegener 9:4

Young SJ, Bertucco M, Sheehan-Stross R et al (2013) Cathodal transcranial direct current stimulation in children with dystonia: a pilot open-label trial. J Child Neurol 28:1238–1244

Zittel S, Helmich RC, Demiralay C et al (2015) Normalization of sensorimotor integration by repetitive transcranial magnetic stimulation in cervical dystonia. J Neurol 262:1883–1889

# Dementia and Cognitive Disorders

# 15

Chunyan Liu, Tao Han, and Wensi Hao

Alzheimer's disease (AD) is the most common form of dementia. With the aging of the population, dementia has become the leading cause of disability in the elderly over 65 years. Drug therapy is currently the dominant option for the treatment of AD; however, available treatment options partially contribute to delaying progression of symptoms, but can't alter the disease course. Over the years, extensive efforts have been made to conduct trials worldwide to develop appropriate drugs. Unfortunately, drugs that have a significant impact on disease progression have not yet been discovered. Based on this fact, many scholars have begun to pay attention to non-medication regimens, including lifestyle intervention, cognitive training, and neuromodulation therapy. Among them, cognitive training related to memory and art has been found to improve cognitive symptoms to a certain extent. In recent years, more and more attention has been paid to the study of neuromodulation therapy for AD.

## 1 Theoretical Basis for Neuromodulation Therapy in AD

Neuromodulation therapy includes noninvasive neuromodulation therapy and invasive neuromodulation therapy. Among them, noninvasive neuromodulation therapy is painless, noninvasive, inexpensive, and easy to operate, which is easier to be widely used. Noninvasive neuromodulation therapy techniques for AD include repetitive transcranial magnetic stimulation (rTMS), transcranial direct current stimulation (tDCS), transcranial photobiomodulation (tPBM), and low-intensity transcranial focused ultrasound stimulation (tFUS).

C. Liu (✉) · W. Hao
Department of Neurology, Xuanwu Hospital, Capital Medical University, Beijing, China

T. Han
Department of Neurology, Shandong Provincial Hospital Affiliated to Shandong First Medical University, Jinan, Shandong, China

AD is characterized by diffuse atrophy of the whole brain, especially the cerebral cortex, with extracellular beta-amyloid plaques and intracellular neurofibrillary tangles (NFT) composed of hyperphosphorylated tau proteins. In addition, recent studies have shown that synaptic transmission abnormalities exist in the brain in the early stage of AD. Some studies have even confirmed that soluble Aβ oligomers have synaptic toxicity and can cause synaptic plasticity disorders. Therefore, before Aβ plaque deposition and formation, the dysfunction of synapses may already be present, and synaptic dysfunction will lead to abnormal information integration in brain networks. High-frequency rTMS increases cerebral blood flow and glucose metabolism in both stimulated and remote brain regions and reduces intracortical inhibition at stimulation sites. Enhanced synaptic plasticity has been considered as a possible mechanism underlying high-frequency rTMS (Robinson 2000). De Taboada et al. conducted a study on photobiomodulation (PBM) in AD animal model in 2011. The 810 nm laser was emitted into the heads of amyloid-β precursor protein (AβPP)-transgenic mice (three times a week for 6 months). The number of Aβ plaques in the brain was found to be significantly reduced. The use of PBM could reduce amyloid protein, soluble Aβ protein, and cerebral inflammatory markers, but increase ATP level, mitochondrial function, and c-fos (Taboada et al. 2011). Several trials and reviews have shown that individuals with AD may have seriously damaged metabolic interaction between neurons and astrocytes due to abnormal glutamate-glutamine (Glx) circulation (Sjogren et al. 2002). The application of high-frequency rTMS in the left dorsolateral prefrontal cortex (DLPFC) area has been shown to increase Glx levels and normalize Glx cycles (Michael et al. 2002).

The cerebral oscillatory activity can cause the rapid synchronous kindling of neurons within or among brain regions. According to the different oscillation frequencies, it is divided into delta oscillation (1–4 Hz), theta oscillation (4–7 Hz), alpha oscillation (8–13 Hz), beta oscillation (13–30 Hz), and gamma oscillation (30–80 Hz). Studies have



Y. Wang (ed.), *Therapeutics of Neural Stimulation for Neurological Disorders*, https://doi.org/10.1007/978-981-99-4538-2_15

shown that gamma oscillation and theta oscillation are significantly related to advanced cognitive function. Among them, slow oscillation builds a framework for fast oscillation. Fast oscillation for rapid information exchange on this basis plays an important role in advanced cognitive processing. Transcranial alternating current stimulation (tACS) delivers a current oscillating above and below zero at a specific frequency and a given stimulus intensity; the current interacts directly with ongoing neuronal activity during cognitive or sensorimotor processes, leading to the desynchronization or synchronization of brain network oscillations, thus resulting in behavioral changes.

It is an important attempt to realize the effective treatment of AD by regulating the oscillations among the core nodes of cognitive network, promoting the recovery of local and overall network functions, improving the functional connection of synapses, and reducing the number of Aβ plaques in the brain through neuromodulation technique.

## 2 Noninvasive Neuromodulation Therapy in AD

### 2.1 Repetitive Transcranial Magnetic Stimulation

#### 2.1.1 Treatment Principle

Repetitive transcranial magnetic stimulation (rTMS) is a noninvasive neuromodulation technique for the superficial cortical neurons. It utilizes a rapidly changing magnetic field to induce cortical excitability changes in the areas covered by the coils. Low-frequency stimulation at ≤1 Hz will cause the decrease of cortical excitability, while high-frequency stimulation at ≥5 Hz will cause the increase of cortical excitability (Brighina et al. 2010).

#### 2.1.2 Clinical Application

At present, in the clinical studies on the treatment of AD with TMS, the therapeutic target is mostly focused on the dorsolateral prefrontal cortex (DLPFC), but the conclusions about the improvement of AD cognitive function by the stimulation in DLPFC are not consistent. Lozano et al. included 19 patients with mild AD and completed a 3-week randomized controlled trial comparing the efficacy of 5 Hz rTMS stimulation in the left dorsolateral prefrontal lobe with that of multi-target stimulation. It was found that there was no significant difference between the two at least 4 weeks after treatment. A meta-analysis showed that the stimulation in the bilateral or right DLPFC was helpful to the recovery of cognitive function of AD patients, but stimulation in the left DLPFC was ineffective. At the same time, other studies have proposed that the stimulation in the bilateral or right DLPFC can improve the performance of memory-related tasks in patients with mild cognitive impairment and can also improve the performance of episodic memory and processing speed tasks in AD patients. The above controversial study conclusions may be the reason why the study on DLPFC has not been recommended in the latest guidelines (Lefaucheur et al. 2020).

Some investigators have performed studies on the affected brain regions (mainly the Papez memory circuit and the posterior core nodes of the default mode network (DMN), including the precuneus and angular gyrus of the inferior parietal lobule) in the early stage of AD patients as stimulation targets. In a study by Koch et al., the left precuneus was located by navigation, 14 patients with mild AD were included and treated with high-frequency rTMS at 20 Hz for 2 weeks, and it was found that the episodic memory function was improved after treatment, while other cognitive functions were not significantly improved. Stephanie et al. conducted a comprehensive analysis for the studies on the treatment of AD with rTMS from 2016 to 2018, and the stimulation targets included multiple targets, DLPFC, and the posterior parietal and posterior temporal cortex. In most studies, a certain cognitive domain or comprehensive cognitive function was partially improved to a certain extent. Lee et al. conducted 10 Hz multi-target stimulation in AD patients in 2016, and the results showed that the ADAS-Cog score was improved compared with that before treatment. Rabey and Nguyen et al. conducted high-frequency magnetic stimulation on several brain regions, and the results showed that the ADAS-Cog and MMSE scores were all improved compared with those before treatment. In addition, there was a clear correlation between the severity of AD and the clinical efficacy of rTMS treatment. The clinical benefit of rTMS treatment in patients with mild to moderate AD is higher than that in patients with severe AD, suggesting that effective rTMS treatment as early as possible in the early stage of AD is a better choice.

We have summarized the above studies on the treatment of AD-related cognitive impairments with rTMS (see Table 15.1). In general, the current stimulation targets for AD mainly include the bilateral DLPFC, bilateral parietal cortex, and multi-target stimulation including both, in combination with Broca and Wernicke regions. So far, however, the stimulation effects of various targets have varied. Expert guidelines for rTMS treatment were published in 2014 by the European Expert Panel of the International Federation of Clinical Neurophysiology and updated in 2020 (Lefaucheur et al. 2020). Currently, rTMS treatment for AD has not obtained grade A or grade B recommendations, no consistency guidance has been formed for stimulation targets and stimulation parameters, and only multi-target rTMS treatment involving six regions, i.e., the bilateral

**Table 15.1** Summary of the clinical application of rTMS in the treatment of AD

| Year | Investigator | Study design | No. of enrolled subjects | Patient type | Stimulation parameters | | | | | Efficacy |
|---|---|---|---|---|---|---|---|---|---|---|
| | | | | | Site | Frequency | Intensity | Time course | Course of treatment | |
| 2016 | Avirame et al. | Self-control | 11 | AD | Bilateral prefrontal lobes | 10 Hz | 120%RMT | Stimulation for 2 s, rest for 20 s<br>20 min | 20 times | Improvement in mindstreams<br>Improvement in ACE |
| 2016 | Rabey et al. | Self-control | 30 | AD | Six brain regions | 10 Hz | 90–110%RMT | Stimulation for 2 s, rest for 28 s<br>Three sites, 1300 pulses | 30 times | Improvement in ADAS-Cog<br>Improvement in MMSE |
| 2017 | Zhao et al. | Parallel-controlled | 30 | AD | Bilateral temporoparietal regions | 20 Hz | – | Stimulation for 10 s, rest for 20 s<br>Three sites, 10 min | 30 times | Improvement in ADAS-Cog, MMSE, and AVLT compared to prior treatment<br>No difference between true and sham stimulations |
| 2017 | Nguyen et al. | Self-control | 10 | AD | Six brain region DLPFC | 10 Hz | 100%RMT | Stimulation for 2 s, rest for 28 s<br>Four sites, 1300 pulses | 25 times | Improvement in ADAS-Cog |
| 2018 | Koch et al. | Cross-over design | 14 | AD | Precuneus | 20 Hz | 100%RMT | Stimulation for 2 s, rest for 28 s<br>1600 pulses | 10 times | Improvement in RAVLT<br>No improvement in MMSE |
| 2018 | Lozano et al. | Parallel-controlled | 19 | AD | Six brain region left DLPFC | 5 Hz | 100%RMT | Stimulation for 10 s, rest for 60 s<br>1500 pulses | 15 times | Improvement in ADAS-Cog and MMSE<br>No difference in efficacy between the two groups |
| 2018 | Padala et al. | Cross-over design | 8 | MCI | Left DLPFC | 10 Hz | 120%RMT | Stimulation for 3 s, rest for 26 s<br>3000 pulses | 10 times | Improvement in AES and MMSE |
| 2020 | Bagattini et al. | Parallel-controlled | 50 | Mild to moderate AD | Left DLPFC | 20 Hz | 100%RMT | Stimulation for 2 s, rest for 28 s<br>2000 pulses | 20 times | Improvement in trained associative memory, add-on effect |
| 2022 | Liu et al. | Parallel-controlled | 37 | Mild AD | Angular gyrus | 40 Hz | 40% maximal output intensity | Stimulation for 2 s, rest for 58 s<br>2400 pulses | 20 times | Improvement in ADAS-Cog, MMSE, and MoCA |

**Six brain regions: bilateral DLPFC, bilateral parietal association cortex, and Broca and Wernicke regions**

*AD* Alzheimer's disease; *ACE* Addenbrooke Cognitive Examination; *ADAS-Cog* Alzheimer's Disease Assessment Scale-Cognitive Subscale; *AES* Apathy Evaluation Scale; *AVLT* Auditory Verbal Learning Test; *DLPFC* dorsolateral prefrontal cortex; *MCI* mild cognitive impairment; *MMSE* Mini-Mental State Examination; *RAVLT* Rey Auditory Verbal Learning Test

DLPFC, bilateral parietal association cortex, and Broca and Wernicke regions, has obtained grade C recommendations. To find out its cause, AD is a disease with a variety of advanced cognitive dysfunction, including memory, language, spatial orientation, and executive control. Therefore, in theory, neuromodulation therapy for multiple targets may provide more help.

## 2.2 Transcranial Direct Current Stimulation

### 2.2.1 Treatment Principle

The tDCS device consists of an adjustable DC stimulator and a pair of stimulation electrodes (one anode and one cathode). The two stimulation electrodes are placed at different positions on the scalp, with a weak current of 0.5–3 mA passing through between the electrodes. A small portion of the current between the electrodes will enter the brain, modulating the activity of neurons in the area covered by the electrodes and even in distant sites. The more consistent view is believed that tDCS does not directly cause neurons to generate action potential, but affects the magnitude and conduction of action potential (Horvath et al. 2015).

### 2.2.2 Clinical Application

In recent years, tDCS has been recognized that it can play an important role in the treatment of various brain functional diseases. A number of study results showed that tDCS had a relatively definite efficacy on neuropathic headache and fibromyalgia, which has obtained grade B recommendation in the expert guidelines issued by the European Expert Panel of the International Federation of Clinical Neurophysiology in 2020 (Fregni et al. 2020). In the field of AD treatment, scholars have carried out beneficial exploration. Ferrucci et al. conducted a double-blind, cross-over clinical study involving ten patients with AD, and the stimulation target was the temporoparietal cortex (TPC). The study confirmed that single anodal tDCS stimulation could significantly improve the performance of patients' memory tasks, while cathodal tDCS would further worsen the performance of memory tasks, and sham stimulation had no effect on the performance of subjects' memory tasks. Boggio et al. also conducted a controlled study of anodal tDCS and sham stimulation in ten patients and found that a single anodal tDCS stimulation in the left DLPFC and temporal lobe could improve memory task performance, while sham stimulation had no effect. Another study involving 15 patients with AD showed that 2 mA anodal stimulation in the temporal cortex improved memory function by 8.99% compared with that before treatment, while sham stimulation caused the decrease of memory function by 2.62%. Celina et al. conducted a systematic review of 12 tDCS studies on cognitive impairment, including a total of 202 patients with AD and mild cognitive impairment (11 studies including patients with AD/a study for patients with mild cognitive impairment). The study suggested that there were positive findings in 10 out of 12 studies, indicating that tDCS is a promising technique for noninvasive neuromodulation therapy in AD. However, the used stimulation parameters and selected stimulation targets in these 12 studies varied greatly. In addition, the physiological state of patients during stimulation and the cognitive level before treatment also had a significant impact on the treatment effect. Stephanie et al. conducted another review study. This study included 11 studies on the treatment of AD with tDCS from 2016 to 2018, 5 of which were conducted in patients with mild cognitive impairment. Most of stimulation targets were the bilateral temporal lobe and prefrontal cortex. The sham stimulation was set in three studies, and the other three adopted the cross-over design of true and sham stimulations. The evaluation scales selected in the 11 studies varied greatly, including comprehensive cognitive function evaluation, verbal memory evaluation, visual memory evaluation, subjective memory evaluation, and verbal function evaluation.

In addition, we summarized non-case-reported studies on the treatment of AD-related cognitive impairment with tDCS for five or more treatment courses (see Table 15.2). Khedr, Cotelli, and Murugaraja et al. used a single anodal tDCS to stimulate the left DLPFC, and the results showed that the visual memory tasks and MMSE scores were improved compared with those before treatment. Suemoto found, however, that the effect of the stimulation of anodal tDCS in the left DLPFC was not significant. There were also reports about the application of tDCS in the left temporal and parietal lobes, with different results. Overall, tDCS has a positive effect on the treatment of cognitive impairment, but it cannot be denied that these studies have a placebo effect.

## 2.3 Transcranial Near-Infrared Stimulation (tNIRS)

### 2.3.1 Treatment Principle

tNIRS is a new noninvasive neuromodulation technique, also known as PBM technique, that stimulates biological response by using near-infrared light with the wavelength of 630–1064 nm. The near infrared light upregulates mitochondrial function by regulating the activity of cytochrome oxidase, thereby increasing local neuronal metabolism. The near-infrared light has advantages of deep penetration; can penetrate the scalp and skull into the brain; easily passes through the brain tissue; is less dispersed by the brain tissue, with easily controlled direction, range, and energy size; and is easy to operate and painless (Tsai and Hamblin 2017).

**Table 15.2** Summary of the clinical application of tDCS in the treatment of AD

| Year | Investigator | Study design | No. of enrolled subjects | Patient type | Stimulation parameters | | | | | | Efficacy |
|---|---|---|---|---|---|---|---|---|---|---|---|
| | | | | | Active electrode site | Reference electrode site | Active electrode polarity | Current intensity | Stimulus duration | Treatment course | |
| 2012 | Boggio et al. | Cross-over design | 15 | AD | Bilateral temporal | Right deltoid | Anode | 2 mA | 30 min | 5 times | Improvement in visual memory task performance<br>No improvement in ADAS-Cog and MMSE |
| 2014 | Cotelli et al. | Parallel-controlled | 36 | AD | Left DLPFC | Right deltoid | Anode | 2 mA | 25 min | 10 times | Improvement in memory task performance |
| 2014 | Suemoto et al. | Parallel-controlled | 40 | AD | Left DLPFC | Right supraorbital | Anode | 2 mA | 20 min | 6 times | Invalid |
| 2016 | Bystad et al. | Parallel-controlled | 25 | AD | Left temporal | Right supraorbital | Anode | 2 mA | 30 min | 6 times | Invalid |
| 2016 | Yun et al. | Parallel-controlled | 16 | MCI | Left DLPFC | Right DLPFC | Anode | 2 mA | 30 min | 9 times | Improvement in subjective memory scale MMQ |
| 2017 | Roncero et al. | Cross-over design | 3 | AD | Left temporoparietal | Right supraorbital | Anode | 2 mA | 30 min | 10 times | Improvement in picture-naming task performance |
| 2017 | Murugaraja et al. | Self-control | 11 | MCI | Left DLPFC | Right supraorbital | Anode | 2 mA | 20 min | 5 times | Improvement in visual memory scale PMIT |
| 2020 | Manenti et al. | Parallel-controlled | 13 | MCI | Left lateral PFC | Right supraorbital | Anode | 1.5 mA | 15 min | 1 time | Improvement in free recall and recognition of memory |
| 2021 | Gangemi et al. | Parallel-controlled | 25 | Advanced AD | Left temporoparietal | Right supraorbital | Anode | 2 mA | 20 min | 10 times/80 times | Improvement in neuropsychological performance, long-term protective effect |

*AD* Alzheimer's disease; *ADAS-Cog* Alzheimer's Disease Assessment Scale-Cognitive Subscale; *DLPFC* dorsolateral prefrontal cortex; *MCI* mild cognitive impairment; *MMQ* Multifactorial Memory Questionnaire; *MMSE* Mini-Mental State Examination; *MoCA* Montreal Cognitive Assessment; *PMIT* Picture Memory Impairment Test

### 2.3.2 Clinical Application

Several animal studies have shown that the use of transcranial photobiomodulation (tPBM) in neurodegenerative disease models (such as AD and PD) of mice has positive effects, so people are interested in treating patients with AD or PD using tPBM. A small-scale preliminary study investigated the effect of the combination of tPBM and intranasal PBM on five patients with mild to moderate AD. After 12 weeks of treatment, the memory and cognitive levels were compared before and after treatment. The memory and cognitive ability were also assessed by the Mini-Mental State Examination (MMSE) and Alzheimer's Disease Assessment Scale (Cognitive Subscale). The MMSE score was 10–24 at baseline. After 12 weeks of active treatment, cognitive ability was significantly improved; sleep quality was improved; anger, anxiety, and wandering decreased; and there were no related adverse events. A recent Monte Carlo simulation study showed that transcranial multi-directional illumination using multiple LED arrays (181-point sources across the scalp) could improve the uniformity of photon flux and photon distribution in the brain without causing temperature rise. Another small-scale placebo-controlled pilot study involved 11 AD patients (MMSE between 15 and 25) treated for 6 min per day using 1072 nm near-infrared light (1100 LEDs in 15 arrays, 70 LEDs per group) for 28 consecutive days (all matched to 1060–1080 nm, pulsed at a frequency of 10 Hz with a 50% duty cycle). The results showed that the cognitive function of patients completing the treatment was improved and the clock drawing, immediate recall, practical memory, visual attention and trails A and B, and EEG amplitude and connection measurement showed an improvement trend.

Therefore, from the overall results, tNIRS has a positive effect on the treatment of cognitive impairment, but the specific study parameters still need to undergo the large-scale validation, so further study is needed on the application of tNIRS to the treatment of cognitive impairment.

## 2.4 Low-Intensity tFUS

### 2.4.1 Treatment Principle of Low-Intensity tFUS

Ultrasound is a pressure wave with a frequency above the range of human hearing (> 20 kHz). High-intensity focused ultrasound (>200 W/cm$^2$) caused significant tissue heating for therapeutic ablation, while low-intensity ultrasound (<100 W/cm$^2$) produced mechanical effects on the tissue without causing tissue heating or damage. Unlike light, magnetic fields, or currents, ultrasound can be focused on solid structures and transmitted at far distances in biological soft tissues with minimal power loss. Stimulation was performed using a short-duration (500 ms) transcranial focused ultrasound waveform with short exposure times and relatively low-intensity pulses not enough for ultrasound stimulation to produce tissue heating. The 500 ms pulsed sound pressure wave can alter the balance of local excitation and inhibition by acting on mechanically sensitive components of the brain, including the cell membrane, ion channels, and synaptic vesicles. In hippocampal slices, low-intensity ultrasound (<300 mW/cm$^2$) and low-frequency ultrasound (<0.65 MHz) can stimulate action potential and synaptic transmission. When a low duty cycle (i.e., 5%) waveform is used, inhibitory interneurons in the cortex are preferentially activated, producing a net inhibition of pyramidal neuron activity, while a higher duty cycle waveform (i.e., 50%) results in the excitation of cortical pyramidal neurons. Ultrasonic neuromodulation has the advantages of noninvasiveness, high spatial resolution, and deep access to action targets as compared with traditional neuromodulation technologies such as deep brain stimulation (DBS), TMS, and transcranial electrical stimulation (tES). Therefore, ultrasonic-based noninvasive neuromodulation is considered to be one of the most promising novel neuromodulation technologies for clinical transformation (Fini and Tyler 2017).

### 2.4.2 Clinical Application

In a small-scale preliminary study, AD patients had continuous improvement in memory after 3 months of ultrasonic brain stimulation. In another study, using low-intensity tFUS based on single ultrashort ultrasound pulses was applied to small areas of the brain. This study with a total of 35 participants was conducted at a research center in Austria and Germany, respectively. During the 2–4-week course of treatment, the investigators used a handheld transducer to apply sonic pulses to the skull of the patient. Participants in the Austrian group were given infrared cameras for the location of AD-related brain regions. The results showed significant improvement from baseline in neuropsychological scores in both groups and remained stable during follow-up. Functional magnetic resonance imaging (MRI) in the Austrian group showed the corresponding upregulation of memory network. These treatments were well tolerated, with no major adverse reactions and no new intracranial lesions on MRI. Roland Beisteiner, the lead author of this study, said that the study proved the rationality of a larger-scale clinical study using transcranial low intensity pulsed ultrasound stimulation technique for AD patients and it can be used in all diseases, in which activating neurons may promote the recovery of brain dysfunction, including PD and stroke. Unlike TMS currently under study and other noninvasive electrophysiological methods, low-intensity tFUS can be adjusted for deep brain regions, but further studies are still needed to clarify the precise action mode and the optimal action parameters of this technique.

# 3 Invasive Neuromodulation Therapy for AD

## 3.1 Deep Brain Stimulation

### 3.1.1 Treatment Principle

At present, it has been proved that DBS can regulate the dysfunctional neural circuits in many neural networks. In addition, the cellular reaction after DBS may directly affect the nutritional effects of local cerebral tissue and may delay the progress of chronic degenerative diseases. The neural basis for the use of DBS in many neurological and psychiatric disorders is the dysfunction of circuits in single or multiple neural networks. Through metabolic, functional, and electrophysiological studies, key abnormal nodes potentially for therapeutic intervention have been identified. The processing of these nodes by DBS can interrupt abnormal circuit activity or connection, thereby rescuing the dysfunctional circuit and restoring normal physiological network activity. In patients with dementia, dysfunction of key functional circuits controlling memory and cognition, such as cholinergic signaling system and Papez circuit, constitutes an attractive therapeutic target. In addition to the direct electrical effect of DBS, the changes of cell nutrition also occur in local or distant interconnected parts. In rodent studies, DBS in the anterior nucleus of the thalamus and entorhinal cortex were examined, and it was found that neurogenesis increased by three times in the subgranular zone of the dentate gyrus and the performance of tasks related to spatial memory was also improved. After behavioral assessment, these cells expressed the cellular oncogene fos (c-fos), which was inhibited by antimitotic agents, indicating that these cells were physiologically active and capable of integrating in the working neural circuit. In humans, similar hippocampal neurogenesis also occurs in adults, but it remains unknown whether this ability can be used by DBS to reverse cell degeneration of various types of dementia.

### 3.1.2 Clinical Application

#### 3.1.2.1 Meynert Basal Nucleus Stimulation

**Basic Principles and Targets**

Cholinergic signals in the hippocampus and related networks play an important role in the encoding and consolidation of new memories. The basal nucleus of Meynert (NbM) is a key cholinergic nucleus with extensive connections projected onto the cerebral cortex and medial temporal lobe. In AD patients, selective cell loss is observed in NbM but not in the adjacent subcortical structures, supporting cholinergic deficiency as a cause of cognitive decline and the role of anticholinesterase inhibitors as medication. Stimulation in NbM has been shown to enhance visual memory and overall cortical acetylcholine signaling. Defining the anatomical structure of NbM is a complicated task. In a histological study, Mesulam and his colleagues defined NbM as the caudal cholinergic cell population (Ch4) in the basal forebrain, which consists of five horizontal subareas, with sizes of 13–14 mm anteroposterior and 16–18 mm medial and lateral, respectively. In practice, due to the large amount of overlap and variation of each cell population, it is impossible to reliably target the specific subarea of NbM. The current strategy is that, on the coronal plane 1–2 mm in front of the mammillary body, the starting point of the target is located below the external medullary lamina, which separates the globus pallidus from the putamen, and the anterior commissure adjacent to the ventral putamen is used as an upper lateral marker. On the axial plane, the target is located on the posteromedial side of the anterior commissure and the anterolateral side of the optic tract, approximately 2 mm from the mammillary body, in order to locate a specific subregion of NbM.

**Clinical Studies**

In 1985, Turnbull et al. published the first report of NBM-targeted therapy for AD. In the report, a patient with early AD had a unilateral electrode placed in the left NBM. No clinical efficacy was reported after 9 months of treatment, but at the 2-month follow-up, fluorodeoxyglucose PET imaging showed metabolic decline rate of the stimulated left temporal lobe and parietal lobe was half that of the contralateral hemisphere. In a phase I clinical trial involving six patients with mild to moderate AD, according to NINCDS-ADRDA criteria, cerebrospinal fluid changes involved tau and amyloid Aβ42 related to AD. All six patients underwent bilateral electrode placement in NBM Ch4i and then participated in a 2-week randomized, double-blind, sham-stimulation control period, in which they were assigned to the treatment group with the deepest contact of unipolar stimulation (2.5 V, 20 Hz, 90 μs) and then switched to sham stimulation or vice versa. All patients were in the open stimulation period within 1 year after surgery. At the end of the follow-up, the ADAS-Cog-13 score decreased in two patients, remained unchanged in three patients, and was improved by nine points in one patient. Longitudinal metabolism PET imaging was performed in four patients, and glucose metabolism in three patients increased overall by 2% to 5% within 1 year after surgery, most notably in the temporal region. No adverse events related to surgery or stimulation were reported. Kuhn et al. later reported NBM stimulation in two additional AD patients who were younger than the six patients in their early cohorts and had lower baseline ADAS-Cog scores. The two patients received open stimulation therapy and were followed for 2 years without adverse events related to surgery or stimulation. At the end of the study, one patient had a 7-point decrease in ADAS-Cog score after a 26-month follow-up, with an improvement in cognitive function, and the other patient had no improvement in cognitive function after 24 months of stimulation.

#### 3.1.2.2 Fornix Stimulation

**Basic Principles and Targets**

The cognitive effects of fornix stimulation were initially discovered incidentally in a 50-year-old obese patient with normal cognition who received bilateral DBS treatment in the hypothalamus. Stimulating any electrode during surgery can cause vivid deja vu experience. Analysis of the location of the track electrodes shows that their ventral leads are adjacent to the fornix column. Three weeks after surgery, the patients show multiple improvements in the California Verbal Learning Test and the spatial association learning test. Standardized low-resolution electromagnetic tomography analysis shows that activation of the electrodes drives electrophysiological activities in the medial temporal lobe structure and the hippocampus. Anatomically, the fornix is the main output channel of the hippocampus, containing more than 1.2 million axonal fibers projected into the Papez circuit. Injury of the fornix has long been thought to lead to anterograde amnesia, while stimulative effects are thought to in turn drive the activities of their associated memory networks, which may play a role in AD patients. Stereotactic targeting of the fornix includes placing the lead 2 mm in front of the fornix column and tangent to the fornix column via the ventricular approach (the approach point is about 2 cm outside the midline, and the depth of most of the ventral contact is near the mammillary body).

**Clinical Studies**

Laxton et al. evaluated six patients who were possibly suffering from early AD (MMSE score higher than 20) and received continuous bilateral fornix stimulation for 12 months. The stimulation parameters included unipolar setting, amplitude of 3–5 v, frequency of 130 Hz, and pulse width of 90 μs. One month after surgery, PET imaging of all patients showed that the intake of fluorodeoxyglucose in the temporal parietal lobe increased, and the postoperative follow-up lasted for 1 year. One year after surgery, the cognitive decline rate of five patients was lower than the clinical course before surgery, and no adverse events occurred during the follow-up. In 2016, Lozano et al. reported the follow-up results of a phase II randomized, multicenter, double-blind, controlled trial (ADvance) evaluating 42 patients with mild AD. Study participants were men and women aged 45–85 years, diagnosed with mild dementia according to NINCDS-ADRDA criteria, with a Clinical Dementia Rating Scale Sum of Boxes (CDR-SB) of 0.5 or 1 and an Alzheimer Disease Assessment Scale-Cognitive Subscale 13 (ADAS-Cog-13) of 12–24. Patients were randomly given monopolar electrical stimulation with pulse width of 3.0–3.5 v, 130 Hz, and 90 μs for 12 months. There were no significant differences in changes in ADAS-Cog-13 and CDR-SB scores between the treatment and sham stimulation groups at the end of the study. In PET imaging, the group "On" showed statistically significant increases in metabolism in several brain regions after randomization for 6 months and did not persist at 12 months. Interestingly, in multiple comparative analyses, different clinical results were related to age. Younger patients (age <65 years old) receiving stimulation had the largest decrease in ADAS-Cog-13 and CDR-SB scale scores in all groups, with the most obvious improvement in the cognitive function. Patients over 65 years old had the least decrease in ADAS-Cog-13 and CDR-SB scale scores at 12 months. The safety results of ADvance showed that 5 of 42 patients (11.9%) developed serious adverse reactions. Of these five patients, two had an infection at the site of the internal pulse generator requiring device implantation, one patient returned to undergo surgery with the electrode repositioned based on a suboptimal position on radiograph, one patient developed bilateral chronic subdural hematoma requiring to be removed, and one patient developed severe headache and nausea requiring additional two nights of hospitalization.

DBS in the treatment of dementia is still in the initial stage. Early studies have shown that treatment response of individual patients can delay disease progress, but the effects of the whole population have not been confirmed. In order to promote future study, we need to further improve patient selection, combine it with other indicators of disease severity, and repeatedly explore different stimulation parameters.

## 3.2 Invasive Vagus Nerve Stimulation

### 3.2.1 Treatment Principle

Vagus nerve stimulation (VNS) regulates brain network activity by stimulating the vagus nerve (cranial nerve X), including direct invasive stimulation and indirect transcutaneous noninvasive stimulation. Invasive VNS (iVNS) system consists of subclavicular implanted pulse generator and lead around the left vagus nerve in the carotid sheath. The pulse generator delivers programmable electrical stimulation to the vagus nerve.

The possible mechanism for cognitive improvement by iVNS is based on the neuroanatomical basis. Studies showed atrophied locus coeruleus (LC) and decreased concentration of norepinephrine (NE) in the temporal cortex of AD patients, while NE can inhibit the inflammatory activation of microglia and act as an endogenous anti-inflammatory agent. In animal models, VNS has been shown to play a role in activating the locus coeruleus and increasing norepinephrine output to the basolateral amygdala and hippocampus. It has also been recently confirmed that VNS can activate the raphe nucleus, recruit serotonin and noradrenergic pathways through the VNS system, and bring additional benefits to the established AD therapy by promoting complementary cognitive and behavioral pathways. While iVNS can increase NE concentration and reduce inflammation, these mechanisms may involve specific cognitive functions of AD, but further large-scale trials are needed to study this hypothesis (Cimpianu et al. 2017).

### 3.2.2 Clinical Application

The relationship between iVNS and AD was investigated in two studies. In Sweden, investigators recruited ten AD patients undergoing pulse generator implantation, which delivered programmable signals to the left cranial nerve X. Initial parameter settings are as follows: frequency 20 Hz, pulse width 500 μs, and current 0.25 mA. Stimulation continued for 30 s, followed by a 5-min pause. After 3 months of treatment, seven of ten patients had improved ADAS-Cog scores (median improvement of 3.0 points), and nine of ten patients had improved MMSE scores (median improvement of 1.5 points). After 6 months of treatment, seven patients had improved ADAS-Cog scores (median improvement of 2.5 points), and seven patients had improved MMSE (median improvement of 2.5 points). The result suggested that VNS was well tolerated with slight side effects. This result suggested that VNS had a positive effect on the cognitive function of patients with AD, but further study is still needed (Sjogren et al. 2002). Based on the findings of this study result, the same study team recruited 7 additional AD patients and followed up these 17 patients for at least 1 year. They found that ADAS-Cog and MMSE improved or did not decrease from baseline in 7 of 17 patients (41.2%) and 12 of 17 patients (70.6%). The two small-scale trials showed that after 3 months and 1 year of treatment, invasive VNS was well tolerated and the specific cognitive functions in MMSE and ADAS-Cog were improved, respectively.

### 3.2.3 Conclusion

An increasing number of studies have shown that neurostimulation techniques as new treatment options for AD have shown some positive therapeutic effects. However, due to lack of large-scale randomized controls in many of these studies, reliable comparisons among these studies remain incomplete. However, stimulation-related improvement related to memory and specific cognitive function improvement are promising. In addition, the stimulation in multiple brain regions or the stimulation combined with other treatments (such as cognitive training) seems to produce more positive effects. Therefore, although the field of brain stimulation is not yet mature, it has a broad development prospect and becomes a new hope for the future treatment of AD and improvement of cognitive function.

## References

Brighina F, Palermo A, Daniele O et al (2010) High-frequency transcranial magnetic stimulation on motor cortex of patients affected by migraine with aura: a way to restore normal cortical excitability? Cephalalgia 30(1):46–52

Cimpianu CL, Strube W, Falkai P et al (2017) Vagus nerve stimulation in psychiatry: a systematic review of the available evidence. J Neural Transm 124:145–158

De Taboada L, Yu J, El-Amouri S et al (2011) Transcranial laser therapy attenuates amyloid-beta peptide neuropathology in amyloid-beta protein precursor transgenic mice. J Alzheimers Dis 23:521–535

Felipe F, El-Hagrassy Mirret M, Kevin P-B et al (2020) Evidence-based guidelines and secondary meta-analysis for the use of transcranial direct current stimulation (tDCS) in neurological and psychiatric disorders. Int J Neuropsychopharmacol 24(4):256–313

Fini M, Tyler WJ (2017) Transcranial focused ultrasound: a new tool for non-invasive neuromodulation. Int Rev Psychiatry 29(2):168–177

Horvath JC, Forte JD, Carter O (2015) Evidence that transcranial direct current stimulation (tDCS) generates little-to-no reliable neurophysiologic effect beyond MEP amplitude modulation in healthy human subjects: a systematic review. Neuropsychologia 66:213–236

Lefaucheur JP, Aleman A, Baeken C et al (2020) Evidence-based guidelines on the therapeutic use of repetitive transcranial magnetic stimulation (rTMS): an update (2014-2018). Clin Neurophysiol 131(2):474–528

Michael N, Gösling M, Reutemann M et al (2002) Metabolic changes after repetitive transcranial magnetic stimulation (rTMS) of the left prefrontal cortex: a sham-controlled proton magnetic resonance spectroscopy (1H MRS) study of healthy brain. Eur J Neurosci 17(11):2462–2468

Robinson SR (2000) Neuronal expression of glutamine synthetase in Alzheimer's disease indicates a profound impairment of metabolic interactions with astrocytes. Neurochem Int 36(4–5):471–482

Sjogren MJ, Hellstrom PT, Jonsson MA et al (2002) Cognition-enhancing effect of vagus nerve stimulation in patients with Alzheimer's disease: a pilot study. J Clin Psychiatry 63:972–980

Tsai SR, Hamblin MR (2017) Biological effects and medical applications of infrared radiation. J Photochem Photobiol B 170:197–207

# Attention Deficit

# 16

Qing Xue

Attention deficit hyperactivity disorder (ADHD) is a neurodevelopmental disorder characterized by persistent attention deficit, hyperactivity, and impulsivity. The disorder tends to develop before the age of 12. The resulting dysfunction interferes with the quality of social, academic, or workplace functioning, which cannot be explained by other mental disorders or personality disorders. The pathophysiology of ADHD is not well-established. Since ADHD is associated with dopamine and norepinephrine abnormalities, typical treatment strategies include drug therapy (methylphenidate, atomoxetine) combined with cognitive behavioral therapy (CBT), parental training, and other treatments. Predominant drug therapies may lead to adverse reactions such as digestive problems and growth restriction and do not produce a response in some cases. As a result, physical therapy is considered as an alternative option.

## 1 Treatment Principle

Some electrophysiological and fMRI researches show the most consistently found dysfunctional regions, including the right inferior frontal cortex (IFC) followed by the right dorsolateral prefrontal cortex (DLPFC), anterior cingulate cortex (ACC), right inferior parietal lobe, or basal ganglia, could potentially be used as targets for neurotherapeutics. High-frequency stimulation applied to the motor cortex (MC) or the DLPFC promotes the release of endogenous dopamine into the caudate nucleus, as a complement to dopamine deficiency in the basal ganglia in patients with ADHD (Rubia et al. 2021).

Q. Xue (✉)
Department of Neurology, Xuanwu Hospital, Capital Medical University, Beijing, China

## 2 Treatment Regimen

High-frequency magnetic or electrical stimulation is mostly targeted to the right prefrontal cortex. The studies on low-frequency stimulation are rare.

## 3 Clinical Application

About 10–30% of patients with ADHD fail to benefit from conventional medication and behavioral therapy or have some side effects. As a result, a noninvasive neurostimulation technique becomes a possible treatment option. Previous studies have confirmed that high-frequency repetitive transcranial magnetic stimulation (rTMS) applied to MC or DLPFC could promote the release of endogenous dopamine into the caudate nucleus. This serves as a complement to dopamine deficiency in the basal ganglia in patients with ADHD.

### 3.1 Repetitive Transcranial Magnetic Stimulation

In most rTMS treatment regimens, the right DLPFC is selected as the target for stimulation, typically at a high frequency. However, no treatment options are recommended in the guidelines. In the existing studies, stimulation was applied to the frontal cortex as the target at a varying frequency.

#### 3.1.1 High-Frequency Therapy

Bloch and his colleagues conducted a double-blind crossover study. The study included 13 patients with ADHD. Stimulation was applied to the right DLPFC (5 cm anterior to the optimal site of a motor twitch) as target, at 20 Hz and 100% motor threshold (MT), in 42 sequences, including an additional sham stimulation group; the duration of treatment was 1



Y. Wang (ed.), *Therapeutics of Neural Stimulation for Neurological Disorders*, https://doi.org/10.1007/978-981-99-4538-2_16

week. Half of the subjects received the true stimulation first and then the sham stimulation, and the other half of the subjects received the opposite stimulation. The symptoms of attention deficit were improved in the true stimulation group but not in the sham stimulation group (Bloch et al. 2010).

Weaver et al. conducted a randomized controlled study in nine adolescents. The right DLPFC was used as the target for true stimulation, at the frequency of 10 Hz, 100% MT, 50 sequences per day, 5 days per week for 2 weeks. The clinical global impression scale and ADHD-IV scale were significantly improved in the true stimulation group (Weaver et al. 2012).

Ustohal et al. reported a 36-year-old male with childhood-onset ADHD, who did not respond to atomoxetine and had concomitant depressive disorder. When the treatment was given in the left DLPFC with 10 Hz frequency and 120% MT intensity, the patient showed improvement of attention symptoms; however, it was also improved after sham stimulation (Ustohal et al. 2012).

Shahar et al. conducted high-frequency stimulation in the right prefrontal lobe of 20 adults with ADHD, with the measurements of Conner's Adult ADHD Rating Scale and stop-signal reaction-time Test, and found that the attention measures were improved (Shahar et al. 2016).

Paz et al. conducted a double-blind placebo-controlled study in 22 adults with ADHD. The magnetic stimulation coil was H-shaped. The frequency was 18 Hz, the intensity was 120% of the motor threshold, and the target was the bilateral prefrontal cortex. Each sequence was 2 s stimulation plus a 20 s interval, 55 sequences per day, 5 days per week, a total of 4 weeks. There was improvement in both the true and sham stimulation groups (Paz et al. 2018).

Cao et al. conducted a study in children with ADHD, who were randomized into three groups: atomoxetine group, rTMS group (10 Hz, right DLPFC), and atomoxetine + rTMS group. The results showed that rTMS and atomoxetine had the same improvement effect on attention deficit, hyperactivity, and oppositional defiance (Cao et al. 2018).

In conclusion, high-frequency stimulation on DLPFC could improve some clinical symptoms in ADHD.

#### 3.1.2 Low-Frequency Therapy

Compared with high-frequency magnetic stimulation, the studies on low-frequency stimulation are rare. Niederhofer described a 42-year-old female with ADHD (hyperactivity-dominated) who did not respond to methylphenidate. Methylphenidate was discontinued 2 months before the trial. The SMA was stimulated at 1 Hz with 1200 pulses per day for 5 days. The patient's hyperactive symptoms were improved, which lasted for at least 4 weeks. But the symptoms of attention deficit were not improved (Niederhofer 2008). Niederhofer then also applied magnetic stimulation with the same frequency and pulse number to a similar patient, but the right MC was selected and treated for 21 days. This patient's hyperactive symptom was improved, but the attention deficit was not obviously improved (Niederhofer 2011). Gomez and colleagues stimulated the left DLPFC with a low frequency of 1 Hz, a 90% MT, and 1500 pulses per day for 5 days in ten boys aged 7–12 years who did not respond to conventional treatment. Evaluations by both parents and teachers showed improvement in hyperactive/impulsiveness symptoms at home and attention deficit symptoms at school (Gómez et al. 2014). In a pilot study in Wang's group, ten patients of ADHD received 0.5 Hz stimulations on the right parietal region and improved better than other patients receiving sham stimulations.

### 3.2 Transcranial Direct Current Stimulation

DLPFC is used as the focal point for tDCS, with the current being between 1 and 2 mA. No guidelines are recommended.

#### 3.2.1 Placement of Anode in the Right Frontal Lobe

Cosmo et al. conducted a randomized controlled treatment study in adult ADHD.

In the true stimulation group, the anode was placed in the right DLPFC, and the cathode was placed in the left DLPFC, with the current of 1 mA; there was no current in the sham stimulation group. The improvements were similar between the two groups (Cosmo et al. 2015).

Soltaninejad et al. conducted a single-blind, crossover, controlled study on ADHD students in 20 universities. Subjects were divided into three groups: anodal stimulation, anode in the left DLPFC, cathode in the right supraorbital, and current 1.5 mA; cathodal stimulation, cathode in the left DLPFC, anode in the right supraorbital, and current 1.5 mA; and sham stimulation, no current. The cathodal stimulation group showed certain improvement in the Go/NoGo tasks (Soltaninejad et al. 2019).

Breitling et al. described a study in 21 male adolescents with ADHD. There were three groups: group 1, anode in the right gyrus frontalis inferior and cathode in the left mastoid, 1 mA of current for 20 min; group 2, cathode in the right gyrus frontalis inferior and anode in the left mastoid; and group 3, sham stimulation group, no current. Interference control was improved in the anodal stimulation group, while the performance was impaired in the cathodal and sham stimulation groups (Breitling et al. 2016).

#### 3.2.2 Placement of Anode in the Left DLPFC

Bandeira and colleagues conducted a study in nine ADHD subjects. The cathode was placed on the right supraorbital region, the anode was placed on the left DLPFC, and the current increased from 1 to 2 mA. Then the subjects would play a game called "Super Lynx." After 5 days of treatment, the selective attention of patients was improved (Bandeira et al. 2016).

Soff and colleagues conducted a double-blind, randomized, crossover, controlled study in 15 adolescents. The anode was placed in the left DLPFC in the true stimulation group, there was no current in the sham stimulation group, and the treatment lasted for 5 days. After a 2-week washout period, the two groups received the alternative treatment. The true stimulation group showed obvious improvement in attention and impulsiveness symptoms (Soff et al. 2017).

Sotnikova and colleagues performed similar trials using the above pattern. The reaction time was significantly improved in the true stimulation group. The fMRI scan showed enhanced activation in the left DLPFC, left PMC, left SMA, and precuneus and enhanced connection between DLPFC and working memory network (Sotnikova et al. 2017).

Nejati et al. conducted a double-blind, randomized, controlled study in 25 children. There were two groups: true stimulation group, anode in the left DLPFC, cathode in the right DLPFC, and current 1 mA, and sham stimulation group, no current. After a 72-h washout period, each group received the other mode of treatment. True stimulation had decreased reaction time in the Stroop task (Nejati et al. 2020).

Allenby et al. conducted a double-blind, crossover, controlled study in 37 adults with ADHD. There were two groups: true stimulation group, anode in the left DLPFC, cathode in the right supraorbital region, and current 2 mA, and sham stimulation group, no current. The two groups were rotated after a 2-week washout period. Finally, the subjects were required to perform continuous performance test (CPT) and stop-signal reaction-time (SSRT) task. The true stimulation group showed less error rate (Allenby et al. 2018).

### 3.2.3 Placement of Anode in the Bilateral DLPFC

Jacoby and Lavidor conducted a double-blind, randomized, controlled study in 20 adults with ADHD. There were two groups: true stimulation group, anode in the bilateral DLPFC, cathode in 1 cm below the occipital eminence, and current 1.8 mA, and sham stimulation group, no current. The two groups were rotated after a 1-week washout period. CPT failed to confirm the clear effect of tDCS due to learning and repetition effects (Jacoby and Lavidor 2018).

In summary, methods of rTMS and tDCS for ADHD are varying and no guidelines are recommended.

## References

Allenby C, Falcone M, Bernardo L et al (2018) Transcranial direct current brain stimulation decreases impulsivity in ADHD. Brain Stimul 11(5):974–981

Bandeira ID, Guimarães RS, Jagersbacher JG et al (2016) Transcranial direct current stimulation in children and adolescents with attention-deficit/hyperactivity disorder (ADHD): a pilot study. J Child Neurol 31(7):918–924

Bloch Y, Harel EV, Aviram S et al (2010) Positive effects of repetitive transcranial magnetic stimulation on attention in ADHD Subjects: a randomized controlled pilot study. World J Biol Psychiatry 11(5):755–758

Breitling C, Zaehle T, Dannhauer M et al (2016) Improving interference control in ADHD patients with transcranial direct current stimulation (tDCS). Front Cell Neurosci 10:72

Cao P, Xing J, Cao Y et al (2018) Clinical effects of repetitive transcranial magnetic stimulation combined with atomoxetine in the treatment of attention-deficit hyperactivity disorder. Neuropsychiatr Dis Treat 14:3231–3240

Cosmo C, Baptista AF, de Araújo AN et al (2015) A randomized, double-blind, sham-controlled trial of transcranial direct current stimulation in attention-deficit/hyperactivity disorder. PLoS One 10(8):e013537

Gómez L, Vidal B, Morales L et al (2014) Low frequency repetitive transcranial magnetic stimulation in children with attention deficit/hyperactivity disorder. Preliminary results. Brain Stimul 7(5):760–762

Jacoby N, Lavidor M (2018) Null tDCS effects in a sustained attention task: the modulating role of learning. Front Psychol 9:476

Nejati V, Salehinejad MA, Nitsche MA et al (2020) Transcranial direct current stimulation improves executive dysfunctions in ADHD: implications for inhibitory control, interference control, working memory, and cognitive flexibility. J Atten Disord 24(13):1928–1943

Niederhofer H (2008) Effectiveness of the repetitive transcranial magnetic stimulation (rTMS) of 1 Hz for attention-deficit hyperactivity disorder (ADHD). Psychiatr Danub 20(1):91–92

Niederhofer H (2011) Additional biological therapies for attention-deficit hyperactivity disorder: repetitive transcranial magnetic stimulation of 1 Hz helps to reduce methylphenidate. Clin Pract 2(1):e8

Paz Y, Friedwald K, Levkovitz Y et al (2018) Randomised sham-controlled study of high-frequency bilateral deep transcranial magnetic stimulation (dTMS) to treat adult attention hyperactive disorder (ADHD): negative results. World J Biol Psychiatry 19(7):561–566

Rubia K, Westwood S, Aggensteiner PM et al (2021) Neurotherapeutics for attention deficit/hyperactivity disorder (ADHD): a review. Cell 10(8):2156

Shahar N, Teodorescu AR, Karmon-Presser A et al (2016) Memory for action rules and reaction time variability in attention-deficit/hyperactivity disorder. Biol Psychiatry Cogn Neurosci Neuroimaging 1(2):132–140

Soff C, Sotnikova A, Christiansen H et al (2017) Transcranial direct current stimulation improves clinical symptoms in adolescents with attention deficit hyperactivity disorder. J Neural Transm (Vienna) 124(1):133–144

Soltaninejad Z, Nejati V, Ekhtiari H (2019) Effect of anodal and cathodal transcranial direct current stimulation on DLPFC on modulation of inhibitory control in ADHD. J Atten Disord 23(4):325–332

Sotnikova A, Soff C, Tagliazucchi E et al (2017) Transcranial direct current stimulation modulates neuronal networks in attention deficit hyperactivity disorder. Brain Topogr 30(5):656–672

Ustohal L, Prikryl R, Kucerova HP et al (2012) Emotional side effects after high-frequency rTMS of the right dorsolateral prefrontal cortex in an adult patient with ADHD and comorbid depression. Psychiatr Danub 24(1):102–103

Weaver L, Rostain AL, Mace W et al (2012) Transcranial magnetic stimulation (TMS) in the treatment of attention-deficit/hyperactivity disorder in adolescents and young adults: a pilot study. J ECT 28(2):98–103

# Autism Spectrum Disorder

# 17

Yingxue Yang

Autism spectrum disorder (ASD), first reported and named by the American psychiatrist Kanner in 1943, is a group of neurodevelopmental disorders that seriously affect children's health and have high etiological heterogeneity. It is characterized by persistent impairment of interactive social communication and social interaction, repetitive stereotyped behaviors, and narrow interests.

With the further understanding of ASD and the improvement of diagnostic criteria, the prevalence of ASD is increasing year by year. Survey data from the Centers for Disease Control and Prevention showed that the prevalence of ASD among children in the United States was 1 in 110 in 2006, rising to 1 in 68 in 2014 and reaching an alarming 1 in 54 in 2020. Currently, the prevalence of ASD is estimated to be 1% worldwide. Statistics showed that the prevalence of ASD was 11.8 per 10,000 in mainland China and the combined prevalence in mainland China, Hong Kong, and Taiwan was 26.6 per 10,000. The first multicenter epidemiological study of ASD in China in 2020 showed that the prevalence of school-age children with ASD in China was 1/142, and about 2/3 of them had at least one neuropsychiatric comorbidity, and nearly half of them were first diagnosed, 90% of which were from ordinary primary schools (Zhou et al. 2020). It is estimated that the number of children with ASD aged 6–12 years in China is about 700,000–1,000,000. The study also found that 50% of these children with ASD at school age were diagnosed with ASD for the first time and nearly 90% of newly diagnosed children with ASD were enrolled in ordinary primary schools. Diagnosis through the assessment of ASD comorbidities revealed that 2/3 of children with ASD had at least one or more neuropsychiatric comorbidities, with the top three most common being attention deficit hyperactivity disorder, specific phobias, and agoraphobia.

The increasing prevalence of ASD is mainly due to the rapid increase in the level of knowledge, alertness, and diagnosis of these disorders, which is also influenced by factors such as changes in diagnostic criteria. However, due to the complex etiology of ASD, which is the result of the combination of genetic variation and environmental factors, many factors may lead to the increased prevalence of ASD, which needs to be further studied.

For the neuropsychological mechanism of ASD, many hypotheses have been put forward by scholars, mainly including theory of mind, executive dysfunction, weak central coherence, and mirror neuron dysfunction. These hypotheses can explain the core symptoms of ASD to some extent, but lack a unified theory that can explain all clinical manifestations of ASD.

With the development of neuroimaging technology, we come to realize that ASD is not the result of structural and functional abnormalities in a single brain region. More and more functional imaging studies have shown that there is an abnormality in functional connectivity between brain regions of ASD. Just et al. proposed the hypothesis of frontal-posterior brain underconnectivity in ASD (Just et al. 2012). The hypothesis suggests that functional connectivity in ASD brain regions is excessive and disorganized, while long-distance (especially between frontal and posterior regions) intercortical or functional networks are asynchronous, weakly responsive, and poorly informative. These abnormalities can lead to the abnormal integration and regulation of information in the brain and reduced modular processing of information in the whole brain, resulting in psychological and behavioral abnormalities and various clinical symptoms of ASD. This hypothesis is consistent with the findings of abnormal white matter connectivity in ASD, and is supported by some studies, and needs to be further verified and substantiated by more evidence.

Because the etiology and pathogenesis of ASD are not clear, there is no specific treatment at present. Treatment is mostly based on parent-mediated interventions and behavioral interventions, supplemented by symptomatic treatment with medication. However, the process of behavioral

Y. Yang (✉)
Department of Neurology, Xuanwu Hospital, Capital Medical University, Beijing, China



Y. Wang (ed.), *Therapeutics of Neural Stimulation for Neurological Disorders*, https://doi.org/10.1007/978-981-99-4538-2_17

interventions is long, requiring parents to invest a lot of time and money, and the efficacy is variable. Meanwhile, the functional impairment of children with ASD prevents most of them from integrating into society, which seriously affects the physical and mental health and quality of life of children and brings a huge burden to the family and society. It is becoming a huge health problem facing the world today. For more than half a century, multidisciplinary efforts have been devoted to explore the etiology and pathogenesis of ASD, which has not been explained so far. Therefore, there is still a lack of biomarkers for diagnosis and of specific clinical treatment methods. It is extremely important to develop new routes, new targets, or new methods for ASD treatment.

## 1 Repetitive Transcranial Magnetic Stimulation

Repetitive transcranial magnetic stimulation (rTMS) is a new neuroelectrophysiological technique developed on the basis of TMS. Different stimulation frequencies can change the excitability of neurons. Low-frequency rTMS (below 1 Hz) can inhibit the local neuronal activity, thereby reducing cortical excitability and metabolism. High-frequency rTMS (5–20 Hz) can facilitate local neurons, which leads to increased cortical excitability and enhanced metabolism. Noninvasive modulation of cortical excitability can be achieved by rTMS.

Dozens of studies have been reported on the application of rTMS for the treatment of ASD, mostly with different parameters of low-frequency stimulation of the prefrontal cortex attempted. Several studies applied low-frequency TMS to stimulate the left, right, and bilateral dorsolateral prefrontal cortex (DLPFC) (6 stimulations on the left, 6 stimulations on the right, and 6 stimulations bilaterally) at 0.5 Hz or 1 Hz, respectively, with a stimulation intensity of 90% motor threshold or resting motor threshold. After treatment, the scores of the Aberrant Behavior Checklist and Repetitive Behavior Scale were improved for patients; especially the patients' irritable behavior, hyperactivity, and social withdrawal behavior were partially improved (see review by Masuda et al. 2019). Stimulation of the left DLPFC with low-frequency TMS (1 Hz) by Gomez et al. improved patients' social interaction and behavior scale scores (Gomez et al. 2017). Panerai et al. stimulated the left premotor cortex with high-frequency TMS (8 Hz) in ASD patients with severe mental retardation and found that the hand-eye coordination was improved and no rTMS-related adverse effects occurred (Panerai et al. 2014). After Abujadi et al. applied intermittent theta-burst stimulation (iTBS) (50 Hz, 100% motor threshold, 8 s of train duration, 2 s of inter-train intervals, 30 trains, and 15 (5 days per week) stimulation sessions in total) to the right DLPFC, the executive function and the scores of the Repetitive Behavior Scale and Obsessive-Compulsive Inventory were improved, and no relevant side effects occurred (Abujadi et al. 2017). A randomized, double-blind, and controlled trial applied deep rTMS stimulation of the bilateral medial prefrontal cortex (mPFC) (each session involved thirty 10 s trains at 5 Hz, with 20 s inter-train interval), and patients' self-rated social relating impairment and socially related anxiety symptoms were improved after treatment (Enticott et al. 2014).

Yang et al. first attempted high-frequency TMS in the parietal lobe of patients with ASD (Yang et al. 2019). A preliminary study of high-frequency TMS (20 Hz) was performed on the left parietal lobe of ASD patients with severe mental retardation. It was found that after high-frequency rTMS stimulation of the left parietal cortex increased the excitability of the parietal cortex, the patients' verbal communication and social communication ability were improved, and the cognitive function and imitative abilities of the patients were also improved to varying degrees. Three patients had transient irritability during or after treatment that resolved spontaneously after 3–5 days.

Unfortunately, at present, rTMS treatment for ASD is mostly for small clinical observations, and there are few RCT studies. Large-scale multicenter controlled clinical trials are needed to further explore and validate the stimulation targets and treatment parameters of rTMS.

## 2 Transcranial Direct Current Stimulation

Transcranial direct current stimulation (tDCS) is a noninvasive technique that utilizes constant, low-intensity direct current (0.5–3 mA) to modulate neuronal activity in the cerebral cortex. Positive stimulation usually enhances the excitability of neurons at the site of stimulation, while negative stimulation decreases the excitability of neurons at the stimulation site. Pseudostimulation is often used as a control stimulus. Repeated stimulation also produces long-term potentiation and long-term depression.

The application of tDCS in the treatment of ASD has been preliminarily explored in recent years. Schneider et al. treated ASD patients with tDCS at a current intensity of 2 mA for 30 min, with anode placed on the left DLPFC, and language acquisition was improved in ASD patients after treatment (Schneider and Hopp 2011). Amatachaya et al. conducted a randomized, double-blind, controlled trial enrolling a total of 20 patients with mild to moderate ASD aged 5–8 years with a current intensity of 1 mA and an anode placed on the left DLPFC for 20 min per day for a total of 5 consecutive days. After treatment, ASD symptom severity according to the Childhood Autism Rating Scale and Autism Treatment Evaluation Checklist global score was statistically significantly improved (Amatachaya et al. 2014). Gomez et al. treated ASD patients with 20 consecutive 20 min cathodal (1 mA) stimulation sessions over the left DLPFC. Autistic symptoms improved after treatment, and the efficacy lasted for 6 months (Gomez et al. 2017). Hadoush et al. performed bilateral hemispheric stimulation on patients with ASD at a current intensity of 1 mA for 20 min duration, with anodal placement in the bilateral prefrontal and motor cortex for a total of ten sessions (five times per week for 2 weeks). After treatment, the social ability and physical conditions of the patients were significantly improved compared with the control group, and no relevant side effects were observed (Hadoush et al. 2020).

At present, the treatment of ASD with tDCS is still in the exploratory stage. In the future, large controlled trials and long-term follow-up observations are needed to validate the efficacy of tDCS in patients with ASD.

## 3 Vagus Nerve Stimulation

There are a few case series and case reports looking at the effects of implantable vagus nerve stimulation for ASD in patients with epilepsy along with the treatment of epilepsy. It was found that implantable vagus nerve stimulation has the potential to improve behavior in ASD patients, and it is possible that this improvement is not dependent on improvements in epileptic seizure frequency and mood of the patient (van Hoorn et al. 2019). Vagus nerve stimulation for ASD is still in the exploratory stage and needs to be validated in further clinical trials.

## References

Abujadi C, Croarkin PE, Bellini BB et al (2017) Intermittent theta-burst transcranial magnetic stimulation for autism spectrum disorder: an open-label pilot study. Rev Bras Psiquiatr 40:309–311

Amatachaya A, Auvichayapat N, Patjanasoontorn N et al (2014) Effect of anodal transcranial direct current stimulation on autism: a randomized double-blind crossover trial. Behav Neurol 2014:173073

Enticott PG, Fitzgibbon BM, Kennedy HA et al (2014) A double-blind, randomized trial of deep repetitive transcranial magnetic stimulation (rTMS) for autism spectrum disorder. Brain Stimul 7(2):206–211

Gomez L, Vidal B, Maragoto C et al (2017) Non-invasive brain stimulation for children with autism spectrum disorders: a short-term outcome study. Behav Sci 7(63):1–12

Hadoush H, Nazzal M, Almasri NA et al (2020) Therapeutic effects of bilateral anodal transcranial direct current stimulation on prefrontal and motor cortical areas in children with autism spectrum disorders: a pilot study. Autism Res 13(5):828–836

Just MA, Keller TA, Malave VL et al (2012) Autism as a neural systems disorder: a theory of frontal-posterior underconnectivity. Neurosci Biobehav Rev 36(4):1292–1313

Masuda F, Nakajima S, Miyazaki T et al (2019) Clinical effectiveness of repetitive transcranial magnetic stimulation treatment in children and adolescents with neurodevelopmental disorders: a systematic review. Autism 23(7):1614–1629

Panerai S, Tasca D, Lanuzza B et al (2014) Effects of repetitive transcranial magnetic stimulation in performing eye-hand integration tasks: four preliminary studies with children showing low-functioning autism. Autism 18:638–650

Schneider HD, Hopp JP (2011) The use of the Bilingual Aphasia Test for assessment and transcranial direct current stimulation to modulate language acquisition in minimally verbal children with autism. Clin Linguist Phon 25:640–654

van Hoorn A, Carpenter T, Oak K et al (2019) Neuromodulation of autism spectrum disorders using vagal nerve stimulation. J Clin Neurosci 63:8–12

Yang Y, Wang H, Xue Q et al (2019) High-frequency repetitive transcranial magnetic stimulation applied to the parietal cortex for low-functioning children with autism spectrum disorder: a case series. Front Psychiatry 10:293

Zhou H, Xu X, Yan W et al (2020) Prevalence of autism spectrum disorder in China: a nationwide multi-center population-based study among children aged 6 to 12 years. Neurosci Bull 36(9):961–971

# Depression and Bipolar Affective Disorder

# 18

Zhong Zheng, Ke Zou, Jiaxi Huang, Tianhao Bao, and Jiaqi Han

The unipolar depression and bipolar disorder depressive episode have different pathogenesis, which are difficult to distinguish in the early clinical stage. Patients with bipolar disorder begin with depressive episode more often than with manic episodes, during which hypomania lasts for a short time with mild symptoms, making them difficult to distinguish from normal emotion changes. Patients with depressive episode at a young age and a family history of bipolar disorder, who once had a history of transient mania and are currently manifested by mixed depression, retardative depression, agitated depression, and psychotic depression, are mostly considered as having bipolar depression. At present, the main complaints are somatic symptoms, with initial insomnia, anorexia, and unipolar depression. Despite the accumulation of numerous data on brain imaging, neuroendocrine, neurotransmitter, neurophysiological, and other biological markers of unipolar and bipolar affective disorders, the differences in the physiopathologic mechanism between the two disorders remain unclear in essence so far. In terms of treatment, both are primarily treated through medication, psychological approach, behavioral approach, and neurostimulation techniques. Drug treatments include selective norepinephrine reuptake inhibitors (SNRIs), selective serotonin uptake inhibitors (SSRIs), and tricyclic antidepressants. Neurostimulation techniques include electroconvulsive (ECT), repetitive transcranial magnetic stimulation (rTMS), transcranial direct current stimulation (tDCS), and transcranial alternating current stimulation (tACS).

Although the pathogenesis of depressive disorder and bipolar affective disorder are thought to be related to neurotransmitter disorders such as 5-HT, norepinephrine, and dopamine, the hypothesis of neurotransmitter disorder cannot fully explain all problems. The clinical application and research of rTMS and tDCS in recent years have gradually expanded the horizon to the relationship between asymmetric changes in hemispheric function of the human brain and the pathogenesis of psychiatric disorders, and there may be a correlation between neurotransmitter changes and asymmetric changes in hemispheric function. Since the first report on rTMS for depression in 1995, the Food and Drug Administration (FDA) approved rTMS for treating treatment-resistant depression in 2008 and for treating migraine in 2013. Deep TMS (deepTMS, dTMS) with H7 stimulation coils has been approved by FDA for treating obsessive-compulsive disorder in 2018 and for treating treatment-resistant depression in 2021. Therefore, the exploration of the mechanisms of neurostimulation and antidepressant treatment is of great clinical and academic-theoretical significance to rediscover the pathogenesis of psychiatric disorders and expand the treatment pathways of psychiatric disorders.

## 1 Functional Asymmetry of Left and Right Hemispheres and Human Evolution

The structure and function of the left and right hemispheres in human are characterized by asymmetry or lateralization. The left hemisphere is known as the dominant hemisphere as

Z. Zheng (✉)
Neurobiological Laboratory, West China Hospital, Sichuan University, Chengdu, Sichuan, China

Center for Neurological Function Test and Neuromodulation, West China Xiamen Hospital, Sichuan University, Xiamen, China

K. Zou
Neurobiological Laboratory, West China Hospital, Sichuan University, Chengdu, Sichuan, China

J. Huang
Center for Neurological Function Test and Neuromodulation, West China Xiamen Hospital, Sichuan University, Xiamen, China

T. Bao
Department of Geriatrics, Yunnan Province Mental Health Center, Kunming Medical University, Kunming, Yunnan, China
e-mail: doctor@kmmu.edu.cn

J. Han
Department of Neurology, Xuanwu Hospital, Capital Medical University, Beijing, China



Y. Wang (ed.), *Therapeutics of Neural Stimulation for Neurological Disorders*, https://doi.org/10.1007/978-981-99-4538-2_18

it is dominant over the right hemisphere. The lateralization of human brain function is the result of the interaction between genes and the environment, which is featured with cross-race, cross-culture, and cross-territory. Symmetry in organisms is the default condition, whereas asymmetric evolution is a genetic mutation that occurs when an organism adapts to nature and corresponds to the evolution of brain asymmetry with the formation of right-handedness, language lateralization, and visual-auditory attentional bias in human.

## 1.1 Biological Evolution and Lateralization of Human Function

### 1.1.1 Genetic-Environmental Factors, Handedness, and Language Lateralization

Hemispheric dominance of handedness and language functions are two of the most important lateralization phenomena in human. The proportion of populations with right-handedness and left-hemispheric language dominance is higher than 90% worldwide, and the left dominance proportion is as high as 70% in left-handed populations, which is consistent across cultural settings. Handedness and language lateralization precede the development of the individual's ability to receive environmental information and the cognitive abilities. Previous studies have found that most individuals show preferential development and movement of the right arm from the seventh week of embryonic development and that most fetal brain structures appear to develop preferentially on the left side. Before the age of 3 years, the development of the right hemisphere slows down, while the left hemisphere continues. Corresponding to the development of language skills in young children, the lateralization of body language, such as gestures, also begins in infancy. It is evident that there is a high association between handedness and language lateralization. The control of hand and language is a highly complex and elaborate motor output activity of individuals acting on the environment. This activity is dominated by the same hemisphere, with a clear ecological advantage, which helps individuals to conserve processing capacity and thus make quick and accurate responses. Many organisms are bilaterally symmetric. This kind of symmetry evolution may be dated back to 40 million to 55 million years ago, and the rightward shift started in prehistoric chimpanzees and primitive humans 1.5 million to 200,000 years ago, which distinguished them from other vertebrates. Multiple studies have shown that asymmetric evolution occurred in the planum temporale and Broca's area in chimpanzees and hominids, suggesting that such structural changes in brain asymmetry associated with language function may have been present long before humans. According to the gene-environment interaction (G × E) theory, the breakthrough of brain symmetry and the formation of handedness are influenced by both environmental factors and genetic mutations. Currently, there are three theories describing the evolution of brain asymmetry based on genes: the right shift theory, the directional-chance allele model, and the recessive X-linkage model. Instead of simply assuming that alleles determine left- and right-handedness, the progressive development and complementarity of the three theories suggest that alleles first determine the lateralization of language function, making the left hemisphere the dominant hemisphere for language. The left hemisphere dominance favors right-handedness, while the lateralization of language and handedness has some chance of being to the left or to the right, as they are influenced by cultural, environmental, and gender factors. Recent studies have explored the influence of genetic and environmental factors on handedness, coming up with different conclusions: A study conducted by Medland et al. (2009) investigated a large sample (54,270 cases) of twins from Australia and the Netherlands by behavioral ratings or self-reports. After conducting maximum likelihood analyses of various genetic and environmental factors (sex, age, and weight at birth) affecting handedness, they found that genetic factors were the most important in the formation of handedness. However, the study conducted by Vuoksimaa et al. (2009) investigated a large sample of twins (30,161 cases) from Finland using the same method and found that environmental factors were more prominent. It can be seen that both genes and environment are important factors influencing human behaviors such as handedness.

## 1.2 Human Visual-Auditory Attentional Bias

Since the first discovery of monaural dominance in humans by the dichotic listening test (DLT) in 1954, the visual asymmetry found in normal and split-brain human has been repeatedly reported and confirmed, suggesting that the human brain is characterized by not only the lateralization of handedness and language but also attentional bias associated with hemispheric dominance.

### 1.2.1 Visuospatial Attentional Bias

Visuospatial attentional bias refers to the fact that normal people do not focus on the center of the stimulus when observing it, but have left visual field dominance. What corresponds to the attentional bias is pseudoneglect, which is a slight right-sided neglect that normal individuals will show steadily in spatial attention tests. Most scholars believe that pseudoneglect is based on the same neural mechanism as spatial hemineglect or unilateral neglect, but it is a different manifestation of the same mechanism in patients and normal people. The line bisection task (LBT) found that normal people were prone to erroneously assume that the midpoint of a line segment is further to the left when determining its location, whereas patients with left-sided visuospatial neglect caused by brain injury significantly underestimated the length of the left segment of the line (Manly et al. 2005). In the visual search task, subjects always proceeded from left to right when searching. Studies on eye-movement paradigm also found that in a scene with one distinct object in each of the left and right regions, it was more likely that the left-side object would first receive attention than the right-side object. Right-sided dominance occurred when texts, words, and letters were used as stimulus. This kind of asymmetric phenomenon may be due to the fact that the left hemisphere dominates language and the corresponding right visual field is more sensitive to verbal information. As the right hemisphere is critical to picture recognition and comprehension, the corresponding left visual field is dominant in image processing. Damage to the right hemisphere is more likely to result in the severe and persistent neglect than that to the left hemisphere. It is inferred that the right hemisphere is probably dominant in the processing of visual attention.

### 1.2.2 Auditory Attentional Bias

Brodbent (1954) performed DLT with language materials targeting left and right ears and found better perception when the stimulation was presented to the right ear. This is known as right ear advantage (REA). Normal subjects with right-handedness all showed REA in listening to numbers, words, letters, and nonsense consonant syllables, but showed left ear advantage (LEA) in listening to melodies, tunes, and familiar environmental sounds. The conduction model proposed by Kimura (1967) emphasizes the inhibition of ipsilateral projection fibers by contralateral projection fibers in the auditory pathway and suggests that the left hemisphere (right ear) is better at processing verbal information to obtain REA, while the right hemisphere (left ear) is better at processing non-verbal information, including space, emotion, music, and intonation to obtain LEA. The attentional bias model proposed by Kinsbourne (1987) suggests that the asymmetry observed in DLT is a spatial attentional bias to one side when one hemisphere is activated Wang (1992). The REA existing during the dichotic listening to verbal information is generated by the activation of the left hemisphere by this information, resulting in an attentional bias toward the right side of space. The opposite phenomenon would take place during the dichotic listening to non-verbal information. Auditory attentional bias is also related to handedness. Performed DLTs to subjects with LEA and REA, respectively, among which two tests were verbal and one was non-verbal. The results revealed that both verbal tests showed REA. The two tests in the REA group were both statistically significant, while only one test in the LEA group was statistically significant. The non-verbal test showed LEA, which was of statistical significance to the REA group. The absolute difference in scores of both ears in the LEA group was smaller than that in the REA group Wang (1992). According to the right shift theory, left-handed individuals may lack right shift genes, and the formation of their dominant hemisphere of language and handedness may be mainly determined by environmental factors. Gandour et al. (2004) believed that both the left and right brains were involved in the processing of pitch information in speech: the right brain (left ear) was responsible for analyzing acoustic phonological features, while the left brain (right ear) comprehended the linguistic significance. Since morphology of Chinese words is very different from that of Latin, the two may correspond to different dominant hemispheric models in the brain mechanisms of cognition. The processing of Chinese characters, as ideographic writing, is associated with both the left and right hemispheres. Chinese is the oldest and the only language in the world that has never died out and also the only language in the world where semantics are identified through tones. Each Chinese character may have its own meaning or may form different semantics with other characters. The identification of tones is especially important. Wang et al. (2001) studied the cognitive processing of tones in Americans with no knowledge of Chinese tones or no experience in learning Chinese tones with dichotic listening techniques and found that 48% of tone recognition errors in Americans came from the left ear (right hemisphere). Based on brain imaging techniques, Zhang et al. (2008) prepared experimental materials with French, a language unfamiliar to Chinese language learners, and a continuous stream of six syllables to test Chinese language learners. The results showed that acoustic phonological processing of pitch (intonation) information in speech by Han Chinese is primarily from the right brain (left ear).

## 1.3 Lateralization of Emotional Information Processing in Human Brain Hemispheres

Emotion is the human brain's reflection of objective reality and also the attitude of human toward the consistency or inconsistency of objective reality with personal needs. Emotions reflect a subject-object relationship, i.e., a relationship between the needs of the person as a subject and objective things. Emotions are not directly and mechanically determined by objective reality, but by the connection of various events in the external world with various human needs. Such connection is established in the cognitive process, related to the previous knowledge and experience accumulated by the individual and influenced by specific environment, time, cultural background, etc. Therefore, emotions are innately acquired and can also be acquired later in life. The cognitive evaluation process, hence, becomes a key factor and prerequisite during the generation of emotions, while the brain's processing of stimulating information is also influenced by emotional features. The emotional process comprises the activities of three subsystems: emotional experience, emotional physiology, and expressive behavior. The "basic emotion theory" suggests that emotions are composed of seven basic emotions, namely, happiness, sadness, expectation, surprise, disgust, fear, and anger. Meng et al. (1985) found that happiness, surprise, sadness, anger, disgust, and fear were the basic emotions of 1-year-old infants by photographing their expressions. It suggested that five emotions other than disgust were already present in infancy. The "dimensional theory" suggests that emotions have three main dimensions: valence/pleasantness (from pleasant to unpleasant), arousal (from excitability to calmness), and dominance (from dominant to dominated). Valence refers to the degree to which an individual experiences negative or positive emotion; arousal refers to the degree of activation of the body energy associated with the emotional state. For example, happiness is of high arousal and high positivity, while sadness is of low arousal and high negativity. Besides, valence is positively correlated with dominance. When emotions are transformed from positive → partially positive → neutral → partially negative → extremely negative, the arousal, pleasure, and dominance of emotions are also transformed smoothly, thus constituting different human emotions. The basic idea of both theories is to study complex emotions with a simple model of emotions, while the dimensional theory is at the base of the emotional system and closely related to an organism's survival-related tendency to seek advantages and avoid harms. The International Affective Picture System (IAPS) created by Lang et al. based on the dimensional theory well represents this three-dimensional space, making it easy to compare various emotion-induced responses. Among them, sexy and blissful pictures are more pleasurable, with higher arousal and dominance, while fearful and disgusting pictures are less pleasurable and less dominant but more arousing; neutral pictures are less arousing and less likely to attract attention although they are relatively highly pleasurable and dominant. The human brain tends to select information consistent with personal emotional state in the process of information input-encoding-memory updating-storage-output, i.e., the "mood-congruent effect." The enhancement in human memory to negative emotional material is considered as the "mood-congruent effect" related to negative mood, which is associated with memories of harm from natural threats and interpersonal interactions.

Emotions have a tendency to be bipolar such as affirmative and denying, positive and negative, proactive and reactive, convergent and distant, and tense and sluggish. Experimental emotions are often induced in emotion studies. This study paradigm of observing emotional effects is called affective priming. EEG activity changes in real time at the time of affective priming. A study on event-related potentials (ERP) conducted by Recio et al. (2014) found that high-intensity emotional pictures induced greater ERP deflections than low-intensity emotional ERP deflections; in the affective priming test of six emotions, there were significant changes in ERP deflections evoked by pictures expressing anger, fear, and surprise, while no significant changes were seen in ERP deflections evoked by those expressing pleasure, disgust, and sadness. It indicated that the emotional intensity of anger, fear, and surprise was greater than that of happiness, disgust, and sadness. The effect of positive and negative emotions on the body must carry energy changes of positive and negative nature. The adaptation degree of the physical and mental functions of human body, an open system in constant exchange of matter, energy, and information with the outside world, can be expressed by the thermodynamics term of energy (e.g., EEG activity). The exposure of human beings to various things in the social environment also causes corresponding emotional changes in the human body, and the emotions in turn act on the organism and thus produce a tangible willpower that increases emotional effects.

## 2 Lateralization of Cerebral Hemispheric Functions and Psychiatric Disorders

### 2.1 Handedness and Psychiatric Disorders

The lateralization of human hemispheric functions has developed in response to natural selective evolution, which on the one hand accelerates the evolution of human brain and on the other hand, as a genetic mutation, increases human susceptibility to various types of psychiatric disorders. It has been shown that patients with epilepsy, schizophrenia, and cerebrovascular disease have a higher rate of left-handedness than normal people. According to systematic evaluations, the incidence of non-right-handedness is significantly higher in preterm infant than in full-term children and may be a potential factor of neurological and psychiatric developmental disorders. In recent years, brain imaging and neurophysiological techniques have revealed variations of lateralization of hemispheric functions, which may be one of the important links in the onset of affective disorders and schizophrenia, with reduced left hemispheric (LH) activity during depressive episodes; hyperactive left hemisphere activation in manic, impulsive, violent, dissocial personality disorders and paranoid schizophrenia; and hyperactive right hemisphere (RH) activation in generalized anxiety and post-traumatic stress disorders (Zheng et al. 2014, 2015a, b, c, d) (Fig. 18.1) (the hypothesis proposed in this figure does not represent all

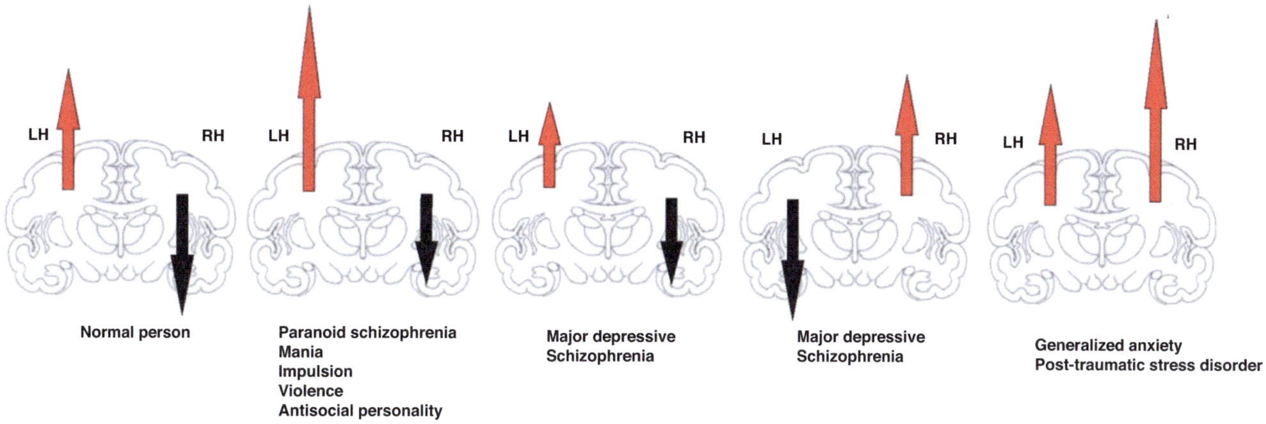


**Fig. 18.1** Hypothesis of the relationship between different types of psychiatric disorders and left/right hemispheric activation. (Modified based on Zheng et al. 2014, 2015a, b, c, d)

types of brain function lateralization in psychiatric disorders). Altered activation balance implies cognitive and adaptive impairment. Therefore, the left and right hemisphere lateralization must be maintained at a certain level, as excessive or inadequate lateralization is associated with the onset of psychiatric disorders.

## 2.2 Lateralization of Cerebral Hemispheric Functions and Psychiatric Disorders

### 2.2.1 Study of Hemispheric Functional Lateralization in Depression

1. *Studies on transcranial magnetic stimulation (TMS)-motor evoked potential (MEP)* The technique of transcranial magnetic stimulation-motor evoked potential (TMS-MEP) initially proposes the concept of motor cortex excitability (MCE). The commonly used MCE of the left and right hemispheres receiving TMS can be measured by resting motor threshold (RMT), cortical silent period (CSP), intracortical inhibition (ICI), and intracortical facilitation (ICF). The relationship between motor cortices can be determined by the ipsilateral silent period (ISP) and interhemispheric inhibition (IHI), and MCE can be determined by the contradiction and balance between cortical inhibition and facilitation (Zheng 2014). Maeda et al. (2000) found by a single-pulse TMS (sp-TMS) study that patients with depression had reduced excitability in the left hemisphere and increased excitability in the right hemisphere. Meanwhile, through the evaluation of intracortical inhibition (ICI) and intracortical facilitation (ICF) by paired-pulse TMS (pp-TMS) with different interstimulus intervals (ISI), it was found that the facilitation in the right hemisphere of depression was enhanced, while the facilitation in the left hemisphere was weakened.

The sp-TMS study conducted by Cui et al. (2008) showed that the left RMT in the depression group was significantly higher than that in the healthy control group ($P < 0.05$), i.e., the cortical excitability in depression patients had no asymmetry, which was in contrast to normal people, manifested by lower excitability in the dominant hemisphere than in the non-dominant hemisphere.

Lefaucheur et al. (2008) used RMT and CSP of the first dorsal interosseous muscle with sp-TMS and tested ICI and ICF with pp-TMS. Studies showed that there was no significant difference between the left and right RMT in the healthy control group, but the left RMT was lower than that in depression group and the left RMT was higher than the right RMT in depression group. There was no significant difference between the left and right CSP in healthy control group, but the left CSP was longer than that in depression group and the left CSP was shorter than the right CSP. There was no significant difference between the left and right ICI in the healthy control group. The left and right ICI in the depression group was lower than that in the control group, and the left ICI was significantly lower than the right ICI in the depression group. The right ICF was significantly lower than the left ICF in the depression group. This degree of abnormal left and right hemisphere functional asymmetry was associated with depression symptoms.

Huang et al. (2022) found that patients with major depression (MDD) demonstrated significantly increased intracortical inhibition at 1-ms and 2-ms interstimulus intervals (ICI-1ms, ICI-2) than healthy controls in the left hemisphere ($P < 0.05$), suggesting impaired cortical inhibition compared with healthy controls. There were no significant group differences in rest motor threshold (RMT), cortical silent period (CSP), and intracortical facilitation at 12-ms and 15-ms (ICF-12ms, ICF-15ms). Among MDD, significant positive correlations were

**Table 18.1** Comparison of left/right motor cortex excitability between the healthy control and depression groups

| | Depression ($n$ = 40) | | | Healthy control ($n$ = 40) | | | Healthy vs. depression | |
|---|---|---|---|---|---|---|---|---|
| | Left hemisphere | Right hemisphere | $P$ | Left hemisphere | Right hemisphere | $P$ | Left hemisphere ($P$) | Right hemisphere ($P$) |
| RMT | 63.42 ± 4.86 | 62.55 ± 5.97 | 0.241 | 62.48 ± 7.14 | 63.58 ± 8.26 | 0.502 | 0.919 | 0.380 |
| ICF-12ms (%) | 87.51 ± 52.11 | 87.13 ± 73.89 | 0.61 | 93.79 ± 51.55 | 86.24 ± 47.97 | 0.558 | 0.532 | 0.491 |
| ICF-15ms (%) | 83.22 ± 50.27 | 99.35 ± 73.72 | 0.259 | 97.00 ± 58.63 | 85.71 ± 47.76 | 0.409 | 0.281 | 0.590 |
| ICI-1ms (%) | 25.20 ± 21.28 | 29.41 ± 51.74 | 0.300 | 15.99 ± 17.44 | 13.97 ± 12.34 | 0.830 | 0.016 | 0.175 |
| ICI-2ms (%) | 37.43 ± 25.26 | 30.47 ± 27.06 | 0.264 | 25.09 ± 21.82 | 21.34 ± 16.28 | 0.408 | 0.012 | 0.227 |
| CSP (ms) | 131.12 ± 38.51 | 112.79 ± 45.03 | 0.014 | 139.47 ± 43.32 | 128.5 ± 39.82 | 0.431 | 0.549 | 0.112 |

Modified based on Huang et al. (2022)

**Table 18.2** Correlation between motor cortex excitability and HDRS-24 scores in patients with major depressive disorder

| | HDRS | |
|---|---|---|
| | $r$ | $P$ |
| RMT (LH) | −0.037 | 0.823 |
| RMT (RH) | −0.11 | 0.498 |
| ICF-12ms (LH) | 0.156 | 0.335 |
| ICF-12ms (RH) | −0.284 | 0.076 |
| ICF-15ms (LH) | 0.197 | 0.223 |
| ICF-15ms (RH) | −0.207 | 0.201 |
| ICI-1ms (LH) | 0.335 | 0.035 |
| ICI-1ms (RH) | −0.034 | 0.837 |
| ICI-2ms (LH) | 0.323 | 0.042 |
| ICI-2ms (RH) | −0.079 | 0.629 |
| CSP (LH) | −0.007 | 0.965 |
| CSP (RH) | −0.258 | 0.107 |

Modified based on Huang et al. (2022)

observed between the 24-item Hamilton Depression Rating Scale (HDRS-24) score and left ICI ($P < 0.05$). Moreover, patients showed a significant hemispheric asymmetry in CSP (Tables 18.1 and 18.2).

Based on the literatures published in PubMed over the last 20 years, Cantone et al. (2017) studied and summarized 25 items of changes in cortical excitability and synaptic plasticity in depression.

(a) *Overall cortical hypoexcitability and abnormal left-right hemispheric asymmetry* RMT studies showed reduced excitability in the left hemisphere or/and increased excitability in the right hemisphere in depression patients. In a study on hemispheric asymmetry in the prefrontal and motor cortices, 1 Hz inhibitory rTMS was applied to the left M1 area, which did not increase right hemispheric excitability while decreasing ipsilateral cortical excitability, suggesting that patients had lost normal interhemispheric modulation. The intervention effect of applying high-frequency rTMS to the left prefrontal region on clinical symptoms was associated with a significant increase in ipsilateral hemispheric cortical excitability. Application of high-frequency rTMS to the left dorsolateral prefrontal cortex (DLPFC) for 1 month produced a reduction in excitability differences between the left and right hemispheres with improvement in clinical symptoms.

(b) *Abnormal changes in cortical inhibition* In contrast to RMT excitability indicators, inhibition indicators such as CSP and SICI showed contradictory results, such as prolonged CSP in patients with unipolar or bipolar depression, while other studies showed reduced CSP and SICI. Shortening of CSP has also been reported in patients previously diagnosed with depression and in patients with treatment-resistant depression. Only one group of studies showed a significant decrease in SICI. The results above are consistent with the "cortical disinhibition" hypothesis that GABAergic systemic transmission is involved in the pathophysiology of depression. Animal experiments and neurochemical and neuroimaging studies also suggest abnormalities in this mechanism.

(c) *Neuroplasticity features* The reduced functioning of ICF after exercise is characterized by an initial exercise-induced facilitation followed by an early return to baseline MEP amplitude, suggesting that the motor cortex is less excitable in depression patients and may be a neurophysiological signal of the imbalance between excitatory and inhibitory mechanisms involved in plasticity. Player et al. (2013) stimulated depression patients receiving drug treatment with paired associative stimulation (PAS) and found significantly lower neuroplasticity in patients than in healthy controls. Kuhn et al. (2016) further confirmed this hypothesis and found that LTP-like plasticity was restored after symptom remission, suggesting that LTP-like plasticity was state-dependent. This pathway involved postsynaptic depolarization of cells due to the activation of receptors of glutamate, NMDA, and α-amino-3-hydroxy-5-methyl-4-isoxazole propionic acid (AMPA).

2. *Studies on transcranial magnetic stimulation-electroencephalography (TMS-EEG)* Sun et al. (2016) assessed the relation between cortical inhibition and suicidal ideation in patients with treatment-resistant depression and the neurological mechanisms of magnetic seizure therapy (MST) and found that a significant reduction in suicidal ideation after MST intervention was associated with negative potential N100 induced by magnetic stimulation and ICI baseline value, suggesting that cortical inhibition may be indicative of suicidal ideation and that these two values are highly specific and sensitive in predicting suicidal ideation (accuracy of 89%, sensitivity of 90%, specificity of 89%). Chung et al. (2016) and Pellicciari et al. (2017) evaluated the effect of ten times of pulse-train stimulations to the bilateral dorsolateral prefrontal cortex (left iTBS and right cTBS) on oscillatory frequency in the frontal lobe. Results showed that the oscillation of dorsolateral prefrontal cortices of depression patients was extremely asymmetric at the baseline. TBS induced a rearrangement of the oscillation on the bilateral dorsolateral prefrontal cortex, corresponding to the improvement in clinical symptoms. Furthermore, Noda et al. (2018) investigated the effect of single PAS to the dorsolateral prefrontal cortex on EEG oscillations in the prefrontal cortex of patients and found that PAS increased the δ, theta, and γ band power spectrum in healthy controls, while no corresponding effect was obtained in the depression group, indicating abnormal changes in prefrontal plasticity in depression patients.
3. *Studies on magnetic resonance imaging and transcranial magnetic stimulation-functional magnetic resonance* Patients of severe depression suffer from asymmetric loss of left and right brain functions, with decreased fractional anisotropy in the left hippocampus and reduced brain white matter volume in the left limbic system. In patients with treatment-resistant depression, the metabolic signal ratio of *N*-acetyl aspartic acid (NAA) to creatine (Cr) (NAA/Cr) is enhanced in the right hippocampus. The disappearance of bilateral hippocampal NAA/Cr asymmetry and the reduction of local coherence in the left brain region (Zou et al. 2007; Li et al. 2010) suggest a lateralized reduction or inversion of neuronal activity and connectivity in the corresponding brain regions in depression patients. Vink et al. (2018) studied the effect of TMS to the left DLPFC of ten healthy subjects with transcranial magnetic stimulation-functional magnetic resonance (TMS-fMRI). The stimulation of the DLPFC triggered activities in many associated brain regions. The subgenual anterior cingulate cortex (sgACC) of the human brain was affected in four of the nine subjects. The normalization of sgACC was associated with the remission of depressive symptoms in the patients. The results suggest that in some individuals, TMS to the left DLPFC can cause activation propagating to sgACC, which is related to the connectivity of brain structures of different individuals. This kind of individual tendency can be used as a predictor of the patient's response to rTMS treatment. Thus, the mode of transmission of TMS-induced activity and the ability of rTMS to modulate specific brain regions may depend on the current state of individual brain networks. Using resting-state (RS)-based fMRI connectivity features, different depression biotypes have been identified in adults with clusters of functional connectivity features associated with different clinical phenotypes. Studies on RS-fMRI connectivity also elucidate the mechanisms of neuromodulation intervention in depression patients. For example, rTMS changes the brain network connectivity, which may be related to the antidepressant effects of rTMS. Effective neurostimulation treatment should target specific brain regions or networks, which may depend on the patient's biotype. Therefore, biomarkers must be used to guide treatment. TMS indirectly measures $GABA_B$ receptor tone through stimulation paradigms such as LICI and CSP, which triggers unique motor responses quantified by MEP. Balzekas et al. (2018) made the first attempt to perform RS-fMRI and TMS measurements simultaneously in adolescents with depression. Preliminary data confirmed the feasibility of the method of biomarkers in depression adolescents, which allowed direct comparison of limited measurements of cortical inhibition with connectivity features in the same subject. In the study, LICI (ISI = 200 ms) showed a greatest correlation with functional connectivity and hence was most likely to serve as a cross-modal biomarker of cortex inhibitory network dysfunction in depression patients.
4. *Hemodynamic study* Zhang et al. (2014) observed functional near-infrared spectroscopy (fNIRS) changes using verbal fluency tasks (VFTs) and found that the mean curves of oxygenate hemoglobin (Oxy-Hb) and deoxygenate hemoglobin (Deoxy-Hb) in the bilateral dorsolateral prefrontal regions of depression patients were lower and flatter than those in the control group. However, studies on single photon emission computed tomography (SPECT) showed significantly lower activation in the left prefrontal cortex in depression patients (Zhao et al. 2000). According to a recent meta-analysis (Ho et al. 2020), almost every fNIRS study found that Oxy-Hb concentration was negatively correlated with the overall severity of depressive symptoms. Pu et al. (2012, 2015) found that the hemodynamic changes in orbitofrontal, prefrontal, and dorsolateral frontal cortex regions were negatively correlated with the degree of suicidal ideation in depressed individuals. The hemodynamic responses in the right gyrus frontalis medius were negatively correlated with the aggression and despair in suicidal motiva-

tion, while not related to the hemodynamic responses in patients in the healthy control group and in patients without suicidal motivation. A study conducted by Nishida et al. (2017) found a negative correlation between changes in mean Oxy-Hb concentration and the Pittsburgh Sleep Quality Index when depression patients were performing tasks, suggesting that the sleep disorder self-rating scale was associated with reduced left prefrontal responsiveness when patients performed VFT, thus indicating that the responsiveness of prefrontal regions was susceptible to sleep disorders. Another study found (Ho 2020) that psychomotor retardation scores on the Hamilton Depression Rating Scale (HDRS) were significantly positively associated with mean Oxy-Hb changes in the right temporal region in depression patients characterized with melancholy, while negatively related with the mean Oxy-Hb changes in the left temporal regions in depression patients with non-melancholic features.

### 2.2.2 Study on Hemispheric Functional Lateralization in Schizophrenia

1. *Studies on transcranial magnetic stimulation-motor evoked potential* Pascual-Leone et al. (2002) used RMT, ICI, and ICF to study the differences in left- and right-side MCE in schizophrenia (SCZ) treated and untreated with medication. The cMEP for ICI and ICF was the TMS of 130% MT and used as the Test-TMS; the Condition-TMS was 80% RMT stimulation. The percentage area difference between Test-MEP and cMEP was measured (△MEP%). It was found that the bilateral hemispheric RMT was 5% higher in the medication group than in the non-medication and health control groups and 10% higher in the right hemisphere than in the left hemisphere in the health control group, while the opposite in the non-medication group. ICI was significantly lower in the medication group compared to the non-medication and health control groups, suggesting that antipsychotics can modulate the patient's MCE.

   Gao et al. (2008) studied the changes and correlation of cortical excitability in first-episode SCZ patients and their first-degree relatives. The results showed that the RMT in the left side was significantly lower than in the right side for the health control group, while there was no significant difference in RMT between the left and right hemispheres in first-episode SCZ patients and their first-degree relatives without the disease, and there was no significant difference in RMT of the left and right sides among the three groups. It suggested that the disappeared asymmetry of cortical excitability in the first-episode SCZ patients and their first-degree relatives may be related to the diminished or absent synaptic transmission facilitation of the dominant hemispheric neurons.

2. *Studies on transcranial magnetic stimulation-EEG* Schizophrenia (SCZ) is characterized by positive symptoms (such as delusions and hallucinations) and negative symptoms (such as apathy, hypoexcitability, and social withdrawal). Although there are few knowledges of the pathophysiology of the disease, animal and postmortem studies suggest the presence of excitatory and inhibitory neuronal dysfunction, such as diminished GABAergic neuron function in DLPFC. Additional evidences for GABAergic inhibitory dysfunction in SCZ come from EEG literatures, which suggest that abnormal spontaneous EEG rhythms are specifically associated with $\beta$ and $\gamma$ oscillations (Uhlhaas et al. 2013). GABAergic inhibitory interneurons are considered as associated with the generation ($GABA_A$ receptors) and modulation ($GABA_B$ receptors) of $\gamma$ oscillations (Uhlhaas et al. 2013).

### 2.2.3 Post-traumatic Stress Disorder

Centonze et al. (2005) applied pp-TMS to the left motor area 1 (M1) of 14 patients with post-traumatic stress disorder (PTSD) and 11 patients with minor head trauma and compared them with 12 healthy controls to study the ICI characteristics of these patients. The results showed that PTSD patients had relatively higher Condition-MEP wave amplitude and higher percentage difference between Test-MEP and Condition-MEP wave amplitude. However, patients with minor head trauma were not significantly different from controls, indicating abnormal intracortical inhibition impairment in patients with PTSD.

Rossi et al. (2009) compared cortical excitability and the difference in left and right hemispheric excitability between 16 normal and 15 non-medicated PTSD patients with 80% MT as Condition-TMS and 120% MT as Test-TMS, analyzed the percentage difference between Test-MEP and cMEP wave amplitudes, and measured the CSP and RMT of the left and right hemispheres. The results showed that ICI was extensively missing in PTSD patients; in half of the patients, ICI was reversed between the left and right hemispheres, ICI in the right hemisphere increased, and ICI and ICF in the left and right hemispheres were correlated with HDRS and Hamilton Anxiety Rating Scale (HARS), while there was no significant difference between RMT and CSP. It suggested that reduced inhibition in the left and right hemispheres and increased facilitation in the right hemisphere in PTSD patients were closely associated with affective disorders.

### 2.2.4 Obsessive-Compulsive Disorder

Wassermann et al. (2001) used MT to study the correlation between ICI/IFC with ISI at 3, 4, 10, and 15 ms and the scores of NEO-Personality Inventory-Revised (NEO PI-R) in patients with obsessive-compulsive disorder (OCD). It was found that the overall ISI and the cMEP/Test-MEP amplitude ratio across time were significantly correlated with neuroticism scores of all study subjects on the NEO PI-R. The correlation of ICI and ICF with neuroticism trait scores in OCD patients was higher than with OCD symptoms themselves, a result that was more pronounced in male patients, suggesting that neuroticism trait scores in OCD are associated with reduced inhibition in the left motor cortex and enhanced cortical facilitation.

## 2.3 Visual-Auditory Attentional Bias and Psychiatric Disorders

### 2.3.1 Visual Attentional Bias and Psychiatric Disorders

In a line bisection task (LBT), Tian et al. (2011) marked the midpoint of line segments of different lengths and found that the points marked by patients with schizophrenia significantly shifted to the left, while no significant difference was found in normal control group, suggesting a rightward visual attention bias. Ozel-Kizil et al. (2012) used LBT to evaluate the difference in visual hemispatial neglect (LBT) between patients with schizophrenia and their normal siblings and found significant left visual field neglect in the patient group, while there was no significant difference between normal controls and their normal siblings, concluding that schizophrenia patients have abnormal left hemispatial recognition. He et al. (2010) compared and studied the differences in visual attentional bias among healthy controls, patients with generalized anxiety disorder (GAD), and patients with treatment-resistant depression (TRD) and found that errors of GAD patients showed more significant left bias than the normal control group. It suggested over-activation of the right hemisphere and/or reduced activity in the left hemisphere in GAD patients. However, this kind of bias was not observed among TRD patients.

### 2.3.2 Auditory Attentional Bias and Psychiatric Disorders

Right-handed subjects showed the right ear advantage (REA) and right-hand motor response dominance during dichotic listening and processing of verbal stimulation. This functional lateralization was achieved mainly through the following neural pathways: stimulation from the right ear reached the left hemisphere dominating speech processing through a stronger afferent pathway while suppressing weaker projections from the left ear. Correspondingly, stimulation from the left ear reached the right hemisphere not dominating speech processing through a stronger afferent pathway while suppressing weaker projections from the right ear. Therefore, the left hemisphere was significantly more efficient and faster in processing verbal stimulation from the right ear. Although verbal information received by the right hemisphere from the left ear could be transmitted through the corpus callosum to the left hemisphere for processing, the accuracy and response speed were both affected. It can be seen that the processing of verbal stimulation from the right ear by right-handed subjects is more accurate and faster than the left ear advantage (LEA) subjects. A study conducted by Romney et al. (2000) performed dichotic listening tests and dichotic viewing tasks to study the response of right-handed patients with paranoid schizophrenia and anxiety disorders to verbal and non-verbal information in comparison with normal right-handed populations. The subjects were asked to respond to visual and auditory information by pressing the "same" and "different" buttons. It was found that the left ear of paranoid schizophrenics was less accurate in response to verbal information and the time of its response to verbal and non-verbal information was prolonged. There was no significant difference between the anxiety patients and the control group. It suggested that the right hemispheric function was reduced, while the left hemispheric function became dominant in patients with paranoid schizophrenia.

## 2.4 Lateralization of Emotional Information Processing in Brain Hemispheres and Psychiatric Disorders

Nerve fibers originating in the prefrontal cortex project to the striatum (putamen and caudatum) and thalamus and then return to the prefrontal cortex. Numerous studies have shown that there may be a functional asymmetry between the left and right PFC in the processing of emotions, with the left DLPFC involved in the generation and regulation of positive emotions and the right DLPFC involved in the generation and regulation of negative emotions. The amygdala plays an important role in the generation of negative emotions as well as in aversive association learning. The prefrontal → amygdala top-down connection and the amygdala → prefrontal bottom-up connection are mutually constrained. Some studies have found that left or right hemispheric damages resulted in a pathological crying response or a pathological laughing response, respectively (Agarwal et al. 1992; Alfredo et al. 2009; Nicholas et al. 2011). Davidson et al. (2000) and Grimm et al. (2008) believed that the weakened function of the left PFC and the relatively activated function of the right prefrontal cortex and amygdala contributed to the enhancement of bottom-up connection related to negative emotion processing and low cognitive control related to top-down connectivity, resulting in a disturbed balance of positive and negative emotion processing functions, which may be one of the biological factors for the development of depression. The lateralization of $N_2$ wave and slow wave in ERP decreased in depression patients during the recognition of pictures expressing positive and negative emotions, while left lateral-

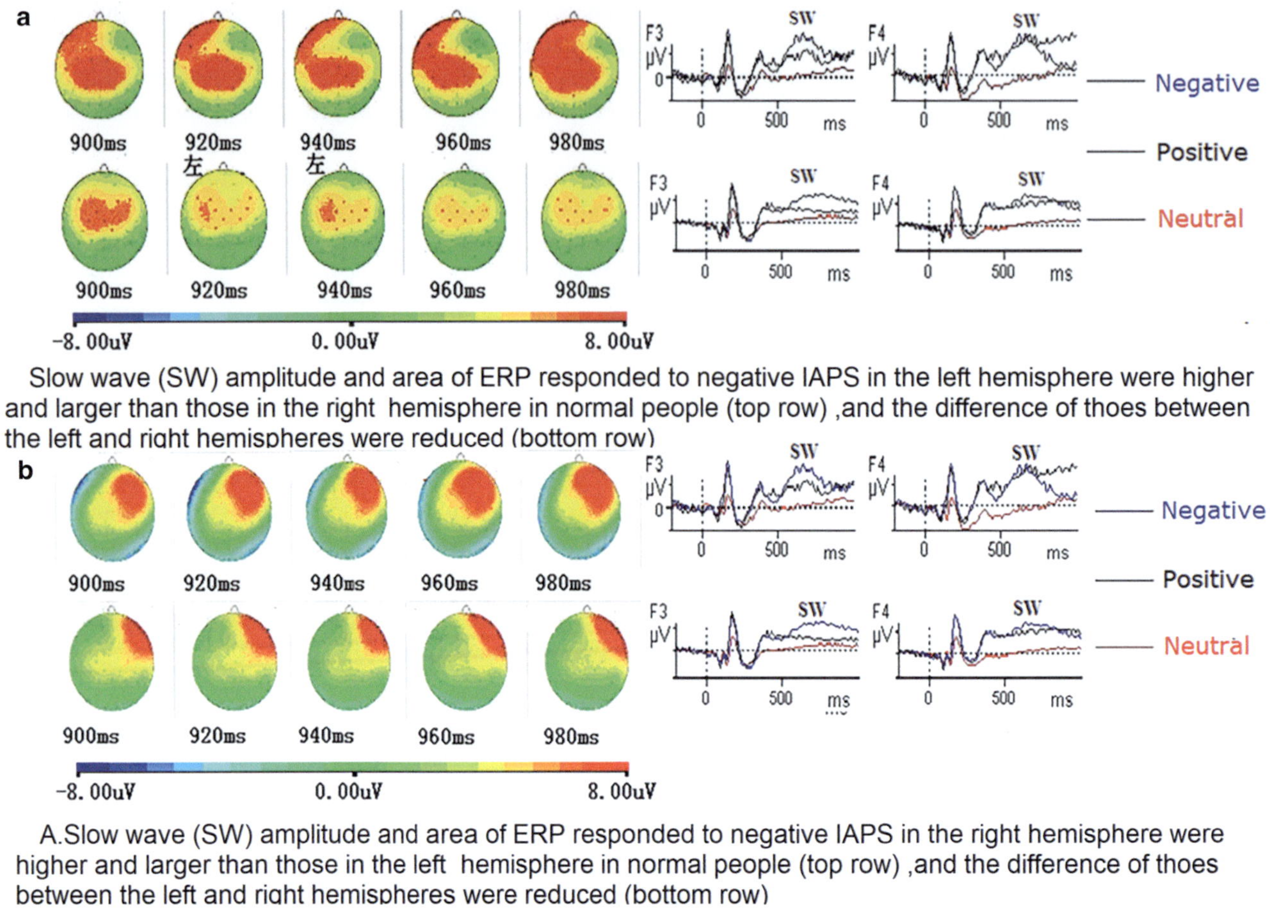


**Fig. 18.2** Topography of positive and negative IAPS-EPR slow wave (900–980 ms) lateralization in normal populations and depression patients (Zheng et al. 2014, 2015a, b, c, d)

ization showed in normal control groups during the recognition of negative pictures and right lateralization occurred during the recognition of positive pictures.

Some studies showed that pictures expressing negative emotions would induce an increase in the emotional slow wave (SW) amplitude of the left hemisphere EEG $\gamma$ band activity and ERP in normal people and positive mood pictures induce an increase in the right hemisphere EEG and ERP in normal people; positive and negative mood pictures induce asymmetrical left and right hemisphere EEG and ERP activity in depression that are opposite to or diminished from normal people (Fig. 18.2). Depression is characterized by the "cognitive triad," i.e., negative self-evaluations, negative interpretations of previous events, and negative perceptions of the future. Studies on emotional experience showed that patients tended to respond negatively to pictures expressing negative, positive, and neutral emotions, which was considered as a "mood-congruent effect" associated with depressive mood. Therefore, patients showed reduced or reversed lateralization in ERP activation when identifying negative (sadness) and positive (happiness) of International Affective Picture System (IAPS) pictures.

## 2.5 Bipolar Disorder

Studies suggest "worry" and "anger" are the main emotional pathological factors of bipolar affective disorder, among which mania is more related to "anger," and the incidence of "anger" is higher than that of "happiness" in mania, and depression is more related to "worry," and the incidence of worry is higher than that of sadness in depressive episodes, which may be the symptomatological characteristics of manic and depressive episodes in China at present (Cui et al. 2009). Corresponding studies found that the asymmetry index (AI) (difference of right hemispheric values minus left hemispheric values) of EEG $\alpha$ power spectrum in patients with type I bipolar affective disorder had different manifestations: AI was significantly biased toward the right hemisphere in patients with hypomanic episode, while there was a tendency to bias toward the left hemisphere in patients with depressive episode, i.e., the activity of right hemisphere decreased and that of left hemisphere relatively increased in hypomanic episode.

## 3 Effects of rTMS on Hemispheric Functional Lateralization in Depression

### 3.1 Effects of rTMS on Hemispheric Motor Cortex Excitability in Depression

Bajwa et al. (2008) studied the interactive effects of 1 Hz low-frequency rTMS to the left primary motor cortex on the excitability of the left and right hemispheres by comparing patients with heavy depression with healthy controls. It was found that low-frequency rTMS increased right hemispheric excitability while decreasing left hemispheric excitability in normal subjects, while it was not the case in patients with depression, suggesting that depression patients lack this kind of interhemispheric interactive response. The correlation between the degree of abnormal interhemispheric interaction modulation with Beck Depression Inventory scores suggested that emotional disorders in depression are associated with a slow information switching mechanism between the hemispheres. This absence of left and right hemisphere interaction also occurred in patients with degenerative and underdeveloped corpus callosum (Meyer et al. 1995). High-frequency rTMS can correct fractional anisotropy, cerebral glucose metabolic rates, and abnormal lateralization of event-related potentials in brain areas of depression patients. Therefore, increasing left hemispheric excitability and decreasing right hemispheric excitability are the mechanisms by which rTMS treats depression. Due to the lack of interactive inhibitory mechanism of the corpus callosum in the left and right hemispheres in patients with depression, excitatory high-frequency stimulation in the left hemisphere and inhibitory low-frequency stimulation in the right hemisphere may be more effective than unilateral stimulation.

MCE may be deemed as a result of the paradoxical action of intracortical facilitation (ICF) and intracortical inhibition (ICI) (Cantone et al. 2017). It is manifested by increasing MCE when ICF is higher than ICI and by decreasing MCE if ICI is higher than ICF. In depression patients, the excitability, facilitation, and inhibition of the left and right hemispheres may be opposite to those in normal subjects. Currently, depression is mostly treated by restoring the normal asymmetry between the hemispheres with high-frequency rTMS to the left dorsolateral frontal cortex or low-frequency stimulation to the right dorsolateral frontal cortex, the separate application of which results in similar effects (Isenberg et al. 2005). It was proved by the study on TMS in combination with PET that ICI and ICF were positively related with regional cerebral blood flow (rCBF) (Strafella and Paus 2001). Speer et al. (2000, 2009) found that the application of 20 Hz high-frequency rTMS to the left DLPFC in patients with depression and bipolar affective disorder improved depressive symptoms and increased rCBF (significantly higher on the left than on the right) in the bilateral prefrontal cortex and gyrus cinguli and rCBF in the left amygdala, bilateral insula, bilateral basal ganglia, and bilateral cerebellum. However, the application of 1 Hz rTMS to the left dorsolateral frontal cortex worsened depressive symptoms and decreased rCBF in the right prefrontal cortex, left medial temporal cortex, left basal ganglia, and left amygdala. Therefore, the method mostly applied is high-frequency stimulation to the patient's left dorsolateral frontal cortex.

### 3.2 The Relationship Between rTMS and Emotions of Depression Patients

Certain regions of the brain intrinsically linked to emotion regulation and the pathogenesis of depression include the middle and dorsolateral prefrontal cortices, the gyrus cinguli, and other structures of the limbic system (e.g., the amygdala, hippocampus, hypothalamus, limbic thalamus, and insula). The limbic system-cortex-striatum-pallidum-thalamus is an important neural circuit for emotion regulation. The frontal lobe is the superlative part that develops at the latest in the brain, which includes the primary motor area, premotor area, and prefrontal lobe. The prefrontal cortex (PFC) is divided into the DLPFC, supraorbital prefrontal cortex, and ventromedial prefrontal cortex. The frontal lobe is anatomically adjacent to the anterior cingulate cortex, where a large number of fibers are connected to the limbic system. It was found by Jorge et al. (2009) in a study on fMRI that the left top-down effective functional connections were significantly lower in depression patients than in healthy controls when the stimulation of happy faces was given, while the bottom-up connections in depression patients did not differ from healthy controls. The same trend of alteration in the left top-down connections was observed when the stimulation of sad faces was given. In female patients, the left top-down connections reduced in response to happy faces, which might be due to the neurological mechanism of extreme low self-esteem, self-blame, and anxiety in depression. Since the DLPFC is part of the PFC and a key brain area for cognition and emotion regulation circuits, it is now considered as an important brain area for the positioned treatment with rTMS. DLPFC is the area 4–5 cm horizontally anterior to the motor area of the abductor pollicis brevis (APB-brain area) for evoking MEP. This method of positioning is also known as the 5 cm rule (Fig. 18.3). Another method to position left/right DLPFC is to select F4/F3 in 10–20 EEG system. Stimulation frequencies are divided into low and high frequencies, with low frequencies being 1 Hz or less and high frequencies being 5 Hz or more. The high frequencies adopted now are usually up to 20 Hz. TMS-MEP studies on left and right MCE asymmetry in psychiatric disorders have demonstrated that patients with

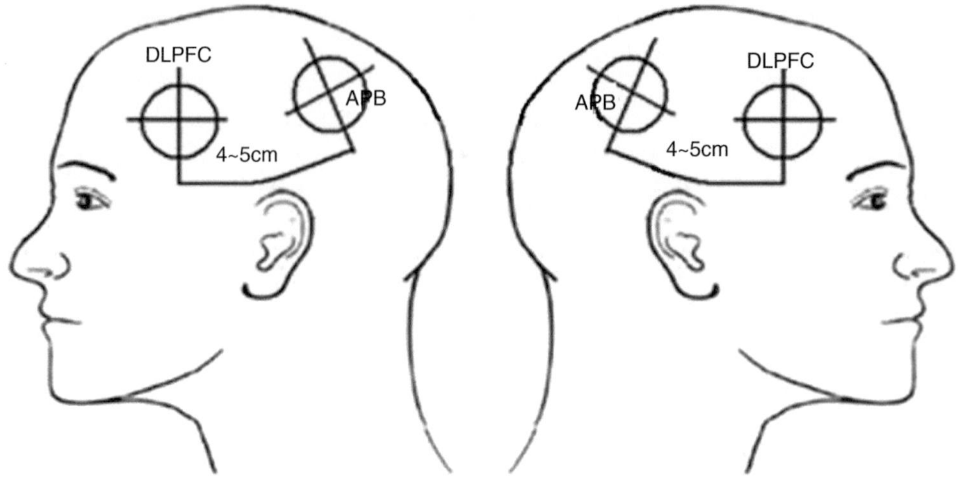


**Fig. 18.3** Schematic diagram of left and right dorsolateral frontal positioning

depression and schizophrenia mostly have a relative decrease in MCE in the dominant hemisphere and a relative increase in MCE in the non-dominant hemisphere or experience non-symmetry disappearance.

## 3.3 Action Mechanism of rTMS and Antidepressants

After summarizing previous studies on the effects of different neurotransmitters on MCE, Herwig et al. (2002) found that neurotransmitters related to the changes in MCE included dopamine, gamma-aminobutyric acid, 5-hydroxytryptamine, glutamic acid, and norepinephrine, while MCE may be independent of acetylcholine (Table 18.3). Dopamine, gamma-aminobutyric acid, and 5-hydroxytryptamine can increase ICI and decrease ICF, while glutamic acid can only increase ICF, and norepinephrine can decrease ICI and increase ICF.

A study conducted by Herwig et al. (2002) found that the selective norepinephrine reuptake inhibitor (SNRI), reboxetine, caused a decrease in RMT but not an increase in AMT in healthy subjects, while reboxetine and the presynaptic α2 receptor blocker, yohimbine, both increased ICF and decreased ICI in the left hemisphere of healthy subjects. In contrast, the SSRI, sertraline, decreased ICF (Ilic et al. 2002). Li et al. (2018) treated persistent somatoform pain disorder (PSPD) is a type of somatoform disorder with rTMS (800 pulses to the left DLPFC in iTBS mode and 800 pulses at 1 Hz low frequency to the right DLPFC) together with sertraline, the results of which showed that the efficacy of sertraline combined with rTMS was better than that of sertraline alone, as sertraline could increase RMT in the right hemisphere but could not affect RMT in the left hemisphere, while sertraline combined with rTMS could increase RMT in the right hemisphere and also decrease RMT in the left hemisphere. It suggested that sertraline acted only on the right hemisphere to decrease its excitability, but not on the left hemisphere to increase its excitability. It can be seen that some SNRIs and SSRIs may have opposite effects on left and right hemispheric MCE, i.e., SNRIs increase left hemisphere excitability, while SSRIs decrease right hemisphere excitability, which is similar to the treatment of high-frequency rTMS to the left DLPFC and low-frequency rTMS to the right DLPFC. To put it in another way, it also indicates the fact that there is abnormal asymmetry between the left and right hemispheres in patients with depressive disorder.

**Table 18.3** Effects of several different neurotransmitters on ICI and ICF (Quoted from Herwig, 2002)

| | ICI | ICF | Effect marker |
|---|---|---|---|
| Dopamine | ↑ | ↓ | Receptor agonists (pergolide and bromocriptine) only resulted in changes of ICF, while the reverse effect of antagonist haloperidol was significant (Ziemann et al. 1996a, 1997) |
| Gamma-aminobutyric acid | ↑ | ↓ | Lorazepam, ethanol, antiepileptic drugs (Ziemann et al. 1996b, c; Werhan et al. 1999) |
| 5-Hydroxytryptamine | ↑ | ↓ | ICF reduced after oral administration of 25 mg of clomipramine (Manganotti et al. 2001) and 100 mg of sertraline (Ilic et al. 2002) |
| Glutamic acid | ←→ | ↑ | Antagonists (riluzole, memantine, dextromethorphan) (Ziemann et al. 1998; Schwenkreis et al. 1999, 2000) |
| Norepinephrine | ↓ | ↑ | Yohimbine, reboxetine (Plewnia et al. 2001a, b) |
| Acetylcholine | ←→ | ←→ | Muscarinic receptor blockers (hyoscyamine) (Di Lazarro et al. 2001) |

Note: ↑ increasing ↓ decreasing, ←→ no effect
Pergolide, bromocriptine, haloperidol, lorazepam, ethanol, clomipramine, sertraline, riluzole, memantine, dextromethorphan, yohimbine, reboxetine

### 3.4 Effects of rTMS and Antidepressants on Brain Networks in Depression

In essence, a human brain is structured as spatially and temporally separable functional networks. In patients with depression, the neuronal activity patterns in at least default mode networks (DMN) and the fronto-parietal central executive network (CEN) among many functional networks are always abnormal. DMN is involved in rumination, self-referential processing, and episodic memory recovery and includes multiple regions in the medial prefrontal cortex, posterior cingulate cortex, and posterior parietal cortex (mainly medial). CEN plays a key role in the regulation of attention, working memory, and decision-making and includes the DLPFC and the frontal cortices of multiple posterior regions (mainly lateral). The activities of CEN and DMN are also closely related. Anatomical tracing studies in non-human primates have shown a strong interconnection among the DLPFC, the subgenual anterior cingulate cortex (sgACC), and the medial prefrontal cortex (mPFC) of DMN. DMN activity is suppressed in cognitive control tasks dependent on DLPFC, a part of CEN. It is found by studies of functional MRI that the DLPFC is always in a state of low activity in patients with depression (Conor et al. 2014). Sometimes, the activities of CEN and DMN activity are inversely related, with depression patients showing abnormally increased functional connectivity in DMN and abnormally reduced functional connectivity in CEN and also abnormally changed connectivity between the CEN and DMN (Li et al. 2013). DMN includes anterior and posterior subnetworks, the connectivity between which excessively increases in depression patients. After 12 weeks of treatment with antidepressants, posterior subnetwork function is recovered, while the functional connectivity within the anterior subnetwork remains abnormal (Li et al. 2013). rTMS to the DLPFC can normalize the subgenual hyperconnectivity of DMN but cannot alter connectivity within the CEN. In addition, rTMS can induce inverse connection between the DLPFC and medial prefrontal DMN nodes (Li et al. 2013). It can be seen that antidepressants and rTMS may each exert different efficacies through different pathways when acting on brain networks, the combination of which may create superimposed effects.

## 4 Progress in Studies on rTMS Treatment Protocols for Bipolar and Unipolar Depression

In 2014, the European Expert Panel of the International Federation of Clinical Neurophysiology Organization published the *Evidence-Based Guidelines on the Therapeutic Use of Repetitive Transcranial Magnetic Stimulation (rTMS)* (Lefaucheur et al. 2014). It introduces treatment protocols for 12 kinds of diseases by hierarchy. On October 30, 2019, the National Health Commission of the People's Republic of China released the "Criteria for the Establishment of National Psychiatric Centers" and "Criteria for the Establishment of Regional Psychiatric Medical Care Centers of National Level," which included rTMS in the list of key technologies. In 2019, the European Expert Panel of the International Federation of Clinical Neurophysiology Organization re-evaluated the studies on rTMS treatment released since 2014 and further updated the guidelines on the application of rTMS treatment and published the *Evidence-Based Guidelines on the Therapeutic Use of Repetitive Transcranial Magnetic Stimulation (rTMS)* (updated from 2014 to 2018), which introduced treatment protocols for the following disease by hierarchy.

In January 2013, the legal entity of the Clinical TMS Society was established by clinical TMS researchers from around the United States and the world to assist members in optimizing TMS management, developing new indications for TMS treatment, expanding insurance coverage, increasing public awareness of TMS, promoting and guiding the application of TMS by individual clinics, etc. In 2016, the association led the preparation of an expert consensus on rTMS for the treatment of depression (Perera et al. 2016), the basic content of which is as follows: patients with primary or recurrent major depressive disorder (depression) meeting the diagnostic criteria of the US *Diagnostic and Statistical Manual of Mental Disorders* (5th Edition) (DSM-V) or an equivalent nosology, who have failed to respond to 1–4 antidepressants, will be treated with a standard protocol of high-frequency rTMS to the left DLPFC for 4–6 weeks. Part of the patient's symptoms only improvement after 6 weeks of treatment and are in a period of significant stagnation as evaluated clinically may be treated for an additional 1–2 weeks. For patients who have not achieved satisfactory efficacy after 6 weeks of treatment, the treatment course may be extended beyond 6 weeks when there is a lag in the antidepressant taking effect in the prior history. There is no evidence showing that continued treatment beyond 6 weeks (up to 10 weeks has been reported in literatures) is neurotoxic. There is no strong clinical evidence showing that rTMS under MRI neuronavigation improves clinical efficacy compared with the rTMS under conventional navigation. Patients benefiting from treatment in the acute phase may still choose rTMS for continuous or maintenance treatment in the subsequent relapse and recrudescence phase. rTMS can be used in combination with antidepressants or antipsychotics or alone.

### 4.1 High-Frequency Stimulation to Left DLPFC and/or Low-Frequency Stimulation to Right DLPFC for Unipolar Depression

So far, there are mainly two approaches to treat major depressive disorder with rTMS, with high-frequency stimulation to the left DLPFC or low-frequency stimulation to the right DLPFC. Lefaucheur et al. (2014) found that eight of the ten positive studies applied high-frequency stimulation to the left DLPFC and two of them applied low-frequency stimulation to the right DLPFC. Based on these studies, they were sure that all ten studies were on unipolar depression and this method was also applicable to the treatment of moderate unipolar depression not responsive to at least one antidepressant. High-frequency stimulation to the left DLPFC was also consistent with the treatment protocol approved by FDA. During 2014–2018, four additional studies were performed for further analysis, with one of them being Class I, one being Class II, and two being Class III. Of these, one study showed that treatment performed twice a day was more effective than that preformed once a day. Thus, the Grade A of evidence indicating the definite efficacy of high frequency to the left DLPFC remained unchanged. A recent meta-analysis (Teng et al. 2017) showed that increasing the number of treatments and the total number of pulses per treatment (1200–1500) was beneficial in increasing antidepressant efficacy. Most studies have shown that low-frequency stimulation to the right DLPFC has an antidepressant effect, while its statistical value is lower than that of high frequency to the left DLPFC. From 2014 to 2018, there were no new large samples or sham stimulation-controlled studies. Therefore, the grade of evidence thereof remained as B (Lefaucheur et al. 2020). Some studies and meta-analyses compared the two types of stimulation and showed similar antidepressant efficacy. Low frequency may be better tolerated than high frequency (Berlim et al. 2013). Among the seven studies analyzed by Lefaucheur et al. (2014), where the combination of left and right stimulations was compared with sham stimulation in terms of efficacy, only three studies showed significantly better efficacy than that of sham stimulation. Considering the highly contradictory results, rTMS to the bilateral DLPFC was not recommended for depression patients. During 2014–2018, a total of 12 studies on combined bilateral stimulation were analyzed (two Class I, six Class II, and four Class III studies) and did not find any superiority of bilateral stimulation compared to unilateral stimulation. However, Blumberger et al. (2016) applied rTMS to the bilateral DLPFC (600 pulses at 1 Hz to the right DLPFC; 1500 pulses at 10 Hz to the left DLPFC), which produced a significant antidepressant response compared to the high frequency to the left DLPFC (2100 pulses at 10 Hz). Therefore, combined bilateral stimulation to DLPFC was considered as Grade B evidence (Lefaucheur et al. 2014).

## 4.2 Combined with Antidepressants

A large multicenter Class I study including 170 patients with depression (Brunelin et al. 2014) showed that low-frequency rTMS to the right DLPFC was as effective as venlafaxine alone or the combination of both treatments. A Class III retrospective study including 32 patients (Verma et al. 2018) showed that high frequency to the left DLPFC in combination with drugs was applicable for treatment-resistant depression. As to the synergy study of rTMS, a Class III study (Dell'Osso et al. 2015) showed that high frequency to the left DLPFC or low frequency to the right DLPFC could be used in the treatment of patients with acute phase and bipolar depression or with a range of adverse drug reactions or resistant to drugs. Therefore, there was no difference between the treatment with rTMS alone and the combination of rTMS with antidepressants (Level III recommendation) (Lefaucheur et al. 2020).

## 4.3 Studies on Bipolar Disorder

For rTMS treatment of bipolar depression, studies conducted by Nahas et al. (2003) concluded a negative outcome. Conclusive results are difficult to make from the other ten literatures on rTMS treatment for bipolar depression analyzed by Lefaucheur et al. (2014) due to factors such as the presence of population heterogeneity and drug uptake in patients. In an open study, Frank et al. (2011) found that the efficacy of rTMS in treating patients with bipolar depression was equivalent to that of rTMS in treating unipolar depression. Similarly, Berlim et al. (2014) did not find a difference in the treatment of samples of primary unipolar depression and bipolar depression with high-frequency rTMS to the left DLPFC. Currently, there is also no evidence of a risk of mild mania caused by rTMS.

Saba et al. (2004) intervened in patients with bipolar disorder in manic phase with rTMS and found a synergy effect of rTMS on drugs: an open study including eight subjects applied high-frequency rTMS (five sequences of 15 s, 10 Hz, 80% MT) to the right dorsolateral prefrontal cortex and evaluated the efficacy before and on the 14th day of treatment with the mania assessment scale (MAS) and clinical global impression (CGI) and found a significant alleviation of manic symptoms.

Zhang et al. (2016) selected 64 bipolar disorder patients with manic episodes and divided them equally into study and control groups. Both groups were treated with lithium carbonate sustained-release tablets. The study group was given rTMS to the right DLPFC at 20 Hz and 80% MT, and the control group was given sham stimulation. Symptom changes were evaluated by the Bech-Rafaelsen Mania Scale (BRMS). The BRMS scores of the study group were significantly lower than those of the control group ($P < 0.05$) 14 days after treatment. The effective rate in the study group was significantly higher than that in the control group ($P < 0.05$).

Cao et al. (2017) selected 30 patients of bipolar disorder with manic episodes and divided them equally into a study group and a control group. Both groups were treated with magnesium valproate sustained-release tablets. The study group was given rTMS to the right DLPFC at 1 Hz and 100% MT, and the control group was given sham stimulation. Symptom changes were evaluated by the Bech-Rafaelsen Mania Scale (BRMS). The left RMT changes were also measured. The results showed that after 10 days of treatment, BRMS scores were lower in both groups than those before treatments. BRMS scores were lower in the study group than those in the control group ($P < 0.01$), and MT was higher in both groups than that before treatments, and MT was higher in the study group than that in the control group ($P < 0.01$).

It can be seen that the rTMS treatments for manic episodes in bipolar disorder and unipolar depression are completely opposed, from which the objectivity of the hypothesis of left and right hemispheric activation equilibrium in psychiatric disorders can be further inferred. To be specific, the activation balance inclines to the right hemisphere during depressive episodes and excessively to the left hemisphere during manic episodes. High-frequency stimulation to the left hemisphere or/and low-frequency stimulation to the right hemisphere or high-frequency stimulation to the right hemisphere or/and low-frequency stimulation to the left hemisphere may correct the abnormal interhemispheric activation balance during depressive and manic episodes, respectively. According to the studies conducted by Gold et al. (2019) and Kazemi et al. (2016), high-frequency rTMS to the left DLPFC could reduce depressive symptoms in bipolar disorder, and 2 Hz to the right DLPFC had a better efficacy than 1 Hz. There was no significant difference in efficacy between low frequency (1 Hz) to the right DLPFC and high frequency (10 Hz) to the left DLPFC. High frequency (10 Hz) to the left DLPFC in combination with low frequency (1 Hz) to the right DLPFC showed a significantly higher proportion of responses than unilateral stimulation (10 Hz to the left DLPFC or 1 Hz to the right DLPFC). Other studies had found that TBS may have a faster, stronger, and longer-lasting effect than conventional rTMS protocols in bipolar and unipolar disorders (Beynel et al. 2014; Bulteau et al. 2017).

Adult bipolar disorder patients with manic episodes may have reduced metabolism in the right hemisphere and increased metabolism in the left hemisphere (Gold et al. 2019), so that high-frequency (20 Hz) rTMS to the right DLPFC had a significant ameliorating effect on manic symptoms compared with sham stimulation (Praharaj et al. 2009). A study on adolescents applied low-frequency stimulation to the left DLPFC together with low-frequency stimulation to the right DLPFC, which showed no significant difference in the efficacy of real and sham stimulation, suggesting that the failure of rTMS treatment in adolescents was due to the different metabolic patterns of the left and right hemispheres between adolescents and adults (Pathak et al. 2015).

### 4.4 iTBS Versus cTBS

Cheng-Ta et al. (2014) conducted a clinically randomized controlled and sham stimulation double-blind controlled study by selecting 60 patients with treatment-resistant depression and dividing them into group A (cTBS at 1800 pulses to the right DLPFC), group B (iTBS at 1800 pulses stimulation to the left DLPFC), group C (randomly alternating treatment with cTBS stimulation to the right DLPFC with iTBS to the left DLPFC at 1800), and group D (coil flipped 90°, alternating treatment with sham stimulation cTBS and iTBS at 1800 pulses), with 15 patients in each group and stimulation intensity of each group being 80% MT. Each group was given ten sequences of stimulations for 2 weeks. The reduction rate of 17 items in the Hamilton Depression Rating Scale (HDRS) at week 2 after treatment compared with that before treatment (week 0) was as follows: group B > group A, group B > group D, group C > group A, and group C > group D. It was believed that the treatment of the left DLPFC with iTBS (group B or C) had better efficacy than the treatment of the right DLPFC with cTBS only (group A) and the alternating treatment with iTBS to the left DLPFC and cTBS to the right DLPFC was slightly more effective than the treatment of the left DLPFC with iTBS alone.

Samuel Bulteau et al. (2017) performed a randomized, sham stimulation-controlled, and double-blind study to compare the efficacy of 10 Hz rTMS and iTBS to the left DLPFC in patients with major depressive disorder. The study consisted of two phases: a treatment phase consisting of 20 courses, followed by a 6-month longitudinal follow-up phase. It was concluded that the efficacy and maintenance effect of iTBS were superior to the 10 Hz standard stimulation during the periods of treatment as well as follow-up observation. Compared to the conventional rTMS approved by FDA, which requires 37.5 min per stimulation, iTBS delivers 600 pulses in 3 min, demonstrating a stronger excitatory effect. There are some clinical trials and two meta-analyses showing that iTBS is superior to sham stimulation for treatment-resistant depression. If the efficacy of 3-min iTBS is not less than standard rTMS, the treatment volume, cost, and implementability of rTMS will be increased several-fold, and its clinical utility will be greatly enhanced. Blumberger et al. (2018) compared the effectiveness of iTBS and 10 Hz rTMS in the treatment of more than 400 depression patients by 17 items in the Hamilton Depression Rating Scale (HDRS). The results showed no significant difference in the reduction rate of HDRS between iTBS and 10 Hz rTMS.

Andrew et al. (2019) conducted a cost-effectiveness analysis from a health economics perspective by a large randomized non-inferiority trial to investigate the benefit-cost of treating depression adults aged 18–65 years with 10 Hz rTMS or iTBS. The cost of achieving alleviation was calculated by verifying directly related healthcare expenses for equipment, coils, physician assessment, and working hours of technicians involved in the treatment. Results showed the mean cost per treatment course for each patient was $1108 ± $166 for iTBS and $1844 ± 04 for 10 Hz rTMS, resulting in net savings of $735 (95% CI: 688–783). The mean cost per remission of iTBS treatment was $3695 ± 552 and of 10 Hz rTMS was $6146 ± 1015, with net mean incremental savings of $2451 (95% CI: 2293–2610), confirming the high efficiency and low cost of TBS protocol for depression patients.

Two Class II and III studies have shown that cTBS to the right DLPFC may have no antidepressant effect (Level III recommendation), while another study found that cTBS to the right combined with iTBS to the left had a better antidepressant response than sham stimulation. Another randomized controlled study on sham stimulation versus real rTMS compared the efficacy of high frequency to the left DLPFC combined with low frequency to the right DLPFC with the efficacy of iTBS to the left DLPFC combined with cTBS to the right DLPFC. The results showed an increasing trend in response rates for both treatments with real TBS and rTMS, with this trend being most prominent in the group of TBS (Lefaucheur et al. 2020).

## 4.5 Accelerated rTMS Protocol

In order to enhance the antidepressant effect and reduce the days of stimulation, it has been suggested that increasing the number of rTMS to more than once per day may be more effective. This is known as the accelerated rTMS protocol. The accelerated rTMS protocol mainly aims to reduce the burden on patients and the operator and free them from traveling and repeating the procedure over several weeks. It seems to be safe and well tolerated in depression patients, even in the elderly (Lefaucheur 2020). Baeken et al. (2013) reported a sham-controlled crossover study of 20 patients treated with 20 times of 20 Hz stimulation to the left DLPFC for 4 days (five times/day). No significant efficacy difference from one stimulation/day was observed. It was also found that the response to accelerated high-frequency rTMS protocol could be predicted according to the higher metabolic activity in sgACC and stronger negative functional connectivity with the left superior medial prefrontal cortex. George et al. (2014) treated 41 patients with suicidal ideation with 9 sequences of 10 Hz rTMS to the left DLPFC for 3 days (3 stimulations per day, 6000 pulses each stimulation). The scores of suicide ideation decreased in both the groups given real stimulation and sham stimulation, while the rate of decrease was faster when rTMS was applied. A recent study compared the efficacy of accelerated high-frequency rTMS (three stimulations per day for 3 weeks, with each stimulation lasting 1–3 days) with daily treatment with 1 stimulation (5 days per week for 4 weeks), in which 58 subjects were given accelerated rTMS and 57 subject were given standard stimulation. It showed no significant differences between the two groups in terms of remission rates, remission rate reduction, or depression score reduction. A sham stimulation group, however, was not set up in this study (Fitzgerald 2018). In a crossover study by Duprat et al. (2016), 47 patients received 20 times of active rTMS or sham stimulation (five times per day) over 4 days. A reduction in depression scores in both treatment groups was observed, while the response and remission rates appeared to increase in the group of active rTMS compared to the sham stimulation group after 2 weeks of treatment. Desmyter et al. (2016) and Duprat et al. (2018) concluded that there was no significant difference in overall response to either suicide risk or benefit for the accelerated iTBS protocol with real and sham stimulation.

An fMRI study revealed that the left DLPFC of health participants correlated least with the subgenual anterior cingulate cortex (sgACC), whereas in MDD the correlation between DLPFC and sgACC was enhanced. A higher inhibitory correlation between the left DLPFC and sgACC detected by fMRI was associated with a better clinical efficacy in MDD (Cole 2022). In 2020, Cole and his colleagues of Stanford University in the United States developed an accelerated intermittent theta-burst stimulation (iTBS) protocol based on neuroscience, namely, Stanford accelerated intelligent neuromodulation therapy (SAINT), for treatment-resistant depression (Cole et al. 2020, 2022). This study recruited 19 patients with MDD and 2 patients with depressive episodes in bipolar II disorder. Each participant received MRI and resting-state fMRI scans. Neuronavigation device was applied to individually localize the region of the left DLPFC (site 1) most anti-correlated with sgACC (site 2) as the stimulated target for rTMS. The iTBS sessions were delivered at 90% MT, 1800 pulses per session with a 50-min inter-session interval. Pulses at 50 Hz were delivered in 2-s trains with an 8-s intertrain interval. Ten sessions were applied per day (cumulative 18,000 pulses per day) for five consecutive days (90,000 pulsed in total). Between treatments, participants seated in a separate waiting area (with no researchers or other patients) to limit their interactions with researchers and other participants and to prevent group effects. Symptoms were evaluated by the Hamilton Depression Rating Scale (HDRS-17), Montgomery-Åsberg Depression Rating Scale (MADRS), Beck Depression Inventory-II (BDI-II), and Columbia Suicide Severity Rating Scale (C-SSRS).

The results showed after 5 days of SAINT treatment, the response rate reached 90.48% (MADRS score reduction rate ≥ 50%). All responders had met the remission criteria

(MADRS score < 11 points) after the treatment. At all follow-up time points, the suicide risk was significantly reduced. After 1 month of SAINT treatment, the suicide risk remission rate was still above 80% on different scale scores. Among the participants, six were retreated with the same SAINT protocol when they failed to meet the remission criteria. The average time to start the retreatment was 20.5 weeks. There were barely differences in the daily 6-item HAM-D scores and the weekly MADRS scores of the six participants between the initial SAINT treatment and retreatment. The study demonstrated that SAINT remarkably alleviated depressive symptoms and suicidal ideation in patients with treatment-resistant depression within 5 days, without cognitive side effects. Meanwhile, the remission rate of SAINT was observed to be higher than the response rates for standard FDA-approved rTMS regimens (37%), ECT (48%), and ketamine (31%). The SAINT scheme enables to identify the most tightly junctional part of the left DLPFC with the deeper sgACC, which is typically hyperactive during the onset of MDD. The emotion regulation is retrievable by precisely stimulating subregions of the DLPFC with high-intensity magnetic pulses to excite neurons, accompanying with a reduced activity of sgACC. Dose-response curves exhibited that SAINT therapy was several-fold higher than the earlier FDA-approved rTMS doses, consequently significant improvements in response rates. Accordingly, the FDA approved SAINT in 2022 for the accelerated treatment of treatment-resistant depression.

On the other hand, this study elucidated the importance of individualized localization by fMRI. The mean distance between the left DLPFC localized by individualized neuronavigation and the F3 guided by the EEG 10–20 system is 25.18 mm, suggesting that individualized targets of fMRI are more precise than the conventional DLPFC localization.

## 4.6 Deep Transcranial Magnetic Stimulation

The H-coil was used in deep transcranial magnetic stimulation (deep TMS, dTMS) is designed to suit the head surface, will generate induced electric fields at different locations surrounding. These electric fields have the same direction and can be superimposed in deep neural regions. Although there are more than a dozen types of H-coil, H1 coil is comparable to eight-shaped coil, which is mainly used to stimulate the left DLPFC. The focusing performance of H1 coil is more concentrated than eight-shaped coil and can reach deeper regions. The action range of H1 coil magnetic lines is larger than those of eight-shaped coil. Moreover, the attenuation of the electric field intensity induced by the H-coil is far weaker than that of eight-shaped coil. The action of eight-shaped coil reaches 0.7 cm below the dura mater, with a range of action of 3 $cm^3$; the action of H1 coil can reach 1.8 cm below the dura mater, with a range of action of 18 $cm^3$; and the action of H7 coil can reach 3 cm below the dura mater, with a range of action of 73 $cm^3$. H7 coil can act directly on the medial prefrontal lobe, anterior cingulate cortex, etc. and was approved by FDA in 2018 for the treatment of obsessive-compulsive disorder. According to literatures, H1 coil produced the same short-term antidepressant efficacy as regular coils and was approved by FDA in 2013 for treating treatment-resistant unipolar depression. In 2021, FDA approved the protocol of iTBS at 600 pulses with H1 coils for treating treatment-resistant depression.

Filipčić et al. (2019) randomly assigned 228 depression patients to the H1 coil group ($n$ = 65), the eight-shaped coil group ($n$ = 72), and the regular medication control group ($n$ = 72). rTMS treatment protocols were performed by apply high-frequency stimulation to the left DLPFC. At the end of treatment in week 4, the H1 coil group showed a remission rate of 1.74 (95% CI: 0.79–3.83) superior to that of the eight-shaped coil group, and the efficacy difference between the methods of applying rTMS was not significant. The remission rates were significantly higher in the H1 and eight-shaped coil groups than in the control group: 60% (95% CI: 48–71%), 43% (95% CI: 31–55%), and 11% (95% CI: 5–20%). The H1 coil group responded significantly better than the eight-shaped coil group: OR = 2.33; 95% CI: 1.04–5.21 ($P$ = 0.040). Reduction rate of HDRS scores: 59% reduction was seen in the H1 coil group, 41% reduction in the eight-shaped coil group ($P$ = 0.048), and 17% reduction in the control group ($P$ = 0.001 compared with the H1 coil group; $P$ = 0.003 compared with the eight-shaped coil group).

## 4.7 Selection and Setting of rTMS Treatment Course

According to the Chinese *Guidelines for the Prevention and Treatment of Depressive Disorders*, depression is a psychiatric disorder with a very high relapse rate, with a 50% relapse rate after treatment in patients with one episode, 75% in patients with two episodes, and up to 90% in patients with three episodes. In a long-term follow-up study of 257 patients treated with rTMS, Dunner et al. (2014) found that after the treatment in acute phase, 93 patients (36.2%) received at least 1 consolidation therapy with rTMS from the second month onward, with the times of consolidation therapy gradually reducing over several weeks. A 1-year follow-up revealed that among the 78 patients in complete remission after rTMS treatment in acute phase, 23 patients (29.5%) had a recurrence of depression within the first 6 months of follow-up and 55 patients (70.5%) experienced no recurrence within 1 year.

1. Course setting for rTMS treatment in the acute phase of depression (Shawn et al. 2018): it is recommended to take rTMS as an acute treatment to alleviate depressive symptoms (Level I recommendation and Grade A study evidence). It is recommended to apply one stimulation a day in the acute phase, five stimulations a week. rTMS treatment reduces depressive symptoms as early as 2 or 3 weeks after the treatment starts. The treatment in acute phase will be completed within 6 weeks in most cases (20–30 stimulations).
2. Depression treatment in remission phase (Shawn et al. 2018): rTMS may be used as maintenance treatment for patients who respond well to rTMS in acute phase (Level I recommendation and Grade A study evidence).

   It is believed by most experts that rTMS maintenance therapy may be considered if other effective maintenance therapies that have been confirmed fail to produce definite efficacies or if the patient has a history of frequent relapses (twice or more per year). The typical frequency of maintenance therapy gradually reduces from once or twice per week to once a month to stopping treatment, depending on the patient's response level.
3. Recommendations for clinical use

   Treatment with eight-shaped coils or H1 coils:

   Clinical recommendation 1: high frequency to the left DLPFC. Level I recommendation, Grade A, B, and C evidence (Lefaucheur 2020)

   Clinical recommendation 2: low frequency to the right DLPFC. Level II recommendation, Grade B and C evidence (Lefaucheur 2020)

   Clinical recommendation 3: low frequency (or cTBS) to the right DLPFC combined with high frequency (or iTBS) to the left DLPFC. Level II recommendation, Grade B and C evidence (Lefaucheur 2020)

## 5 Studies on Application of Transcranial Direct Current Stimulation to Depression

### 5.1 Progress in Studies on Transcranial Direct Current Stimulation

The European Expert Panel of the International Federation of Clinical Neurophysiology Organization collected studies on treatments with transcranial direct current stimulation (tDCS) published before September 2016 for evidence-based analysis and developed guidelines for the application of tDCS treatment by exploring the efficacy for pain, Parkinson's disease, motor dysfunction disorders, stroke, post-stroke aphasia, multiple sclerosis, epilepsy, disorders of consciousness, Alzheimer's disease, tinnitus, depression, schizophrenia, and addiction. Study evidence showed that there were no Level I recommendations for tDCS in the treatment of related disorders, but only Level II and III recommendations (Lefaucheur 2017). Felipe et al. (2021) summarized the English literature on all tDCS studies since the 2010s and published updated guidelines for the treatment of tDCS in 2021.

The treatment of depressive disorders with tDCS has been based on previous studies of functional and structural abnormalities in the left and right dorsolateral and ventral medial frontal cortices, amygdala, and hippocampus in depression patients. For example, it was found by fNIRS and EEG techniques that tDCS could normalize neuronal activity and interhemispheric imbalances between the two DLPFC areas. As with high-frequency rTMS to the left DLPFC or/and low-frequency rTMS to the right DLPFC adopted for treating depression, the current approach is to enhance the activity of the left DLPFC with anodal stimulation and/or to decrease the activity of the right DLPFC or orbitofrontal cortex (OFC) with cathodal stimulation. tDCS largely affects deeper brain structures such as the amygdala, hippocampus, and other subcortical structures, while it is unclear how changes in resting-state brain networks affect the antidepressant effects of tDCS and how each electrode, respectively, acts on brain network modulation.

Antidepressant studies showed that the combination of tDCS with sertraline hydrochloride (50 mg/day) was superior to tDCS or placebo alone, suggesting a mutually synergy effect of tDCS and antidepressant drugs. It has been shown that pharmacological modulation of neurons by 5-hydroxytryptaminergic and noradrenergic in deep brain structures may mediate the effects of tDCS, although they are not directly influenced by the superficial current flow generated by tDCS. Currently, it is recognized to place anodes at the left DLPFC, while more studies have shown that placing cathodes at the right OFC is more effective than the right DLPFC as it is believed that the intracerebral currents created by cathodes in these two brain areas are different.

As to the studies on tDCS cognitive effect, beneficial effects have been reported in four Class II studies, while negative outcomes have been reported one Class I study and three Class II studies. Therefore, the studies remain contradictory at present. tDCS cannot be recommended to improve cognitive symptoms in patients with depression.

### 5.2 Recommendations for Clinical Treatment

Place anodes at the left DLPFC and cathodes at the right OFC; apply at least ten stimulations with an intensity of 2 mA. Maintain each stimulation for 20–30 min. It is identified as Level I recommendation (with Grade A, B, and C evidence) and can be applied either in combination with drugs or alone (Fregni et al. 2021).

## 6 Studies on Application of Deep Brain Stimulation to Depression

Deep brain stimulation (DBS) has been widely studied over the past four decades. Emerging evidence suggests that DBS is a prospective therapeutic option for psychiatric disorders including treatment-resistant depression (TRD) with an average of 60% response rates in chronically TRD patients (Figee et al. 2022). The initial study was based on the experience of DBS in Parkinson's disease. Selected targets for TRD include the subcallosal cingulate cortex (SCC), the ventral capsule and ventral striatum (VC/VS), the nucleus accumbens (NAc), the medial forebrain bundle (MFB), the lateral habenula (LHb), and the inferior thalamic peduncle (ITP).

Among these targets, the SCC was the first and the most studied target in the practice of DBS for TRD. The SCC is involved in the processing of emotions with its reciprocal connections to the medial prefrontal cortex, cingulate cortex, NAc, anterior thalamus, and others (Riva-Posse et al. 2014). Multiple studies have demonstrated the efficacy and safety of DBS for TRD in the SCC. Mayberg et al. applied DBS to the SCC firstly in six patients with TRD. Four of six patients gained remission of depression in the 6-month follow-up period. Antidepressant effect was associated with a reduction of cerebral blood flow in SCC and downstream limbic and cortical regions (Mayberg et al. 2005). Later studies also confirmed the effectiveness of DBS treatment in SCC. In long-term studies, Kennedy et al. and Crowell et al. have shown that the effect of DBS in SCC can last for several years with good tolerance underlying long-term therapeutic prospects (Kennedy et al. 2011; Crowell et al. 2019).

The NAc plays the critical role in the reward circuitry. Functional activity of the NAc is associated with the experience of pleasure underlying its significance in the treatment of major depression. Bewernick et al. demonstrate antidepressant and antianhedonic effects of DBS to NAc in five TRD patients with a 50% reduction of the Hamilton Depression Rating Scale (Bewernick et al. 2010). Long-term therapeutic effect was also shown in a subsequent study with a response rate of 45% at 4 years (Bewernick et al. 2012).

The VC/VS is concerned with the anterior limb of the internal capsule was revealed by a previous tractography study to have widespread projections to the frontal pole, medial temporal lobe, NAc, thalamus, etc. (Gutman et al. 2009). All the connected regions have been implicated in depression, making the VC/VS a potential therapeutic target. The first open-label trial of DBS for VC/VS conducted by Malone et al. demonstrated its safety and efficacy, with a response rate of 40% at 6 months and 53.3% at 4 years (Malone et al. 2009).

The LHb was involved in the brain's anti-reward system, mediating negative motivations. Chronic deep brain stimulation of the LHb alleviated depressive symptoms in rats (Meng et al. 2011). Sartorius et al. firstly demonstrated the therapeutic potential of DBS for LHb (Sartorius et al. 2010). A recent open-label study conducted by Zhang et al. found that DBS for LHb gained 56%, 46%, and 64% reduction for depression at 3-month, 6-month, and 12-month follow-up, respectively (Zhang et al. 2022).

The MFB projects from the ventral tegmental area to the NAc and to the prefrontal cortex. DBS of the supero-lateral branch of the medial forebrain bundle (slMFB) in TRD is associated with both acute and long-term antidepressant effects. Bewernick et al. applied DBS to the slMFB, six of eight patients (75%) were responders at 12 months, and four patients reached remission. Long-term results revealed a stable effect up to 4 years (Bewernick et al. 2017).

Convergent evidence indicates that DBS is a promising neurostimulation approach in TRD. The existing studies have shown efficacy and safety for various brain targets in both short-term and long-term follow-up. However, studies of DBS in TRD are still in progress, since many studies were small-scale single-center studies or case studies. Future research should focus on large-scale, multicenter, randomized, double-blind studies. In addition, the search for optimal DBS targets in TRD is still needed. Therapeutic outcomes may vary with regard to different DBS targets. Standard and individualized DBS treatment protocols are supposed to be established. Brain network connectome research and animal studies may help understand the nature of DBS treatment in TRD.

## References

Agarwal A, Nag D, Jain AK (1992) Pathological laughter after right hemispheric infarction. Indian J Psychiatry 34(4):385–387
Alfredo B, Giuliana L, Andrea M et al (2009) Asymmetries of the human social brain in the visual, auditory and chemical modalities. Philos Trans R Soc Lond B Biol Sci 364(1519):895–914
Almeida JR, Versace A, Mechelli A et al (2009) Abnormal amygdala-prefrontal effective connectivity to happy faces differentiates bipolar from major depression. Biol Psychiatry 66(5):451–459
Andrew B. Mendlowitz, Alaa Shanbour, et al (2019). Implementation of intermittent theta burst stimulation compared to conventional repetitive transcranial magnetic stimulation in patients with treatment resistant depression: a cost analysis. PLoS One 14(9):e0222546
Bajwa S, Bermpohl F, Rigonatti SP, Pascual-Leone A et al (2008) Impaired interhemispheric interactions in patients with major depression. J Nerv Ment Dis 196(9):671–677
Baeken C, Vanderhasselt MA, Remue J et al (2013) Intensive HF-rTMS treatment in refractory medication-resistant unipolar depressed patients. J Affect Disord 151(2):625–631
Balzekas I, Lewis CP, Shekunov J et al (2018) A pilot study of GABAB correlates with resting-state functional connectivity in

five depressed female adolescents. Psychiatry Res Neuroimaging. 30(279):60–63

Berlim MT, Van den Eynde F, Daskalakis ZJ (2013) High-frequency repetitive transcranial magnetic stimulation accelerates and enhances the clinical response to antidepressants in major depression: a meta-analysis of randomized, double-blind, and sham-controlled trials. J Clin Psychiatry 74(2):122–129

Berlim MT, Eynde F, Tovar-Perdomo S et al (2014) Response, remission and drop-out rates following high-frequency repetitive transcranial magnetic stimulation (rTMS) for treating major depression: a systematic review and metaanalysis of randomized, double-blind and sham-controlled trials. Psychol Med. 44(2):225–239

Bewernick BH, Hurlemann R, Matusch A et al (2010) Nucleus accumbens deep brain stimulation decreases ratings of depression and anxiety in treatment-resistant depression. Biol Psychiatry 67:110–116

Bewernick BH, Kayser S, Sturm V et al (2012) Long-term effects of nucleus accumbens deep brain stimulation in treatment-resistant depression: evidence for sustained efficacy. Neuropsychopharmacology 37:1975–1985

Bewernick BH, Kayser S, Gippert SM et al (2017) Deep brain stimulation to the medial forebrain bundle for depression - long-term outcomes and a novel data analysis strategy. Brain Stimul 10(3):664–671

Beynel L, Chauvin A, Guyader N et al (2014) What saccadic eye movements tell us about TMS-induced neuromodulation of the DLPFC and mood changes: a pilot study in bipolar disorders. Front Integr Neurosci 8:65

Blumberger DM, Vila-Rodriguez F, Thorpe KE et al (2018) Effectiveness of theta burst versus high-frequency repetitive transcranial magnetic stimulation in patients with depression (THREE-D): a randomised non-inferiority trial. Lancet 391(10131):1683–1692

Brodbent DE (1954) The role of auditory localization in attention and memory span. J Exp Psychol. 47(3):191–196

Brunelin J, Jalenques I, Trojak B, Attal J et al (2014) The efficacy and safety of low frequency repetitive transcranial magnetic stimulation for treatment-resistant depression: the results from a large multicenter French RCT. Brain Stimul. 7(6):855–863

Bulteau S, Sébille V, Fayet G et al (2017) Efficacy of intermittent theta burst stimulation (iTBS) and 10-Hz high-frequency repetitive transcranial magnetic stimulation (rTMS) in treatment-resistant unipolar depression: study protocol for a randomised controlled trial. Trials 18:17. https://doi.org/10.1186/s13063-016-1764-8

Cantone M, Bramanti A, Lanza G et al (2017) Cortical plasticity in depression: a neurochemical perspective from transcranial magnetic stimulation. ASN Neuro 9(3):175909141771151

Cao C, Gu D, Yu L (2017) Controlled study of low-frequency repetitive TMS adjuvant for manic episodes. J Psychiatry 30(2):137–139

Centonze D, Palmieri MG, Boffa L et al (2005) Cortical hyperexcitability in post-traumatic stress disorder secondary to minor accidental head trauma: a neurophysiologic study. J Psychiatry Neurosci 30(2):127–132

Cheng-Ta Li, Mu-Hong Chen, Chi-Hung Juan, et al (2014) Efficacy of prefrontal theta-burst stimulation in refractory depression: a randomized sham-controlled study. Brain 137(Pt 7):2088–2098

Chung Y, Hancock KE, Delgutte B (2016) Neural Coding of Interaural Time Differences with Bilateral Cochlear Implants in Unanesthetized Rabbits. J Neurosci 18;36(20):5520-5531

Cole EJ, Stimpson KH, Bentzley BS et al (2020) Stanford accelerated intelligent neuromodulation therapy for treatment-resistant depression. Am J Psychiatry 177(8):716–726

Cole EJ, Phillips AL, Bentzley BS et al (2022) Stanford neuromodulation therapy (SNT): a double-blind randomized controlled trial. Am J Psychiatry 179(2):132–141

Conor L, Chen AC, Zebley BD et al (2014) Default mode network mechanisms of transcranial magnetic stimulation in depression. Biol Psychiatry 76(7):517–526

Crowell AL, Riva-Posse P, Holtzheimer PE et al (2019) Long-term outcomes of subcallosal cingulate deep brain stimulation for treatment-resistant depression. Am J Psychiatry 176:11. https://doi.org/10.1176/appi.ajp.2019.18121427

Cui X, Zhang L, Gao Z et al (2008) The research on motor threshold of patients with depression. Chin J Health Psychol 16(4):421–422

Cui J, Zou Y, Wang M (2009) Etiology of traditional Chinese medicine in bipolar disorder. LiShiZhen Med Materia Medica Res 20(5):1251–1252

Davidson RJ, Marshall JR, Tomarken AJ et al (2000) While a phobic waits: regional brain electrical and autonomic activity in social phobics during anticipation of public speaking. Biol Psychiatry 47(2):85–95

Dell'Osso B, Oldani L, Camuri G et al (2015) Augmentative repetitive Transcranial Magnetic Stimulation (rTMS) in the acute treatment of poor responder depressed patients: a comparison study between high and low frequency stimulation. Eur Psychiatry 30(2):271–276

Desmyter S, Duprat R, Baeken C et al (2016) Accelerated intermittent theta burst stimulation for suicide risk in therapy-resistant depressed patients: a randomized, sham-controlled trial. Front Hum Neurosci 10:480. eCollection 2016

Dunner DL, Aaronson ST, Sackeim HA et al (2014) A multisite, naturalistic, observational study of transcranial magnetic stimulation for patients with pharmacoresistant major depressive disorder: durability of benefit over a 1-year follow-up period. J Clin Psychiatry. 75(12):1394–1401

Duprat R, Desmyter S, Rudi de R et al (2016) Accelerated intermittent theta burst stimulation treatment in medication-resistant major depression: a fast road to remission? J Affect Disord 200:6–14

Duprat R, Wu GR, De Raedt R et al (2018) Accelerated iTBS treatment in depressed patients differentially modulates reward system activity based on anhedonia. World J Biol Psychiatry 19:497–508

Felipe Fregni, Mirret M El-Hagrassy, Kevin Pacheco-Barrios el al (2021) Evidence-Based Guidelines and Secondary Meta-Analysis for the Use of Transcranial Direct Current Stimulation in Neurological and Psychiatric Disorders. Int J Neuropsychopharmacol. 1;24(4):256–313

Figee M, Riva-Posse P, Choi KS et al (2022) Deep brain stimulation for depression. Neurotherapeutics 19:1229–1245

Filipčić I, Filipčić I, Milovac Z et al (2019) Efficacy of repetitive transcranial magnetic stimulation using a figure-8-coilor an H1-Coil in treatment of major depressive disorder; a randomized clinical trial. J Psychiatr Res 114:113–119

Fitzgerald PB, Hoy KE, Elliot D et al (2018) Accelerated repetitive transcranial magnetic stimulation in the treatment of depression. Neuropsychopharmacology. 43(7):1565–1572

Frank E, Eichhammer P, Burger J et al (2011) Transcranial magnetic stimulation for the treatment of depression: feasibility and results under naturalistic conditions: a retrospective analysis. Eur Arch Psychiatry Clin Neurosci 261(4):261–266

Fregni F, El-Hagrassy MM, Pacheco-Barrios K et al (2021) Evidence-based guidelines and secondary meta-analysis for the use of transcranial direct current stimulation in neurological and psychiatric disorders. Int J Neuropsychopharmacol 24(4):256–313

Gandour J, Tong Y, Wong D et al (2004) Hemispheric roles in the perception of speech prosody. Neuroimage. 23(1):344–357

Gao Zhiqin, Yu Haiying, Sun Jianel al (2008) A controlled study of cortical excitability in patients with first-episode schizophrenia and their first-degree relatives. Journal of Psychiatric. 21(5):341-343

George MS, Raman R, Benedek DM et al (2014) A two-site pilot randomized 3 day trial of high dose left prefrontal repetitive transcranial magnetic stimulation (rTMS) for suicidal inpatients. Brain Stimul. 7(3):421–431

Gold AK, Ornelas AC, Cirillo P et al (2019) Clinical applications of transcranial magnetic stimulation in bipolar disorder. Brain Behav 9(10):e01419

Grimm S, Beck J, Schuepbach D et al (2008) Imbalance between left and right dorsolateral prefrontal cortex in major depression is linked to negative emotional judgment: an fMRI study in severe major depressive disorder. Biol Psychiatry 63(4):369–376

Gutman DA, Holtzheimer PE, Behrens TEJ et al (2009) A tractography analysis of two deep brain stimulation white matter targets for depression. Biol Psychiatry 65:276–282

He W, Chai H, Zhang Y et al (2010) Line bisection performance in patients with generalized anxiety disorder and treatment-resistant depression. Int J Med Sci. 7(4):224–231

Herwig U, Bräuer K, Connemann B et al (2002) Intracortical excitability is modulated by a norepinephrine-reuptake inhibitor as measured with paired-pulse transcranial magnetic stimulation. Psychopharmacology 164(2):228–232

Ho CSH, Lim LJH, Lim AQ et al (2020) Diagnostic and predictive applications of functional near-infrared spectroscopy for major depressive disorder: a systematic review. Front Psych 11:378. https://doi.org/10.3389/fpsyt.2020.00378. eCollection

Huang J, Li J, Zou K et al (2022) Cortical excitability and inhibition in patients with major depression and its correlation with depression severity as measured by transcranial magnetic stimulation. J Psychiatry 35(1):14–18

Ilic TV, Korchounov A, Ziemann U (2002) Complex modulation of human motor cortex excitability by the specific serotonin re-uptake inhibitor sertraline. Neurosci Lett 319(2):116–120

Isenberg K, Downs D, Pierce K et al (2005) Low frequency rTMS stimulation of the right frontal cortex is as effective as high frequency rTMS stimulation of the left frontal cortex for antidepressant-free, treatment-resistant depressed patients. Ann Clin Psychiatry 17:153–159

Jorge J, Silvester C, Xiao H et al (2009) MRI-derived measurements of human subcortical, ventricular and intracranial brain volumes: reliability effects of scan sessions, acquisition sequences, data analyses, scanner upgrade, scanner vendors and field strengths. NeuroImage 46(1):177–192

Kazemi R, Rostami R, Khomami S et al (2016) Electrophysiological correlates of bilateral and unilateral repetitive transcranial magnetic stimulation in patients with bipolar depression. Psychiatry Res 240:364–375

Kennedy SH, Giacobbe P, Rizvi SJ et al (2011) Deep brain stimulation for treatment-resistant depression: follow-up after 3 to 6 years. Am J Psychiatry 168:502–510. https://doi.org/10.1176/appi.ajp.2010.10081187

Kimura (1967) Functional asymmetry of the brain in dichotic listening. Cortex. 3:163–178

Kinsbourne M (1987) Mechanisms of unilateral neglect. In: Jeannerod M, ed. Neurophysiological and neuropsychological aspects of spatial neglect. Amsterdam: North-Holland, p69–86

Kuhn M, Mainberger F, Feige B et al (2016) State-Dependent partial occlusion of cortical LTP-Like Plasticity in Major Depression. Neuropsychopharmacology. 41(6):1521–1529

Lefaucheur JP, Lucas B, Andraud F et al (2008) Inter-hemispheric asymmetry of motor corticospinal excitability in major depression studied by transcranial magnetic stimulation. J Psychiatr Res 42:389–398

Lefaucheur JP, André-Obadia N, Antal A et al (2014) Evidence-based guidelines on the therapeutic use of repetitive transcranial magnetic stimulation (rTMS). Clin Neurophysiol 125:2150–2206

Lefaucheur JP, Antal A, Ayache SS et al (2017) Evidence-based guidelines on the therapeutic use of transcranial direct current stimulation (tDCS). Clin Neurophysiol. 128(1):56–92

Lefaucheur JP, André-Obadia N, Baeken C et al (2020) Evidence-based guidelines on the therapeutic use of repetitive transcranial magnetic stimulation (rTMS): an update. Clin Neurophysiol 131(2):474–528

Li D, Huang X, Wu Q et al (2010) Brain functions in major depressive disorder: a resting-state functional magnetic resonance imaging study. J Biomed Eng 27(1):16–19

Li B, Liu L, Friston KJ et al (2013) A treatment-resistant default mode subnetwork in major depression. Biol Psychiatry 74(1):48–54

Li B, Jiang G, Jin R et al (2018) The mechanism study of treating persist somatoform pain disorder with Sertraline combined with rTMS based on the human brain function asymmetry theory. Chin J Rehabil Med 33(8):934–939

Maeda F, Keenan JP, Pascual-Leone A (2000) Interhemispheric asymmetry of motor cortical excitability in major depression as measured by transcranial magnetic stimulation. Br J Psychiatry 177:169–173

Malone DA, Dougherty DD, Rezai AR et al (2009) Deep brain stimulation of the ventral capsule/ventral striatum for treatment-resistant depression. Biol Psychiatry 65:267–275

Manganotti P, Bortolomasi M, Zanette G et al (2001) Intravenous clomipramine decreases excitability of human motor cortex. A study with paired magnetic stimulation. J Neurol Sci 184:27–32

Manly T, Dobler VB, Dodds CM et al (2005) Rightward shift in spatial awareness with declining alertness. Neuropsychologia 43(12):1721–1728

Mayberg HS, Lozano AM, Voon V et al (2005) Deep brain stimulation for treatment-resistant depression. Neuron 45:651–660

Medland SE, Duffy DL, Wright MJ et al (2009) Genetic influences on handedness: data from 25, 732 Australian and Dutch twin families. Neuropsychologia 47(2):330–337

Meng H, Wang Y, Huang M et al (2011) Chronic deep brain stimulation of the lateral habenula nucleus in a rat model of depression. Brain Res 1422:32–38

Meng Z-L (1985) Human emotions. Shanghai People's Publishing House, Shanghai, p 152–156

Meyer BU, Röricht S, Gräfin von Einsiedel H et al (1995) Inhibitory and excitatory interhemispheric transfers between motor cortical areas in normal humans and patients with abnormalities of the corpus callosum. Brain 118(2):429–440

Nahas Z, Kozel FA, Li X et al (2003) Left prefrontal transcranial magnetic stimulation (TMS) treatment of depression in bipolar affective disorder: a pilot study of acute safety and efficacy. Bipolar Disord 5(1):40–47

Nicholas TO, Madeleine SG, Catherine LH et al (2011) Behaviour, physiology and experience of pathological laughing and crying in amyotrophic lateral sclerosis. Brain 134(12):3455–3466

Nishida M, Kikuchi S, Matsumoto K et al (2017) Sleep complaints are associated with reduced left prefrontal activation during a verbal fluency task in patients with major depression: a multi-channel near-infrared spectroscopy study. J Affect Disord 207:102–109

Noda Y, Barr MS, Zomorrodi R et al (2018) Reduced short-latency afferent inhibition in prefrontal but not motor cortex and its association with executive function in schizophrenia: a combined TMS-EEG study. Schizophr Bull. 44(1):193–202

Ozel-Kizil ET, Baskak B, Gunes Eet al (2012) Hemispatial neglect evaluated by visual line bisection task in schizophrenic patients and their unaffected siblings. Psychiatry Res. 200 (2-3):133–136

Pascual-Leone A, Manoach DS, Birnbaum R et al (2002) Motor cortical excitability in schizophrenia. Biol Psychiatry 52(1):24–31

Pathak V, Sinha VK, Praharaj SK (2015) Efficacy of adjunctive high frequency repetitive transcranial magnetic stimulation of right prefrontal cortex in adolescent mania: a randomized sham-controlled study. Clin Psychopharmacol Neurosci 13(3):245–249

Perera T, George MS, Grammer G et al (2016) The Clinical TMS Society consensus review and treatment recommendations for TMS therapy for major depressive disorder. Brain Stimul 9(3):336–346

Pellicciari MC, Veniero D, Miniussi C (2017) Characterizing the Cortical Oscillatory Response to TMS Pulse. Front Cell Neurosci. 11:38. https://doi.org/10.3389/fncel.2017.00038. eCollection 2017

Player MJ, Taylor JL, Weickert CS et al (2013) Neuroplasticity in depressed individuals compared with healthy controls. Neuropsychopharmacology 38(11):2101–2108

Plewnia C, Bartels M, Cohen L, Gerloff C (2001a) Noradrenergic modulation of human cortex excitability by the presynaptic alpha (2)-antagonist yohimbine. Neurosci Lett 307:41–44

Plewnia C, Cohen L, Bartels M, Gerloff C (2001b) Human cortex excitability is modulated by central norepinephrine. World J Biol Psychiatry [Suppl] 2:240S

Praharaj SK, Ram D, Arora M (2009) Efficacy of high frequency (rapid) suprathreshold repetitive transcranial magnetic stimulation of right prefrontal cortex in bipolar mania: a randomized sham controlled study. J Affect Disord 117(3):146–150

Pu S, Yamada T, Yokoyama K et al (2012) Reduced prefrontal cortex activation during the working memory task associated with poor social functioning in late-onset depression: multichannel near-infrared spectroscopy study. Psychiatry Res 203(2–3):222–228

Pu S, Nakagome K, Yamada T et al (2015) Suicidal ideation is associated with reduced prefrontal activation during a verbal fluency task in patients with major depressive disorder. J Affect Disord 181:9–17

Recio G, Conrad M, Hansen La B et al (2014) On pleasure and thrill: the interplay between arousal and valence during visual word recognition. Brain Lang 134:34–43

Riva-Posse P, Choi KS, Holtzheimer PE et al (2014) Defining critical white matter pathways mediating successful subcallosal cingulate deep brain stimulation for treatment-resistant depression. Biol Psychiatry 76(12):963–969

Romney DM, Mosley JL, Addington DE (2000) Hemispheric processing deficits in patients with paranoid schizophrenia. J Genet Psychol. 161(1):99–114

Rossi S, De Capua A, Tavanti M et al (2009) Dysfunctions of cortical excitability in drug-naïve posttraumatic stress disorder patients. Biol Psychiatry. 66:54–61

Saba G, Rocamora JF, Kalalou K et al (2004) Repetitive transcranial magnetic stimulation as an add-on therapy in the treatment of mania: a case series of eight patients. Psychiatry Research. 128(2):199–202

Sartorius A, Kiening KL, Kirsch P et al (2010) Remission of major depression under deep brain stimulation of the lateral habenula in a therapy-refractory patient. Biol Psychiatry 67:e9–e11. https://doi.org/10.1016/j.biopsych.2009.08.027

Schwenkreis P, Liepert J, Witscher K et al (2000) Riluzole suppresses motor cortex facilitation in correlation to its plasma level. A study using transcranial magnetic stimulation. Exp Brain Res 135:293–299

Schwenkreis P, Witscher K, Janssen F et al (1999) Influence of the N-methyl-d-aspartate antagonist memantine on human motor cortex excitability. Neurosci Lett 270:137–140

Shawn M. McClintock, Irving M. Reti, Linda L. Carpenter, et al (2018) Consensus recommendations for the clinical application of repetitive transcranial magnetic stimulation (rTMS) in the treatment of depression. J Clin Psychiatry 79(1): https://doi.org/10.4088/JCP.16cs10905

Speer AM, Kimbrell TA, Wassermann EM et al (2000) Opposite effects of high and low frequency rTMS on regional brain activity in depressed patients. Biol Psychiatry 48(12):1133–1141

Speer AM, Benson BE, Kimbrell TK et al (2009) Opposite effects of high and low frequency rTMS on mood in depressed patients: relationship to baseline cerebral activity on PET. J Affect Disord 115(3):386–394

Strafella AP, Paus T (2001) Cerebral blood-flow changes induced by paired-pulse transcranial magnetic stimulation of the primary motor cortex. J Neurophysiol 85(6):2624–2929

Sun Y, Farzan F, Mulsant BH et al (2016) Indicators for remission of suicidal ideation following magnetic seizure therapy in patients with treatment-resistant depression. JAMA Psychiatry. 73(4):337–345

Teng S, Guo Z, Peng H et al (2017) High-frequency repetitive transcranial magnetic stimulation over the left DLPFC for major depression: Session-dependent efficacy: A meta-analysis. Eur Psychiatry. 41:75–84

Tian Y, Ling W, Wang C et al (2011) Dissociation between visual line bisection and mental number line bisection in schizophrenia. Neurosci Lett. 491(3):192–195

Uhlhaas PJ, Singer W (2013) High-frequency oscillations and the neurobiology of schizophrenia. Dialogues Clin Neurosci. 15(3): 301–313

Verma R, Kumar N, Kumar S (2018) Effectiveness of adjunctive repetitive transcranial magnetic stimulation in management of treatment-resistant depression: A retrospective analysis. Indian J Psychiatry. 60(3):329–333

Vink JJT, Mandija S, Petrov PI et al (2018) A novel concurrent TMS-fMRI method to reveal propagation patterns of prefrontal magnetic brain stimulation. Hum Brain Mapp. 39 (11):4580-4592

Vuoksimaa E, Koskenvuo M, Rose RJ et all (2009) Origins of handedness: a nationwide study of 30, 161 adults. Neuropsychologia 47(5):1294-1301

Wang Y, Jongman A, Sereno JA (2001) Dichotic perception of Mandarin tones by Chinese and American listeners. Brain Lang. 78(3):332–348

Wang Y, Wang S (1992) Binaural split hearing test and its clinical application. Fore Medl Neuro Neurosurg 19(6): 289–292

Wassermann EM, Greenberg BD, Nguyen MB et al (2001) Motor cortex excitability correlates with an anxiety-related personality trait. Biol Psychiatry 50(5):377–382

Zhang L-J, Zhou F-Y, Wang X-Y (2008) Brain lateralization of acoustic phonetics processing of pitch information in speech. Chin J App Psychol. 114(14):330–335

Zhang X, Bin G, Shi L et al (2014) Brain activation pattern during a verbal fluency task in depression patients: a near-infrared spectroscopy imaging study. Chin J Psychiatry 47(1):7–11

Zhang X, Ouyang J, Yang X (2016) Effects of repetitive transcranial magnetic stimulation combined on the manic episode of bipolar disorder. China Med Pharm 6(3):199–201

Zhang C, Zhang Y, Luo H et al (2022) Bilateral Habenula deep brain stimulation for treatment-resistant depression: clinical findings and electrophysiological features. Transl Psychiatry 12:52. https://doi.org/10.1038/s41398-022-01818-z

Zhao J, Lin X, Jiang K et al (2000) Hypoperfusion in baseline and cognitively activated brain SPECT imaging of adult and elderly patients with depression. Chin J Nucl Med 20(4):165–168

Zheng Z, Zou K, Yan T (2014) Neurophysiological study of transcranial magnetic stimulation. In: Wang X, Lu L (eds) Transcranial magnetic stimulation and neuropsychiatric disorders, 1st edn. Peking University Medical Press, Beijing, pp 54–111

Zheng Z, Zou K, Jin R et al (2015a) Discussing Qi activity ascending/descending the theory of TCM basing on cerebral asymmetry variation in psychiatric disorders(2). J Chengdu Univ TCM 38(2):96–100, 108

Zheng Z, Zou K, Wang C et al (2015b) Discussing Qi activity ascending/descending the theory of TCM basing on cerebral asymmetry variation in psychiatric disorders(1). J Chengdu Univ TCM 38(1):95-98,108

Zheng Z, Zou K, Yang C et al (2015c) Discussing Qi activity ascending/descending the theory of TCM basing on cerebral asymmetry variation in psychiatric disorders(3). J Chengdu Univ TCM 38(3): 88–93

Zheng Z, Zou K, Yang C et al (2015d) Discussing Qi activity ascending/descending the theory of TCM basing on cerebral asymmetry variation in psychiatric disorders(4). J Chengdu Univ TCM 38(4): 72–79

Ziemann U, Bruns D, Paulus W (1996a) Enhancement of humanmotor cortex inhibition by the dopamine receptor agonist pergolide:

evidence from transcranial magnetic stimulation. Neurosci Lett 208:187–190

Ziemann U, Lonnecker S, Steinhoff BJ, Paulus W (1996b) The effect of lorazepam on the motor cortical excitability in man. Exp Brain Res 109:127–135

Ziemann U, Lonnecker S, Steinhoff BJ, Paulus W (1996c) Effects of antiepileptic drugs on motor cortex excitability in humans: a transcranial magnetic stimulation study. Ann Neurol 40:367–378

Ziemann U, Tergau F, Bruns D et al (1997) Changes in human motor cortex excitability induced by dopaminergic and anti-dopaminergic drugs. Electroencephalogr Clin Neurophysiol. 105(6):430–437

Ziemann U, Tergau F, Wischer S et al (1998) Pharmacological control of facilitatory I-wave interaction in the human motor cortex. A paired transcranial magnetic stimulation study. Electroencephalogr Clin Neurophysiol. 109(4):321–330

Zou K, Sun X, Huang X et al (2007) A proton magnetic resonance spectroscopy research on hippocampus in patients with treatment-resistant depression. Chin J Psychiatry 40(2):74–77

Zou K, Huang X, Li T et al (2008) Alterations of white matter integrity in adults with major depressive disorder: a magnetic resonance imaging study. J Psychiatry Neurosci 33(6):525–530

# Anxiety Disorders

# 19

Kun Wang, Wensi Hao, Shiyu Wang, and Xiaona Dai

## 1 Post-traumatic Stress Disorder

Post-traumatic stress disorder (PTSD) is a chronic anxiety disorder that an individual develops in response to one or more traumatic events. PTSD is characterized by flashbacks of past traumatic events, avoidance of reminders of trauma, and hypervigilance. In terms of the pathogenesis, PTSD is considered as a consequence of interactions between genetic susceptibility and environment factors. The dysregulation of hypothalamic-pituitary-adrenal (HPA) axis in the neuroendocrine system, molecular abnormalities in the glucocorticoid receptor stress signaling pathway, and structural and/or neuronal activity abnormalities in the prefrontal cortex, anterior cingulate gyrus, amygdala, and hippocampus are reported to be associated with the development of PTSD. PTSD has complex clinical manifestations, with different symptom combinations and variable mechanisms specific to neural circuit abnormalities.

Psychological intervention and medication, as the current dominant therapies for PTSD, fail to produce rapid and effective responses in PTSD patients, with limited efficiency rate and cure rate. Treatments for PTSD include exposure therapy and acupuncture. Modern neuromodulation strategies are developed based on functional neuroimaging techniques and have been applied to the treatment of PTSD, including transcranial magnetic stimulation (TMS), electroconvulsive therapy, transcranial direct current stimulation (tDCS), vagus nerve stimulation, percutaneous nerve stimulation, deep brain stimulation, etc. (Koek et al. 2019).

K. Wang
Department of Neurology, Beijing Puren Hospital, Beijing, China

W. Hao · X. Dai
Department of Neurology, Xuanwu Hospital, Capital Medical University, Beijing, China

S. Wang (✉)
Department of Pediatric, Xuanwu Hospital, Capital Medical University, Beijing, China

Since no FDA-approved neuromodulation therapies for PTSD are available, all such interventions for PTSD are mostly used in clinical trials. Studies have shown that the electroconvulsive therapy (ECT) affects the possibility of reactivating traumatic memories in PTSD patients. In most studies on tDCS for PTSD, the dorsolateral prefrontal cortex (DLPFC) and ventromedial prefrontal cortex (VMPFC) were targeted to reduce the expression of fear in patients. This section will focus on the TMS therapy.

### 1.1 Treatment Principle of rTMS

The neurobiological hypothesis for the efficacy of repetitive TMS (rTMS) in the treatment of PTSD is based on dysfunction in brain regions (including the amygdala, frontal lobe, and hippocampus) associated with the processing of threatening and fearful stimuli. rTMS exerts its therapeutic effect by regulating abnormal activity in brain regions associated with PTSD symptoms and modulating multiple neural circuits, including dopaminergic, serotonergic, and noradrenergic neurons, and to some extent by potentially reversing pathological and molecular abnormalities in the brain of PTSD patients. A study in mice showed that rTMS normalized the HPA axis after stress while exerting neuroprotective effects by increasing brain-derived neurotrophic factors (BDNFs) (Czeh et al. 2002).

### 1.2 Treatment Regimen of rTMS

No optimal parameter options of rTMS for PTSD are available. Generally, increasing stimulation intensity and number of pulses and prolonging treatment duration within a safe range may contribute to successful treatment. However, due to the lack of a unified standard for rTMS parameters, large samples are required for double-blind controlled studies.



Y. Wang (ed.), *Therapeutics of Neural Stimulation for Neurological Disorders*, https://doi.org/10.1007/978-981-99-4538-2_19

#### 1.2.1 Stimulation Target

Without guidance from a navigation system, the therapist places a magnetic stimulation coil (parallel to the skull) on the stimulated area of the patient's head. With the development of neuronavigation, an optical tracking device allows for more accurate and stable positioning of the stimulation coil. The DLPFC has been the most often targeted site in studies with PTSD. The rTMS treatment options for PTSD include excitation of the left DLPFC, excitation of the left DLPFC combined with the inhibition of the right DLPFC, excitation of the right DLPFC, inhibition of the right DLPFC, inhibition of the right DLPFC, and bilateral DLPFC (Koek et al. 2019).

#### 1.2.2 Stimulation Intensity

In practical application, the stimulation intensity is individualized, depending on the stimulation effect on the brain; the stimulation intensity is generally set to 80–110% MT, based on 100% motor threshold (MT). Although high-intensity stimulation may be more effective than low-intensity stimulation, suprathreshold intensity stimulation may induce epilepsy, localized scalp pain, and muscle contractions in the hands. In the elderly, due to significant prefrontal atrophy and increased distance between the skull and prefrontal cortex, the stimulation intensity may be appropriately increased to achieve a therapeutic effect.

#### 1.2.3 Stimulation Frequency

The optimal TMS frequency for the treatment of PTSD remains unclear: generally, low frequency ≤1 Hz and high frequency ≥5 Hz.

#### 1.2.4 Stimulation Time

Typical duration is about 20 min per treatment.

#### 1.2.5 Number of Treatment Sequences

Current clinical studies are generally designed to continuously treat once daily for 5 days, followed by suspension for 2 days, or daily continuous treatment (i.e., continuous treatment for 7 days every week). The duration of treatment is typically 2 weeks, ranging from 1 to 8 weeks. For a treatment with a duration of more than 2 weeks, re-determination of motor thresholds is often required.

### 1.3 Clinical Application of rTMS

There are many randomized, placebo-controlled clinical studies of rTMS in the treatment of PTSD. The results suggested that rTMS was effective in relieving PTSD and rTMS induced on the right DLPFC may have the potential to treat PTSD. In addition, both low-frequency and high-frequency rTMS alleviated PTSD symptoms (Yan et al. 2017); low-frequency rTMS alleviated PTSD and depression symptoms, and high-frequency rTMS led to improvement in the main symptoms and related symptoms of PTSD. High-frequency rTMS seemed to be superior to low-frequency rTMS; however, another meta-analysis and systematic review in 2014 revealed that rTMS applied to the right DLPFC may be more effective than to the left DLPFC, while no difference was noted between high-frequency rTMS and low-frequency rTMS.

The Evidence-Based Guidelines on the Therapeutic Use of rTMS released by the International Federation of Clinical Neurophysiology (IFCN) in 2014 recommended that high-frequency TMS applied to the right DLPFC may be effective in patients with PTSD (Level C recommendation) (Lefaucheur et al. 2014). The efficacy of this therapy was further confirmed in 2018 (Grade of Evidence B) (Lefaucheur et al. 2020).

In summary, we found that TMS was clinically and statistically superior to sham TMS in the treatment of PTSD. However, given the limited number of available studies and the heterogeneity of these studies, it is crucial to further investigate the potential effects of TMS in the treatment of PTSD in clinical practice.

### 1.4 Treatment Principle of tDCS

The amygdala (AMG), hippocampus (HPC), and prefrontal cortex (PFC) are mostly involved in the fear network that is associated with PTSD. PTSD is characterized by a persistent maladaptation to fear responses, in which the PFC is unable to regulate fear signals from the amygdala and dorsal anterior cingulate gyrus (Koch et al. 2016). The PFC, which consists of the VMPFC and the DLPFC, inhibits the fear response. Numerous studies have revealed that overactivity in the AMG and DLPFC is positively connected with PTSD symptom severity, whereas decreased activity in the VMPFC in PTSD is inversely correlated with symptom severity (Diehl et al. 2018). Most studies involving tDCS in the treatment of PTSD have targeted DLPFC or VMPFC, which may modify the amygdala's reactivity. In addition, research has demonstrated that the lateral temporal cortex (LTC) and insula are hyperactivated in PTSD and that LTC and insula may interact with fear circuits, resulting in maladjustment in PTSD (Simmons and Matthews 2012).

tDCS modulates cortical excitability and spontaneous neuronal activity in target regions by modifying the resting potential of neurons, which may last for up to an hour. Anodal tDCS hyperpolarizes the axon terminal membrane, whereas cathodal tDCS depolarizes it. tDCS can also cause low-level neural oscillations that affects functional connections in local and distant brain regions.

## 1.5 Treatment Regimen of tDCS

### 1.5.1 Stimulation Target

The DLPFC has been the most often targeted site for PTSD with the anode over the F3 and cathode over the F4. Additionally, the VMPFC is a typical application target, with the anode over AF3 and the cathode over PO8 or the anode over AF3 and the cathode over M1. Studies have also been conducted on the right temporal cortex, with the HD-tDCS central electrode (cathode) placed at T8 and the ring anode placed at F8, C4, P8, and EX10, respectively.

### 1.5.2 Stimulation Intensity

The current intensity is generally set to 2.0 mA.

### 1.5.3 Stimulation Time

Typical duration is about 20–25 min per treatment.

### 1.5.4 Number of Treatment Sequences

Current clinical studies are generally designed to be treated once a day for 10 consecutive days.

## 1.6 Clinical Application of tDCS

Jungwon Han and Ahmadizadeh et al. (Han et al. 2022; Ahmadizadeh et al. 2019) discovered that tDCS in the bilateral DLPFC can considerably alleviate symptoms of PTSD, depression, and anxiety. Van't Wout-Frank et al. (2019) found that anodal VMPFC tDCS paired with exposure to virtual reality decreased psychophysiological arousal and clinical symptoms of PTSD. A 2015 pilot study (Saunders et al. 2015) demonstrated that anodal VMPFC tDCS combined with computerized working memory training significantly improved working memory and PTSD symptoms in four veterans, along with normalization of neurophysiological markers (qEEG in P3a). During the consolidation of extinction memory, anodal VMPFC tDCS alleviated PTSD symptoms much more than when applied during fear extinction (Van't Wout et al. 2017). Preliminary research (Hampstead et al. 2020) with HD-tDCS in the right temporal cortex revealed that it is feasible to treat PTSD symptoms, which are accompanied by alterations in the connectivity of the right LTC and the fear network.

In summary, TMS and tDCS were clinically and statistically superior to sham TMS and tDCS in the treatment of PTSD. However, given the limited number of available studies and the heterogeneity of these studies, it is crucial to further investigate the potential effects of TMS and tDCS in the treatment of PTSD in clinical practice.

# 2 Panic Disorder

## 2.1 Definition

Panic disorder (PD) is a functional disorder in the category of neurosis, also known as acute anxiety attacks. It is clinically characterized by repeated episodes of severe panic, often accompanied by repeated palpitations, chest pain, sweating, choking, tremor, dizziness, and even a strong sense of impending death or loss of control; the episodes may last for a few minutes or tens of minutes, with a self-limiting disease course. PD typically occurs in young and middle-aged adults and is more common in women than in men. The lifetime prevalence of PD is 1–5%, and the annual prevalence in the USA and UK is 1.8% and 1.7% (Kessler et al. 2005; Skapinakis et al. 2011); according to surveys in China, the prevalence of PD is 4.3% (3.6–4.1%) (Huang 2008; He et al. 2012). The risk of suicide in patients with PD is twice that of patients suffering from other psychiatric disorders and almost 20 times that of those without psychiatric disorders. Early diagnosis and treatment have satisfactory long-term outcomes in PD patients, resulting in symptomatic relief in more than 50% of patients (CampbeH and Abbass 2007; Wu 2010). Patients who fail to receive timely and effective early treatment may develop a chronic fluctuating course or even become incapacitated. PD is associated with psychiatric comorbidities and related social medical costs and may lead to significantly impaired quality of life, daily activities, and work performance in patients with PD; therefore, it is particularly important to strengthen the diagnosis and treatment of PD.

## 2.2 Therapeutic Principle of Transcranial Magnetic Stimulation

TMS is a safe and noninvasive neuromodulation technique, which has been widely used in the treatment of neurological and psychiatric diseases due to few side effects. rTMS transiently modulates neural excitability/neuroplasticity in measured and connected regions. It has been widely used in the research and treatment of various neurological and psychiatric disorders such as depression, epilepsy, stroke, and Parkinson's disease.

Some studies suggest that TMS may affect the plasticity of the cerebral cortex; its main mechanism is related to the plasticity of synapses, and to a certain extent, it is also related to various factors such as ion channels and membrane potential. TMS is a noninvasive neurostimulation technique that works by modulating the activity of neurons in the cerebral cortex. rTMS mainly affects the synaptic plasticity of neurons, especially the long-acting potentiation/depression mechanisms. The modulation of cortical excitability by rTMS is influenced by stimulus intensity/frequency, number of pulses, stimulus duration, and coil position and type; in particular, frequency is most closely related. Current studies suggest that low-frequency rTMS (≤1 Hz) reduces motor cortex excitability, while high-frequency rTMS (≥5 Hz) improves motor cortex excitability. Extensive TMS experiments reported rTMS also had effects on intermediate links in synaptic plasticity, including gene initiation, protein expression, BDNFs, amino acids, monoamine neuromodulation, secretion of neurotransmitters (e.g., norepinephrine, dopamine, serotonin), and migration of NMDA and AMPA receptor synthesis. In addition to the regulation of synaptic plasticity, the changes in neural excitability induced by rTMS may involve other potential mechanisms, such as changes in cell membrane excitability, changes in resting membrane potential, modification of ion channels, hyperpolarization, depolarization of membrane potential, threshold potential, or regulation of neurotransmitters, and metabolism of amino acids.

Maladaptive structural and functional neuroplasticity in the prefrontal and limbic regions has now been identified as an important pathological mechanism in anxiety disorders. Studies have revealed that anxiety disorders are associated with hypoactivation in the left DLPFC and increased activation in the amygdala. On the other hand, evidence suggests that activation in the right DLPFC is higher in patients with anxiety disorders (e.g., panic disorder). These functional activation patterns may also be associated with structural abnormalities in each of these brain regions, which may contribute to the emergence of cognitive/affective deficits (i.e., exaggerated fear responses/threat perception). Based on the above findings, rTMS may be an effective treatment for anxiety disorders by modulating the pathological hypoactivation/hyperactivation in the DLPFC to counteract the induced dyskinesia and maladaptive neuroplasticity.

## 2.3 Clinical Recommendations for Neuromodulation in the Treatment of PD

Low-frequency rTMS applied to the right DLPFC and temporoparietal region for panic attacks and generalized anxiety disorder. Grade C consistent evidence

## 2.4 Clinical Applications

A total of four English literatures on neuromodulation therapy for PD were found by searching the database (Prasko et al. 2007; Mantovani et al. 2013; Deppermann et al. 2017; Kumar et al. 2018).

Prasko conducted a randomized, double-blind controlled rTMS study in 15 patients with PD. Eligible patients who met the ICD-10 diagnostic criteria were enrolled. Fifteen patients received drug therapy for at least 6 weeks prior to enrollment and were refractory to such drug therapy. The stimulation parameters were as follows: stimulation frequency 1 Hz, right DLPFC as stimulation site, 110% RMT, and 1800 pulses/day, five times per week for 2 weeks, for a total of ten times. Symptomatic assessments of panic symptoms, anxiety symptoms, adverse events, and cognitive function were performed before treatment, at 2 weeks after treatment, and at follow-up visits. The outcomes were as follows: Anxiety improvement was noted in the true stimulation group and sham stimulation group, without statistically significant differences between the groups; treatment with rTMS did not cause cognitive impairment or other adverse events. No subjects dropped out during the study. No serious adverse events (SAEs) were reported during the study.

Mantovani conducted a randomized, double-blind controlled rTMS study in 25 patients with PD (aged 18–65 years), of whom 13 received true stimulation and 12 received sham stimulation. All patients met the DSM-IV diagnostic criteria for PD and depression. The stimulation parameters were as follows: stimulation frequency 1 Hz, right DLPFC as stimulation site, 110% RMT, and 1800 pulses/day, five times per week for 4 weeks, for a total of 20 times. The outcomes were as follows: Before treatment and at the end of 4-week treatment, the true stimulation group had better PD response than the sham stimulation group (50% vs. 8%, Fisher's test $P = 0.005$) based on symptomatic assessments of PD (PDSS), anxiety (HAMA), and depression, without significant improvement in depression. During the study, four patients dropped out due to the inconvenience of daily rTMS treatment. No SAEs such as epilepsy, neurological complications, or memory or attention impairment occurred during the study. There were no statistically significant differences in common AEs between the true stimulation group and the sham stimulation group.

Deppermann conducted a randomized, controlled, double-blind study of iTBS. A total of 44 patients (aged 18–65 years) with PD were included, all of whom met DSM-IV diagnostic criteria for PD. Patients were randomized into 2 groups, including 22 patients in the true stimulation group and 22 patients in the sham stimulation group; patients were treated with iTBS combined with 9-week cognitive behavioral therapy (CBT). There was no dose adjustment of SRIs within 3 weeks of enrollment. iTBS stimulation parameters were as follows: the stimulation site was the left DLPFC, and intensity was 80% RMT, 15 sessions for a total of 3 weeks. There is no significant difference between the true and the sham stimulation groups in the main outcomes, but there was a significant improvement in the plaza fear avoidance score compared with the sham stimulation group.

Conducted an open-label rTMS study which included a total of 13 patients (aged 28–52 years) with PD and depression (diagnosed based on ICD-10 diagnostic criteria by psychiatrists) who developed resistance to at least 2 different classes of drugs at optimal doses and for at least 8–12 weeks. No sham stimulation control group was set for the study. The stimulation parameters were as follows: stimulation frequency was 20 Hz, stimulation site was the left DLPFC, intensity was 110% RMT, and stimulation time was 5 s with 20 s interval and a total of 10 sessions of stimulation, 1000 pulses/day, five times per week for 4 weeks for a total of 20 times. Compared with pre-treatment, patients treated with 20 sessions of rTMS had significantly lower scores based on the Panic Disorder Severity Scale (PDSS) and Hamilton Depression Rating Scale (HDRS), with PDSS decreasing by 38% and HDRS by 40%. The main side effects were one case of headache and two cases of local scalp discomfort, respectively. All patients had no serious adverse reactions during and after treatment and were well tolerated.

Most of the studies described above suggested the potential beneficial effects of rTMS for panic attack treatment. Studies have shown that inhibitory stimulation of the right DLPFC and excitatory stimulation of the left DLPFC both have significant improvement. The results of most studies (albeit uncontrolled) were consistent with those of the randomized, double-blind, sham-controlled study by Mantovani et al., suggesting that clinical improvement in these subjects may not be a placebo effect. In addition, it is not clear whether improvement in depressive comorbidities panic symptoms is associated with improvement in depressive symptoms. Another source of variability was the number of stimuli, which varied from a minimum of 9 to a maximum of 40 in the above studies. The results of Prasko et al. (2007) are in contrast to those of Mantovani et al. (2013) and might be due to the different number of stimuli. In the study by Mantovani et al. (2013), the author used 40 stimuli, compared with 10 stimuli in the study by Prasko et al. (2007). Therefore, it is speculated that a sufficient number of stimuli may play a key role in the clinical effect.

## 3 Generalized Anxiety Disorder

### 3.1 Overview

#### 3.1.1 Disease Manifestation and Current Situation

Generalized anxiety disorder (GAD) is an uncontrollable chronic anxiety disorder, which is often accompanied by physiological symptoms such as sleep disorders, muscle tightness, and difficulty concentrating. According to the ICD-10 diagnostic criteria, this anxiety is characterized by excessive worrying and false predictions about neutral or threatening events, the negative effects of which the patient views as catastrophic. Anxiety can spread to multiple areas of daily life, including health, family relationships, occupation, or financial situation (Bandelow et al. 2013). The lifetime prevalence was about 5.7%, with the highest prevalence in the age group of 45–59 years (7.7%), and reduced in people aged 30–44 years and older than 60 years. The prevalence was twice as high in women (7%) as in men (4%). The International Study of Disabilities in Mental Disorders found that 38% of GAD patients had moderate to severe occupational role impairment, with an average of 6.3 days of absence or loss of function per month (Hoge et al. 2012).

#### 3.1.2 Pathogenesis

The pathogenesis of GAD is not clear; both genetic and social factors may contribute to the onset of the disease. The incidence of first-degree relatives is significantly higher than that of the normal population, and twin studies showed a moderate genetic impact; Spanish researchers examined more than 1000 patients attending primary care centers and found that polymorphic variants in the serotonin 1A receptor gene are associated with common clinical manifestations in anxiety comorbidities with depression (Molina et al. 2011). Social factors also have a certain impact on the onset of patients with GAD; some patients show personality characteristics of easy tension, fear, sensitivity, and high vigilance before the onset of the disease. Large epidemiological survey studies have shown that in adult patients, separation, widowhood, divorce, unemployment, and long-term housewife may be risk factors of the disease. In adolescent patients, risk factors include low socioeconomic status, abuse, traumatic events, and self-regulation disorders (Ma and Mao 2016). Neurobiological factors such as neurotransmitter disturbances may also play a role in the pathogenesis of GAD. In recent years, with the gradual progress of neural network theory, network dysfunction is also considered to cause the onset of GAD.

#### 3.1.3 Treatment

Current guidelines recommend psychotherapy and drug therapy as first-line treatment regimens. Psychotherapy mainly refers to cognitive behavioral therapy, the response rate of patients is 47–75%, and it not only has short-term effect but also has positive efficacy in long-term follow-up studies. First-line antidepressants of selective serotonin reuptake inhibitors (SSRIs) and serotonin-norepinephrine reuptake inhibitors (SNRIs) as well as pregabalin are preferred, and benzodiazepines and buspirone can be selected for second-line drugs, with a response rate of patients ranging 44–81%. However, about 50% of GAD patients do not respond to current treatments, and their symptoms are protracted, resulting in more severe impairment of social functioning. As a new therapeutic method, neuromodulation technology has been gradually applied in the treatment of GAD due to its safety and effectiveness.

## 3.2 Treatment Principle

The underlying mechanism of GAD indicates that patients have persistent activity in brain regions associated with introspective thinking and mental behavior (De Lima et al. 2019). Recent evidence from neuroimaging studies suggests that emotion regulation is controlled by specific neural circuits and that GAD patients may have abnormalities in prefrontal-limbic network connectivity. In the process of inducing anxiety, it is found that GAD patients show stronger connectivity between prefrontal and limbic lobes than the normal population. In addition, GAD patients are unable to inhibit anxiety-related neural activity once they complete the task of inducing anxiety; when they complete the emotion regulation task, GAD patients show insufficient activation of the prefrontal cortex and/or anterior cingulate cortex compared with the normal population, while the connection between the medial prefrontal cortex and other frontal regions (especially the dorsolateral prefrontal cortex) and limbic lobe (such as the insula) is weakened; in the task of emotional conflict requiring latent emotion regulation, GAD patients are unable to mobilize the anterior cingulate cortex (a brain region that successfully resolves conflict and emotional control) compared with the normal population (Diefenbach et al. 2016). Neuroimaging studies suggest that GAD patients have hyperactivation of the amygdala (a brain region implicated in the regulation of fear responses) and abnormal activity in this region may be associated with abnormalities in neurotransmitters. Imbalances in neurotransmitters may lead to altered connectivity and abnormal network function between the amygdala and prefrontal regions (Dilkov et al. 2017). The dorsolateral prefrontal cortex plays a key role in emotion regulation, especially in cognitive and emotional control, by influencing decision-making and flexibly guiding behavior to achieve situational goals, and clinical studies have found that inhibition of right dorsolateral prefrontal cortex activation can improve emotional inhibition during decision-making, while increasing the activity in the left dorsolateral prefrontal cortex can mobilize the intensity of emotional regulation. Simultaneous alteration of dorsolateral prefrontal cortex activity can also alter its functional connectivity with the ventromedial prefrontal cortex, a region critical for successful emotion suppression (Diefenbach et al. 2016). GAD patients have a negative stimulation bias that makes it difficult to divert attention from negative stimuli. The posterior parietal cortex is involved in attentional bias. Electroencephalography, magnetoencephalography, and functional magnetic resonance studies have shown hyperactivation of the parietal cortex (e.g., the superior parietal lobule and intraparietal sulcus) in anxiety patients. There are structural and functional connections between the dorsolateral prefrontal cortex and parietal cortex, and both participate in the top-down attentional control network. Studies on functional magnetic resonance imaging have shown abnormalities in the frontoparietal executive control network (including the dorsolateral prefrontal cortex and posterior parietal cortex) in GAD patients (Lin et al. 2020). At the same time, GAD patients may also have functional abnormalities in the coracoid part of the left anterior cingulate cortex, the left orbitofrontal cortex, and the right cingulate isthmus.

Overall, the activity of the left dorsolateral prefrontal cortex in GAD patients is negatively correlated with the degree of anxiety, and excitatory stimulation by tDCS or TMS can upregulate positive responses to positive emotional stimuli and downregulate negative responses to emotional stimuli while enhancing prefrontal cortex regulation of the limbic lobe in the processing of negative emotions. In addition, due to excessive activation of the right dorsolateral prefrontal cortex in GAD patients, inhibition of its activation by tDCS or TMS can improve emotional control during decision-making, thereby improving patient symptoms.

## 3.3 Treatment Options and Clinical Applications

There are few clinical studies on physical therapy for GAD up to now, and there is no consensus on treatment regiments. Existing clinical studies mainly use tDCS and TMS for treatment, which will be described in detail below.

### 3.3.1 tDCS Therapy

Ana Lucia de Lima et al. conducted a randomized controlled study of tDCS in 30 GAD patients. In the true stimulation group, the anode placement was placed at the left dorsolateral prefrontal cortex (corresponding to F3 in 10/20 electroencephalograms), and the cathode placement was placed above the contralateral supraorbital region (corresponding to Fp2 in 10/20 electroencephalograms). The direct current intensity was 2 mA, and the electrode sheet area was 35 cm$^2$ (5 cm × 7 cm). Treatment was continued for 5 consecutive days from Monday to Friday, 20 min per day, for a total of five treatments. The electrodes of the sham stimulation group were placed in the same position as the true stimulation group, and the current parameters were set consistently, but the current was turned off after 30 s of stimulation. The results showed that there was a significant decrease in anxiety levels in the test and control groups and the mean HAMA score was reduced by 48.7% in the true stimulation group and 33.6% in the sham stimulation group compared with the baseline period. None of the patients who completed the treatment had complications, and very few patients had adverse reactions such as headache and drowsiness (De Lima et al. 2019).

Farzad Nasiri et al. conducted a randomized controlled study of the Unified Protocol for Transdiagnostic Treatment of Emotional Disorders (UP) combining with tDCS in 43 GAD depression patients. Fifteen patients were treated with UP for 60 min every week for 12 weeks; 13 patients were treated with tDCS in the last 2 weeks of UP treatment, with the cathode placed in the right dorsolateral prefrontal cortex (corresponding to F4 in 10/20 electroencephalogram) and the anode placed in the deltoid muscle. The direct current intensity was 2 mA, and the electrode sheet area was 25 $cm^2$ (5 cm × 5 cm). Treatment was continued for 5 consecutive days, 30 min a day, for 2 weeks from Monday to Friday, for a total of 10 treatments; 15 patients underwent blank control. The anxiety and depression scores of the treatment group were significantly decreased compared with the control group after the end of treatment and at the 3-month follow-up, and the scores of patients with UP therapy and transcranial electrical stimulation (tES) were significantly decreased, as were the remission rates (69% vs. 60% vs. 0% at the end of treatment and 77% vs. 60% vs. 0% at the 3-month follow-up). No complications were observed in all patients who completed the treatment (Nasiri et al. 2020).

Shiozawa Pedro et al. treated one GAD patient with tDCS. The cathode was placed at the right dorsolateral prefrontal cortex (corresponding to F4 in 10/20 electroencephalograms), and the anode was placed at the deltoid muscle. The direct current intensity was 2 mA, and the electrode patch area was 35 $cm^2$ (5 cm × 7 cm). Treatment was continued for 15 consecutive days, 30 min per day, for a total of 15 treatments. The patient's anxiety symptoms relieved significantly after 15 days, and there were no complications or adverse reactions (Shiozawa et al. 2014).

### 3.3.2 TMS Therapy

Alexander Bystritsky et al. enrolled ten GAD patients treated six times (two times per week) with low-frequency stimulation (1 Hz) of the right dorsolateral prefrontal cortex at an intensity of 90% resting motor threshold (RMT) for 15 min per treatment (900 pulses). Patients' anxiety symptoms and HAMA scores decreased significantly after treatment, with six patients achieving complete remission after treatment and two patients' HAMA scores decreasing by more than 50% (Bystritsky et al. 2008).

Gretchen J. Diefenbach et al. conducted a randomized controlled trial of 25 GAD patients. The right dorsolateral prefrontal cortex was stimulated at low frequency (1 Hz) at 90% RMT with each treatment lasting 15 min (900 pulses) for 30 sessions (five times per week). The sham stimulation group had the same sites and parameters as the true stimulation group, but the sham stimulation group used a false coil. The results suggested that the anxiety symptoms were significantly improved in patients in the true stimulation group after treatment and during the follow-up period, but no effect was observed in the sham stimulation group (Diefenbach et al. 2016).

Dancho Dilkov et al. conducted a randomized controlled trial of 42 GAD patients. The stimulation target is on the right side of the dorsolateral prefrontal cortex, using the high-frequency stimulation (20 Hz), with 110% intensity of RMT; each session consisted of 20 trains (9 s per train and 51 s between trains) for a total of 25 sessions (5 sessions per week in the first 4 weeks, 3 sessions in the fifth week, and 2 sessions in the sixth week). The stimulation site and parameters in sham group were consistent with the true stimulation group, but the coil of the sham stimulation group remained 90° with the skull. Scores on the HAMA scale in the true stimulation group decreased from moderate to severe (25–38 points) to mild (3–10 points), with an average decrease of 79%, and all patients met the response criteria (score decrease ≥50%). There was no significant change in sham group scores (Dilkov et al. 2017).

Zhaoyang Huang et al. conducted a randomized controlled trial in 36 patients with GAD on insomnia. The right parietal cortex (corresponding to the P4 site of the EEG, containing the superior parietal lobule and the intraparietal sulcus) was stimulated with low frequency (1 Hz) at an intensity of 90% RMT with three trains of 500 pulses per train separated by 10 min, for a total of ten sessions. The coil used in the sham group was identical in shape and sound to the true stimulation group, but did not perform active stimulation. Anxiety scores of patients in the true stimulation group were significantly decreased, while those in the sham group were basically unchanged. Only a small percentage of patients experienced mild headache and neck pain (Huang et al. 2018).

Yicong Lin et al. conducted a randomized controlled trial of 29 GAD patients. Patients in the three groups were stimulated in the right dorsolateral prefrontal cortex for 10 ms, 20 ms, and 50 ms followed by stimulation of the right inferior parietal lobule. Each session consisted of 15 trains (100 pulses per train, train interval of 30 s) for a total of 10 sessions using low-frequency stimulation at 1 Hz and 90% RMT. There was a significant decrease in HAMA scores in the second and third groups after treatment and during the follow-up period, and there was a significant difference between the third and first groups (Lin et al. 2020).

## 3.4 Conclusion

An increasing number of studies on neural network imbalance in GAD have been conducted in recent decades. A variety of neuromodulation techniques have been applied in the clinical studies of GAD, but these studies are currently in the exploratory stage. Although the parameter settings and stimulation targets are not completely consistent in the treatment of GAD patients, both TMS and tES can significantly improve the clinical symptoms of patients with few side effects, and the efficacy and safety have been preliminarily verified. In the future, it is necessary to conduct more clinical studies to develop precise neuromodulation protocols for GAD.

## References


Ahmadizadeh MJ, Rezaei M, Fitzgerald PB (2019) Transcranial direct current stimulation (tDCS) for post-traumatic stress disorder (PTSD): a randomized, double-blinded, controlled trial. Brain Res Bull 153:273–278

Bandelow B, Boerner RJ, Kasper S et al (2013) The diagnosis and treatment of generalized anxiety disorder. Dtsch Arztebl Int 110(17):300–310

Bystritsky A, Kaplan JT, Feusner JD et al (2008) A preliminary study of fMRI-guided rTMS in the treatment of generalized anxiety disorder. J Clin Psychiatry 69:1092–1098

CampbeH SG, Abbass AA (2007) Chest pain: consider panic disorder. Can Faro Physician 53(5):807–808

Czeh B, Welt T, Fischer AK et al (2002) Chronic psychosocial stress and concomitant repetitive transcranial magnetic stimulation: effects on stress hormone levels and adult hippocampal neurogenesis. Biol Psychiatry 52(11):1057–1065

De Lima AL, Braga FM, Da Costa RM et al (2019) Transcranial direct current stimulation for the treatment of generalized anxiety disorder: a randomized clinical trial. J Affect Disord 259:31–37

Deppermann S, Vennewald N et al (2017) Neurobiological and clinical effects of fNIRS-controlled rTMS in patients with panic disorder/agoraphobia during cognitive-behavioural therapy. Neuroimage Clin 16:668–677

Diefenbach GJ, Assaf M, Goethe JW et al (2016) Improvements in emotion regulation following repetitive transcranial magnetic stimulation for generalized anxiety disorder. J Anxiety Disord 43:1–7

Diehl MM, Bravo-Rivera C, Rodriguez-Romaguera J et al (2018) Active avoidance requires inhibitory signaling in the rodent prelimbic prefrontal cortex. Elife 7:e34657

Dilkov D, Hawken ER, Kaludiev E et al (2017) Repetitive transcranial magnetic stimulation of the right dorsal lateral prefrontal cortex in the treatment of generalized anxiety disorder: a randomized, double-blind sham controlled clinical trial. Prog Neuropsychopharmacol Biol Psychiatry 78:61–65

Hampstead BM, Mascaro N, Schlaefflin S et al (2020) Variable symptomatic and neurophysiologic response to HD-tDCS in a case series with posttraumatic stress disorder. Int J Psychophysiol 154:93–100

Han J, Choi KM, Yang C et al (2022) Treatment efficacy of tDCS and predictors of treatment response in patients with post-traumatic stress disorder. J Affect Disord 318:357–363

He Y, Zhang L et al (2012) Prevalence of anxiety disorders among outpatients in general hospitals. Chin Ment Health J 26(3):165–170

Hoge EA, Ivkovic A, Fricchione GL (2012) Generalized anxiety disorder: diagnosis and treatment. BMJ 345:e7500

Huang Y (2008) Current status of epidemiological research on mental disorders in China. China Prev Med 9(5):445–446

Huang Z, Li Y, Bianchi MT et al (2018) Repetitive transcranial magnetic stimulation of the right parietal cortex for comorbid generalized anxiety disorder and insomnia: a randomized, double-blind, sham-controlled pilot study. Brain Stimul 11(5):1103–1109

Kessler RC, Chiu WT et al (2005) Prevalence, severity, and comorbidity of 12 month DSM-IV disorders in the National Comorbidity Survey Replication. Arch Gen Psychiatry 62(6):617

Koch SB, van Zuiden M, Nawijn L et al (2016) Aberrant resting-state brain activity in posttraumatic stress disorder: a meta-analysis and systematic review. Depress Anxiety 33(7):592–605

Koek RJ, Roach J, Athanasiou N et al (2019) Neuromodulatory treatments for post-traumatic stress disorder (PTSD). Prog Neuropsychopharmacol Biol Psychiatry 92:148–160

Kumar S, Singh S et al (2018) Effect of high-frequency repetitive transcranial magnetic stimulation (rTMS) in patients with comorbid panic disorder and major depression. Australas Psychiatry 26(4):398–400

Lefaucheur JP, Andre-Obadia N, Antal A et al (2014) Evidence-based guidelines on the therapeutic use of repetitive transcranial magnetic stimulation (rTMS). Clin Neurophysiol 125(11):2150–2206

Lefaucheur JP, Aleman A, Baeken C et al (2020) Evidence-based guidelines on the therapeutic use of repetitive transcranial magnetic stimulation (rTMS): an update (2014–2018). Clin Neurophysiol 131(2):474–528

Lin Y, Chen P, Yang K et al (2020) Efficacy of repetitive dual-site paired associative Transcranial magnetic stimulation in the treatment of generalized anxiety disorder. Brain Stimul 13(5): 1170–1172

Ma X, Mao FQ (2016) Psychiatry. Peking University Medical Press, Beijing

Mantovani A, Aly M et al (2013) Randomized sham controlled trial of repetitive transcranial magnetic stimulation to the dorsolateral prefrontal cortex for the treatment of panic disorder with comorbid major depression. J Affect Disord 144:153–159

Molina E, Cervilla J, Rivera M et al (2011) Polymorphic variation at the serotonin 1-A receptor gene is associated with comorbid depression and generalized anxiety. Psychiatr Genet 21:195–201

Nasiri F, Mashhadi A, Bigdeli I et al (2020) Augmenting the unified protocol for transdiagnostic treatment of emotional disorders with transcranial direct current stimulation in individuals with generalized anxiety disorder and comorbid depression: a randomized controlled trial. J Affect Disord 262:405–413

Prasko J, Zalesky R et al (2007) The effect of repetitive transcranial magnetic stimulation (rTMS) add on serotonin reuptake inhibitors in patients with panic disorder: a randomized, double blind sham controlled study. Neuro Endocrinol Lett 28:33–38

Saunders N, Downham R, Turman B et al (2015) Working memory training with tDCS improves behavioral and neurophysiological symptoms in pilot group with post-traumatic stress disorder (PTSD) and with poor working memory. Neurocase 21(3):271–278

Shiozawa P, Leiva AP, Castro CD et al (2014) Transcranial direct current stimulation for generalized anxiety disorder: a case study. Biol Psychiatry 75:e17–e18

Simmons AN, Matthews SC (2012) Neural circuitry of PTSD with or without mild traumatic brain injury: a meta-analysis. Neuropharmacology 62:598–606

Skapinakis P, Lewis G et al (2011) Panic disorder and subthreshold panic in the UK general population: epidemiology, comorbidity and functional limitation. Eur Psychiatry 26(6):354–362

Van't Wout M, Longo SM, Reddy MK et al (2017) Transcranial direct current stimulation may modulate extinction memory in posttraumatic stress disorder. Brain Behav 7(5):e00681

van't Wout-Frank M, Shea MT, Larson VC et al (2019) Combined transcranial direct current stimulation with virtual reality exposure for posttraumatic stress disorder: feasibility and pilot results. Brain Stimul 12(1):41–43

Wu W (2010) Guidelines for the prevention and treatment of anxiety disorders. People's Medical Publishing House, Beijing, pp P82–P114

Yan T, Xie Q, Zheng Z et al (2017) Different frequency repetitive transcranial magnetic stimulation (rTMS) for posttraumatic stress disorder (PTSD): a systematic review and meta-analysis. J Psychiatr Res 89:125–135

# Schizophrenia

# 20

Zhong Zheng, Ke Zou, Jiaxi Huang, Junlan Yang, Jingshu Zhou, Ruicai Xiong, and Yingxuan Li

Schizophrenia (SCZ) is a serious psychiatric disorder of unknown etiology, with an acute or subacute onset, and is a leading cause of disability in young adults. The disease is often manifested as perceptual, cognitive, emotional, and behavioral disorders. Schizophrenia is characterized by positive and negative symptoms and cognitive impairment. Positive symptoms refer to abnormal or hyperactive mental functions, including abnormal sensations (visual and auditory hallucinations), delusions, confused thoughts, disorganized speech and behavior, etc. Negative symptoms refer to reduced or absent mental functions, including lack of motivation, lack of pleasure, social withdrawal, reduced willpower, etc. Cognitive deficits include impaired attention, decline in executive function and working memory, etc. Although the use of chlorpromazine in 1950 cured or alleviated the psychiatric symptoms of most SCZ patients, there were still 20% of the patients who were drug-resistant or dissatisfied with psychotropic treatments. So far, SCZ treatment mainly includes medication, electroconvulsive therapy (ECT), psychotherapy, and rTMS treatment.

Z. Zheng (✉)
Neurobiological Laboratory, West China Hospital, Sichuan University, Chengdu, Sichuan, China

Center for Neurological Function Test and Neuromodulation, West China Xiamen Hospital, Sichuan University, Xiamen, China

K. Zou
Neurobiological Laboratory, West China Hospital, Sichuan University, Chengdu, Sichuan, China

J. Huang · J. Zhou
Center for Neurological Function Test and Neuromodulation, West China Xiamen Hospital, Sichuan University, Xiamen, China

J. Yang
Department of Geriatrics, The Fourth People's Hospital (Mental Health Centre) of Pengzhou City, Chengdu, Sichuan, China

R. Xiong
Department of Psychiatry, The Fourth People's Hospital (Mental Health Centre) of Pengzhou City, Chengdu, Sichuan, China

Y. Li
Department of Neurology, Xuanwu Hospital, Capital Medical University, Beijing, China

## 1 Research on the Pathogenesis of Schizophrenia

### 1.1 Research Progress on the Mechanism of Auditory Hallucinations in Schizophrenia

Phenomenological research on the correlation between subjective perception and the function of inner speech network (Barber et al. 2021) identified three independent symptom dimensions in auditory hallucinations: language complexity (left temporal cortex and Broca's area), self-other misattribution (cingulate and left temporal cortex), and spatial location (right temporoparietal junction) (Plaze 2011). The cognitive model of auditory-verbal hallucinations (AVH) asserts that AVH in SCZ results from the misperception of an individual's own inner speech as others' overt speech. Cognitive theory suggests that AVH derives from one's inner speech that has been misidentified as an external source due to defective speech monitoring of the nervous system. Imaging studies have found that inner speech can activate brain regions in language-dominant hemisphere such as inferior frontal cortex, insula, and supplementary motor area, as well as brain areas associated with speech monitoring such as superior temporal cortex and anterior cingulate. The inner speech of SCZ patients with AVH is different from that of patients without AVH and normal individuals, particularly in the abnormalities of monitoring circuits. Inner speech and AVH share the same neural circuit, thus competing with each other. There are differences in the perception of speech between the two. The loudness of AVH is the quantification of sensation, while the reality of AVH depends on its heterogeneity. The metrical stress evaluation task is an effective way to study the relationship between AVH heterogeneity



Y. Wang (ed.), *Therapeutics of Neural Stimulation for Neurological Disorders*, https://doi.org/10.1007/978-981-99-4538-2_20

and inner speech through competition for shared neurophysiological resources. In order to investigate the relationship between subjective perception and the neural activation of inner speech network, Vercammen et al. (2011) used the metrical stress evaluation task to dissociate and evaluate the differences between patients' speech production and perception. The study showed that the louder the AVH, the lower the tasked-related activity in bilateral angular gyrus, anterior cingulate gyrus, left inferior frontal gyrus, left insula, and left temporal cortex, suggesting that the loudness of AVH was associated with a competition for neural resources. The reality of AVH was related to reduced language lateralization, while one's inner speech production might contribute to the loudness of AVH.

### 1.1.1 Auditory Hallucination and Sensory Gating

Under normal circumstances, information is regularly transmitted from the low-level center to the high-level center in the human brain. The normal human brain in response to external stimulus involves a selection and filtering process called sensory gating (SG) (Braff et al. 1992). SG is a normal brain function by which the brain inhibits the inputs of indifferent sensory stimulus, so that higher functions of the brain are not overloaded with such stimulus. SG deficit is one of the underlying etiologies of psychiatric disorders, which is associated with indifferent stimulus overload and may lead to various attention-related psychiatric symptoms such as pressure of thought, inability to calm the mind, and looseness of thought. The etiology of SG involves responses to novel stimulus or changes in successive stimulus and minimization or cessation of responses to incoming indifferent stimulus.

SG can be assessed by neuroelectrophysiological measures, that is, a prepulse inhibition (PPI) evoked by condition-test paradigm or paired-stimulus paradigm. The paradigm involves S1 (condition stimulus) and S2 (test stimulus), with an interstimulus interval (ISI) between S1 and S2. The potential evoked by S2 within a certain time period after S1 will be suppressed by the after-effect of S1, demonstrating that S1 suppresses S2. The time difference between S1 and S2 is called interstimulus interval (ISI). The auditory sensory gating P50 (SG-P50) normally adopts S1–S2 paired-click stimulations with a 500 ms ISI. The interval between each acoustic pair is 10 s. SG potential is defined as the maximum positivity between 30 and 90 ms post-stimulus with a peak latency between 50 and 60 ms. So, it is called P50. The P50 signal is averaged superimposed for 16–32 times to shape a pair of S2-P50 and S1-P50 waveform (Fig. 20.1). The 256-channel auditory P50 studies found that the P50 response of Cz lead was most prominent. Thus, the recording electrode was mostly connected to Cz, while the reference electrode was placed behind left and right ears connected in parallel. We adopted paired-click stimulations containing a short click and a pure tone, with the short click lasting 0.1–0.2 ms and the pure tone lasting

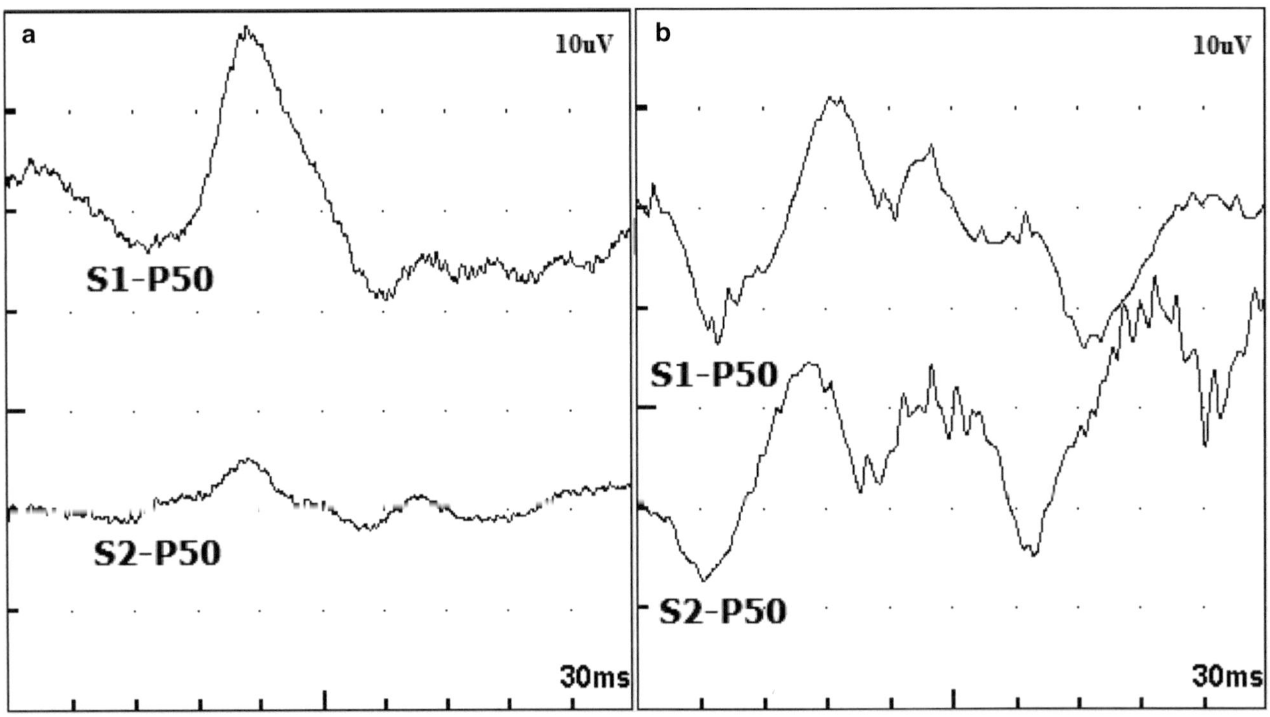


**Fig. 20.1** Comparison between the P50 mean waveform of healthy subjects (**a**) and SCZ patients (**b**)

40 ms. The sound stimulation intensity was 50 dB above the hearing threshold. The ISI between S1 and S2 is 500 ms. The S1 and S2 stimulation responses are averaged for 16–32 times. The patient is not required to respond to the stimulus during the test. To control confounding variables such as individual differences, the amplitude ratio of the test stimulus to the condition stimulus ($S_2/S_1$), the difference between the condition stimulus and the test stimulus ($S_1 - S_2$), and 100 ($1 - S_2/S_1$) were used to express abnormal P50 activity. Besides SG-P50, SG-N100 and SG-P200 can also be used as a neurophysiological measure to evaluate auditory gating function.

SG is an automatic, unlearned response. SG-P50 is an auditory middle-latency response. The effect of repeated auditory stimulation of the P50 amplitude is considered to reflect an automatic suppression mechanism of the brain to filter indifferent stimulus. S1 recurs after a long cycle and is a novel component during successive stimulation changes. The sensory gate of a normal human brain is open to S1 stimulation, and all the sensory information can enter higher-order cortical center. However, S2 is repeated in a short time after S1, and the two have the same characteristics. As a result, S2 does not transmit any new information and becomes an indifferent stimulus which is suppressed. This phenomenon is manifested in the test as the inhibition of S2-P50 response (amplitude decrease) and a smaller S2-P50/S1-P50 amplitude ratio. Compared with the P50 amplitude caused by S1, the P50 amplitude caused by S2 in healthy adult is reduced by 60–80% (at least >50%). But in patients with SCZ and depression, the absolute amplitude of S2-P50 and S2-P50/S1-P50 amplitude ratio increase (Chen et al. 2006; Zheng et al. 2012).

SG functions may be related to dopamine, acetylcholine (α7-nicotinic acetylcholine), γ-aminobutyric acid (GABA-B), dopamine (D2), norepinephrine, serotonin ($5\text{-}HT_3$) receptors, and adenosine. Currently, a consistent view is that P50 suppression deficits are genetically linked to the α7-nicotinic receptor subtype (α7-AchR) of cholinergic receptors (Freedman et al. 1997). The α7-cholinergic receptor gene (cHRNA7), located at 15q13-14, is one of the three most well-documented candidate susceptibility genes for SCZ (Waternor et al. 2002). The α7-AchR, encoded as cHRNA7, belongs to the nicotinic acetylcholinergic receptor of the central nervous system, which is distributed in the hippocampus, striatum, thalamus, and cortex of the brain. The hippocampal pyramidal cells are the brain regions where auditory-evoked potentials occur. Neurobiological studies have found that α7-AchR activates cholinergic neurons that project to the hippocampus. When cholinergic nerve fibers are excited, GABA is released, which excites inhibitory interneurons that suppress the response of hippocampal pyramidal cells to the second stimulus. If the hippocampal α7-AchR function is defective, the GABA release is decreased, which is not enough to excite inhibitory interneurons causing SG deficits. The phenomenon of P50 suppression is formed during the auditory process. After analyzing three models of auditory sensorimotor gating disorders in SCZ, researchers believe that the inferior colliculus in the midbrain is the first relay station of the auditory neural pathway and is critical in the auditory PPI. The deep layer of the superior colliculus sends projection fibers to the pedunculopontine tegmental nucleus, which is also an important midbrain structure responsible for the PPI formation. The balance of the mutual regulation between various neurotransmitters is important for maintaining normal PPI (Li et al. 2007). However, auditory P50 suppression is not just a physical property of the brain's response to stimulation. Craniotomy studies have found that P50 suppression also exists in the cortex and neocortex of the temporal lobe, olfactory area, and hippocampus (Boutros et al. 2005, 2008). Paired-pulse stimulation applied to moss fibers in rats can record paired-pulse inhibitory potentials in the pyramidal neurons of the hippocampal CA3 region. Norepinephrine (NE) can eliminate this inhibition, while the antagonist of NE, such as haloperidol and phentolamine, can eliminate the disinhibition effect of NE, thus supporting that the SG-P50 originates from the CA3 region (Xu et al. 1992, 1997). Amphetamine or phencyclidine can also eliminate SG suppression (Stevens et al. 1991). Paired-click stimulation induces not only P50 activity but also subsequent N100 (SG-N100) and P200 (SG-P200) potentials (Picton et al. 1974; Marijn et al. 2009).

Hirano et al. (2010) used vowel instead of pure tones to measure the SG-evoked magnetoencephalography P50m inhibition ratio (S2-P50m/S1-P50m) and N100m inhibition ratio (S2-N100m/S1-N100m). The study showed that the speech-auditory gating deficits in the left hemisphere of SCZ patients were more pronounced and were positively correlated with the severity of auditory hallucinations. The SG in the right hemisphere was associated with negative symptoms in SCZ. Recent studies suggest that the human left temporal cortex is essential in noise filtering. For example, humans can selectively focus on more important voices while screening out the interference of other sounds in a noisy environment.

Robert et al. (2017) compared the SG function differences between AVH-on state and AVH-off state. The results showed that the inhibition ratios of SG-P50, SG-N100, and SG-P200 during AVH-on were significantly higher than that of AVH-off. The Psychotic Symptom Rating Scales (PSYRATS) AVH total score was negatively correlated with SG-N100 inhibition ratio, positively correlated with S2-N100, and positively correlated with S1-N100.

### 1.1.2 Auditory Hallucinations and Cognitive Function

Kim et al. (2012) suggested that salience-related regions are involved in particular stimuli that trigger hallucinations in SCZ patients. Event-related potential (ERP) studies have shown that late-onset SCZ auditory hallucinations may be associated with abnormal P300 of negative emotion (Li et al. 2007). The amplitude of left temporal auditory P300 in the auditory hallucination group was significantly lower than that in the non-auditory hallucination group ($P < 0.05$). The latency of each electrode in the auditory hallucination group was significantly longer than that in the non-auditory hallucination group ($P < 0.05$) (Ni and Guo 2001). The fMRI study suggests that SCZ patients with auditory hallucinations have increased activity in the amygdala and parahippocampus when they receive implicit emotional words (Escartí et al. 2010). It indicates that patients with auditory hallucinations are accompanied by cognitive impairment and emotional experiences, such as anxiety and depression.

Oh and Kim (2011) applied high-frequency rTMS to the left DLPFC along with low-frequency rTMS to the left temporoparietal cortex (TPC) when treating patients with drug-resistant SCZ. They found that both Positive and Negative Symptom Scale (PANSS) scores and Depressive Symptom Scale scores were significantly reduced and short-term semantic memory was enhanced. The case study reported by Fitzgerald and Daskalakis (2011) also demonstrated that the effect of multi-brain combination therapy was better than that of single-brain therapy, with improvements in auditory hallucinations and reductions in depression and anxiety scores. Therefore, while rTMS treats auditory hallucinations in SCZ, other related symptoms other than auditory hallucinations should receive more attention.

## 2 rTMS Therapeutic Mechanism and Treatment Protocol Study

### 2.1 Effects of rTMS on Motor Cortex Excitability in SCZ

Oxley et al. (2004) used 1 Hz low frequency to stimulate the motor cortex of health control and schizophrenic patients and found that low-frequency rTMS reduced the normalized scalp area for MEP (percentage difference from baseline MEP) and motor threshold (MT) in health control, while low-frequency stimulation caused the MEP area and MT to increase in patients. This indicated that patients with SCZ have abnormal neuroplasticity responses and further suggested that patients have abnormal GABAergic neuron activity.

### 2.2 rTMS Treatment Study of Auditory Hallucinations in SCZ

### 2.2.1 Low-Frequency rTMS Treatment in the Left Temporoparietal Lobe

Based on basic research, the current treatment of auditory hallucinations in SCZ uses low-frequency rTMS (≤ 1 Hz) to stimulate the left temporoparietal lobe, which is the midpoint of T3-P3 connection line in the international 10-20 EEG system (Hoffman et al. 1999). This is the temporoparietal cortex (TPC), which includes the superior temporal gyrus (STG) and temporoparietal junction (TPJ). Hoffman et al. used low-frequency rTMS to treat intractable auditory hallucinations in SCZ patients with four consecutive trials. In 2000, in a crossover experiment with sham stimulation as the control group in 12 SCZ patients, 8 people's intractable hallucinations were significantly improved, but other symptoms did not significantly change between real and sham stimulation (Hoffman et al. 2000). Hoffman et al. (2003) also used a double-blind controlled trial, in which 24 patients were randomized to receive real and sham stimulation at 1 Hz for 9 days (Hoffman et al. 2003). Results showed that there were significant differences between rTMS group and sham stimulation group. Hoffman et al. (2005) recruited 50 patients with auditory hallucinations at least 5 times a day as subjects. Using a double-blind parallel design, 50 subjects were randomly assigned to the study group and the control group. The study group received a frequency of 1 Hz in the left prefrontal cortex, with 90% MT intensity. The control group received sham stimulation and were assessed with the clinical general impression (CGI) scale before and after stimulation. The results suggested that the CGI score of the study group was significantly improved and the number of auditory hallucinations was significantly reduced. The study also showed that 52% of patients maintained treatment effects for 15 weeks or longer. Holi et al. (2004) conducted a crossover experiment. The patient group was stimulated by 10 Hz, 100% MT, and 20 series/day of rTMS, while the control group was stimulated by sham stimulation. There was no significant difference between the treatment group and the control group. Aleman et al. (2007) systematically reviewed the efficacy of 1 Hz rTMS in the left TPC regarding the treatment of auditory hallucinations, which includes 10 studies and 212 patients. The results showed that rTMS was effective in improving auditory hallucinations, compared with sham stimulation, with an effect size of 0.76 (95% CI = 0.36–1.17). When nine studies with continuous stimulation therapy were included, the effect size increased to 0.88, and the heterogeneity disappeared, suggesting that rTMS could improve auditory hallucinations, but not general psychotic symptoms.

West China Hospital of Sichuan University cooperated with the Fourth People's Hospital (Mental Health Centre) of Pengzhou City Sichuan Province to study the variance of

SG-P50 in SCZ patients with AVH and the effect of low-frequency stimulation of the left TPC on symptoms and SG-P50. Symptoms were assessed by PANSS and the Auditory Vocal Hallucination Rating Scale (AHRS); SG-P50 was measured by three-lead recording. The recording electrodes were located in the left temporoparietal lobe (TP3) (the midpoint of the T3-P3 connection line), the right temporoparietal lobe (TP4) (the midpoint of the T4-P4 connection line), and the parietal lobe (Cz). The reference electrodes were placed in parallel on the left and right mastoids. The SG-P50 studies found that, compared with the healthy control group, the S1-P50 amplitude in Cz decreased, while the S2-P50 amplitude and S2/S1 in TP3, TP4, and Cz increased ($P < 0.05$–$0.001$). The S2-P50 in TP3 increased ($P < 0.05$–$0.001$). The increasing trend of S2-P50 in TP3 is more significant. Based on clinical routine drug treatment, the observation group was treated with 1 Hz, 110% MT, stimulation interval of 15 s, and 1600 pulses of rTMS for 15 times (2 weeks), while the control group was only given routine clinical drug treatment. Symptoms and SG-P50 were measured before treatment and 2 weeks and 4 weeks after treatment. The results showed that PANSS and AHRS in the observation group before treatment were significantly lower than those at 2 weeks and 4 weeks after treatment ($P < 0.05$) and there was no significant difference in symptom scores in the control group before and after treatment ($P > 0.05$). Two weeks after treatment in the observation group, only the S2/S1 in TP3 and Cz were significantly lower than those before treatment ($P < 0.05$), and there was no significant difference in SG-P50 in TP4 before and after treatment ($P > 0.05$). There was no significant difference between the two groups in the parameters before treatment ($P > 0.05$). The study suggests that although rTMS can reduce symptom scores and the symptom improvement can be maintained until the fourth week, the improvement of S2/S1 ratio only exists at the end of treatment, which has a transient effect.

The therapeutic mechanism of low-frequency rTMS may be related to the reduction of cortical excitability of language processing circuits. Giesel et al. (2012) showed that rTMS can improve auditory hallucination symptoms and reduce the excitability of primary auditory cortex. Low-frequency rTMS can also normalize the functional connectivity between the frontal temporoparietal lobes by improving the functional connectivity between the left temporoparietal brain regions (Mishra et al. 2011; Wang et al. 2008). The study found that rTMS can significantly improve auditory hallucination symptoms in patients and it is better than the improvement effect of drug treatment. A possible underlying mechanism is that stimulation increases the excitability of the bilateral cortex, enhances the SG function, and improves cognition and negative emotions, which facilitates patient's filtering ability to indifferent sensory stimulus and reduces the interference of indifferent stimulus, thereby improving the patient's auditory hallucinations and positive and negative symptoms.

Lefaucheur et al. (2014) summarized the previous research results and suggested that low-frequency rTMS of the left temporoparietal cortex (TPC) may be effective in the treatment of auditory hallucinations in SCZ. They recommended rTMS as a Level III clinical treatment for SCZ auditory hallucinations. Since 2014, there are only three Grade B evidence-based studies reporting randomized controlled trials of real and sham stimulation, but they still cannot change the Level III recommendation for low-frequency stimulation of the left TPC (Lefaucheur et al. 2020).

### 2.2.2 High-Frequency rTMS Treatment in the Left Temporoparietal Lobe

In addition to low-frequency stimulation, some scholars have also reported the use of 20 Hz high-frequency stimulation of the left TPC. Dollfus et al. (2018) used the rTMS in individualized fMRI neuronavigation technique to stimulate fMRI activation region, left superior temporal sulcus, activated by language tasks and used the rTMS automatic positioning device to make the stimulation site always located at the intersection between the projection of the ascending ramus of the left lateral fissure and the left superior temporal sulcus. But the classical location of T3-P3 based on the 10–20 EEG system showed great anatomical variability between subjects. By comparing the 20 Hz real stimulation and the sham stimulation after 4 weeks, it was found that rTMS could significantly reduce AHRS, but there was no significant difference between the real and sham stimulation AHRS after treatment.

Several studies have evaluated the clinical value of using inhibitory cTBS interventions. Koops et al. (2016) conducted a study with a large sample of 71 patients (34 real stimulation, 37 sham stimulation) who received 10 treatments (2 times/day). The results showed that real stimulation can significantly reduce AVHS and PSYRATS. However, another study in which 1 Hz stimulation and cTBS were applied to the left TPC separately (Kindler et al. 2013) did not find any statistical difference. Lefaucheur et al. (2020) summarized two Grade B evidence-based studies, which demonstrated that there was insufficient evidence for the efficacy of temporal and parietal high-frequency stimulation of auditory hallucinations.

## 2.3 rTMS Treatment Research on Negative Symptoms of SCZ

The current consensus is that cognitive impairment is one of the core symptoms of SCZ, mainly including deficits in attention, memory (including working memory, instant memory, short-term memory, and long-term memory), abstract thinking (mainly manifested in executive function), and information integration. Cognitive impairment is the primary symptom of SCZ rather than a consequence of disease.

Therefore, correct evaluation of cognitive function has crucial guiding significance for clinical work. Liu et al. (2008) treated 12 patients with high-frequency rTMS in the left DLPFC with 110% MT, 10 Hz, and 1500 pulses per day. After 20 treatments, compared with 11 cases of sham stimulation in the control group, there was no effect on the indicators of attention network test ($P > 0.05$). However, rTMS could shorten the total time, incorrect response time, error rate, and correct response time of the Wisconsin Card Sorting Test ($P < 0.05$). The test time of the sham stimulation group was also shortened ($P < 0.01$). Studies have shown that rTMS has a tendency to improve some cognitive impairment in patients, but the conventional rTMS mode may not be effective in treating SCZ. Zhang et al. (2010) used iTBS mode to stimulate the left DLPFC in 15 patients in the treatment group, with a stimulation volume of 80% MT, 2400 pulses/day. The efficiency was 46.7%, while the effective rate was 8.3% in the sham stimulation group. PANSS total score, positive factor score, negative factor score, and general psychopathology score and Anderson Negative Symptom Scale (Scale for the Assessment of Negative Symptoms, SANS) total score were significantly reduced. The conventional rTMS mode has a weak effect on synaptic plasticity. The TBS mode can induce long-term excitability changes in the human cerebral cortex with a shorter stimulation duration and lower stimulation frequency.

Feinsod et al. (1998) found that 1 Hz low-frequency rTMS of the left prefrontal lobe was effective for patients with anxiety, tension, and restlessness. Cohen et al. (1999) found that 1 Hz low-frequency rTMS of the left prefrontal lobe was effective in patients with negative symptoms. These results suggest that the left prefrontal lobe may be the treatment area for negative symptoms. Compared with sham stimulation, Liu (2011) studied the effect of 1 Hz low-frequency rTMS in the left temporoparietal lobe on serum brain-derived neurotrophic factor (BDNF) in SCZ patients and its relationship with antipsychotic drugs. They found that BDNF levels were significantly increased after 10 and 20 days of atypical antipsychotics combined with rTMS treatment ($P < 0.001$) and BDNF levels were significantly increased after typical antipsychotics combined with rTMS and atypical antipsychotics ($P < 0.05$). There was no significant change after treatment with typical antipsychotics alone. BDNF level at the beginning of the treatment was negatively correlated with the Clinical Global Impression Scale-Severity (CGI-S) and the Clinical Global Impression Scale-Global Improvement (CGI-GI) at the end of treatment ($P < 0.05$) and was also negatively correlated with the Positive and Negative Symptom Scale (PANSS) before and after treatment ($P < 0.001$), suggesting that low-frequency rTMS of the left temporoparietal lobe could increase serum BDNF concentration and alleviate clinical symptoms in SCZ patients.

Clinical EEG studies have shown that the decrease in alpha wave frequency is partially related to negative symptoms of SCZ. Yi Jin et al. (2006) selected 37 SCZ patients with negative symptoms and randomized them to receive 2 weeks of αTMS (8–13 Hz), 3 Hz, 20 Hz rTMS and sham stimulation with 80% MT loop coils to stimulate the left and right frontal areas (10–20% of the system's F3, F4). Negative symptom reduction rate and alpha EEG power spectrum analysis in F7, F3, Fz, F4, and F8 were evaluated at the end of treatment (2 weeks) and 2 weeks later (4 weeks). The results showed that the negative symptom reduction rate (29.6%) in the αTMS mode stimulation group was significantly higher than that in the other two rTMS groups and sham stimulation group (both <9%) ($P = 0.007$). The improvement of clinical symptoms was highly correlated with the increased percentage (34%) of frontal alpha wave amplitude ($r = 0.86$, $P = 0.001$).

Liu and Shen (2016) selected 80 SCZ patients and randomly assigned them to sham stimulation group, TBS group, 10 Hz group, and 20 Hz group. The left DLPFC was selected as the stimulation site. The stimulation intensity was 80% MT. The total number of stimulation pulses was 1200 every time. The four groups were treated once a day for 2 months. All patients were treated with olanzapine after admission, with a dosage of 10 mg/time, once a day. Visuospatial working memory and vocabulary fluency tests were used to evaluate patients' cognitive function. PANSS was used to evaluate patients' psychiatric symptoms. The results showed that the letter fluency test of the TBS group was higher than the other three groups and the PANSS negative symptom score was lower than the other three groups ($P < 0.05$). The PANSS general psychopathological score of the 10 Hz group was lower than the other three groups ($P < 0.05$). The visuospatial working memory of the 20 Hz group was higher than that of the other three groups, and the general psychopathological score of PANSS was lower than that of the other three groups ($P < 0.05$). They believed that the 10 Hz stimulation cannot improve the cognitive dysfunction of patients. The TBS stimulation can better improve the negative symptoms of SCZ. The 20 Hz stimulation can not only improve the general psychiatric symptoms of patients but also improve their visuospatial working memory. Research suggests that the 20 Hz stimulation can improve the cognitive function of SCZ patients. Studies have shown that 20 Hz rTMS can increase gamma resonance activity, which is a major physiological mechanism associated with cognitive activity (Barr et al. 2009).

Zheng et al. (2012) recruited 80 SCZ patients and randomly assigned them to 4 rTMS intervention groups. Sham, TBS, 10 Hz, and 20 Hz stimulation were used. The stimulation site was the left DLPFC, the stimulation intensity was 80% MT, and the total number of stimulation pulses was 1200/day. The stimulation was continued for 5 days. During the intervention, the original antipsychotic treatment was not changed. Cognitive function was assessed by VSWM and VFT, and psychiatric symptoms were assessed by PANSS. The results showed that 20 Hz stimulation could improve patients' visuospatial working memory ($t = -2.469$, $P = 0.024$). TBS led to improvement in vocabulary fluency test ($t = -4.538$, $P = 0.000$). Both TBS and 10 Hz stimulation led to improvement in the negative symptoms of patients (TBS, $t = 5.373$, $P = 0.000$; 10 Hz, $t = 2.272$, $P = 0.036$), while 10 Hz stimulation and 20 Hz stimulation led to improvement in general psychopathological symptoms (10 Hz, $t = 2.725$, $P = 0.014$; 20 Hz, $t = 3.632$, $P = 0.002$).

Lefaucheur et al. (2014) summarized previous research results and believed that high-frequency rTMS in the left DLPFC may be effective for SCZ negative symptoms. They recommended rTMS as a Level II clinical treatment. Between 2014 and 2018, two additional class II and III studies with "positive" results were added, while another class I and one class III study had "negative" results. There is also a wide heterogeneity in the patients participating in these studies. Therefore, the recommendation level will be lowered from B to C in 2020.

## 2.4 Recommendations for Clinical Treatment

### 2.4.1 Verbal Auditory Hallucinations

Clinical recommendation: low-frequency stimulation in the left temporoparietal lobe (the midpoint of the T3-P3 connection line of the 10–20 system). Level III recommendation, Grade B and C evidence (Lefaucheur et al. 2020)

### 2.4.2 Negative Symptoms

Clinical recommendation: high-frequency stimulation in the left DLPFC. Level III recommendation, Grade B and C evidence (Lefaucheur et al. 2020)

# 3 Transcranial Direct Current Stimulation in the Treatment of SCZ

## 3.1 Research Progress of Transcranial Direct Current Stimulation in the Treatment of SCZ

### 3.1.1 SCZ Auditory Verbal Hallucination

The treatment of auditory hallucinations is based on left prefrontal anodal stimulation and left temporoparietal cathodal stimulation. Brunelin et al. (2012) used real and sham stimulation RCTs and found that after 1 week of tDCS treatment with real stimulation twice a day, the AHRS score in the real stimulation group decreased significantly by 31% and then decreased by 38% after 3 months. Mondino et al. (2016a, 2016b) found that after real tDCS treatment twice a day, patients made fewer misattributions of covert and overt speech errors and the frequency of auditory hallucinations decreased by 46%, which theoretically supports the hypothesis that SCZ auditory hallucinations related to internal events are mistakenly thought to occur in the external spaces. In addition, the investigators found that improvement in auditory hallucinations after tDCS treatment was associated with decreased resting-state functional connectivity between the left temporoparietal junction and the left anterior insula. There is a significant reduction in the right inferior frontal gyrus (active during auditory hallucination episodes) and resting-state functional connectivity in the left temporoparietal lobe, but a significant increase in the resting-state functional connectivity in the left DLPFC, left angular gyrus, and precuneus. In another study targeting clozapine-resistant refractory SCZ or schizoaffective disorder, researchers found that increasing the number of treatments might improve clinical outcomes, with a greater decline at week 4 than at week 2 (Fregni et al. 2021).

### 3.1.2 SCZ Positive and Negative Symptoms

The studies on positive and negative symptoms were mainly focused on left DLPFC anodal stimulation and right DLPFC cathodal stimulation. Two class II studies showed no significant change in positive or negative symptoms with bilateral DLPFC stimulation. There was no significant result in clinical multicenter research (Jeon et al. 2018) or treatments with higher current density, duration, twice-a-day therapy combined with cognitive therapy (Shiozawa et al. 2016). However, a class I RCT ($n = 100$) study by Valiengo et al. (2019) found that tDCS with anodal stimulation of the left prefrontal lobe and cathodal stimulation of the left temporoparietal junction area resulted in significant and relative benefits for negative symptoms, respectively. This led to a Level II recommendation of tDCS for the treatment of SCZ negative symptoms.

## 3.2 Recommendations for Clinical Treatment

Verbal auditory hallucinations and negative symptoms in SCZ: a 2-mA current stimulation is applied by placing the anode over the left DLPFC and the cathode over the left temporoparietal junction, twice daily for more than 10 days. Level II recommendation (A, B, C evidence) (Fregni et al. 2021; Valiengo et al. 2019)

## 4 Deep Brain Stimulation in the Treatment of SCZ

### 4.1 Research Progress of Deep Brain Stimulation in the Treatment of SCZ

It is estimated that 10–30% of patients with SCZ show little or no response to first-line, non-clozapine antipsychotic medications, which is collectively termed as treatment-resistant schizophrenia (TRS). TRS is defined as the persistence of positive symptoms, despite two or more adequate trials of antipsychotic treatment with adequate dose, duration, and adherence. Clozapine is the gold standard antipsychotic for TRS, but approximately 40% of TRS patients do not respond even to clozapine (Wada et al. 2022).

Since the results of augmenting clozapine with other medications have been modest, researchers have been exploring the utility of brain stimulation procedures in TRS patients (Nucifora et al. 2019). Deep brain stimulation (DBS), which was originally applied in the 1980s (Benabid et al. 1987) for the successful treatment of tremor in Parkinson's disease (PD), is an invasive brain stimulation technique. Directly stimulating deep brain structures, DBS is thought to attenuate clinical symptoms by balancing dysfunctional networks in neuropsychiatric disorders (Nucifora et al. 2019). DBS now serves as an effective therapy for treatment-refractory movement disorders and PD, and it has been used in growing numbers of patients with treatment-resistant psychiatric disorders (Corripio et al. 2022), particularly in obsessive-compulsive disorder (OCD) and major depressive disorder (MDD). But there are only a small number of DBS studies in SCZ, partly due to uncertainties over where to site the electrodes (Corripio et al. 2020).

#### 4.1.1 DBS in Regions Within Cortico-Basal Ganglia Circuits

Previous studies indicate that basal ganglia structures may be morphology altered in SCZ patients, and functional neuroimaging studies support that SCZ is a brain disorder of abnormal functional connectivity within several brain structures, including those in the basal ganglia (Macpherson and Hikida 2019). As summarized in a recent review (Wada et al. 2022), the key neurological basis of SCZ is thought to be located in cortico-basal ganglia circuits, including the anterior cingulate cortex (ACC), medial prefrontal cortex (mPFC), ventral subiculum of the hippocampus (vSub), substantia nigra (SN), ventral pallidum (VP), nucleus accumbens (NAcc), nucleus reuniens of the midline thalamus (RE), and ventral tegmental area (VTA). And neuroimaging findings of TRS also showed increased glutamate levels in ACC as well as lower dopamine synthesis in the associative striatum. In other words, there might be abnormal neurotransmission from the medial prefrontal cortex (mPFC) to the associative striatum.

The substantia nigra pars reticulata (SNr) functions as a major basal ganglia output of limbic and associative cortico-striatal-thalamic circuits (CSTs) via ascending projections to the mediodorsal nucleus of the thalamus and limbic and associative cortices. Fann (2021) reported the first 6-month outcome of DBS of SNr in TRS (ClinicalTrials.gov, NCT02361554). They postulated that the SNr may serve as a common node, modulating limbic and associative cortices through a projection to the mediodorsal nucleus of the thalamus, and this would make it a potentially effective target for DBS in TRS. The subject is a 35-year-old Caucasian woman with TRS, paranoid type, with comorbid OCD. Her positive symptoms consisted of persistent auditory and visual hallucinations, thought broadcasting, and persecutory delusions. Antipsychotic medications, including clozapine, failed to significantly reduce her symptoms. Two DBS leads were implanted in bilateral SNr, and the parameters of monopolar stimulation were set as 60 μs of pulse width, 130 Hz of frequency, and 1.0 V in the left while 0.8 V in the right SNr. The subject reported immediate and complete resolution of chronic hallucinations in an amplitude dose-responsive fashion upon selective SNr stimulation initiation, and remission of delusions occurred within 12 weeks. After 24 weeks of stimulation, her baseline Brief Psychiatric Rating Scale (BPRS) for unusual thought content and hallucinations had both decreased to 1 (not present). Suspiciousness up to a delusional level was also down to 1 from a baseline score of 6 (severe). Her total BPRS score decreased from 42 at baseline to 20. Her Scale for the Assessment of Negative Symptoms anhedonia/asociality and avolition/apathy ratings were improved slightly. And the subject experienced no significant complications or adverse events related to the DBS device implantation except for increased appetite for the first 3 months. Now the patient remains stably improved at 1-year follow-up. This is the first reported case of SNr DBS for TRS, indicating that SNr DBS could be safely used in TRS and provide relief from symptoms that did not respond to pharmacologic treatment.

Corripio et al. (2020) reported the outcome of the first completed randomized crossover clinical trial of DBS in TRS patients (ClinicalTrials.gov, NCT02377505). They selected eight TRS patients who were also resistant to/intolerant of clozapine and randomly assigned them to NAcc or sgACC group. An open stabilization phase lasting at least 6 months was followed by a randomized double-blind crossover phase lasting 24 weeks in those who met symptomatic improvement criteria. The parameters of DBS were as follows: 130–210 Hz of frequency, 2.0–2.5 V of voltage, and 60–210 μs of pulse width. The primary endpoint was a 25% improvement in the Positive and Negative Symptom Scale (PANSS) total score. One patient did not receive DBS due to complications of surgery. Of the remaining seven patients, four patients (2/3 with NAcc and 2/4 with sgACC electrode placements) showed ≥ greater than 25% improvement in PANSS in their symptoms, thus meeting symptomatic improvement criteria. Three of these four patients entered the crossover phase, and all showed worsening when the stimulation was discontinued. The fourth patient worsened after the current was switched off accidentally without her or the investigators' knowledge. Physical adverse events were uncommon, but two patients developed persistent psychiatric adverse effects (negative symptoms/apathy and mood instability, respectively). Although the sample size was small and the variability in treatment response made it difficult to generalize, there were suggestions that the NAcc placement had greater beneficial effects than that in the sgACC. NAcc-targeted DBS also seemed to be particularly effective in patients with a clinical picture characterized mainly by delusions and hallucinations. Meanwhile, six patients (three patients received DBS in the NAc and three patients in the sgACC) underwent 18F-FDG PET at baseline and after at least 6 months of stimulation (Roldan et al. 2020). In this first series of cases examining metabolic changes after DBS treatment in TRS patients, metabolic increases were observed after NAcc stimulation but less clearly so with sgACC stimulation. It was also indicated that greater clinical improvement was associated with increases in activity in areas of the cortico-striatal-thalamic-cortical circuit.

Other structures defined as the neurobiological basis of SCZ including the hippocampus and VTA would also be the potential targets of treatment for TRS, though it remains untried (Corripio et al. 2022). Structural imaging studies have indicated that the hippocampus is the site of one of the largest volume reductions in SCZ, while functional imaging studies examining resting hippocampal activity in SCZ have had decidedly mixed results, including increased, equivocally increased, and normal activity. Another research proposed to examine the effect of VTA-targeted and NAcc-targeted DBS in SCZ patients with dominated negative symptoms. It is postulated that negative symptoms were a result of reduced function in these regions, according to the revised or extended dopamine hypothesis. It should be noted that this proposal assumes excitatory rather than inhibitory effects of DBS. Unfortunately, it was abandoned due to a lack of recruitment.

#### 4.1.2 DBS in the Habenula

Located in the epithalamus near the pineal gland, the habenula (Hb) is a set of well-conserved nuclei in all vertebrates that is divided into two anatomically and functionally distinct regions: the medial (MHb) and the lateral (LHb) habenula (Fakhoury 2017). The Hb is thought to play a key role in not only sleep and wakefulness but also the processing of aversive stimulus events, including pain, stress, and fear-provoking stimuli. Given its role in emotional processing, motivation, and dopaminergic function, it may be hypothesized that Hb dysfunction contributes to the pathophysiology of SCZ (Wang et al. 2020).

Wang et al. (2020) explored the utility of bilateral DBS of Hb in the treatment of two young adult male patients with TRS. The primary clinical outcome was the severity of the patients' positive and negative symptoms, as assessed by using the PANSS total and scale scores. The DBS parameters used for patient 1 were 60 μs, 60 Hz, and 2.0 V on the left and 2.5 V on the right side. Initial DBS parameters for patient 2 were 60 μs, 60 Hz, and 2.5 V on the left and 2.0 V on the right side, while it took approximately 8 months to adjust and identify the DBS parameters most suitable for patient 2 (3.15 V, 80 μs, and 135 Hz on the left side; 3.2 V, 60 μs, and 135 Hz on the right side). Although both patients experienced improvements in their clinical symptoms in the first 6 months of Hb DBS, only patient 2 reached a favorable clinical outcome at 12-month follow-up. By contrast, the symptoms of patient 1 subsequently worsened and became so severe and disruptive that he had to be hospitalized at 10-month follow-up and could no longer participate in the study. These results make it difficult to reach a general and robust conclusion about the clinical effectiveness of Hb DBS for TRS. The safety profile and long-term efficacy of this approach are poorly understood and need further investigation.

### 4.2 Recommendations for Future Clinical Trials

Bilateral DBS to restore normal function in separate, but connected, regions within cortico-basal ganglia circuits may be necessary for the efficacious treatment of the different types of SCZ (Macpherson and Hikida 2019). NAcc may represent the strongest candidate for DBS electrode placement in SCZ, with SNr also showing promise (Corripio et al. 2022). The DBS parameters should be among the following range: 60–210 Hz of frequency (mostly 60 Hz), 0.8–7.5 V of voltage (mostly 2.0–2.5 V), and 60–210 μs of pulse width (mostly 60 μs). And the follow-up duration should be 6–12 months.

## References

Aleman A, Sommer IE, Kahn RS (2007) Efficacy of slow repetitive transcranial magnetic stimulation in the treatment of resistant auditory hallucinations in schizophrenia: a meta-analysis. J Clin Psychiatry 68(3):416–421

Barber L, Reniers R, Upthegrove R (2021) A review of functional and structural neuroimaging studies to investigate the inner speech model of auditory verbal hallucinations in schizophrenia. Transl Psychiatry 11(1):582. https://doi.org/10.1038/s41398-021-01670-7

Barr MS, Farzan F, Rusjan PM et al (2009) Potentiation of gamma oscillatory activity through repetitive transcranial magnetic stimulation of the dorsolateral prefrontal cortex. Neuropsychopharmacology 34(11):2359–2367

Benabid AL, Pollak P, Louveau A et al (1987) Combined (thalamotomy and stimulation) stereotactic surgery of the VIM thalamic nucleus for bilateral Parkinson disease. Appl Neurophysiol 50(1–6):344–346

Boutros NN, Trautner P, Rosburg T et al (2005) Sensory gating in the human hippocampal and rhinal regions. Clin Neurophysiol 116(8):1967–1974

Boutros NN, Mears R, Pflieger ME et al (2008) Sensory gating in the human hippocampal and rhinal regions: regional differences. Hippocampus 18(3):310–316

Braff DL, Grillon C, Geyer MA (1992) Gating and habituation of the startle reflex in schizophrenic patients. Arch Gen Psychiatry 49(3):206–215

Brunelin J, Mondino M, Gassab L et al (2012) Examining transcranial direct-current stimulation (tDCS) as a treatment for hallucinations in schizophrenia. Am J Psychiatry 169:719–724

Chen X, Zhang M, Lou F et al (2006) Follow-up study on sensory gating in first-episode schizophrenia. Chin J Psychiatry 39(4):193–196

Cohen E, Bernardo M, Masana J et al (1999) Repetitive transcranial magnetic stimulation in the treatment of chronic negative schizophrenia: a pilot study. J Neurol Neurosurg Psychiatry 67(1):129–130

Corripio I, Roldan A, Sarro S et al (2020) Deep brain stimulation in treatment resistant schizophrenia: a pilot randomized cross-over clinical trial. EBioMedicine 51:102568

Corripio I, Roldan A, McKenna P et al (2022) Target selection for deep brain stimulation in treatment resistant schizophrenia. Prog Neuropsychopharmacol Biol Psychiatry 112:110436

Dollfus S, Jaafari N, Guillin O et al (2018) High-frequency Neuronavigated rTMS in auditory verbal hallucinations: a pilot double-blind controlled study in patients with schizophrenia. Schizophr Bull 44(3):505–514

Escartí MJ, de la Iglesia-Vayá M, Martí-Bonmatí L (2010) Increased amygdala and parahippocampal gyrus activation in schizophrenic patients with auditory hallucinations: an fMRI study using independent component analysis. Schizophr Res 117(1):31–41

Fakhoury M (2017) The habenula in psychiatric disorders: more than three decades of translational investigation. Neurosci Biobehav Rev 83:721–735

Fann JR (2021) Deep brain stimulation of the substantia nigra pars reticulata for treatment-resistant schizophrenia: a case report. Biol Psychiatry 90(10):e57–e59

Feinsod M, Kreinin B, Chistyakov A et al (1998) Preliminary evidence for a beneficial effect of low-frequency, repetitive transcranial magnetic stimulation in patients with major depression and schizophrenia. Depress Anxiety 7(2):65–68

Fitzgerald PB, Daskalakis ZJ (2011) Concurrent treatment of depression and auditory hallucinations in a patient with schizophrenia. Aust N Z J Psychiatry 45(8):681–683

Freedman R, Coon H, Myles-Worsley M et al (1997) Linkage of a neurophysiological deficit in schizophrenia to a chromosome 15 locus. Proc Natl Acad Sci U S A 94(2):587–592

Fregni F, El-Hagrassy MM, Pacheco-Barrios K et al (2021) Evidence-based guidelines and secondary meta-analysis for the use of transcranial direct current stimulation in neurological and psychiatric disorders. Int J Neuropsychopharmacol 24(4):256–313

Giesel FL, Mehndiratta A, Hempel A et al (2012) Improvement of auditory hallucinations and reduction of primary auditory area's activation following TMS. Eur J Radiol 81:1273–1275

Hirano Y, Hirano S, Maekawa T et al (2010) Auditory gating deficit to human voices in schizophrenia: a MEG study. Schizophr Res 117(1):61–67

Hoffman RE, Boutros NN, Berman RM et al (1999) Transcranial magnetic stimulation of left temporoparietal cortex in three patients reporting hallucinated "voices". Biol Psychiatry 46(1):130–132

Hoffman R (2000) Transcranial magnetic stimulation in clinical psychiatry. In: George M, Belmaker R (eds) Transcranial magnetic stimulation of schizophrenia. American Psychiatric Publishing, Inc, p 189

Hoffman ER, Hawkins AK, Gueorguieva R et al (2003) Transcranial magnetic stimulation of left temporoparietal cortex and medication-resistant auditory hallucinations.Arch Gen Psychiatry 60(1): 49–56

Hoffman R (2006) Transcranial magnetic stimulation in clinical psychiatry. In: George M, Belmaker R (eds) Transcranial magnetic stimulation of schizophrenia. American Psychiatric Publishing, Inc, pp189

Holi MM, Eronen M, Toivonen K et al (2004) Left prefrontal repetitive transcranial magnetic stimulation in schizophrenia. Schizophr Bull 30(2):429–434

Jeon DW, Jung DU, Kim SJ et al (2018) Adjunct transcranial direct current stimulation improves cognitive function in patients with schizophrenia: a double-blind 12-week study. Schizophr Res 197:378–385

Jin Y, Potkin SG et al (2006) Therapeutic effects of individualized alpha frequency transcranial magnetic stimulation (αTMS) on the negative symptoms of schizophrenia. Schizophr Bull 32(3):556–561

Kim JJ, Ku J, Lee H et al (2012) Distinct neural responses used to gain insight into hallucinatory perception in patients with schizophrenia. J Psychiatr Res 46(10):1318–1325

Kindler J, Homan P, Flury R, Strik W et al (2013) Theta burst transcranial magnetic stimulation for the treatment of auditory verbal hallucinations: results of a randomized controlled study. Psychiatry Res 209(1):114–117

Koops S, Dellen E, Schutte M et al (2016) Theta burst transcranial magnetic stimulation for auditory verbal hallucinations: negative findings from a double-blind-randomized trial. Schizophr Bull 42(1):250–257

Lefaucheur JP, André-Obadia N, Antal A et al (2014) Evidence-based guidelines on the therapeutic use of repetitive transcranial magnetic stimulation (rTMS). Clin Neurophysiol 125:2150–2206

Lefaucheur JP, André-Obadia N, Baeken C et al (2020) Evidence-based guidelines on the therapeutic use of repetitive transcranial magnetic stimulation (rTMS): an update (2014–2018). Clin Neurophysiol 131:474–528

Li N, Ji W, Zhang D et al (2007) Study on late schizophrenia and negative feelings and auditory event-related potential. Chin J Mod Med 17(5):613–615

Liu X (2011) Clinical control study about the effect of slow repetitive transcranial magnetic stimulation (rTMS) on serum levels of brain-derived neurotrophic factor (BDNF) in schizophrenia. J Psychiatry 24(3):169–199

Liu X, Shen J (2016) Effects of different repetitive transcranial magnetic stimulation modes on cognitive function and psychiatric symptoms in patients with schizophrenia. J Psychiatry 29(1):58–60

Liu R, Wang J, Liu H et al (2008) Comparative study for influence of rTMS on cognitive function in schizophrenia. Shanghai Arch Psychiatry 20(5):257–260, 307

Macpherson T, Hikida T (2019) Role of basal ganglia neurocircuitry in the pathology of psychiatric disorders. Psychiatry Clin Neurosci 73(6):289–301

Marijn Lijffijt, F. Gerard Moeller, Nash N. Boutros, et al (2009) The Role of Age, Gender, Education, and Intelligence in P50, N100, and P200 Auditory Sensory Gating. J Psychophysiol. 23(2):52–62

Mishra BR, Sarkar S, Praharaj SK et al (2011) Repetitive transcranial magnetic stimulation in psychiatry. Ann Indian Acad Neurol 14:245–250

Mondino M, Jardri R, Suaud-Chagny M et al (2016a) Effects of Fronto-temporal transcranial direct current stimulation on auditory verbal hallucinations and resting-state functional connectivity of the left

Temporo-parietal junction in patients with schizophrenia. Schizophr Bull 42(2):318–326

Mondino M, Haesebaert F, Poulet E et al (2016b) Fronto-temporal transcranial direct current stimulation (tDCS) reduces source-monitoring deficits and auditory hallucinations in patients with schizophrenia. Schizophr Res 161(2–3):515–516

Ni M, Guo Q (2001) Relationship between auditory hallucinations and P300 schizophrenia. Patients. Chin J Mod Med 11(7):45–47

Nucifora FC, Woznica E, Lee BJ et al (2019) Treatment resistant schizophrenia: clinical, biological, and therapeutic perspectives. Neurobiol Dis 131:104257

Oh SY, Kim YK (2011) Adjunctive treatment of bimodal repetitive transcranial magnetic stimulation (rTMS) in pharmacologically non-responsive patients with schizophrenia: a preliminary study. Prog Neuropsychopharmacol Biol Psychiatry 35(8):1938–1943

Oxley T, Fitzgerald PB, Brown TL et al (2004) Repetitive transcranial magnetic stimulation reveals abnormal plastic response to premotor cortex stimulation in schizophrenia. Biol Psychiatry 56:628–633

Picton T, Hillyard S, Krausz HI et al (1974) Human auditory evoked potentials: 1. Evaluation of components Electroencephalogr Clin Neurophysiol 36:179–190

Plaze M, Paillère-Martinot ML, Penttilä J, et al(2011) "Where do auditory hallucinations come from?"--a brain morphometry study of schizophrenia patients with inner or outer space hallucinations. Schizophr Bull 37(1):212–221

Robert JT, Andrew M, Jon H et al (2017) Diminished auditory sensory gating during active auditory verbal hallucinations. Schizophr Res 188:125–131

Roldan A, Portella MJ, Sampedro F et al (2020) Brain metabolic changes in patients with treatment resistant schizophrenia treated with deep brain stimulation: a series of cases. J Psychiatr Res 127:57–61

Shiozawa P, Gomes JS, Ducos DV et al (2016) Effect of transcranial direct current stimulation (tDCS) over the prefrontal cortex combined with cognitive training for treating schizophrenia: a sham-controlled randomized clinical trial. Trends Psychiatry Psychother 38:175–177

Stevens KE, Fuller LL, Rose GM (1991) Dopaminergic and noradrenergic modulation of amphetamine-induced changes in auditory gating. Brain Res 26;555(1):91–98. https://doi.org/10.1016/0006-8993(91)90864-r

Valiengo LDCL, Stephan G, Pedro CG et al (2019) Efficacy and safety of transcranial direct current stimulation for treating negative symptoms in schizophrenia: a randomized clinical trial. JAMA Psychiat 77:121–129

Vercammen A, Knegtering H, Bruggeman R et al (2011) Subjective loudness and reality of auditory verbal hallucinations and activation of the inner speech processing network. Schizophr Bull 37(5):1009–1016

Wada M, Noda Y, Iwata Y et al (2022) Dopaminergic dysfunction and excitatory/inhibitory imbalance in treatment-resistant schizophrenia and novel neuromodulatory treatment. Mol Psychiatry 27(7):2950–2967

Wang X, He J, Ren Y et al (2008) Research progress of transcranial magnetic stimulation in the treatment of schizophrenia. Shanghai Arch Psychiatry 20:115–117

Wang Y, Zhang C, Zhang Y et al (2020) Habenula deep brain stimulation for intractable schizophrenia: a pilot study. Neurosurg Focus 49(1):E9

Waternor DM, Bassett AS, Brzustowicz LM et al (2002) Recent advances in the genetics of schizophrenia. Cell Mol. Life Sc. 59(2):331–348

Zhang Z, Zhang X, Li H et al (2010) Double-blind randomized controlled trial of repetitive transcranial magnetic stimulation in the treatment of the negative symptom s of schizophrenia. Shanghai Arch Psychiatry 22(5):262–265

Zheng L, Guo Q, Li H et al (2012) Effects of repetitive transcranial magnetic stimulation with different paradigms on the cognitive function and psychotic symptoms of schizophrenia patients. J Peking Univ (Health Sciences) 44(5):732–736

# Obsessive-Compulsive Disorder

# 21

Yulian Niu and Xiaotong Yang

Obsessive-compulsive disorder (OCD) is a psychiatric disorder characterized by intrusive thoughts accompanied by anxiety, persistent obsessive thinking, and/or compulsive behaviors with repetitive actions to relieve this distress. The lifetime prevalence is 0.8–3.0% worldwide. The China Mental Health Survey reported an annual prevalence of 1.63% for OCD. Medication and cognitive behavioral therapy (CBT) remain as predominant first-line therapies for OCD in clinical practice; however, patients with refractory OCD (40–60% of OCD patients) fail to respond to such interventions. In recent years, physiotherapy has been introduced in clinical practice and has produced certain therapeutic effects. Numerous studies have confirmed that functional and structural abnormalities of the cortico-striatal-thalamic-cortical (CSTC) circuit are leading etiologies of OCD (Milad and Rauch 2012). This circuit includes the dorsolateral prefrontal cortex (DLPFC), orbitofrontal cortex (OFC), supplementary motor area (SMA), anterior cingulate cortex (ACC), caudate nucleus, putamen, pallidum, and multiple regions in the thalamus. These brain regions are involved in processing various information such as emotion, motivation, cognition, and movement, which are the key areas for individual behavioral control. Based on the research of the CSTC circuit, physical interventions for the treatment of OCD at this stage are mainly carried out on the aforementioned brain regions. The methods mainly include deep brain stimulation, transcranial magnetic stimulation, transcranial direct current stimulation, modified electric convulsive therapy, focused ultrasound, and vagus nerve stimulation (Compiling group of the Guidelines for the Prevention and Treatment of OCD in China 2016).

Y. Niu (✉)
Department of Neurology, Daxing District People's Hospital, Beijing, China

X. Yang
Department of Neurology, The First Hospital of Hebei Medical University, Shijiazhuang, Hebei, China

## 1 Deep Brain Stimulation

Deep brain stimulation (DBS) is a reversible, modifiable, invasive neuromodulation technique. DBS electrodes are precisely implanted at specific target sites within the skull through stereotactic technology, using weak, sustained electrical pulses from the electrodes to modulate cerebral nuclei or tissues. The exact mechanism of the therapeutic action of DBS is unclear. It is generally accepted that it is similar to lesion therapy, inducing neuronal inhibition by stimulating and depolarizing neurons near the electrode. Synaptic inhibition, synapse pruning, depolarization block, and stimulus-generated modulation of pathological network activity are the putative mechanisms of the DBS.

The main targets selected in existing studies exploring the intervention of DBS in refractory OCD include anterior limb of the internal capsule (ALIC), ventral capsule/ventral striatum (Vc/Vs), nucleus accumbens (NAc), subthalamic nucleus (STN), and inferior thalamic peduncle (ITP). These nuclei or white matter are involved in the composition of the CSTC circuit, and any abnormality may lead to the development of OCD.

ALIC was the first therapeutic target chosen based on the beneficial effects of ALIC lesion in patients with treatment-refractory OCD. This target was first explored by Nuttin et al. in 1999, with four subjects, three of whom showed improvement in clinical symptoms and >35% reduction in the Yale-Brown Obsessive-Compulsive Scale (Y-BOCS) scores (Nuttin et al. 1999). Thereafter, Vc/Vs started to be explored as another target. Greenberg et al. first reported that after ten patients with treatment-refractory OCD received DBS with bilateral Vc/Vs, complete responses (>35% reduction in Y-BOCS scores) and partial responses (25–35% reduction in Y-BOCS scores) were achieved in 40% and 20% of patients, respectively. In 2010, Greenberg et al. implemented a multi-center study, in which a total of 26 patients with treatment-refractory OCD received DBS with bilateral ALIC-Vc/Vs with a response rate of 62% (>35% reduction



Y. Wang (ed.), *Therapeutics of Neural Stimulation for Neurological Disorders*, https://doi.org/10.1007/978-981-99-4538-2_21

in Y-BOCS score) (Greenberg et al. 2010). In 2015, Alonso et al. conducted a comparative meta-analysis of 116 subjects involved in 31 relevant studies, including 83 subjects treated with DBS applied to striatal regions (including ALIC, Vc/Vs, NAc, and ventral caudate nucleus), 27 subjects treated with DBS applied to STN, and 6 subjects treated with DBS applied to ITP (Alonso et al. 2015). The results showed that the overall reduction rate of Y-BOCS was 45.1% and the overall response rate of treatment reached 60%, indicating that DBS was effective in the treatment of refractory OCD, but there was no significant difference in the efficacy between the targets. In 2018, Borders et al. conducted a retrospective analysis of 153 patients involved in 32 relevant reports comparing the relationship between different treatment targets and improvement in OCD symptoms and found that the effectiveness of each target treatment was higher than 50% and that Y-BOCS scores decreased consistently over a certain time window (Borders et al. 2018).

Benefit from advances in imaging, neuro-navigation, and localization techniques, the precision and safety of DBS treatment have improved considerably and may provide benefit to patients with treatment-refractory OCD. However, numerous randomized controlled clinical studies are still needed to explore the optimal targets and clinical efficacy.

**Recommendation** DBS is effective in treatment-refractory OCD. However, the treatment is invasive and requires surgical transcranial surgery, and long-term efficacy is controversial. Neurosurgical treatments for mental illness have not been approved in China, and if physical therapy is necessary, it is recommended to strictly comply with the relevant medical and legal regulations in China (Compiling group of the Guidelines for the Prevention and Treatment of OCD in China 2016).

## 2 Repetitive Transcranial Magnetic Stimulation

Transcranial magnetic stimulation (TMS) induces intracranial pulsed currents under a varying magnetic field. TMS regulates cortical excitability by stimulating synapses and altering receptor function and induces the electrophysiological activity of excitatory or inhibitory neurons. Repetitive transcranial magnetic stimulation (rTMS) produces a cumulative effect, resulting in higher levels of excitation in neurons, which not only affects the function of the distant brain associated with the local stimulation, achieving a regional reconstruction of cortical function, but also produces a long-term effect, allowing the biological effects produced to persist for some time after the stimulus has ceased. In 1997, rTMS was first reported as a treatment for OCD by Greenberg et al. (1997). In August 2018, TMS was approved by the US FDA for the treatment of OCD. In 2018, the Chinese Medical Doctor Association Neuromodulation Specialty Committee gave clinical recommendation Level II and III evidence for high-frequency or low-frequency rTMS stimulation of bilateral DLPFC for OCD (Xu et al. 2018).

Because of the hyperactivation of the supplementary motor area (SMA), dorsolateral prefrontal cortex (DLPFC), and orbital frontal cortex (OFC) in OCD patients (Del Casale et al. 2011), the stimulation targets selected in the current study are mainly in these brain areas, among which the DLPFC is the most clinically evident target (Zhou et al. 2017; Rehn et al. 2018). Because the OFC is more difficult to stimulate directly, rTMS has been less studied at this target, and the results obtained are controversial. SMA, on the other hand, has a clear effect on the basal ganglia and is also easier to stimulate compared to OFC, so more studies have been done on SMA. In a multi-center randomized controlled trial involving 22 patients with treatment-refractory OCD, the rTMS group (10 patients) was given bilateral SMA low-frequency rTMS for 6 weeks, and 80% of patients achieved a 40% mean reduction in Y-BOCS scores from baseline, significantly improving the clinical symptoms of OCD (Hawken et al. 2016). Berlim et al. screened 10 randomized controlled trials with a total of 282 subjects for meta-analysis and found that the response rates in the true and pseudo-stimulation groups were 35% and 13%, respectively (Berlim et al. 2013). The recommended treatment frequency was low frequency, and the recommended stimulation targets were SMA and OFC. In recent years, the pre-supplementary motor area (pre-SMA) has also attracted the interest of researchers. A study using pre-SMA as a target for OCD intervention using LF-rTMS found that after 4 weeks of intervention, the Y-BOCS reduction rate was 25% and 12% in the true and pseudo-stimulation groups, respectively, while after 8 weeks of intervention, the reduction rate was more than 40% in the true stimulation group (Mantovani et al. 2013). The findings suggest that pre-SMA may be another effective stimulation target for rTMS treatment. There is no consistent standard for the duration of rTMS treatment, and most current studies choose a treatment duration of 2–6 weeks. Some studies have suggested that the longer the course of treatment, the better the clinical outcome is likely to be.

Deep transcranial magnetic stimulation (dTMS), which modulates not only cortical excitability but also the activity of deep neural circuits (up to 6 cm in depth), has become a new mode of stimulation that has received much attention in recent years. Modirrousta et al. used low-frequency dTMS stimulation of the medial prefrontal cortex (mPFC) to treat OCD, and after ten consecutive treatments, all patients showed improvement in OCD symptoms, with a mean Y-BOCS score reduction of 39.4% and a maximum of 76.5%, suggesting that low-frequency dTMS stimulation of the mPFC may be effective in improving patients' clinical symptoms and that this improvement still be maintained after 1 month (Modirrousta et al. 2015). In the study by Carmi et al., subjects were divided into high-frequency (20 Hz) and low-frequency (1 Hz) dTMS and pseudo-stimulation groups with stimulation sites of mPFC and ACC and treated for 5 weeks. 43.75% of patients in the high-frequency group had significantly lower Y-BOCS scores at the end of the intervention, and this effect lasted for 1 month, while there was no significant improvement in either the low-frequency or pseudo-stimulation groups (Carmi et al. 2018). This result suggests that treatment of OCD with HF-dTMS may be preferable. However, it is important to note that high-intensity stimulation may lead to cortical hyperactivation, which may be at risk for seizures. Therefore, it is recommended to clarify the action potential threshold before starting treatment to prevent cortico-spinal hyperexcitability.

In summary, rTMS is a safe, effective, and well-tolerated technique, but there are no agreed-upon optimal treatment parameters. The exploration of optimized stimulation parameters and optimal target sites will be an important research direction for rTMS in the treatment of OCD.

**Recommendation** Low-frequency rTMS stimulation of the supplementary motor area and orbitofrontal cortex can significantly improve OCD symptoms, with relatively good safety and no significant effects on normal cognitive function, perception and endocrine, etc. The main risk is the induction of epileptogenesis, especially with high-frequency rTMS treatment (Compiling group of the Guidelines for the Prevention and Treatment of OCD in China 2016).

## 3 Transcranial Direct Current Stimulation

Transcranial direct current stimulation (tDCS) is also a noninvasive transcranial stimulation technique. tDCS is a neuromodulation technique that can reshape or restore neuronal function by changing the resting membrane potential of neurons by applying weak direct current to the cerebral cortex through two electrode pads, which can produce neuroelectrophysiological and biochemical responses in brain neurons with good safety.

At present, the research on the use of tDCS for OCD is still in the primary exploration stage, and the targets are mostly DLPFC, OFC, SMA, or pre-SMA, and there is no uniform standard for the treatment parameters. The current intensity is mostly 2 mA, the treatment time is mostly 20–30 min, the frequency is 1 time/day or 2 times/day, and the number of treatments is mostly 10–20 times. In 2019, Gowda et al. used a randomized, double-blind method to divide 25 cases of treatment-refractory OCD into a stimulation group (12 cases) and a pseudo-stimulation group (13 cases), in which the anode was placed on the left pre-SMA and the cathode on the right supraorbital region, and the treatment was given twice a day for 20 min per intervention for 5 days. The results showed that four individuals in the stimulation group had ≥35% Y-BOCS reduction, whereas no one in the pseudo-stimulation group showed improvement (Gowda et al. 2019). This study suggests that tDCS is potentially effective for the treatment of treatment-refractory OCD.

At this stage, the studies of tDCS for OCD are mostly case reports and open-ended studies, so it is not possible to determine whether there is any confounding of trial conclusions due to placebo effects and individual differences, and optimal treatment parameters are not yet available. Further exploration of optimal protocols and clinical efficacy will be required by expanding the sample size and setting up controlled studies.

## 4 Modified Electric Convulsive Therapy

Modified electric convulsive therapy (MECT) is a therapeutic method that modifies the activity of the cerebral cortex by applying a brief, moderate amount of electric current to the brain, causing extensive firing of neurons in the cerebral cortex.

There are limited reports on ECT therapy for OCD. Fontenelle et al. summarized 50 relevant papers involving 279 patients, and the analysis suggested that 60% of patients were effectively treated with ECT. However, because most of the studies are case reports and lack the evidence-based basis of randomized controlled studies, they are not a basis for the effectiveness of ECT treatment.

**Recommendation** Because of the lack of evidence-based reports, there is still a great deal of controversy about whether ECT is effective in treating patients with OCD. With major depressive symptoms, psychotic symptoms, or other psychiatric disorders such as comorbid Tourette syndrome, ECT is effective. Common adverse reactions are cognitive dysfunction and poor circulatory system, the former generally recovered within 6 months (Compiling group of the Guidelines for the Prevention and Treatment of OCD in China 2016).

## 5 Focused Ultrasound

Transcranial focused ultrasound stimulation (tFUS), especially MR-guided focused ultrasound (MRgFUS), has become a new research hotspot because of its noninvasiveness and focus on deep targets (Kubanek 2018). Current clinical research on tFUS is still dominated by trials, involving movement disorders, obsessive-compulsive disorder (OCD), chronic pain, and brain tumors.

In the study conducted by Jung et al., four patients with refractory OCD were subjected to bilateral thermal capsulotomy with MRgFUS in the anterior limb of internal capsule (ALIC) (Jung et al. 2015). The results showed that the Y-BOCS scores gradually decreased (mean 33%) over the 6-month follow-up period; all patients experienced near-immediate and sustained improvement in depression (mean 61.1% reduction) and anxiety (mean 69.4% reduction), while none of the patients experienced any surgery-related side effects. A recent meta-analysis by Kumar et al. showed the potential advantages of MRgFUS over radiofrequency ablation and radiosurgery in the treatment of refractory OCD, with preliminary data demonstrating effectiveness (reduction in Y-BOCS score of 56.6% on average) and safety (complication rate of 16.2%) (Kumar et al. 2019).

MRgFUS can also be applied using lower frequency (LIFU, low-intensity focused ultrasound), which is reversible and allows identifying the best stimulation target to improve symptoms. Because of its novelty, no study published has so far employed LIFU in OCD. However, given its ability to noninvasively modulate deep brain regions with precise spatial resolution, LIFU may be a promising technology for treating and further understanding the aberrant neural circuits of OCD. Two clinical trials are currently underway to assess the safety and feasibility of LIFU in patients with OCD targeting the ventral striatum and the caudate of the basal ganglia, respectively.

In the foreseeable future, lesion using MRgFUS is expected to become one of the most effective treatments for refractory OCD, but more studies are needed to further optimize the stimulation pattern and to verify the safety and durability of its efficacy. At the same time, the large size and relatively high cost of MRgFUS equipment are also important factors limiting the development of tFUS, which requires continuous research exploration and enhancement in the future.

## 6 Vagus Nerve Stimulation

Vagus nerve stimulation (VNS) is the electrical stimulation of the vagus nerve by means of spiral electrodes applied to the vagus nerve. The signals transmitted to the cerebral cortex are widely projected through the nucleus tractus solitarius and ascending reticular system to the deeper nuclei and cortex, thereby modulating brain function. The treatment of OCD with VNS is still in the early stage of experimentation and lacks evidence-based basis, so its specific mechanism and therapeutic effects remain to be proven.

**Recommendation** At this stage, VNS can only be used as an adjunct to OCD treatment or as an exploratory treatment for those with poor outcomes.

OCD is a psychiatric disorder that keeps being recognized, and physical interventions to treat OCD will become the forefront of cognitive neuroscience research in the future as the mechanisms of brain circuits continue to be explored and new technologies are applied. It is believed that the technology will become more and more established in the future and will benefit more OCD patients as the treatment plan is optimized.

## References


Alonso P, Cuadras D, Gabriëls L et al (2015) Deep brain stimulation for obsessive-compulsive disorder: a meta-analysis of treatment outcome and predictors of response. PLoS One 10(7):e0133591

Berlim MT, Neufeld NH, Van DEF (2013) Repetitive transcranial magnetic stimulation (rTMS) for obsessive-compulsive disorder (OCD): an exploratory meta-analysis of randomized and sham-controlled trials. J Psychiatr Res 47(8):999–1006

Borders C, Hsu F, Sweidan AJ et al (2018) Deep brain stimulation for obsessive compulsive disorder: a review of results by anatomical target. Ment Illn 10(2):7900

Carmi L, Alyagon U, Barnea-Ygael N et al (2018) Clinical and electrophysiological outcomes of deep TMS over the medial prefrontal and anterior cingulate cortices in OCD patients. Brain Stimul 11(1):158–165

Chinese guidelines for the prevention and treatment of OCD 2016 (refined edition) (2016) Chin J Psychiatry 6:353–366.

Del Casale A, Kotzalidis G, Rapinesi C et al (2011) Functional neuroimaging in obsessive-compulsive disorder. Neuropsychobiology 64(2):61–85

Gowda SM, Narayanaswamy JC, Hazari N et al (2019) Efficacy of pre-supplementary motor area transcranial direct current stimulation for treatment resistant obsessive compulsive disorder: a randomized, double blinded, sham controlled trial. Brain Stimul 12:922–929

Greenberg B, George M, Martin J et al (1997) Effect of prefrontal repetitive transcranial magnetic stimulation in obsessive-compulsive disorder: a preliminary study. Am J Psychiatry 154(6):867–869

Greenberg BD, Gabriels LA, Malone DA et al (2010) Deep brain stimulation of the ventral internal capsule/ventral striatum for obsessive-compulsive disorder: worldwide experience. Mol Psychiatry 15(1):64–79

Hawken E, Dilkov D, Kaludiev E et al (2016) Transcranial magnetic stimulation of the supplementary motor area in the treatment of obsessive-compulsive disorder: a multi-site study. Int J Mol Sci 17(3):420

Jung HH, Kim SJ, Roh D et al (2015) Bilateral thermal capsulotomy with MR-guided focused ultrasound for patients with treatment-refractory obsessive-compulsive disorder: a proof-of-concept study. Mol Psychiatry 20(10):1205–1211

Kubanek J (2018) Neuromodulation with transcranial focused ultrasound. Neurosurg Focus 44(2):E14

Kumar KK, Bhati MT, Ravikumar VK et al (2019) MR-guided focused ultrasound versus radiofrequency capsulotomy for treatment-refractory obsessive compulsive disorder: a cost-effectiveness threshold analysis. Front Neurosci 13:66

Mantovani A, Rossi S, Bassi BD et al (2013) Modulation of motor cortex excitability in obsessive-compulsive disorder: an exploratory study on the relations of neurophysiology measures with clinical outcome. Psychiatry Res 210(3):1026–1032

Milad MR, Rauch SL (2012) Obsessive-compulsive disorder: beyond segregated cortico-striatal pathways. Trends Cogn Sci 16(1): 43–51

Modirrousta M, Shams E, Katz C et al (2015) The efficacy of deep repetitive transcranial magnetic stimulation over the medial prefrontal cortex in obsessive compulsive disorder: results from an open-label study. Depress Anxiety 32(6):445–450

Nuttin CP, Demeulemeester H et al (1999) Electrical stimulation in anterior limbs of internal capsules in patients with obsessive-compulsive disorder. Lancet 354(9189):1526

Rehn S, Eslick G, Brakoulias V (2018) A meta-analysis of the effectiveness of different cortical targets used in repetitive transcranial magnetic stimulation (rTMS) for the treatment of obsessive-compulsive disorder (OCD). Psychiatry Q 89(3):645–665

Xu Y, Li D, Tan LW et al (2018) Expert consensus on repetitive transcranial magnetic stimulation (2018 China edition). J Transl Med 01:4–9

Zhou D, Wang W, Wang G et al (2017) An updated meta-analysis: short-term therapeutic effects of repeated transcranial magnetic stimulation in treating obsessive-compulsive disorder. J Affect Disord 215:187–196

# Addiction

# 22

Yiran Duan

There are three main techniques that stimulate specific areas of the brain for the treatment of addiction, including repetitive transcranial magnetic stimulation (rTMS), transcranial direct current stimulation (tDCS), and deep brain stimulation (DBS). The underlying principle of stimulation therapy is to suppress addiction by restoring normal brain function in the target area. rTMS exerts a therapeutic effect on addiction by generating repetitive magnetic field pulses under specific parameters, altering the excitability of neurons, with few side effects and higher cost-effectiveness. tDCS is another noninvasive stimulation technique that applies low-intensity direct current to a specific target scalp area through two or more electrodes to depolarize or hyperpolarize the neurons in the target area and modulate the resting membrane potential and cortical excitability of the target brain area, with few side effects and higher cost-effectiveness; however, it is necessary to further investigate the efficacy of tDCS. DBS is an invasive stimulation technique that applies stimulation through a bipolar electrode (connected to a pulse generator) implanted in specific brain regions. The stimulation parameters are programmable. Existing studies support the effectiveness of DBS in the treatment of addiction, with more side effects and higher costs.

## 1 rTMS

### 1.1 Treatment Principle

rTMS is a noninvasive brain stimulation technique that is applied to treat various neurological and psychiatric disorders, including Parkinson's disease, schizophrenia, obsessive-compulsive disorder, and chronic pain. rTMS generates a repetitive pulsed magnetic field applied to a specific brain region through an electromagnetic coil placed on a specific scalp region. This process has been demonstrated to modulate cortical excitability by inducing transient electrical currents in the local cortex and to produce lasting neuroplastic changes in target brain regions. Studies revealed that decreased activity in reward circuits during addiction withdrawal was associated with craving, relapse, and sustained nicotine intake. rTMS is a potential tool for manipulating this circuit, which induces the release of dopamine and leads to sustained changes in neuronal excitability. In particular, high-frequency rTMS (>3 Hz) increases cortical excitability and promotes neuroplasticity. rTMS has been used in the treatment of various neuropsychiatric disorders associated with abnormal dopamine activity and altered cortical excitability (Dinur-Klein et al. 2014).

### 1.2 Treatment Regimen

One problem with TMS is limited accessible depth in brain tissue. The rapid attenuation of the electric field results in TMS being largely restricted in superficial cortical targets. Therefore, TMS was applied to the dorsolateral prefrontal cortex (DLPFC) in most studies on TMS for psychiatric disorders including addiction. Diversity rTMS parameters may have different impact on the neuron excitability. Studies demonstrated that low-frequency rTMS (<1 Hz) led to reduced neuronal firing rate and cortical excitability, while high-frequency rTMS (10–20 Hz) exhibited the opposite effect. In addition, theta-burst stimulation (TBS) has been widely used in recent years, which involves three repetitions of pulsed stimulation at a frequency of 50 Hz every 200 ms. The first option is intermittent theta-burst stimulation (iTBS), which is applied to the cortex for facilitation. This parameter includes a 2-s train of TBS that is repeated every 10 s to give a total time of 190 s (600 pulses). Another technique using continuous burst frequencies, also known as continuous theta-burst stimulation (cTBS), produces temporary and

Y. Duan (✉)
Department of Neurology, Beijing Friendship Hospital, Capital Medical University, Beijing, China



Y. Wang (ed.), *Therapeutics of Neural Stimulation for Neurological Disorders*, https://doi.org/10.1007/978-981-99-4538-2_22

long-term inhibitory effects. This technique consists of a non-stop 40-s train of TBS (600 pulses) (Huang et al. 2005; Hanlon et al. 2015).

### 1.3 Clinical Application

rTMS is a promising treatment for neurological and psychiatric diseases, as it is well tolerated by most patients with few side effects and is painless. Additionally, rTMS is a cost-effective alternative to other expensive therapy options (such as electroconvulsive therapy). Although rTMS is a promising potential technique for treating addiction, current studies are still in the developmental stage. Many studies have shown that 1–2 periods of high-frequency rTMS may be effective in reducing addiction compared to the sham group, the most powerful evidence for that is in smoking-related diseases. Randomized controlled studies show that high-frequency rTMS can reduce smoking, and matching rTMS with smoking triggers can improve the efficacy of rTMS. Additionally, these studies indicate that multiple treatments are needed to keep the effect (Johann et al. 2003).

## 2 tDCS

### 2.1 Treatment Principle

tDCS is a noninvasive brain stimulation technique where two or more electrodes are placed on the scalp to deliver a low-intensity direct current stimulation to the brain's target areas. The stimulation modes of tDCS vary depending on the sizes and quantities of electrodes used, the number of contact mediators and the duration of stimulation, and the distribution of current delivered to the brain, thereby stimulating the brain differently. There are two mechanisms suggesting that tDCS modulates brain activities: (1) by changing neuronal resting membrane potential to depolarize the neurons underlying the anode electrode and hyperpolarize neurons underlying the cathode electrode and (2) by modulating synaptic activities in a manner similar to long-term potentials (at the anode) and long-term depression (at the cathode). Thus, the neuromodulation effects are thought to be dependent on the intensity, duration, and direction of the current, where anodal tDCS can increase excitability and cathodal tDCS can contribute to hyperpolarization and inhibition. Craving is a complicated behavior and one of the key factors for addiction. As the same as the behaviors induced by other stimuli, strong cravings come after stimuli such as smoking. This response is associated with a large-scale activation of a wide range of neural circuits, including the prefrontal region. The DLPFC is closely associated with processing the craving of smoking and drugs like cocaine; specifically, craving is associated with the increased activity of the DLPFC. DLPFC activation by noninvasive brain stimulation, which thus may mimic its rewardrelated activation, has been demonstrated to be effective in decreasing craving symptoms in cigarette smokers. The rationale for the effect of tDCS on craving is that tDCS is a powerful but simple technique and it can change spontaneous neuronal activities, thereby modifying cognitive processing (Boggio et al. 2009).

### 2.2 Treatment Regimen

Much of what we know about tDCS comes from studies on the motor cortex; it remains unclear whether the same principles also apply to other brain regions, such as the prefrontal cortex. Furthermore, because the organization of neurons in the whole brain may be different, it may result in different effects when the same type of stimulation is applied. However, some experiments have demonstrated that tDCS has a long-lasting functional effect on the activities of the non-motor cortex, which can last 1 h after the stimuli have ceased (Philip et al. 2017). tDCS stimulation generates a low-intensity current (0.5–2.0 mA); it can modulate the resting membrane potential and cortical excitability of target brain regions by depolarization or hyperpolarization mechanisms based on the difference of stimulus parameters. The anode is proven to be increasing cortical excitability, whereas the cathode has the opposite role (Coles et al. 2018).

### 2.3 Clinical Application

tDCS is a cost-effective, accessible, and painless stimulation technique, with few side effects, occasionally itching or tingling on the scalp, and is well tolerated by patients. The effects of tDCS on substance abuse disorders remain unclear. Only three studies measured the actual drugs or alcohol intake from subjects who have accepted tDCS treatment, and these studies didn't show a beneficial effect. One of these investigated alcohol intake addiction-related tDCS treatments and reported that although cravings for alcohol in the case group significantly decreased, tDCS resulted in a trend for relapse of using alcohol (da Silva et al. 2013). The other two studies investigated smoking addiction and found a decrease in smoking based on the subjects' self-report (smoking diary). However, when objective measures were used, no obvious differences were seen between the tDCS group and the sham group (Fecteau et al. 2014). Many addiction-related studies had small sample sizes, and different standards have been used to measure the results.

## 3 DBS

### 3.1 Treatment Principle

DBS is a surgical operation where bipolar electrodes are placed into specific brain regions and stimulated through implanted pulse generators. The stimulation parameters are programmable and dependent on the target brain region, disease, and patient response. Stimulation induces an electrical field up to 1 cm depending on specific neuron tissue density; the mechanism of action is still unclear. As stimulation has various actions on cellular physiology, each stimulating process may be depolarized or hyperpolarized depending on the distance from the electrode. Furthermore, many stimulation variables including brain region, frequency, intensity, and state of neuron synchronization can influence the direction and duration of its effects on neuronal activities. Although the specific neuronal effect of DBS remains largely debated, DBS may act through multiple mechanisms (Herrington et al. 2016). Like normal pleasurable events, drinking alcohol triggers the release of more dopamine in the nucleus accumbens (NAc), which is the core region of the brain reward system. The release of dopamine activates the reward system and leads to pleasing feelings. The reward system is a part of the dopamine system; its neurons are located in the ventral tegmental areas and project to the NAc, septum, and amygdala. Chronic alcohol intake can lead to dysfunction of this pathway, cravings for alcohol, and thus addiction. Accordingly, NAc, one of the major regions of the brain reward system, plays a key role in addiction treatment. Studies on rodents showed that stimulating NAc reduced alcohol, morphine, and cocaine abuse and similar behaviors. These studies also showed the effect on addictive behaviors when the prefrontal cortex, lateral hypothalamus, subthalamic nucleus, and lateral habenula are stimulated. A major limitation of these studies is that the human brain structure is hard to mimic that of rodents, and non-human primate studies have not been attempted.

### 3.2 Treatment Regimen

For severe alcohol-dependent subjects, bilateral DBS was performed on the ventral striatum (nucleus accumbens); the stimulation parameters are 130 Hz, 120 μs, and 4.5 or 5.5 V (Kuhn et al. 2011). Cocaine-dependent subjects received bilateral DBS on the ventral striatum; the stimulation parameters are 130–140 Hz, 90 or 120 μS, and 0–5 V (Kuhn et al. 2014). Nicotine-dependent subjects received bilateral DBS on the ventral striatum; the stimulation parameters are 130–145 Hz, 90–180 μS, and 3–6.5 V (Herrington et al. 2016).

### 3.3 Clinical Application

DBS has not been used extensively in addiction studies, but there are some preliminary studies. Previous case studies showed, for patients who received DBS treatment for other symptoms (such as anxiety or mood disorder), stimulation to the ventral striatum/nucleus accumbens can reduce the consumption of substances of abuse, such as alcohol, nicotine, and heroin. In one study, bilateral DBS was performed on the ventral striatum (nucleus accumbens) in five severely alcohol-dependent subjects; all subjects experienced improvement in addiction, and two of them achieved complete abstinence. In another case report, subjects who improved error processing on cognitive testing reduced addictive behaviors. A case report of two heroin-dependent subjects who received bilateral DBS to the ventral striatum showed their depressive symptoms and anxiety were improved and a reduction in their drug use, though not cessation. Three of ten nicotine-dependent subjects who received DBS to the ventral striatum were able to quit smoking (Kuhn et al. 2009). These preclinical studies have supported the application of DBS in addictive behaviors.

## References


Boggio PS, Liguori P, Sultani N (2009) Cumulative priming effects of cortical stimulation on smoking cue-induced craving. Neurosci Lett 463(1):82–86

Coles AS, Kozak K, George TP (2018) A review of brain stimulation methods to treat substance use disorders. Am J Addict 27(2):71–91

da Silva MC et al (2013) Behavioral effects of transcranial direct current stimulation (tDCS) induced dorsolateral prefrontal cortex plasticity in alcohol dependence. J Physiol Paris 107(6):493–502

Dinur-Klein L, Dannon P, Hadar A et al (2014) Smoking cessation induced by deep repetitive transcranial magnetic stimulation of the prefrontal and insular cortices: a prospective, randomized controlled trial. Biol Psychiatry 76(9):742–749

Fecteau S, Agosta S, Hone-Blanchet A (2014) Modulation of smoking and decision-making behaviors with transcranial direct current stimulation in tobacco smokers: a preliminary study. Drug Alcohol Depend 140:78–84

Hanlon CA, Dowdle LT, Austelle CW et al (2015) What goes up, can come down: novel brain stimulation paradigms may attenuate craving and craving-related neural circuitry in substance dependent individuals. Brain Res 1628(Pt A):199–209

Herrington TM, Cheng JJ, Eskandar EN (2016) Mechanisms of deep brain stimulation. J Neurophysiol 115(1):19–38
Huang YZ, Edwards MJ, Rounis E et al (2005) Theta burst stimulation of the human motor cortex. Neuron 45(2):201–206
Johann M, Wiegand R, Kharraz A et al (2003) Transcranial magnetic stimulation for nicotine dependence. Psychiatr Prax 30(Suppl 2):S129–S131
Kuhn J, Bauer R, Pohl S et al (2009) Observations on unaided smoking cessation after deep brain stimulation of the nucleus accumbens. Eur Addict Res 15(4):196–201
Kuhn J, Gründler TO, Bauer R et al (2011) Successful deep brain stimulation of the nucleus accumbens in severe alcohol dependence is associated with changed performance monitoring. Addict Biol 16(4):620–623
Kuhn J, Möller M, Treppmann JF et al (2014) Deep brain stimulation of the nucleus accumbens and its usefulness in severe opioid addiction. Mol Psychiatry 19(2):145–146
Philip NS, Nelson BG, Frohlich F et al (2017) Low-intensity transcranial current stimulation in psychiatry. Am J Psychiatr 174(7):628–639

# Sleep Disorders

# 23

Jianghong Liu, Jingwen Zhang, Li Wang, and Bingqi Guo

Neurostimulation for the treatment of sleep disorders is one of the research hotspots in recent years. However, electroconvulsive therapy (ECT) and transcranial electrical stimulation (tES), which have been used for decades, remain clinically relevant. Transcranial magnetic stimulation (TMS), deep brain stimulation (DBS), and tES have continued to evolve over the past two decades. This section summarizes the fundamental and clinical studies of neurostimulation in the treatment of sleep disorders in recent years, so as to provide a reference for clinical practice and research on sleep disorders by neurostimulation techniques.

## 1 Insomnia

Insomnia disorder, one of the most common sleep disorders, is a subjective experience that affects daytime functioning as poor sleep quality or short sleep duration cannot satisfy normal physiological needs due to difficulties in falling asleep and/or maintaining sleep. Chronic insomnia is considered if the disorder persists for at least 3 months at a frequency of at least three times a week (Sateia et al. 2017). Short-term insomnia is considered when the disorder meets the diagnostic criteria for insomnia disorder but persists for less than 3 months. According to the statistics from the National Health and Family Planning Commission, the prevalence of insomnia disorder in China is as high as 10–20%. Chronic insomnia has many adverse effects on both physical function and quality of life. Epidemiological studies have shown that the physical functional status is significantly impaired in patients with chronic insomnia. It is also an important risk factor for psychiatric disorders, especially emotional disorders. Recent studies have shown that chronic insomnia is associated with an increased risk of cardiovascular and cerebrovascular diseases and dementia. Insomnia with short objective sleep time is particularly an important risk factor for hypertension (Zhang et al. 2008).

Treatment options for insomnia include drug and non-drug therapies. Drug therapy has limited efficacy and some adverse effects associated with long-term use. Non-medication treatment for insomnia includes cognitive behavioral therapy for insomnia (CBT-I), physical therapy, and Chinese medicine therapy. CBT-I is currently recognized as the first line of treatment for insomnia disorders. CBT-I is a multi-mode treatment that combines many behavioral and cognitive interventions, including sleep hygiene, stimulus control, sleep restriction, relaxation training, and cognitive restructuring. However, CBT-I is not commonly performed due to the lack of volunteers trained in this field. In addition, in many countries, CBT-I is expensive for patients, and clinicians find it difficult for patients to maintain CBT-I economically due to health insurance limitations. Thus, although the efficacy of CBT-I for insomnia disorders is basically not controversial, issues caused by its feasibility and cost-effectiveness inhibit many patients with chronic insomnia from benefiting from this treatment. A host of recent researches suggested that different types of neurostimulation can modulate sleep. Neurostimulation is a noninvasive stimulation technique, the efficacy of which in treating sleep disorders has been validated by many studies.

### 1.1 Transcranial Electrical Stimulation

Transcranial electrical stimulation is a noninvasive method of applying low-intensity electric currents to the head. It evolved from the concept of "electrosleep," which was first studied in the early twentieth century, with most of the early researches and applications in Russia. Since the 1960s, the concept of "electrosleep" has become increasingly popular in the United States. The name was changed from "electrosleep" to "transcranial electrical stimulation" as it was

J. Liu (✉) · J. Zhang · L. Wang · B. Guo
Department of Neurology, Xuanwu Hospital, Capital Medical University, Beijing, China
e-mail: Liujh@xwhosp.org



Y. Wang (ed.), *Therapeutics of Neural Stimulation for Neurological Disorders*, https://doi.org/10.1007/978-981-99-4538-2_23

believed that electrical stimulation did not actually induce sleep directly, but rather facilitated sleep by the auxiliary effect of relaxing caused by stimulation. Transcranial electrical stimulation applies a weak electric current to the scalp, which is not strong enough to directly induce neural discharge. In contrast, it is able to cause local changes in cortical excitability and may generate action potential, which can be extended to interconnected neural network (Song et al. 2019). The common transcranial electrical stimulations are transcranial direct current stimulation (tDCS) and transcranial alternating current stimulation (tACS).

### 1.1.1 Transcranial Direct Current Stimulation (tDCS)

#### 1.1.1.1 Mechanisms of tDCS

tDCS acts on the corticothalamic loop regulating sleep and wakefulness and hence can regulate cortical excitability. The cathodal tDCS reduces the resting membrane potential and hyperpolarizes neurons, while the anodal stimulation leads to depolarization (Valiengo et al. 2017). This kind of effect on cortical excitability can further regulate sleep duration and cycle. For example, it was found by some studies that bifrontal anodal tDCS could significantly shorten the total sleep time of subjects, while bifrontal cathodal stimulation could prolong the total sleep time. In addition, tDCS with slow-wave oscillation can realize resonance with the frequency of sleep slow waves, thus complementing the missing slow-wave activity. In conclusion, the mechanism of tCDS for the treatment of insomnia disorder has not been established and remains to be further explored.

#### 1.1.1.2 tDCS for Insomnia

The recent research hotspots of tDCS for insomnia include exploring the mechanism and efficacy of tDCS based on the pathogenesis of insomnia. In current studies, bifrontal tDCS is mainly applied to induce changes in the neuronal excitability of the cerebral cortex, with anodal stimulation applied to increase cortical excitability and cathodal stimulation to decrease cortical excitability. The hypothesis that total sleep time (TST) could be regulated by tDCS was experimentally verified in a study in which 19 healthy participants were screened and subjected to repeated tDCS interventions in the laboratory over a period of 5 nights. Bifrontal electrodes and biparietal current-return electrodes linked the tDCS device to subjects before sleep. The results revealed that bifrontal anodal tDCS significantly decreased the subjects' TST compared to cathodal and sham stimulation, while bifrontal cathodal tDCS increased TST but in a minor way (Frase et al. 2019). Their study demonstrated that anodal tDCS reduced TST in healthy populations, while cathodal tDCS increased TST in healthy populations.

It was found by another study that the lack of slow-wave activity may play a fundamental role in the pathogenesis of insomnia. Some researchers applied slow oscillating tDCS during the second stage of nonrapid eye movement sleep (NREM) in insomnia patients to resonate with the frequency of the sleep slow wave, thus complementing the missing slow-wave activity. Six insomnia patients with difficulty in maintaining sleep were included in this study. After polysomnogram (PSG) recording, patients were randomized to receive tDCS or sham stimulation on D3 and D4 of the study. The results showed that slow oscillating tDCS increased the duration of NREM stage 3 by 33 ± 26 min after termination of stimulation, which reduced the duration of NREM stage 1 and wake-up time after sleep duration by 55.4 ± 51 min together. It was also found that the stimulation also improved sleep efficiency and increased the probability of transition from stage 2 to stage 3 of NREM while also reduced the transition from NREM stage 2 to wakefulness (Saebipour et al. 2015). These results suggest that slow oscillating tDCS has the effect of stabilizing sleep, similar to that of sleep slow wave-enhancing drugs.

### 1.1.2 Transcranial Alternating Current Stimulation

#### 1.1.2.1 Mechanisms of tACS

tACS is a noninvasive transcranial low-frequency electrical stimulation method that regulates brain activity. tACS enhances the consolidation of motor memory, improves human physiological-psychological state, alters the connectivity of default mode network (DMN), and regulates neuroplasticity. It has been shown by some studies that tACS was effective in relieving anxiety, depression, stress, and insomnia. Moreover, it has been proven to be effective in regulating pain, emotion, and aggressive behavior. However, studies on the action mechanism of tACS are restricted. The ability of tACS to improve sleep is still controversial. It depends on the location of electrodes, the treatment methods for different patients, or tACS intensity or density. A study investigating the effect of alternating current stimulation on cortical excitability demonstrated that tACS applied to the motor cortex (0.80 $A/m^2$, 15 Hz, 20 min) could result in measurable changes in cortical excitability, most likely due to the effect of tACS acting on local circuit neurons. Several studies on tACS showed that different current intensities and frequencies produced conflicting results with respect to cortical excitability. In one of these studies, current density of 25 $mA/cm^2$ was applied at 1 Hz, 10 Hz, 15 Hz, 30 Hz, and 45 Hz, respectively, with a stimulation duration of 5 min. The results showed that alternating current stimulation did not cause significant changes in cortical excitability. A different conclusion was reached in another study in which the investigators tested the effect of tACS on cortical excitability at a frequency of 15 Hz for a duration of 20 min under a lower current density (0.08 mA/cm$^2$). The results showed changes in cortical excitability. In several other studies, current density of 25 mA/cm$^2$ was applied at 1 kHz, 2 kHz, and 5 kHz, respectively, for 10 min of stimulation (Chaieb et al. 2011). It was found that the amplitude of motor evoked potential (MEP) increased at all frequencies, suggesting that tACS at high frequency may be associated with increased excitability of the motor cortex. These studies suggest that the effect of tACS on cortical excitability is frequency- and intensity-dependent.

#### 1.1.2.2 tACS for Insomnia

Most of previous randomized controlled tests on the effects of tACS on insomnia provided limited information due to the limitations of experimental designs. Other studies reported conflicting results of tACS improving sleep efficiency, which may be associated with the differences in electrode positions, patient management procedures, and tACS current frequencies and/or intensities. When the frequency of tACS is 77.5 Hz, the current is delivered through electrodes placed in the prefrontal and mastoid regions, thereby altering β-endorphin levels and neurotransmitters, including 5-hydroxytryptamine, in the cerebrospinal fluid, brainstem, hypothalamus, and cortex. The neurobiological mechanisms of the hypothalamus and the serotonergic neurons involved in chronic insomnia are well known, while the mechanism of tACS on chronic insomnia in adults is unknown.

The existing literatures on the analysis of the effects of tACS on insomnia are limited to a few randomized controlled tests, with unsatisfactory results in most cases. Most importantly, these randomized controlled tests even produced inconsistent results, and this kind of inconsistency seems to be related to the differences in stimulation sites and parameters including the treatment frequency and current intensity and duration. A study completed a double-blind, randomized, parallel-group, placebo-controlled test of tACS targeting the prefrontal and mastoid regions at a frequency of 77.5 Hz, which examined the efficacy of tACS to chronic insomnia in adults by assessing response and remission rates, sleep severity, sleep onset latency (SOL), TST, sleep efficiency, and sleep quality. According to its results, patients who received an active tACS intervention within 8 weeks had lower insomnia severity, shorter SOL, longer TST, improved sleep efficiency and sleep quality. A strict and reproducible approach was adopted to study patients with chronic insomnia. It was found that the tACS protocol applied could treat chronic insomnia, but more similar studies were needed to validate the clinical effects of tACS in adults with chronic insomnia at a frequency of 77.5 Hz (Wang et al. 2019) and a current intensity of 15 mA.

In conclusion, tES had certain effect in the treatment of insomnia. Both tDCS and tACS have been demonstrated in corresponding studies. tDCS can improve sleep by increasing TST as well as slow-wave activity through a rational stimulation protocol. However, the current results are still restricted by insufficient number of studies. Due to the complex and diverse pathophysiology of different types of insomnia, the electrophysiological characteristics, the degree of response to stimulation, etc., future studies should consider the various factors, such as the severity of insomnia, age, and gender. In addition, more studies are also needed to improve the stimulation protocols. The most common adverse effects of tDCS are mild tingling sensation at the stimulation site, followed by moderate fatigue, mild discomfort, mild pain sensation under the electrodes, headache, etc. Occasionally, patients feel nervous or overexcited during stimulation. Currently, although tDCS does not cause serious adverse effects or irreversible brain damage, there are potential side effects. In contrast to tDCS, the application of tACS to the treatment of insomnia is seldom studied. The above tests demonstrated that tACS is promising as a new research target for the treatment of insomnia by CES. However, due to the limitation of the number of these studies and the uncertainty of the treatment parameters, tACS cannot yet be concluded as a good clinical treatment tool. In addition, the action mechanism of tACS is still unclear, which should be further explored by more basic researches.

## 1.2 Transcranial Magnetic Stimulation

### 1.2.1 Mechanisms of Repetitive Transcranial Magnetic Stimulation

TMS is a painless, noninvasive physiological technique. Repetitive transcranial magnetic stimulation (rTMS) is defined as the continuous application of magnetic stimulation of a fixed frequency and appropriate intensity to a brain region. That with a frequency equal to or below 1 Hz is low-frequency TMS (LF-rTMS), which suppresses motor cortex excitability; that with a frequency above 5 Hz is high-frequency TMS (HF-rTMS), which usually increases cortical excitability (Binghu Jiang et al. 2019). The regulation effect can last for several minutes, while its mechanism is unclear and may be related to long-term potentiation (LTP) and long-term depression (LTD) of the central nervous system (CNS). Besides affecting brain electrical activity, rTMS may also affect cerebral blood flow, blood-brain barrier permeability, neural metabolism, and protein signaling transduction. rTMS has been widely applied in the treatment of neurological and psychiatric disorders. In recent years, rTMS has been found to be effective in improving sleep quality, optimizing sleep structure, and maintaining the efficacy of drug treatment and CBT-I for insomnia, and is safe for the treatment of insomnia in pregnant or lactating women, the elderly, and other special populations. However, the most effective rTMS parameters have not been determined.

### 1.2.2 LF-rTMS for Insomnia

Hyperarousal is considered to be the ultimate common pathway in the pathophysiology of insomnia. Studies on TMS for insomnia showed that insomnia was characterized by altered intracortical excitability. Therefore, reducing the hyperexcitability of insomnia patients can effectively improve their sleep. The potential mechanisms of rTMS improving insomnia may include suppressing the hyperexcitable state of the cerebral cortex, affecting metabolic activity and sleep-related hormones, and promoting hippocampal neurogenesis. Recently, neuroimaging shows that insomnia patients indicate higher excitability in the dorsolateral prefrontal cortex (DLPFC) than healthy individuals. In addition, studies also showed cortical hyperexcitability in healthy individuals with restless legs syndrome, chronic insomnia, and sleep deprivation. In consideration of the inhibitory effect of LF-rTMS on cortical excitability, it is inferred that continuous bilateral LF-rTMS on DLPFC may be effective in treating patients with insomnia.

Meanwhile, it is widely believed that rTMS exerts therapeutic effects on different disorders by regulating cortical excitability and altering the levels of neurotransmitters. Increasing evidences demonstrated that rTMS regulates brain plasticity by promoting the synthesis and release of brain-derived neurotrophic factor (BDNF). rTMS also promotes the release of gamma-aminobutyric acid (GABA) from neurons. The results suggest that both BDNF and GABA are involved in sleep regulation, while BDNF promotes the function of GABAergic neurons. Investigators recruited 32 patients with primary insomnia consistent with diagnostic criteria of the International Classification of Sleep Disorders (ICD-3) and applied 1 Hz rTMS to DLPFC for 10 consecutive days. The severity of their sleep disorders was assessed by the Pittsburgh Sleep Quality Index (PSQI) before and after the whole rTMS process. Also, serum concentrations of BDNF and GABA in patients were measured. In addition, the amplitude of MEPs was used to reflect the cortical excitability. In the end, Pearson correlation analysis was performed to evaluate the correlation between these variables. Their findings revealed that after treatment with rTMS, PSQI was significantly lower than that before rTMS treatment, the serum concentrations of BDNF and GABA increased significantly, and the amplitude of MEPs decreased significantly. Pearson correlation analysis showed that changes in PSQI were positively correlated with changes in serum concentrations of BDNF and GABA, changes in MEPs amplitude were negatively correlated with changes in serum concentrations of BDNF and GABA, and increases in serum concentration of GABA were positively correlated with that of BDNF. These findings suggest that LF-rTMS on DLPFC may significantly improve insomnia symptoms by increasing the levels of BDNF and GABA in the brain and decreasing cortical excitability (He et al. 2019).

In addition to the above stimulation of the DLPFC, LF-rTMS on the right dorsolateral prefrontal cortex is common at present. A recent study investigated the effect of LF-rTMS on the right DLPFC combined with acupuncture for treating chronic PI. Seventy-eight patients with chronic PI were randomized into the treatment group and control group. Patients assigned to the treatment group received acupuncture with rTMS, while the control group was given acupuncture alone. PSQI, Insomnia Severity Index (ISI), TST, and sleep efficiency (SE) were mainly evaluated. The scores of both groups were improved after treatment, while more significantly in the treatment group than in the control group (Nardone et al. 2020). Thus, LF-rTMS combined with acupuncture can effectively improve the sleep quality and hence improve the quality of life of patients. In addition, other researchers applied continuous 1 Hz rTMS low-frequency magnetic stimulation to posterior parietal cortices of 20 insomnia patients for 14 consecutive days. A 20-minute TMS-EEG was immediately performed before and after treatment. The time-varying EEG network was analyzed with adaptive directed transfer function. The results of patient EEG network were compared with those of 20 healthy subjects. The results revealed that PSQI, ISI, and SE of insomnia patients significantly reduced after treatment with rTMS ($P < 0.05$), and the improvement in scores lasted for 1 month (Song et al. 2019). It was concluded that LF-rTMS targeting the posterior parietal cortices had a significantly positive effect on PI treatment, which could last for at least 1 month (Raffaele Nardone et al. 2020).

### 1.2.3 HF-rTMS for Insomnia

Compared with LF-rTMS, studies and application of HF-rTMS in treating insomnia are not sufficient. However, there are some studies on HF-rTMS in treating sleep-related disorders, such as depression, restless legs syndrome (RLS), narcolepsy, sleep apnea, etc.

In a study, 160 patients with depressive sleep disorders were evenly randomized into an observation group and a control group, with 80 patients in each group. The control group was treated with conventional methods. The observation group was treated with rTMS at 20 Hz besides the conventional treatment. The efficacy was observed after 14 days of treatment and followed up after 2 months. The results showed that the TST, delayed time to early awakening, and number of wakefulness were significantly improved in the observation group compared with the control group and the sleep disorder factors (difficulty in falling asleep, poor sleep, and early awakening) in Hamilton's Depression Scale were significantly reduced. The follow-up after 2 months showed that the observation group had better sleep quality than the control group (Du et al. 2011).

There were also other investigators aiming at investigating whether HF-rTMS could have any beneficial effect in

RLS, as patients with RLS could develop symptomatic insomnia (secondary insomnia).

Another study enrolled 120 hospitalized male patients with drug-dependent insomnia disorder and randomized them into the invention group of 10 Hz rTMS, the sham stimulation group, and the control group, with 40 patients in each group. Patients in the intervention group were given 5 times of 10 Hz rTMS for 6 consecutive weeks, with a total of 180,000 stimulation pulses. No inventions were applied to the control group. Patients were assessed before and after the intervention with PSQI, Self-Rating Anxiety Scale (SAS), and Self-Rating Depression Scale (SDS), respectively. The results of this study showed that 6 weeks of treatment with HF-rTMS could significantly improve sleep quality ($P < 0.001$) and alleviate depression ($P < 0.001$) and anxiety status ($P < 0.001$) in patients hospitalized with drug-dependent insomnia (Lin et al. 2019). It suggests that chronic HF-rTMS has a positive impact on the treatment of drug-dependent inpatients with insomnia disorders.

#### 1.2.4 Summary and Outlook

TMS has been studied for nearly 30 years and is undoubtedly one of the most widely applied techniques of noninvasive brain stimulation. Its efficacy and safety in a wide range of nerve diseases have been proven. rTMS is the most recognized technique, with many relevant reports both at home and abroad. LF-rTMS, in particular, has been thoroughly studied in terms of the action mechanism and efficacy. Most studies investigated the action mechanisms and the effects of LF-rTMS in inhibiting hyperexcitability of the cerebral cortex and regulating neurotransmitters. Compared with LF-rTMS, there are not many studies and applications of HF-rTMS for insomnia. Therefore, the mechanism of HF-rTMS in improving sleep is not yet clear. More studies are needed to verify the clinical utility of HF-rTMS for insomnia. We are also concerned about the safety of rTMS. For the general population, severe side effects of rTMS include seizures, hearing impairment, and cognitive impairment (Taylor et al. 2008). However, the incidence of seizures is low, and with adequate hearing protection measures, a course of rTMS will not cause hearing changes. In addition, studies have shown that rTMS does not affect cognition. This guarantees the application of rTMS to the general population. Furthermore, rTMS has its unique advantage in treating the elderly, not only with relatively few side effects but also without damages caused by drug-drug interactions. Since the side effects of rTMS in children are similar to those in adults, it is recommended that the latest guidelines for adult safety be followed when rTMS is used in this population. rTMS is well tolerated in pregnant patients, with usually mild and transient adverse effects. It has been demonstrated that rTMS is particularly effective and safe for pregnant women with prenatal depression. Based on these studies, we believe that the safety of rTMS for clinical use is trustworthy.

### 1.3 Deep Brain Stimulation

#### 1.3.1 Mechanisms of DBS

Electrical stimulation of brain has been proven to affect various mechanisms involved in neuronal function and signaling. The mechanisms of how DBS acts on brain neurons and affects their function are still being investigated. DBS has been proven to have therapeutic effects on Parkinson's disease (PD), dystonia, and tremor. Currently, most of the studies on DBS involve its application in treating PD. Several studies have shown that DBS can improve the insomnia symptoms in PD. There are few studies on how DBS can improve non-motor symptoms, especially sleep disorders, probably because the effect of DBS in improving the latter has not been fully validated.

#### 1.3.2 DBS for Insomnia

The basal ganglia region is a deep subcortical network structure that plays a crucial role in movement, sleep, and mental behavior. There is a hypothesis that dopamine acts on D2 receptors at the ends of the striatum, enhancing the activity of the pallidum and improving sleep. To test it, the investigators explored the efficacy of pallidum DBS in improving sleep for the first time in rats. It was found by this study that unilateral DBS to the rat pallidum significantly increased nonrapid eye movement sleep and rapid eye movement.

In a study exploring the effects of deep brain stimulation of the subthalamus (DBS-ST) on the sleep and other non-motor symptoms in patients with PD, 36 patients with advanced PD were enrolled. Among them, 24 PD patients were evaluated with PSG of 2 nights before surgery and 6 months after DBS treatment. In addition, the motor symptoms, non-motor symptoms (especially sleep disorders), and quality of life of all patients were assessed before DBS treatment and 6 and 12 months after treatment, respectively. The results of this study showed that according to the assessment of PSG, objective sleep indicators significantly deteriorated after DBS-STN treatment, mainly in terms of TST, SE, N1 and N2 sleep time, wakefulness after sleep onset, and sleep latency. In contrast, subjective sleep quality improved significantly in the first 6 months after DBS-STN treatment. All clinical results of sleep remained good 12 months after DBS-STN treatment. Another investigator conducted a prospective, non-blinded clinical test to evaluate sleep in five patients with PD who had DBS implanted in the pallidum. An effective sleep survey was conducted, and PSG was used to assess the sleep quality before and 6 months after implantation. The assessment items included the Epworth sleep score, the PD sleep score, the ISI, and the RLS severity scale. According to PSG, most patients showed significant improvements in sleep quality, indicated by sleep efficiency (SE) and sleep onset latency (SOL), which, however, did not have statistical significance. Most of the findings displayed improvements in the ISI scale, suggesting that pallidum DBS is promising for improving insomnia. It was concluded that while the pallidum DBS did not show statistically significant improvement in sleep, the PSG did show trends in sleep improvement in terms of sleep efficiency, sleep onset latency, and sleep survey scores. A larger, blinded clinical study is therefore required to more definitively determine whether pallidum DBS affects sleep (Dulski et al. 2019).

### 1.3.3 Summary and Outlook

The recent studies of DBS on insomnia are mainly related to the efficacy of DBS in sleep-related diseases, such as Parkinson's disease (PD) and Tourette syndrome (Martinez-Ramirez et al. 2018). As to the pathogenesis of PD, we find that DBS in the basal ganglia region can improve the sleep disorder in PD patients. Both animal and clinical studies have validated this idea. However, the action mechanism of DBS in this aspect has not been fully explored, and there are no reports on the application of DBS alone for the treatment of insomnia. Such studies are expected in the future to more adequately validate the beneficial effects of DBS on sleep disorders.

In conclusion, neurostimulation is expected to become a major direction for the treatment of insomnia in the future, with many possibilities to be explored.

## 2 Neuromodulation for Restless Legs Syndrome

RLS is a common neurological sensorimotor disorder that involves cortical, subcortical, spinal cord, and peripheral nerves, resulting in increased excitability or decreased inhibition. RLS is clinically characterized by an irresistible urge to move the legs accompanied by unpleasant sensations in the lower extremities. Symptoms of RLS usually worsen at night and when an individual is resting and can be relieved by lower extremity movement. The onset of RLS frequently occurs in the evening or at night, with relative relief during the daytime. Periodic limb movements (PLM) affect 80% of RLS patients, i.e., periodic mechanical and involuntary movements of the lower extremities unilaterally or bilaterally during sleep or wakefulness, typically characterized by toe rhythmic dorsiflexion and ankle dorsiflexion and occasionally hip and knee flexion. RLS is associated with severely reduced sleep and quality of life and may cause fatigue, blood pressure fluctuation, memory loss, and low mood.

RLS is a lifelong disease, with unclear pathogenesis. So far, it is believed that the pathogenesis of RLS is associated with a variety of factors, including genetic factors, abnormal iron metabolism, dysfunction of central dopaminergic neurons, dysfunction of cortical sensorimotor integration, and pathological changes in spinal cord signaling pathways. RLS is a treatable disorder. Whenever possible, etiological treatment should be given to patients with secondary RLS with an established etiology. The overall goal of treatment is to reduce or eliminate the clinical symptoms of RLS, including reducing the severity of symptoms, reducing the number of nocturnal leg movements, and shortening the duration of nocturnal awakenings, thereby improving daytime function, sleep, and quality of life. Current clinical treatments are performed both pharmacologically and non-pharmacologically, with pharmacological treatment (including dopaminergic drugs, alpha-2-delta calcium channel ligands, benzodiazepines, and opioids) remaining the primary treatment at present. Usually, there are efficacies when drugs are taken for the first time. However, there are still 15–40% of patients having no respond to dopaminergic drugs. There is growing concern about the long-term use, decreased efficacy, increased dosage, and "augmentation" of the current pharmacological treatment; hence, the non-pharmacological treatment for RLS has become a hotspot of research. Non-pharmacological treatments mainly include exercise (aerobic exercise), rTMS, transcutaneous spinal direct current stimulation, and special devices.

According to recent studies on the pathogenesis of RLS, the dysfunction of cortical sensorimotor integration is one of the possible causes (Zeng et al. 2020). The decline in the inhibitory mechanism of excitability in the motor cortex leads to the increase of the excitability in patients. Some investigators also propose that the dysfunction of the downstream inhibitory pathway of the central nervous system can be the cause, which leads to hyperexcitability of spinal cord neurons. Based on these new insights, the neuromodulation therapy targeting the excitability regulation of cortical and spinal cord neurons is gradually gaining attention.

### 2.1 Repetitive Transcranial Magnetic Stimulation

rTMS is a novel noninvasive, safe, and effective neuromodulation technique, which has been applied to the treatment of RLS in recent years. In the study conducted by Lin et al., 14 patients with primary RLS were treated with high-frequency rTMS (15 Hz at resting motor threshold) on bilateral primary motor cortex in the leg motor areas for 14 days (Lin et al.

2015). Scale scores at different time points were compared by one-way analysis of variance (ANOVA). The results showed that the scale scores at different time points were compared. The results showed that significant improvements in RLS severity, PSQI, and HAMA scores were noted at the end of the 14-day treatment, with effects lasting at least 2 months. The studies revealed that high-frequency rTMS in the supplementary motor area (SMA) and lower extremity representative area M1 was effective in relieving RLS symptoms. In addition, rTMS at a frequency of 5 Hz in the supplementary motor area appeared to be a safe, noninvasive method that could be considered as a potential treatment for restless legs syndrome.

There are no recommended treatment parameters for rTMS treating RLS. The treatment parameters used in the neuromodulation laboratory of Xuanwu Hospital of Capital Medical University are presented below. rTMS for RLS stimulates the leg motor representative area in cortex with a high-frequency and moderate-intensity magnetic field. Patients wore earplugs to reduce noise stimulation and were seated in a comfortable chair. Their heads were held in a fixed position. The coil holder was adjusted to a fixed position, with the coil parallel to the vertex and the tangent surface pressed against the surface of the patient head. Stimulation site: The stimulation was applied directly to the primary motor cortex (M1) of the lower extremities, with the midpoint of the coil placed in the cortical representative area corresponding to the tibialis anterior muscle of the lower extremities. It was usually located 0–2 cm lateral from the vertex and then 1–2 cm backward. The stimulation intensity was 120% of the active motor threshold (AMT) of the contralateral tibialis anterior muscle. The stimulation frequency was 5 Hz. A series of stimulation was 5 Hz stimulation for 10 s, with an interval of 50 s. The daily dose was 20 series of rTMS given to the motor cortex corresponding to the tibialis anterior muscle of lower right and left extremities or each 10 series of rTMS on the right and left sides and another 10 series of rTMS after an interval of 10 min. Usually, 10 days of treatment was one cycle.

Safety considerations: The safety of TMS devices lies in the high voltage, high current, strong magnetic field, and high-temperature protection. The coil is equipped with a temperature sensor. During operation, the device will immediately shut down when the temperature is exceeded, so as to protect head from high-temperature burns. Strong electromagnetic field: The electromagnetic field around the coil will have an effect on the surrounding metal and magnetic substances. Metal and magnetic objects should be removed from the body before operation. The most serious risk of TMS to the human body is the induction of epilepsy. In patients with epilepsy or a family history of epilepsy, epilepsy can be induced even with conventional high-frequency treatment protocols. The only absolute contraindication is the presence of metal and electronic devices near the coil stimulation site, especially in vivo medical implants, such as cochlear implants, pacemakers, and intracranial electrodes.

## 2.2 Transcranial Direct Current Stimulation

tDCS is a noninvasive method of brain stimulation that induces persistent and polarity-specific changes in human cortical excitability and therefore may also be useful for RLS. Koo et al. (2015) randomized 33 females with RLS into 3 groups that received 5 sessions of tDCS (2 mA, 20 min) using cathodal, anodal, or sham stimulation. The active electrode was placed on the vertex (Cz scalp position of the international 10–20 system), covering the bilateral medial aspect of the primary motor cortex involving the leg areas. Neurophysiological changes were assessed using event-related desynchronization/synchronization (ERD/ERS) of electroencephalography. The results showed that the changes in IRLS scores, as well as the responder rate in CGI-I scale, did not differ significantly among the groups. There was also no significant difference in any of the secondary outcome measures (including PGI, PSQI, MOS sleep subscales, and BDI) and ERD/ERS among the groups. The negative results could be the effect of a strong placebo effect, as seen in the study where there was also a significant decrease in scores in the sham stimulation group. In addition, the small sample size could have led to insignificant results. In conclusion, there are too few studies and limited evidence for tDCS for RLS. Further studies of tDCS with larger numbers of participants and different tDCS parameters such as electrode location, session duration, number of sessions, and the duration of the follow-up period should be encouraged.

Safety considerations: Research on the side effects of tDCS is ongoing, but so far, the established side effects are minor and restricted to the electrode location. They include temporary skin redness, itching, burning sensations, and tingling. Other suggested side effects of tDCS include headache, nausea, and dizziness. Nonetheless, it should be noted that all of the tolerability and safety data on tDCS comes from controlled human trials using specialized equipment and strictly controlled protocols (e.g., limiting current duration, number of sessions).

## 2.3 Deep Brain Stimulation

Deep brain stimulation (DBS) is a well-established neuromodulatory therapy for various movement disorders leading to a significant reduction in motor and non-motor symptoms. Several case reports have reported that DBS of the globus pallidus internus (GPi) improved RLS symptoms, indicating that DBS could be a feasible approach to treat refractory RLS.

Casoni et al. (2020) reported a case of an 80-year-old patient with severe RLS refractory to all medical treatments which significantly improved after bilateral GPi DBS, with reduction in the IRLS score and an objective improvement of polysomnographic parameters. Ondo et al. (2017) report two female patients (onset RLS in childhood) with an IRLS of 36/40 and 40/40 at initial visit who attempted >10 therapies and then improved after bilateral GPi DBS. Although DBS is effective in relieving symptoms of refractory idiopathic RLS, none has achieved complete remission nor eliminates RLS medications. In addition, multiple studies (Dulski et al. 2022; Klepitskaya et al. 2018; Chahine et al. 2011; Driver-Dunckley et al. 2006) have shown that STN DBS significantly decreased RLS symptoms in patients with PD despite a decrease in dopaminergic treatment. Two studies (Kedia et al. 2004; Marques et al. 2015) reported the emergence of RLS after STN DBS, but none reported that GPi DBS worsened RLS.

Safety considerations: As a procedure, DBS carries the risk of surgical complications. Also, the brain stimulation itself can cause side effects. Complications of surgery or after surgery may include misplacement of leads, bleeding in the brain, stroke, infection, breathing problems, nausea, seizure, headache, confusion, and hardware complications, such as an eroded lead wire. Possible side effects included numbness or tingling sensations, tightness of the muscles on the face or arms, speech or balance disorders, lightheadedness, vision problems such as diplopia and mood changes such as anger or depression.

## 2.4 Spinal Cord Stimulation

In studies on the pathogenesis of RLS, it is believed by some investigators that abnormalities in the focal dopamine transmission pathway of the spinal cord reduce the inhibitory effect of upper motor neurons on the spinal cord, resulting in the hyperexcitability of spinal cord neurons and thus the symptoms of RLS (Rijsman et al. 2005). Therefore, the generation and regulation of RLS symptoms can be performed in the spinal cord.

Electrical neurostimulation is a functional electrical stimulation technique that acts on the nervous system by distributing pulsed current through an external pulse generator. In the last decades, direct current neurostimulation has been applied to the scalp as a noninvasive technique to regulate the excitability of the brain. Drawing on the therapeutic approach of transcranial direct current stimulation, some investigators have applied direct current stimulation, a safe tool causing no damages to any issues, to the spinal cord. Direct current stimulation can also regulate the excitability of spinal cord neurons. The techniques of spinal cord stimulation include transcutaneous spinal direct current stimulation (tsDCS) and spinal cord stimulation (SCS).

### 2.4.1 Transcutaneous Spinal Direct Current Stimulation

Transcutaneous spinal direct current stimulation (tsDCS) is a local stimulation with surface electrodes placed directly on the skin of the corresponding spinal segment, which affects the excitability of the upstream and downstream spinal cord pathways and spinal cord reflexes through direct current stimulation on the skin. The basic mechanism of tsDCS regulating spinal cord excitability is the change of hyperpolarization or depolarization of resting membrane potential caused by the different polarity of the stimulation. tsDCS can regulate the excitability of spinal conduction tracts and can also exert an inhibitory effect on spinal conduction pathways by directly activating interneurons at the level of spinal segments or by regulating information transmission between the posterior funiculus and posterior horn of the spinal cord to affect the reflex activity of spinal cord and the electrical activity of the spinal circuit. The polarization of membrane is the main mechanism of the immediate post-tsDCS effect. In addition to the immediate effect, tsDCS also has a post-stimulation effect. The change in neural excitability can persist after successive treatments, and this phenomenon cannot be explained by the potential polarization of neuronal membrane alone. Studies conducted by Wang Li et al. showed that tsDCS could regulate the functional connectivity and plasticity of the cortex. This change may be involved in the post-tsDCS effect, the mechanism of which is similar to the long-time course of synaptic facilitation (Wang et al. 2020). Further studies confirmed that tsDCS could alter the activity of NMDA receptors or GABA to regulate the synaptic microenvironment, thus acting as a regulator of synaptic plasticity.

A study was conducted to use tsDCS in the treatment of RLS, where its efficacy was observed and the mechanism was investigated (Heide et al. 2014). This was done by soaking thick rectangular sponge electrode pads (5 × 7 cm, 6 mm thick) in saline to increase conductivity and avoid skin damage. Anodal electrodes were placed on the back at the T11 level, and cathodal electrodes were placed on the right shoulder. The stimulation was given once at a current intensity of 2.5 mA for 15 min. The results showed significant symptom relief in patients receiving anodal stimulation and a significant reduction in VAS scores after treatment.

The largest limitation of the technique of spinal direct current stimulation was that the follow-up effect could last only a few hours or even minutes. Repeated stimulation was one way to maintain the follow-up effect. To apply transcutaneous spinal direct current stimulation to RLS, it is necessary to explore optimal treatment parameters, including stimulation intensity, single stimulation time, number of stimulations per day, and days of stimulation, so as to maintain the efficacy for a longer time, and even completely reverse the spinal hyperexcitability in patients with RLS, with a view to achieving a complete cure.

Safety considerations: Although transcutaneous spinal direct current stimulation is a noninvasive neuromodulation technique, attention should still be paid to safety issues. The safety of tsDCS depends on the current density during stimulation and the total amount of electricity received per unit area. Studies have shown that tissues are not damaged when current density is <25 mA/cm$^2$ and total electricity is <216 C/cm$^2$. The stimulation parameters adopted by the current clinical study are much lower than the criteria and only cause local numbness for a short time at the beginning or end of the treatment. In addition, neuronal nonspecific enolase (a sensitive and specific indicator of neuronal damage) in serum did not change significantly immediately after tsDCS stimulation, suggesting that direct current stimulation does not cause nerve damages.

### 2.4.2 Spinal Cord Stimulation

Spinal cord stimulation usually refers to implantable spinal cord electrical stimulation, a technique in which stimulating electrodes are surgically placed directly outside the spine canal dura of the corresponding segment, adjacent to the posterior column of the spinal cord. Electrical stimulation is delivered through a pulsed stimulator to act on the conduction tracts of the posterior column of the spinal cord and the sensory neurons of the posterior horn of the spinal cord. Spinal cord stimulation (SCS) has been developed based on the "gate control" theory of pain signaling. SCS has been used clinically for 50 years in Europe and the United States as an effective treatment for chronic pain and now is an important technique in clinical pain treatment. The action mechanism of spinal cord stimulation includes:

1. Stimulating $A_\beta$ fibers in the posterior column of the spinal cord to activate inhibitory interneurons, thus closing the gate of pain signal afferents
2. Stimulating spinal cord interneurons to release inhibitory neurotransmitters, inhibit the release of excitatory amino acids, reduce the transmission of nociceptive information, and activate the downstream pain inhibitory pathway of the central nerves of the upper spinal cord
3. Regulating sympathetic nerves to increase the release of vasodilatory substances and increase the blood flow of local tissues

In the pathogenesis of RLS, increased excitability of spinal cord neurons causing increased sensitivity of spinal flexor reflexes and microvascular dysfunction of the lower extremities and reduced blood flow leading to distal tissue hypoxia are all possible causes. It is even believed by some investigators that restless legs syndrome is a dysregulation of pain control due to impaired downward regulation of upper spinal cord neurons and that the disorder itself is a chronic pain. According to a case report (Holland et al. 2016), a patient with chronic back pain combined with severe RLS experienced complete relief of back pain along with complete relief of RLS symptoms and a reduction in IRLS score from 33 to 0 after receiving spinal cord stimulation ($T_{10}$ to $T_{11}$ levels). Patients only turned on the stimulator before going to bed to receive approximately 30 min of electrical stimulation. Oral medication was not required. The follow-up evaluation 2 years after implantation showed that this effect persisted. Case reports of the application of SCS for the treatment of RLS have been continuously presented in recent years. The remarkable efficacy of SCS reported suggests that the therapeutic effect of spinal cord stimulation in RLS warrants further investigation, so as to provide a new treatment for patients with refractory RLS.

Since there are few studies on the use of spinal cord stimulation in the treatment of RLS, further studies and multicenter trials are needed to observe the efficacy as well as to explore the optimal treatment parameters. This section gives a brief introduction to the spinal cord stimulation technique. The system of spinal cord stimulation consists of a neurostimulator that delivers electrical pulses, electrodes that deliver the pulses to the spinal cord, wires that connect the electrodes to the stimulator, and an external programmer. After the electrodes are implanted in the epidural cavity, a current is delivered from the stimulator and reaches the electrodes via an extension lead to stimulate the spinal cord nerves for therapy. Stimulation electrodes can be implanted by percutaneous puncture and surgical operation. The positioning of electrodes in surgery is based on the area of discomfort in the lower extremity caused by numbness or tingling sensation produced by the stimulated spinal cord segment. After the electrode position is adequately confirmed, an external temporary stimulator is attached. After the stimulation parameters are adjusted, effects will be observed from patients for 7–10 days. If the symptoms are relieved by more than 50%, the treatment is considered effective. Then a permanent stimulator can be implanted under the skin in the upper abdomen. At discharge, the physician will set the original stimulation parameters based on the data from the testing period, and the patient can turn on the stimulator on his or her own for regular treatment during the daily routine. Patients will return regularly for the assessment of effects and the adjustment of treatment parameters.

Safety considerations: The technique of spinal cord stimulation has been widely used in the field of pain treatment, with mature technology and high safety. However, there are some possible complications:

1. Cerebrospinal fluid leakage: The electrode penetrating the dura mater can result in cerebrospinal fluid extravasation and hence trigger intracranial hypotension headache.
2. Displacement of electrodes: It usually occurs within a few days after implantation. Strenuous physical activity should be avoided in the early stage of electrode implantation.
3. Local infection: It is mostly superficial. Prophylactic antimicrobial agents are usually used after SCS surgery to reduce the incidence of postoperative infection.
4. Dural hematoma and local injury: It mostly occurs after electrode implantation by laminectomy, which is a more serious but rare complication of SCS.

Therefore, in order to improve the accuracy of electrode implantation position and reduce complications such as injury, electrodes should be implanted by percutaneous puncture whenever possible.

In conclusion, there are relatively few studies on the application of neuromodulation technique for the treatment of RLS, and the evidence provided is not always reliable. It seems that there are not sufficient evidences for the neuromodulation of RLS, while there are sufficient preliminary data that support further high-quality randomized controlled studies.

## 3 Other Sleep Disorders

### 3.1 Effects of Neurostimulation Therapy in Narcolepsy

#### 3.1.1 Narcolepsy

Narcolepsy is a chronic sleep disorder of unknown etiology. It is mainly characterized by recurrent episodes of irresistible daytime sleepiness, cataplexy, and nocturnal sleep disturbance. It is also accompanied by a range of other symptoms, including sleep paralysis, hypnagogic hallucination, and cataplexy, which usually begin in adolescence. The International Classification of Sleep Disorders (2014 Edition) reclassifies narcolepsy into Type 1 and Type 2 according to clinical manifestations and the level of orexin in the cerebrospinal fluid. Patients with Type 1 narcolepsy experience cataplexy and/or have low levels of orexin in the cerebrospinal fluid, whereas patients with Type 2 narcolepsy do not experience cataplexy and have normal levels of orexin in the cerebrospinal fluid. Orexin-related neurons are related with the excitability in several areas of the brain that promote awakening, such as areas with adrenergic (locus coeruleus), dopaminergic (ventral tegmental area and periaqueductal gray matter), 5-hydroxytryptaminergic (raphe nuclei), and histaminergic (nodular papillary nucleus). Thus, the missing of stimulation in these areas due to lack of orexin may lead to a general low-arousal state, causing low threshold required for falling asleep. The current treatment for narcolepsy alleviates excessive daytime somnolence (EDS) and nervousness and partially reduces symptoms such as nocturnal sleep disruption, hypnagogic hallucination, and sleep paralysis. It is performed mainly by medication targeting three aspects, that is, CNS stimulants for daytime sleepiness, antidepressants for improving symptoms of cataplexy, and sedative-hypnotics for nocturnal sleep disturbances.

#### 3.1.2 Neurostimulation for Narcolepsy

Fundamentally, drug therapy for narcolepsy is generally symptomatic and emphasizes lifestyle adjustments. In view of this, new therapeutic strategies with good compliance, efficacy, and tolerability are in urgent need. Some recent studies assess the cortical excitability in patients with narcolepsy with TMS and find reduced cortical excitability, which is believed to be a cause of EDS. In addition, there are also several studies examining the therapeutic effects of TMS on narcolepsy. There are more studies on the assessment of cortical excitability in narcolepsy by rTMS. It is reported that reduced cortical excitability in narcolepsy may contribute to excessive daytime somnolence. It is also found that Type 1 narcolepsy is divided into four phases: trigger phase (CA1), resistance phase (CA2), dystonia phase (CA3), and recovery phase (CA4). With the method of rTMS-MEP, it is found that the MEP wave amplitudes in the quiet wakefulness state (QW) and the cataplexy period were highly fluctuating. The wave amplitude in CA1 and CA2 phases was significantly higher compared to the quiet wakefulness state, while that of the CA3 phase was below the baseline level and gradually recovers in the CA4 phase. The increase in wave amplitude is significant in CA1 and CA2, and the latency of MEP is prolonged in CA3. However, there are relatively few studies on rTMS in the treatment of narcolepsy.

In a study, a 14-year-old girl experiencing EDS, cataplexy, and hypnagogic hallucination received 25 times of high-frequency (10 Hz) rTMS in the left dorsolateral prefrontal cortex (LPFC). The follow-up 1 year after showed that the symptoms of EDS and cataplexy seizure improved significantly after rTMS treatment. In the meanwhile, her Epworth Sleep Score (ESS) decreased significantly. In the first year after the outpatient follow-up, the relief of her symptoms was maintained. This case suggests that rTMS may be a safe and effective alternative strategy to alleviate symptoms of narcolepsy. However, it is only one case. Studies with a large sample are required.

HF-rTMS to the left dorsolateral prefrontal cortex is effective for the symptoms of narcolepsy. Narcolepsy also affects attention, emotion, and memory. The left DLPFC is chosen as a target because this site regulates emotion and executive function. The mechanism of rTMS to the left

DLPFC may be the modulation of the anterior cingulate cortical connections in the corticolimbic circuit (Martin Tik et al. 2017).

Some investigators established a mouse model to perform deep brain stimulation to the medial prefrontal cortex and the arousal center of the lateral hypothalamus (LH) to investigate whether DBS could benefit the mouse model of Type 1 narcolepsy. The results showed that continuous stimulation of LH resulted in an 86–100% reduction in the time consumed by a cataplexy and was able to increase the wake time by approximately 15–27% compared to that before treatment. Stimulation also reduced cataplexy, indicating that DBS at the LH site was specific and effective in relieving the symptoms of narcolepsy. However, this is a study on an animal model. Studies in humans are still needed (Anna A. Rogers et al. 2020).

### 3.1.3 Summary and Outlook

The case report above is the first to report that HF-rTMS in DLPFC is effective in alleviating the symptoms of narcolepsy, suggesting that rTMS may be able to regulate sleep. It is a breakthrough study that provides a new direction for the treatment of narcolepsy, and more such studies are needed in the future to provide a more adequate experimental and theoretical basis for its application. The basic study above also provides evidence for the effectiveness of neurostimulation for narcolepsy, while it is necessary to further explore stimulation parameters, additional targets, and long-term effects. Their findings provide the first evidence that deep brain stimulation can function as a new therapeutic approach for the treatment of NT1.

## 3.2 Application of Neurostimulation in Obstructive Sleep Apnea Syndrome

### 3.2.1 Obstructive Sleep Apnea Syndrome

Obstructive sleep apnea syndrome (OSAS) is characterized by recurrent upper airway stenosis or obstruction during sleep, and manifested as snoring, apnea, and excessive daytime somnolence, and is associated with reduced oxyhemoglobin saturation, which leads to intermittent hypoxemia, sympathetic arousal, and sleep fragmentation. OSAS is also associated with various disorders including hypertension, obesity, type 2 diabetes, depression, and psychological and cognitive defects. Clinically, it is defined as symptoms of at least 5 obstructive respiratory events (hypopnea, apnea, respiratory effort-related arousals, etc.) per hour of sleep accompanied by excessive daytime somnolence, loud snoring, apnea, or awakening due to gasping for air or 15 obstructive respiratory events per hour without accompanying signs and symptoms. The prevalence of OSAS ranged from 7.0% to 49.7% in male and 3.0% to 23.4% in female. Untreated OSAS can lead to cardiovascular disease (hypertension, coronary artery disease, congestive heart failure, arrhythmias, and stroke), metabolic disorders, and diabetes. Currently, the gold standard treatment for OSAS is continuous positive airway pressure (CPAP), which has been proved to be able to reduce mortality and morbidity in OSAS. However, many patients fail to maintain the treatment. Other treatment strategies such as behavioral strategies, oral appliances, and adjunctive therapies have also been reported, while the value of their application remains to be explored.

### 3.2.2 Application of TMS in OSAS

Studies on the application of TMS in OSAS can be divided into two major categories: those that use TMS as an exploratory tool to describe the pathophysiology and mechanisms of OSAS and those that use it as a therapeutic tool to improve OSAS.

#### 3.2.2.1 Application of TMS in the Exploration of OSAS Pathophysiology

Abnormalities in cortical excitability are considered as the pathophysiological basis of various neurocognitive manifestations of OSAS. TMS is a noninvasive method for monitoring cortical function. TMS can be used for in vivo physiological studies to assess intracerebral excitatory and inhibitory circuits. TMS uses a brief magnetic field to induce currents in the cerebral cortex, producing a motor evoked potential (MEP), followed by a period of inhibition of muscle activity, known as cortical silent period (CSP). The threshold of MEP reflects the excitability of neuron center, whereas CSP reflects the excitability of neurons.

There were investigators using TMS to assess motor cortical excitability and investigate the function of the motor cortex during the day in OSAS patients. They recruited ten OSAS patients and ten healthy volunteers and divided them into an active group and a control group. TMS was applied to the motor cortex in the active group, and the MEP of the first interosseus of the dominant hand was recorded to assess the CSP threshold, and CSP and RMT induced by five stimulation intensities, and Stanford Sleepiness Scale (SSS) were also measured. The results showed that ESS and SSS were significantly higher in OSAS patients than in the control group. The longer duration of CSP in OSAS patients receiving 95%, 100%, and 105% stimulation intensities proved the changes in motor cortical excitability. In addition, the increment of CSP was related to the stimulation intensity, i.e., CSP increment was greater at low-intensity stimulation (95% RMT, 100% RMT, 105% RMT), while the TMS stimulation intensity was higher, the CSP increased only in the morning.

#### 3.2.2.2 Application of TMS as a Therapeutic Tool in OSAS

The activity of the upper airway dilator muscle is controlled by the brainstem and output by the cortex, both of which are transmitted by peripheral motor neurons to the upper airway dilator muscle to make movements. In the treatment of OSAS, TMS can improve upper airway kinetics without waking up patients. Stimulation of the genioglossus with TMS during the ascendant portion of the inspiratory flow of airflow-limited breaths maintains the recruitment of the upper airway dilator muscles, thus improving upper airway kinetics without affecting the patient sleep. The use of rTMS in the inspiratory or expiratory phase produces a different kinetic response in the upper airway. It has been shown that TMS can activate upper airway dilator muscles in healthy subjects. However, the ability of TMS to activate upper airway dilator muscles in OSAS patients has not been demonstrated. A study collected 14 patients with moderate-to-severe OSAS and examined the MEP of the upper airway dilator muscle in these patients during wakefulness and sleep, respectively. TMS-induced muscle activity was applied during the period of stable nonrapid eye movement sleep, with stimulation intensity ranging from that reaching muscle motor threshold to the maximum intensity under which patients are not awaken. It assessed the maximum inspiratory flow, mean inspiratory volume, EEG frequency changes, and pulse rate variability. It was found that after TMS stimulation, the maximum inspiratory flow increased with the gradual increase of stimulation and the inspiratory volume increased in almost all patients. It should be noted that cortical autonomic arousal was observed in only 13.8% of patients throughout the study. It was thus concluded that TMS could activate the submental muscle during sleep without awakening sleeping OSAS subjects and that TMS caused the expansion of the upper airway dilator muscle, which led to an increase in both maximal inspiratory flow and mean inspiratory volume.

### 3.2.3 DBS for OSAS

In addition to the role of TMS in OSAS, there are also reports describing other neurostimulation methods that may alleviate sleep symptoms in patients with OSAS. According to a case report, in a patient with idiopathic PD with OSAS receiving the implantation of a deep brain stimulator in the subthalamic nucleus (DBS-STN), the sleep symptoms were alleviated. The patient was a 68-year-old patient with advanced PD who was previously treated with medication until 2011, when the patient was implanted with a deep brain stimulator in the subthalamic nucleus. Motor symptoms (muscle stiffness, mild bradykinesia, mild cervical dystonia) were alleviated significantly after the surgery, and the dose of dopaminergic drugs was reduced. Improvements in sleep were also found, such as a decrease in snoring, a decrease in obstructive respiratory time, and improvements in polysomnogram as well as ESS. This case report suggests the possibility of improving sleep by implanting DBS-STN in PD patients.

### 3.2.4 Summary and Outlook

Studies on the assessment of motor cortical excitability with TMS have shown that TMS provides new insights for the pathophysiology of OSAS, which indicates the irregularities of cortical excitability in OSAS patients, mainly referring to the increase in RMT and CSP. The studies instruct future researches and treatment strategies. In addition, while different TMS modes have shown positive effects on treatment, it is clear that this tool needs further validation before it can be clinically accepted and applied. The increase in inspiratory flow in OSAS patients by TMS also offers new targets for the treatment of OSAS. However, due to the limitation of sample size and the lack of similar trials, further studies are needed to address issues such as proposing optimal protocols to improve dynamic upper airway flow during sleep and to prevent upper airway obstruction. In addition, the above case report provides a new target for DBS in the treatment of OSAS. More studies may be conducted in the future to explore DBS in the treatment of OSAS.

## 3.3 Application of Neurostimulation in Rapid Eye Movement Sleep Behavior Disorder

### 3.3.1 Introduction of Rapid Eye Movement Sleep Behavior Disorder (RBD)

RBD is characterized by dream-related violent behavior that occurs when muscular flaccidity disappears during rapid eye movement (REM) sleep, often manifested as abnormal sleep behavior interpreted by dreams, such as punching and kicking at night, abnormal vocalizations, and, in severe cases, injury to self or bed partner. RBD is divided into primary and secondary depending on the etiology. Secondary RBD is associated with neurological diseases, including Parkinson's disease (PD), multiple system atrophy (MSA), and other diseases involving brainstem structures such as tumors and demyelination. RBD is now considered a potential harbinger of the development of neurodegenerative diseases, particularly PD, MSA, dementia with Lewy bodies (DLB), and autonomic failure. RBD may also present with other symptoms such as upright hypotension and depression. RDB occurs mostly in patients with PD, and in some cases, RBD can harm themselves or even their bed partners due to dream-enacting violence. Such injuries have been reported to occur in 38–59% of patients and also occasionally caused by hitting a wall or nightstand or even falling out of bed when the disorder attacks. Minor bruises, lacerations, and sprains are most common, while there are also a few cases of subarachnoid hemorrhage.

### 3.3.2 Application of Neurostimulation in RBD

RBD has been found to be closely associated with a group of neurodegenerative diseases called synuclein diseases, including PD, DLB, and MSA. The pathogenesis of RBD and the neurobiological significance of cognitive defects in RBD are not known, but the cholinergic system is known to play a key role in cognitive function. TMS has been recently introduced to assess certain cholinergic circuits in the human brain. TMS may produce information about the function of certain cholinergic circuits in the human brain. It is a technique that relies on short-interval afferent inhibition (SAI) in the motor cortex, which significantly reduces in the cholinergic system of patients with dementia, such as Alzheimer's disease and dementia with Lewy bodies, whereas SAI is normal in patients of dementia with abnormalities in non-cholinergic systems, such as frontotemporal dementia. It should be noted that SAI also reduces in patients with mild cognitive impairment of the type of multi-domain amnesic.

There is a study using TMS to test the hypothesis on the association of the cognitive function and cortical activation in RBD patients with the dysfunction of cholinergic system. This study applied the SAI technique to 10 patients with idiopathic RBD (iRBD) and compared them with 15 age-matched healthy subjects. All iRBD patients and controls received an extensive neuropsychological assessment. The results revealed that the mean SAI was significantly lower in patients with iRBD compared to controls. It was shown by the neuropsychological assessment that six of the ten patients with iRBD experienced mild cognitive impairment. SAI assessment was closely related to tests of verbal episodic memory and executive function. These results support the hypothesis that some patients with iRBD suffering from cognitive impairment have cholinergic dysfunction. One possible explanation is that the expression involving the cholinergic circuit in the brainstem of RBD patients alters the balance of the entire cholinergic system or its balance with other neurotransmitter systems. Thus, studies on TMS may also contribute to the identification of individuals with iRBD who develop cholinergic degeneration. Because of this, it is possible to predict whether iRBD individuals with abnormal SAI are at risk for cognitive impairment.

Subthalamic nucleus (STN) deep brain stimulation (DBS) (STN-DBS) seemed to have a negative effect on RBD, which was probably related to the diagnosis of RBD. However, some studies also showed that the RBD of some patients with PD and RBD was mitigated after DBS. PD patients who developed RBD symptoms within 1 year after DBS were older than those who did not develop RBD after stimulation. A study showed that clinical RBD increased after bilateral STN-DBS. The stimulation site of STN-DBS is the subthalamic nucleus, whereas RBD is associated with the pons oblongata. Therefore, they do not have a definite relationship in terms of anatomical sites.

The RMT, central motor conduction time (CMCT), short-latency intracortical inhibition (SICI), intracortical facilitation (ICF), and SAI were measured by TMS in patients with PD combined with RBD and patients with PD but without RBD. It was found that there was a high correlation between neuropsychological scores and SAI in patients with PD combined with RBD and the SAI of patients with PD combined with RBD was significantly lower than that of patients with PD but without RBD. In addition, this correlation was not high in patients with PD but without RBD. SAI was affected by GABAergic drugs, such as benzodiazepines, and also affected by dopaminergic drugs. In contrast, monoaminergic drugs did not affect SAI.

### 3.3.3 Summary and Outlook

There are few reports on the application of neurostimulation in RBD and only a few studies on the pathogenesis. Studies above suggest that TMS may be useful to explain the pathophysiological mechanisms underlying impaired cortical activation in patients with iRBD. The combined results of neurophysiological and neuropsychological studies support the hypothesis that some patients with RBD with cholinergic dysfunction may develop dementia.

## References


Casoni F, Galbiati TF, Ferini-Strambi L et al (2020) DBS in restless legs syndrome: a new therapeutic approach? Sleep Med 76:155–157

Chahine LM, Ahmed A, Sun Z (2011) Effects of STN DBS for Parkinson's disease on restless legs syndrome and other sleep-related measures. Parkinsonism Relat Disord 17(3):208–211

Chaieb L, Antal A, Paulus W (2011) Transcranial alternating current stimulation in the low kHz range increases motor cortex excitability. Restor Neurol Neurosci 29(3):167–175

Driver-Dunckley E, Evidente VGH, Adler CH et al (2006) Restless legs syndrome in Parkinson's disease patients may improve with subthalamic stimulation. Mov Disord 21(8):1287–1289

Du Z, Liu Y, Wang L et al (2011) Efficacy of transcranial magnetic stimulation therapy in the treatment of depression sleep disorders. Shandong Med 51(25):48–49

Dulski J, Schinwelski M, Konkel A et al (2019) The impact of subthalamic deep brain stimulation on sleep and other non-motor symptoms in Parkinson's disease. Parkinsonism Relat Disord 64:138–144

Dulski J, Wąż P, Konkel A et al (2022) The impact of subthalamic deep brain stimulation on restless legs syndrome in Parkinson's disease. Neuromodulation 25(6):904–910

Frase L, Selhausen P, Krone L et al (2019) Sulamith Tsodor differential effects of bifrontal tDCS on arousal and sleep duration in insomnia patients and healthy controls. Brain Stimul 12(3):674–683

He Y, Sun N, Wang Z et al (2019) Effect of repetitive transcranial magnetic stimulation (rTMS) for insomnia: a protocol for a systematic review. BMJ Open 9(7):e029206

Heide AC, Winkler T, Helms HJ et al (2014) Effects of transcutaneous spinal direct current stimulation in idiopathic restless legs patients. Brain Stimul 7(5):636–642

Holland MT, Rettenmaier LA, Flouty OE et al (2016) Epidural spinal cord stimulation: a novel therapy in the treatment of restless legs syndrome. World Neurosurg 92:582.e15–582.e18

Jiang B, He D, Guo Z et al (2019) Efficacy and placebo response of repetitive transcranial magnetic stimulation for primary insomnia. Sleep Med 63:9–13

Kedia S, Moro E, Tagliati M et al (2004) Emergence of restless legs syndrome during subthalamic stimulation for Parkinson disease. Neurology 63(12):2410–2412

Klepitskaya O, Liu Y, Sharma S et al (2018) Deep brain stimulation improves restless legs syndrome in patients with Parkinson disease. Neurology 91(11):e1013–e1021

Koo YS, Kim SM, Lee C et al (2015) Transcranial direct current stimulation on primary sensorimotor area has no effect in patients with drug-naïve restless legs syndrome: a proof-of-concept clinical trial. Sleep Med 16(2):280–287

Lin Y, Feng Y, Zhan S et al (2015) Repetitive transcranial magnetic stimulation for the treatment of restless legs syndrome. Chin Med J 128(13):1728–1731

Lin J, Liu X, Li H et al (2019) Chronic repetitive transcranial magnetic stimulation (rTMS) on sleeping quality and mood status in drug dependent male inpatients during abstinence. Sleep Med 58:7–12

Marques A, Fantini ML, Morand D et al (2015) Emergence of restless legs syndrome after subthalamic stimulation in Parkinson's disease: a dopaminergic overstimulation? Sleep Med 16(5):583–588

Martinez-Ramirez D, Jimenez-Shahed J, Leckman JF et al (2018) Efficacy and safety of deep brain stimulation in Tourette syndrome. JAMA Neurol 75(3):353

Nardone R, Sebastianelli L, Versace V et al (2020) Effects of repetitive transcranial magnetic stimulation in subjects with sleep disorders. Sleep Med 71:113–121

Ondo W (2017) Deep brain stimulation (DBS) for severe restless legs syndrome: therapeutic and physiologic considerations. Sleep Med 31:93–94

Rijsman RM, Stam CJ, de Weerd AW (2005) Abnormal H-reflexes in periodic limb movement disorder; impact on understanding the pathophysiology of the disorder. Clin Neurophysiol 116:204–210

Rogers AA, Aiani LM, Blanpain LT et al (2020) Deep brain stimulation of hypothalamus for narcolepsy-cataplexy in mice. Brain Stimul 13(5):1305–1316

Saebipour MR, Joghataei MT, Yoonessi A et al (2015) Slow oscillating transcranial direct current stimulation during sleep has a sleep-stabilizing effect in chronic insomnia: a pilot study. J Sleep Res 24(5):518–525

Sateia MJ, Buysse DJ, Krystal AD et al (2017) Clinical practice guideline for the pharmacologic treatment of chronic insomnia in adults: an American Academy of Sleep Medicine Clinical Practice Guideline. J Clin Sleep Med 13(02):307–349

Song P, Lin H, Li S et al (2019) Repetitive transcranial magnetic stimulation (rTMS) modulates time-varying electroencephalography (EEG) network in primary insomnia patients: a TMS-EEG study. Sleep Med 56:157–163

Taylor R, Galvez V, Loo C (2008) Transcranial magnetic stimulation (TMS) safety: a practical guide for psychiatrists. Australas Psychiatry 26(2):189–192

Tik M, Hoffmann A, Sladky R et al (2017) Towards understanding rTMS mechanism of action: stimulation of the DLPFC causes network-specific increase in functional connectivity. NeuroImage 162:289–296

Valiengo LC, Goulart AC, de Oliveira JF et al (2017) Transcranial direct current stimulation for the treatment of post-stroke depression: results from a randomised, sham-controlled, double-blinded trial. J Neurol Neurosurg Psychiatry 88(2):170–175

Wang HX, Li W, Zhang WR et al (2019) Effect of transcranial alternating current stimulation for the treatment of chronic insomnia: a randomized, double-blind, parallel-group, placebo-controlled clinical trial. Psychother Psychosom 89(1):38–47

Wang L, Liu C, Hou Y et al (2020) Altered cortical gray matter volume and functional connectivity after transcutaneous spinal cord direct current stimulation in idiopathic restless legs syndrome. Sleep Med 74:254–261

Zeng M, Wang L, Cheng B et al (2020) Transcutaneous spinal cord direct-current stimulation modulates functional activity and integration in idiopathic restless legs syndrome. Front Neurosci 14:873

Zhang Y, Liao W, Xia W (2008) Effect of acupuncture cooperated with low-frequency repetitive transcranial magnetic stimulation on chronic insomnia. Curr Med Sci 38(3):491–498

# Pain Disorders

# 24

Hongwei Zhu, Bing Ni, Zhexue Xu, Nuo Yang, and Huicong Wang

## 1 Chronic Neuropathic Pain

In 2020, the International Association for the Study of Pain (IASP) revised the definition of pain; the updated definition of pain is "an unpleasant sensory and emotional experience associated with, or resembling that associated with, actual or potential tissue damage." Pain is regarded as a symptom or a specific condition known as pain syndrome. According to etiology, pain can be generally divided into two types: nociceptive pain (NCP) elicited by stimulation of receptors in tissues and neuropathic pain (NPP) elicited by primary lesions or dysfunctions in the nervous system.

NPP mainly manifests as spontaneous pain with or without sensory or motor dysfunction in the painful area. NPP can be divided into two main types, peripheral and central, depending on the part of nervous system involved. NPP, upon the analysis of its occurrence mechanism, has one common feature—that the lesion interferes with the normal transmission of pain information at a certain level of the nervous system, and is therefore called deafferentation pain. The common causes of NPP include vascular diseases (e.g., hemorrhage, infarction, vascular malformation), demyelinating diseases (e.g., multiple sclerosis, Guillain–Barre's syndrome), physical factors (e.g., violent trauma, hyperosteogeny, hypertrophy of ligament), infections (e.g., herpes zoster virus, HIV, EB virus, *Mycobacterium tuberculosis*), and metabolic abnormalities (e.g., gout, diabetes, heavy metal intoxication).

The nature of NPP varies greatly among individuals, generally manifested as pricking, burning, lightning, and cutting pains; the duration of attacks is characterized by explosive exacerbations with persistent, intermittent, or persistent background pain; pain intensity varies from "severe" to "insufferable" among individuals, and some sufferers are complicated with significant anxiety and depression. Pain progresses to chronic NPP if the primary cause of pain cannot be eradicated or if the pain persists for more than 3 months after the primary cause is eradicated.

Based on common etiologies, approximately 30% of patients with spinal cord injury, 20% with multiple sclerosis, and 1.5% with stroke may have NPP. Painful areas vary greatly among individuals with chronic NPP who generally have poor responses to conventional Nonsteroidal Anti-inflammatory Drugs (NSAIDs), anticonvulsants, opioids, tricyclic antidepressants, nerve block, rehabilitation physiotherapy, spinal canal decompression surgery, etc. (Schaefer et al. 2014). Neuromodulatory stimulation for chronic NPP includes noninvasive and invasive approaches. Noninvasive stimulation mainly includes two types: transcranial direct current stimulation (tDCS) and transcranial magnetic stimulation (TMS). Invasive stimulation mainly includes motor cortex stimulation (MCS), spinal cord stimulation (SCS), and peripheral nerve stimulation (PNS). MCS, SCS, and PNS, as non-destructive procedures, are emerging treatment options for chronic NPP in the past 20 years and have become the main procedures for neuromodulation in the treatment of chronic pain relaying on the advantages of minimal invasive, reversibility, and adjustability (Bremer 2016, Deer 2017, Kallewaard 2019).

H. Zhu (✉) · B. Ni
Department of Functional Neurosurgery, Beijing Institute of Functional Neurosurgery, Xuanwu Hospital, Capital Medical University, Beijing, China
e-mail: zhuhongwei@xwh.ccmu.edu.cn; nibing@xwh.ccmu.edu.cn

Z. Xu
Department of Neurology, Xuanwu Hospital, Capital Medical University, Beijing, China

Department of Neurology, Dongfang Hospital Beijing University of Chinese Medicine, Beijing, China

N. Yang · H. Wang
Department of Neurology, Xuanwu Hospital, Capital Medical University, Beijing, China


Y. Wang (ed.), *Therapeutics of Neural Stimulation for Neurological Disorders*, https://doi.org/10.1007/978-981-99-4538-2_24

## 1.1 Noninvasive Stimulation

tDCS may relieve NPP in some patients by direct current stimulation of specific targets in the brain. Common stimulation sites include dorsolateral prefrontal cortex (DLPFC), primary motor cortex (M1), and right orbitofrontal cortex. tDCS may have satisfactory analgesic effects, especially in patients with lower extremity pain after spinal cord injury. Anodal stimulation is generally applied to targets for about 20 days, with 35 $cm^2$ of the electrode cross-sectional area, 2 mA of the stimulation intensity, and 10–20 min of the daily duration.

Repetitive transcranial magnetic stimulation (rTMS) is a noninvasive tool for clinical treatment and research, which generates localized electrical currents in the cerebral cortex based on the electromagnetic induction principle, thus stimulating the central nervous system, changing the excitability of neurons, changing the plasticity of synapses, and achieving the purpose of treatment. Early stage rTMS application in the management of chronic pain was designed for the preexperiment before successful electrode placement of motor cotex stimulation (MCS). With the advancement of science and technology, the role of rTMS has gradually changed from testing to treatment; in particular, the revolutionary change of stimulation coils is the most significant, and various types have emerged so far (e.g., 8-shaped focusing coils) based on the original circular non-focusing coils. 8-shaped focusing coils can provide more focused focal stimulation than non-focusing coils, resulting in more favorable clinical outcomes. In addition, in combination with navigation technology, precise stimulation of specific areas, including specific cortical and subcortical structures, as well as spinal cords and nerves, can be achieved. Generally, the M1 hand part in the cerebral hemisphere M1 opposite to the pain is selected as the stimulation site, and the effects of rTMS on cortical excitability are frequency- and intensity-dependent. Several studies have shown that rTMS with higher frequency may be more effective, and most of the current studies have selected the stimulation frequency of 10–20 Hz. The intensity of rTMS is generally 80–110% of the resting motor threshold (RMT). High-frequency stimulation ($\geq$10 Hz) is applied. A series of stimulation for several seconds to tens of seconds is followed by a pause for several seconds to tens of seconds. Dozens-hundreds of stimulations, generally at 2000–3000 pulses/day, are given in dozens of minutes in a single day, for 5–10 days (continuous stimulation for 10 s per minute + rest for 50 s, 2000 stimulations/time, for a total of 20 min). It applies to central post-stroke pain, spinal cord injury pain, postherpetic neuralgia, etc. (Hamid 2019).

## 1.2 Motor Cortex Stimulation

In 1991, Tsubokawa et al. examined the therapeutic effect of cortical stimulation on thalamic pain, and unexpectedly found that direct MCS in the precentral gyrus can effectively inhibit pain. Thalamic pain originates from the hemorrhage, infarction or trauma in the thalamus, and manifests as contralateral numbness, pain, burning sensation, creeper sensation, etc.; such pain, generally severe, may not be relieved even after recovery of motor function in most patients, and is intractable as a typical central neuropathic pain. Following the discovery that stimulation of M1 on the lesion side of thalamus was beneficial for pain relief, it was also found that stimulation of the ipsilateral sensory cortex usually exacerbated pain in many patients, and that the best area for pain control was precentral gyrus, where muscle contractions in the pain sites could be induced (Maarrawi 2007).

There is also a DPM system in the central nervous system in addition to the ascending transmission system that produces pain from peripheral to central nervous systems. Normally, the two systems are in equilibrium. However, in the presence of inflammation or pathological changes in peripheral tissue or nervous system, the two systems will become unbalanced by diminished DPM function and diminished or even lost analgesia, ultimately leading to or facilitating pain. M1 is considered to be the highest pain neuroregulatory target in the nervous system. However, so far, the analgesic mechanism of MCS has not been fully clarified, and the neural pathway that reduces pain signals by stimulating M1 remains unclear. Several scholars have suggested that MCS may improve pain by affecting DPM (Fig. 24.1). Studies have shown that stimulation of M1 may result in direct inhibition of brain regions involved in emotional responses to pain, triggering the functioning of descending inhibition pathways at the dorsal horn level. It suggests the presence of top-down inhibitory signals in central nervous regulation, providing persistent clinical effects that last beyond the stimulation phase. MCS enhances the release of endogenous opioids in different regions of the brain. Electromagnetic stimulation of the central nervous system not only increases the secretion of endogenous opioids, but also modulates neurotransmitters such as glutamate, dopamine, and GABA aminobutyric acid.

### 1.2.1 Indications for MCS Treatment

Indications for MCS treatment included: (1) Chronic unilateral pain following thalamic, or medulla lesions (medulla oblongata lesions may manifest as contralateral pain if located at the rostral of the pyramidal chiasma, or as chiasma or ipsilateral pain if located at the pyramidal chiasma or its caudal end), requiring that patients have normal or basically normal motor function in the painful area; (2) Chronic facial pain following peripheral trigeminal branch injury (different from typical trigeminal neuralgia, it is often found after radiofrequency or balloon compression of the trigeminal nerve, or after oral and maxillofacial surgery, and mainly manifests as hypoesthesia in the pain area combined with pain, numbness, creeper sensation and burning sensation, also known as anes-

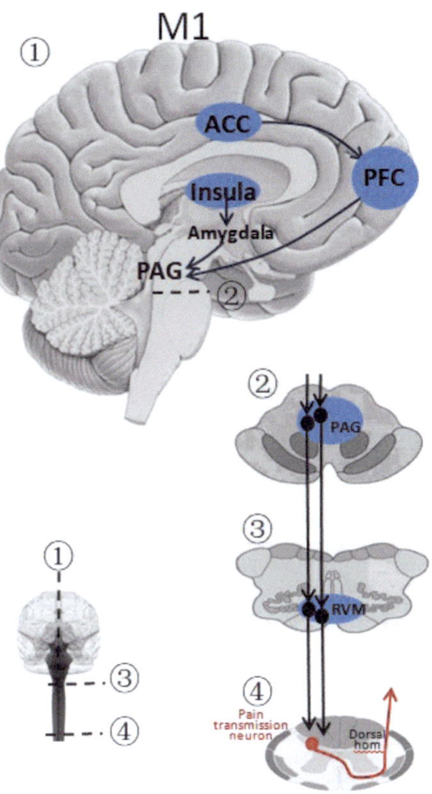


**Fig. 24.1** Possible relationship between M1 area and descending pain modulation (DPM) system. *M1* motor cortex, *ACC* anterior cingulate cortex, *PFC* prefrontal cortex, Insula; amygdala, *PAG* periaqueductal gray matter, *RVM* rostral ventromedial medulla; Dorsal horn

thesia dolorosa); (3) Chronic postherpetic neuralgia involving the head and face (in most cases, varicella-zoster virus activates the ganglion of the trigeminal nerve along its peripheral branches and causes skin lesions and pain; in a few cases, the virus activates from the geniculate ganglion along the intermediate nerve, causing skin lesions in the lateral ear canal, parotid gland and retroauricular area, pain and dizziness, known as Hunt syndrome; in some patients, the virus involves the C1–C3 area, causing pain in the head and face); (4) Phantom limb pain, a pain syndrome in which chronic phantom limb pain occurs after limb amputation due to trauma, infection, gangrene, ischemic infarction, and other reasons.

### 1.2.2 Preoperative Evaluation

Imaging examination includes Magnetic Resonance Imaging (MRI), Diffusion Weighted Imaging (DWI), and Multiplanar reconstruction (MPR), aiming at locating the original lesion in the brain, observing the degree of brain atrophy, locating the position of the central sulcus, and providing the necessary volumetric data for navigation. Pain assessment includes pain location, characteristics of pain episode, VAS pain score, assessment of pain area depth sensation and motor function, anxiety, and depression scale.

Repetitive transcranial magnetic stimulation (rTMS) can be used as a preoperative pre-assessment for MCS surgery. The response rate of the therapy for various pain syndromes is about 50–60%. However, this therapy is an invasive and expensive surgical technique. If the efficacy of rTMS is confirmed, it is worthwhile to try MCS again. Since the hand, head, and face have the largest corresponding area in M1, the hand position in the primary motor cortex is generally determined upon navigation, and then the adjacent face region is selected as the stimulation site. The stimulation intensity is 70% of the motor threshold, with a frequency of 10 Hz; continuous stimulation is performed for 10 s/min, with an interval of 50 s and at 2500 pulses/day, and for 5 days of treatment per week. The stimulation therapy is considered effective with a 50% reduction in VAS score, and MCS surgery may be performed for responsive patients.

### 1.2.3 Surgical Method

MCS surgery can be performed in one or two stages. One-stage operation was performed under general anesthesia with intraoperative waken-up, and no muscle relaxants were used intraoperatively. The scalp projection point of the gyrus precentralis was determined preoperatively by imaging plus navigation or robotics. Bone flap craniotomy or drill craniotomy may be performed, and a U-shaped or linear surgical incision is designed according to the projection point, and bone flap craniotomy is usually the main method. Continuous limb EMG monitoring is performed on the painful side prior to anesthesia.

Adequate local anesthesia should be performed on the scalp to avoid restlessness during awakening. After the bone flap was opened, the dura mater was dissected towards the midline along the U-shaped incision to reduce the anesthesia depth, and the stimulation site was determined to be the M1 area according to the anatomical markings by using 2 Hz square wave stimulation with concentric electrodes and gradually increasing the stimulation voltage threshold until the corresponding myoelectric response on the contralateral side was induced. Experimental electrical stimulation can also be performed by directly connecting the pre-implanted electrode to the stimulation generator. In this way, both the electrode position can be determined and the stimulation threshold for causing muscle spasms or twitches in the contralateral limb can also be measured. The electrodes can be placed in the convex epidural or subdural area for patients with upper extremity pain, or in the subdural paracentral lobule area for patients with lower extremity pain, with the electrodes positioned as parallel to the sulcus centralis as possible (Fig. 24.2a, b). For the one-stage complete placement patients, the implantable pulse generator (IPG) was placed under ipsilateral clavicle, which was connected with the electrode via a subcutaneous tunnel. For staged patients, electrodes were connected to an extension cord and then

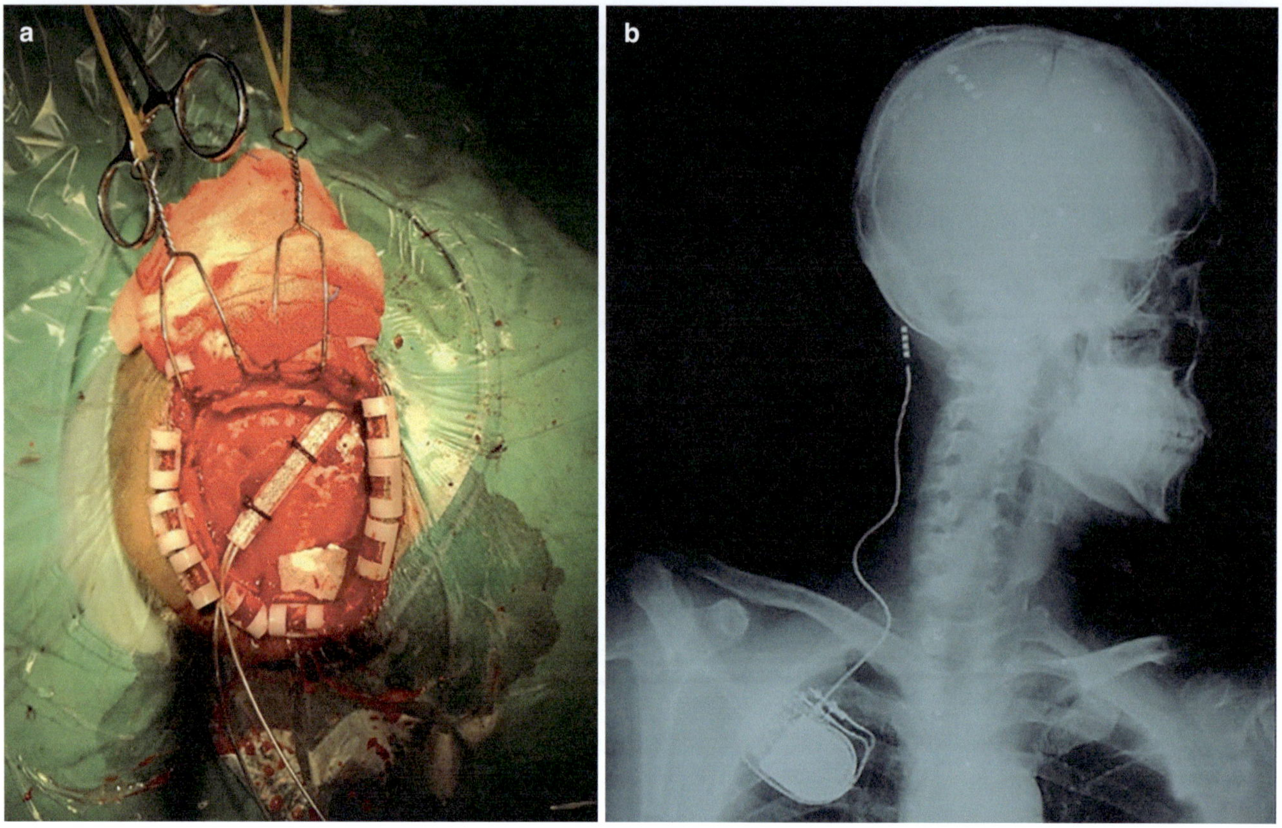

**Fig. 24.2** Intraoperative and postoperative photos of MCS. (**a**) Bone flap craniotomy, after epidural electrical stimulation, the fixed electrodes parallel to the gyrus precentralis; (**b**) X-ray image after the full set of MCS device placed

pierced through the skin behind the ear. After returning to the ward, stimulation was initiated with an external temporary stimulator after the intracranial situation was stabilized, generally at a wave width of 180–400 μs and a frequency of about 40 Hz. The stimulation intensity did not exceed 70% of the threshold for inducing muscle convulsions to avoid causing seizures, during which the improvement of pain was observed. The implantable pulse generators were placed under general anesthesia in the second-stage test for the responsive patients, and the stimulation electrodes and extension cords were removed for the non-responsive patients in the second-stage test to find other treatment options.

### 1.2.4 Judgment of Stimulation Effect

Responsive: (1) After implantation of MCS stimulating electrode, the in vitro test system is well connected, and no hardware faults such as short circuit and circuit break were found; (2) Ideal stimulation sites and parameters can be found, the pain improvement rate is >50%, the amount of oral analgesic drugs is significantly reduced, and further implantation of implantable pulse generator is required to continue treatment; (3) The pain improvement rate is >50%, and the test treatment is terminated later due to short circuit, circuit break, displacement, hardware failure, loose connection, and other reasons of the test system; (4) The pain improvement rate is >50%, and patients cannot continue IPG implantation therapy due to economic reasons; (5) The pain improvement rate is >50%, and patients are forced to discontinue test treatment due to infection or rupture at the implantation site, cerebrospinal fluid leakage and hematoma; (6) The first test during hospitalization has poor results, and the pain improvement rate is >50% after adjusting the electrode position by reoperation.

Non-responsive: (1) After implantation of MCS stimulating electrode, the external test system is well connected, and there is no hardware failure such as short circuit and circuit break; (2) After repeated adjustment of stimulation contacts, wave width, frequency and voltage, or reoperation adjustment of electrode position, patients always complain of poor pain relief and pain improvement rate of <50%, and ask for the termination of treatment.

### 1.2.5 MCS Efficacy and Complications

Long-term follow-up of 17 patients in one group revealed that the original pain after MCS was relieved to varying degrees, the analgesic efficacy was satisfactory within 1 month, and the VAS score was significantly lower than that before MCS. It also revealed that the analgesic efficacy fluctuated from time to time, and most patients could still obtain analgesic efficacy after adjusting the stimulation parameters for multiple times, and the pain was relieved by 10–90% compared with that before surgery; patients with post-stroke pain had better long-term analgesic efficacy than those with post-spinal cord injury pain and phantom limb pain, and patients with good motor function in the pain area had better efficacy than those with motor dysfunction. However, the overall response rate of MCS gradually decreased with the extension of follow-up time.

Surgical complications can be divided into two categories: craniotomy-associated complications, such as incision infection, intracranial hematoma (e.g., epidural hematoma, subdural hematoma, intracerebral hematoma), seizure and subdural effusion; MCS hardware-associated complications, such as electrode displacement, electrode short circuit, electrode fracture, early depletion of pulse generator power; both types of complications have a low overall incidence.

## 1.3 Spinal Cord Stimulation

Since the first clinical application of spinal cord stimulation (SCS) by Shealy et al. (1967), SCS has been used in the field of chronic pain treatment for more than half a century. The development of SCS electrodes and devices has gradually matured, and there has been much experience in the treatment of chronic neuropathic pain, with more than 50,000 SCS system implants worldwide in 2017, a significant proportion of which were used for the treatment of chronic neuropathic pain (Jensen & Brownstone 2019).

In 1965, Melzak and Wall put forward the famous gate theory, which laid the theoretical foundation of SCS. The theory suggests that the "electro-chemical" information of peripheral pain is transmitted to the spinal cord through thin unmyelinated C fibers and a small amount of myelinated $A_\delta$ fibers, which terminates in the spinal dorsal horn, while fine touch and vibration sensations are also transmitted to the spinal dorsal horn from thick $A_\beta$ fibers. After receiving the fine touch and vibration sensations transmitted by the coarse fibers, the body will close the "door" for receiving the information of the fine fibers, which means, electrical stimulation of thick fibers in the posterior column of the spinal cord can reversibly inhibit the transmission of pain information by thin fibers. However, it was later found that the "gate theory" could not fully explain the analgesic principle of SCS. At present, there is no complete theoretical system for the analgesic mechanism of SCS, and some scholars believe that the mechanism of SCS in treating pain with different etiologies is also different. SCS is much more extensively used than MCS due to its convenience in taking care of bilateral pain, overall low surgical risk and low surgical difficulty (Sivanesan 2019, Velásquez 2018).

### 1.3.1 Indications for SCS Treatment

Indications for SCS treatment included: (1) Failed back surgery syndrome (FBSS), manifested as the original waist or lower limb pain does not relieve or the new occurrence of chronic persistent pain in the lower back after open surgery of the lower back; (2) complex regional pain syndrome (CRPS), manifested as severe pain disproportionate to the degree of injury after minor injury to peripheral extremities, with marked sweating abnormalities, rough nail bed, hair loss, and thin skin and other denervated vegetative nerve symptoms of pain syndrome, with severe pain and marked local hyperalgesia; (3) pain after spinal cord and peripheral nerve injury, including spinal cord injury segmental pain and subsegmental pain, pain after brachial plexus/lumbosacral plexus nerve injury or avulsion, pain after cauda equina injury, stump pain or phantom limb pain after amputation, etc.; (4) peripheral vascular pain, found in limb atherosclerotic stenosis, thromboangiitis obliterans, Raynaud's disease, and other limb ischemic pain diseases; (5) diabetic peripheral neuropathy (DPN), manifested as a syndrome of stocking and glove sensory loss in the extremities with pruritus, pricking, burning sensation, etc.; (6) postherpetic neuralgia below the neck, manifested as a clinical syndrome in which the pain in the original skin lesions is not relieved after the herpes zoster skin lesions are relieved, combined with obvious hyperalgesia; (7) craniofacial pain, as reported in some literature, high cervical SCS can treat craniofacial neuropathic pain. In addition, there are chronic neuropathic pain with relatively fixed pain areas caused by other etiologies.

A 76-year-old male patient who underwent T9–10 intraspinal tumor surgery presented with pain in the abdominal wall below the navel and in the anterior thighs of both lower extremities, normal defecation and lower extremity activities, and mild decreased sensation in the pain area. T2 sagittal view (Fig. 24.3a) showed intramedullary hyperintensity signal at the T10 vertebral segment (yellow arrow), considered to be an intramedullary gliosis lesion. Axial T2 of T10 vertebral segment (Fig. 24.3b) showed disorganization of the spinal cord structures in the spinal canal. Segmental and subsegmental pain of spinal cord injury was considered in combination with medical history.

### 1.3.2 SCS Preoperative Evaluation

All patients are subjected to anteroposterior and lateral X-ray of the thoracolumbar spine or cervical spine, sagittal + axial MRI (Fig. 24.3), and plain CT scan +3D reconstruction depending on the pain sites, in order to understand the bone substance and sequence of vertebral body and spinal canal, possible spinal internal fixation site and segment, spinal cord cone or cervical enlargement site and the proposed SCS electrode placement segment with or without severe spinal cord atrophy and to exclude intravertebral canal tumor, disc herniation, fracture fragment, hematoma, or

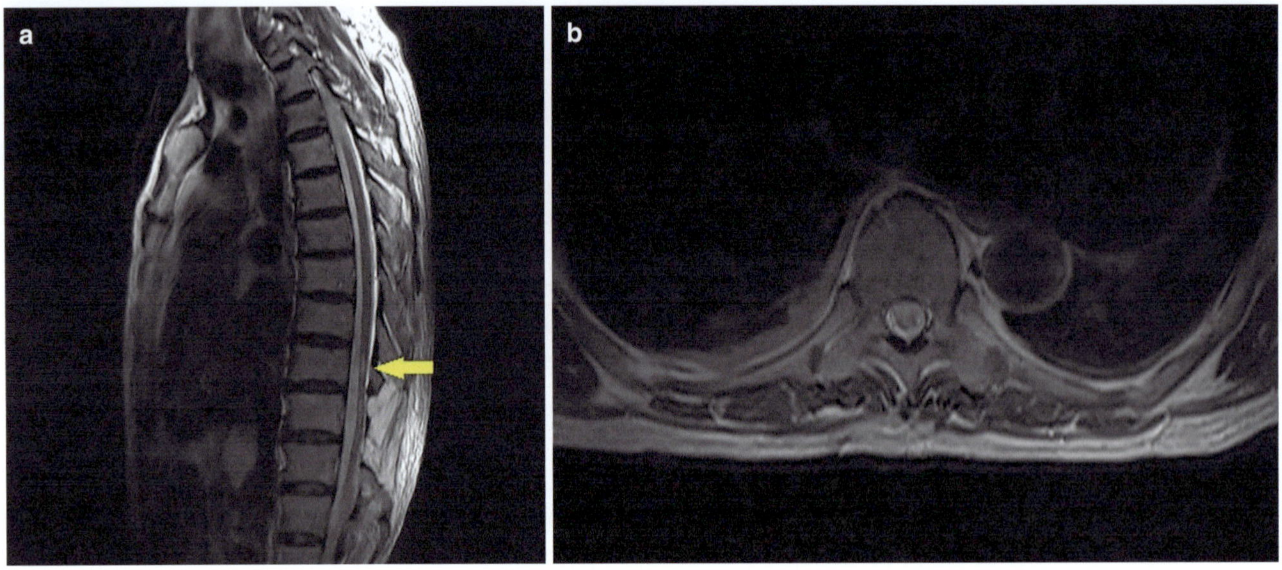


**Fig. 24.3** Preoperative MRI findings of thoracic vertebrae in patients with pain of lower extremities after intraspinal tumor surgery. (**a**) sagittal view (**b**) axial view

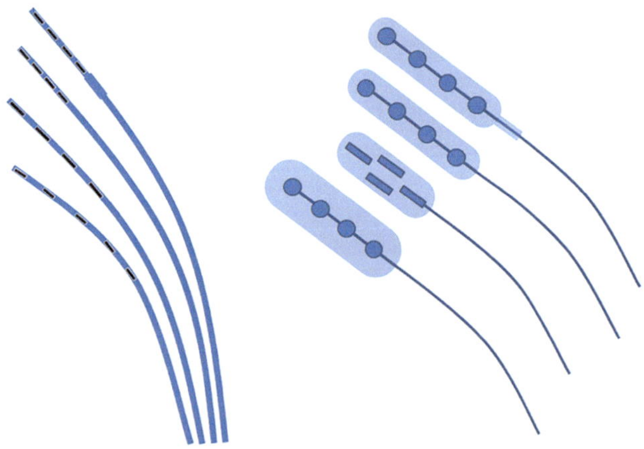

**Fig. 24.4** SCS puncture electrodes and surgical electrodes of different sizes and models

calcification. Pain assessment includes the determination of the upper and lower bounds of the pain area, the characteristics of pain episode, VAS pain score, assessment of pain area depth sensation and motor function, and evaluation of anxiety and depression scale.

### 1.3.3 SCS Surgical Approach

The SCS surgery is generally performed in two stages. The one-stage SCS surgery is performed under local anesthesia plus neuroleptanalgesia (thoracolumbar) or general anesthesia (cervical). There are currently two main types of SCS electrodes available, i.e., puncture (columnar) electrodes and surgical (plate) electrodes (Fig. 24.4). For surgical electrode implantation, patients are generally required to take the prone position, and the corresponding spinal cord and spinal segment are selected according to the upper bound of dermatomes in the painful area. Generally, the spinal cord segments corresponding to the uppermost margin of the dermatome towards the cephalic end of the painful area are taken as the stimulation area, and surgical sites are selected according to the level of spinal vertebral lamina space corresponding to different spinal segments. Taking lower limb pain as an example, a T11–T12 interlaminar approach is adopted. After posture fixation, a C-arm X-ray is positioned positively and laterally, and a longitudinal midline incision is made from the lower edge of the T10 vertebral body to the T12 spinous process. The T11 and T12 spinous processes are exposed by cutting the skin and separating the paravertebral muscles on both sides and then retracting them, and the T11–T12 interlaminar space was revealed by removing about one third of the bone substance in the lower part of the T11 spinous process. After slightly enlarging the T11–T12 interlaminar space with tools such as Kerrison biting forceps or ultrasound bone knife, paying attention to maintaining the midline position, and after the yellow ligament opening of about 1.0 cm, the electrode is slowly placed from the dorsal epidural space of the spinal cord towards the cephalic side, during which the patient mostly experiences more obvious back discomfort and needs to pay attention to the presence of lower limb paresthesia, and then the electrode is connected to the external stimulator through the extension cord for testing, with intraoperative attention to the site, side and threshold of the abnormal sensation induced by the stimulation, the presence of lower limb or thoracic back twitching, etc., combined with intraoperative C-arm X-ray fluoroscopy, after judging that the electrode position is satisfactory, the electrode tail is fixed in the paravertebral muscle, and the electrode extension cord is led out through another poked hole next to the incision to connect to the external stimulator.

The mean test duration is 7 days, and program control is performed by a designated physician. Combined with the

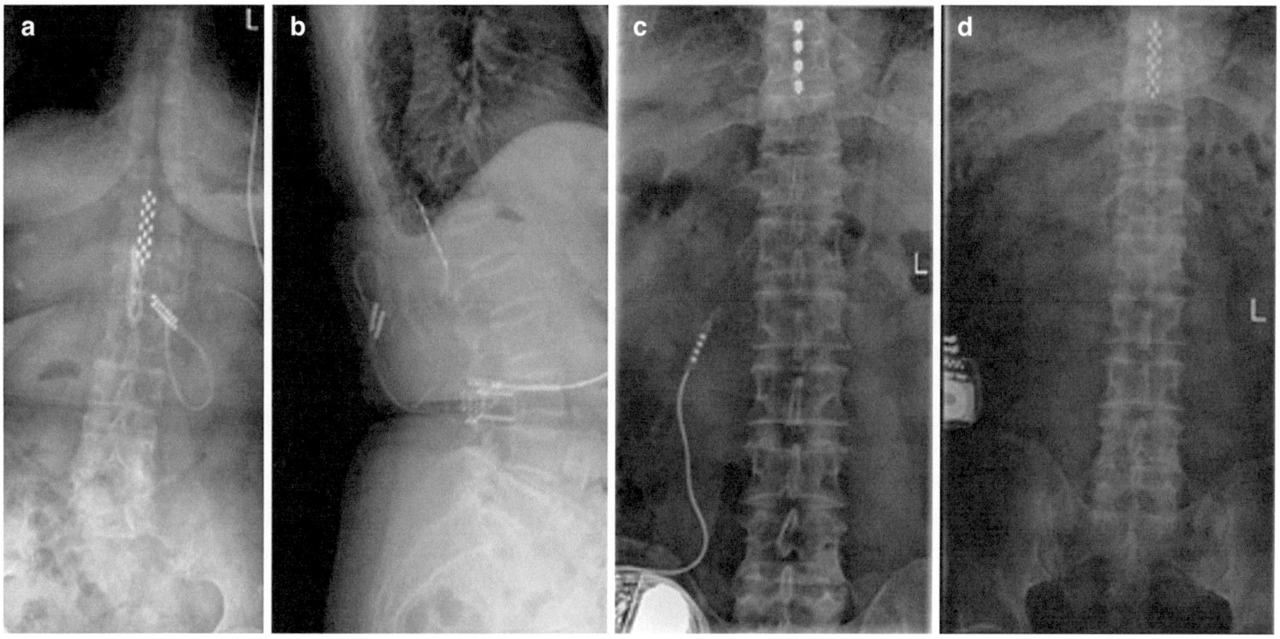


**Fig. 24.5** Postoperative radiographic findings of SCS in anterior and lateral positions. (**a**, **b**) Anteropositive and lateral lumbar vertebrae showed the implantation of 3 rows of 16-contact SCS electrodes connected with extension wires; (**c**) the patient received 4-contact electrode implantation in another hospital and had poor pain relief 1 year after surgery; (**d**) after admission, the original electrodes were removed and replaced with 3 rows of 16-contact 5–6–5 electrodes. The pulse generator was implanted after good results were tested

electrode positions determined by the X-ray electrode locator (Fig. 24.5a–d), the bipolar mode with positive and negative electrode contacts are selected, and square wave stimulation is adopted with a frequency of 30 60 Hz, wave width of 180–260 μs, and voltage of 0.5–2 V. The stimulation intensity is determined according to the voltage threshold of stimulation-elicited abnormal sensation (vibration and numbness sensation commonly described by patients), and it is generally preferable that the abnormal sensations are obvious when the patient is in the supine position without discomfort. During this period, patients experience at least three different sets of electrodes contact combinations and a period of time is selected for comparative observation with subthreshold stimulation. In the early stage of testing, the incision pain is severe, and the main evaluation reference is based on the results after 3 days of surgery. The followings are recorded, including the electrode position of each patient, the range and intensity of abnormal sensations produced by each combination of electrode contacts, the presence or absence of limb or trunk muscle twitching, the patient's chief complaint VAS score and oral drug changes. See Sect. 24.4 of Part I of this chapter for the judgment of test effects.

### 1.3.4 Analysis of SCS Efficacy and Complications

The overall efficiency of the phase I test of SCS is about 60–70%. The analysis of the test response rates in 81 patients found 56.3% of patients with failed back surgery syndrome and complex regional pain syndrome, 58.6% of patients with pain after spinal cord injury, and 52.8% of patients with pain after peripheral nerve injury. There was no statistical difference in the response rate of SCS test among neuropathic pain caused by different pain etiologies; the response rates of SCS in different pain sites in phase I test were 37.5% in the medial axis, 46.2% in the upper limb, and 68.2% in the lower limb. The therapeutic effect of SCS on pain in the midline of trunk was inferior to that in extremities (Fig. 24.5). SCS treatment is basically ineffective for patients with complete loss of superficial sensation in the painful area. SCS treatment also suffers from an overall decrease in efficacy with prolonged postoperative time.

Therapeutic coverage of painful areas (therapeutic coverage), which is considered a necessary factor for a valid SCS test, is defined as the area where the stimulation induces abnormal sensations needs to include painful areas. It is difficult to cover the medial axis areas using SCS treatment such as the trunk (especially the back and the midline of the waist), head and neck, while it is easy to cover extremities. High cervical SCS have also been reported for the treatment of facial trigeminal neuropathic pain, but cases are rare because of the wide gap between the dorsal spinal cord and the dura of C1–C2 and the difficulty of stimulating the spinal cord when placed in the epidural. Therefore, SCS treatment should be disregarded as much as possible for patients with NP in the trunk, neck, and head and face alone, and even if barely tested, the failure rate should be high, especially in patients with simple low back pain after multiple back surgeries.

Most data support the concept of $T_8$ (spinal cord segment) mute region, i.e., the small area covered by electrode stimulation treatment for $T_8$ spinal cord segment, so SCS electrodes should avoid the $T_8$ spinal cord segment as much as possible (roughly located in the $T_5$–$T_6$ spinal cord segment). Overall, the incidence of electrode displacement due to placement of surgical electrodes is lower than that of puncture electrodes, and poor therapeutic coverage due to electrode displacement is an important factor in ineffective stimulation. Single-row 1 × 4 contact or double-row 4 × 2 contact electrodes were mostly used previously, the electrode contacts were few and the spacing was wide, and there were fewer electrode contact combinations available for program control mode. In recent years, after using three rows of 5–6–5 size 16-contact electrodes, regardless of unilateral and bilateral symptoms, the electrodes are placed in the median position, so that the stimulation in the left column covers the left side, the stimulation in the right column covers the right side, and the stimulation in the middle column covers both sides. There is a wide choice for postoperative program control (Eldabe 2016).

In some patients, the painful area is covered well during the test, and pain relief is still not significant. These patients are involved in multiple pain etiologies and sites. It is possible that the production of such pain is involved in the centers above the spinal cord, but the SCS itself is also involved in the regulation of the circuitry above the spinal cord. Therefore, there are many reasons for the ineffectiveness of the test. For these patients, high-frequency (10 kHz) SCS may be effective if the painful area is extensive (Salmon 2019).

The surgical complications of SCS are similar to those of MCS and can be divided into two categories: (1) Complications related to vertebral canal surgery, such as incision infection, intraspinal hematoma, incision effusion, and cerebrospinal fluid leakage; (2) Complications associated with hardware system, such as electrode displacement, electrode short-circuit, electrode breakage, and premature depletion of pulse generator power. SCS puncture electrodes are more prone to electrode breakage and displacement (Petraglia 2016).

## 1.4 Other Treatment Options

In addition to MCS and SCS, there are also surgical methods for the treatment of chronic pain with electrical stimulation, including peripheral nerve electrical stimulation (PNS), and field stimulation (Fig. 24.6). Similar to MCS and SCS, stimulation electrodes connected by pulse generators can continuously stimulate a single nerve or a pain area to achieve long-term pain relief. It is indicated for neuropathic pain after chronic peripheral nerve injury with limited pain site, not close to important blood vessels and organs, and belonging to single innervation area. Generally, before performing peripheral nerve electrical stimulation treatment, the governing nerves can be judged according to the anatomical characteristics of the pain area, and local anesthetics are used for diagnostic block. For patients with effective diagnostic block, stimulation devices can be implanted in stages or in one stage.

Since 2011, electrical stimulation of dorsal root ganglion (DRG) has gradually become a research hotspot in Europe and

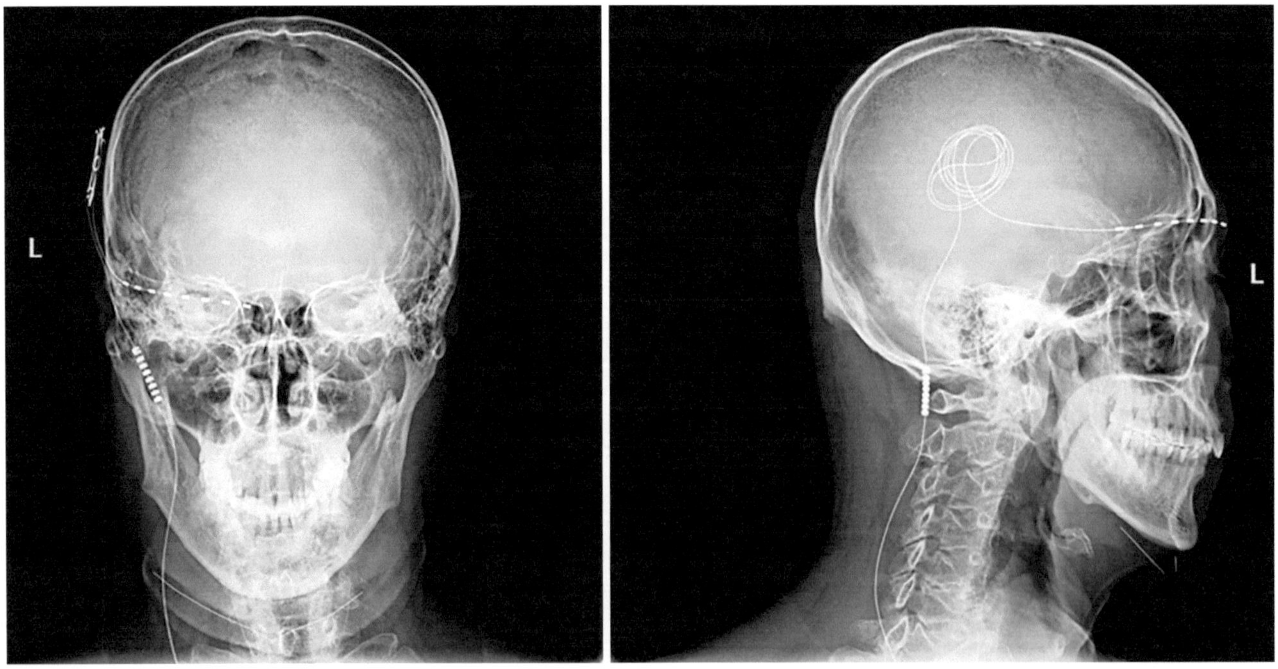


**Fig. 24.6** Frontal and lateral head X-ray view of a patient with chronic intractable frontal pain following frontal trauma after PNS electrode implantation. Electrodes are delivered subcutaneously under tempus through a skin tunnel to the painful area of the forehead

America; in particular, DRG electrical stimulation has shown unique advantages for some patients with chronic NPP who have unsatisfactory therapeutic effects with SC. DRG is an inflatable nodule of the dorsal root of spinal nerves, which is the first stage of somatosensory input and plays an important role in the generation and modulation of chronic pain, making DRG an important target in the field of chronic pain treatment. Compared with spinal cord stimulation, DRG electrical stimulation has the advantages of precise stimulation site, less power consumption, and less influence of body position on the stimulation effect. DRG electric stimulation has its uniqueness and adapts to a wide range of people, which is worth developing.

## 2 Chronic Musculoskeletal Pain

### 2.1 Overview

Chronic musculoskeletal pain (CMP) refers to pain that occurs in muscles, bones, joints, tendons, or soft tissues for more than 3 months, excluding pain elicited by malignant tumors. Although the disease is not life-threatening, long-term pain will lower the patients' quality of life, and promote the occurrence of anxiety and depression, placing a great burden on families and society. Epidemiological studies have shown that CMP is the leading cause of disability worldwide, with 20–30% of the world's population suffering from musculoskeletal diseases, and the disease increases with age. The incidence of CMP is increasing year by year due to factors such as the aging of China's population and the increase of various traumas. It occurs in an increasingly young population. Up to now, there is no uniform standard classification for this disease. The latest revision of the *International Classification of Diseases* (ICD-11) divides CMP into primary and secondary. Primary chronic pain occurs mainly in the muscles, bones, joints, or tendons with marked affective disorder (anxiety or depression) or functional impairment (impairing daily life and social interaction). Secondary chronic pain is divided into inflammatory response, structural changes, and pain elicited by nervous system according to etiologies. Also, according to pathogenic sites, it is divided into bone pain, muscle pain, tendon and ligament pain, joint pain, spinal pain, and multi-site pain.

### 2.2 Mechanisms of Chronic Musculoskeletal Pain

**At present, the mechanism of this disease has not been fully clarified, and the possible mechanisms include the following:**

1. Inflammatory Response: After the body is infected, the activity of inflammatory cells and the level of mediators increase locally and systemically, thus activating peripheral pain receptors, causing sensitization of the receptors, activating primary afferent nerve fibers, and producing abnormal discharge, which leads to pain.
2. Fibrosis: Inflammatory response can lead to localized fibrotic scarring, causing further damage and creating a vicious cycle of pain.
3. Massive Release of Neurotransmitters and Abnormal Changes in Neuroimmunity: Substance P, stimulating kinin and calcitonin gene-related peptide (CGRP), and *N*-methyl-D-aspartate (NMDA) in the tendon, dorsal root ganglion (DRG), and spinal dorsal horn are elevated.
4. Peripheral and Central Sensitization: Endogenous serotonin, norepinephrine, and endogenous opioids and other substances secreted by the glands lead to a decrease in the pain threshold of patients and the development of pain hypersensitivity.

### 2.3 Evaluation

After diagnosis, a detailed medical history will be collected, and careful physical examination as well as laboratory and imaging tests are performed to rule out infection, tumor, and other causes. Once the diagnosis is confirmed, it is important to assess pain-related conditions by selecting the appropriate assessment method based on the patient's specific circumstances (numeric rating scale (NRS), visual analog scale (VAS), verbal rating scale (VRS), faces pain scale, etc.). The evaluation of pain properties includes Nerve Pain Questionnaire (NPQ), DN4 questionnaire, ID Pain Scale and Leeds Assessment of Neuropathic Symptoms and Signs to evaluate whether there is concomitant neuropathic pain, etc. The pain questionnaires include the short form MPQ, McGill Pain Questionnaire, and Brief Pain Inventory, etc., which are used for comprehensive evaluation of the nature and intensity of pain. In terms of psychological assessment, scales such as GAD-7, PHQ-9, and PHQ-15 are used to evaluate psychological disorders. In terms of function, the SF-36 and sleep assessment, WOMAC pain score, Nordic Musculoskeletal Questionnaire (NMQ), and other related disease assessment forms are used. Other relevant measurements include physiologically relevant parameters: heart rate, blood pressure, electrodermal activity, electromyogram, and cortical evoked potentials.

### 2.4 Clinical Therapy

#### 2.4.1 Conventional Medication

Analgesia is accompanied by active treatment of the primary disease and selection of drugs according to the etiology, with particular attention to differentiating between nociceptive pain and neuropathic pain, and pain caused by multiple biological, psychological, and social factors, requiring attention to the assessment of psychological status, etc. Current oral medications include nonsteroidal anti-inflammatory drugs and opiates. For pain caused by central and peripheral nerve injury, appropriate treatment should be provided for the damaged nerves, including administration of antiseizure medication, neurotrophic medicines, etc., in addition to the treatment of peripheral musculoskeletal lesions.

### 2.4.2 Noninvasive Neuromodulation Therapy

#### 2.4.2.1 Central Nervous System Stimulation Therapy

Lefaucheur et al. used high-frequency TMS with 1000 pulses at 10 Hz to intervene in patients with neuralgia. After one treatment, pain scores then decreased, but the pain relief was short-lived. Hirayama treated chronic neuralgia patients with 500 pulses of 5 Hz 90% RMT in the M1 area. The results showed that the pain scores and McGill scores of patients in the M1 area decreased significantly, and subjective pain was significantly relieved. Its mechanism of action may be related to synaptic LTP/LTD, and rTMS therapy in motor cortex may be realized by modifying the central pain modulatory system. Current studies on cortical stimulation in the treatment of chronic pain have shown modest pain relief, but most of them show significant improvement relative to the placebo group, which is a new option for patients with neuralgia who cannot tolerate drugs.

Harvey et al. treated older patients suffering from chronic musculoskeletal pain with anodal tDCS over primary motor cortex, contralateral to the most painful site. Anodal stimulation was applied to targets for five consecutive days, with sponge electrodes (5 × 7 cm), 2 mA of the stimulation intensity, and 20 min of the daily duration. The results showed that pain intensity reduced during treatment and follow-up period (Harvey et al. 2022). Kim et al. used tDCS combined with physical therapy (PT) in older adults with chronic musculoskeletal pain. Anode electrodes were placed over the left DLPFC, and cathode electrodes were placed over the right DLPFC. The stimulation intensity was set to 2 mA for 30 min, three times per week for 8 weeks. The study suggested a beneficial effect in reduction of pain sensation and depression incidents (Kim et al. 2022). The anodal tDCS in the left M1 at 2 mA for 20 min with 35 $cm^2$ electrodes on five consecutive days may relieve pain and improve sleep quality in patients with fibromyalgia (Deus-Yela et al. 2017). Furthermore, another study showed that transcranial random noise stimulation (tRNS) of M1 on 5 days a week (Monday–Friday) for 2 weeks induced a significant reduction in musculoskeletal pain in fibromyalgia (Curatolo et al. 2017). In the future, transcranial electrical stimulation (tES) has the potential to be an effective treatment for patients with CMP.

#### 2.4.2.2 Peripheral Nerve Stimulation Therapy

##### 2.4.2.2.1 Transcutaneous Electrical Nerve Stimulation (TENS)

In terms of peripheral nerve stimulation, transcutaneous electrical nerve stimulation (TENS) is the most widely used method. There are currently two recognized hypotheses: (1) Gate control theory suggests that glial cells in the spinal dorsal horn contain a neurological "gate" that either attenuates or enhances nerve impulses from the periphery to the central system. The passage and upward transmission of pain signals is determined by the opening and closing of the gate, thus enhancing or attenuating peripheral nerve impulses to the central system, triggering pain sensations and responses. A-δ and C neurons release the excitatory neurotransmitter substance P, which opens the gate and allows the pain impulses to travel upward; the thicker and faster-transmitting A-β neurons release inhibitory neurotransmitters, closing the gate, and preventing pain impulses from traveling upward; TENS regulates pain sensation by activating coarse fibers and inhibiting fine fibers. This theory emphasizes a dynamic balance between excitatory and inhibitory activities. (2) Another hypothesis is that the therapy can causes the central release of various analgesic substances, among which endogenous opioid peptides are the most important; endogenous opioid peptides coexist with other neuropeptides or neurotransmitters and neuromodulators, and can act as neurotransmitters, neuromodulators, or neurohormones together with opioid receptors to form a powerful endogenous pain-modulating system to regulate pain. In addition, various neurotransmitters, such as 5-hydroxytryptamine and norepinephrine, also contribute to the analgesic effects of TENS.

The therapy often involves applying electrodes to the skin surface for pain management and rehabilitation. The difference between TENS and traditional electrical stimulation is that the latter mainly stimulates motor fibers, while TENS mainly stimulates sensory fibers. However, sensory and motor fibers can be stimulated simultaneously when the stimulation parameters or thresholds are changed. Sensory level stimulation: high-frequency, low-intensity electrodes are mostly placed in the nerve trunk. Kinesthetic level stimulation: Low-frequency, high-intensity electrodes are mostly placed at the motor points. In clinical application, frequency conversion and different current intensities are generally used so that TENS can produce a variety of stimuli, and the central nervous system secretes enough analgesic substances to achieve better therapeutic effects.

Despite the large number of studies reporting on the treatment of CMP with TENS, the inclusion of patients with varying etiologies of chronic pain makes it difficult to determine the efficacy of the therapy for specific origins of pain. In addition, the stimulation parameters of TENS vary in different studies, even in different subjects of the same study.

##### 2.4.2.2.2 Transcutaneous Spinal Direct Current Stimulation

Transcutaneous spinal direct current stimulation (tsDCS) is another electrical stimulation method for low back pain, similar to TENS, which is also derived from sensory gating theory. This method modulates the electrical activity of spinal cord conduction tracts and spinal cord circuits, thus affecting the activity of innervation at and above the level of the spinal cord. The main idea of the epidural stimulation of the spine is to activate low-threshold primary afferent fibers ascending to the dorsal column nucleus, but the physiological basis for pain relief from dorsal column stimulation is uncertain as such stimulation may activate other neural tissues.

Truini et al. used tsDCS to stimulate the foot and perioral region of healthy subjects with laser stimulation at the position of the T10 spinous process. The reference electrode was placed on the right shoulder (electrode plate size 5 × 7 cm), and the current intensity was 2.5 mA for 20 min. The results showed that anodal tcDCS significantly reduced the amplitude of LEPs produced by laser stimulator stimulation of the foot, whereas no similar effect was seen with cathodal stimulation; however, neither cathodal nor anodal stimulation changed the stimulated perioral LEPs. From the above experiments, it was concluded that direct current stimulation induced changes at the spinal cord level.

Meyer-friessem et al. conducted a sham-controlled, single-blind trial involving 24 healthy subjects, in which the anode was placed at position of the T11 spine process, the reference electrode was placed at the back of the left shoulder (electrode plate size 5 × 7 cm), and the current intensity was 2.5 mA for 15 min. Anodes could significantly reduce the pain sensation induced by high and mild mechanical stimulation, which lasted for 1 h, but had no effect on the pain sensation induced by single-pulsed current stimulation.

In addition to the suppressive effect on pain, the therapy had an effect on motor function after CNS injury, such as muscle tone and walking function in patients recovering from stroke.

### 2.4.3 Other Therapies

#### 2.4.3.1 Rehabilitation Exercise and Adjunctive Therapy

Common examples include strength, stretching, endurance, resistance training, and aerobic exercise. Other forms of exercise include Tai Chi, yoga, etc.; auxiliary braces are often used as a combination therapy to effectively relieve pain and improve function.

#### 2.4.3.2 Psychological Intervention Therapy

Positive psychological assessment and intervention are beneficial for chronic pain control, as an alternative or adjunctive therapy to poorly controlled analgesics, or even as a first-line treatment option. It includes cognitive behavioral therapy and guided imagery.

#### 2.4.3.3 Self-Management of Pain

Educational methods are used to teach patients about the knowledge and skills of pain management, and to enable them to adopt the right behaviors to control the active process of pain. These include pain cognition, pain assessment, ways of reporting pain, knowledge and beliefs about pharmacological analgesia, and non-pharmacological analgesia.

#### 2.4.3.4 Traditional Chinese Medical Treatment

Traditional therapies are used for pain control, such as acupuncture, acupotomology, silver acupuncture, bone setting massage, and traditional Chinese medicine (oral decoction, topical plaster), especially acupuncture has some short-term efficacy in pain control.

#### 2.4.3.5 Minimally Invasive Interventional Therapy and Surgery

If none of the above methods can relieve pain, minimally invasive interventional therapy is one of the methods. In addition, surgery is the ultimate treatment for some CMP patients if the primary disease meets the indications for surgical treatment.

# 3 Chronic Visceral Pain

## 3.1 Overview

Chronic visceral pain (CVP) refers to pain from the internal organs, usually described as blunt, diffuse, poorly localized, and characterized by hypersensitivity to stimuli such as organ distention. CVP is often the result of multiple etiologies, which may include acute or chronic inflammation, obstruction or interruption of normal mechanical processes, benign or malignant tumor compression, visceral nerve ischemia and conduction changes, psychological factors, etc. Visceral pain is characterized by extremely inaccurate localization, difficult to describe in nature, and often accompanied by referred pain. Most patients have suffered from mental disorders, especially anxiety and depression, because the disease has not been effectively treated for a long time. Therefore, this kind of pain seriously affects the quality of life of patients and increases the medical burden on patients and their families as well as the whole society.

## 3.2 Mechanism

Compared with somatic pain study, there are relatively a few studies on CVP, and the understanding of the pathological mechanisms of CVP is relatively lagging behind.

### 3.2.1 Brain–Gut Axis

Environmental factors, especially stress, have a huge impact on the function of the gastrointestinal tract. In this process, brain–gut axis plays an important role in bidirectional signaling between central nervous system and gastrointestinal tract. Dysfunction of this axis can lead to imbalance of intestinal environment, gastrointestinal dysfunction, intestinal inflammation, chronic abdominal pain syndrome, and eating disorders (Olesen 2016).

### 3.2.2 Enteric Microorganisms

It is currently recognized that enteric microorganisms themselves are crucial information mediators in dual signaling pathways. The regulation of enteric microorganisms through the use of probiotic therapy has shown that a mixture of eight probiotics can hinder the development of visceral pain hypersensitivity induced by maternal-infant separation during the neonatal period.

### 3.2.3 Immune System

There are direct or indirect connections between the immune system and the HPA axis, and between the autonomic nervous system (ANS) and enteric nervous system (ENS), which are involved in the pathophysiology of visceral pain.

### 3.2.4 Sex Hormone, Hydrogen Sulfide, Ion Channel/Receptor, Genetic and Epigenetic Regulation All Contribute to Visceral Pain

The fluctuation of estrogen in the menstrual cycle may cause changes in neurotransmitter systems such as 5-hydroxytryptamine and glutamate, thereby affecting the occurrence of stress-related diseases and the efficacy of pain treatment; hydrogen sulfide can promote the function of Cav3.2 and TRPA1 (transient receptor potential ankyrin 1) to induce pain sensation, and can enhance the function of voltage-gated sodium channels; ion channels/receptors include glutamate receptors, TRPV1 cation channels (which can be activated by noxious thermal stimuli, low pH, and capsaicin, etc., and are an integrator of pain stimuli), and P2X receptors (which are involved in the occurrence and development of inflammatory pain and neuropathic pain), and the regulation of $\gamma$-aminobutyric acid spinal GABAergic circuit plays an important role in pain process; epigenetics influences pain-related genes, resulting in the release of pain.

### 3.2.5 Abnormalities in the Visceral and Sympathetic Nerves

The viscera are innervated by the visceral nerves and sympathetic nerves, both of which are connected to the spinal cord through the gray-white communicating branches, so abnormalities in the visceral and sympathetic nerves can also cause chronic visceral pain. These include cervical and thoracic sympathetic disorder syndromes, chronic visceral pain elicited by celiac plexus stimulation, and chronic perineal pain induced by abnormalities in ganglion impar, etc.

## 3.3 Evaluation

### 3.3.1 Test Method for Visceral Pain Response

At present, the evaluation of visceral pain mainly focuses on the evaluation of gastrointestinal tract, including colorectal and gastric distension. Unlike surface tissues, the gastrointestinal tract is an empty organ whose proper or natural stimulation is dilation or traction.

### 3.3.2 Monitoring and Recording of Visceral Pain

Abdominal muscle contraction elicited by the above stimulation methods is commonly referred to visceral motor response, which is the most widely used methods for detecting visceral pain, including abdominal withdrawal reflex (AWR) and abdominal EMG; the postdilated response of stomach can also be observed by electromyography of the neck muscles; heart rate variation reflects the intensity of visceral pain; in recent years, functional imaging techniques have also been used to assess the response of higher brain centers that are highly sensitive to visceral pain. In addition, the visual analog scale and McGill questionnaire for pain assessment have also been used to measure visceral pain.

## 3.4 Clinical Therapy

### 3.4.1 Medication

The previously commonly used analgesics, antispasmodics, or antidepressants are used to treat visceral pain. With the advancement of science and technology, some drugs targeting peripheral signal transduction and sensitization have been developed, including voltage-gated ion channel blockers, ligand-gated cation channel blockers, G-protein coupled receptor blockers, etc.

### 3.4.2 Noninvasive Neuromodulation Therapy

Transcranial magnetic stimulation is a noninvasive technique that has been used to treat various chronic pains. For visceral pain disorders, its role has been studied for functional pain as well as organic pain. Irritable bowel syndrome (IBS) is a typical disease of chronic visceral pain.

Melchior et al. conducted a randomized, placebo-controlled study to observe the effect of TMS on rectal sensitivity in irritable bowel syndrome. The study enrolled 21 IBS patients with a crossover design of true and false stimulation, 5 days per week for 2 months, with a stimulation frequency of 20 Hz, 80% resting motor threshold, 20 sequences per session, 5 s stimulation, and 55 s stimulation interval, for a total of 2000 pulses. Primary outcome measures were elevated pressure pain threshold after rTMS stimulation, secondary measures were rectal volume tolerance and rectal compliance, and mean pain intensity between baseline and end of treatment. The results showed that there was no difference between true and false stimulation in terms of pressure pain threshold, maximum volume tolerance, and rectal compliance, but the volume threshold was significantly improved in the true stimulation group.

Algladi et al. investigated the regulatory effects of transcranial magnetic stimulation on the modulation of visceral sensitivity in healthy people and individuals with irritable bowel syndrome. A total of 16 normal subjects and 10 IBS patients were enrolled in this study. The single-pulse lumbosacral magnetic stimulation or transcranial magnetic stimulation was used to evaluate the excitability of the spinal cord root and cerebral cortex and the effect of nerve stimulation on anorectal sensation and pain induced by local electrical stimulation. In normal subjects, a variety of stimulation parameters were screened, and only 1 Hz repetitive lumbosacral nerve magnetic stimulation (rLSMS) increased the response of the spinal cord and changed the excitability of the healthy anus. Both 1 Hz rLSMS and 10 Hz repetitive transcranial magnetic stimulation increased rectal pain thresholds in healthy subjects within 1 h after the intervention. In IBS patients, the rectal pain threshold for this parameter was elevated at all post-stimulation time points compared with placebo.

In general, transcranial magnetic stimulation can be used for the control of visceral pain in individual centers with sufficient experience although it is at a preliminary stage and further research is required.

#### 3.4.3 Other Therapies

The development of future psycho-behavioral therapy based on Pavlov's classical theory of conditioned reflex can significantly improve functional gastrointestinal disorders. Traditional Chinese medicine acupuncture and moxibustion also play a certain role in relieving visceral pain. Minimally invasive interventions for patients with refractory visceral pain can be performed by blocking the corresponding innervated nerve according to the general localization of the pain, and if the block is effective, nerve modulation, and nerve destruct treatment can be attempted.

## 4 Chronic Migraine

### 4.1 Overview

Migraine is a widespread neurological disorder clinically characterized by recurrent headache attacks lasting 4–72 h. Typical migraines are unilateral, throbbing and exacerbated by physical activity, often accompanied by nausea, vomiting, photophobia, and fear of sound. Migraine attacks are often accompanied by reversible focal neurological symptoms, including visual or sensory disturbances, such as flashes, blurred vision, and limb numbness; common visual auras include scotoma, amaurosis, and flashes. Comorbid conditions include depression, epilepsy (Bauer et al. 2021), stroke, and myocardial infarction.

### 4.2 Mechanism

#### 4.2.1 Vasogenic Hypothesis

Studies have confirmed that the occurrence of migraine may be of vascular origin: pre-attack constriction ischemia of the intracranial arteries causing visual disturbance aura, followed by dilation of the external carotid artery system, which produces vasoactive polypeptides and non-infectious inflammation leading to migraine. Migraine attacks are vascularly pulsatile; the attack locations also coincide with the craniofacial branches of the external carotid artery. Vasodilating substances have been shown to induce migraine-like attacks in people with migraine (Ferrari et al. 2022). But common migraines often start with general sensory symptoms such as fatigability and altered mood. The regional cerebral blood flow tomography of classical migraines found that decreased blood flow during attack occurred only in the cerebral cortex, such as the occipital lobe, while blood flow to the deep brain was usually normal. In addition, inadequate inflow can also persist for hours to migraine onset; even after the aura disappears; late delayed hyperinflow (reactive hyperemia) is not associated with the onset.

#### 4.2.2 Neurogenic Hypothesis

Some studies have suggested that migraine may be of neuropathic origin, with persistent vasomotor changes: migraine aura is neurologically induced by previous behavioral changes, changes in state of mind, changes in eating habits, etc. Symptoms include fear, anxiety, insomnia, yawning, fatigability, attention deficit, body temperature instability, etc. Pulse headaches are also not elicited by blood vessels. The main etiologies of migraine are all related to the nervous system: mental factors, hunger, sleep factors, female physiological factors and sensory stimulation. Besides, EEG, brain metabolism and other studies support this hypothesis.

#### 4.2.3 Trigeminal Neurovascular Reflex Theory

As a widely accepted theory, it states that nociceptive signals from the trigeminovascular system are transmitted to the brain to produce migraines. Activation of trigeminal neurons releases vasoactive peptides and induces local inflammatory responses, a process that sequentially sensitizes and discharges secondary neurons in the brainstem and tertiary neurons in the thalamus until the final injurious impulse reaches somatosensory and other cortical areas associated with pain perception, a process that involves various transcription factors, inflammatory factors, c-fos genes, etc. Retrospective evaluations are limited by recall bias and misattribution, and studies support high misattribution rates of migraine triggers (e.g., stress, sleep disorders, specific foods and non-eating).

## 4.3 Evaluation

Migraine is divided into migraine with aura and migraine without aura. The evaluation of migraine can be scored by Numerical Rating Scale (NRS) scale (Soler et al. 2010), recording the frequency of weekly attacks, the duration of pain each time, the intensity of pain, etc. Migraine attacks are often accompanied by a state of anxiety and depression, which can also be assessed using the Anxiety and Depression Scale. The EEG may also be altered and there may be an increase in focal waves, which may appear as spike waves, sharp waves, spikes and slow waves, or there may be an increase in widespread slow waves.

## 4.4 Clinical Therapy

### 4.4.1 Transcranial Magnetic Stimulation

#### 4.4.1.1 Clinical Application

Repetitive transcranial magnetic stimulation (rTMS) modulates cerebral cortical excitability through depolarization or hyperpolarization. Low frequencies (e.g., <1 Hz) decrease cortical excitability, and high frequencies (5–20 Hz) increase cortical excitability. rTMS can modulate the excitability of the visual cortex and correct the electrophysiological abnormalities of the visual cortex, so it can be used in the treatment of migraine. Experts in the treatment of repetitive transcranial magnetic stimulation reach a consensus on the clinical recommendation that low-frequency stimulation of the occipital lobe by rTMS has Grade A consistent evidence, and two other major rTMS targets, M1 and dorsolateral prefrontal cortex, have also been studied (Johnson & Martinson 2007, Lefaucheur 2020).

#### 4.4.1.2 Therapeutic Effect

A meta-analysis of 16 randomized clinical trials of single-pulse and repetitive TMS indicated that single-pulse transcranial magnetic stimulation was effective in treating migraine with aura after the first attack (Lan et al. 2017). In a randomized, double-blind, parallel-group, two-phase, sham-controlled investigation at 18 centers in the USA, 164 patients with at least one seizure were treated with TMS ($n = 82$) or sham stimulation ($n = 82$). A handheld, portable, single-pulse TMS device was placed under the occipital bone, and treatment consisted of two pulses with an interval of approximately 30 s (the peak value: 0.9 T, and the rise time: 180 μs. The total pulse length is less than 1 ms). Pain-free response rates at 2 h were significantly higher with TMS (32/82 [39%]) than with sham stimulation (18/82 [22%]), for a therapeutic gain of 17% (Reed et al. 2010). In the randomized, double-blind, parallel-group, sham-controlled trial of 42 migraine patients using tabletop clinic-based sTMS over the occiput with two pulses and 30-s interval, 69% reported mild or no pain for 2 h in the active group, while 48% reported in the sham group (Lipton and Pearlman 2010). A report on 14 patients who received 12 sessions of high-frequency (10 Hz) rTMS over the left motor cortical area (MC, M1) compared with 15 patients who received botulinum toxin A revealed that rTMS was comparable to BTX-A injections in the treatment of chronic migraine, but had a poorer sustained effect (Shehata et al. 2016). Data from a group of patients repeatedly reporting false controls supported the efficacy of three HF-rTMS administrations for left-sided M1 migraines, which was similar to the classical treatment regimen for NPP (Misra et al. 2013; Kalita et al. 2016). There were two conflicting results in the comparison of true and false rTMS stimulation on the dorsolateral prefrontal cortex (DLPFC) for migraine (Brighina et al. 2004; Conforto et al. 2014).

#### 4.4.1.3 Mechanism of Action

Based on the trigeminal nerve dysfunction theory, TMS at the onset of migraine attacks can reduce the sensitivity of afferent fibers in the trigeminal nerve vessels, thereby alleviating headache. Based on the cortical diffusion mechanism, TMS can reduce the depolarization or excitability of neurons during the aura attack, thus interfering with the spreading depression in the cortex and relieving headache. Studies have shown that TMS can modulate a variety of neurotransmitters, and transcranial magnetic stimulation can often significantly alter the content of various receptors such as serotonin receptors. Furthermore, clinical improvements were associated with increased plasma β-endorphin levels (Misra et al. 2017).

### 4.4.2 Transcranial Electrical Stimulation

#### 4.4.2.1 Clinical Application

tDCS modulates the resting potential and cortical excitability of brain cells through weak currents and modulates the functional connection of cortico–striatal–thalamocortical circuits, so it can be used for migraine treatment. In general, tDCS targets for migraine are M1, DLPFC, or primary visual cortex (V1).

#### 4.4.2.2 Therapeutic Effect

The first positive result was reported with cathode tDCS using V1 (Antal et al. 2011). Other positive results were reported for anodal tDCS using M1 in two studies. A study involving more than 10 patients who received active tDCS showed a reduction in pain intensity, episodes, and medication at the end of the intervention and within 8 weeks of active tDCS (Auvichayapat et al. 2012). Using different parameters, pain reduction was greater after left DLPFC than M1 tDCS in the other study (Andrade et al. 2017). However, another study was a preliminary report of results in 8 patients, in which reductions in pain intensity and migraine attack duration were observed only at follow-up (Dasilva et al. 2012). The current level of evidence for the analgesic effect of TDCS on M1 stimulation is weaker than that for rTMS, which has a level of evidence of B (probably effective) (Fregni et al. 2021).

#### 4.4.2.3 Mechanism of Action

tDCS delivers electrical currents to brain tissue through scalp electrodes to induce long-lasting brain modulatory effects. Anodal tDCS facilitates cortical excitability, whereas cathodal tDCS suppresses it. A large body of research has shown that stimulation of the primary motor cortex (M1) inhibits the activity of the thalamus, which is associated with pain perception (Kurt et al. 2017). Another area often targeted for pain-alleviating strategies is the dorsolateral prefrontal cortex (Lorenz et al. 2003). tDCS may reduce pain by activating different neural circuits in the precentral gyrus, which would be afferents or efferents connecting structures involved in the sensory or emotional component of pain processing, such as the thalamus or DLPFC (Meissner et al. 2013). Neurophysiological studies show that the visual cortex is hyperactive during the interictal periods of migraine, and thus the suppression of the excitability may help to reduce migraine pain. Cathodal stimulation of V1 is thought to reduce its hypersensitivity or hyperresponsiveness at the origin of headache (Lefaucheur et al. 2020).

### 4.4.3 Remote Electrical Neuromodulation

#### 4.4.3.1 Clinical Application

Remote electrical neuromodulation (REN) is a novel nonpharmacological, noninvasive FDA-approved acute treatment for migraine. Peripheral nerves in the upper arm are stimulated to induce conditioned pain modulation, an endogenous analgesic mechanism in which conditioning stimulation inhibits pain in remote regions of the body (Nie et al. 2020). Recent systematic review and meta-analysis of randomized controlled trials found REN to be the only migraine neuromodulation intervention with sufficient published high-quality research (Ashina 2020).

#### 4.4.3.2 Therapeutic Effect

A post-hoc, within-subject analysis of data from 78 adults with chronic migraine enrolled in a clinical trial of REN showed no statistical difference from medication on any of the efficacy measures, and REN provides an effective treatment alternative that may reduce acute medication use (Grosberg et al. 2022). A multicenter study of 91 subjects had a 4-week treatment with REN found pain relief and pain disappearance at 2 h was achieved by 59.3% and 20.9% respectively, and showed that REN has a favorable effect on nausea, photophobia, and phonophobia and improves functional ability (Grosberg et al. 2021). A double-blind, randomized, pseudo-control study of 71 patients (299 treatments) confirmed that 64% of participants based on 200-μs, 150-μs, and 100-μs pulse width stimulation obtained 50% pain reduction, while the sham stimulation was 26%. For patients starting with severe or moderate pain, reduction to mild or no pain occurred in 58% (25/43) of participants (66/134 treatments) for the 200-μs stimulation protocol and 24% (4/17; 8/29 treatments) for placebo (Yarnitsky et al. 2017). According to a randomized, double-blind, sham-controlled, multicenter study, 252 migraineurs were applied for 30–45 min stimulation on the upper arm within 1 h of attack onset, the result demonstrated that active stimulation was more effective than sham stimulation in achieving pain relief (Yarnitsky et al. 2019).

#### 4.4.3.3 Mechanism of Action

REN is a novel acute migraine treatment that stimulates C and Aδ noxious brachial peripheral nerve fibers to induce conditioned pain modulation, an endogenous analgesic mechanism whereby conditioning inhibits pain in remote body regions (Nir, Supportive, and Care 2015). Presumably, the noxious information reaches the brainstem via the ascending pain pathway and activates descending inhibitory pathways originating in the periaqueductal gray and rostral ventromedial medulla, which globally inhibit incoming pain messages in the trigeminal cervical complex that occur during headache by releasing serotonin and norepinephrine (Yarnitsky et al. 2019).

### 4.4.4 Other Therapies

Peripheral nerve stimulation (PNS) treats pain in the area innervated by the nerve by electrically stimulating the peripheral nerve (Bushnell 1991). PNS is widely used in the treatment of chronic pain syndromes such as neuropathic pain or complex regional pain syndrome, as well as headaches such as occipital neuralgia, migraine, and cluster headaches (Schoenen 2013). It includes transcutaneous supraorbital neurostimulation, vagus nerve stimulation, occipital nerve stimulation, pterygopalatine ganglion stimulation, etc.

## References

Andrade SM, de Brito Aranha REL, de Oliveira EA et al (2017) Transcranial direct current stimulation over the primary motor vs prefrontal cortex in refractory chronic migraine: a pilot randomized controlled trial. J Neurol Sci 378:225–232

Antal A, Kriener N, Lang N et al (2011) Cathodal transcranial direct current stimulation of the visual cortex in the prophylactic treatment of migraine. Cephalalgia 31:820–828

Ashina M (2020) Migraine. N Engl J Med 383(19):1866–1876

Auvichayapat P, Janyacharoen T, Rotenberg A et al (2012) Migraine prophylaxis by anodal transcranial direct current stimulation, a randomized, placebo-controlled trial. J Med Assoc Thail 95:1003–1012

Bauer S, Hodl M, Eglseer D. (2021). Association between malnutrition risk and pain in older hospital patients. Scand J Caring Sci, 35(3): 945–951

Bremer N, Ruby J, Weyker PD et al (2016) Neuromodulation: a focus on dorsal root ganglion stimulation. Pain Manag 6(3):205–209

Brighina F, Piazza A, Vitello G et al (2004) rTMS of the prefrontal cortex in the treatment of chronic migraine: a pilot study. J Neurol Sci 227:67–71

Bushnell MC, Marchand S, Tremblay N (1991) Electrical stimulation of peripheral and central pathways for the relief of musculoskeletal pain. Can J Physiol Pharmacol 69(5):697

Conforto AB, Amaro E Jr, Gonçalves AL et al (2014) Randomized, proof-of-principle clinical trial of active transcranial magnetic stimulation in chronic migraine. Cephalalgia 34:464–472

Curatolo M, La Bianca G, Cosentino G et al (2017) Motor cortex tRNS improves pain, affective and cognitive impairment in patients with fibromyalgia: preliminary results of a randomised sham-controlled trial. Clin Exp Rheumatol 105(3):100–105

Dasilva AF, Mendonca ME, Zaghi S et al (2012) tDCS-induced analgesia and electrical fields in pain-related neural networks in chronic migraine. Headache 52:1283–1295

Deer TR, Levy RM, Kramer J et al (2017) Dorsal root ganglion stimulation yielded higher treatment success rate for complex regional pain syndrome and causalgia at 3 and 12 months: a randomized comparative trial. Pain 158(4):669–681

Deus-Yela J, Soler MD, Pelayo-Vergara R et al (2017) Transcranial direct current stimulation for the treatment of fibromyalgia: a systematic review. Rev Neurol 65(8):353–360

Eldabe S, Buchser E, Duarte RV (2016) Complications of spinal cord stimulation and peripheral nerve stimulation techniques: a review of the literature. Pain Med 17(2):325–336

Ferrari S, Vanti C, Giagio S, et al. (2022). Low back pain and sexual disability from the patient's perspective: a qualitative study. Disabil Rehabil 44(10):2011–2019

Fregni F, El-Hagrassy MM, Pacheco-Barrioset K et al (2021) Evidence-based guidelines and secondary meta-analysis for the use of transcranial direct current stimulation in neurological and psychiatric disorders. Int J Neuropsychopharmacol 24(4):256–313

Grosberg B, Rabany L, Lin T, et al. (2021). Safety and efficacy of remote electrical neuromodulation for the acute treatment of chronic migraine: an open-label study. Pain Rep 6(4):966

Grosberg B, Rabany L, Vizel M et al (2022) Effectiveness comparison of remote electrical neuromodulation and standard-care medications for acute treatment of chronic migraine: a post-hoc analysis. Pain Manag 12(7):837–844 https://doi.org/10.2217/pmt-2022-0053

Hamid P, Malik BH, Hussain ML (2019) Noninvasive transcranial magnetic stimulation (TMS) in chronic refractory pain: a systematic review. Cureus 11(10):e6019

Harvey MP, Martel M, Houde F et al (2022) Relieving chronic musculoskeletal pain in older adults using transcranial direct current stimulation: effects on pain intensity, quality, and pain-related outcomes. Front Pain Res (Lausanne) 3:817984

Jensen MP, Brownstone RM (2019) Mechanisms of spinal cord stimulation for the treatment of pain: still in the dark after 50 years. Eur J Pain 23(4):652–659

Johnson M, Martinson M (2007) Efficacy of electrical nerve stimulation for chronic musculoskeletal pain: a meta-analysis of randomized controlled trials. Pain 130(1–2):157–165

Kalita J, Laskar S, Bhoi SK et al (2016) Efficacy of single versus three sessions of high rate repetitive transcranial magnetic stimulation in chronic migraine and tension-type headache. J Neurol 263:2238–2246

Kallewaard JW, Nijhuis H, Huygen F et al (2019) Prospective cohort analysis of DRG stimulation for failed back surgery syndrome pain following lumbar discectomy. Pain Pract 19(2): 204–210

Kim S, Salazar Fajardo JC, Seo E et al (2022) Effects of transcranial direct current stimulation on physical and mental health in older adults with chronic musculoskeletal pain: a randomized controlled trial. Eur Geriatr Med 13(4):959–966

Kurt E, Henssen D, Steegers M et al (2017) Motor cortex stimulation in patients suffering from chronic neuropathic pain: summary of expert meeting and premeeting questionnaire, combined with literature review. World Neurosurg 108:254–263

Lan LH, Zhang XN, Li XP et al (2017) The efficacy of transcranial magnetic stimulation on migraine: a meta-analysis of randomized controlled trails. J Headache Pain 18:86

Lefaucheur JP, Aleman A, Baeken C et al (2020) Evidence-based guidelines on the therapeutic use of repetitive transcranial magnetic stimulation (rTMS): an update (2014–2018). Clin Neurophysiol 131:474–528

Lorenz J, Minoshima S, Casey KL (2003) Keeping pain out of mind: the role of the dorsolateral prefrontal cortex in pain modulation. Brain J Neurol 126(Pt 5):1079–1091. https://doi.org/10.1093/brain/awg102

Maarrawi J, Peyron R, Mertens P et al (2007) Motor cortex stimulation for pain control induces changes in the endogenous opioid system. Neurology 69(9):827–834

Meissner K, Fassler M, Rücker G et al (2013) Differential effectiveness of placebo treatments: a systematic review of migraine prophylaxis. JAMA Intern Med 173(21):1941–1951

Misra UK, Kalita J, Bhoi SK (2013) High-rate repetitive transcranial magnetic stimulation in migraine prophylaxis: a randomized, placebo-controlled study. J Neurol 260:2793–2801

Misra UK, Kalita J, Tripathi G et al (2017) Role of β endorphin in pain relief following high rate repetitive transcranial magnetic stimulation in migraine. Brain Stimul 10:618–623

Nie ZB, Rao YF, Chen JP et al (2020) Interpretation of expert consensus on drug treatment of chronic musculoskeletal pain: clinical application of antidepressants in chronic musculoskeletal pain. Chin J Pain Med 26(3):169–173

Olesen AE, Farmer AD, Olesen SS et al (2016) Management of chronic visceral pain. Pain Manag 6(5):469–486

Petraglia FW, Farber SH, Gramer R et al (2016) The incidence of spinal cord injury in implantation of percutaneous and paddle electrodes for spinal cord stimulation. Neuromodulation 19(1):85–90

Reed KL, Black SB, Banta CJ et al (2010) Combined occipital and supraorbital neurostimulation for the treatment of chronic migraine headaches: initial experience. Cephalalgia 30(3):260–271

Salmon J (2019) High-frequency spinal cord stimulation at 10 kHz for widespread pain: a retrospective survey of outcomes from combined cervical and thoracic electrode placements. Postgrad Med 131(3):230–238

Schaefer C, Mann R, Sadosky A et al (2014) Burden of illness associated with peripheral and central neuropathic pain among adults seeking treatment in the United States: a patient-centered evaluation. Pain Med 15(12):2105–2119

Schoenen J, Jensen RH, Lanteri MM et al (2013a) Stimulation of the sphenopalatine ganglion (SPG) for cluster headache treatment. Pathway CH-1: a randomized, sham-controlled study. Cephalalgia 33(10):816–830

Shealy CN, Mortimer JT, Reswick JB. (1967). Electrical inhibition of pain by stimulation of the dorsal columns: preliminary clinical report. Anesth Analg 46(4):489–491

Shehata HS, Esmail EH, Abdelalim A et al (2016) Repetitive transcranial magnetic stimulation versus botulinum toxin injection in chronic migraine prophylaxis: a pilot randomized trial. J Pain Res 9:771–777

Sivanesan E, Maher DP, Raja SN et al (2019) Supraspinal mechanisms of spinal cord stimulation for modulation of pain: five decades of research and prospects for the future. Anesthesiology 130(4):651–665

Soler MD, Kumru H, Pelayo R et al (2010) Effectiveness of transcranial direct current stimulation and visual illusion on neuropathic pain in spinal cord injury. Brain J Neurol 133(9):2565–2577

Tsubokawa T, Katayama Y, Yamamoto T et al (1991) Chronic motor cortex stimulation for the treatment of central pain. Acta Neurochir Suppl (Wien) 52:137–139

Lipton RB, Pearlman SH (2010) Transcranial magnetic simulation in the treatment of migraine. Neurotherapeutics 7(2):204-212 https://doi.org/10.1016/j.nurt.2010.03.002.

Velásquez C, Tambirajoo K, Franceschini P et al (2018) Upper cervical spinal cord stimulation as an alternative treatment in trigeminal neuropathy. World Neurosurg 114:641–646

Yarnitsky D, Volokh L, Ironi A, et al. (2017). Nonpainful remote electrical stimulation alleviates episodic migraine pain. Neurology 88(13):1250–1255

Yarnitsky D, Dodick DW, Grosberg BM, et al. (2019). Remote Electrical Neuromodulation (REN) Relieves Acute Migraine: A Randomized, Double-Blind, Placebo-Controlled, Multicenter Trial. Headache 59(8):1240–1252

# Epilepsy

# 25

Dongju Yang, Tao Yu, Xueyuan Wang, and Wenwen Shen

## 1 Overview of Application of Neurostimulation in the Treatment of Epilepsy

Epilepsy is a common disease of the central nervous system. It is estimated that the number of existing patients with epilepsy in China is up to nine million, of which about two to three million are patients with refractory epilepsy that are still difficult to control despite regular application of antiseizure medication therapy. Surgery and neuromodulation therapies are important alternative therapies for drug-refractory epilepsy. However, surgical treatment has strict indications, which is suitable for patients with clear epileptogenic foci and better demarcation from functional areas. The surgery is invasive and expensive. Neuromodulation, by contrast, might be noninvasive, reversible, inexpensive, and easy to popularize, which has certain advantages.

With the rapid development of epilepsy discipline, the definition of epilepsy, epileptic seizures, and classification of epileptic diseases have been updated rapidly in the past 30 years, and an in-depth understanding of epilepsy is the basis of neuromodulation in the treatment of epilepsy. Therefore, this part introduces the diagnosis of epilepsy, epileptic seizures, and classification of epileptic diseases. Epilepsy is traditionally defined as a clinical syndrome of highly synchronized abnormal discharges of brain neurons due to multiple etiologies, with characteristic clinical manifestations of seizures, transience, repetition, and stereotypy. The latest definition of epilepsy published by the International League Against Epilepsy (ILAE) in 2014 is a brain disease with pathological tendency to have recurrent seizures. More specifically, epilepsy should be considered in case of: (1) at least two unprovoked seizures or reflex seizures at an interval of more than 24 h, (2) a single unprovoked or reflexive seizure, but comparable to the overall risk of recurrence (at least 60%) of two unprovoked seizures over the next 10 years, and (3) diagnosis of epilepsy syndrome. The new definition emphasizes the early diagnosis of epilepsy and provides better operability for clinical practice.

In recent years, ILAE has also updated the epileptic seizure classification and epilepsy classification framework, and in the classification, framework published in 2017, seizures are classified into three categories: focal onset, generalized onset, and unknown onset (Fisher et al. 2017). The concepts of "focal" and "generalized" are clearly elaborated in the report published by ILAE in 2010 (Berg et al. 2010). "Focal epileptic seizures" are defined as "seizures originating in a unilateral hemispheric network," may originate in very focal brain areas, in slightly more widely distributed areas, or even in the subcortex. For the same patient, although focal seizures may spread to the contralateral hemisphere, the origin point and origin side are constant, stable across seizures with a dominant propagation pathway; although the same patient may suffer from different seizure types, each type has its own relatively constant origin and dominant propagation pathways. "Generalized epileptic seizures" are defined as "seizures that originate at one site within a bilater ally distributed abnormal network but rapidly engage the whole network." The abnormal network of generalized seizures may not cover the entire brain. The origin of some seizures may seem limited, but for the same patient, its origin points or side is not constant or stable in different seizures. For the definitions of "focal" and "generalized," more consideration is given to the origin and propagation characteristics of within-networks. The operational classification of ILAE epileptic seizures in 2017 was based on the above

D. Yang
Department of Neurology, Xuanwu Hospital, Capital Medical University, Beijing, China

T. Yu (✉) · X. Wang
Department of Functional Neurosurgery, Beijing Institute of Functional Neurosurgery, Xuanwu Hospital, Capital Medical University, Beijing, China
e-mail: wangxueyuan@xwhosp.org

W. Shen
Department of Pediatric, Xuanwu Hospital, Capital Medical University, Beijing, China



Y. Wang (ed.), *Therapeutics of Neural Stimulation for Neurological Disorders*, https://doi.org/10.1007/978-981-99-4538-2_25

theoretical basis, with seizures of focal origin, generalized origin, and unknown origin. Focal epilepsy presented as awareness or impairment of awareness depending on the degree of perceptual involvement and was further divided into motor and non-motor symptoms based on the main symptoms. Instead of emphasizing the classification of the degree of awareness impairment in generalized epilepsy, it was divided into motor symptoms and non-motor symptoms based on the main semiology. The new classification emphasized the symptoms that appeared first.

In the same year, ILAE classified epilepsy into focal epilepsy, generalized epilepsy, combined generalized, and focal epilepsy and unknown epilepsy (Scheffer et al. 2017). The classification and comprehensive determination of epilepsy is based on all clinical information such as seizure classification, electrophysiology, symptomatology, imaging, genetics, laboratory tests, and complications. All seizure information should be taken into account in making the judgment. Focal epilepsy refers to epilepsy of definite and limited origin, including unifocal epilepsy and multifocal epilepsy, whose EEG epileptiform abnormal discharges are mostly limited. Generalized epilepsy refers to epilepsy with extensive origins, and its EEG manifestations are of epileptiform abnormal discharges, which may have various seizure types, such as absence, myoclonus, and atonia. Combined generalized and focal epilepsy is a new type proposed in 2017, which refers to epilepsy with both focal and generalized seizures. Common examples are Dravet syndrome and Lennox-Gastaut Syndrome.

Neurostimulation techniques can be simply divided into two categories: noninvasive stimulation and implantable stimulation. Common noninvasive stimulation techniques for treating epilepsy include transcranial magnetic stimulation, transcranial electrical stimulation, etc. Implantable stimulation techniques mainly include vagal nerve stimulation, deep brain stimulation, and responsive neurostimulation. Despite the different principles of the different techniques, since the mechanism of epilepsy is believed to be increased excitability/synchronization of the cortical/subcortical epileptic network, the common principle of different techniques for treating epilepsy is to reduce or inhibit cortical excitability or synchronization, only the methods and parameters used are different.

# 2 Implantable Device Stimulation Protocol for Focal Epilepsy

About 30% of patients have seizures that cannot be effectively controlled by medication and are drug-refractory epilepsy, and a significant number of these patients have focal epilepsy. For these patients, the selection of appropriate surgical excision is effective. However, there are also patients who are not eligible for surgical excision, or whose surgery fails, and patients who are not willing to undergo surgery. Neuromodulation therapy is precisely adapted to this part of the clinical needs. Neuromodulation therapy for epilepsy is divided into implantable stimulation and noninvasive stimulation, wherein implantable stimulation includes vagus nerve stimulation, deep brain stimulation, and responsive neurostimulation. This section briefly introduces the current neuromodulation methods for the treatment of focal refractory epilepsy.

## 2.1 Vagus Nerve Stimulation

### 2.1.1 Mechanism

The vagus nerve originates in the brain stem and travels through the neck to the chest and abdomen. It is one of the longest nerves in the human body and affects multiple functions. The left and right vagus nerves are distributed on both sides of the body, and the right vagus nerve is involved in the innervation of the sinoatrial node. Stimulation of the right vagus nerve is associated with the risk of cardiac arrhythmia, so stimulation of the left cervical vagus nerve trunk is usually adopted clinically.

At present, the mechanism of vagus nerve stimulation (VNS) is still not completely clear, and the nucleus tractus solitaries is considered to be closely related to the mechanism of action of VNS. The nucleus tractus solitaries is the relayed nucleus of the visceral primary afferent fibers in the dorsal medulla oblongata, which has close fiber connections with many nuclei and regions in the brain in the ascending pathway of vagus nerve. The nucleus tractus solitaries receive a wide range of somatic and visceral sensory projections, as well as extensive projections from other parts of the brain. After a large amount of information is processed internally, it sends out motor and autonomic efferent fibers, which are projected to the parabrachial nuclei, the locus coeruleus, and other structures, producing diffuse effects on the brain. Therefore, stimulation of the vagus nerve through the nucleus tractus solitaries can increase the threshold of epileptic seizures, thereby reducing seizures. In addition, VNS also reduces seizures by increasing the concentration of inhibitory transmitters and decreasing the concentration of excitatory transmitters in the brain.

### 2.1.2 Indications

In 1997, the Food and Drug Administration (FDA) approved the vagus nerve stimulator as the first implantable device for the treatment of refractory epilepsy, and it has been widely used in the treatment of focal epilepsy and generalized epilepsy. VNS was initially approved for the treatment of refractory epilepsy in patients over 12 years of age, but with the observation of clinical efficacy, VNS treatment became more commonly used in children with epilepsy younger than 12 years of age as it demonstrated better efficacy and clinical tolerability. Compared with antiseizure medication and surgical resection, VNS has fewer complications and less impact on children's neurodevelopment, and early treatment with VNS in children with refractory epilepsy has been observed to improve their quality of life and cognitive ability.

In 2014, American Academy of Neurology (AAN) proposed several recommendations on VNS for the treatment of epilepsy, and recommended VNS as a safe and reliable treatment comparable to new antiseizure medication, including: (1) For children with drug-refractory epilepsy (including focal and generalized seizures), VNS can be considered as an alternative treatment if it is clear that the child does not meet the indications of resective surgery; (2) VNS may have additional therapeutic effects in improving mood and controlling depressive symptoms for some patients; (3) The therapeutic effect produced by VNS can be maintained for a long time, which can make up for the defect of the honeymoon effect of drug treatment; (4) Stimulation intervention with the VNS device may stop the subsequent seizure process if activated immediately when the patient has an aura; (5) VNS therapy may reduce the incidence of sudden unexpected death in epilepsy and is particularly recommended as an adjunctive therapy for patients with epilepsy combined with sleep apnea; (6) VNS can be used as a treatment option for refractory epilepsy in children and epileptic encephalopathy such as LGS.

It should be noted that patients with arrhythmia, abnormal cardiac conduction function, dysphagia, dyspnea, etc. should avoid VNS therapy.

### 2.1.3 Implantation Surgery

The hardware of vagus nerve stimulator consists of stimulation pulse generator, stimulation electrode, and extension cord, and the external neuromodulation device includes external parameter programmer and magnet.

The procedures of vagus nerve stimulator implantation are as follows: (1) The surgery is performed under general anesthesia during which tracheal intubation or laryngeal mask airway (LMA) can be applied; (2) The patient is in the supine position, with the shoulders elevated so that the head is mildly tilted back and to the right, to fully expose the left sternocleidomastoid muscle; (3) Disinfect the whole left upper chest, left neck and left maxillofacial region; (4) Cut the skin 4–6 cm horizontally along the natural lines at the level of the thyroid cartilage in the middle of the medial margin of the left sternocleidomastoid as the cervical incision, separate the platysma, bluntly separate the medial margin of the sternocleidomastoid, enter the carotid sheath while noting carotid pulsation, find the vagus nerve between the carotid artery and the jugular vein, separate and fully free the vagus nerve by about 3 cm, and surround it with a flexible rubber-insulated wire for later use; (5) Incise the skin 5 cm from the left anterior axillary line to the clavicle direction along natural skin lines, separate the gap between subcutaneous tissue and muscle towards the clavicle direction, and separate the "skin bag" for placing the pulse generator; (6) Penetrating the subcutaneous tunnel: use a special subcutaneous tunnel cleaning rod to penetrate the subcutaneous tunnel between the cervical incision and the thoracic incision and introduce electrodes from the cervical incision and export from the thoracic incision through the cleaning rod; (7) Fully expose the vagus nerve, then place a soft rubber pad under the nerve, and sequentially wind three sets of spiral electrodes on the vagus nerve trunk under a microscope, and fix the electrodes on the fascia by suture with special buckles; (8) Connect the pulse generator and detect the pulse generator with the programmer. If the impedance is good, it indicates that the installation is successful; (9) Put the pulse generator into the chest "skin bag," fix it with sutures, and suture the cervical and thoracic incisions, respectively.

### 2.1.4 Adverse Reactions

Common side effects and adverse reactions of VNS are generally mild (common in high-intensity stimulation), including voice alteration or hoarseness, cough, dyspnea, sore throat, paresthesia, headache, etc. A few patients may also have dyspepsia, nausea and vomiting, tinnitus, hiccups and other symptoms, and with the passing of time, the symptoms gradually relive or disappear. If adverse reactions occur, they can generally be alleviated by reducing the output current and/or pulse width. The incidence of severe complications produced by stimulation is less than 1%, such as bradycardia, permanent vocal cord paralysis, and dysphagia.

### 2.1.5 Postoperative Program Control

Postoperative program control is a crucial link for implantable stimulation devices to exert therapeutic effects. The current VNS stimulation parameters are also empirically improved and will be continually improved and revised. The postoperative program control of VNS is roughly divided into two stages. The goal of the first stage is to increase the output current to the target range. At this stage, the stimulation starts with a low current (0.25 mA) and gradually increases. During the initial modulation, patients may frequently receive treatment adjustments until the target level of appropriate seizure control and minimal side effects is achieved. The objective of the second phase is to gradually increase duty cycle. The commonly used stimulation parameters for VNS stimulation are as follows: frequency 20–30 Hz, pulse width 250–500 μs, stimulation time 30 s, and intermittent time 3–5 min. The output current of the pulse generator is gradually adjusted from 0.25 mA to 1.0–2.0 mA but does not exceed 3.0 mA. To a certain extent, different patients respond differently to different combinations of parameters, which requires the debugging physician to try to find the best parameters for the patient in order to achieve satisfactory efficacy and minimal side effects. If patients have achieved the expected results at any time, the adjustment can be stopped. In principle, the parameters are no longer changed in patients with good treatment results, unless the patient has intolerance.

In addition, when the patient senses an aura or frequent seizures, the magnet (external modulation device) stimulation can be activated immediately, and it is possible to terminate the seizures in time. At present, the Vagus Nerve Stimulation Model 106 (AspireSR) is characterized by automatic stimulation when a patient has a seizure with increased heart rate or even tachycardia. With the deepening of understanding of the antiepileptic mechanism of VNS, it will also lead to new stimulation modalities and individualized stimulation protocols for different types of epilepsy.

#### 2.1.6 Efficacy

Extensive clinical studies on VNS showed that about 30% of patients treated with VNS had a reduction of >75% in epileptic seizure frequency, and approximately more than 50% had a reduction of >50%. Moreover, the severity of seizure was usually lessened and the postoperative quality of life (QoL) was notably improved in the patients. The therapeutic effect may be further improved with the prolongation of stimulation time year by year. Elliott et al. reported follow-up results for a group of 436 patients, which showed that 7.5% of patients reached Engel Grade I, 13.0% achieved Grade II, and 43.3% achieved Grade III after surgery (Elliott and Morsi 2011a); among the 65 patients who had been followed up for more than 10 years, 24.5% reached Engel Grade I, 15.6% achieved Grade II, and 46.2% achieved Grade III, and the efficacy was further improved over time (Elliott and Morsi 2011b). In 2016, Englot et al. conducted VNS efficacy study with the largest sample size to date. 5554 patients were enrolled based on the Cyberonics database, and 2869 epilepsy patients from 78 studies were covered by literature review (Englot and Rolston 2016). The results showed that the patient response rate and the mean epilepsy seizure reduction rate were all 63.0%, and seizures of 8.2% patients were completely controlled within 24–48 months of treatment with VNS.

### 2.2 Deep Brain Stimulation

Deep Brain Stimulation (DBS) has achieved success in the treatment of dyskinesia, and its safety and efficacy are widely recognized. The spectrum of diseases treated with this approach is expanding due to its advantages of adjustability and reversibility. In recent decades, DBS has made great progress in the treatment of refractory epilepsy. Researchers have conducted clinical research and exploration on nuclei such as the anterior thalamic nucleus, subthalamic nucleus, centromedian nucleus, hippocampus–amygdala, of which the anterior nucleus of thalamus (ANT) DBS for the treatment of epilepsy has been certified in many countries and has become a clinical treatment method.

#### 2.2.1 Mechanism

Despite the increasing clinical application of DBS, the mechanism of its control of epileptic electrical activity remains unclear, and previous studies have mostly been based on animal experiments. There are different viewpoints or hypotheses about the mechanism of DBS controlling epilepsy, for example: (1) electrical stimulation directly obstructs neuronal activity or inactivates neurons near the electrode by inactivating the sodium channels; (2) electrical stimulation affects the membrane potential of neurons by releasing electrons, promoting the generation of cellular action potentials, excitation of the corresponding neuron, and conduction along the axon; (3) electrical stimulation results in selective release of inhibitory neurotransmitters. Studies have shown that at the level of the cortical–subcortical network, several pathways involving structures in the cortex and basal ganglia regions play a role in suppressing epileptic discharges. Therefore, electrical stimulation of an anatomical relay station in these network structures can be regarded as a method to remotely control the generation and spread of epilepsy.

However, DBS differs from VNS in that it is more targeted. By regulating specific targets (nuclei), the corresponding neural networks or neural circuits can be affected, and thus the effect may be more significant for certain specific types of epilepsy. This is where the complexity of DBS for epilepsy comes in, but on the other hand, it is also the advantage and potential of DBS to achieve better seizure control. To achieve effective improvement and control of epileptic seizures, more in-depth and detailed research is needed on the mechanism of DBS in controlling epilepsy.

##### 2.2.1.1 Electrical Stimulation on Anterior Nucleus of Thalamus

The ANT is the key node in the Papez circuit. The mammillary body projects intensively to ANT through the mammillothalamic tract, and ANT forms direct and indirect connections with the hippocampus through the cingulate tract and intercortical pathways. In terms of function, ANT converges and integrates the hippocampal and mammillary body input information, allowing a functional complex to be formed in the anterior thalamus and closely related to the fornix structures, making ANT the most important target for the regulation of epilepsy, especially effective for temporal lobe epilepsy. Previous studies on the mechanism of ANT electrical stimulation to control epileptic seizures mostly focused on animal experiments. The study by Stypulkowski et al. showed that electrical stimulation of ANT had an inhibitory effect on the local field potential (LFP) of the hippocampus in large animals, and it was related to the stimulation frequency, and the inhibitory effect of high-frequency stimulation on LFP was more significant (Stypulkowski and Stanslaski 2013). The clinical study of Yu et al. showed that high-frequency stimulation of ANT in patients with epilepsy originating from the hippocampus resulted in the weakening and desynchronization of neuronal activity and epileptic electrical activity in the epileptogenic foci (Yu and Wang 2018). Moreover, when ANT was electrically stimulated, the hippocampus showed a frequency-dependent stimulation effect, i.e., low-frequency stimulation led to an increase in the synchronization of hippocampal neuronal activity; stimulation at frequencies higher than 50 Hz results in desynchronization of abnormal hippocampal discharge, which is likely to inhibit seizures of hippocampal origin. In addition, high-frequency stimulation also resulted in a decrease in large-scale cortical network connectivity, which may be related to the inhibition of epileptic seizures.

#### 2.2.1.2 Hippocampus–Amygdala Electrical Stimulation

Chronic hippocampus–amygdala electrical stimulation is another well-characterized treatment. Medial structures of the temporal lobe may be both seizure-generating brain regions and important structures in epilepsy networks. In the hippocampal slice model, high-frequency stimulation induces negative slow potential transformation, increases extracellular potassium accumulation, and results in decreased neuronal excitability. On the one hand, hippocampal–amygdala DBS regulates the activity of brain regions that trigger seizures by inhibiting neurons adjacent to the electrodes and reduces the occurrence of seizures; on the other hand, DBS acts on the neural projection network of several brain regions connected to the hippocampus and produces a regulatory effect.

#### 2.2.1.3 Electrical Stimulation of Subthalamic Nucleus

Subthalamic nucleus (STN) is a potentially attractive stimulation target, as tens of thousands of patients with dyskinesia have been treated with STN stimulation. Given the important role of STN in the motor cortex-basal ganglia circuit, Ren Liankun et al. studied the propagation path of epileptic seizure discharges originating from the motor cortex to STN and found that the seizure discharges originating from the motor cortex can quickly spread to STN, suggesting the specific propagation mode of motor cortex-STN in motor area epilepsy (Ren and Yu 2020). When the STN-cortex evoked potential technique was used to apply single-pulse stimulation to the STN, significant cortical evoked potentials were recorded in the motor cortex, suggesting a specific upward propagation between the subthalamic nucleus and the motor cortex, providing further evidence for the classic motor cortical-basal ganglia circuit from an electrophysiological perspective. Similar to ANT stimulation, STN stimulation was found to be frequency-dependent. Low-frequency subthalamic nucleus stimulation could specifically activate the abnormal discharge of motor cortex, while high-frequency stimulation could specifically inhibit abnormal discharge of motor cortex.

### 2.2.2 Implantation Surgery

ANT-DBS is a clinically approved and most commonly used procedure for deep brain stimulation in the treatment of refractory epilepsy, and the procedure is as follows:

(1) Multiplanar Reformat (MPR) scan: whole-head MPR scans with slice thickness 1 mm are performed preoperatively, including T1 and T2 sequences. The coordinates of the anterior (AC) and posterior commissures (PC) and commissural midpoints are calculated based on the MRI images. (2) Surgical plan: typical coordinates of ANT are as follows: 6 mm lateral to the middle point of the commissure, 8 mm anterior to the posterior commissure, and 12 mm above the AC–PC line. Most corrections are made currently based on the visual target of the image. The angular trajectory is approximately 60° in the front-rear direction. The entry point is preferably on the cerebral gyrus and avoids the cerebral sulcus and veins. (3) Frame placement: a stereotaxic frame was installed under local anesthesia on the day of surgery. The frame is oriented as parallel as to the line between the lateral canthus and the tragus to align with the anterior–posterior commissure. (4) Surgical operation: fix the patient's head frame on the operating bed, routinely disinfect the towel, adjust the guide arch to set the target point and angle, mark the entry point on the scalp, incise the scalp, and drill the skull. Install the Stimlock and cross the dura mater to minimize CSF loss and avoid calculating changes in the relative position of the coordinates. (5) Intraoperative microelectrode recording: once the ventricle wall is punctured to reach the thalamus, the characteristic action potential can be encountered, and recording is usually started 10 mm above the target and 5–10 mm beyond the target. (6) Electrode implantation: place the four-contact DBS electrode in the expected target. The size of the electrode is larger than that of the anterior thalamic nucleus (the thickness of the anterior nucleus is about 6 mm), so the bottom contact (contact 0) is usually located in the DM nucleus. (7) Pulse generator implantation: after the DBS electrodes are placed and fixed, the electrodes are connected to the pulse generator through a subcutaneous tunnel, which is completely internalized. The operation is mostly performed under general anesthesia.

### 2.2.3 Postoperative Program Control

Electrostimulation of the anterior thalamic nucleus has the potential for therapeutic success after failed excisional procedures, which cannot be achieved without successful postoperative program control, but the process of program control is complex. The stimulation parameters documented in the literature are listed below: Kerrigan et al. used bipolar stimulation, with 100 Hz, 90 μs, and 1–10 V, in a cyclic mode (1 min "on" and 9 min "off"). Osorio et al. (2007) used double cathode electrodes, with 145–170 Hz, 90 μs, and 1.5–5 V, in a cyclic mode (1 min "on" and 5 min "off"). The stimulus parameters in the SANTE study were for unipolar stimulation, with 145 Hz, 90 μs, and 5 V, in cyclic mode (1 min "on" and 5 min "off"). Lee et al. used unipolar or bipolar stimulation, with 100–185 Hz, 1.5–3.1 V, and 90–150 μs, in continuous mode. Krishna et al. used 100–185 Hz, 2.4–7 V, 90 μs, and in cyclic mode (1 min "on" and 5 min "off"). Lehtimaki et al. used single cathode, double cathode electrodes, or bipolar electrode stimulation, 140 Hz, 90 μs, 5 V, in a cyclic mode (1 min "on" and 5 min "off").

In our central parameter settings, the stimulation electrode points located in the anterior thalamic nucleus are first selected for unipolar or bipolar stimulation, where the range of electric fields produced in the unipolar mode is closer to a sphere, while an ellipsoidal electric field can be produced by stimulation in the bipolar mode with a higher current density than the former. For patients who are not satisfied with the stimulation effect only by using the electrode points in the anterior thalamic nucleus, an attempt could be made to selectively increase the electrode points below the anterior thalamic nucleus to form a longer-span stimulating electric field (1–3+ is an alternative if the original option of 2–3+ is not satisfactory). The frequency of stimulation is selected from 130 to 145 Hz; theoretically, above 60 Hz can inhibit epileptiform discharges. A pulse width of 90 μs is mostly used. For those with poor effect, an attempt can be made to increase the pulse width, but that means more power consumption. The stimulation voltage is selected to start at 2 V, which can be tolerated by most patients, but there are also patients with increased seizures. At this point, titration modulation is required from a smaller voltage. For safety of long-term stimulation, a cyclic mode is recommended.

### 2.2.4 Clinical Efficacy

Before 2010, studies on ANT-DBS reported a small number of cases. In 2010, a multi-center randomized controlled trial (SANTE) organized by Fisher et al. was published in Epilepsia, which was also the largest clinical study of DBS in the treatment of epilepsy to date. The study included 17 epilepsy centers in North America, and 110 cases were followed up for 2 years (Fisher and Salanova 2010). The results showed a 41–56% reduction in the frequency of seizures. During the double-blind phase, 36.3% reduction in the stimulus group and 12.1% reduction in the control group, respectively. Follow-up results in the unblinding phase showed that seizures were reduced by 44% at 13 months and 57% at 25 months; the proportion of patients who improved more than 50% was 43% at 13 months, 54% at 25 months, and 67% at 37 months; seizures disappeared for at least 6 months in 12.7% of patients, for at least 1 year in 7.3% of patients, and for 16% of patients, seizures were reduced by an average of more than 90% at 2 years after surgery. In this study, temporal lobe origin was found to be more effective, followed by diffuse, frontal, parietal, or occipital origin. The 5-year follow-up of the study showed an average reduction of 69% in seizures (Salanova and Witt 2015).

Further studies have been reported since SANTE's study. For example, Lee and Shon (2012) reported an average 70.4% reduction in seizures in 15 patients with drug-resistant epilepsy treated with ANT-DBS (mean follow-up: 27 months) (Lee and Shon 2012); Oh et al. reported an average 57.9% reduction in seizures in 9 cases (1-year follow-up) (Oh and Kim 2012).

With the progress of image reconstruction technology, studies on the relationship between clinical efficacy and stimulation targets in patients with ANT-DBS began to emerge. In 2016, Wu et al. analyzed the image data of some patients in SANTE study, and showed that only a few contacts of stimulation electrodes were located in ANT, and the rest electrodes were located outside ANT (Wu and D'Haese 2016). Lehtimaki et al. analyzed the optimal placement of ANT electrodes on 30 sides of 15 patients with epilepsy, and the results showed that the contacts located in the anterior and upper parts had the greatest antiepileptic effect (Lehtimaki and Mottonen 2016). The Canadian research team also reached similar conclusions and found that 9 of 16 patients had mild damage lasting for 2–4 months, and the longest one lasted for 3 years (Krishna and King 2016).

#### 2.2.4.1 Hippocampus–Amygdala Electrical Stimulation

A total of nine patients with chronic hippocampal electrical stimulation were enrolled in a double-blind controlled study conducted by Velasco et al. with follow-up duration ranging from 18 months to 7 years. Five patients with normal MRI had a 95% reduction in seizures, while four patients with hippocampal sclerosis had a seizure reduction of approximately 50–70% (Velasco and Velasco 2007). In the study by Vonck et al., a total of seven patients received bilateral hippocampal stimulation, four of which were responsive, and one of them lasted 1.5 years without seizure (Vonck and Boon 2005). In 2007, the study group also reported 13 patients who received either unilateral or bilateral hippocampal stimulation. Of the ten patients followed up for more than 12 months, five had a >50% reduction in seizures, one had no seizure for 2 years, one had only nocturnal seizure, two had a <25% reduction in seizure, and one had no change in seizure frequency. Continuous follow-up of these patients showed a mean follow-up of 8.5 years in 11 patients, with >90% reduction in seizures in six cases, including three cases seizure-free for more than 3 years, 40–70% reduction in seizures in three cases, <30% reduction in seizures in two cases, and further reduction in three patients after replacement with bilateral stimulation (Boon and Vonck 2007). Boex et al. reported that eight patients with medial temporal lobe epilepsy underwent hippocampal DBS, and two patients with combined hippocampal sclerosis had 65–75% reduction in seizures (Boex and Seeck 2011). Of the six patients with normal MR, two had no seizures, one of which experienced the event of not turning on, the other two had a 65–70% reduction in seizures, and the remaining two had no significant changes in seizures. Yu et al. reported a study of five cases of hippocampal DBS with a mean reduction in seizures of >75% in one case, >50% in three cases, and no significant improvement in one case at 14–24 months postoperative follow-up (Yu and Zhang 2015).

#### 2.2.4.2 Electrical Stimulation of Subthalamic Nucleus

Benabid et al. first treated a child with refractory epilepsy and left central parietal focal cortical dysplasia with STN as a stimulation target in 1998, manifested by convulsions of the right limb, which could have a persistent state, and chronic stimulation using 0 and 1 contacts. At 30 months of follow-up, seizures were reduced by 80% (Benabid and Minotti 2002). Subsequently, STN-DBS therapy was performed on four other patients with refractory epilepsy. The seizures in three patients were reduced by 67–80.7%, and the lesions in patients with seizure reduction >50% were all located in the central region. Loddenkemper et al. reported five patients with drug-resistant epilepsy, two of whom had seizures reduced by 80% after 10 months and 60% after 16 months, and the other three had no obvious effect (Loddenkemper and Pan 2001). Handforth et al. reported the results of STN-DBS treatment in two cases of refractory epilepsy, in which the frequency of seizures of bilateral temporal origin was reduced by 50%, but drop attacks were still present. Another patient with a left frontal encephalomalacia had a one-third reduction in seizure frequency and, more significantly, a reduction in seizure severity (Handforth and DeSalles 2006). Lee et al. reported that three patients receiving STN-DBS treatment had an average reduction in onset frequency of 49.1%, which was slightly worse than the three patients receiving ANT-DBS at the same time (Lee and Jang 2006). Although the number of case reports of STN-DBS for partial seizures is limited, it still shows that it may be more effective for epileptogenic foci in the functional area, which is also consistent with our study on the mechanism of STN-DBS in controlling epilepsy.

### 2.3 Responsive Neurostimulation

Responsive neurostimulation (RNS) is a quite creative idea to inhibit the spread of seizure propagation by triggering stimulation when neurophysiological changes are detected in the pre-seizure or electrical seizure period, which is a new option for those patients with well-localized epileptic foci. Morrell et al. published a controlled clinical study on RNS in the journal *Neurology*, enrolling 191 patients with partial seizures (Morrell and Group RNSSiES 2011). The blind period was 3 months. The results showed that the treatment group reduced seizures by 38% and the control group by 17%, with a significant difference. During subsequent long-term follow-up, the frequency of epileptic seizures decreased by 48–66% within 3–6 years. Later, longer follow-up showed that 37% patients had no seizures for more than 3 months, 23% for more than 6 months and 13% for more than 1 year. It has been proved to be safe and effective in clinical practice and has been approved for clinical use by some countries.

However, RNS treatment of refractory epilepsy still faces some obvious challenges, such as the accuracy of epileptogenic foci location, the accuracy and timeliness of real-time EEG detection in judging epileptic seizures, the success rate of seizure termination with pulsed current stimulation, battery life. RNS modulation protocols are also being improved and refined, and new RNS systems have been integrated with the DBS system to provide feedback stimulation of deep tissues such as the hippocampus or anterior thalamic nucleus to improve seizure control. Although it is still far from perfect, it is undeniable that the closed-loop neuroregulation represented by RNS is a promising treatment. The combination of RNS and DBS or responsive DBS may further improve the effect of epilepsy modulation treatment.

## 3 Noninvasive Stimulation Techniques for Focal Epilepsy

At present, mature noninvasive stimulation techniques for treating epilepsy include auricular vagus nerve stimulation, transcranial magnetic stimulation, and transcranial direct current stimulation, while other techniques, such as transcranial alternating current stimulation, transcranial ultrasonic stimulation, and transcranial light stimulation, are still being explored.

### 3.1 Transcranial Magnetic Stimulation

The basic technical principle of TMS is Faraday's law of electromagnetic induction, that is, the application of the energized coil to generate a short (100–400 μs) magnetic pulse, with the magnetic field strength of up to 1.5–2 T, can penetrate the skull and other tissues to act on the nervous system, generating the corresponding induced electric field to produce various effects. The key components of transcranial magnetic stimulation (TMS) equipment include: capacitors, coils, and control components. The conventional coil generates magnetic pulses at a depth of about 2 cm, and H coil is a special coil for stimulating deep brain, which can be used for stimulating deep brain structure. The effects of TMS are associated with the stimulation frequency, intensity, time, mode, target, brain structure, and even the direction of the coil (the relationship between the induced electric field and the arrangement of cortical neurons). The stimulation intensity of TMS is usually expressed by the relative intensity of resting motor threshold (RMT) of a single patient, such as 70% RMT. RMT is expressed by the maximum output relative value of each device. Various TMS stimulation modes are commonly used in clinic, including single pulse (mainly for examining motor cortex function), paired-pulse (for measuring cortical inhibitory and excitatory functions), as well as repetitive stimulation and burst stimulation (usually for clinical treatment).

Repetitive transcranial magnetic stimulation (rTMS) is the most commonly used stimulation mode for treatment. Its clinical application is mainly based on the principle that low frequency transcranial magnetic stimulation (LF-TMS, frequency $\leq$1 Hz) reduces cortical excitability and high frequency transcranial magnetic stimulation (HFTMS, frequency above 5 Hz) increases cortical excitability. The basic mechanisms by which TMS changes cortical excitability are long-term depression (LTD) effect of low-frequency stimulation and long-term potentiation (LTP) effect of high-frequency stimulation. Since the basic feature of epilepsy is the hyperexcitability of cortical neurons, when rTMS is applied to focal epilepsy, the main therapeutic strategy is to use low-frequency stimulation to reduce the cortical excitability of the epileptogenic foci, thereby reducing seizures (Lefaucheur et al. 2020). Theodore et al. first reported the results of a controlled study in 2002 in which only a reduction in transient seizures was observed in patients with focal epilepsy dominated by temporal lobe epilepsy who were treated with TMS at 1 Hz, 120% RMT, 900 pulses, 2 pulses per day, for 7 consecutive days (Theodore et al. 2002). Fregni et al. conducted a control study with more pulses in 2006, using 1 Hz, 70% maximum output and 1200 pulses for 5 consecutive days of TMS treatment and found that epileptic seizures decreased within 2 months after treatment, and the reduction rates were 72%, 53%, and 58% at 2, 4, and 8 weeks after treatment; epileptiform abnormal discharges were also significantly reduced in the treatment group, with the reduction rates of 31% immediately after treatment, 16% at week 4 and 14% at week 8, respectively (Fregni et al. 2006a). Sun et al. also reported the results of a controlled study in 2012 which applied a longer treatment course: 0.5 Hz, 90% RMT, 1500 pulses, and 14 consecutive days of TMS treatment, significantly reducing seizures by 80% and epileptiform abnormal discharges by 65.8%. In addition, the study also found that post-treatment mental status scores of epileptic patients also showed significant improvement (Sun et al. 2012). Based on the above results, the TMS clinical application guidelines issued by the International Federation of Clinical Neurophysiology in 2014 recommended that low-frequency repetitive transcranial magnetic stimulation targeting the epileptogenic area may be effective in the treatment of focal epilepsy (Grade of Evidence: C) (Lefaucheur et al. 2014). Also, some researchers have explored the non-targeted low-frequency repetitive transcranial magnetic stimulation targeting the parietal region, but the efficacy was uncertain, and the 2014 guidelines did not make recommendations for this treatment regimen. In 2020, the International Federation of Clinical Neurophysiology updated its guidelines and still recommended low-frequency repetitive transcranial magnetic stimulation for the treatment of epilepsy (Grade of Evidence: C) (Lefaucheur et al. 2020). In addition to the commonly used low-frequency transcranial magnetic stimulation, theta rhythm magnetic stimulation is also a newer treatment option in recent years. Continuous theta rhythm magnetic stimulation can decrease cortical excitability, while intermittent theta rhythm magnetic stimulation can increase cortical excitability. However, the application of this protocol in the field of epilepsy treatment is currently limited.

## 3.2 Transcranial Electrical Stimulation

Transcranial electrical stimulation includes transcranial direct current stimulation (tDCS), transcranial alternating current stimulation (tACS), and transcranial random noise stimulation (tRNS).

At present, tDCS is a relatively mature technique in the field of epilepsy, and its technical principle is relatively simple, as follows: constant current stimulation (0.5–2 mA) is performed with the corresponding device, and due to various electrical conductivities of skin, skull, cerebrospinal fluid, gray matter and white matter, the current output of tDCS can eventually enter the skull, resulting in a dispersed electric field and affecting the function of neurons. tDCS is a quite safe technique that uses weak direct current stimulation and does not directly cause action potentials. However, a weak constant stimulation can have long-lasting effects. As early as in 2000, Nitsche et al. reported that placing a cathode in M1 region for only 4 s could reduce the amplitude of TMS motor-evoked potential, reflecting its decreased excitability (Nitsche and Paulus 2000). Earlier in 1964, Bindman et al. reported that neuronal firing rates decreased with cathodal tDCS but increased with anodal tDCS, and the effect could last for several hours after a few minutes of stimulation

(Bindman et al. 1964). The main components of tDCS include current generators and electrodes. The main effects of tDCS on nerve tissue are associated with electrode size, shape, conductive medium, target, electrode distribution pattern, and polarity. The most common effect is that anodal stimulation increases cortical excitability and cathodal stimulation decreases it, mainly because the former produces membrane depolarization effect, while the latter produces cell membrane hyperpolarization.

The main strategy of transcranial direct current stimulation in the treatment of epilepsy is to apply cathodal electrical stimulation to reduce the abnormally increased cortical excitability in epileptic foci. Liebetanz et al. found through animal experiments that cathodal tDCS could increase epilepsy threshold, and the effect was associated with current intensity and duration (Liebetanz et al. 2006). Focal epilepsy has a definite origin. Therefore, the common treatment is to place the cathode at the scalp position corresponding to the brain region of epileptic origin. In 2006, Fregni et al. enrolled ten adult patients with malformation of cortical development as the treatment group and nine patients as the control group to receive tDCS for 20 min in 1 day at 1 mA and found that epileptic discharges were reduced, with a tendency to minimizing the risk of seizures (Fregni et al. 2006b). San-Juan et al. reported in 2016 that 28 patients with medial temporal lobe epilepsy and hippocampal sclerosis were enrolled and treated with tDCS for 30 min at 2 mA. The study found that true stimulation treatment could significantly reduce seizures, and the efficacy of 5-day treatment was superior to that of 3-day treatment, with reduction rates of 54.6% and 43.4%, respectively; epileptic abnormal discharges were also significantly reduced (San-Juan et al. 2017). Similarly, in 2016, Assenza et al. also found that a single tDCS treatment for 20 min at 1 mA could reduce temporal lobe seizures, but no reduction of epileptiform discharges was reported (Assenza et al. 2017). In addition, Auvichayapat et al. (2013) reported that a single tDCS treatment for 20 min at 1 mA could reduce the epileptiform abnormal discharges in children with focal epilepsy; however, a significant reduction in the number of seizures might not be found due to the short-term course (Auvichayapat et al. 2013). Importantly, in 2020, Yang et al. reported the results of a multicenter randomized controlled study in which the investigators enrolled 78 patients with refractory focal epilepsy treated with cathodal tDCS at 2 mA for 20 min targeting the epileptogenic foci for 2 weeks and found that tDCS significantly reduced seizures, and the efficacy of a 40-min daily treatment was significantly superior to that of a 20-min daily treatment (Yang et al. 2020). In 2017, the International Federation of Clinical Neurophysiology reported guidelines for the clinical application of tDCS, but due to insufficient research evidence at the time of guideline preparation, no recommendations had been made for tDCS for the treatment of focal epilepsy (Lefaucheur et al. 2017). As of 2020, when the Neuromodulation Center Working Group updated its clinical guidelines for tDCS, it was recommended that tDCS for epilepsy was likely to be safe and effective (Level II recommendations) and that the effect of cathodal tDCS reduction in epileptic seizure frequency was 0.7 (95% confidence interval 1.38–0.02) (Fregni et al. 2021). Transcranial alternating current stimulation (tACS), transcranial random noise stimulation (tRNS) and transcranial pulsed current stimulation (tPCS) have not been systematically explored in the field of epilepsy, and more research evidence is required.

## 3.3 Transcutaneous Vagus Nerve Stimulation

The main mechanism of implantable vagal nerve stimulation for epilepsy is associated with the neuroendocrine of vagus nerve projection area. Transcutaneous vagus nerve stimulation (t-VNS) technique was proposed by Canadian scholar Ventureyra in 2000 inspired by traditional Chinese acupuncture and moxibustion. The auricular branch of the vagus nerve is the only branch of the vagus nerve on the body surface, distributed in the auricular concha cavity and cymba conchae on the outer side of the auricle. It is a mixed nerve of facial nerve, glossopharyngeal nerve, and vagus nerve and contains general somatosensory and general visceral sensory fibers (Butt et al. 2020). Stimulation of the skin in the distribution area of the vagus auricular branch can activate the auricular branch of vagus nerve and produce parasympathetic excitatory effect similar to stimulation of cervical vagus nerve trunk, which may be transmitted upward to the nucleus tractus solitaries and projected to the locus coeruleus to produce antiepileptic effect. Existing treatment parameters include stimulation intensity, frequency, pulse width, duration, and course of stimulation (Yap et al. 2020). The current international study on t-aVNS is presented in several case series and one randomized controlled study: Stefan et al. reported in 2012 the application of t-aVNS in the treatment of 10 adults with drug refractory epilepsy at a frequency of 10 Hz, pulse width of 300 ms, voltage of 25 V, and 3 h/day for 9 months; five out of seven patients had fewer seizures without specific discomfort (Stefan et al. 2012). He et al. reported in 2013 that the application of therapeutic regimen at 20 Hz and 0.4–1.0 mA for 30 min/day could reduce the frequency of refractory epileptic seizures in children, and the efficacy accumulated with the treatment time: seizures were reduced by 31.83% after 2 months of treatment, 54.13% after 4 months and 54.21% after 6 months; the response rates of seizures reduced by half in the same period were 23.57%, 53.85%, and 53.85%, respectively (He et al. 2013). Bauer et al. reported a randomized controlled study of t-aVNS at 25 Hz and 1 Hz in

2016, which applied a simulation protocol with a pulse width of 250 ms, a cycle with 30 s-on and 30 s-off, and 4 h per day for five consecutive months; when the stimulation intensity was above the sensory threshold but below pain threshold, it was found that stimulation at 25 Hz significantly reduced seizures by about 34%, while under 1 Hz stimulation, seizures were not significantly reduced, but increased by about 2.9% (Bauer et al. 2016). Due to the limited number of studies, no guidelines or consensus have been formed, and overall, a longer course of treatment, for example, 6–9 months, or longer, is required. Because of its delayed cumulative effect, it can be used as a third or fourth line treatment for epilepsy, mainly in drug-refractory epilepsy, where the epileptic focus is unclear, multifocal, or overlaps with functional areas, or where resective surgical treatment is ineffective and has contraindications.

### 3.4 Other Simulation Protocols

Transcranial focused ultrasound stimulation refers to mechanical acoustic stimulation above the human hearing threshold (20 kHz), with high spatial resolution of millisecond level, and can penetrate the skull to a depth of 10 cm. Previous studies have reported that low-intensity ultrasound stimulation can excite or inhibit neurons. Chen et al. reported that transcranial focused ultrasound pulse stimulation could inhibit epileptic abnormal discharges elicited by pentylenetetrazol, and preliminarily explored the effects of mechanical pressure, exposure time and other parameters, and found the optimal efficacy at 0.75 MI (mechanical stress index) and 600 s of exposure (Chen et al. 2020). However, all current results of transcranial ultrasound stimulation for epilepsy are derived from animal experiments, and its efficacy in humans has not been verified. Chaieb et al. reported that 10 min of near-infrared light stimulation at a wavelength of 810 nm reduced cortical excitability in healthy subjects with an effect lasting 30 min (Chaieb et al. 2015), which, however, was not explored in epilepsy.

## 4 Implantable Device Stimulation Protocol for Generalized Epilepsy

### 4.1 Vagus Nerve Stimulation

Although the therapeutic mechanism of VNS for epilepsy is still not fully understood, the neuroanatomy-based hypothesis, neurotransmitter hypothesis, immunological hypothesis, or brain network hypothesis have all revealed that VNS may have a control effect on generalized epilepsy. In most clinical studies, the efficacy of VNS for focal and generalized epilepsy has not been rigorously distinguished.

Englot et al. conducted the VNS efficacy study with the largest sample size and found that the efficacy of >50% reduction in seizures was significantly associated with negative preoperative MRI results, and patients with generalized onset seizures after the age of 12 were more likely to achieve optimal control of seizure (Englot and Rolston 2016). All these suggested that VNS had a better effect on generalized epilepsy. A study conducted by Englot and Chang (2011) revealed that the reduction rates were 58% in patients with generalized seizures, 43% in those with partial seizures, and 54% in those with combined seizures and that VNS had a good control effect on generalized seizures (Englot and Chang 2011).

In recent years, VNS has also attracted attention in the treatment of some special epilepsy syndromes. According to the evidence-based guideline-Grade C evidence for epilepsy treatment with VNS provided by the American Academy of Neurology (AAN), about 55% of patients with Lennox-Gastaut syndrome can achieve more than 50% seizure relief with VNS. Several studies showed that VNS had a favorable efficacy in LGS patients, with follow-up results showing a reduction in epileptic seizure frequency and severity, and an improvement in quality of life. Lancman et al. and Cukiert et al. suggested that the efficacy of VNS for tonic seizures, tonic–clonic seizures, myoclonic seizures, and atypical absence seizure in patients with LGS did not differ significantly from that of corpus callosotomy. The study conducted by Konstantin also reflected favorable effects of VNS on atonic seizures, with a reduction rate of more than 80%. Besides, some studies have also found that VNS has a certain effect on Dravet syndrome (e.g., reducing seizures by more than 50% in 55.9% of patients, with an average reduction of 55%); other studies have revealed that 69% of patients experienced a 50% or above reduction in the seizure frequency, with an average reduction of 70%. Few studies have also shown that VNS has also shown significant efficacy in Doose syndrome, and although the number of cases is small, it is still a treatment that deserves attention.

### 4.2 Deep Brain Stimulation

#### 4.2.1 Centromedian Nucleus (CM) Electrical Stimulation

The effect of CM electrical stimulation on generalized epilepsy is noteworthy. Experiments have confirmed that CM is associated with arousal and epilepsy thresholds, and the epilepsy threshold decreases if arousal is inhibited. In 1987, Velasco et al. first reported the clinical results of CM electrical stimulation and found that the frequency of complex partial seizures was reduced by 60–100%, while the frequency of generalized tonic–clonic seizures was reduced by 80–100% (Velasco and Velasco 1987). It was found that the

EEG showed rhythmic activities at 2–3 Hz after 2 days of CM stimulation, which might be a "beneficial igniting" of the inhibitory process and could explain the long-term effects of such stimulation; the arousal and rhythmical changes elicited by CM stimulation seemed to suggest a possible effect on generalized epilepsy. Velasco et al. conducted a follow-up study and reported the efficacy of CM electrical stimulation in 13 LGS patients with generalized tonic–clonic seizures and atypical absence seizures (Velasco and Velasco 2000). Seizures began to decrease in the first month of stimulation, and the optimal efficacy was achieved in the 6th month; two youngest patients had stimulator removal due to skin breakage, three had poor control, two had no seizures, and eight experienced gradual improvement. Velasco pointed out that CM electrical stimulation was most effective in the treatment of atypical absence seizures of nonfocal GTCS and LGS, and the most effective target was the parvocellular part of CM. Cukiert et al. reported the results of thalamic CM-DBS in four patients with generalized epilepsy who had undergone enlarged corpus callosotomy, with a 65–95% reduction in seizure frequency and a 25–95% reduction in the frequency of spike waves, although their morphology remained the same (Cukiert and Burattini 2009). Valentin et al. recently reported the results of two children with generalized epilepsy who were followed up for 12–36 months after CM-DBS; one case had a 90% reduction in seizures, while the other had no improvement (Sa and Singh 2019).

#### 4.2.2 Subthalamic Nucleus

The efficacy of STN electrical stimulation on myoclonic seizures deserves attention. Alaraj et al. reported the results of bilateral STN-DBS in a patient with Lennox-Gastaut syndrome manifested as myoclonic, convulsive seizure and generalized tonic–clonic seizures, with complete disappearance of generalized tonic–clonic seizures and 75% reduction in myoclonic and absence seizures after surgery (Chabardes and Kahane 2002). Wille et al. performed surgery plus ventral intermediate nucleus (Vim)-DBS in four patients with progressive myoclonic epilepsy, and after a mean follow-up of 24 months, SNr/STN DBS achieved the optimal clinical outcome in all patients, with a 30–100% reduction in myoclonic seizures, whereas Vim stimulation was ineffective and induced myoclonic emergence (Wille and Steinhoff 2011).

## 5 Noninvasive Stimulation Techniques for Generalized Epilepsy

### 5.1 Transcranial Electrical Stimulation

tDCS is a noninvasive technique that modulates neuronal excitability changes in the cerebral cortex by applying a constant weak current to the scalp. There are two types of stimulation, namely anodal stimulation and cathodal stimulation. The anodal stimulation enhances the excitability of neurons at the stimulation site by depolarization of cell membrane, while cathodal stimulation reduces the excitability of neurons at the stimulation site by hyperpolarization.

Liebetanz et al. found in young mice that long-term cathodal tDCS treatment could reduce the count of dead hippocampal cells in status epilepticus and the sprouting of granular cells and CA3 area in hippocampus, thereby improving cognitive impairment and reducing the occurrence of epilepsy. Kamida et al. also found through studies in young mice that long-term effects of tDCS might have neuroprotective effects on the immature hippocampus, thus improving cognitive level and antiepileptic effects. Nitsche et al. found that serotonin reuptake inhibitors (SSRI) (e.g., citalopram) could enhance the facilitation of anodal tDCS, eliminate the inhibitory effects of cathodal tDCS, and turn it into facilitation.

At present, the tissue damage threshold of tDCS used clinically to stimulate cortex is as follows: current density 0.05 mA/cm$^2$ and mean electric quantity 0.06 C/cm$^2$. The safety parameters of tDCS are set as follows: electrode plate 5 cm × 7 cm or circular electrode plate 3–6 cm in diameter, duration of stimulation 20–30 min, current 1.0–2.0 mA (no more than 4 mA). The sponge mats for the electrodes are soaked in saline; one is placed above the skull in the cortical area stimulated by the electrodes, and the other is placed on the orbital skin or mastoid on the opposite side. Reference electrodes are placed on the shoulder or elsewhere outside the skull to ensure minimal interference between the two stimulation plates. The effect of tDCS is associated with current density, electrode polarity, duration of each treatment, treatment days, and other factors.

tDCS was applied in animal experiments in the 1890s to study the increase or decrease of excitability and activity of cerebral cortex immersed in fluid by direct current stimulation. Later, this technique was applied to humans as a noninvasive stimulation. The inhibitory effects of tDCS on epileptic activity and seizures have been confirmed in animal experiments. Zobeiri et al. reported that a rat model of epilepsy induced by genetic defects showed changes in cortical excitability and slow spike and waves after tDCS stimulation. As a result, the number of slow spike and waves was reduced during cathodal stimulation and the mean duration was affected after stimulation, suggesting that cathodal direct current stimulation may be an effective noninvasive tool for reducing cortical excitability.

San-Juan et al. reviewed the literature on the efficacy and safety of tDCS in the treatment of epilepsy in animals and humans from January 1969 to October 2013. Animal experiments showed that direct current stimulation in vivo and in vitro can lead to a reduction in seizures without neurological damage, and 67% of clinical studies suggested that tDCS could significantly reduce the epileptic seizure frequency and was well tolerated by patients. Varga et al. treated children with 25 $cm^2$ electrode pad at 1 mA current intensity for 20 min to study the effect of single cathodal tDCS stimulation on children with refractory persistent spikes during slow-wave sleep. The electrode position was determined according to the visualized three-dimensional voltage distribution diagram, with the cathode in the negative peak region and the anode in the positive peak region. There was no significant decrease in epileptic activity during slow-wave sleep after treatment. This may be related to the fact that tDCS only improves resting excitability, but it remains unclear.

As a noninvasive physical therapy, tDCS has been proved to be safe in clinical trials. Nitsche et al. observed whether subjects had brain lesions 30 min and 1 h after tDCS stimulation in the generally accepted safe mode under T-weighted imaging and DWI imaging of MRI. The results revealed no tissue edema, structural changes of brain tissue, and imbalance of the blood–brain barrier in subjects. Poreisz et al. used questionnaires to study the safety status of tDCS in both healthy people and patients, and 102 people were enrolled in this experiment. After 567 sessions of tDCS, it was found that: according to current safety guidelines for tDCS, its application in motor and non-motor cortices both resulted in relatively minor side effects, such as mild tingling during stimulation (70.6%), itching (30.4%), mild fatigue (35.3%), mild headache after stimulation (11.8%), nausea (2.9%), and insomnia (0.98%). Other studies also found no seizures or other adverse events after tDCS, except transient tingling, itching, and skin redness under electrode stimulation only when the current was switched on, but all of them were transient and reversible.

Preliminary data from a recent randomized single-blind controlled trial for the treatment of generalized epilepsy were available: 22 patients with epilepsy were randomly divided into active group and control group; in the active group, the current intensity was 2 mA, the duration was 20 min, and the treatment lasted for 5 days. The negative electrode patches for patients with focal epilepsy were positioned on epileptogenic foci according to MRI or EEG, while the electrodes patches for patients with idiopathic epilepsy were positioned on the vertex (Cz zone) by a 10–20 computer system. Intergroup comparison suggested that after tDCS stimulation, the scores of auditory verbal learning test (AVLT) in the active group decreased, but there was no significant change; the results of other cognitive tests [e.g., Symbol Digit Modalities Test (SDMT) and Stroop Color and Word Test (SCWT)] were also not statistically different between groups. Therefore, it was believed that tDCS stimulation might have transient negative effects on short-term memory, while there was insufficient evidence for the effect of tDCS stimulation on other cognitive function changes (Luo 2017).

A total of 22 children with drug-resistant Lennox-Gastaut syndrome were enrolled in a randomized, controlled, double-blind trial and randomly assigned to an active group or a control group. The active group included 15 patients who received 5 days of tDCS (with current intensity of 2 mA and duration of 30 min) on the basis of antiseizure medication; the negative electrode patch was located in the left M1 region and the positive electrode patch in the right shoulder region. The control group was given sham stimulation and antiseizure medication. The results showed that the number of epileptic seizures and abnormal interictal discharges in the active group were significantly reduced. Subsequently, the investigators classified seizures into five categories, namely tonic, myoclonic, atonic, absence, and partial seizures. The results of intergroup analysis suggested that tDCS treatment was only effective for tonic, atonic, and absence seizures (Auvichayapat 2016).

A total of eight epileptic patients with epileptic spasm received tDCS treatment, with a treatment target cathode of P3/P4 for 14 consecutive days as a course of treatment, and antiseizure medication remained unchanged during the treatment. Among the 8 epileptic subjects, 7 had symmetrical epileptic spasm, and only 1 had asymmetrical epileptic spasm, who received 1–6 courses of tDCS treatment, respectively. The results revealed that tDCS had sustained effect in 3 cases, transient effect in 1 case, aggravation in 2 cases, and no effect in 2 cases. The more total sessions, the better the outcome for patients with predominantly symmetrical convulsive seizures (Yang et al. 2019).

Existing studies suggest that tDCS can significantly improve idiopathic generalized epilepsy.

## 5.2 Transcranial Magnetic Stimulation

rTMS regulates neuronal activity by generating a certain intensity of magnetic field through a specially shaped coil, which produces a sustained excitation or inhibition effect on neurons, so that the abnormal neural network can be improved or even restored to normal for a period of time. Local stimulation of magnetic pulses can even affect distant regions in a way that is not yet known. The LTD of TMS can reduce the excitability of epileptic focal cortex. Its effects are known to last for more than an hour after treatment. In case of low-frequency repetitive stimulation, the arrival of a second pulse induces a longer period of synaptic inhibition, assuming that stimulation occurs during the inhibitory period produced by the previous pulse. In addition, factors such as enhanced GABAergic inhibition, regulation of neuropeptide-Y expression levels, and influence on early gene expression may also be part of the antiepileptic mechanism of rTMS.

Current studies have found that common side effects of TMS include mild headache, neck pain, acute mania, syncope, tinnitus, anxiety, and scalp irritation, but these side effects are often transient and can be tolerated by most patients. Of greatest concern is the potential risk of TMS-induced seizures. In 2009, Rossi et al. searched 143 TMS papers related to epileptic seizures and conducted a systematic review of thousands of cases. Only 16 TMS-induced seizures were identified, 7 of which occurred before the safety guidelines for rTMS application were developed in 1998, and in most cases with risk factors such as a history of epilepsy, taking medications that lower the seizure threshold, or having other disorders affecting cortical excitability. Bae et al. retrospectively analyzed the safety of rTMS in 280 patients with epilepsy, only 4 (1.4%) had seizures, all of which were non-persistent, and the possibility of using drugs to reduce excitability could not be ruled out. Overall, rTMS-induced epileptic seizures were only individual cases with a low probability.

Studies on the efficacy of rTMS were not inconsistent. On the one hand, some studies believed that rTMS had a relatively positive effect on the number of epileptic seizures and epileptiform discharges, and the psychological status was significantly improved and could be maintained for at least 2 months. On the other hand, other studies suggested that rTMS had weak and short-term antiepileptic effects; although abnormal discharges were significantly reduced in the treatment group and patients with neocortical epilepsy might benefit more, the results were not statistically different. In 2011, Hsu et al. included 11 trials of a total of 164 patients with epilepsy and systematically analyzed the effect of standard transcranial magnetic stimulation on epilepsy and found that overall rTMS had only moderate antiepileptic effects. After classification according to epilepsy etiologies, it was found that rTMS had a strong inhibitory effect on patients with neocortical epilepsy and cerebral cortical dysplasia, but a weak effect on other types of epilepsy (Hsu 2011).

In conclusion, rTMS had a favorable effect on focal epilepsy, but a weaker effect on generalized epilepsy.

### 5.3 Others

Transcranial light stimulation, transcranial ultrasound stimulation, electromagnetic synchronous brain stimulation and electro-optical synchronous brain stimulation are currently emerging clinical noninvasive neuromodulation techniques, which have the advantages of being painless, inexpensive, adjustable stimulation mode and repeatable operation, etc. Some trials have confirmed that they are effective for focal epilepsy with definite foci, but there is a lack of relevant trial basis for generalized epilepsy.

## References


Assenza G, Campana C, Assenza F et al (2017) Cathodal transcranial direct current stimulation reduces seizure frequency in adults with drug-resistant temporal lobe epilepsy: a sham controlled study. Brain Stimul 10:333–335

Auvichayapat N (2016) Transcranial direct current stimulation for treatment of childhood pharmacoresistant Lennox-Gastaut syndrome: a pilot study. Front Neurol 7:66

Auvichayapat N, Rotenberg A, Gersner R et al (2013) Transcranial direct current stimulation for treatment of refractory childhood focal epilepsy. Brain Stimul 6:696–700

Bauer S, Baier H, Baumgartner C et al (2016) Transcutaneous vagus nerve stimulation (tVNS) for treatment of drug-resistant epilepsy: a randomized, double-blind clinical trial (cMPsE02). Brain Stimul 9:356–363

Benabid AL, Minotti L (2002) Antiepileptic effect of high-frequency stimulation of the subthalamic nucleus (corpus luysi) in a case of medically intractable epilepsy caused by focal dysplasia: a 30-month follow-up: technical case report. Neurosurgery 50:1385–1391

Berg AT, Berkvoic SF, Brodie MJ et al (2010) Revised terminology and concepts for organization of seizures and epilepsies: report of the ILAE Commission on Classification and Terminology, 2005–2009. Epilepsia 51(4):676–685

Bindman L, Lippold O, Redfearn J (1964) The action of brief polarizing currents on the cerebral cortex of the rat (1) during current flow and (2) in the production of long-lasting after-effects. J Physiol 172:369–382

Boex C, Seeck M (2011) Chronic deep brain stimulation in mesial temporal lobe epilepsy. Seizure 20:485–490

Boon P, Vonck K (2007) Deep brain stimulation in patients with refractory temporal lobe epilepsy. Epilepsia 48:1551–1560

Butt MF, Albusoda A, Farmer AD et al (2020) The anatomical basis for transcutaneous auricular vagus nerve stimulation. J Anat 236:588–611

Chabardes S, Kahane P (2002) Deep brain stimulation in epilepsy with particular reference to the subthalamic nucleus. Epileptic Disord 4(Suppl 3):S83–S93

Chaieb L, Antal A, Masurat F et al (2015) Neuroplastic effects of transcranial near-infrared stimulation (tNIRS) on the motor cortex. Front Behav Neurosci 9:147

Chen S-G, Tsai C-H, Lin C-J et al (2020) Transcranial focused ultrasound pulsation suppresses pentylenetetrazol induced epilepsy in vivo. Brain Stimul 13:35–46

Cukiert A, Burattini JA (2009) Centro-median stimulation yields additional seizure frequency and attention improvement in patients previously submitted to callosotomy. Seizure 18:588–592

Elliott RE, Morsi A (2011a) Vagus nerve stimulation in 436 consecutive patients with treatment-resistant epilepsy: long-term outcomes and predictors of response. Epilepsy Behav 20:57–63

Elliott RE, Morsi A (2011b) Efficacy of vagus nerve stimulation over time: review of 65 consecutive patients with treatment-resistant epilepsy treated with VNS >10 years. Epilepsy Behav 20:478–483. https://doi.org/10.1016/j.yebeh.2010.12.042

Englot DJ, Chang EF (2011) Vagus nerve stimulation for epilepsy: a meta-analysis of efficacy and predictors of response. J Neurosurg 115:1248–1255

Englot DJ, Rolston JD (2016) Rates and predictors of seizure freedom with vagus nerve stimulation for intractable epilepsy. Neurosurgery 79:345–353

Fisher R, Salanova V (2010) Electrical stimulation of the anterior nucleus of thalamus for treatment of refractory epilepsy. Epilepsia 51:899–908

Fisher RS, Cross JH, French JA et al (2017) Operational classification of seizure types by the international league against epilepsy: position paper of the ILAE Commission for Classification and Terminology. Epilepsia 58(4):522–530

Fregni F, Otachi PTM, Do Valle A et al (2006a) A randomized clinical trial of repetitive transcranial magnetic stimulation in patients with refractory epilepsy. Ann Neurol 60:447–455

Fregni F, Thome-Souza S, Nitsche MA (2006b) A controlled clinical trial of cathodal DC polarization in patients with refractory epilepsy. Epilepsia 47:335–342

Fregni F, El-Hagrassy MM, Pacheco-Barrios K et al (2021) Evidence-based guidelines and secondary meta-analysis for the use of transcranial direct current stimulation in neurological and psychiatric disorders. Int J Neuropsychopharmacol 24:256–313

Handforth A, DeSalles AA (2006) Deep brain stimulation of the subthalamic nucleus as adjunct treatment for refractory epilepsy. Epilepsia 47:1239–1241

He W, Jing X, Wang X et al (2013) Transcutaneous auricular vagus nerve stimulation as a complementary therapy for pediatric epilepsy: a pilot trial. Epilepsy Behav 28:343–346

Hsu WY (2011) Antiepileptic effects of low frequency repetitive transcranial magnetic stimulation: a meta-analysis. Epilepsy Res 96:231–240

Krishna V, King NK (2016) Anterior nucleus deep brain stimulation for refractory epilepsy: insights into patterns of seizure control and efficacious target. Neurosurgery 78:802–811

Lee KJ, Jang KS (2006) Chronic deep brain stimulation of subthalamic and anterior thalamic nuclei for controlling refractory partial epilepsy. Acta Neurochir Suppl 99:87–91

Lee KJ, Shon YM (2012) Long-term outcome of anterior thalamic nucleus stimulation for intractable epilepsy. Stereotact Funct Neurosurg 90:379–385. https://doi.org/10.1159/000339991

Lefaucheur J-P, André-Obadia N, Antal A et al (2014) Evidence-based guidelines on the therapeutic use of repetitive transcranial magnetic stimulation (rTMS). Clin Neurophysiol 125:2150–2206

Lefaucheur J-P, Antal A, Ayache SS et al (2017) Evidence-based guidelines on the therapeutic use of transcranial direct current stimulation (tDCS). Clin Neurophysiol 128:56–92

Lefaucheur J-P, Aleman A, Baeken C et al (2020) Evidence-based guidelines on the therapeutic use of repetitive transcranial magnetic stimulation (rTMS): an update (2014–2018). Clin Neurophysiol 131:474–528

Lehtimaki K, Mottonen T (2016) Outcome based definition of the anterior thalamic deep brain stimulation target in refractory epilepsy. Brain Stimul 9:268–275

Liebetanz D, Klinker F, Hering D et al (2006) Anticonvulsant effects of transcranial direct-current stimulation (tDCS) in the rat cortical ramp model of focal epilepsy. Epilepsia 47:1216–1224

Loddenkemper T, Pan A (2001) Deep brain stimulation in epilepsy. J Clin Neurophysiol 18:514–532

Luo WY (2017) Preliminary results of efficacy and cognitive affects of cathodal transcranial direct current stimulation for the treatment of epilepsy. Brain Stimul 10(2):501–502

Morrell MJ, Group RNSSiES (2011) Responsive cortical stimulation for the treatment of medically intractable partial epilepsy. Neurology 77:1295–1304

Nitsche MA, Paulus W (2000) Excitability changes induced in the human motor cortex by weak transcranial direct current stimulation. J Physiol 527(3):633–639

Oh YS, Kim HJ (2012) Cognitive improvement after long-term electrical stimulation of bilateral anterior thalamic nucleus in refractory epilepsy patients. Seizure 21:183–187

Osorio I, Overman J, Giftakis J, et al (2007) High frequency thalamic stimulation for inoperable mesial temporal epilepsy. Epilepsia 48(8):1561–1571

Ren L, Yu T (2020) Subthalamic nucleus stimulation modulates motor epileptic activity in humans. Ann Neurol 88:283–296

Sa M, Singh R (2019) Centromedian thalamic nuclei deep brain stimulation and Anakinra treatment for FIRES-two different outcomes. Eur J Paediatr Neurol 23:749–754

Salanova V, Witt T (2015) Long-term efficacy and safety of thalamic stimulation for drug-resistant partial epilepsy. Neurology 84:1017–1025

San-Juan D, Espinoza López DA, Vázquez Gregorio R et al (2017) Transcranial direct current stimulation in mesial temporal lobe epilepsy and hippocampal sclerosis. Brain Stimul 10:28–35

Scheffer IE, Berkovic S, Capovilla G et al (2017) ILAE classification of the epilepsies: position paper of the ILAE Commission for Classification and Terminology. Epilepsia 58(4):512–521

Stefan H, Kreiselmeyer G, Kerling F, Kurzbuch K, Rauch C, Heers M et al (2012) Transcutaneous vagus nerve stimulation (t-VNS) in pharmacoresistant epilepsies: a proof of concept trial: transcutaneous vagus nerve stimulation. Epilepsia 53:e115–e118

Stypulkowski PH, Stanslaski SR (2013) Chronic evaluation of a clinical system for deep brain stimulation and recording of neural network activity. Stereotact Funct Neurosurg 91:220–232

Sun W, Mao W, Meng X et al (2012) Low-frequency repetitive transcranial magnetic stimulation for the treatment of refractory partial epilepsy: a controlled clinical study: rTMS for the treatment of refractory partial epilepsy. Epilepsia 53:1782–1789

Theodore WH, Hunter K, Chen R et al (2002) Transcranial magnetic stimulation for the treatment of seizures: a controlled study. Neurology 59:560–562

Velasco F, Velasco M (1987) Electrical stimulation of the centromedian thalamic nucleus in the treatment of convulsive seizures: a preliminary report. Epilepsia 28:421–430

Velasco F, Velasco M (2000) Predictors in the treatment of difficult-to-control seizures by electrical stimulation of the centromedian thalamic nucleus. Neurosurgery 47:295–304

Velasco AL, Velasco F (2007) Electrical stimulation of the hippocampal epileptic foci for seizure control: a double-blind, long-term follow-up study. Epilepsia 48:1895–1903

Vonck K, Boon P (2005) Long-term deep brain stimulation for refractory temporal lobe epilepsy. Epilepsia 46(Suppl 5):98–99

Wille C, Steinhoff BJ (2011) Chronic high-frequency deep-brain stimulation in progressive myoclonic epilepsy in adulthood—report of five cases. Epilepsia 52:489–496

Wu C, D'Haese PF (2016) Variations in thalamic anatomy affect targeting in deep brain stimulation for epilepsy. Stereotact Funct Neurosurg 94:387–396

Yang D, Du Q, Huang Z et al (2019) Transcranial direct current stimulation for patients with pharmacoresistant epileptic spasms: a pilot study. Front Neurol 10:50

Yang D, Wang Q, Xu C et al (2020) Transcranial direct current stimulation reduces seizure frequency in patients with refractory focal epilepsy: a randomized, double-blind, sham-controlled, and three-arm parallel multicenter study. Brain Stimul 13:109–116

Yap JYY, Keatch C, Lambert E et al (2020) Critical review of transcutaneous vagus nerve stimulation: challenges for translation to clinical practice. Front Neurosci 14:284

Yu T, Wang X (2018) High-frequency stimulation of anterior nucleus of thalamus desynchronizes epileptic network in humans. Brain 141:2631–2643

Yu T, Zhang GJ (2015) Therapeutic efficacy of chronic unilateral hippocampus electric stimulation for temporal lobe epilepsy. Chin J Minim Invasive Neurosurg 7:296–299

# Stroke

# 26

Haiqing Song, Zu Wang, Weiqun Song, Zhiyuan Shen, Xin Guo, and Shujuan Tian

## 1 Post-Stroke Motor Rehabilitation

Stroke is the leading cause of disability worldwide, while post-stroke motor disturbance is often persistent and disabling. Motor rehabilitation is currently dominated by active exercise training; nevertheless, there are many alternative or auxiliary motor rehabilitation approaches in clinical practice, which are promising in the rehabilitation of patients with post-stroke motor disturbance.

### 1.1 Post-Stroke Motor Disturbance

With advancements in medical techniques, mortality from stroke has declined over the past few decades. Despite the increased incidence of stroke, higher survival rates are driven by an aging population, leading to more stroke survivors now than before. Nowadays, stroke is the leading cause of disability worldwide. Studies have shown that stroke is also the third most common cause of death internationally, after cancer and heart disease.

Post-stroke impairments, such as motor paralysis, sensory disturbance, decreased cognitive function, can negatively affect independent function, with motor paralysis being the most important factor affecting independent function. In 55–75% of patients with stroke, functional disturbance persists even 3–6 months after onset, affecting their quality of life and making it difficult for them to return to their normal occupation or activities of daily living.

Observational studies suggest that within 3 months of stroke is a plateau in motor recovery. Animal models reveal that improvements in the first few weeks are related to the nature and timing of the neurobiological mechanisms underlying spontaneous recovery. While patients in the chronic phase benefit from rehabilitation, earlier initiation of rehabilitation leads to better results, probably because this approach maximizes the interaction between treatment and natural recovery.

### 1.2 Evaluation of Post-Stroke Motor Function

Despite improvements in acute treatment and new tools applied to promote rehabilitation, most patients still have post-stroke dysfunction, frequently resulting in post-stroke disability. However, motor disturbance varies greatly from patient to patient. In patients with severely impaired function, it is difficult to predict recovery in the acute or subacute phase using clinically common assessment methods alone. Accurate early rehabilitation predictions will help rationalize rehabilitation goals and meliorate the design of trial testing strategies to facilitate rehabilitation. In fact, although both initial motor impairment and long-term motor outcome depend on the severity of corticospinal tract (CST) injury and the integrity of alternative motor fibers (e.g., the CST and the cortico-tectospinal tract), the inter-individual variability during rehabilitation is large for patients with initially moderate to severe injury. Mounting evidence highlights the limitations of using clinical scores alone to predict motor recovery, especially in patients with initially severe injuries. Therefore, setting realistic rehabilitation goals for individual patients requires applying other methods.

H. Song · Z. Wang
Department of Neurology, Xuanwu Hospital, Capital Medical University, Beijing, China
e-mail: songhq@xwhosp.org

W. Song (✉)
Department of Rehabilitation, Xuanwu Hospital, Capital Medical University, Beijing, China

Z. Shen · X. Guo · S. Tian
Department of Neurology, Neuromedical Technology Innovation Center of Hebei Province, Brain Aging and Cognitive Neuroscience Laboratory of Hebei Province, Hebei Hospital, Xuanwu Hospital of Capital Medical University, The First Hospital of Hebei Medical University, Shijiazhuang, Hebei, China



Y. Wang (ed.), *Therapeutics of Neural Stimulation for Neurological Disorders*, https://doi.org/10.1007/978-981-99-4538-2_26

Neurophysiological and structural imaging studies demonstrate that motor outcome is strongly dependent on the integrity of motor fibers, whereas the degree of CST injury limits performance and recovery from movement. Indeed, the CST constitutes the major motor efferent pathway. Voluntary limb movements originate mainly in the contralateral motor cortex and receive fiber convergence from frontal and parietal regions, and these regions play an important role in superior sensory motor processing. The motor cortex is divided into primary motor cortex (M1), premotor cortex (PMC), cingulate motor area (CMA), and supplementary motor area (SMA). The CST is composed of large pyramidal neurons of M1 that received the converge with fibers in the SMA, PMC, somatosensory cortex, and posterior parietal cortex (Davidoff 1990). Stroke-related CST injury is difficult to be detected on conventional magnetic resonance imaging (MRI) but can be clearly reflected on diffusion tensor imaging (DTI).

The recovery of motor function may be a remodeling of the CST itself or by spontaneous and therapeutic induction of highly plastic nerve fibers that can replace the motor tract. DTI indicators represent promising clinical biomarkers for predicting motor recovery, monitoring, and predicting response to neurorehabilitation interventions.

## 1.3 Traditional Post-Stroke Motor Rehabilitation

Exercise therapy is a key factor in post-stroke rehabilitation. Motor exercises performed after stroke may vary in their orientation or in their technical characteristics. Training orientation can be divided into goal orientation, task orientation, and repetitive task training orientation. Technical features can focus on duration, training load, and feedback type. There are a variety of therapeutic approaches to post-stroke motor rehabilitation, and discussions are still underway regarding the relative effectiveness of these approaches on post-stroke motor function.

### 1.3.1 Muscle Strengthening Training

Muscle strengthening training is a progressive active exercise for resistance in the paralyzed limb. These exercises can be performed by the therapist with manual resistance or with weight-bearing devices. Muscle strengthening and endurance training in post-stroke limb motor rehabilitation have long been condemned for its so-called evoked spasticity, but has now been regarded as an important part of rehabilitation programs used for patients with brain injury (Patten et al. 2004). Big data studies have demonstrated the superiority of intensive muscle training, which is valuable and can be part of post-stroke motor rehabilitation strategies in order to improve dyskinesia.

### 1.3.2 Constraint-Induced Movement Therapy

Constraint-induced movement therapy (CIMT) is a treatment method that applies the principle of motor skill learning to post-stroke rehabilitation and belongs to task-oriented training methods. CIMT is promising for the less affected limb with limited movement. The specific treatment strategy is to induce motor learning, enhance neuroplasticity, and strengthen training. Patients improve specificity through exercise training and have brain changes after repeated exercises, which in turn increase the complexity, motivation, and reward of exercise to enhance neuroplasticity. The dyskinetic limb must undergo tasks and exercises tailored to individual abilities. The initial high-intensity treatment program of CIMT emphasized: (1) Repeated task-oriented hemiplegic limb exercises for 6 h per day for 10 consecutive working days; (2) Transformation of skills acquired in the clinical environment into real-world movements of patients per day; (3) Restriction of the use of the non-paralyzed limb to promote the use of the hemiplegic limb during 90% of awake time; and (4) Shaping and improving neuroplasticity through certain rewards. Currently, the dose regimen of CIMT used is 0.5–6 h per day.

Evidence shows that either high-intensity or modified CIMT is superior to standard rehabilitation methods in terms of upper extremity impairment and disability, and therefore can be recommended to patients with stroke after 3 months of onset (Hatem et al. 2016).

#### 1.3.2.1 Mirror Therapy

Mirror therapy was initially reported as a treatment method to relieve pain in amputees' phantom limbs. For treatment, the mirror is placed in the patient's median sagittal plane and reflects the non-paralytic limb, simulating the affected limb. With this setup, movement of the non-hemiplegic limb creates a visual illusion of normal movement of the hemiplegic limb.

One of the advantages of mirror therapy is that it is easy for patients to manage and perform self-managed home therapy and is suitable for patients with severe motor disturbance. The mechanism underlying the effect of mirror therapy should be related to the activity of mirror neurons. Mirror neurons fire both in action execution and in action observation. However, the exact mechanism of mirror therapy in patients with stroke remains speculative. Studies have shown that mirror illusions can prevent or reverse the habituation of the paralytic limb not to be used because the visual image of the paralytic limb is similar to the patient's own motor limb sensation (Liepert et al. 1995). Furthermore, mirror therapy directly stimulates motor recovery by modulating cortical excitability. Mirror therapy, with responses lasting up to 6 months after treatment, can be integrated into post-stroke rehabilitation strategies, which is promising in improving motor disturbance or disability.

## 1.4 New Techniques for Post-Stroke Motor Rehabilitation

Although physical therapy is a necessary tool for motor recovery, when used for patients in chronic recovery stage, the impact of physical therapy on post-stroke motor recovery is often limited. That is why more effective methods of stroke rehabilitation are needed. Nowadays, there are several new techniques for alternative and adjunctive treatment. The latest technology interventions in stroke rehabilitation include noninvasive brain stimulation, robot-aided training, and virtual reality immersion therapy. These therapies largely overcome the limitations of traditional therapies and provide additional benefits to patients.

### 1.4.1 Noninvasive Brain Stimulation

Repetitive transcranial magnetic stimulation (rTMS) and transcranial direct current stimulation (tDCS) affect the function of the corticospinal tract by modulating the excitability of the motor cortex. rTMS and tDCS have been studied in stroke rehabilitation as noninvasive neuromodulation therapies.

#### 1.4.1.1 Transcranial Magnetic Stimulation

The transcranial magnetic stimulation (TMS) therapy has been recommended for a variety of neurological and psychiatric disorders, including stroke, chronic pain, and Parkinson's disease. In a safety study of high-frequency rTMS in patients with chronic stroke, a single session of rTMS at a subthreshold intensity of 20 Hz was found to be safe and beneficial for recovery of motor function (Yozbatiran et al. 2009). There is a competing model of balance between the two hemispheres, and this balance is affected after stroke. The excitability of the contralateral hemisphere is enhanced, while the interhemispheric inhibition to the affected hemisphere is abnormally increased. These changes in excitability may be an important reason for the impaired recovery of motor function. Therefore, a possible strategy for stroke rehabilitation is to modulate the plasticity of interhemispheric balance by rTMS, thus restoring normal activity patterns.

Theta burst stimulation (TBS) is a form of rTMS, which is effective in altering cortical excitability. According to the stimulation pattern, TBS can be divided into intermittent theta burst stimulation (iTBS) and continuous TBS. While iTBS affects motor evoked potentials and produces long-term potentiation, continuous TBS can inhibit the brain activity for up to 1 h. Interestingly, continuous TBS can maintain the excitability of the motor cortex. The effects also depend on the behavior caused by changes in pre-stimulus network state and post-stimulus network interaction. In addition, continuous TBS requires lower stimulation intensities to produce more lasting aftereffects compared to conventional low-frequency rTMS (Hellriegel et al. 2012). In addition to improving motor function, the use of continuous TBS on the left posterior parietal cortex can significantly improve cognitive disorder in patients with cerebral trauma, manifested by improved visuospatial ability and attention.

Despite the lack of standard operating procedures and coordination, rTMS has been demonstrated to be a promising tool for stroke rehabilitation. More researches and consensus on the fundamental mechanisms and standardization of rTMS are likely to be available in the future.

#### 1.4.1.2 Transcranial Direct Current Stimulation

tDCS, a noninvasive application, acts on brain tissue through weak currents. It can be used to manipulate membrane potential and regulate the spontaneous discharge rate of neurons in animals and humans. The neuromodulation by tDCS in stroke patients with hemiplegia is intended to reduce interhemispheric imbalances and improve brain plasticity. tDCS can be applied in three ways: (1) Anodal stimulation, where anodal electrodes are placed on the affected side; (2) Cathodal stimulation, where cathodal electrodes are placed on the healthy side; and (3) Dual hemisphere stimulation (dual tDCS), where anodal and cathodal stimulation are applied to the affected and healthy sides, respectively.

With thousands of subjects receiving tDCS, no serious adverse events have ever been reported. Besides, the minor side effects of tDCS are well defined, including tingling sensation or rash at the electrode site at the beginning of stimulation or temporarily, and erythematous rash due to capillary dilation. Therefore, tDCS can be considered as a relatively safe aid for post-stroke motor recovery.

The combination of tDCS with various rehabilitation therapies can enhance the efficacy of rehabilitation treatments for motor disturbance alone. A number of evidences demonstrate the superiority of transcranial direct current stimulation and the current value of tDCS as an adjunctive therapy for integration into stroke rehabilitation strategies, taking into account safety guidelines and the different effects of stimulation protocols.

### 1.4.2 Peripheral Electrical Stimulation

Initial damages to neuronal pathways and subsequent functional reorganization limit the recovery of motor function. In traditional studies on stroke, potential treatment strategies targeting motor function recovery emphasize the combination of motor stimulation with massive exercise training to restore motor skills. However, many stimulations have shown potential therapeutic benefits for improving post-stroke functional recovery. Post-stroke electrical stimulation can be divided into two types: (1) sensory electrical stimulation and (2) motor electrical stimulation.

#### 1.4.2.1 Somatosensory Electrical Stimulation

Transcutaneous electrical nerve stimulation (TENS) is the somatosensory electrical stimulation of peripheral nerves through the use of skin electrodes. Based on the stimulation effects, TENS is divided into high-frequency TENS and low-frequency TENS, corresponding to stimulation frequencies of 80–100 Hz and 1–5 Hz, respectively. High-frequency TENS triggers sensory responses, while low-frequency TENS triggers motor contractions.

Afferent motor stimulation can improve nerve function, but the degree of functional recovery varies. The effect of increased sensory input on motor outcome is relatively neglected. To be specific, improvement in motor performance and effective motor learning can be facilitated by afferent information in the skin and proprioception, as the method above increases the excitability of the cortical motor in areas representing the stimulated body part. Clinical studies have shown that sensory input improves motor function in rehabilitation after spinal nerve injury (Gomes-Osman and Field-Fote 2015).

The exact action mechanism of TENS on post-stroke motor recovery is unclear. Most likely, the mechanism of long-term potentiation of excitatory glutamatergic connections between primary sensory and motor cortex mediates the direct effects of repetitive transcutaneous electrical stimulation on corticospinal excitability and motor ability. High-frequency TENS is currently valuable for post-stroke motor recovery and can be integrated into stroke rehabilitation strategies as an adjunctive therapy.

#### 1.4.2.2 Motor Electrical Stimulation

Electrical stimulation via electrodes on skin surface can induce muscle contraction. Low-frequency TENS to peripheral nerves induces muscle contraction at a stimulation frequency of 1–5 Hz. Neuromuscular electrical stimulation (NMES) on neuromuscular end plates induces muscle contraction at a stimulation frequency of 10–50 Hz. NMES may be applied as a passive technique to induce simple muscle contractions or can be actively triggered by electromyographic activity (EMG-NMES) or by limb position (position-triggered NMES). Both forms of electrical stimulation can improve the enthusiasm of stroke patients in task-based limb training.

EMG-triggered electrical stimulation combines EMG biofeedback with the delivery of electrical stimulation. When a stroke patient attempts to move the affected limb, and the EMG signal of the active contraction exceeds a preset threshold, electrical stimulation will be delivered to the target muscle, thus spreading the movement throughout the limb. This treatment is appropriate for stroke patients as the signal can actively activate the paralyzed muscle, but does not generate enough electrical information to activate the muscle to move. The working principle of positional feedback stimulation is same as that of EMG feedback, but it relies on limb position to trigger stimulation rather than EMG signals.

NMES is a therapeutic method that uses low-frequency current to stimulate nerves and muscles, causing muscle contraction. Clinically, low-frequency electric pulses is commonly used to stimulate motor nerves and make muscles contract. Usually, the stimulation frequency is set between 12 and 50 Hz, and the intensity of muscle contraction is modulated by changing the pulse amplitude (usually 0–100 mA) or the pulse width (usually 0–300 μs). The main principles are as follows: (1) NMES stimulates motor nerve fibers and afferent sensory nerve fibers at the same time and promotes the plasticity of the brain; (2) NMES causes the reverse activation of motor nerve fibers and leads to the coupling activity of presynaptic membrane and postsynaptic membrane; (3) The process of functional recovery of neural network is similar to that of motor learning. Repeated functional movements under the action of NMES can cause the redistribution of motor learning areas in the cortex; (4) Visual feedback information of stimulating actions will further enhance the ability of motor learning.

NMES treatment can increase the neuroplasticity of spinal cord pathways. In a meta-analysis article published in 2017, 29 randomized clinical trials were included, including 940 subjects, NMES can reduce spasticity after stroke and can increase the range of limb movement, it is better when combined with other treatment methods. However, chronic tissue changes caused by immobilization, such as atrophy, muscle transformation into connective tissue, and shortening of muscle resting length, the loss of motor units of hemiplegic arm caused by secondary transsynaptic degeneration may affect the effect of NMES. At present, there is no final conclusion on the setting of NMES stimulation parameters, and there are differences not only between different studies, but also among different individuals in inducing optimal muscle contraction and avoiding discomfort, pain, and skin irritation. Meta-analysis shows that the frequency of NMES use is between 30 and 50 Hz, the pulse width is between 0.1 and 0.5 ms, 5 times a week, 30 min each time, lasting 3–4 weeks, and the effect is better.

### 1.4.3 Robot-Aided Therapy

The robot in robot-aided therapy is defined as a reprogrammable and multi-functional manipulator designed to move materials, parts, or specialized devices through variable programmed movements for the performance of a variety of tasks (Pignolo 2009). Today an increasing number of robotic devices are available for post-stroke motor rehabilitation. For example, robotic therapy for upper limb rehabilitation requires three basic components: (1) A motorized mechanical component attached to the hand that provides hand-targeted passive, active assisted or active resistance movement; (2) Performance-related visual feedback displayed by a screen; and (3) An interactive computer program that monitors and incrementally advances training to motivate the stroke patients.

The main advantage of robot-aided therapy is the provision of high-dose and high-intensity training. Most robotic devices are tailored for elbow and shoulder movements. There is a lack of robotic training devices for finger and wrist movements. The current robotic systems for upper limbs can be divided into passive systems for limb stabilization, active systems for limb movement, and interactive systems that are for both.

### 1.4.4 Virtual Reality Immersion Method

The computerization techniques of virtual reality focus on the interaction between subjects with virtual environment, manifested by task-oriented training with robotic devices. This type of training is usually based on interaction with a two-dimensional virtual environment presented on a computer or television screen.

Virtual reality immersion techniques consist primarily of a computer-generated 3D graphical environment in combination with visual, auditory, or haptic devices. Operators are supposed to experience the computer-generated environment as if it were part of the real world. Users can interact with the virtual environment by using standard input devices (e.g., keyboard and mouse) or through multi-modal devices (e.g., wired gloves). As the training is fun, subjects tend to be more motivated in a virtual reality environment than in a traditional rehabilitation setting.

Playing so-called serious games can improve patient compliance and self-management, contribute to the recovery of their physical and mental health and, to a certain extent, improve the knowledge of patients and clinicians. When stroke patients play specific games, they can potentially increase their patience, exercise motor skills through repetitive learning in a rich environment, gain confidence through reinforcement and immediate feedback, and increase motivation through achievements and social interaction.

In summary, many clinical and research interventions are feasible to promote the motor function of upper limbs in stroke patients. In addition, interventions can be made together in order to maximize the recovery of motor function in each patient. Currently, the field of stroke rehabilitation is faced with the challenge of how to tailor training to the needs of individual stroke patients. Although the effectiveness of some interventions is still under debate, some specific rehabilitation approaches bring hope to the prognosis of post-stroke limb movement.

## 2 Aphasia

Language processing is accomplished by the interaction of complex language-specific and domain-general networks. Because aphasia is caused by brain injury that damages certain nodes or connections in language processing networks, recovery from aphasia is a process of dynamic reorganization of linguistic and nonlinguistic networks, reflecting the brain adaptation and plasticity for language processing.

The most commonly used noninvasive brain stimulation (NIBS) techniques to promote recovery from aphasia are rTMS and tDCS. Initially, experimental studies and clinical treatments tend to use inhibitory or excitatory stimulation to observe the change of language performance to confirm its effectiveness, with stimulation sites mostly located in Broca's area and Wernicke's area in the left hemisphere for excitatory stimulation and in the homologous regions of Broca's area in the right hemisphere for inhibitory stimulation. With the development of neuroimaging techniques, clinical studies and treatments have focused more on the alteration of language function and its associated neural networks.

Virtual lesion is made to healthy subjects by local inhibitory stimulation for the purpose of understanding the compensatory mechanisms of the language network. Excitatory or inhibitory stimulation is applied to different sites in aphasic patients for the purpose of understanding the effect of NIBS on the aphasia recovery, interactions, and compensatory potential of specific language functional networks and other language-related networks in post-stroke aphasia.

### 2.1 Noninvasive Brain Stimulation in Normal Subjects

In a study applying functional magnetic resonance imaging (fMRI) combined with continuous theta burst stimulation (cTBS), the reorganization of the parietal-frontal network related to lexical comprehension in healthy subjects was explored (Hartwigsen et al. 2017). When inhibitory cTBS was applied to the left angular gyrus, a critical area for semantic processing, the inhibitory effect was enhanced not only at the stimulated site but also in the anterior part of the left inferior frontal gyrus, another semantically critical node in the large semantic network, suggesting that virtual lesion of the left angular gyrus during semantic processing increased the inhibitory effect on the anterior part of the left inferior frontal gyrus. In addition, the supramarginal gyrus neighboring phonological areas showed semantic-related upregulation, possibly reflecting partially compensation for disruptive cTBS effects and contributing to the maintenance of task processing. In contrast, following the suppression of the phonological area of the left supramarginal gyrus, a delay in phonological decision was observed without any compensatory upregulation of the semantic area. These results suggested that neighboring low-level task (e.g., phonological) processing areas may compensate to some extent for higher-level semantic processing impairments, but the higher-level semantic processing areas could not compensate for lower-level phonological processing impairments.

In one study, cTBS was applied to suppress activity in the left middle temporal gyrus of 16 healthy subjects while EEG was recorded. The task was to present a picture after the subject read the opening sentence and ask the subject to name it (a context-oriented speech production task). cTBS increased the error rates of naming and caused a decrease in the α-β (8–24 Hz) oscillatory power in the left temporoparietal cortex, which extended to left prefrontal cortex associated with domain-specific and domain-general control mechanisms. This study showed that the lexical-semantic network could reconfigure rapidly after the left middle temporal gyrus was disturbed, while also recruiting both domain-specific and domain-general control networks (Klaus et al. 2019).

What will occur in the interaction between the homologous regions of the two hemispheres when a region in the left language areas is suppressed? When inhibitory cTBS was applied to the left posterior inferior frontal gyrus in healthy subjects, the activity of stimulation target was reduced, while the activity in the contralateral homologous region was increased and the response became faster during pseudoword repetition (Hartwigsen et al. 2013). These results indicated a dynamic modulation of interhemispheric interactions, and increased activation of the right posterior inferior frontal gyrus when the left posterior inferior frontal gyrus was disrupted may support early aphasia recovery.

A study of semantic memory in healthy subjects found reduced target area activity after perturbation of the left ventral lateral anterior temporal lobe using cTBS. While increased compensatory activity in the right ventral anterior temporal lobe was observed. It indicates there is an effective connection between the left and right anterior temporal lobes (Jung and Ralph 2016). This may reflect the involvement of alternative language networks, so as to adapt to changing demands or unilateral damage, which provided experimental evidences for the compensatory mechanism after perturbation.

## 2.2 Noninvasive Brain Stimulation in Aphasia

### 2.2.1 Excitatory Stimulation of the Left Hemisphere

What effect does excitatory stimulation of the left inferior frontal gyrus have on the language network of the post-stroke aphasics? Excitatory intermittent theta burst stimulation (iTBS) was applied 10 times to residual left Broca's area in 8 patients with moderate to severe chronic aphasia, 6 of which showed improvement in semantic fluency (Szaflarski et al. 2011). fMRI showed a significant increase in activation of the language network in the left frontal-temporoparietal lobe. Excitatory stimulation to a region in the left hemisphere could not only increase cortical excitability but also enhance the connectivity of white matter, thus improving language function in patients with aphasia. It was also reported that rTMS at 10 Hz in the left inferior frontal gyrus may normalize activity in both hemispheres by altering functional connectivity.

In 9 patients with chronic post-stroke aphasia, the anodal tDCS was placed on the left inferior frontal gyrus and the cathode on the right inferior frontal gyrus. After 3 weeks of tDCS coupled with speech therapy, their repetition accuracy of syllables and words and the correct rate of noun and verb naming significantly increased. The pattern of functional connectivity changes was found in the left hemispheric premotor cortex, anterior cingulate cortex, anterior cuneus, and cerebellar. These left hemispheric structures were associated with speech planning, maintenance, and execution, demonstrating that the bilateral tDCS determines functional connectivity changes within the lesioned hemisphere, enhancing the language recovery process (Marangolo et al. 2016).

To determine at which stimulation site in Broca's area or Wernicke's area better therapeutic effects can be obtained, a 5-day randomized crossover experimental study on the treatment of anodal tDCS combined with speech therapy in 20 patients with aphasia in the subacute phase of left frontal-temporal or frontal-temporoparietal stroke was conducted. The improvement in picture naming showed superior tDCS in left Wernicke's area than in Broca's area (Wu et al. 2015).

Can stimulation to the left representation of hand in primary motor cortex promote language function after left language network is impaired? In 16 patients with chronic post-stroke aphasia, with mild naming disorder and without lesions affecting the hand representation of the left motor cortex and underlying white matter, anodal tDCS was applied to the left representation of the hand M1 during fMRI scanning (Darkow et al. 2017). Investigators selected pictures that patients could reliably name when choosing language materials, so as to find out the effect of electrical stimulation independent of behavioral therapy. When anodal tDCS was applied, aphasic patients showed improvement in picture naming and reduced activity in domain-general brain regions associated with high-level cognitive control. The independent component analysis further revealed increased activity in larger language-related networks, as well as increased connectivity between these brain regions. Compared to unstimulated healthy controls, tDCS resulted in normalized network activity and connectivity in patients with aphasia. The overall reduction in task-related brain activity reflected increased neural efficiency and reduced processing efforts, suggesting that excitatory tDCS to the left representation of hand M1 improved picture naming ability in patients with aphasia.

### 2.2.2 Inhibitory Stimulation of the Right Hemisphere

There is a general consensus on the dominance of the left hemisphere for most elements of language processing. Activity in the right-hemispheric homologs of the perisylvian language areas may interfere with the left hemispheric language mechanisms and reduce performance in post-stroke aphasic patients. rTMS stimulation at 1 Hz was applied to three sites in the right inferior frontal gyrus of aphasic patients, and the stimulation sites were selected according to the patient's naming response activity in fMRI (Harvey et al. 2017). 8 patients with mild to moderate non-fluent chronic aphasia received stimulations to pars triangularis (BA45) and 1 received stimulations to pars orbitalis (BA47) for 10 times. 6 of them received fMRI studies. Re-naming evaluation was performed 2 and 6 months after treatment. The naming accuracy reached the maximum improvement at 6 months. This long-term effect was accompanied by a shift in the recruitment from the more anterior pars triangularis (BA45) of the right inferior frontal gyrus to the posterior BA44, 6 and 46. Besides, the naming improvement was accompanied by recruitment in the right frontal operculum, suggesting that the pars triangularis was ineffective in compensating for language processing. Many brain regions in the left hemisphere (left medial frontal lobe, cingulate gyrus, supplementary motor areas, and fusiform gyrus) were also recruited, the mechanism of which involved dynamic reorganization of the bilateral neural network for language processing.

In summary, increased bilateral neural activation associated with language improvement has also been found in studies on the application of NIBS to the left hemisphere. The brain regions recruited among patients with aphasia, even within the language domain, vary considerably, involving the activation of left perisylvian and extra-sylvian and homologous regions in the right hemisphere for various language processing. Therefore, the mechanisms of left and right hemispheric compensation may be complex, as the recruitment of neural tissue in the left and right hemispheres at different phases of recovery may reflect the complex dynamic interactions between the two hemispheres.

The application of many new treatments for hyperacute ischemic stroke, such as intravenous thrombolysis and mechanical thrombectomy, leads to the diversity of brain injury sites, making the classical types of aphasia less and less. In addition, the size of the brain injury, the integrity of the subcortical fiber tracts, the anatomical differences between individuals, and the differences in language function compensation in the adjacent and remote areas of the injury influence the effectiveness of stimulation. To sum up, stimulation targets are often selected based on the neuroimaging examination and language characteristics of patients in clinical practice. Identifying impaired nodes in language network and choosing individualized stimulus targets may contribute to satisfactory clinical outcome.

## 3 Unilateral Neglect

Unilateral neglect (UN) is also known as hemispatial neglect or unilateral spatial agnosia. It is a syndrome in which patients with hemispheric injury are unable to report or localize the contralateral objects or stimuli of the lesion. This disorder does not result from damage to the sensory or motor systems, but is more often considered as a specific type of attentional disorder. Jackson first reported the syndrome in 1876, while a multicenter study conducted by Ringman et al. found a 20% prevalence of UN after injury to the left cerebral hemispheres, compared to 43% for the right cerebral hemisphere. In addition, the follow-up after 3 months revealed 5% and 17% of neglect symptoms remaining in patients with left and right hemisphere injury, respectively (Buxbaum et al. 2004). The study showed that the severity of UN was positively correlated with age, and there was no clear correlation with gender or handedness. The most common after right hemisphere injury was left hemispatial neglect.

In the past 20 years, with the advent and development of TMS and tDCS, studies on neglect treatment with both have been frequently reported.

### 3.1 Studies on TMS Associated with Neglect

The common TMS includes rTMS and TBS. Treatment protocols include low-frequency rTMS and continuous TBS to the healthy hemisphere and high-frequency rTMS and intermittent TBS to the affected hemisphere. However, differences exist in stimulation parameters, stimulation position, number of treatments, stimulation intensity, number of pulses per treatment, duration, sequence interval, motor threshold (MT), and evaluation methods. In 2016, the U.S. *Guidelines for Adult Stroke Rehabilitation and Recovery* included neglect symptom improvement with rTMS as Level IIb recommendation with Grade of Evidence: B. In 2017, the Australian *Clinical Guidelines for Stroke Management* included neglect treatment with noninvasive brain stimulation as a weak recommendation for clinical treatment. In 2018, the Chinese *Expert Consensus on Repetitive Transcranial Magnetic Stimulation Therapy* of China clinically included post-stroke UN treatment with sequential cTBS to the left posterior parietal cortex as Consistency of Evidence III.

#### 3.1.1 rTMS

So far, several studies have applied rTMS to improve symptoms of visuospatial neglect by reducing interhemispheric competitive inhibition. The number of rTMS sequences ranged from 1 to 20, with 10 sequences (5 per week) being a common pattern. The intervention duration ranged from 1 day to 4 weeks. Among these studies, the total number of pulses and duration/sequence interval of each treatment varied considerably. The intensity of rTMS also varied among studies, but, in most studies, it ranged from 80 to 110% of MT. Three different control groups were commonly set up, i.e., high-frequency rTMS (10 or 20 Hz), sham stimulation, and blank control. More than 40 different tests were performed in these studies to assess cognition.

The effect of rTMS on post-stroke cognitive disorder is similar to that of motor disturbance, both of which may change at different time points after stroke. This is related to the post-stroke repair of brain tissue. The first phase of brain repair begins in the first few hours after stroke and lasts for several weeks. Early molecular and synaptic remodeling is essential to induce brain plasticity and regulate brain excitability. The second phase: the chronic phase begins approximately 90 days after stroke, when recovery slows down but still exists. Extensive cortical reorganization during these two phases is considered to be the most important for the behavioral recovery.

##### 3.1.1.1 Acute Phase—Subacute Phase (Onset—3 Months)

Song et al. applied 0.5 Hz-rTMS to patients with a mean onset time of 38.4 days of post-stroke visuospatial neglect at the site of P3 (location of the 10–20 EEG system) on the healthy side (Song et al. 2009). The control group was given only conventional rehabilitation training without sham stimulation. Patients showed significant improvement in both the line bisection and line cancelation tests at the end of treatment. This behavioral improvement persisted 2 weeks after TMS treatment.

In an open-label study conducted in 2010, treatment with low-frequency rTMS to the left posterior parietal cortex (PPC) resulted in a significant improvement in line bisection test in the stimulation group, suggesting that rTMS was an adjunctive strategy during the cognitive rehabilitation period and that it may improve post-stroke visuospatial neglect symptoms.

##### 3.1.1.2 Chronic Phase (More Than 3 Months)

In 2003, Brighina et al. applied low-frequency rTMS in the left hemisphere to three patients with right hemisphere ischemic stroke and found that this parameter improved visuospatial neglect (Brighina et al. 2003). The improvement in visuospatial function with this treatment lasted 15 days after the end of treatment. A pilot study revealed that low-frequency rTMS applied to the left PPC could significantly reduce visuospatial neglect in stroke patients for at least 6 weeks. The study showed that the treatment was dose dependent, with 10 treatments being more effective than 1 treatment alone.

#### 3.1.2 Theta Burst Stimulation

##### 3.1.2.1 Acute phase—Subacute Phase (Onset—3 Months)

Cazzoli et al. applied inhibitory cTBS (continuous theta burst stimulation) to the left hemisphere of patients with early post-stroke spatial neglect and treated them with 8 cTBS sequences at the left PPC for 2 days, which resulted in satisfactory results. Patients showed a 37% improvement in the quantitative assessment of daily activities for spatial neglect symptoms at the end of treatment, which persisted at least 3 weeks after stimulation (Cazzoli et al. 2012).

Later, Koch et al. designed a randomized, double-blind, sham stimulation controlled study and found that 2 weeks of cTBS to the left PPC was effective in improving the results of assessment on neglect behavior and attention disorder. In addition, cTBS acting on the left PPC could reduce the excitability of the frontoparietal pathway in the left hemisphere (Koch et al. 2012).

Fu randomized 12 patients with post-stroke visuospatial attention disorder in the subacute phase into two groups and applied cTBS to the left posterior parietal cortex of 80% MT (cTBS group) or 40% MT (sham stimulation group), respectively, for 10 days. Both groups received resting-state fMRI and behavioral paper-and-pencil tests at baseline examination and after treatment. After treatment, the functional connectivity was weak between the right temporoparietal junction (TPJ) and the right anterior insula, and between the right sulcus temporalis superior and the right anterior insula in group cTBS. Both groups showed improvement in behavioral tests, with greater changes in the cTBS group compared to the baseline examination. This suggested that cTBS to the left posterior parietal cortex could cause changes in RSFC between right ventral attention network regions in patients with post-stroke visuospatial neglect, which may play a key role in recovery from post-stroke visuospatial neglect (Fu et al. 2017).

##### 3.1.2.2 Chronic Phase (More Than 3 Months)

In 2009, Nyffeler et al. first observed the potential function of TBS in visuospatial neglect after recruiting and treating right hemispheric stroke patients with left visuospatial neglect. The study performed cTBS stimulation on the left PPC of patients and found that the use of 2 cTBS sequences significantly increased the number of perceived left visual targets compared to sham stimulation and blank controls, which was maintained for up to 8 h. This effect could be maintained for up to 32 h with 4 TBS sequences (Nyffeler et al. 2009).

Although there is a large number of literatures suggesting the benefit of TMS for recovery from post-stroke UN, there is no standard clinical protocol for TMS in the UN rehabilitation. When to stimulate? Should excitatory stimulation or inhibitory stimulation be selected? Or both? What is the optimal duration, number of repetitions, and stimulation intensity? Combined with what cognitive training protocols will TMS be more effective? These are all the questions needing to be considered at present.

### 3.2 Studies on tDCS Associated with Neglect

In recent years, studies have emerged on the generation, improvement mechanism, and therapeutic effects of tDCS for visuospatial neglect. Studies on healthy subjects found that tDCS could modulate posterior parietal cortex excitability in this population, and that anodal tDCS could increase the detection accuracy of contralateral spatial visual stimulation at the stimulation site, while cathodal tDCS would decrease the contralateral spatial visual stimulation at the stimulation site (Sparing et al. 2009). Another study found that cathodal tDCS to the right PPC in healthy subjects shifted visuospatial placing to the right, inducing neglect-like changes, while dual tDCS (left anode + right cathode) induced a stronger attentional shift toward the right than single tDCS (right cathode). Recent studies suggest that tDCS for parietal cortex may also have an effect on visuospatial neglect. Ko et al. found that 1 session of anodal transcranial direct current stimulation to the right posterior parietal cortices improved visuospatial scanning function, and Sparing et al. found that the inhibitory effect of 1 session of cathodal tDCS on the non-injured side or the facilitative effect of 1 session of anodal tDCS on the injured side could both relieve the symptoms of visuospatial neglect. Yi et al. further found that the alleviation effect of post-stroke visuospatial neglect could be maintained for at least 1 week after increasing the times of anodal tDCS stimulations on the injured side of the posterior parietal cortex or cathodal tDCS stimulations on the uninjured side of the posterior parietal cortex (Ko et al. 2008). Sunwoo et al. found that unimodal tDCS (anodal tDCS to the posterior parietal cortices) or bimodal tDCS (anodal tDCS to the right posterior parietal cortices together with cathodal tDCS to the left posterior cortices) could both improve the behavioral performance in line bisection test of patients with left visuospatial neglect, while the effect of bimodal tDCS was better than that of unimodal tDCS (Yi et al. 2016). Whereas, Bang et al. found that tDCS combined with feedback training was better than feedback training alone in improving visuospatial neglect.

The current international application of noninvasive brain stimulation is mainly focused on the posterior parietal cortex at the healthy side and injured side of patients with UN, but not all neglect patients can achieve significant results. The reason for this is that the posterior parietal cortex is not the only key node of the attention network, in addition to differences in treatment parameters and the course of the disease in the enrolled patients. Is it possible that one or several core encephalic regions may act as a link, or in other words, a "functional regulator," between the ventral and dorsal attention networks? If effective neuromodulation techniques are available, rebalancing the interhemispheric excitability at the key node will facilitate the recovery of neglect and the exploration of its functional remodeling mechanisms.

## 4 Dysphagia

Dysphagia refers to the process that food cannot enter the stomach from the mouth due to various reasons, mainly manifested as choking when drinking or eating, hoarseness or aphonia, etc. The swallowing function of patients is obviously impaired when they swallow autonomously and consciously, but the automatic and unconscious swallowing function is relatively preserved (Yuan et al. 2013). Dysphagia is one of the common complications after stroke. Stroke ranks second among the causes of death in the world, posing a serious threat to human health. Domestic studies reported that the incidence of dysphagia after stroke was 30–50%, while foreign reports were 51–73% (Baijens et al. 2016). Patients with dysphagia after stroke suffer from food intake difficulties, reduced food intake, and prolonged eating time, leading to weight loss, malnutrition, and electrolyte disturbances. When food choking into the respiratory tract can also cause complications such as aspiration pneumonia and repeated fever. It seriously affects the nutritional intake, disease recovery, and quality of life of stroke patients, increases the burden on families and society, and seriously endangers the life safety of patients. Although a large proportion of patients recover spontaneously in a short period of time, more than 10% of patients have residual dysphagia after conventional therapy. Therefore, it is particularly important for the recovery of patients to more effectively promote the improvement of swallowing function and reduce the risk of complications in the early stage of stroke.

NIBS techniques for post-stroke dysphagia involve TMS and tDCS. As a noninvasive neuromodulation technology, TMS uses a changing magnetic field to generate an induced electric field to change the action potential of neurons in the cerebral cortex, that is, to achieve the purpose of neuromodulation by regulating the excitability of the cerebral cortex. And with the deepening of research, different pulse stimulation modes were gradually developed and experimentally applied to the treatment of dysphagia after stroke. At present, the commonly used modes in clinical transcranial magnetic stimulation therapy are rTMS and TBS, and the two treatment modes have different characteristics. tDCS uses weak electrical currents to enhance or inhibit neuronal excitability. This article mainly discusses the stimulation site, stimulation parameters, and stroke stages.

## 4.1 TMS

### 4.1.1 rTMS

rTMS is a series of TMS pulses that continuously act on the cerebral cortex with constant stimulation intensity. rTMS has been shown to induce changes in brain excitability with long-term effects. Khedr et al. (2009) first reported the clinical effect of using rTMS to treat dysphagia after stroke in 2009. Some scholars have studied the effects of low-frequency and high-frequency rTMS on the swallowing disorder after stroke in the contralateral hemisphere and found that the swallowing function of both groups of patients was improved, and the therapeutic effect of high-frequency rTMS was more significant in scoring performance. Park et al. tried to use the treatment scheme of bilateral high-frequency rTMS, and the results suggested that bilateral mandibular hyoid muscle motor cortex stimulation could recover swallowing function faster than unilateral rTMS. Therefore, although the swallowing muscle tissue has representative regions in both hemispheres and is controlled bilaterally, in the treatment of swallowing disorder after stroke, the hemispheric reorganization after exciting the contralateral hemisphere may be more critical than inhibiting the cortical excitability of the contralateral hemisphere. A recent meta-analysis in China showed that rTMS therapy was more effective than standard swallowing training or placebo in patients with dysphagia after stroke. There were no significant differences between the groups in the stimulation of the affected, unaffected, and bilateral hemispheres. However, a study from Korea showed that bilateral hemispheric high-frequency rTMS (10 Hz) was more effective than unilateral high-frequency rTMS (10 Hz) in improving post-stroke dysphagia. The swallowing motor cortex is distributed in both cerebral hemispheres, so stimulation of both the affected and contralateral cerebral hemispheres may improve dysphagia after stroke. However, it is now generally accepted that one of the cerebral hemispheres may be dominant, so dysphagia caused by a unilateral hemisphere stroke may result from damage to the "dominant hemisphere" for swallowing. Some scholars believe that the related connections between the swallowing cortical network in the affected cerebral hemisphere may be reduced after stroke, so the recovery of swallowing function may depend on the unaffected side. In addition, Sasegbon et al. conducted a study on healthy volunteers and found that high-frequency rTMS in the cerebellar vermis had an inhibitory effect on pharyngeal motor cortex activity and swallowing behavior. Similarly, a study in China showed that high-frequency rTMS applied to mylohyoid cortical representation of the cerebellum can effectively improve the swallowing function of stroke patients. Others have placed rTMS coils on the left mastoid of stroke patients to stimulate their vagus nerve and found that it also significantly improved the patients' swallowing function. In conclusion, rTMS applied to the affected side, unaffected side, bilateral cerebral hemispheres, cerebellum, and left mastoid all helped to improve swallowing function. It is necessary to further study the influence of different stimulation sites on the therapeutic effect to find the best stimulation site and provide a basis for clinical treatment.

As rTMS is increasingly used to treat dysphagia after stroke, the current study began to look at whether different stimulation frequencies would affect treatment outcomes. It is generally believed that high-frequency stimulation is used on the affected side to increase cortical excitability, while low-frequency stimulation is used on the unaffected side to reduce its excitability to balance the cortical excitability of the bilateral hemispheres, thereby improving swallowing function. Stimulation frequencies of 1, 3, 5, and 10 Hz were used in previous studies. 5 Hz is the most commonly used stimulation frequency. Previous studies have shown that both high-frequency ($\geq$5 Hz) and low-frequency rTMS (1 Hz) can effectively promote the recovery of swallowing function after stroke. A recent meta-analysis in China showed that high-frequency stimulation has a larger effect size than low-frequency stimulation. In order to study the optimal stimulation frequency for rTMS, a subgroup analysis was conducted, but there was no significant difference between the groups at 1, 3, and 5 Hz.

Using 1 Hz-rTMS with different stimulation intensities in healthy volunteers, Fitzgerald et al. found that stimulation intensities of 115% of the resting motor threshold (RMT) were more effective in inhibiting cortical excitability than 85% of RMT. Mistry et al. used 1 Hz-rTMS with different stimulation intensities in patients with dysphagia and found that the stimulation intensity of 120% RMT was more effective than 80% RMT in improving patients' swallowing function. Both studies were given low-frequency rTMS and compared the effects of different stimulation intensities on cortical excitability. There is a lack of research on the effects of different stimulation intensities on high-frequency rTMS and cortical excitability at different stimulation sites.

#### 4.1.2 TBS

TBS uses short high-frequency (50 Hz) pulses with lower stimulation intensity and shorter stimulation time in burst-like or clump-like stimulation patterns. TBS is divided into continuous TBS (continuous theta burst stimulation, cTBS) and intermittent TBS (intermittent theta burst stimulation, iTBS), which can have opposite effects on cortical excitability. cTBS produces long-term inhibition of motor cortex excitability, and iTBS produces long-term potentiation of motor cortex excitability. A study of the effects of motor cortex excitability in the suprahyoid muscle group in healthy subjects found that cTBS inhibited the excitability of the ipsilateral suprahyoid muscle group motor cortex while activating the contralateral suprahyoid muscle group motor cortex area, whereas iTBS promoted the excitability of the motor cortex of the ipsilateral suprahyoid muscle group, but did not affect the contralateral side, and the effect on excitability lasted for at least 30 min. In addition, iTBS could reverse the inhibitory effect of cTBS on the motor cortex of the suprahyoid muscle group. Cosentimo et al. applied iTBS in the right pharyngeal motor cortex for 5 days in 42 patients with neurogenic primary or secondary dysphagia including stroke. The dysphagia outcome and severity scale (DOSS) was used to evaluate the swallowing function of the patients, respectively, before stimulation, 1 month and 3 months after stimulation, and the results suggested that iTBS could effectively improve the dysphagia in the long term. cTBS can inhibit the excitability of the motor cortex of the ipsilateral suprahyoid muscle group and activate the motor cortex of the contralateral suprahyoid muscle group; while iTBS promotes the excitability of the motor cortex of the suprahyoid muscle group on the ipsilateral side, but does not affect the contralateral side, the effect on the excitability lasts for at least 30 min. In addition, iTBs can reverse the inhibitory effect of cTBS on the motor cortex of the suprahyoid muscle group (Tai et al. 2021).

At present, there are few studies on TBS in dysphagia after stroke, and most of the studies are on the mechanism of healthy people. In conclusion, TBS, as a stimulation mode with low stimulation intensity, short stimulation time and strong stimulation effect, has a good application prospect in the treatment of dysphagia after stroke. However, further research is needed to verify its effect and determine the optimal stimulation regimen and suitable population.

### 4.2 tDCS

The theoretical basis of tDCS therapy is the phenomenon of cellular "electrotaxis," that is, cells move in a certain direction toward or away from an electrode in response to a potential gradient signal. tDCS can affect cell migration speed and direction, metabolism, and differentiation through cellular "electrotaxis." Usually, tDCS anodal stimulation can increase cell excitability, and cathodal stimulation can reduce cell excitability.

Current studies have shown that tDCS is effective and safe in improving neurological function in chronic stroke patients, but there are few data on the efficacy and safety of tDCS in acute stroke patients. Galovic et al. used survival analysis to estimate the recovery time of oral intake in stroke patients to establish a clinical map of the recovery of swallowing function after stroke and found that the trajectory of spontaneous recovery of swallowing function depends on the time after stroke. In the first few weeks after stroke onset, the slope of recovery of swallowing function was steeper, suggesting that early rehabilitation after stroke may be more conducive to the recovery of swallowing function. Suntrup et al. found that the time after stroke had a significant impact on the efficacy of tDCS and proposed that early stimulation (acute phase, but more than 24 h after onset) could better improve post-stroke dysphagia. The spontaneous recovery process in the acute phase is thought to be associated with cortical reorganization and increased pharyngeal motility in the contralateral motor cortex, so it may be advantageous to use neuromodulation techniques such as tDCS to facilitate this process. However, the current study was not able to determine which period after stroke has a larger effect size of tDCS.

In the literature on tDCS for the treatment of dysphagia after stroke, most investigators have used anodal tDCS stimulation. There is currently no consensus on whether the affected or the contralateral cerebral hemisphere of patients with dysphagia after anodal stimulation has a greater effect on the recovery of swallowing function. Studies on the recovery of swallowing function after stroke have shown that remodeling of the swallowing motor cortex on the contralateral side can promote the improvement of swallowing function, and anodal stimulation of the contralateral cerebral hemisphere can increase its compensation, thereby contributing to the recovery of swallowing function. However, since swallowing has corresponding functional areas in the cortex of both cerebral hemispheres, the remodeling of the affected cerebral hemisphere may also play an important role in the recovery of swallowing function. However, it has been suggested that tDCS applied to the contralateral cerebral hemisphere may have some inherent advantages compared to the affected side, as the current density distribution is not disturbed by tissue inhomogeneity, topographic abnormalities, or impaired intracortical connectivity. Ahn et al. used the bihemispheric anode tDCS model. The results showed that the bihemispheric anode tDCS combined with conventional dysphagia treatment can more effectively improve the swallowing function of chronic stroke patients compared with conventional dysphagia treatment alone.

The application of tDCS in the treatment of dysphagia after stroke is mainly based on weak current stimulation of 1–2 mA, but the most effective intensity of current stimulation still lacks the support of large samples and multicenter high-quality research. In the study of Kumar et al. anodal stimulation was performed on patients with dysphagia after stroke with a current strength of 2 mA. The DOSS scores of the patients were improved after stimulation compared with those before stimulation, and the difference was statistically significant compared with the control group. Suntrup et al. used a 1 mA current to anodically stimulate patients with dysphagia at least 24 h after stroke onset, and 1 month later, patients treated with tDCS showed greater improvement in fiberoptic endoscopic dysphagia severity scale (FEDSS) than the sham group. Intuitively, it would be tempting to assume that an increase in stimulus intensity (i.e., higher current intensities, current densities, and longer durations) would have a greater impact on the results. This notion is also supported by some earlier studies that found approximately linear effects of stimulus intensity and duration. Du et al. gave 1.0 mA, 1.5 mA, and 2.0 mA anodic stimulation to patients with dysphagia after stroke. After treatment, the scores of aspiration degree of the three groups all increased, but only the improvement of high-intensity current stimulation group had significant difference. It shows that under the same stimulation time, high-intensity tDCS can effectively improve the excitability of the swallowing cortex and improve the dysphagia after stroke. However, this study has shortcomings such as a small sample size (Du et al. 2020). Further research should expand the sample size and explore the mechanism of action of tDCS to provide theoretical reference for clinical application.

## 5 Apraxia

Apraxia was first proposed by Steinthal in 1871 (Wheaton and Hallett 2007). It means that patients cannot perform purposeful movements without motor and sensory impairment (Heilman and Watson 2008). Clinically, apraxia is divided into conceptual apraxia, conceptual motor apraxia, limb motor apraxia, structural apraxia, and dressing apraxia (Cantagallo et al. 2021). Some studies have shown that the incidence rate of apraxia after left and right brain injury is 51.3% and 6%, respectively, of which the incidence rate of conceptual apraxia in acute phase is 28% and 13%, respectively, and the incidence rate of conceptual motor apraxia is 57% and 34%, respectively (Zwinkels et al. 2004).

When stroke patients having apraxia, they often show obstacles to the daily skilled tools, specific gestures, and musical instrument playing skills, such as speech apraxia, limb apraxia, and swallowing apraxia. The treatment of apraxia can be divided into behavior training and rehabilitation treatment (Hu et al. 2008). At present, after systematic behavioral training and rehabilitation treatment, the symptoms of apraxia will be significantly improved, but its long-term effect is not ideal. Noninvasive brain stimulation techniques, such as rTMS or tDCS, have been increasingly used to intervene the brain activity of healthy people and stroke patients.

### 5.1 rTMS

TMS is based on the application of a very short strong current to generate a rapidly changing high-intensity magnetic field through the coil. The magnetic field will induce vertical currents in the brain, which are strong enough to directly depolarize neurons and affect the excitability of the cortex, so as to regulate the network state of the local or whole brain and promote the remodeling of brain function.

#### 5.1.1 rTMS and Speech Apraxia

Speech apraxia is characterized by motor speech impairment, which is thought to be caused by impairment in motor planning and/or programming (Yang and Wang 2014). A large number of studies have found that rTMS can promote the recovery of language function after stroke. Some studies have adopted the routine rehabilitation combined with low-frequency rTMS for the patients with speech apraxia after stroke and found that the speech apraxia score was significantly increased after low-frequency rTMS (Ren et al. 2016). This treatment scheme is mainly based on the theory of brain remodeling. It is believed that the dominant hemisphere suppresses the non-dominant hemisphere through the corpus callosum, and the non-dominant hemisphere also has a certain influence on the dominant hemisphere, so as to achieve a dynamic balance. When stroke occurs in the dominant hemisphere, this balance will be broken, the inhibitory effect of the dominant hemisphere on the contralateral hemisphere will be weakened, and the inhibitory effect of the contralateral hemisphere on the dominant hemisphere will be strengthened (Otal et al. 2015). Generally speaking, the brain function localization of aphasia is thought to overlap with Broca area, that is, the inferior frontal gyrus of the left hemisphere. Therefore, the stimulation site of rTMS is selected in Broca mirror area of the right hemisphere. Low-frequency rTMS (1–4 Hz) stimulation can reduce nerve cell excitability and inhibit cortical activity. In this study, low-frequency rTMS was used to stimulate the Broca mirror area of the right hemisphere in patients with post-stroke aphasia, reduce the excitability of the Broca mirror area of the right hemisphere, eliminate the inhibitory function of the Broca mirror area of the left hemisphere, promote the excitability of the bilateral cerebral hemispheres to a balanced state, and rebuild the language function network, so as to promote the recovery of the patient's speech function. After high-frequency (>5 Hz) transcranial magnetic stimulation for left dorsolateral prefrontal cortex was performed on patients with primary progressive speech apraxia, it was found that high-frequency transcranial magnetic stimulation can improve the language apraxia of patients by stimulating the language functional area and its adjacent areas (Shpiner et al. 2019). In addition, intermittent theta burst stimulation (iTBS) is an excitatory rTMS mode, which can improve the language function of aphasia patients after stroke. A double-blind randomized controlled trial showed that the application of 20 Hz excitatory rTMS to the Broca area of the affected side and 1 Hz inhibitory rTMS to the Broca area of the healthy side could significantly improve the language ability (Liang et al. 2021). The iTBS scheme is a unique rTMS pattern because it is similar to the firing pattern of hippocampal neurons and may promote long-term enhancement.

### 5.1.2 TMS and Limb Apraxia

Limb apraxia is the loss of the ability to perform random and skilled movements (Li 2013). Although this loss cannot be attributed to the basic sensorimotor loss, it is a kind of movement disorder with skilled and purposeful movements in clinical practice. This loss cannot be explained by hypotonia, dystonia, tremor, or chorea. It is the motor damage of the higher cortical centers rather than the primary motor control damage. A recent study compared the contralateral low-frequency rTMS regimen alone or in combination with the contralateral high-frequency rTMS regimen for 15 consecutive days. It was found that the upper limb motor function of stroke patients could be effectively improved, but the bilateral combined application was more effective than the contralateral low-frequency rTMS regimen alone. Continuous theta burst stimulation (cTBS) is an inhibitory rTMS scheme. When applied to the left inferior frontal gyrus (IFG) and the left inferior parietal lobe (IPL), the gesture behavior will be temporarily interrupted. Gesture behavior may depend not only on different regions of the left and right hemispheres, but also on their interactions (Bologna et al. 2016). iTBS can enhance the excitability of the affected cortex and promote the plasticity of the brain, while cTBS can inhibit the excitability of the cortex through continuous stimulation of the contralateral cortex. Both cTBS and iTBS can restore the balance of excitability between the two hemispheres by regulating the excitability of the cerebral cortex, so as to improve the limb function of the hemiplegic side of the patients after stroke. It has recently been shown that the integrity of the corpus callosum is critical for the interhemispheric effects of cTBS.

## 5.2 tDCS

tDCS technology uses low amplitude continuous current applied through scalp electrodes and can regulate brain activity by affecting resting membrane potential. Cathodic polarization inhibited cortical excitability, while anodic polarization increased cortical excitability. It has been demonstrated that changes in cortical excitability are related to changes in potential cortical neuronal activity and local cerebral blood flow.

### 5.2.1 tDCS and Speech Apraxia

In recent years, tDCS can effectively restore different cognitive abilities of aphasia patients, including language ability, such as speech intelligibility and behavioral verbs. Some studies have used anode tDCS to treat the left lip area of primary motor cortex (M1) in patients with speech apraxia after stroke and examined the activation changes of speech related areas after treatment with EEG. It was found that tDCS can activate the dorsolateral prefrontal cortex and Broca area of the left cerebral hemisphere in addition to the stimulation site when stimulating M1 area, revealing that tDCS on primary motor cortex can improve the speech function of patients with apraxia of speech (AOS) after stroke, and can stimulate and recruit more motor speech network areas (Wang et al. 2019). Some scholars used double hemispheric tDCS to stimulate the left inferior frontal gyrus and inhibit the right inferior frontal gyrus. Combined with repetitive language training, it can not only improve the accuracy of patients in expressing therapeutic stimuli, but also improve the performance of patients in different language tasks (picture description, naming of nouns and verbs, word repetition, word reading). Recently, it has been proposed that spinal cord stimulation increases somatosensory activity from the spinal cord to the brain through the current transmitted through the thoracic spine, inducing neurophysiological changes on the cortex (such as the sensorimotor cortex). There is evidence that spinal cord stimulation can improve the correlation between language and sensorimotor schema in post-stroke AOS patients.

### 5.2.2 tDCS and Limb Apraxia

No matter in the acute period (<2 weeks), recovery period (2 weeks–6 months), and sequelae period (>6 months), the recovery of upper limb motor function can be better promoted through anode tDCS (anode electrode piece is placed in the primary motor cortex of the affected side of the upper limb, cathode electrode piece is placed in the orbitofrontal cortex of the healthy side) or cathode tDCS (cathode is placed in the healthy side, anode is placed in the orbitofrontal cortex of the affected side) combined with other rehabilitation physiotherapy (Sattler et al. 2015). However, not all rehabilitation physiotherapy methods combined with tDCS can restore the original upper limb motor function and enhance the efficacy. For example, tDCS combined with wrist robot assisted therapy can improve the motor function of some upper limbs in stroke patients in convalescence period, but there is no significant difference between tDCS and wrist robot assisted therapy alone (Mazzoleni et al. 2017). At any time after stroke, tDCS alone can promote the recovery of upper limb motor function. For example, in the sequela period of stroke, neither anode tDCS nor cathode tDCS can significantly improve the upper limb motor function. However, after stratification of stroke severity, cathode tDCS can improve the upper limb motor function of patients with moderate to severe stroke, but there is also a risk of reducing the upper limb motor function of the healthy side. In addition, a recent meta-analysis found that the therapeutic effect of tDCS in patients with sequelae stroke was better than that in patients with acute and convalescent stroke (Bai et al. 2019).

### 5.2.3 tDCS and Dysphagia

The use of anode tDCS in the swallowing motor cortex of healthy subjects increased pharyngeal excitability in the stimulated hemisphere, reversed the neurophysiological and behavioral effects of cortical focal inhibition on swallowing, and increased the activation of bilateral swallowing related cortical networks. Many studies have found that anode tDCS can improve the swallowing function of patients with dysphagia after stroke. Moreover, with the significant increase in the amplitude of esophageal motor evoked potentials evoked by both hemispheres, there is a decrease in cortical inhibition and an increase in facilitation in both hemispheres (Jefferson et al. 2019). It was found that before tDCS treatment, NDA showed that cortical excitability in the central parietal temporal lobe region of the affected hemisphere increased during reflexive swallowing, but there was no significant increase in excitability in the monitored region during volitional swallowing. However, after tDCS treatment, cortical excitability in more regions of the affected or unaffected cerebral hemispheres increased during reflexive swallowing, and cortical excitability in the central parietal temporal region of the affected or unaffected hemispheres also increased during volitional swallowing. This finding may indicate that the improvement of swallowing function is related to the level of cortical activation after tDCS, and that the size and/or location of the lesion also affect the prognosis of swallowing apraxia (Yuan et al. 2011). The spontaneous recovery process in the acute phase of stroke is thought to be related to the remodeling of cortical function and the compensatory reorganization of the uninjured cerebral hemisphere, so it may be advantageous to apply tDCS to promote this process in the acute phase. Existing studies have shown that tDCS combined with traditional swallowing training can significantly improve the swallowing function of post-stroke patients with swallowing disorders. However, it is still uncertain whether the effects of TDCS on swallowing disorders at different times after stroke are different, and further research is still needed.

## 6 Post-Stroke Depression

Post-stroke depression (PSD) is an important neuropsychiatric complication of stroke patients, involving about 1/3 of stroke survivors, with a cumulative prevalence of 55% (Goldstein et al. 2016). The main clinical manifestations of PSD patients include low mood, sleep disorder, and decreased volitional activity, and even self-injuring behaviors (Bartoli et al. 2018). PSD deterioration has a great impact on the motor function rehabilitation and quality of life of stroke patients. At the same time, depression may have a negative impact on the existing health status, thus creating a vicious circle and greatly increasing the economic burden of the family and society. Therefore, it is very important to deal with depression correctly during the rehabilitation treatment of stroke (Huang et al. 2018).

The pathophysiological mechanism of PSD is still unclear, and its causes may be multifactorial, such as lesion location, genetic susceptibility, and imbalance of amine transmitters (Liu et al. 2019). At present, the main treatment of PSD is drugs, but the treatment period is usually long, and there are many adverse reactions such as dizziness, nausea and sexual dysfunction. In addition, depression subtype, sex, stroke location, and other factors may affect the treatment results of PSD (Vahid-Ansari and Albert 2018). Therefore, researchers continue to explore new methods of PSD treatment to better improve post-stroke affective disorder.

Noninvasive neuroelectrophysiological techniques, such as rTMS, tDCS, tACS, and so on, are safe, efficient, painless, easy to operate, and noninvasive. At present, they have been applied to stroke rehabilitation research as a treatment method, but it has also carried out some research in PSD treatment.

## 6.1 rTMS

Currently, the rTMS used in clinic are mainly divided into high-frequency rTMS and low-frequency rTMS. The former has an exciting effect on cerebral cortex, while the latter is considered to have an inhibitory effect. Pascual-Leone et al. first reported the antidepressant effect of rTMS in 1996. PSD as a special form of depression, rTMS provides a new idea for its treatment, which has attracted extensive attention of researchers (Pascual-Leone et al. 1996). As early as 2004, Jorge et al. found that high-frequency rTMS could improve the depression of stroke patients in the study of refractory PSD patients. Thereafter, Gu and Chang demonstrated once again that high-frequency rTMS is essential for the treatment of PSD by implementing high-frequency rTMS (10 Hz) intervention on the left dorsolateral prefrontal cortex of patients with chronic stroke. Similarly, low-frequency rTMS (1 Hz) also has a positive effect on the relief of bad emotions in post-stroke patients (Gu and Chang 2017). Although the current research indicates that rTMS may have some effects on improving the outcome of PSD, it is not yet clear whether there is a difference in the efficacy between high-frequency and low-frequency rTMS, and it may be related to the regulation of dopaminergic secretion, cortical excitability, and synaptic plasticity. Subsequently, some scholars explored the long-term efficacy of rTMS in PSD patients after stopping treatment. The results showed that both excitatory (10 Hz) and inhibitory (1 Hz) rTMS could significantly improve depressive symptoms during the treatment period, and its efficacy persisted for at least 1 month after stopping the treatment. Several other studies have also confirmed the long-term effectiveness of rTMS in most patients with major depression, but the antidepressant effect shown during the follow-up period is small, and the antidepressant effect gradually decreases with time (Kedzior et al. 2015). In addition, the efficacy of bilateral rTMS for PSD has also attracted extensive attention. However, it has not been found that bilateral rTMS has additional benefits compared with unilateral stimulation, and its efficacy is equivalent to that of unilateral high-frequency and low-frequency rTMS. Existing studies have shown that the effect of rTMS on PSD is significantly better than that of traditional treatment, regardless of the specific implementation scheme. Therefore, the future research of rTMS in the treatment of PSD should pay more attention to the efficacy of rTMS combined with traditional rehabilitation methods (Best et al. 2019). For example, previous studies have confirmed that rTMS combined with aerobic exercise therapy may be effective in patients with mild PSD.

## 6.2 Theta Burst Stimulation

Recently, a new type of theta rhythmic stimulation TBS, which can be divided into two different methods, iTBS and cTBS. iTBS not only has the characteristics of noninvasive, painless and less side effects, but also can easily regulate cortical excitability with weak stimulation and short time (Bulteau et al. 2017). A randomized sham-controlled study by Li et al. showed that TBS has good antidepressant effects, especially for some patients with refractory depression (Li et al. 2014). A randomized, double-blind, sham-controlled crossover study by Duprat et al. showed that the clinical remission rate of patients with drug-resistant major depression could reach 30% after 2 weeks of accelerated (20 times) iTBS treatment. iTBS can induce stable long-term excitation in a short time, regulate the activity of neurons in the brain, affect the glucose metabolism in the brain, increase the transmission of brain-derived neurotrophic factors and neurotransmitters, thereby enhancing brain plasticity and regulating cortical excitability. The mechanism of action is similar to rTMS. Its advantage is that it can promote the improvement of depression in patients in a shorter time and improve the clinical treatment efficiency (Duprat et al. 2016). These results indicate that TBS is safe and effective in the treatment of PSD, with good tolerance and recurrence prevention. Standard rTMS treatment takes about 30 min, while TBS treatment takes only 3 min, which can reduce the waiting time of patients and reduce the cost required to relieve depression (Blumberger et al. 2018). However, the exact effect of TBS on PSD needs further clinical validation.

## 6.3 tDCS

The research of tDCS in PSD is not sufficient, and the evidence supporting clinical application is not perfect. Some scholars studied 62 patients with PSD and randomly divided them into the combination group and the control group, 31 cases each. The patients in both groups were given droperidol tablets orally. The combination group was treated with transcranial electrical stimulation therapy. The nonfixed frequency and intensity were set to medium, 30 min each time, twice a day, for 5 weeks. After treatment, the scores of HAMD and SDS in the two groups were significantly lower than those before treatment, and the combination group was better than the control group, indicating that therapy is helpful to alleviate and improve the depressive symptoms of patients (Gan et al. 2015). Bueno et al. treated one patient with tDCS (2 mA, 30 min) and fluoxetine for 10 days. The anode of tDCS was placed on the left DLPFC, and the cathode was placed on the right orbitofrontal part. As a result, the depressive symptoms of the patient were significantly improved (Bueno et al. 2011). Valiengo et al. treated 4 patients with post-stroke aphasia and depression with tDCS (2 mA, 30 min). The anode was placed on the left DLPFC, and the cathode was placed on the right DLPFC. After 10 days of continuous treatment, the patients were treated once every 2 weeks and 4 weeks, a total of 12 times. The results showed that the depressive symptoms of all patients were improved. The scores of aphasia depression scale and stroke aphasia depression scale decreased by 65.7% and 47.5%, respectively. The effect was still maintained after 4 weeks of treatment, and all patients tolerated it, no adverse reaction (Valiengo et al. 2016). After that, Valiengo et al. conducted a randomized, sham stimulation controlled, double-blind clinical trial on 48 PSD patients and adopted the same tDCS intervention scheme. The results showed that the tDCS group was significantly better than the sham stimulation group at the end of treatment, and the effective rate (37.5%) and remission rate (20.8%) were higher than those of the sham stimulation group (Valiengo et al. 2017). These clinical studies showed that tDCS improved the depressive symptoms of PSD. However, the long-term efficacy and follow-up effects of tDCS are not clear. Different research results may also be related to different tDCS treatment schemes. In the future, more standardized and reasonable large-scale clinical randomized controlled studies should be designed to further explore the efficacy of tDCS on PSD patients and provide theoretical basis for clinical application.

## 6.4 tACS

tACS provides a new treatment for adult PSD patients. Compared with tDCS, the adverse reactions of tACS are less, and the adverse reactions are mild and transient. Regarding the mechanism by which tACS regulates brain activity, previous studies have shown that tACS may target α, γ, and θ oscillation plays an important role in regulating brain function. Therefore, more and more attention has been paid to the effect of tACS on regulating brain activity of different subjects. Preclinical experiments of tACS in the treatment of depression at a frequency of 77.5 Hz and a current of 15 mA showed that tACS had therapeutic effects by stimulating the forehead and two mastoid regions. Similarly, in order to evaluate the effectiveness and safety of tACS in the treatment of PSD, some scholars carried out a double-blind, randomized, controlled prospective study. It was found that tACS can reduce the severity of depression in PSD patients with 77.5 Hz frequency and 15 mA current, and it is safe and effective.

## 6.5 Other Electrical Stimulation

### 6.5.1 Bionic Electrical Stimulation of EEG

The cerebellar fastigial nucleus is considered to play an important role in the regulation of cerebral blood flow. As a therapeutic instrument that simulates the low-frequency biological current of EEG through direct digital frequency synthesis technology, the bionic electrical stimulator of EEG mainly uses the combined waveform of electromagnetic pulse formed by biological information simulation technology and computer software technology through the electrodes attached to the mastoid process, temple, and Fengchi point of both ears, so that it can penetrate the skull barrier without traumatic effects and directly act on the fastigial nucleus of cerebellum. Electrical stimulation of the cerebellar fastigial nucleus can improve the blood flow of the frontal lobe and auxiliary motor functional areas, restore and improve the neural control system, change the excitability and functional characteristics of visual, somatosensory, and prefrontal cortical neurons, and then improve the motor function of PSD patients (Gan et al. 2015). Electrical stimulation of the cerebellar fastigial nucleus may also promote the release of 5-HT from thalamic and cerebral cortical cells, thus alleviating the depressive symptoms of PSD patients. In addition, 5-HT can induce synaptic connections, promote the growth of sensory motor synapses, enhance the excitability of spinal cord movement, regulate and improve purposeful motor responses, and be conducive to the functional recovery of paralyzed limbs (Bao and Shao 2019).

### 6.5.2 Transcutaneous Acupoint Electrical Stimulation

Transcutaneous acupoint electrical stimulation (TAES) is a new type of acupuncture treatment at present. It not only uses the physical treatment method of pulse electrical stimulation, but also combines the specific treatment of acupoints (He et al. 2016). Some studies suggest that electrical stimulation at relevant acupoints can accelerate the blood circulation in the brain, promote the release of 5-hydroxytryptamine and norepinephrine in the brain, reduce the concentration of plasma endothelin, and reduce the damage to tissues. Electrical stimulation can also stimulate the afferent nerve fibers and project them to the higher center through the lower center of the spinal cord to promote the activation of the frontal, parietal, and temporal lobes. Different frequency electrical stimulation may activate different neural pathways to achieve the purpose of treating post-stroke depression (Wang et al. 2014). Many studies have found that TAES combined with antidepressants or routine nursing can effectively reduce the depression score of post-stroke patients, improve the degree of depression of stroke patients, and have a good therapeutic effect on depression (Zhang et al. 2016).

### 6.5.3 Neuromuscular Electrical Stimulation

Neuromuscular electrical stimulation (NMES) is a new treatment method that combines neuromuscular electrical stimulation with myoelectric biofeedback technology. Electrical stimulation with preset program is used to innervate muscles, induce muscle contraction, and improve muscle motor function (Zheng et al. 2013). Passive and active functional training induces information transmission from deep and shallow receptors in the skin and joints, which can accelerate the establishment of collateral circulation in the brain tissue, promote the reorganization and compensation of brain cells around the focus, and is conducive to the "plasticity" of brain tissue. Neuromuscular electrical stimulation therapy can synchronize with the stimulated conduction of the central nervous system and promote the functional recovery of the central nervous system. In the flaccid period of stroke patients, the paralyzed muscles are stimulated by low-frequency pulse current to make the muscles no longer atrophy and maintain the contraction ability. During the spasm period, the nerve impulses are transmitted to the spinal cord center through the centripetal pathway through the neuromuscular electrical stimulation, which can effectively inhibit the muscle spasm, make the subcortical center compensate, and improve the walking ability of the patient (Liu et al. 2014). Some studies have shown that the combination of NMES therapy on the basis of conventional rehabilitation therapy for post-stroke patients with depression has significant clinical efficacy, comprehensively and effectively improve the ability of daily living, improve the depressive state, reduce the occurrence of complications, and has high safety. It is worthy of further promotion and use in the clinic.

## References


Bai X, Guo Z, He L et al (2019) Different therapeutic effects of transcranial direct current stimulation on upper and lower limb recovery of stroke patients with motor dysfunction: a meta-analysis. Neural Plast 2019:1372138

Baijens LW, Clavé P, Cras P et al (2016) European Society for Swallowing Disorders—European Union Geriatric Medicine Society white paper: oropharyngeal dysphagia as a geriatric syndrome. Clin Interv Aging 11:1403–1428

Bao S, Shao B (2019) Effect of electroencephalobionic electrical stimulation combined with sertraline on post-stroke depression and its effect on serum 5-HT, NE and BDNF levels. J Math Med Pharm 32(11):1662–1664

Bartoli F, Di Brita C, Crocamo C et al (2018) Early post-stroke depression and mortality: meta-analysis and meta-regression. Front Psychiatry 9:530

Best SRD, Pavel DG, Haustrup N (2019) Combination therapy with transcranial magnetic stimulation and ketamine for treatment-resistant depression: a long-term retrospective review of clinical use. Heliyon 5(8):e02187

Blumberger DM, Vila-Rodriguez F, Thorpe KE et al (2018) Effectiveness of theta burst versus high-frequency repetitive transcranial magnetic stimulation in patients with depression (THREE-D): a randomised non-inferiority trial. Lancet 391(10131):1683–1692

Bologna M, Paparella G, Fabbrini A et al (2016) Effects of cerebellar theta-burst stimulation on arm and neck movement kinematics in patients with focal dystonia. Clin Neurophysiol 127(11):3472–3479

Brighina F, Bisiach E, Oliveri M et al (2003) 1 Hz repetitive transcranial magnetic stimulation of the unaffected hemisphere ameliorates contralesional visuospatial neglect in humans. Neurosci Lett 336(2):131–133

Bueno VF, Brunoni AR, Boggio PS et al (2011) Mood and cognitive effects of transcranial direct current stimulation in post-stroke depression. Neurocase 17(4):318–322

Bulteau S, Sebille V, Fayet G et al (2017) Efficacy of intermittent Theta Burst Stimulation (iTBS) and 10-Hz high-frequency repetitive transcranial magnetic stimulation (rTMS) in treatment-resistant unipolar depression: study protocol for a randomised controlled trial. Trials 18(1):17

Buxbaum LJ, Ferraro MK, Veramonti T et al (2004) Hemispatial neglect: subtypes, neuroanatomy, and disability. Neurology 62(5):749–756

Cantagallo A, Maini M, Rumiati RI (2021) The cognitive rehabilitation of limb apraxia in patients with stroke. Neuropsychol Rehabil 22(3):473–488

Cazzoli D, Müri RM, Schumacher R et al (2012) Theta burst stimulation reduces disability during the activities of daily living in spatial neglect. Brain 135(Pt 11):3426–3439

Darkow R, Martin A, Wurtz A et al (2017) Transcranial direct current stimulation effects on neural processing in post-stroke aphasia. Hum Brain Mapp 38:1518–1531

Davidoff RA (1990) The pyramidal tract. Neurology 40(2):332–339

Du Y, Li X, Liu W et al (2020) Intensities of transcranial direct current stimulation for dysphagia after cerebral infarction. Chin J Rehabil Theory Pract 26(5):583–587

Duprat R, Desmyter S, Rudi de R et al (2016) Accelerated intermittent theta burst stimulation treatment in medication-resistant major depression: a fast road to remission? J Affect Disord 200:6–14

Fu W, Cao L, Zhang Y et al (2017) Continuous theta-burst stimulation may improve visuo-spatial neglect via modulating the attention network: a randomized controlled study. Top Stroke Rehabil 24(4):236–241

Gan X, Liu Q, Wang X et al (2015) Clinical observation of transcranial electrical stimulation combined with delisin in the treatment of post-stroke depression. Chin J Clin 43(02):61–62

Goldstein LB, Adams R, Alberts MJ et al (2016) Primary prevention of ischemic stroke: a guideline from the American Heart Association/American Stroke Association Stroke Council: cosponsored by the Atherosclerotic Peripheral Vascular Disease Interdisciplinary Working Group; Cardiovascular Nursing Council; Clinical Cardiology Council; Nutrition, Physical Activity, and Metabolism Council; and the Quality of Care and Outcomes Research Interdisciplinary Working Group. Circulation 113(24):e873–e923

Gomes-Osman J, Field-Fote EC (2015) Cortical vs. afferent stimulation as an adjunct to functional task practice training: a randomized, comparative pilot study in people with cervical spinal cord injury. Clin Rehabil 29(8):771–782

Gu SY, Chang MC (2017) The effects of 10-Hz repetitive transcranial magnetic stimulation on depression in chronic stroke patients. Brain Stimul 10(2):270–274

Hartwigsen G, Sau D, Price CJ et al (2013) Perturbation of the left inferior frontal gyrus triggers adaptive plasticity in the right homologous area during speech production. Proc Natl Acad Sci 110:16402–16416

Hartwigsen G, Bzdok D, Klein M et al (2017) Rapid short-term reorganization in the language network. Elife 6:1–18
Harvey DY, Podell J, Turkeltaub PE et al (2017) Functional reorganization of right prefrontal cortex underlies sustained naming improvements in chronic aphasia via repetitive transcranial magnetic stimulation. Cogn Behav Neurol 30(4):133–144
Hatem SM, Saussez G, Della Faille M et al (2016) Rehabilitation of motor function after stroke: a multiple systematic review focused on techniques to stimulate upper extremity recovery. Front Hum Neurosci 13(10):442
He Y, Zheng H, Chen J et al (2016) Clinical study of acupuncture in the treatment of post-stroke depression. J Pract Tradit Chin Med 32(08):820–821
Heilman KM, Watson RT (2008) The disconnection apraxias. Cortex 44(8):975–982
Hellriegel H, Schulz EM, Siebner HR et al (2012) Continuous theta-burst stimulation of the primary motor cortex in essential tremor. Clin Neurophysiol 123(5):1010–1015
Hu P, Pang B, Wang K (2008) Clinical study of apraxia. J Clin Neurol 21(05):394–396
Huang J, Zhou FC, Guan B et al (2018) Predictors of remission of early-onset poststroke depression and the interaction between depression and cognition during follow-up. Front Psychiatry 9:738
Jefferson S, Mistry S, Michou E et al (2019) Reversal of a virtual lesion in human pharyngeal motor cortex by high frequency contralesional brain stimulation. Gastroenterology 137(3):841–849
Jung JY, Ralph MAL (2016) Mapping the dynamic network interactions underpinning cognition: a cTBS-fMRI study of the flexible adaptive neural system for semantics. Cereb Cortex 26:3580–3590
Kedzior KK, Reitz SK, Azorina V et al (2015) Durability of the antidepressant effect of the high-frequency repetitive transcranial magnetic stimulation (rTMS) in the absence of maintenance treatment in major depression: a systematic review and meta-analysis of 16 double-blind, randomized, sham-controlled trials. Depress Anxiety 32(3):193–203
Khedr EM, Abo-Elfetoh N, Rothwell JC (2009) Treatment of post-stroke dysphagia with repetitive transcranial magnetic stimulation. Acta Neurol Scand 119(3):155–161
Klaus J, Schutter DJLG, Piai V (2019) Transient perturbation of the left temporal cortex evokes plasticity-related reconfiguration of the lexical network. Hum Brain Mapp 9:1–11
Ko MH, Han SH, Park SH et al (2008) Improvement of visual scanning after DC brain polarization of parietal cortex in stroke patients with spatial neglect. Neurosci Lett 448(2):171–174
Koch G, Bonnì S, Giacobbe V et al (2012) θ-Burst stimulation of the left hemisphere accelerates recovery of hemispatial neglect. Neurology 78(1):24–30
Li J (2013) Research progress of limb apraxia. Chin J Geriatric Cardio Cerebrovasc Dis 15(05):557–558
Li CT, Chen MH, Juan CH et al (2014) Efficacy of prefrontal theta-burst stimulation in refractory depression: a randomized sham-controlled study. Brain 137(Pt 7):2088–2098
Liang J, Chen Z, Li N et al (2021) Effects of rTMS with different frequencies on speech function recovery in patients with moderate-severe motor aphasia after stroke and its mechanism. New Med 52(03):175–181
Liepert J, Tegenthoff M, Malin JP (1995) Changes of cortical motor area size during immobilization. Electroencephalogr Clin Neurophysiol 97(6):382–386
Liu M, Miao L, Ying G et al (2014) Effect of different frequency of neuromuscular stimulation on dysphagia in patients with stroke. Shandong Med 54(26):31–32
Liu C, Wang M, Liang X et al (2019) Efficacy and safety of high-frequency repetitive transcranial magnetic stimulation for post-stroke depression: a systematic review and meta-analysis. Arch Phys Med Rehabil 100(10):1964–1975
Marangolo P, Fiori V, Sabatini U et al (2016) Bilateral transcranial direct current stimulation language treatment enhances functional connectivity in the left hemisphere: preliminary data from aphasia. J Cognit Neurosci 28:724–738
Mazzoleni S, Tran VD, Iardella L et al (2017) Randomized, sham-controlled trial based on transcranial direct current stimulation and wrist robot-assisted integrated treatment on subacute stroke patients: intermediate results. IEEE Int Conf Rehabil Robot 2017:555–560
Nyffeler T, Cazzoli D, Hess CW et al (2009) One session of repeated parietal theta burst stimulation trains induces long-lasting improvement of visual neglect. Stroke 40(8):2791–2796
Otal B, Olma MC, Floel A et al (2015) Inhibitory non-invasive brain stimulation to homologous language regions as an adjunct to speech and language therapy in post-stroke aphasia: a meta-analysis. Front Hum Neurosci 9:236
Pascual-Leone A, Rubio B, Pallardo F et al (1996) Rapid-rate transcranial magnetic stimulation of left dorsolateral prefrontal cortex in drug-resistant depression. Lancet 348(9022):233–237
Patten C, Lexell J, Brown HE (2004) Weakness and strength training in persons with poststroke hemiplegia: rationale, method, and efficacy. J Rehabil Res Dev 41(3A):293–312
Pignolo L (2009) Robotics in neuro-rehabilitation. J Rehabil Med 41(12):955–960
Ren Y, Guo G, Li Q (2016) Effect of low-frequency repetitive transcranial magnetic stimulation on speech apraxia combined with Broca's aphasia after cerebral infarction. J Clin Med 36(02):42–43
Sattler V, Acket B, Raposo N et al (2015) Anodal tDCS combined with radial nerve stimulation promotes hand motor recovery in the acute phase after ischemic stroke. Neurorehabil Neural Repair 29(8):743–754
Shpiner DS, McInerney KF, Miller MA et al (2019) High frequency repetitive transcranial magnetic stimulation for primary progressive apraxia of speech: a case series. Brain Stimul 12(6):1581–1582
Song W, Du B, Xu Q et al (2009) Low-frequency transcranial magnetic stimulation for visual spatial neglect: a pilot study. J Rehabil Med 41(3):162–165
Sparing R, Thimm M, Hesse MD et al (2009) Bidirectional alterations of interhemispheric parietal balance by non-invasive cortical stimulation. Brain 132(Pt 11):3011–3020
Szaflarski JP, Vannest J, Wu SW et al (2011) Excitatory repetitive transcranial magnetic stimulation induces improvements in chronic post-stroke aphasia. Med Sci Monit 17:132–139
Tai J, Wu J, Wang T et al (2021) Research progress of different modes of transcranial magnetic stimulation in the rehabilitation of dysphagia after stroke. Acta Rehabil Sin 31(03):252–264
Vahid-Ansari F, Albert PR (2018) Chronic fluoxetine induces activity changes in recovery from poststroke anxiety, depression, and cognitive impairment. Neurotherapeutics 15(1):200–215
Valiengo L, Casati R, Bolognini N et al (2016) Transcranial direct current stimulation for the treatment of post-stroke depression in aphasic patients: a case series. Neurocase 22(2):225–228
Valiengo LC, Goulart AC, Oliveira JF et al (2017) Transcranial direct current stimulation for the treatment of post-stroke depression: results from a randomised, sham-controlled, double-blinded trial. J Neurol Neurosurg Psychiatry 88(2):170–175
Wang Y, He J, Sun C et al (2014) Clinical study of percutaneous electrical acupoint stimulation and acupuncture in the treatment of post-stroke depression. Chin J Rehabil Med 29(08):751–753
Wang J, Wu D, Cheng Y et al (2019) Effects of transcranial direct current stimulation on apraxia of speech and cortical activation in patients with stroke: a randomized sham-controlled study. Am J Speech Lang Pathol 28(4):1625–1637
Wheaton LA, Hallett M (2007) Ideomotor apraxia: a review. J Neurol Sci 260(1–2):1–10

Wu D, Wang J, Yuan Y (2015) Effects of transcranial direct current stimulation on naming and cortical excitability in stroke patients with aphasia. Neurosci Lett 589(4):115–120

Yang H, Wang S (2014) Advances in diagnosis and treatment of speech apraxia. Chin J Rehabil Med 29(09):893–897

Yi YG, Chun MH, Do KH et al (2016) The effect of transcranial direct current stimulation on neglect syndrome in stroke patients. Ann Rehabil Med 40(2):223–229

Yozbatiran N, Alonso-Alonso M, See J et al (2009) Safety and behavioral effects of high-frequency repetitive transcranial magnetic stimulation in stroke. Stroke 40(1):309–312

Yuan Y, Wang J, Li Y et al (2011) Application of nonlinear electroencephalogram analysis to observe the electrical activity of the cortex in aphagia. Chin J Rehabil Med 26(10):915–920

Yuan Y, Wang J, Wu D (2013) Clinical research progress of aphagia. Chin J Rehabil Med 28(07):680–683

Zhang Y, Hui X, Ni W (2016) Effect of swallowing training combined with transcutaneous electrical acupoint stimulation on dysphagia. Clin Med Pract 25(9):651–654

Zheng C, Xia W, Zhang Y et al (2013) Effect of neuromuscular electrical stimulation combined with swallowing training on dysphagia after stroke. Chin J Phys Med Rehabil 03:201–204

Zwinkels A, Geusgens C, Sande P et al (2004) Assessment of apraxia: inter-rater reliability of a new apraxia test, association between apraxia and other cognitive deficits and prevalence of apraxia in a rehabilitation setting. Clin Rehabil 18(7):819–827

# Disorders of Consciousness

# 27

Jianghong He and Yuanyuan Dang

Disorders of consciousness (DoC) is a state where consciousness has been affected by severe damage to the brain, including acute or prolonged DoC (pDoC). The pDoC means that the loss of consciousness lasts for more than 28 days, with the first cause being cerebral trauma, followed by stroke and hypoxic encephalopathy. pDoC is classified into vegetative state (VS) and minimally conscious state (MCS) according to the level of residual consciousness. VS refers to a state in which basic brainstem reflexes and sleep-wakefulness cycles are preserved, with spontaneous or stimulated eye-opening, but with no consciousness. MCS refers to the presence of clear signs of consciousness with discontinuity and fluctuation in patients after severe brain injury, and MCS− refers to the clinical presence of visual object tracking, pain stimulation localization, and directed voluntary movements, but the inability to complete the activities as required. MCS+ is a state in which patients can move eyes, open or close eyes, and stably move limbs as required while fail in functional communication with others or intentionally using objects. It is more difficult for patients in VS to recover consciousness, while MCS patients have good recovery potential (Giacino et al. 2018). The key of pDoC patient assessment is to determine the consciousness level of patients by identifying whether the response to stimulation is reflexive or comes from active behavior with partial perceptual involvement. The JFK Coma Recovery Scale-Revised (CRS-R) is the standard clinical scale for the examination and assessment of pDoC (Giacino et al. 2004). The CRS-R and the Glasgow Outcome Scale-Extended (GOS-E) are the primary prognostic evaluation tools for DoC (Group of Disorders of Consciousness and Conscious-promotion, Professional Committee of Neurorepair of Chinese Medical Doctor Association 2021).

The treatment option for DoC matters. New clinical and imaging data suggest that some pDoC patients may still benefit from therapeutic interventions several years after the onset of the disease. Most studies aimed at improving the level of consciousness and functional recovery of patients are open trials or case reports about behavior or brain imaging. The results of a few randomized controlled trials, especially on noninvasive neurostimulation treatments, have been published. It suggests that new therapies such as neurostimulation are valuable for the treatment of DoC, while the stimulation method and parameters still need further testing and validation (Bourdillon et al. 2019).

As patients are unable to express themselves, DoC treatment should follow the four basic principles of medical ethics: autonomy, non-maleficence, beneficence, and justice. The recommendation on treatment abandonment in the early stage of injury (within 28 days) needs to be made with great caution, and the treatment strategy after the stabilization period should take the state of consciousness, physical condition, pre-morbid wishes, and family opinions of the patients into account. When attempting new clinical interventions, adequate supportive evidence and relevant study reports, sound study protocols and risk control measures should be provided, and ethical approval should be provided. Families should be adequately informed of the evidence, limitations, potential risks, and harms of treatments. Patients in the prolonged phase need to be systematically evaluated in an experienced center, and the patient's family should be informed of the patient's prognosis and the need for a long-term assisted care plan in case of permanent disability, if possible (Kondziella et al. 2020).

J. He (✉)
Department of Neurosurgery, Beijing Tiantan Hospital, Capital Medical University, Beijing, China

Y. Dang
Department of Neurosurgery, The First Medical Center, Chinese PLA General Hospital, Beijing, China



Y. Wang (ed.), *Therapeutics of Neural Stimulation for Neurological Disorders*, https://doi.org/10.1007/978-981-99-4538-2_27

# 1 Noninvasive Neurostimulation Technique

## 1.1 Repetitive Transcranial Magnetic Stimulation

**Stimulation Principle** Repetitive transcranial magnetic stimulation (rTMS) is based on the principle of electromagnetic induction to create an electric field in the brain, which induces neural depolarization to achieve the effect of modulating the excitability of the cortex. Primary motor cortex (M1) stimulation may enhance the excitability of the motor cortex, and the stimulation to dorsolateral prefrontal cortex (DLPFC) may induce stronger connectivity between the prefrontal cortex and the thalamus (Thibaut et al. 2019).

**Stimulation Protocols** There is no consensus on the parameters of rTMS for pDoC. It is recommended to apply rTMS at 5–20 Hz with an intensity of 90–100% motor threshold to DLPFC, parieto-occipital junction, or motor area M1 for a course of 1–20 days, with 300–1500 pulses in total (O'Neal et al. 2021). Multiple treatment courses may also be considered based on the recovery characteristics of the disease. It may be implemented as early as possible after the primary condition is stabilized and the cerebral edema has resolved. It is not recommended for patients with unstable lesions in the target area, a history of epilepsy, skull defects at the treatment site, or metal implants in the body.

**Clinical Application** A single-blind, uncontrolled study conducted in China applied 20 times rTMS at 10 Hz to DLPFC (11 min per stimulation) in 16 patients with DoC (3–35 months post-injury) (Xia et al. 2017). All 5 MCS patients and 4/11 VS patients showed an improvement in the total score of CRS-R, and the improvement was more significant in MCS patients. The prefrontal lobe may be a better target for stimulation than the motor cortex. Several randomized controlled trials (RCT) tests of high-frequency rTMS to the M1 region of DoC patients have been completed and have found that rTMS affects cerebral blood flow and subsequent electroencephalogram (EEG) power spectrum in MCS patients (He et al. 2018; Liu et al. 2018). Although none of the current studies on RCT of rTMS have shown significant effects at the group level, there are many parameters (e.g., stimulation type, frequency, or duration) that can be optimized to improve its efficacy.

## 1.2 Transcranial Direct Current Stimulation

**Stimulation Principle** To modulate the excitability and connectivity of the cortex with weak direct current; the cumulative effect of long-course transcranial direct current stimulation (tDCS) modulation can reshape the consciousness network (Xia et al. 2018). The prefrontal cortex can be a target for stimulation because it is extensively connected to the striatum. The connectivity between thalamus and cortex can be strengthened by stimulating the striatum and disinhibiting the thalamus. MCS is more likely to benefit from treatment (Barra et al. 2022).

**Treatment Protocols** There is no unified standard regarding the stimulation site, time, parameters, and treatment course of tDCS for pDoC patients up to now. The recommended stimulation site is DLPFC or posterior parietal cortex, at 10–20 min per stimulation and 1–2 mA for 10–20 days. It should be applied with caution in patients with a history of epilepsy or intracranial metal implants.

**Clinical Application** A double-blind randomized controlled trial tested the effects of tDCS on the prefrontal lobe (i.e., continuous stimulation to the left DLPFC at 2 mA for 20 min) in 55 patients with acute and prolonged DoC (1 week–26 years after injury). At the group level, behavioral improvements (as measured by the CRS-R) were observed in MCS patients, but not in VS patients. At the individual level, 13/30 MCS (43%) showed improvements associated with tDCS (i.e., recovery of clinical signs of awareness not observed prior to tDCS or during sham treatment). Importantly, no patients reported side effects associated with tDCS (Thibaut et al. 2014).

## 1.3 Other Neurostimulation Therapies

Transcutaneous auricular vagal nerve stimulation (taVNS) increases cerebral blood flow, enhances brain electrical activity, affects the secretion of neurotransmitters, and increases wakefulness and awareness levels. taVNS may be applied early to the right side with a current of 10–20 mA and a frequency of 40–70 Hz at 1 time/day or 30 min–8 h per stimulation for 7–30 days. taVNS enters the nucleus tractus solitaries of the brainstem through the auricular branch of the vagus nerve and joins the ascending reticular activating system to participate in the modulation of the consciousness loop (Briand et al. 2020). There are no studies with large sample sizes, and the mostly recommended stimulation sites include the margin middle of bilateral auricular concha (AT2.3.4i) and the acupuncture points on brainstem (AT3.4i) (Yu et al. 2021). Stimulation with an intensity of 6 mA is applied continuously at 20 min per time, with 10 days per course. In addition, there are a few case reports on treating early post-traumatic disorders of consciousness and improving the language comprehension and spatio-temporal orientation of patients with low-intensity focused ultrasound pulsations (Cain et al. 2022).

## 2 Implantable Device Stimulation Technique

Implantable neuromodulation therapy for DoC has been studied for nearly 50 years. Several clinical studies have shown that it can significantly improve consciousness and behavior. However, due to the complexity and diversity of injury mechanisms and conditions in disorders of consciousness, systematic studies with large sample size are lacking. As there are many defects in terms of key scientific issues such as case screening, implementation, program control paradigm, and efficacy evaluation of nerve electrical stimulation for arousal, it has not become a definite therapy method. Therefore, neuromodulation surgery for pDoC should be taken as a complementary tool to conventional therapy. Before surgical evaluation, patients should be recommended to receive conventional rehabilitation and arousal treatment preferentially. The results of the evaluation should be fully explained to the family prior to surgery, and the possible efficacies should be clearly communicated. Recommended indications for surgery: (1) Patients with sudden onset of DoC and consistent with the diagnosis of MCS; (2) The duration of disease must be more than 3 months, with no progressive improvement or deterioration of consciousness for more than 4 consecutive weeks; for traumatic injuries, it is recommended to extend the surgery to be 6 months after the injury with no improvement of consciousness for 8 consecutive weeks; (3) Patients do not have serious complications or contraindications to surgery (Functional Neurosurgery Group of the Chinese Neurosurgery Medical Association, Professional Committee of Neurorepair of Chinese Medical Doctor Association, Chinese Society of Consciousness and Disorders of Consciousness 2019).

### 2.1 Deep Brain Stimulation

**Treatment Principle** The implantation target of deep brain stimulation (DBS) for improving the state of consciousness in DoC patients is mainly at the centrothalamic region with the center median-parafascicular complex (CM-pf) as the core, which aims to improve the low neural activity level after brain injury and enhance wakefulness and consciousness based on the central loop mechanism of consciousness by stimulating the key nodes of the loop bilaterally in centrothalamus and activating the consciousness-related neural network of patients. If the connectivity within this large network or between the thalamus and this network is disrupted, DBS is unlikely to significantly affect this network via the thalamus. Traumatic lesions are often numerous but scattered, thus allowing the presence of more neural connections. Therefore, DBS may be more effective in traumatized patients.

The central thalamus, a key integration center for consciousness generation and maintenance, can induce incremental recruiting, augmenting response (Ward 2011), and amplitude modulation of EEG after receiving sustained low-frequency stimulation, thus activating and enhancing consciousness-related brain network activity (Schiff et al. 2007). Therefore, DBS is applied to compensate for the activity of centrothalamic neurons and thus restore cognitive functions based on these networks (e.g., attention, memory, language, and executive functions), and finally improve the level of consciousness (Baker et al. 2016). Animal studies have shown similar results. DBS can also repair neural cells and effectively control the progression of the disease. As for the neural remodeling that may be involved in the sustained effects of treatment, further observations and studies are needed.

**Treatment Protocols** The basic surgical principles and methods are the same as those of other DBS procedures. Patients with severe destruction of intracranial structures or significant brain atrophy are not suitable for DBS. After general anesthesia, the stimulation is usually applied to the centromedian nucleus-parafascicular nucleus of the thalamus (CM-pf). The anatomical coordinates are $X = 7–9$ mm, $Y = 8$ mm (after the midpoint of anterior commissure—posterior commissure line), and $Z = 0–3$ mm. The planning of the electrode implantation path needs to avoid the lateral ventricles. The application of anesthetic or strong sedative drugs that affect cellular activity should be stopped after drilling to obtain satisfactory microelectrode records. According to preliminary observations, the discharge of CM-pf is significantly weaker than that of single cell in the adjacent nucleus mass, while studies with a large sample size are still required for verification. Implanted electrodes with long-pitch contacts are recommended. DBS is mainly unipolar, and the stimulation parameters of the program set should be set as follows: frequency: 25–100 Hz, wave width: 100–240 μs, voltage: 1.0–4.0 V. Cyclic stimulation mode is adopted, with daytime stimulation and nighttime power-off, to correspond to the normal wakefulness-sleep cycle (Functional Neurosurgery Group of the Chinese Neurosurgery Medical Association, Professional Committee of Neurorepair of Chinese Medical Doctor Association, Chinese Society of Consciousness and Disorders of Consciousness 2019).

**Clinical Application** Clinical studies can be generally divided into three phases: (1) Early studies confirmed that DBS has an enhancing effect on the arousal system. Early animal studies found a relationship between the midbrain reticular formation and the arousal level, and stimulation caused an EEG desynchronization manifestation similar to the wakefulness state. In 1968, McLardy first applied DBS to the left medial interlaminar nucleus and midbrain reticular formation of a 19-year-old patient in a vegetative state 8 months after cerebral trauma (McLardy et al. 1968). Other subsequent studies took the intralaminar nucleus and the nucleus of medulla oblongata as DBS targets. Due to the limitations of the times, these studies all applied short-term stimulation lasting a few weeks, finding no clear evidence that the stimulation targets were structurally compatible with the arousal system. However, it was noted by all of these studies that stimulation could enhance behavioral response and lead to EEG desynchronization. (2) Finding targets and indications compliant with the physiological basis. In the 1980s, a multi-center study jointly conducted by France, Japan, and the United States was conducted on a cohort of 25 patients in a vegetative state with a course over 3 months, where DBS was applied to the parafascicular nucleus and the centromedian nucleus of the thalamus. 13 cases showed significant improvement in the level of consciousness after 1–3 weeks of treatment. Tsubokawa et al. evaluated eight patients in a vegetative state with a course of 2–3 months, three of which showed behavioral improvement (response to commands and phonation/verbal expression) and one showed partial behavioral improvement (resumption of oral feeding and expression of emotions) (Tsubokawa et al. 1990). In 2010, Yamamoto et al. evaluated 21 patients in VS, 8 of which awakened (38.1%) (Yamamoto et al. 2010). In 2013, Yamamoto et al. reported DBS or spinal cord stimulation for 36 patients in VS and MCS again, with 15 patients in the DBS group (15/26) regaining consciousness. Some other investigators also reported similar improvements (Yamamoto et al. 2013). This suggests that electrical neurostimulation can significantly improve the consciousness of patients. (3) Individualized DBS treatment protocols consistent with physiological mechanisms. In 2007, *Nature* reported a study published by Schiff et al. that was highly informative in determining the efficacy of DBS. A case report of a patient in minimally conscious state 6 years after trauma, treated with DBS to the intralaminar nucleus of thalamus, demonstrated significant improvement of symptoms after neurostimulator implantation. This was manifested by the comprehensible verbal expression and the correct use of objects that occurred in the early parameter titration phase, followed by compliance activity, voluntary limb movements and oral feeding, and in particular, the restoration of functional communication with others. Continuous DBS produced sustained effects on behavior, and behavioral improvements were maintained even during the DBS off period (Schiff et al. 2007). Animal studies have shown similar results. As for the neural remodeling that may be involved in the sustained effects of treatment, further observations and studies are needed.

## 2.2 Spinal Cord Stimulation

**Treatment Principle** Spinal cord stimulation (SCS) is performed by surgically placing stimulation electrodes in the middle epidural region at C2–C4 level of the cervical cord. The electrical stimulation travels up the high cervical spinal cord to the brainstem and reaches the cerebral cortex through the ascending reticular activating system and the subthalamic activating system so that (1) the impulse stimulation at the initiation site of conscious impulses enhances the activity of conscious impulses, improves the state of nerve conduction and increases EEG activity via the pathway of ascending reticular activating system-thalamus-cortex (Della Pepa et al. 2013). (2) Modulation of sympathetic ganglion activity in the neck enhances CBF and improves cerebral metabolism. (3) It can increase the release of excitatory transmitters in the loop as well as the neurotransmitters (e.g., catecholamines, dopamine, and norepinephrine), and activate partial proteinase to enhance the modulation of biological signals.

**Treatment Protocols** General anesthesia is generally performed in the prone or lateral position. Keep the neck in anterior flexion and incise it along the midline. Take out the C5 spinous process and the vertebral plate and fully loosen the C2–4 epidural space along the midline with a special dilator before delivering the surgical electrodes upward to the middle of the epidural region at the C2–4 level. Connect the wire and implantable pulse generator (IPG). Bipolar stimulation is commonly used for postoperative programmed SCS, with a stimulation frequency of 5–70 Hz for the program group. It is recommended to use 70 Hz as the preferred stimulation frequency, with a pulse width of 100–240 μs and a voltage of 1.0–5.0 V. The mode of cyclic stimulation is adopted, with daytime stimulation and nighttime power-off, to correspond to the normal wakefulness-sleep cycle (Functional Neurosurgery Group of the Chinese Neurosurgery Medical Association, Professional Committee of Neurorepair of Chinese Medical Doctor Association, Chinese Society of Consciousness and Disorders of Consciousness 2019).

**Clinical Application** Komai first reported SCS for VS in the early 1980s (Kanno et al. 1989), after which Kanno and Momose et al. evaluated changes in cerebral glucose metabolism and cerebral blood flow in patients with DoC after SCS and confirmed that the local cerebral glucose metabolic rate and cerebral blood flow increased significantly before and after stimulation. They found that consciousness was restored in 56 (43%) of 130 cases of VS patients. In 2012, Yamamoto performed SCS on 10 MCS patients, 7 of whom showed significant improvement in consciousness (Yamamoto et al. 2013), Yamamoto et al. reported SCS for 10 patients in MCS again, with 5 of them restoring consciousness. The efficacy of SCS is now considered definite, with an overall response rate of 20–40% (Mattogno et al. 2017), while the arousal rate and effectiveness of VS for post-traumatic brain injury are even higher. It was first reported in China by Wang Peidong et al. in 2001 that 6 patients in VS received SCS treatment, 2 of whom were awakened. In 2019, Xia Xiaoyu and He Jianghong reported 110 patients with DoC treated with electrical neurostimulation, and the rate of arousal was about 30% (Xia et al. 2019).

## 2.3 Other Stimulation Modes

There are only few case reports on the arousal by cortical electrical stimulation, vagus nerve stimulation (VNS), and baclofen pump implantation, with unclear efficacy. For example, it was reported that implanted VNS was adopted in an uncontrolled case study on a VS patient with 15 years of onset. The patient improved from VS to MCS and showed enhanced brain connectivity (i.e., increased activity in the frontoparietal-occipital and basal ganglia regions) (Corazzol et al. 2017). As with all uncontrolled studies, this case report needed to be treated with caution. Nevertheless, it suggested the possibility of this approach for patients with DoC.

## 2.4 Major Complications of the Procedure and the Corresponding Management

### 2.4.1 Deviation in the Position of Electrodes

The actual position of DBS electrodes implanted may deviate from the intraoperative plan due to brain atrophy and deformation, cerebral ventriculomegaly, and intraoperative brain shift. Therefore, except for patients with severe brain atrophy and deformation who are not considered for DBS surgery, the procedure of DBS electrode implantation in patients with indications needs to be carefully checked and operated. In addition to the use of surgical planning system and electronic brain nucleus spectrum to improve localization accuracy, it is recommended to select stimulation electrodes with contacts separated by a long distance to cover as many nuclei of the central thalamus as possible, providing sufficient space for postoperative program control and adjustment. For SCS, preoperative X-ray positioning must be performed to ensure accurate implantation of electrodes, which should be placed in the midline of the spinal cord as much as possible to avoid discomfort on one side caused by low stimulation intensity that may otherwise affect the setting of program control parameters.

### 2.4.2 Incision and Skin Breakage

Patients with DoC are generally malnourished and therefore at high risk for skin breakage. Breakage at the IPG implantation site is often caused by postoperative skin pulling because the capsular bag is excessively small and shallow. When an incision split occurs, it should be cleared early, and the capsular bag should be enlarged and deepened. The incision should be sutured without tension. The subcutaneous effusion under IPG is uncommon, and most of the effusion can be self-absorbed under local pressure.

### 2.4.3 Stimulation-Related Convulsion or Epilepsy Seizure

If the SCS electrode deviates from the midline, stimulation is very likely to lead to convulsion and discomfort on one limb of the patient, resulting in the forced reduction of stimulation intensity or even the abandonment of stimulation contact, which ultimately affects the surgical outcome. Neuromodulation has the potential to induce epileptic seizures, which are not uncommon in post-DoC stimulation. When the stimulation parameters are not set properly or the intensity is too high, it can lead to EEG hyperactivity or even epilepsy seizure in patients, mostly generalized seizures. Once it occurs, stimulation should be stopped immediately. If there is no seizure in the following 1–2 weeks, the stimulation intensity can be adjusted downward and then turned on, and the intensity can be adjusted to the appropriate level if there is no abnormality in the subsequent observation.

Although neuromodulation surgery has become one of the major research hotspots and directions for the treatment of DoC, studies on surgical treatment are limited by the limited understanding of DoC, actual modulation ability, and clinical experience. In addition, there are many bottlenecks and difficulties in the selection of patients, determination of treatment target sites, setting of program control parameters and scientific verification of efficacy. Therefore, it should be performed cautiously and scientifically before becoming a common clinical practice, and its clinical efficacy and adverse events should be recorded in detail.

## References

Baker JL, Ryou JW, Wei XF et al (2016) Robust modulation of arousal regulation, performance, and frontostriatal activity through central thalamic deep brain stimulation in healthy nonhuman primates. J Neurophysiol 116:2383–2404

Barra A, Rosenfelder M, Mortaheb S et al (2022) Transcranial pulsed-current stimulation versus transcranial direct current stimulation in patients with disorders of consciousness: a pilot, sham-controlled cross-over double-blind study. Brain Sci 12:429

Bourdillon P, Hermann B, Sitt JD et al (2019) Electromagnetic brain stimulation in patients with disorders of consciousness. Front Neurosci 13:223

Briand M-M, Gosseries O, Staumont B et al (2020) Transcutaneous auricular vagal nerve stimulation and disorders of consciousness: a hypothesis for mechanisms of action. Front Neurol 11:933

Cain JA, Spivak NM, Coetzee JP et al (2022) Ultrasonic deep brain neuromodulation in acute disorders of consciousness: a proof-of-concept. Brain Sci 12:428

Corazzol M, Lio G, Lefevre A et al (2017) Restoring consciousness with vagus nerve stimulation. Curr Biol 27:R994–R996

Della Pepa GM, Fukaya C, La Rocca G et al (2013) Neuromodulation of vegetative state through spinal cord stimulation: where are we now and where are we going? Stereotact Funct Neurosurg 91:275–287

Functional Neurosurgery Group of the Chinese Neurosurgery Medical Association, Professional Committee of Neurorepair of Chinese Medical Doctor Association, Chinese Society of Consciousness and Disorders of Consciousness (2019) Neurosurgery treatment of chronic disorders of consciousness Chinese Expert Consensus. Chin J Neurosurg 35:5

Giacino JT, Kalmar K, Whyte J (2004) The JFK Coma Recovery Scale-Revised: measurement characteristics and diagnostic utility. Arch Phys Med Rehabil 85:2020–2029

Giacino JT, Katz DI, Schiff ND et al (2018) Practice guideline update recommendations summary: Disorders of consciousness: Report of the Guideline Development, Dissemination, and Implementation Subcommittee of the American Academy of Neurology; the American Congress of Rehabilitation Medicine; and the National Institute on Disability, Independent Living, and Rehabilitation Research. Neurology 91:450–460

Group of Disorders of Consciousness and Conscious-Promotion, Professional Committee of Neurorepair of Chinese Medical Doctor Association (2021) Diagnoses and treatments of prolonged disorders of consciousness: an experts consensus. Chin J Neuromed 19:977–982

He F, Wu M, Meng F et al (2018) Effects of 20 Hz repetitive transcranial magnetic stimulation on disorders of consciousness: a resting-state electroencephalography study. Neural Plast 2018:5036184

Kanno T, Kamel Y, Yokoyama T et al (1989) Effects of dorsal column spinal cord stimulation (DCS) on reversibility of neuronal function-experience of treatment for vegetative states. Pac Clin Electrophysiol 12:733–738

Kondziella D, Bender A, Diserens K et al (2020) European Academy of Neurology guideline on the diagnosis of coma and other disorders of consciousness. Eur J Neurol 27:741–756

Liu X, Meng F, Gao J et al (2018) Behavioral and resting state functional connectivity effects of high frequency rTMS on disorders of consciousness: a Sham-controlled study. Front Neurol 9:982

Mattogno PP, Barbagallo G, Iacopino G et al (2017) Recovery from chronic diseases of consciousness: state of the art in neuromodulation for persistent vegetative state and minimally conscious state, vol 124. Springer International Publishing, Cham, pp 19–25

McLardy T, Ervin F, Mark V et al (1968) Attempted inset-electrodes-arousal from traumatic coma: neuropathological findings. Trans Am Neurol Assoc 93:25–30

O'Neal CM, Schroeder LN, Wells AA et al (2021) Patient outcomes in disorders of consciousness following transcranial magnetic stimulation: a systematic review and meta-analysis of individual patient data. Front Neurol 12:694970

Schiff ND, Giacino JT, Kalmar K et al (2007) Behavioural improvements with thalamic stimulation after severe traumatic brain injury. Nature 448:600–603

Thibaut A, Bruno M-A, Ledoux D et al (2014) tDCS in patients with disorders of consciousness: Sham-controlled randomized double-blind study. Neurology 82:1112–1118

Thibaut A, Schiff N, Giacino J et al (2019) Therapeutic interventions in patients with prolonged disorders of consciousness. Lancet Neurol 18:600–614

Tsubokawa T, Yamamoto T, Katayama Y et al (1990) Deep-brain stimulation in a persistent vegetative state: follow-up results and criteria for selection of candidates. Brain Injury 4: 315–327

Ward LM (2011) The thalamic dynamic core theory of conscious experience. Conscious Cogn 20:464–486

Xia X, Bai Y, Zhou Y et al (2017) Effects of 10 Hz repetitive transcranial magnetic stimulation of the left dorsolateral prefrontal cortex in disorders of consciousness. Front Neurol 8:182

Xia X, Yang Y, Guo Y et al (2018) Current status of neuromodulatory therapies for disorders of consciousness. Neurosci Bull 34: 615–625

Xia XY, Yang Y, Dang YY et al (2019) Therapeutic effect of spinal cord stimulation on chronic disorders of consciousness after brain injury: a report of 110 cases. Chin J Neurosurg 35:5

Yamamoto T, Katayama Y, Kobayashi K et al (2010) Deep brain stimulation for the treatment of vegetative state: DBS for the treatment of VS. Eur J Neurosci 32:1145–1151

Yamamoto T, Katayama Y, Obuchi T et al (2013) Deep brain stimulation and spinal cord stimulation for vegetative state and minimally conscious state. World Neurosurg 80:S30.e1–S30.e9

Yu Y, Yang Y, Gan S et al (2021) Cerebral hemodynamic correlates of transcutaneous auricular vagal nerve stimulation in consciousness restoration: an open-label pilot study. Front Neurol 12:684791

MIX
Papier aus verantwortungsvollen Quellen
Paper from responsible sources
FSC® C105338

If you have any concerns about our products,
you can contact us on
**ProductSafety@springernature.com**

In case Publisher is established outside the EU,
the EU authorized representative is:
**Springer Nature Customer Service Center GmbH**
**Europaplatz 3, 69115 Heidelberg, Germany**

Printed by Libri Plureos GmbH
in Hamburg, Germany